TRUMAN L. K

Statistical Analysis in Chemistry and the Chemical Industry

WILEY PUBLICATIONS IN STATISTICS

Walter A. Shewhart, Editor

Mathematical Statistics

BLACKWELL and GIRSHICK · Theory of Games and Statistical Decisions (*in press*)
HANSEN, HURWITZ, and MADOW · Sample Survey Methods and Theory, Volume II
DOOB · Stochastic Processes
RAO · Advanced Statistical Methods in Biometric Research
KEMPTHORNE · The Design and Analysis of Experiments
DWYER · Linear Computations
FISHER · Contributions to Mathematical Statistics
WALD · Statistical Decision Functions
FELLER · An Introduction to Probability Theory and Its Applications, Volume I
WALD · Sequential Analysis
HOEL · Introduction to Mathematical Statistics, *Second Edition in press*

Applied Statistics

FRYER · Elementary Statistical Analysis
BENNETT and FRANKLIN · Statistical Analysis in Chemistry and the Chemical Industry
COCHRAN · Sampling Techniques
WOLD and JURÉEN · Demand Analysis
HANSEN, HURWITZ, and MADOW · Sample Survey Methods and Theory, Volume I
CLARK · An Introduction to Statistics
TIPPETT · The Methods of Statistics, *Fourth Edition*
ROMIG · 50–100 Binomial Tables
GOULDEN · Methods of Statistical Analysis, *Second Edition*
HALD · Statistical Theory with Engineering Applications
HALD · Statistical Tables and Formulas
YOUDEN · Statistical Methods for Chemists
MUDGETT · Index Numbers
TIPPETT · Technological Applications of Statistics
DEMING · Some Theory of Sampling
COCHRAN and COX · Experimental Designs
RICE · Control Charts
DODGE and ROMIG · Sampling Inspection Tables

Related Books of Interest to Statisticians

ALLEN and ELY · International Trade Statistics
HAUSER and LEONARD · Government Statistics for Business Use

Statistical Analysis in Chemistry and the Chemical Industry

CARL A. BENNETT

Chief Statistician, General Electric Company
Richland, Washington

NORMAN L. FRANKLIN

Lecturer in Chemical Engineering
University of Leeds, England

Sponsored by
The Committee on
Applied Mathematical Statistics
The National Research Council

New York · John Wiley & Sons, Inc.
London · Chapman & Hall, Limited

Members of the Committee on
Applied Mathematical Statistics
National Research Council

PRINTED IN THE UNITED STATES OF AMERICA

Committee's Foreword

DURING THE PAST FIVE OR SIX YEARS THERE HAS BEEN AN ACCELERATED interest in the use of modern statistical methods in chemistry, particularly industrial chemistry. As an early instance of the amount of interest it might be mentioned that the American Chemical Society sponsored a session on statistical methods, at its Atlantic City meeting in April, 1947, which was attended by approximately 500 persons. Since then many meetings and special conferences have been devoted to the application of statistical methods in industrial chemistry.

This rapidly growing interest was discussed at a meeting of the Committee on Applied Mathematical Statistics of the National Research Council in June, 1949. It was the opinion of Dr. H. M. Smallwood and Dr. W. A. Shewhart, members of the Committee representing chemistry and quality control, that a comprehensive book on applied mathematical statistics with illustrative examples and material from chemistry and the chemical industry should be prepared under the direction of the Committee. The Committee, in consultation with members of the Mathematics branch of the Office of Naval Research, decided to sponsor such a book. It was furthermore decided that the manuscript should be prepared jointly by two authors: a mathematical statistician with some knowledge of chemistry and a chemist with some knowledge of mathematical statistics, both persons to have had considerable experience in the application of modern statistical methods to problems in the field of chemistry. Messrs. Shewhart, Smallwood, and S. S. Wilks, also of the Committee, were authorized to implement this plan which finally resulted in the present book.

The Committee was fortunate in being able to arrange for Carl A. Bennett, Chief Statistician, General Electric Company, Richland, Washington, and Norman L. Franklin, Lecturer in Chemical Engineering, University of Leeds, England, to prepare the manuscript. These two authors worked together in the Statistical Research Group at Princeton University during the summers of 1950 and 1951 preparing a draft of the manuscript. Dr. Bennett did a considerable amount of preliminary work in the spring of 1950 in Princeton, before Dr. Franklin arrived.

The Committee believes that the authors have done an excellent job of preparing a book along the general lines mapped out by the Committee and its advisors. Whatever value the book may have in assisting those in chemistry and the chemical industry to a better understanding of the theory and use of modern statistical methods, will be due primarily to the authors' painstaking preparation and presentation of the material.

The Committee is indebted to many persons for general guidance in the early stages of outlining the scope and over-all character of this book. It acknowledges, particularly, the advice of Miss Besse B. Day, Statistician, U.S. Naval Engineering Experimental Station; R. H. Noel, Executive Director of Control, Bristol Laboratories; George W. Thomson, Senior Chemical Mathematician, Ethyl Corporation; John W. Tukey, Princeton University; W. J. Youden, Statistician, National Bureau of Standards; and G. L. Wernimont, Eastman Kodak Company.

The Committee also wishes to express its appreciation to the Statistical Research Group at Princeton for providing the facilities for the preparation of the manuscript; to Dean H. S. Taylor of Princeton University for putting the Committee in touch with Dr. Franklin and helping to arrange for him to come to Princeton to join Dr. Bennett; to the officials of the General Electric Company and the University of Leeds for granting leaves of absence, respectively to Dr. Bennett and Dr. Franklin to work on the book. A special word of appreciation goes to the Office of Naval Research for its interest in and financial support of the project through a contract with Princeton University.

The Committee is indebted to Messrs. Shewhart and Smallwood not only for originally proposing this book but for their continuous help and advice during the planning and preparation of the manuscript. We deeply regret that Dr. Smallwood, who contributed so much to the present project, did not live to see this book in final published form. He died on April 27, 1953, after the manuscript itself had been completed.

L. P. Eisenhart, Chairman
Committee on Applied
Mathematical Statistics
National Research Council

January, 1954

Authors' Preface

THE CONSIDERATIONS WHICH LED TO THE COMMISSIONING OF THIS BOOK are more properly the concern of the Committee on Applied Mathematical Statistics of the National Research Council, and have been reported in the Committee's Foreword. This Committee suggested certain general principles which they felt should be followed in its development; but the responsibility for all but the broadest aspects of the presentation rests entirely upon the authors.

We attempted to write at a level between that of the well-known works on mathematical statistics and other texts which deal almost entirely with the application of statistical methods. Some introductory material of a purely mathematical nature and many proofs and derivations have been included because they were considered necessary to the understanding and judicious application of the techniques presented. The mathematical standard has been selected to exclude material which would be beyond the range of the engineering or chemistry graduate. Examples have been taken from work in the chemical and allied industries to illustrate the application of the methods. It has not been possible to find an actual instance of every technique, nor to include examples from all the fields of applied chemistry to which statistical methods may with advantage be applied.

The techniques which are presented represent only a fraction of those now available. The selection has been based upon the frequency with which the various methods are used in the chemical industry, and the degree of correspondence between the practical situations from which data are obtained and the mathematical models on which inferences from these data must be based. In cases where more than one method of achieving the same end have been discussed, we have attempted to present the theoretical, computational, and practical considerations on which the selection of the appropriate technique should be based, and to illustrate, if only by reference to original papers or more complete works, the consequences of errors in the assumptions on which a particular inference rests.

For those who are interested in a more complete or more rigorous treatment of some topic, we have endeavoured to provide a selection of references, which are not intended to be complete. For those interested in the details of the practical application of the methods to particular industries, references have been given to the original papers from which examples were taken. In a fair number of cases the examples represent unpublished work, and in a few, particularly in the early chapters, simple examples have been fabricated.

We hope that the book will be of use to those in the chemical industry whose interest in this subject has quickened in recent years. The first five chapters are intended to develop the basic principles of statistical inference in a logical sequence, although the non-mathematical reader may wish to refer to Chapter 4 only when this is essential to the understanding of the use of tabulated test ratios. In Chapter 6 through 11 the more specialized methods which have been of greatest importance in industrial applications are developed, and the order in which they are considered will depend upon the interest of the individual. Although this was not the original intention, it is felt that this book could be made the basis for a one-year (two-semester or three-quarter) course in applied statistics for upper class students in statistics or graduate students in the fields of application. For this purpose it might be desirable to consider the material in Chapters 9, 10, and 11 (with the exception of Serial Correlation) immediately following Chapter 5, reserving Chapters 6, 7, and 8 for the second semester of study.

In producing the book the authors are indebted to the Committee on Applied Mathematical Statistics of the National Research Council for their sponsorship, and in particular to Dr. H. M. Smallwood and Professor S. S. Wilks for encouragement and advice. We also appreciated the many comments received on the first draft, which aided immeasurably in its revision; in this connection we should mention in particular the careful reading and detailed comments of Miss Besse B. Day, W. L. Gore, G. W. Thomson, and Dr. Smallwood. Professor J. W. Tukey not only read the first draft in detail but also in the course of many discussions contributed much to its preparation and revision. Our thanks are due to G. L. Wernimont, who provided extensive bibliographical information, and to the many workers and organizations who have published examples of the application of statistical methods, or who provided unpublished data for use in the text. We have acknowledged the source of the many tables individually as they appear, but special mention should be made of the extensive reproduction permitted by Professor E. S. Pearson from publications of the Biometrika Office, and of the many individual tables which originally appeared in the *Annals of Mathematical Statistics*. The task of coordinating the preparation of the first draft and final manuscript was in the hands of Mr. D. M. G. Wishart, to whom we wish to express our gratitude. The authors are indebted to Princeton University for providing office and library facilities during the preparation of this work and to the General Electric Company and the University of Leeds for permitting us extensive leaves of absence in order that it might be completed.

C. A. BENNETT
N. L. FRANKLIN

January, 1954

Contents

CHAPTER

1 INTRODUCTION 1

1.1 The Place of Statistical Methods 1

1.2 The Nature of Statistical Methods 2

1.3 Some Problems to Which Statistical Methods can be Applied . . . 3

1.31 Reliability of Measurements 3

1.32 Estimation and Tests of Significance 4

1.33 Relationships between Factors 4

1.34 Danger of Misapplication of Methods 5

2 DESCRIPTIVE STATISTICS 6

2.1 Introduction 6

2.2 Organization of Data 6

2.21 Raw Data 6

2.22 Grouping of Data 10

2.3 Some Simple Statistics 16

2.31 Statistics 16

2.32 Statistics Used to Determine Location 16

2.33 Statistics Used to Determine Dispersion 18

2.34 Relative Merits of the Statistics Given 20

2.35 Numerical Calculation 21

2.36 Interpretation of Mean and Standard Deviation 26

2.4 Limits for Means and Medians 26

2.41 Variance of Means 26

2.42 Degrees of Freedom 27

2.43 Limits for the Mean 28

2.44 Limits for the Median 30

2.5 Associated Measurements 31

2.51 Pairs of Measurements. Scatter Diagrams 31

2.52 Covariance 35

2.53 Regression and Correlation 36

2.54 Limits for the Slope 38

3 PROBABILITY AND SAMPLES 41

3.1 Introduction 41

3.2 Probability and Random Variables 42

3.21 Population and Random Variables 42

3.22 Probability 43

3.23 Laws of Probability, Independence 45

CHAPTER

3.3 Average Values and Variances . . . 46
3.31 One Random Variable . . . 46
3.32 Two Random Variables . . . 48
3.33 Average Value and Variance of a Linear Combination . . . 49
3.34 Approximate Average Value and Variance of an Arbitrary Function . . . 51
3.35 Population and Sample . . . 54
3.4 Samples . . . 55
3.41 Random Sampling . . . 55
3.42 Average Value and Variance of a Sample Mean . . . 55
3.43 Average Value of the Sample Variance . . . 57
3.44 Sampling from Finite Populations . . . 57
3.5 Stratified Sampling . . . 61
3.51 Estimating the Mean . . . 61
3.52 Minimizing the Variance . . . 62
3.53 The Case of Infinite Subpopulations . . . 64

APPENDIX 3*A* PERMUTATIONS AND COMBINATIONS . . . 66
3*A*.1 Permutations . . . 66
3*A*.2 Combinations . . . 66

4 MATHEMATICAL MACHINERY . . . 68
4.1 Introduction . . . 68
4.2 Distribution Functions . . . 68
4.21 Cumulative Distribution Function . . . 68
4.22 Relationships between Cumulative Distribution Function and Probabilities . . . 70
4.23 Discrete Probability Distribution. Probability Density Function 71
4.24 Joint Cumulative Distribution Function of Two Variables . . . 73
4.25 Discrete Probability Distributions and Probability Density Function of Two Variables . . . 74
4.26 Distributions in More than Two Variables . . . 76
4.3 Moments, Semi-Invariants, and Generating Functions . . . 76
4.31 Moments and Moment Generating Functions . . . 76
4.32 Semi-Invariants . . . 79
4.33 Calculation and Interpretation of Semi-Invariants . . . 80
4.4 The Normal Distribution . . . 83
4.41 The Normal Distribution Function . . . 83
4.42 Distribution of a Linear Combination of Normally Distributed Variables . . . 88
4.43 The Central Limit Theorem . . . 89
4.44 Transformations to Approximate Normality . . . 91
4.45 Tests for Non-Normality . . . 92
4.46 The χ^2-Distribution . . . 95
4.5 Sampling from the Normal Distribution . . . 99
4.51 Distribution of Sample Mean . . . 99
4.52 Independence of Linear Combinations . . . 101

CHAPTER

4.53 Distribution of the Sample Variance 102
4.54 The "Student" t-Distribution 105
4.55 The Distribution of the Ratio of Two Variances 108
4.56 Distribution of Other Sample Statistics 110
4.6 Discrete Probability Distributions 111
4.61 Binomial Distribution 111
4.62 Multinomial Distribution 114
4.63 Poisson Distribution 115
4.64 Limiting Forms of the Binomial and Poisson Distributions . . 118
4.65 Hypergeometric Distribution 120

APPENDIX 4*A* GAMMA AND BETA FUNCTIONS 122
4*A*.1 Gamma Functions 122
4*A*.2 Beta Functions 125
4*A*.3 Incomplete Gamma Function 125
4*A*.4 Incomplete Beta Function 126

APPENDIX 4*B* THE NORMAL BIVARIATE DISTRIBUTION 128

5 STATISTICAL INFERENCE 133
5.1 Introduction 133
5.2 General Principles of Statistical Inference 134
5.21 Point Estimation and Efficiency 134
5.22 Confidence Intervals 136
5.23 Example of the Determination of Confidence Limits . . . 137
5.24 Tests of Significance 140
5.25 Example of a Test of Significance 142
5.26 Confidence and Significance Levels 148
5.3 Inferences about Averages 149
5.31 Point Estimates 149
5.32 Point Estimates for Linear Combinations 152
5.33 Confidence Intervals and Tests of Significance Using the t-Distribution 153
5.34 Confidence Intervals and Tests of Significance Based on Order Statistics 157
5.35 Distribution-Free Tolerance Limits 162
5.4 Inferences about the Variance 164
5.41 Point Estimates 164
5.42 Joint Estimates from Several Samples 166
5.43 Confidence Intervals and Tests of Significance for the Variance 172
5.5 Comparison of Means 176
5.51 Comparison of Two Means 176
5.52 Paired Comparisons 180
5.53 Comparison of More than Two Means 185
5.6 Comparison of Variances 192
5.61 Comparison of Two Variances 192
5.62 Tests for the Homogeneity of Several Variances 196
5.7 Sequential Tests 200
5.8 Measurement Errors 206

CHAPTER

APPENDIX 5A MAXIMUM LIKELIHOOD ESTIMATION 209
5A.1 Method of Maximum Likelihood 209
5A.2 Examples of Determination of Maximum Likelihood Estimates 211
5A.3 Relation of the Principle of Least Squares to Maximum Likelihood 214

6 RELATIONSHIPS BETWEEN VARIABLES 216
6.1 Introduction 216
6.11 Pairs of Measurements of Observations 216
6.12 Ways in which Pairs of Measurements can Occur 220
6.2 Linear Regression with One Dependent Variable 223
6.21 Statistical Model for Linear Regression 223
6.22 Estimation of α and β 223
6.23 Confidence Limits and Tests of Significance for β . . . 227
6.24 Confidence Limits and Tests of Significance for Predicted Values 228
6.25 Estimation of β when $\alpha = 0$ 232
6.26 Transformation of Variables to Obtain Linear Regressions . . 234
6.27 Unequal Residual Variances in Regression Analysis . . . 243
6.3 Multiple Regression 245
6.31 Linear Regression with More than One Independent Variable 245
6.32 Polynomial Regression on Equally Spaced Observations . . 255
6.33 Orthogonal Combinations in Multiple Regression . . . 265
6.4 Correlation 273
6.41 Estimation of the Correlation Coefficient 273
6.42 Estimation of the Regression Lines 280
6.43 Rank Correlation 283
6.44 Correlations of Three or More Variables 286
6.45 Discriminant Functions 288

APPENDIX 6A THE SOLUTION OF REGRESSION EQUATIONS . 296
6A.1 Introduction 296
6A.2 Exact Solution of Linear Equations. Evaluation of Determinants 296
6A.3 Method of Single Division. The Abbreviated Doolittle Method 302
6A.4 Solution of Sets of Equations. Computation of Inverse Elements 315

7 ANALYSIS OF VARIANCE 319
7.1 Introduction 319
7.2 Basic Analysis of a One-Way Classification 319
7.21 Example 319
7.22 General Model for One-Way Classification with Equal Numbers 321
7.23 Computations 323
7.24 Unequal Numbers of Observations in Each Class 327
7.25 Interpretation of Low F-Ratios 329

CHAPTER

7.3 Other Considerations in One-Way Classification 330
7.31 Interpretation of the Effects ξ_i in the General Model . . . 330
7.32 Estimation of the Effects ξ_i 331
7.33 Comparison of Class Effects 333
7.34 General Linear Comparisons Based on Single Degrees of Freedom 335
7.35 Confidence Limits, Comparisons and Classifications of Means . 339
7.36 Testing Hypotheses Concerning the Overall Mean . . . 344
7.4 Models and Populations 348
7.41 Introduction 348
7.42 One-Way Classifications 348
7.43 Two-Way Classifications 349
7.431 Simple Crossed Classifications with No Interaction . . 349
7.432 Nested Classifications 350
7.433 Many-Way Classifications 350
7.44 The Effect of Failure in the Assumptions Concerning the Model Parameters 351
7.45 Transformation of Data 355
7.5 Two-Way Classifications 358
7.51 Types of Two-Way Classifications 358
7.52 Analysis of Two Nested Classifications 358
7.53 Analysis of Two Crossed Classifications 368
7.54 Missing Data 379
7.6 Greater Numbers of Classifications 385
7.61 General Consideration of Three Crossed Classifications . . 385
7.62 General Considerations for Nested Classifications. . . . 402
7.63 Analyses Involving Both Types of Classifications 410
7.7 Application of the Analysis of Variance to Regression 427
7.71 Simple Linear Regression 427
7.72 Analysis of Variance for Linear Regression in More than One Variable 429
7.73 Analysis of Variance for Polynomial Regression 431
7.74 Subdivision of Scale Classifications 436
7.8 Analysis of Covariance 441
7.81 Introduction 441
7.82 Differences between Slopes. 442
7.83 Treatment Comparisons 446
7.84 Covariance with Multiple Classification 451
7.85 Analysis of Covariance Involving Multiple Regression . . . 457
7.86 Comments on the Application of Analyses of Covariance . . 461
7.9 Relationship between Variates Each Including a Random Component 463
7.91 Simple Regression with Known Slope 463
7.92 Simple Regression with Slope Unknown 465
7.93 The Use Instrumental Variates 465

APPENDIX 7*A* THE AVERAGE VALUES OF VARIANCE COMPONENTS IN MODELS OF TYPE III 470
7*A*.1 Nested Classifications 470
7*A*.2 Crossed Classifications 474

CHAPTER

8 THE DESIGN OF EXPERIMENTS 478

8.1 Introduction 478
8.2 General Terminology 479
8.3 Collection of Data 480
8.31 General Considerations 480
8.32 Sampling of Materials. 481
8.33 Sampling of Discrete Articles 481
8.34 Stratified Sampling 482
8.35 Sampling of Aggregates 484
8.36 Sample Reduction 485
8.37 Sampling Bias 486
8.38 Replication 487
8.39 Randomization 491

8.4 Factorial Design 493
8.41 Classical and Factorial Designs. 493
8.42 The 2^3 Factorial Experiment 496
8.43 The 2^k Factorial Experiment 501
8.44 Designs Involving Factors at Other than Two Levels 502
8.441 Simplified Analysis when Some Factors are at Two Levels 503
8.442 Single Degrees of Freedom in the 3^2 and 3^3 Design . . 506
8.443 Dummy Comparisons 512

8.5 Elimination of Experimental Error 514
8.51 General Considerations 514
8.52 Replication in Blocks 515
8.53 Latin Squares 521
8.54 Graeco-Latin Squares 526
8.55 Hyper Graeco-Latin Squares 533
8.56 Other Uses of Latin and Graeco-Latin Squares 535

8.6 Confounding 538
8.61 Introduction 538
8.62 Split Plots 539
8.63 Confounding of Interactions 545
8.64 Confounding in 3^k and Other Factorial Designs 564
8.65 Incomplete Blocks 570
8.66 Incomplete Latin Squares or Youden Squares 575

8.7 Limiting the Number of Experimental Units 580
8.71 Introduction 580
8.72 Fractional Replication 581
8.73 The Use of Latin Squares and Hypersquares 587
8.74 Other Types of Design 592

8.8 Combined Experiments 597
8.81 General Discussion 597
8.82 Raw Materials 598
8.83 Time Effects and Sequential Experiments 599

CHAPTER

9 ANALYSIS OF COUNTED DATA 601
9.1 Introduction 601
9.2 Transformation of Counted Data 601
9.3 Inferences from Counted Data 603
9.31 Estimation of the Probability p of Success 603
9.32 Estimation of the Expected Value λ 607
9.33 Comparison of Counts 611
9.34 Analysis of Variance for Counted Data 615
9.4 Chi-Square Tests 620
9.41 Chi-Square Test of Goodness of Fit 620
9.42 Contingency Tables 623
9.43 Tests of Homogeneity of Observed Counts 626
9.5 Acceptance Sampling by Attributes 627

10 CONTROL CHARTS 631
10.1 Introduction 631
10.2 Statistical Control 631
10.21 The Concept of Statistical Control 631
10.22 The Concept of Control Limits 633
10.3 Control Charts for Variables 634
10.31 Rational Subgroups 634
10.32 Control Charts for Specified Quality Levels 635
10.33 Control Charts when No Levels are Specified 638
10.34 Discussion of the Use of Control Charts 642
10.4 Control Charts for Attributes 644
10.41 Control Charts 644
10.42 Control Charts for Number Defective 646
10.5 Control Charts for Errors of Measurement 647
10.51 The Need for the Control of Measurement Errors 647
10.52 Control Charts for Accuracy 648
10.53 Control Charts for Precision 650
10.54 Special Control Charts for Duplicate Analysis 654
10.55 Control Charts for Counting Instruments 659

11 SOME TESTS FOR RANDOMNESS 663
11.1 The Concept of Randomness 663
11.2 Extreme Variations 665
11.21 Use of Control Charts to Detect Extreme Variations 665
11.22 Rejection of Observations 666
11.3 Use of Runs to Detect Non-Randomness 667
11.31 Relationship of Runs to Non-Randomness 667
11.32 Distribution of Number of Runs 669
11.33 Other Uses of the Theory of Runs 677
11.34 Use of Mean Square Successive Difference to Detect Non-Randomness 677
11.35 Use of the Serial Correlation to Detect Non-Randomness 684
11.36 Choice of Tests for Non-Randomness 688

CHAPTER

APPENDIX 689

Table I Ordinates and Areas of the Normal Distribution . . . 689
Table II Percentage Points of the χ^2-Distribution 694
Table III Percentage Points of the t-Distribution 696
Table IV Percentage Points of the F-Distribution 698

BIBLIOGRAPHY 713

INDEX 719

CHAPTER 1

Introduction

1.1. The Place of Statistical Methods

Interest in the application of statistical methods to all types of problems has grown rapidly since the late 1920's. Although the initial development of these methods was to a large extent connected with biological and agricultural experimentation (with several notable exceptions, such as the work of "Student,"* who was a chemist interested, among other things, in routine control of analyses), they are at present being applied in almost every field of scientific investigation. Procedures for the statistical control of quality of production, and the development and use of statistical techniques for acceptance sampling, were introduced in the 1920's, and were the first statistical methods to gain any general recognition in industrial work. These applications came to the forefront during the era of high production accompanying World War II, and served to focus attention on the application of other statistical methods to problems of industrial research and development as well as to industrial production. Traditionally statistical methods have been employed by workers concerned with those branches of science or commerce where methods of measurement or physical relationships are usually inexact, and many of the journals in which developments in statistical methods are published are associated with subjects such as biology, agronomy, psychology, and economics. More recently chemists, physicists, and engineers have come to appreciate that some aspects of their basic research and much of the industrial application and development of their work can profit from the use of appropriate statistical techniques.

Broadly speaking, statistical methods may be of use wherever conclusions are to be drawn or decisions made on the basis of experimental evidence. However, they are most likely to be of use where the experimental data are subject to fluctuations, or "errors," which are of the order of magnitude of the differences which we wish to study. In this case there are likely to be

* Further reference to "Student" is made in Section 4.54.

two advantages to a statistical approach: (1) we may be able to design the experiment in such a fashion as to avoid some of the sources of experimental error, and make the necessary allowances for that portion which is unavoidable; (2) we can, on the basis of more or less restrictive assumptions concerning the nature of the fluctuations present, present the results in terms of probability statements which express their reliability, or indicate the probability of an incorrect conclusion. Furthermore, a statistical approach will frequently force a more complete and comprehensive evaluation of the experimental aims and difficulties, and lead to a more definitive experiment than would otherwise have been performed; in many, if not in most, cases, this type of benefit is the greatest gain from the use of statistical methods.

1.2. The Nature of Statistical Methods

Statistical methods can conveniently be thought of in two classes: (1) those methods which are used to summarize, or describe, data in such a manner as to make them meaningful; (2) those methods which are used to generalize about a large amount of possible data from a smaller amount of available data. The first class includes the many types of statistical graphs, tables, charts, and indices which we see every day in the newspapers and magazines; stockmarket averages, birth and death rates, production graphs, and census tables are familiar to almost everyone. The methods of the second class are not so commonly known; they go beyond those of the first class in that they involve inferences, or predictions, which are based on the observations at hand but are to be applied to a larger mass of similar data. It is with this second class of methods that this book will be primarily concerned, although many of the methods which are used for descriptive purposes are also used as a basis for inferences, and there is no well-marked division.

Most statistical methods are based on a mathematical foundation, and, especially in the second class of problem, the theory of probability plays a major role, although alone it is not sufficient to solve the problem of generalization of observed results. The development of statistical methods through the use of mathematical tools forms the subject matter of mathematical statistics, and the adaptation and application of these methods to practical problems is the field of applied mathematical statistics. As in other situations in which mathematics is used as a tool, assumptions will be required and the statistical inferences obtained will be valid only within the framework of the assumed structure of the variation present in the observations. With descriptive methods these restrictions are rarely present since no attempt is being made to generalize the results obtained. It is extremely important that we constantly keep in mind the assumptions which are necessary to the validity of the various methods, and the effect on our results of departures from these assumptions.

1.3. Some Problems to which Statistical Methods can be Applied

1.31. Reliability of Measurements

One of the first problems with which the experimenter is likely to be faced concerns the *reliability* of individual measurements, and their relation to the quantity which it is required to measure. This requires an investigation of the manner in which the individual measurements vary, and it is recognized that there are many types of distributions to which experimental measurements may conform. However, these distributions have certain common properties which may be used for descriptive and predictive purposes.

Frequently when chemical measurements are made we wish to infer from the results information concerning some property of a batch of material much larger than that on which the measurement was performed. In determining the fixed nitrogen content of a sample of a synthetic fertilizer we might well be interested in the average fixed nitrogen content of a shipment representing the output of a number of production units. In order to ensure that the analytical result represents as closely as possible the the quantity which we wish to measure it is necessary to examine the variation introduced in the measurement by the sampling method by which a quantity of material suitable for analysis is obtained from the shipment, as well as that introduced by the analytical procedure and by the analyst by whom the test is carried out. This examination may be based upon previous experience with the test, or upon the results of an investigation specifically designed to provide suitable information. In either event a statistical approach to the problem will be necessary in order to achieve the best allocation of effort between the sampling and the analytical procedures.

The knowledge gained from this type of investigation enables us to formulate specifications for chemical materials in terms of the measurements we can afford to make, rather than in terms of some property of the shipment which could never be economically determined. Only by examining the distribution of sample measurements can we determine the proportion of cases in which our specification will cause us to reject a satisfactory shipment, or accept a bad one. If it is possible to place an economic penalty upon these incorrect decisions and to determine the cost of sampling procedures, we shall then be able to frame our specification so that the total cost of the control procedure is minimized.

1.32. Estimation and Tests of Significance

From a given set of experimental data we may wish to answer a relatively concise question such as: "Will this type of denuder enable caustic of 68% strength to be produced without loss of current efficiency?" or a more general question such as: "What type of relationship exists between tray loading and plate efficiency in a fractionating column?" In either situation economic factors would require that the decision be made on the basis of limited information, since the reduced risk of the economic consequences of an incorrect decision must be set against the additional cost of securing this information. We may wish to decide, for example, whether a series of pilot runs on a denuder which produces caustic concentrations of 68.5%, 69.2%, 68.2%, 68.6%, 68.8%, and 68.9% provide sufficient evidence to justify the expense of installing and operating a single full-scale unit on a mercury cell for further investigation. To aid in making such a decision, we might be interested in knowing the answers to such questions as: "What can be inferred from these data as to the long-term average caustic concentration this process will produce?" or "What are the chances that this average concentration is as low as 68%?" The answer to the first question involves the *estimation* from the available data of some unknown constant; the answer to the second is provided if we *test* the *significance* of the variations of the observed data from the hypothetical average concentration of 68%. In the latter case, if we are prepared to make certain general assumptions of the nature discussed above, we can show that the decision that such a unit is capable of preparing caustic of 68% average concentration will be correct in about 99% of similar cases. This second type of approach is likely to be the more valuable in cases where an economic penalty can be set on an incorrect decision, but it must be noted that statistical methods determine only the statistical significance of variations or differences. An effect may be statistically significant and yet so small as to be of no economic importance.

1.33. Relationships between Factors

We can attack the problem of determining the relationship between or the effect of a number of factors by means of sample measurements in a similar fashion. In cases where the factors influencing the final measurement can be controlled at a series of fixed levels, or subdivided into classes, the technique of *analysis of variance*, and the associated methods for obtaining estimates and making significance tests, will often enable the experimenter to determine the factors which are of importance in influencing the value of the final measurement and also the nature of the relationship between these factors and the measurements obtained. The appreciation of the value of this technique has led to the *design of experiments* in such a manner that the analysis of variance can be employed. In this instance the application of statistical theory precedes the collection of data, indicating the pattern in which the information should be collected so that the conclusions drawn may have as wide an application as possible, and so that the assumptions made in the statistical analysis may be reasonably correct.

From the statistical viewpoint the above method of investigation is the most efficient and is usually favored. However, when consideration is given to the economic penalties of this approach, such as reduction in output in large-scale investigation, or the cost of maintaining some factor at a series of constant levels, it may be commercially more attractive to forgo the advantages of a designed experiment and infer the relationships or effects required from the fluctuations in the various factors during normal plant operation. Under these circumstances *regression* methods frequently enable the relationships to be inferred and the effects of the various factors to be determined. In the intermediate case the technique of *analysis of covariance* may be employed.

1.34. Danger of Misapplication of Methods

The above remarks serve to illustrate some of the potential uses of statistical methods, as well as their limitations. The proper use of statistical methods, supplemented by, or as a supplement to, personal judgement, will ensure the optimum use of information contained in experimental data, and a minimum of incorrect decisions based on limited information. However, dependence on statistical methods in situations where the mathematical representation of the true situation is doubtful and possibly incorrect, and where the necessary conditions for the application of the mathematical theory of probability are not present, may not only make the results obtained useless but completely misleading. The development of techniques designed to check the validity of the assumptions made in analyzing practical data is an important branch of applied statistics. Some of these techniques are considered in later chapters but the final test of the application of statistical methods is their performance in practice. The indications are that the methods are useful and provide a valuable discipline for the examination of experimental data.

CHAPTER 2

Descriptive Statistics

2.1. Introduction

This chapter will be devoted to some elementary methods of describing and qualifying measurements. When a considerable number of measurements are available it is convenient to process them by rearranging, tabulating, and graphing the data in order to emphasize the salient facts which they contain. The term statistics is commonly applied to such a collection of data or to the operations of collection, tabulation, and graphing. It is often desirable to calculate from the original data a series of numbers which summarize the information contained in these data. Averages and percentages are common examples of numbers of this type which describe data, and which should be considered under this heading.

If we are to obtain from the data information which is applicable to a larger number of observations which could conceivably have been obtained, but which we are prevented from examining, our data must be qualified in a manner which reflects the reliability of the inference involved. Although we shall consider in this chapter some simple examples of methods of qualifying data, the problem of inference in general, which is our primary concern, is deferred until later.

The latter part of the chapter is devoted to the consideration of associated measurements, a subject which will be developed more fully in Chapter 6.

2.2. Organization of Data

2.21. Raw Data

Data recorded in the order in which they are obtained or in some other arbitrary fashion are generally referred to as ***raw data***. An example of such a set of data is given in Table 2.1, which gives the daily average coke yields, rounded to the nearest 0.1 %, of a coke oven plant for 52 consecutive weeks. Certainly these data as individual measurements represent desirable information as to the day-by-day functioning of the plant.

We can also consider means of tabulation and presentation which will reflect the information contained in the data as a whole.

One simple procedure is the retabulation of the raw data in order of increasing magnitude. Table 2.2 shows such a rearrangement, frequently called an array, for the data of Table 2.1. A simple graphical presentation of the data is provided by a *dot frequency diagram* such as that shown in Figure 2.1. Another useful method of "picturing" the data is the construction of a *cumulative graph*, such as that given in Figure 2.2, where the ordinates are the number, or frequency, of values less than or equal to a given abscissa. Both of these figures are most conveniently

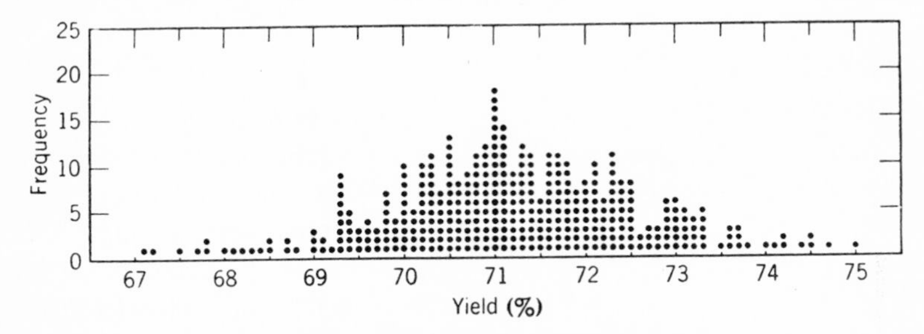

FIG. 2.1. Dot frequency diagram for data of Table 2.2.

prepared from the array in Table 2.2. The ordinates of the cumulative graph are sometimes expressed in terms of fractions (or, more usually, percentages) of the total number of observations. The frequencies expressed in this form will be referred to as relative frequencies. Such a scale is given on the left-hand side of Figure 2.2.

Let us see what we have gained by these simple procedures. From either the array or the dot frequency diagram, the spread of the data, as expressed by the highest and lowest observed values and their difference, the *range*, are easily noted; also it is easy to distinguish those values which occur most frequently. From the cumulative graph it is possible to determine the number or relative frequency of observations less than or equal to (or, by simple subtraction, greater than) any given value; conversely, it is possible to determine the value below which the observations fall with approximately a given relative frequency. Such values are called *quantiles*, or *fractiles*. Those given for relative frequencies in intervals of 0.01 (1%) are called *percentiles*, for intervals of 0.10 (10%) *deciles*, for intervals of 0.25 (25%) *quartiles*, etc. Values of this sort have been extensively used in the fields of education and psychology to describe the results obtained from large-scale testing programs.

Thus from Table 2.2 and Figure 2.1 we may observe that the highest and lowest yields during the year were 75.0% and 67.1% respectively,

TABLE 2.1*

Week	Days						
	1	2	3	4	5	6	7
1	69.4	70.5	73.3	70.3	70.8	70.0	71.6
2	67.5	72.1	70.9	71.1	71.1	72.3	72.2
3	69.7	71.5	72.2	67.1	70.1	73.0	70.4
4	69.2	72.6	70.8	71.7	73.1	69.8	71.0
5	68.3	72.9	72.0	69.1	70.7	71.0	68.8
6	72.2	70.2	71.7	70.4	69.3	72.3	71.6
7	72.4	72.7	71.6	71.2	70.4	70.0	67.2
8	68.1	71.0	71.1	74.5	70.1	70.3	73.0
9	70.0	70.5	70.9	69.7	69.5	71.0	71.8
10	69.3	69.9	69.6	70.0	71.0	70.9	69.5
11	72.8	70.1	71.8	73.2	72.6	70.7	72.4
12	72.3	72.0	70.1	70.6	67.7	72.5	69.8
13	70.5	69.7	71.1	71.6	72.0	67.8	73.6
14	71.3	71.4	71.9	70.5	70.2	72.1	71.7
15	69.4	70.3	72.2	69.8	72.5	71.9	73.1
16	69.3	70.8	68.7	72.1	72.5	73.3	69.8
17	69.8	71.0	68.5	70.3	72.4	69.4	71.4
18	72.7	72.3	71.0	69.3	70.9	71.3	70.7
19	70.5	71.3	70.0	70.6	71.0	70.5	71.3
20	70.0	69.3	70.2	70.9	71.9	70.9	71.2
21	69.3	72.8	72.3	70.5	72.4	70.8	72.3
22	71.2	71.4	72.5	72.5	71.4	67.8	68.2
23	71.7	71.1	70.0	70.8	71.7	71.1	74.0
24	71.4	71.3	68.5	71.1	72.9	69.0	72.3
25	74.7	71.9	70.5	71.2	72.9	68.4	70.0
26	71.1	72.8	71.4	72.3	71.0	70.7	72.2
27	71.3	69.1	72.5	72.9	71.0	69.3	71.3
28	70.6	70.2	73.5	71.7	70.8	70.9	71.1
29	70.9	71.4	75.0	73.0	72.3	70.7	71.2
30	71.3	69.9	71.8	71.1	73.2	71.1	71.8
31	71.6	70.8	72.7	71.9	71.6	74.2	71.0
32	71.0	71.5	71.0	73.6	70.8	73.6	73.1
33	70.2	73.3	73.0	71.7	72.4	71.1	71.0
34	69.4	71.6	70.9	72.3	74.4	72.9	72.5
35	69.8	71.6	71.2	72.4	70.3	71.5	72.1
36	70.8	71.6	70.3	70.6	72.0	71.8	71.3
37	69.9	74.2	70.6	70.9	70.7	70.3	70.2
38	69.6	70.3	69.5	72.1	70.1	73.1	73.3
39	70.8	73.2	71.7	71.2	72.9	71.9	72.5
40	70.2	68.7	71.6	72.0	72.1	70.6	72.1
41	70.4	71.5	71.4	71.2	70.7	71.5	72.0
42	70.9	69.4	73.7	71.6	71.0	70.5	69.3
43	71.0	70.2	70.5	71.7	71.3	70.5	71.5
44	70.7	69.0	70.9	71.0	71.8	71.3	73.8
45	70.2	74.5	71.8	70.4	72.1	70.3	71.8
46	68.0	70.2	70.6	71.3	69.3	71.1	72.3
47	70.4	70.6	72.0	70.5	69.8	73.1	72.2
48	70.7	70.3	69.6	71.7	70.0	70.4	73.7
49	69.0	70.8	72.0	73.7	71.2	73.0	71.8
50	70.3	73.0	70.5	69.9	70.0	69.6	71.1
51	71.8	71.0	71.7	72.1	73.3	72.4	71.9
52	73.2	71.4	72.1	74.1	71.4	71.4	72.4

* Coded data by courtesy of the Director, British Coke Research Association.

TABLE 2.2

67.1	69.8	70.5	70.9	71.3	71.8	72.4	73.6
67.2	69.8	70.5	70.9	71.3	71.8	72.4	73.7
67.5	69.8	70.5	70.9	71.3	71.8	72.4	73.7
67.7	69.9	70.5	71.0	71.3	71.8	72.4	73.7
67.8	69.9	70.5	71.0	71.3	71.8	72.4	73.8
67.8	69.9	70.5	71.0	71.3	71.9	72.5	74.0
68.0	69.9	70.5	71.0	71.4	71.9	72.5	74.1
68.1	70.0	70.5	71.0	71.4	71.9	72.5	74.2
68.2	70.0	70.5	71.0	71.4	71.9	72.5	74.2
68.3	70.0	70.5	71.0	71.4	71.9	72.5	74.4
68.4	70.0	70.5	71.0	71.4	71.9	72.5	74.5
68.5	70.0	70.5	71.0	71.4	71.9	72.5	74.5
68.5	70.0	70.5	71.0	71.4	72.0	72.5	74.7
68.7	70.0	70.6	71.0	71.4	72.0	72.6	75.0
68.7	70.0	70.6	71.0	71.4	72.0	72.6	
68.8	70.0	70.6	71.0	71.4	72.0	72.7	
69.0	70.0	70.6	71.0	71.4	72.0	72.7	
69.0	70.1	70.6	71.0	71.5	72.0	72.7	
69.0	70.1	70.6	71.0	71.5	72.0	72.8	
69.1	70.1	70.6	71.0	71.5	72.0	72.8	
69.1	70.1	70.6	71.0	71.5	72.1	72.8	
69.2	70.1	70.7	71.1	71.5	72.1	72.9	
69.3	70.2	70.7	71.1	71.5	72.1	72.9	
69.3	70.2	70.7	71.1	71.6	72.1	72.9	
69.3	70.2	70.7	71.1	71.6	72.1	72.9	
69.3	70.2	70.7	71.1	71.6	72.1	72.9	
69.3	70.2	70.7	71.1	71.6	72.1	72.9	
69.3	70.2	70.7	71.1	71.6	72.1	73.0	
69.3	70.2	70.7	71.1	71.6	72.1	73.0	
69.3	70.2	70.7	71.1	71.6	72.1	73.0	
69.3	70.2	70.8	71.1	71.6	72.2	73.0	
69.4	70.2	70.8	71.1	71.6	72.2	73.0	
69.4	70.3	70.8	71.1	71.6	72.2	73.0	
69.4	70.3	70.8	71.1	71.6	72.2	73.1	
69.4	70.3	70.8	71.1	71.7	72.2	73.1	
69.4	70.3	70.8	71.2	71.7	72.2	73.1	
69.5	70.3	70.8	71.2	71.7	72.3	73.1	
69.5	70.3	70.8	71.2	71.7	72.3	73.1	
69.5	70.3	70.8	71.2	71.7	72.3	73.2	
69.6	70.3	70.8	71.2	71.7	72.3	73.2	
69.6	70.3	70.8	71.2	71.7	72.3	73.2	
69.6	70.3	70.9	71.2	71.7	72.3	73.2	
69.6	70.3	70.9	71.2	71.7	72.3	73.3	
69.7	70.4	70.9	71.2	71.7	72.3	73.3	
69.7	70.4	70.9	71.3	71.7	72.3	73.3	
69.7	70.4	70.9	71.3	71.8	72.3	73.3	
69.8	70.4	70.9	71.3	71.8	72.3	73.3	
69.8	70.4	70.9	71.3	71.8	72.4	73.5	
69.8	70.4	70.9	71.3	71.8	72.4	73.6	
69.8	70.4	70.9	71.3	71.8	72.4	73.6	

representing a range of 7.9% between the maximum and minimum reported daily yields. From Figure 2.2 it is easily seen that the yield was less than or equal to 68% on 7 days, 69% on 19 days, and 70% on 67 days, or that the yield was greater than 73% on 364 — 333 = 31 days, and 74% on 364 — 356 = 8 days. Similarly we can see that approximately 1 yield in 10 falls below 69.5%, i.e., the relative frequency of

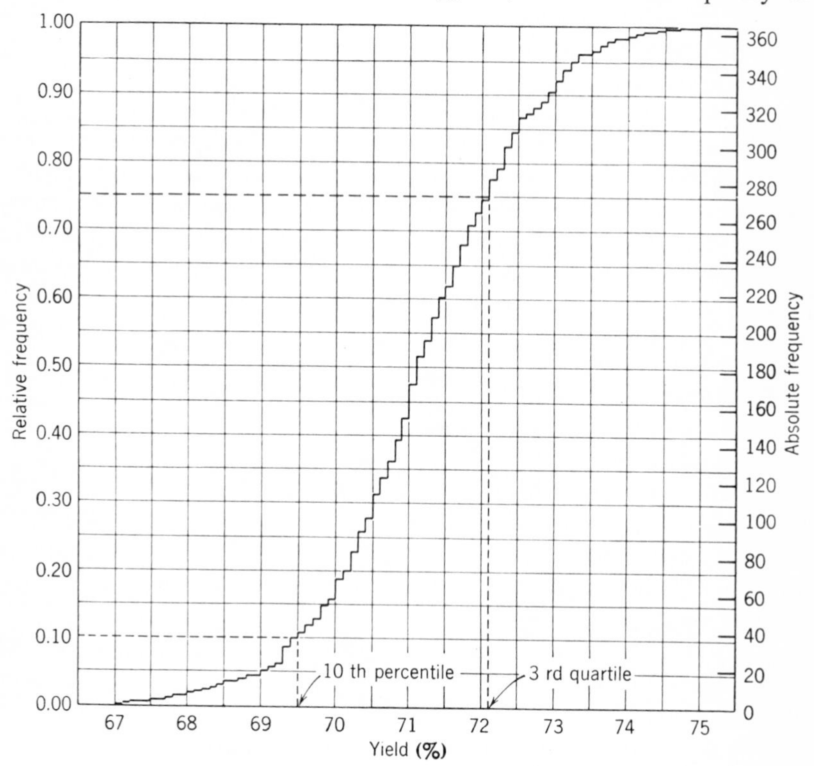

FIG. 2.2. Cumulative graph for data of Table 2.2.

observations below 69.5% is approximately 0.10, and approximately 1 yield in 4 was above 72.1%. Similar observations could be made for any desired value, or any relative frequency.

In the above procedure we have considered the data as a whole, without taking into account the natural order of observation. Frequently a study of the observations in their natural order will yield additional information, and this subject will be discussed at greater length in Chapters 10 and 11.

2.22 Grouping of Data

It will be noted that even in the comparatively dense central portion of the above data the behavior of the frequencies is rather erratic and that the frequency of the individual values is comparatively small. This may be expected when the number of observations and the range of possible values of the observations are both large. In this case a clearer picture of the data may be obtained by tabulating the frequency of occurrence of groups of adjacent values rather than the individual values themselves. Such a tabulation is usually referred to as a *frequency distribution*. The intervals corresponding to the values grouped together are called cells, or classes. In almost every instance it is preferable that the cells be of equal length, since cells of unequal length frequently lead to misinterpretations, and increase the complexity of future calculations (see Section 2.35). The division points between the cells are called *cell boundaries*; any boundary will be the upper cell boundary of the cell below and the lower cell boundary of the cell above. In order to avoid confusion in tabulation, it is often advisable to choose the cell boundaries midway between possible values of the raw data. This is always possible in practice, since data are usually recorded, or can be conveniently rounded, to a fixed number of decimal places. Such a choice of cell boundaries will also ensure that their midpoint is representative of the values which can fall in a particular cell. This is important when calculations are made from the grouped data.

The process of grouping is analogous to the process of "smoothing" data with which many readers will be familiar. The problem is that of removing the erratic behavior of small sections of the data while preserving the larger and more dominant variation. With this in mind, the choice of cell length and number of cells must be made carefully; the use of too few or too large cells may cover up important variations in frequency. Usually the cell length should be chosen so that the number of cells necessary to include all the data will be between 10 and 20, although with small amounts of data less than ten cells may be sufficient, and with very large amounts of data more than twenty cells may be advisable.

Table 2.3 shows the preparation of a grouped frequency distribution directly from the data of Table 2.1. The cell boundaries, cell midpoints, tabulation, and frequencies are given in columns (*a*), (*b*), (*c*), and (*d*), respectively. In this case equal cells of length 0.5 have been chosen, with cell boundaries at the points 66.75, 67.25, 67.75, etc. The form of tabulation shown is that most frequently used, but any other sufficiently accurate method would suffice. If an array or a dot frequency diagram has been prepared, the grouped frequency distribution can easily be obtained by counting, or addition of the frequencies of the values grouped; however, in most instances where there are sufficient data to justify a grouped frequency distribution, it is obtained directly, since the preparation of an array or a dot frequency diagram would involve too much detail.

TABLE 2.3

(*a*) Cell Boundaries	(*b*) Cell Midpoint	(*c*) Tabulation	(*d*) Frequency	(*e*) Relative Frequency	(*f*) Cumulative Frequency	(*g*) Cumulative Relative Frequency
66.75–67.25	67.0	//	2	0.005	2	0.005
67.25–67.75	67.5	//	2	0.005	4	0.011
67.75–68.25	68.0	𝍸	5	0.014	9	0.025
68.25–68.75	68.5	𝍸 /	6	0.016	15	0.041
68.75–69.25	69.0	𝍸 //	7	0.019	22	0.060
69.25–69.75	69.5	𝍸 𝍸 𝍸 𝍸 ////	24	0.066	46	0.126
69.75–70.25	70.0	𝍸 𝍸 𝍸 𝍸 𝍸 𝍸 𝍸 /	36	0.099	82	0.225
70.25–70.75	70.5	𝍸 𝍸 𝍸 𝍸 𝍸 𝍸 𝍸 𝍸 𝍸 ///	48	0.132	130	0.357
70.75–71.25	71.0	𝍸 𝍸 𝍸 𝍸 𝍸 𝍸 𝍸 𝍸 𝍸 𝍸 𝍸 𝍸 ////	64	0.176	194	0.533
71.25–71.75	71.5	𝍸 𝍸 𝍸 𝍸 𝍸 𝍸 𝍸 𝍸 𝍸 𝍸 /	51	0.140	245	0.673
71.75–72.25	72.0	𝍸 𝍸 𝍸 𝍸 𝍸 𝍸 𝍸 𝍸 /	41	0.113	286	0.786
72.25–72.75	72.5	𝍸 𝍸 𝍸 𝍸 𝍸 𝍸 //	32	0.088	318	0.874
72.75–73.25	73.0	𝍸 𝍸 𝍸 𝍸 ////	24	0.066	342	0.940
73.25–73.75	73.5	𝍸 𝍸 //	12	0.033	354	0.973
73.75–74.25	74.0	𝍸	5	0.014	359	0.986
74.25–74.75	74.5	////	4	0.011	363	0.997
74.75–75.25	75.0	/	1	0.003	364	1.000

In the same manner as that used in the preparation of a cumulative graph of the ungrouped frequency distribution, we can prepare a cumulative *grouped frequency distribution* by adding all the frequencies up to and including a given cell. This distribution is given in column (f) of Table 2.3. The relative frequencies for the grouped frequency distribution and the cumulative grouped frequency distribution are given in columns (e) and (g). It should be noted that, whereas the individual cell frequencies are usually associated with the cell midpoints, the cumulative frequencies are associated with the upper cell boundaries. Thus, while

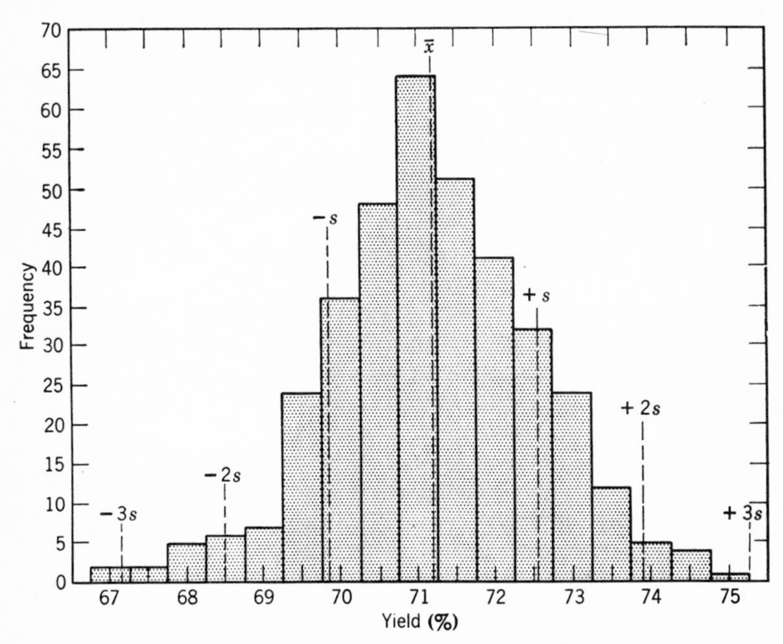

FIG. 2.3. Frequency histogram for data of Table 2.3.

the class frequency of 24 would be associated with the class midpoint of 69.5, the corresponding cumulative frequency of 46 would represent those values less than or equal to 69.75.

The graphical representation of a frequency distribution obtained by erecting on each cell a rectangle whose area is proportional to the cell frequency is called a *frequency histogram*. The corresponding graphical representation of a cumulative frequency distribution is called a *cumulative polygon*; it is obtained by plotting at each cell boundary a point whose ordinate is proportional to the frequency below that value, and connecting these points by straight lines. The frequency histogram and cumulative polygon for the data of Table 2.3 are shown in Figures 2.3 and 2.4. Note that the total area under the histogram below any given value of the yield is proportional to the ordinate of the cumulative polygon at that value; also that in both cases we assume graphically that all values within a class are equally likely.

From the cumulative frequency distribution or from the cumulative polygon the quantiles may be determined in the same manner as they

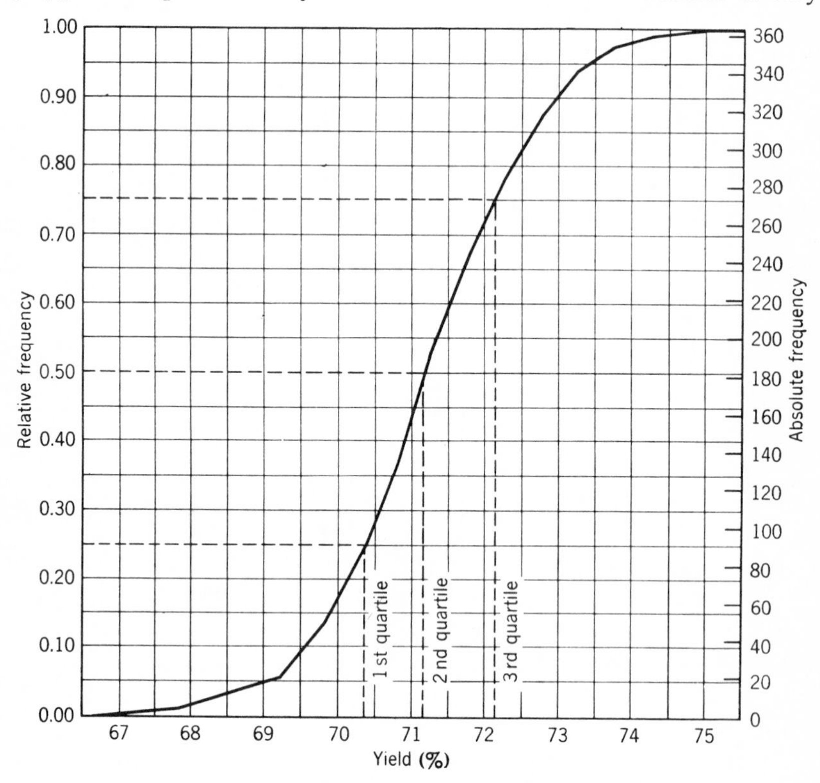

FIG. 2.4. Cumulative polygon for data of Table 2.3.

were from the cumulative graph of ungrouped frequency distribution. If it is so desired, linear interpolation can be used to obtain quantiles which are between those actually occurring at the cell boundaries. Thus, to determine the twenty-fifth percentile, or first quartile, of the distribution of Table 2.3, we proceed as follows: Since 82 of the observations fall below 70.25, this is the 22.5th percentile; since 130 of the observations fall below 70.75, this is the 35.7th percentile; thus the desired value for the 25th percentile would be $70.25 + 0.5 \cdot \dfrac{25 - 22.5}{35.7 - 22.5} = 70.34$. This is equivalent to determining the abscissa on the cumulative polygon corresponding to a relative frequency of 0.25, since the joining of the points at the cell boundaries by straight lines is equivalent to linear interpolation. Similar graphical determinations are shown for the second and third quartiles.

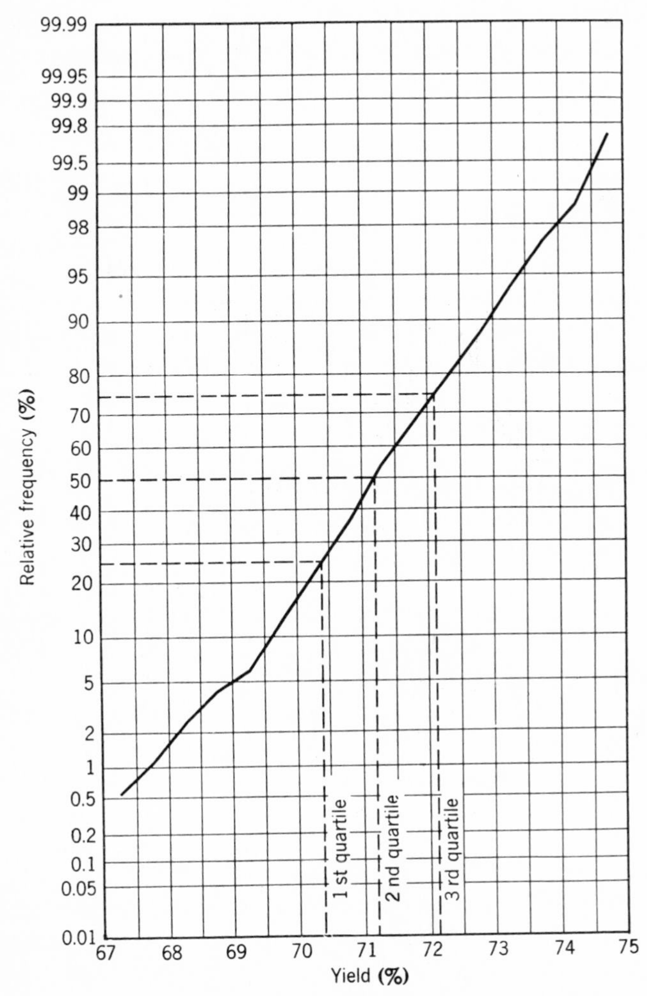

FIG. 2.5. Cumulative polygon on probability paper.

Cumulative polygons are sometimes plotted on so-called "probability paper," on which the ordinate scale is graduated according to the area under a "normal" distribution function (see Section 4.4). If the shape of the grouped frequency distribution is very nearly that of this theoretical distribution, the cumulative polygon so obtained will be very close to a straight line. Graphical interpolation for the various quantiles is more nearly linear, and may be more exact when this type of paper is used. The cumulative polygon for the distribution of Table 2.3 on this paper is shown in Figure 2.5, and the graphical determinations of the first, second, and third quartiles are again indicated. In this instance relative frequencies only can be used unless a special scale of actual frequencies is computed.

2.3. Some Simple Statistics

2.31. Statistics

Single values (such as the quantiles already discussed) which can be computed from a series of observations, are called *statistics.* In the present section we shall present some of the statistics that are useful in describing the characteristics of a series of observations. These and other similar statistics will be used extensively in the more difficult problem of drawing inferences from the observations concerning the larger number of similar observations which might have been obtained.

2.32. Statistics Used to Determine Location

Most series of observations have a tendency to cluster at some particular location on the scale of measurement used. Some typical value, which reflects the location of the data as a whole, may be more informative in this respect than the original data themselves. The most common statistic used for this purpose is the *arithmetic mean*, or simply the *mean*, of the observations; another less common statistic of this type is the *median*. In some cases a more symmetrical distribution of the data is obtained by considering the logarithms of the observations; in such cases the *geometric mean* of the original measures may be an appropriate statistic to use in determining a typical value.

The arithmetic mean of a series of numbers is defined as the sum of the observations divided by the total number of observations; i.e., it is the well-known "average" value. If we denote the n observations by the symbols $x_1, x_2, \cdots, x_n$, and the mean by $\bar{x}$, we can write the simple formula

$$\bar{x} = \frac{\sum_{i=1}^{n} x_i}{n}$$

where the Σ is the usual symbol for the sum of the values indicated, i.e.,

$$\sum_{i=1}^{n} x_i = x_1 + x_2 + \cdots + x_n$$

If a frequency distribution has been prepared, the mean may be computed *approximately* by the formula

$$\bar{x} = \frac{\sum_{i=1}^{k} x_i f_i}{n}$$

where the f_i are the cell frequencies of the cells with midpoints x_i, k is the number of cells, and

$$n = \sum_{i=1}^{k} f_i$$

is the total number of observations. This formula for the mean corresponds to the exact formula for the mean of a series of observations consisting of each midpoint x_i repeated f_i times. If the number of observations is large and the cells are suitably chosen, this approximation is quite good, and the savings in computation, especially if the frequency distribution is to be prepared, are appreciable. Note that the algebraic sum of the deviations of the observations from the mean will always be zero; for, since we defined $\bar{x} = \sum_{i=1}^{n} x_i/n$, or $n\bar{x} = \sum_{i=1}^{n} x_i$, we have

$$\sum_{i=1}^{n} (x_i - \bar{x}) = \sum_{i=1}^{n} x_i - \sum_{i=1}^{n} \bar{x}$$

$$= \sum_{i=1}^{n} x_i - n\bar{x}$$

$$= 0$$

The median is defined as the "middle" value when the given observations are arranged in order of magnitude. If the number of observations is odd, a unique "middle" value exists; if the number of observations is even, a unique middle value does not exist, and the median will be defined as the midpoint of the middle pair of values. The median can easily be determined by simple counting from an array of original data, or it can be approximated from the grouped frequency distribution by interpolation in a manner similar to that used to determine the first quartile in Section 2.22. The median is identical with the second quartile, or the fifth decile, or the fiftieth percentile.

The median is one of the most important of a special group of statistics which are known as *order statistics* because they are obtained by arranging the observations in order of magnitude. The largest value and smallest value of the observations, mentioned in Section 2.21, are also statistics of this type. Other functions of order statistics (frequently called systematic statistics) can be used to obtain typical values. A particular example is the *midrange*, which is defined as the mean of the largest and smallest values.

The geometric mean of a group of n observations is defined as the nth root of the product of all the observations. Its logarithm is the arithmetic mean of the logarithms of the observations, i.e.,

$$\log [\text{geometric mean}] = \frac{\sum_{i=1}^{n} \log x_i}{n}$$

or, for a grouped frequency distribution,

$$\log [\text{geometric mean}] = \frac{\sum_{i=1}^{k} (\log x_i) f_i}{n}$$

where $n = \sum_{i=1}^{k} f_i$, and the x_i are the class midpoints as before.

2.33. Statistics Used to Determine Dispersion

The intuitive reaction of most people to the typical values of the last section will be, "How typical are they?" Intuitively, the answer to this question will be found in the degree to which the observations are clustered, or conversely, the amount of *scatter* or *dispersion* of the observations. Several common statistics used to describe this dispersion are the *range*, *mean deviation*, and the *standard deviation*, whose square, the *variance*, is also common. The first of these is based on order statistics, the last three are based on the deviations of the observed values from some typical value.

The *range* has already been defined as the difference between the largest and smallest observed values. Note that the taking of additional observations can increase the range or leave it fixed, but not decrease it; hence the ranges of two different sets of observations are only comparable if the number of observations is the same, or if suitable correction factors are applied (see Section 5.41).

The *mean deviation* is defined as the mean, or average, of the absolute value of the deviations of the observations from the median, i.e.,

$$d_m = \frac{\sum_{i=1}^{n} |x_i - x_m|}{n}$$

where x_m is the median. The mean deviation has until recently been almost universally defined in terms of deviations from the mean, but deviations from the median are both theoretically and computationally more convenient to use. When the above definitive form is rearranged into the computationally more convenient form given in Section 2.35, it is easy to see that the mean deviation is a systematic statistic, i.e., a function of the ordered observations.

The *variance* is defined as the sum of squares of the deviations from the mean divided by one less than the number of observations. Denoting the variance by s^2, we can write

$$s^2 = \frac{\sum_{i=1}^{n} (x_i - \bar{x})^2}{n - 1}$$

or, in the case of a grouped frequency distribution,

$$s^2 = \frac{\sum_{i=1}^{k} (x_i - \bar{x})^2 f_i}{n - 1}$$

where x_i, f_i, and n are as defined in the previous section. The square root of the variance, s, is called the *standard deviation*. For the purpose of describing the scatter of the observations, the standard deviation, which is in the same units as the original data and has a direct interpretation in terms of the frequency distribution of the observations, will be of more interest than the variance. However, because it arises more naturally in problems of inference and has the desirable property of direct additivity, the variance will be used to a much greater extent as a measure of sample variability in later chapters.

It should be noted that the sum of the squared deviations is a minimum when the deviations are taken from the arithmetic mean. For, if we let

$$Q = \sum_{i=1}^{n} (x_i - m)^2$$

and minimize Q with respect to m, we have

$$\frac{\partial Q}{\partial m} = 2 \sum_{i=1}^{n} (x_i - m) = 0$$

or

$$\sum_{i=1}^{n} x_i - \sum_{i=1}^{n} m = 0$$

Hence

$$\sum_{i=1}^{n} x_i - nm = 0$$

and $m = \bar{x}$ is the value which minimizes Q.

2.34. Relative Merits of the Statistics Given

In the preceding sections we have given several methods of obtaining typical values for a series of observations and of computing their dispersion or scatter. Just as under ordinary circumstances there would be no reason for analyzing for the same element in a sample by two different methods, there is no reason why we should compute more than one typical value or more than one measure of dispersion for a given set of data. Carrying the analogy slightly further, our choice of a statistic will be dictated by its precision, accuracy, and its behavior in the presence of interferences, which is much the same basis as that on which a suitable chemical analysis would be chosen. Although a complete discussion of the relative merits of the statistics will be given in Chapter 5, we shall now indicate briefly some of the reasons why one of these statistics might be used in preference to another.

For data whose distribution does, or would if a sufficient number of observations were made, approximate the normal distribution previously mentioned, the mean and variance give the most information concerning the location and scatter of the data. In addition, both are unbiased; i.e., they tend to reflect accurately the location and scatter (again measured by the mean and variance) of any larger set of observations of which the given observations can be considered a representative part. The median and mean deviation, though not quite so precise as the mean and standard deviation, are much less influenced by the occasional gross error which may affect one of a series of measurements, and the presence of a very small amount of such "contamination" may make their use preferable. Statistics such as the midrange and range are frequently used with small numbers of observations where their precision is not appreciably less than that of the other statistics available, and where we may be willing to sacrifice a little precision for ease of computation. This last situation is somewhat analogous to the use of a quick but less precise colorimetric method for, let us say, plant process control, rather than the more accurate gravimetric or volumetric methods which might be used in a research laboratory.

It should be noted that the measures of dispersion so far given are not necessarily comparable. The sample variance is an accurate estimate of the variance which would have been obtained if all possible measurements of a similar nature had been available, but the sample standard deviation, the mean deviation, and the range are all biased estimates of the square root of this quantity. In Chapter 5 factors depending on the number of observations will be given which correct for this bias and make direct comparisons of these estimates possible.

2.35. Numerical Calculation

The numerical calculation of the variance (and hence the standard deviation) is considerably simplified by changing the definitive form given in Section 2.33 to an algebraically equivalent form more suitable for computational purposes. We have*

$$(a)\qquad s^2 = \frac{\Sigma(x_i - \bar{x})^2}{n - 1}$$

$$(n - 1)s^2 = \Sigma(x_i - \bar{x})^2$$

Expanding the right-hand side we obtain

$$(n - 1)s^2 = \Sigma(x_i^2 - 2x_i\bar{x} + \bar{x}^2)$$

$$= \Sigma x_i^2 - 2\bar{x}\,\Sigma x_i + n\bar{x}^2$$

Since $\Sigma x_i = n\bar{x}$,

$$(n - 1)s^2 = \Sigma x_i^2 - 2n\bar{x}^2 + n\bar{x}^2$$

or

$$= \Sigma x_i^2 - n\bar{x}^2$$

$$(b)\qquad s^2 = \frac{\Sigma x_i^2 - n\bar{x}^2}{n - 1}$$

This form involves the number of observations, the mean, and the sum of squares of the observations themselves, but not their deviations from the mean. It can also be written, since $\bar{x} = \Sigma x_i/n$,

$$(c)\qquad s^2 = \frac{\Sigma x_i^2 - \bar{x}\,\Sigma x_i}{n - 1}$$

or

$$(d)\qquad s^2 = \frac{1}{n - 1}\left[\Sigma x_i^2 - \frac{(\Sigma x_i)^2}{n}\right]$$

* In this section and in the remainder of the book where the range of the summation is clear from the context, we shall frequently write Σx_i or $\Sigma(x_i - \bar{x})^2$ instead of the more cumbersome expressions $\sum_{i=1}^{n} x_i$ or $\sum_{i=1}^{n} (x_i - \bar{x})^2$, or, where more than one subscript is involved, simply indicate the subscript or subscripts over which summation is intended, such as $\Sigma_i x_{ij}$ for $\sum_{i=1}^{n} x_{ij}$, or $\Sigma_{ij} x_{ij}$ for $\sum_{i=1}^{r} \sum_{j=1}^{s} x_{ij}$.

The latter form no longer involves the mean, but only the sum of the observations. If the quantity in the brackets is cleared of fractions, we obtain

$$(e) \qquad s^2 = \frac{n\,\Sigma x_i^2 - (\Sigma x_i)^2}{n(n-1)}$$

This last form is most accurate in the computational sense, since only one final division is involved, although the form (*c*) involves the least operations when the computation of the mean is made in any event.

The quantities that must be obtained from the original data to use the above formulae are n, Σx_i, and Σx_i^2. The last two can be obtained simultaneously on any modern computing machine by accumulating multipliers and products simultaneously in their respective dials, clearing only the keyboard between operations. In case the computations are to be made from grouped data, we can replace n, Σx_i, and Σx_i^2 by Σf_i, $\Sigma x_i f_i$, and $\Sigma x_i^2 f_i$.

The computation of the mean deviation can also be greatly simplified by an algebraic rearrangement of the definitive form given in Section 2.33. Let us designate by $x_{(1)}, x_{(2)}, \cdots, x_{(n)}$ the original n observations arranged in order of increasing magnitude (i.e., $x_{(1)}$ is the smallest observation, $x_{(2)}$ the next smallest, etc.). Then we have

$$x_{(n)} - x_{(1)} = |x_{(n)} - x_m| + |x_{(1)} - x_m|$$

$$x_{(n-1)} - x_{(2)} = |x_{(n-1)} - x_m| + |x_{(2)} - x_m|$$

etc. If n is even, the left-hand side of the last equality will be the difference of the two middle values, whereas, if n is odd, it will be the difference of the two values adjacent to the median. In either case, since $x_m - x_m = 0$, the sum of the right-hand sides is the sum of the absolute values of the deviations from the median, and hence

$$(f) \qquad d_m = \frac{\Sigma(\text{largest half of the observations}) - \Sigma(\text{smallest half})}{n}$$

the median itself being excluded when the number of observations is odd.

The numerical calculations can frequently be simplified in either the ungrouped or the grouped case by the subtraction of an arbitrary constant from all the observations. This does not affect the measures of dispersion since the scatter of the observations is not affected by a change of origin, but the arbitrary constant must be added to any typical value determined from the reduced data. Thus the mean and standard deviation of the numbers 251.4, 252.6, 251.9, 251.7, 252.0, $\cdots$ could be determined from the numbers 1.4, 2.6, 1.9, 1.7, 2.0, $\cdots$, and 250 added to the mean;

the standard deviation of the two sets would be identical. Similarly, all the observations could be increased or reduced by a constant factor, and the results computed from the new values. In this case measures of dispersion would also be affected and would have to be properly adjusted. Data changed in either, or both, of the above fashions, are said to be *coded*. This procedure corresponds to a linear transformation of the data.

In the grouped case, if the cell lengths are equal and one of the cell midpoints is chosen as the arbitrary origin x_0, all the differences $x_i - x_0$ will be divisible by the cell length w. Thus we may in this instance reduce the computational labor involved to a minimum by replacing the x_i by coded values

$$z_i = \frac{x_i - x_0}{w}$$

The mean and standard deviation computed using the z_i would have to be multiplied by w, and the constant x_0 added to the mean to obtain the mean and standard deviation of the original data.

EXAMPLE 1. As an example of the direct calculation of the sample statistics for a set of observations, let us consider the 7 yields in Table 2.1 obtained during the first week. Arranged in order of magnitude they are 69.4, 70.0, 70.3, 70.5, 70.8, 71.6, 73.3. The median is immediately seen to be 70.5, and the range 73.3 — 69.4 = 3.9. To compute the mean, variance, and standard deviation we obtain

$$\Sigma x_i = 69.4 + 70.0 + \cdots + 73.3 = 495.9$$

$$\Sigma x_i^2 = (69.4)^2 + (70.0)^2 + \cdots + (73.3)^2 = 35{,}140.79$$

from which, since $n = 7$,

$$\bar{x}_i = \frac{495.9}{7} = 70.84$$

and, using equation (*e*),

$$s^2 = \frac{7(35{,}140.79) - (495.9)^2}{(7)\ \ (6)} = \frac{68.72}{42} = 1.6362$$

$$s = 1.279$$

By subtracting the constant 70 from each of the observations, we obtain the values—0.6, 0.0, 0.3, 0.5, 0.8, 1.6, 3.3, from which we have

$$\Sigma x_i = (-0.6) + (0.0) + \cdots + (3.3) = 5.9$$

$$\Sigma x_i^2 = (-0.6)^2 + (0.0)^2 + \cdots + (3.3)^2 = 14.79$$

$$\bar{x} = \frac{5.9}{7} = 0.84$$

$$s^2 = \frac{7(14.79) - (5.9)^2}{(6)\ (7)} = \frac{68.72}{42} = 1.6362$$

$$s = 1.279$$

By adding 70 to the mean, we obtain the previous result, and the variance and standard deviation are identical with those previously obtained. When more observations are involved, it may be preferable to choose a constant which is less than all the observations to avoid the confusion of negative numbers.

It is difficult to give a hard and fast rule as to how many significant figures, or decimals, should be retained in the computation of the mean and standard deviation. One procedure is to record the standard deviation to at least 3 significant figures, possibly four in the case of large numbers of observations, and the mean to the number of decimal places thus obtained in the standard deviation. If the data are, as is advisable, recorded to at least the first completely uncertain figure, this will usually imply that the mean should be recorded to at least one and, in the case of large numbers of observations, possibly two more decimal places than the original data. Thus for the above data we should obtain

$$\bar{x} = 70.84$$

$$s^2 = 1.6362$$

$$s = 1.28$$

the retention of the fourth decimal in the variance being questionable.

The mean deviation computed using (f) is found to be

$$d_m = \frac{73.3 + 71.6 + 70.8 - 70.3 - 70.0 - 69.4}{7} = \frac{6.0}{7} = 0.86$$

EXAMPLE 2. Let us consider the calculation of the mean and standard deviation of the complete data of Table 2.1 from the grouped frequency distribution of Table 2.3. Choosing as an arbitrary origin x_0 the mid-point 71.0, and using the class width 0.5, we obtain the coded values

$$z_i = \frac{x_i - 71.0}{0.5}$$

which will be used in place of the x_i. These values, and the necessary computation to obtain the sums $\Sigma f_i = n$, $\Sigma z_i f_i$, and $\Sigma z_i^2 f_i$, are shown in Table 2.4. Note that the column $z_i^2 f_i$ can be obtained simply by multiplying the preceding column of values of $z_i f_i$ by the values of z_i.

TABLE 2.4

	x_i	f_i	z_i	$z_i f_i$	$z_i^2 f_i$
	67.0	2	−8	−16	128
	67.5	2	−7	−14	98
	68.0	5	−6	−30	180
	68.5	6	−5	−30	150
	69.0	7	−4	−28	112
	69.5	24	−3	−72	216
	70.0	36	−2	−72	144
	70.5	48	−1	−48	48
	71.0	64	0	0	0
	71.5	51	1	51	51
	72.0	41	2	82	164
	72.5	32	3	96	288
	73.0	24	4	96	384
	73.5	12	5	60	300
	74.0	5	6	30	180
	74.5	4	7	28	196
	75.0	1	8	8	64
Totals		364		141	2703
		$n = \Sigma f_i$		$\Sigma z_i f_i$	$\Sigma z_i^2 f_i$

Using the totals (frequently called power sums) so obtained, we have for the mean and variance of the coded values

$$\bar{z} = \frac{141}{364} = 0.3874$$

$$s_z^2 = \frac{(364)(2703) - (141)^2}{(364)(363)}$$

$$= \frac{964{,}011}{132{,}132}$$

$$= 7.295818$$

$$s_z = 2.701$$

and thus for the original values

$$\bar{x} = x_0 + w\bar{z}$$

$$= 71.0 + (0.5)(0.3874)$$

$$= 71.194$$

$$s_x = ws_z = 0.5(2.701) = 1.351$$

2.36. Interpretation of Mean and Standard Deviation

If the number of observations is fairly large (say, ≥ 50) and the distribution of the observations is approximately normal, i.e., the cumulative distribution is approximately a straight line on probability paper, the following interpretation of the mean and standard deviation is possible with respect to the frequency of occurrence of the various values of the observations:

The interval $\bar{x} \pm s$ will contain approximately $^2/_3$ of the observations.

The interval $\bar{x} \pm 2s$ will contain approximately $^{19}/_{20}$ of the observations.

The interval $\bar{x} \pm 3s$ will contain approximately $^{997}/_{1000}$ of the observations.

As an example of the first situation, let us consider the observations of Table 2.1. We have already noted in preparing the cumulative polygon on probability paper that the distribution is roughly normal. The mean and standard deviation are 71.19 and 1.35, respectively. Thus the interval $\bar{x} \pm s$ is from 69.84 to 72.54, $\bar{x} \pm 2s$ from 68.49 to 73.89, and $\bar{x} \pm 3s$ from 67.14 to 75.24. The relative frequencies of observations within these intervals as determined from the ungrouped frequency distribution of Table 2.2 are 0.714, 0.945, and 0.997, respectively, in good agreement with the figures given. The mean $\bar{x}$ and these limits are indicated on the histogram of Figure 2.3.

2.4. Limits for Means and Medians

2.41. Variance of Means

Let us suppose that an analytical chemist has analyzed a sample for nitric acid 5 times and obtained the following results (g/l): 41.03, 39.61, 39.91, 40.03, 39.54. Following the line of thought in the preceding sections, he might compute the mean of these data, obtaining $\bar{x} = 40.02$, or he might compute the median and get $x_m = 39.91$. These results in themselves mean nothing except as they reflect the thing which he really wants to know, that is, the true nitric acid content of the sample. As we have indicated, his next step would be to compute some measure of the scatter of the results, since he feels intuitively that this scatter will be indicative of the closeness of his mean or median to the true value. He then comes to the final question: How does one use this measure of the scatter to compute limits within which the true value might reasonably be expected to fall?

Two things should be noted at this point. The first is that we can have no assurance that the typical values obtained will really reflect the "true value" of the nitric acid concentration in the sample. Whether this is true for the case in question cannot be determined from these data. The best we can do statistically is to use the available information to set limits which may reasonably be expected to contain the "average value" that would have been obtained if the chemist had had sufficient time, money, and sample to make a very large number of similar analyses. For any particular method of analysis this "average value" is not necessarily the true value desired. The second point is that, in trying to place limits on the "average value" which would have been obtained from a large number of similar analyses, we are making a statistical inference; that is, we are using our knowledge of the data at hand to draw conclusions about the properties of a larger mass of data of which it is representative.

Let us consider the results from the large number of similar analyses which might have been performed. Had these values been available, we could have computed their mean and variance, which we will designate by μ and σ^2, respectively. These quantities will be defined in more detail in Section 3.3. We wish our limits to have some reasonable chance of containing the quantity μ. Now let us go back to the value of $\bar{x}$ which we obtained from the observations. Since we know that the individual observations varied from the value of μ to an extent indicated by the value of σ^2, we should expect that the values of $\bar{x}$ computed from these observations would also differ from μ. However, we should expect intuitively that under most circumstances this variation would be less than that of the original observations. This is shown to be true in Section 3.42, where we find that, under certain conditions concerning the manner in which the observations are selected, the variance of the means of a large number of similar groups each containing 5 observations would be expected to be $\sigma^2/5$, or, more generally, σ^2/n if each group contained n observations. Given this knowledge of the expected scatter of means, we are one step closer to being able to place the limits desired about our typical value.

2.42. Degrees of Freedom

We have now established the variance of the mean of the observations, but in terms of the variance σ^2 rather than the estimate s^2 computed from the 5 observations which are available. We did not expect the value of $\bar{x}$ to be equal to μ, nor should we expect the value of s^2 to be exactly equal to σ^2.

It should be noted that we defined σ^2 as the variance which would have been computed if a very large number of similar analyses had been available. In this case the deviations would have been computed from the average value μ of all these analyses. However, for a given series of observations we do not know μ, but only $\bar{x}$; hence in computing s^2 the deviations were measured from $\bar{x}$, and not from μ. This use of $\bar{x}$ in place of μ has two consequences. The first is due to the fact that, as we showed in Section 2.33, the sum of the squares of the deviations of the observations from $\bar{x}$ is a minimum. Hence if the deviations had actually been taken about μ, their sum would have been greater. This is exactly compensated, on the average, by dividing the sum of the deviations about $\bar{x}$ by $n - 1$ in defining s^2, rather than by n, as might have been expected; it is shown in Section 3.43 that the average value of a large number of s^2's computed using this definition will be σ^2.

A second consequence is more complicated and is fundamental to all problems of statistical inference in which measures of scatter such as s^2 play a part. As we have seen, the degree to which $\bar{x}$ reflected the average value μ as measured by its variation depended directly upon the number of observations which were used in computing it. In a similar fashion the degree to which s^2 will reflect the variance σ^2 will depend on the number of deviations of the observations from μ, used in computing σ^2, which are contained in the computation of s^2. It will be shown in detail in Section 4.53 that the sum of squares of the deviations from $\bar{x}$ used in computing s^2 really contains only 4 deviations of the type used in computing σ^2. Hence the estimate s^2 in this case is only as reliable as it would be if it had been based on the deviations of 4 observations from μ, and is said to be based on 4 degrees of freedom. The variance s^2 computed from n observations as in Section 2.33 will always have $n - 1$ degrees of freedom.

Another way of looking at this situation is to note that the 5 deviations about the mean $\bar{x}$ are not independent since, as we have seen in Section 2.33, the sum of these deviations must be zero. This is equivalent to saying that we have imposed a linear restriction on the observations. In general, any linear restriction which we impose on the observations, or any linear function of the observations which is used in determining the deviations from which s^2 is to be computed, will result in the loss of a degree of freedom. In this sense the use of the term degrees of freedom is similar to its use in connection with the phase rule of physical chemistry. However, it is preferable for problems of statistical inference to consider the number of degrees of freedom simply as a measure of the reliability of s^2 when used as a replacement for σ^2.

2.43. Limits for the Mean

We are now in a position to determine limits which may reasonably be expected to contain the value μ, provided that we define what we mean by the term "reasonably expected." Statistical inferences are usually made in terms of probability, which will be the subject of the next chapter.

In this case we shall substitute for "reasonably expected" the fact that the chances are about 5 in 100 that the value of μ will fall outside the limits given, and about 95 in 100 that it will fall within them. Such limits are called 95% confidence limits, i.e., we shall be correct about 95% of the time if we say that μ falls within these limits. Alternatively, we can think in terms of taking a 5% risk of making an incorrect statement. As we shall see later we could just as easily have defined "reasonably expected" to mean that we would be right about 99 times in 100 and wrong about 1 time in 100.

Having defined the chances we are willing to take, the limits are then given by $\bar{x} \pm ts/\sqrt{n}$, where t is a constant depending on the number of degrees of freedom on which s^2 is based and the risk we wish to take. Actually two steps are involved here: (1) We obtain an estimate of the standard deviation of the mean $\bar{x}$ by replacing σ^2 by s^2, dividing the latter by n to obtain an estimate of the variance of $\bar{x}$ and taking the square root of this quantity. (2) We multiply this by a constant which has been obtained in such a fashion that it gives the appropriate limits. This constant (called the "Student" t) takes into account both the possible variation of the value of $\bar{x}$ from μ based on its expected variance of $\sigma^2/5$, and the reliability of the use of s^2 in place of σ^2. The details of the derivation of the "Student" t will be given in Section 4.54; particular values for various degrees of freedom and risks (expressed as fractions rather than %) are given in Table III in the Appendix in the back of the book. The value of t, as might be expected, depends on the degree of confidence which we wish to place on our limits and on the reliability of s^2 as a replacement for σ^2. In determining these values it is necessary to assume that the large number of possible observations from which μ and σ^2 would have been computed have the normal frequency distribution previously referred to. Fortunately moderate departures from this form do not seriously affect our probability statements, and so long as we are content if our actual chances are 6 or 7, instead of 5 in 100, this assumption presents no serious difficulty.

EXAMPLE 1. For the data given in Section 2.41 we have $\bar{x} = 40.02$, $s^2 = 0.3577$, $s = 0.60$. From Table III we obtain for 4 degrees of freedom and a probability of 0.05 (5% risk) a value $t = 2.78$. Hence in this case the desired limits are given by

$$40.02 \pm (2.78)\frac{0.60}{\sqrt{5}} = 40.02 \pm 0.75$$

EXAMPLE 2. (Wernimont [74].) Given below in order of magnitude are 10 determinations by the Karl Fischer method of the percentage of water in a methanol solution: 0.50, 0.52, 0.53, 0.54, 0.55, 0.55, 0.56, 0.56, 0.57, 0.64.

For these data $\bar{x} = 0.552$, and $s = 0.037$ based on 9 degrees of freedom. Using in this case 1% risk (probability of 0.01) we obtain from Table III $t = 3.25$. Hence 99% confidence limits are given in this case by

$$0.552 \pm (3.25)\frac{0.037}{\sqrt{10}} = 0.552 \pm 0.038$$

2.44. Limits for the Median

In the preceding sections we have developed limits about the mean of the observations which will, with a certain degree of confidence, contain the mean μ which would have been obtained if a large number of similar analyses had been available. This has been done at some length to show the nature of the problems that arise in making inferences of this type, and the assumptions that must be made in order to obtain a reasonable answer. In this section we shall give briefly another type of argument which will lead to limits containing, with a certain confidence, the median value of the larger number of similar analyses which might have been obtained.

Let us suppose that we take a coin which is equally likely to come up heads or tails and toss it 5 times. The chances of obtaining various numbers of heads and tails in such an experiment are given by the binomial probability distribution, to be discussed in Section 4.61; it is sufficient for our present purposes to state that there are 32 possible outcomes (as the reader can easily confirm by writing down all possible sequences of heads and tails for the 5 tosses) of which 2 correspond to "all heads" or "all tails." Thus the chances are 1 in 16, or about 6.3%, of obtaining all heads or all tails in 5 tosses, assuming every outcome to be equally likely.

Now let us see how we can proceed from coin tossing to the desired limits. The median value we wish our limits to contain is such that any possible observation has an equal chance of falling above or below it. Hence the 5 analyses actually made can be considered as 5 tosses, where "heads" implies "above the median of possible similar analyses," and "tails," "below the median of possible similar analyses." Obtaining 5 analyses above the median value would correspond to tossing 5 heads, and obtaining 5 below the median to 5 tails. Thus the chances that the median value is below the lowest observed analysis, or above the highest observed analysis, are 1 in 16; it follows that the chances that the median value is *between* the highest and lowest values of the 5 observations are 15 in 16. We can therefore state, with a confidence of about 93.7%, or a risk of 6.3%, that the median is between 39.61 and 41.03 g/l.

Several things should be noted about the above procedure. The first is that it did not involve the sample median, as defined in Section 2.32, at all; only the "median of a large number of similar analyses." The distinction between these, which is similar to the distinction between $\bar{x}$ and μ in the preceding section, will be discussed more fully in Chapter 3. The second is that the median value which these limits have a given chance of containing is not necessarily the same as the mean value μ for which limits were previously obtained; which of these values the chemist is really interested in depends on his previous knowledge or insight concerning the distribution of similar analyses. We shall see later that if he is willing to assume that this distribution is symmetric, and the possible number of similar analyses sufficiently large, these values will coincide. The third point is that the risk is in this case predetermined by the number of analyses, or tosses, rather than freely chosen by the chemist determining the limits. This difficulty can be resolved to some extent as shown in the following example. A more detailed discussion will be given in Section 5.34.

EXAMPLE. Let us consider the 10 analyses of Example 2 of the preceding section. In 10 tosses of a coin, the chances of obtaining all heads or all tails can be determined to be 2 in 1024, or about 0.2%. Thus we would have about 99.8% confidence that the median analysis was between 0.50 and 0.64. The chances of 9 or more heads or 9 or more tails are about 22 in 1024, or about 2.1%. By an argument identical with that given above, this means that we have about 97.9% confidence that the median analysis will fall between the next to smallest and next to largest values, i.e., between 0.52 and 0.57. In this case approximate 99% limits could be obtained by interpolating between the smallest and next to smallest, and largest and next to largest, values on the basis of the confidence we have in the limits that they determine, but the practical value of such a procedure is questionable.

2.5. Associated Measurements

2.51. Pairs of Measurements. Scatter Diagrams

In many instances one or more additional measurements or observations may be associated with a given series of observations. For example, along with a series of heights of individuals we might record their weights, reach, etc., or, along with the yield in a series of experiments, we might record the temperature at which the reaction was carried out, the concentration of the reactants used, and other similar data. These sets of associated measurements may be of interest in themselves, but we are more frequently interested in the relationships between them. The study of these relationships will be treated in full in Chapter 6; we shall only give a brief introduction to the problem here, restricting ourselves to pairs of measurements.

An example of a set of raw data consisting of pairs of measurements is given in Table 2.5. These data, reported by Tippett [73], are the percentages of pig iron in the charge to an open hearth furnace during a number of heats, together with the lime consumption during each operation, the latter being recorded as cwt (112 lb) per cast. Since the lime additions are made, in part, to control the oxidation of metalloids in the charge, and since the latter are introduced almost exclusively by the pig iron, it is reasonable to suppose that the two factors, lime additions and percentage pig, are related. On the other hand, the pig iron may vary in composition from cast to cast, and the lime is required to serve a number of other purposes so that no exact type of relationship would be anticipated.

TABLE 2.5

PERCENTAGE PIG AND LIME CONSUMPTION, CWT PER CAST, IN STEEL MAKING WITHOUT SLAG CONTROL

Pig	Lime	Pig	Lime	Pig	Lime	Pig	Lime
43	210	43	212	30	182	50	196
38	145	37	182	42	246	44	215
42	182	37	170	42	194	34	161
36	146	38	175	35	176	38	164
50	158	40	223	37	194	33	228
50	195	45	219	32	185	49	204
40	138	47	220	36	180	48	241
37	126	47	274	46	205	36	195
48	170	36	124	45	187	37	198
38	157	47	310	35	146	53	188
35	156	47	197	29	156	32	159
45	176	33	138	37	176	30	165
37	140	42	213	30	184	33	170
33	155	49	193	26	140	53	219
35	165	36	132	47	218	52	262
48	242	33	192	47	206	53	240
49	206	25	141	31	172	52	208
40	241	30	178	44	176	53	193
42	166	39	201	45	216	53	170
41	212	48	205	47	218	36	176
30	177	35	193	45	219	37	191
40	200	38	281	35	133	50	198
46	235	35	194	47	193	36	201
37	216	39	190	23	164	45	184
45	194	43	207	52	219	38	225

One of the first ways of treating measurements of this type is to prepare a scatter diagram such as that given in Figure 2.6. This merely consists of a graph on which are plotted the points (x, y) corresponding to the pairs of measurements. The resulting picture gives us a graphical indication of the range of the measurements in both the x and the y

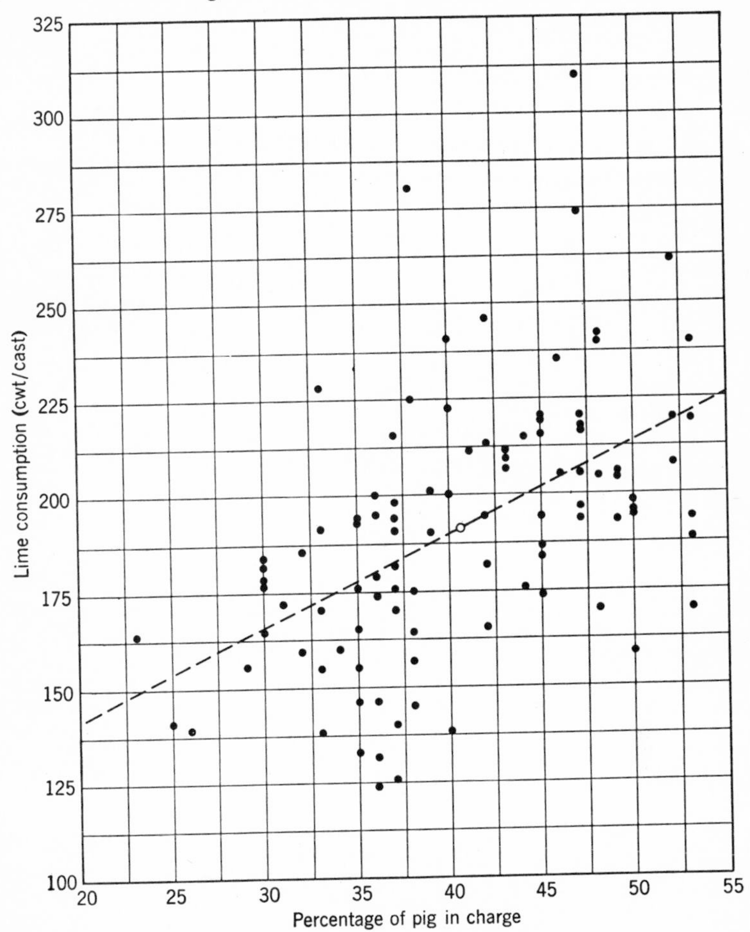

FIG. 2.6. Scatter diagram for data of Table 2.5.

directions, the point or points of concentration of the measurements, and of any dependence which may exist. In the case of pairs of measurements there is no direct parallel to the preparation of an array, although in some instances the pairs are most conveniently arranged in ascending order of one of the two measurements. Thus, for instance, if one of the pairs of observations were the time at which a given event occurred, it would be

TABLE 2.6

Lime Consumption \ % Pig	Class Boundaries	22.5–27.5	27.5–32.5	32.5–37.5	37.5–42.5	42.5–47.5	47.5–52.5	52.5–57.5	Total
Class Boundaries	Class Midpoints	25	30	35	40	45	50	55	
287.5–312.5	300					1			1
262.5–287.5	275				1	1			2
237.5–262.5	250				2		3	1	6
212.5–237.5	225			2	3	8	1	1	15
187.5–212.5	200			8	5	8	8	2	31
162.5–187.5	175	1	7	8	4	4	1	1	26
137.5–162.5	150	2	2	7	3		1		15
112.5–137.5	125			4					4
Total		3	9	29	18	22	14	5	100

natural to record the data in time order, or, if we were deliberately studying the effect on product yield of increased temperature, it would be natural to record the pairs of measurements in order of increasing temperature.

Where it may be desirable, we can prepare a frequency distribution for pairs of measurements in a similar fashion to that used for one measurement. Such a tabulation for the data of Table 2.5 is shown in Table 2.6, where the observed percentages of pig (x) have been grouped into equal cells of length 5 and the corresponding measurements of lime consumption (y) have been grouped into equal cells of length 25. In this case the frequency in any particular square cell denotes the number of pairs of measurements whose x measurement is one of a given group of x values, and whose y measurement is one of a given group of y values. The columns marked "Total" at the right-hand side and the bottom of the table represent the total of all pairs whose x measurement falls in a given x cell and whose y measurement falls in a given y cell, respectively. They represent the distributions that would have been obtained if we had considered the x measurements and the y measurements separately, and are referred to as marginal distributions. Because such tables were frequently prepared in order to compute the correlation coefficient, to be defined in the following section, Table 2.6 may be called a correlation table.

2.52. Covariance

Although the statistics previously discussed are applicable to either or both series of measurements separately, they give no information concerning the possible relationship between pairs. The statistic most frequently computed to reflect this information is the *covariance*, which is defined as

$$(a) \qquad s_{xy} = \frac{\Sigma(x_i - \bar{x})(y_i - \bar{y})}{n - 1}$$

where a particular pair of measurements is designated by (x_i, y_i) and $\bar{x}$ and $\bar{y}$ are the means of the x measurements and y measurements, respectively. This differs very little in form from the definition of the variance, the only change being that the cross product of the deviations of the paired measurements from their respective means is used in place of the square of the deviations. It is easily seen from the definition how this statistic might reflect the presence of a relationship since the cross product of the deviations will tend to be large in absolute value whenever large x deviations are either directly or inversely related to large y deviations.

For numerical computation of the covariance, it is easier to use the equation

$$(b) \qquad s_{xy} = \frac{1}{n - 1}\left[\Sigma x_i y_i - \frac{(\Sigma x_i)(\Sigma y_i)}{n}\right]$$

or

(*c*) $$s_{xy} = \frac{n\,\Sigma x_i y_i - (\Sigma x_i)(\Sigma y_i)}{n(n-1)}$$

which can be obtained algebraically from (*a*) by methods similar to those used for the variance in Section 2.35. The cross products $x_i y_i$ can be accumulated on a modern calculating machine almost as easily as the squares. If a frequency distribution has been prepared, we can simply replace Σx_i, Σy_i, and $\Sigma x_i y_i$ by the approximate quantities $\Sigma x_i f_i$, $\Sigma y_j f_j$, and $\Sigma_{ij} x_i y_j f_{ij}$, where the x_i and y_j are the midpoints of the x and y cells, f_i and f_j their respective marginal frequencies, and f_{ij} the frequency of a particular cell. The first two of these can be computed from the marginal frequency distributions by the methods given in Section 2.35. The last can most easily be computed by first obtaining the $\Sigma_{ij}(x_i - y_j)^2 f_{ij}$, i.e., multiplying each cell frequency by the square of the difference of the x and y midpoints. Then we can obtain the desired cross product from the algebraic identity

(*d*) $$\Sigma_{ij}(x_i - y_j)^2 f_{ij} = -2\Sigma_{ij} x_i y_j f_{ij} + \Sigma_i x_i^2 f_i + \Sigma_j y_j^2 f_j$$

$$\Sigma_{ij} x_i y_j f_{ij} = \tfrac{1}{2}\,[\Sigma_i x_i^2 f_i + \Sigma_j y_j^2 f_j - \Sigma_{ij}(x_i - y_j)^2 f_{ij}]$$

since the quantities $\Sigma_i x_i^2 f_i$ and $\Sigma_j y_j^2 f_j$ will have been obtained from the marginal distributions in order to compute the variances of x and y.

2.53. Regression and Correlation

The use which we make of this measure of the relationship between the pairs of observations depends on what we wish to know. From the data of Table 2.5, for example, we might be interested in determining the increase (or decrease) in lime consumption which we would expect to accompany a given increase in the percentage of pig. To obtain this we can compute the *linear regression coefficient* of y on x, which is defined as

(*a*) $$b_{y\cdot x} = \frac{s_{xy}}{s_x^2}$$

$$= \frac{n\,\Sigma x_i y_i - (\Sigma x_i)(\Sigma y_i)}{n\,\Sigma x_i^2 - (\Sigma x_i)^2}$$

the latter form being the more convenient for direct computation. This is actually an estimate of the best (in a sense which will be considered in more detail in Chapter 6) linear relationship between the deviations of y and x from their respective means; the line

(*b*) $$y - \bar{y} = b_{y\cdot x}(x - \bar{x})$$

is called the *regression line* of y on x, and is in the same sense the best straight line for the purpose of computing the y value corresponding to a given x. Since the regression coefficient in this case is the slope of this line, it is frequently referred to merely as the *slope*. The use of the term regression in this connection is due to an article published in 1886 by Sir Francis Galton in which he used the term "regression toward mediocrity" to describe the fact that in a certain study the deviations in stature of sons from the mean of all sons were less than the deviations of their fathers from the mean of all fathers.

We may also be interested in using the covariance as a measure of the degree of relationship between the two variables. Again using the data of Table 2.5, we can ask the question: "To what extent are the variations in percentage of pig the cause of the variations in lime consumption?" To help answer questions such as these, Karl Pearson first used as a measure of the degree of linear dependence of two variables the coefficient of correlation, which is defined by the equation

$$r_{xy} = \frac{s_{xy}}{s_x s_y}$$

$$(c) \qquad = \frac{n\,\Sigma x_i y_i - \Sigma x_i\,\Sigma y_i}{\sqrt{\{n\,\Sigma x_i^2 - (\Sigma x_i)^2\}\,\{n\,\Sigma y_i^2 - (\Sigma y_i)^2\}}}$$

Actually, this is simply the covariance converted to a relative measure by division by the appropriate scale factors. For our purposes, the square of the correlation coefficient, r_{xy}^2, is more directly applicable, since it reflects that part of the variation in one set of measurements which can be explained by their dependence on the other.

Problems involving regression are most common in the physical sciences, since the form of the equation relating several variables is frequently known, and the problem is that of determining the necessary constants. In this section we have considered only simple linear regression; Chapter 6 will discuss this in more detail, and extend the ideas to curvilinear regressions, regression on more than two variables, and the study of more complicated relationships. The simple linear correlation coefficient defined above, and various other types of correlation coefficients which can be defined in the case of more than two variables, are much used in psychological and educational work. Here the degree of relationship, or the existence of a relationship, between two variables such as intelligence and reading ability may be the only matter in question.

EXAMPLE. For the data of Table 2.5, we obtain

$$n = 100$$

$$\Sigma x_i = 4054 \qquad \Sigma y_i = 19{,}104$$

$$\Sigma x_i^2 = 169{,}430 \qquad \Sigma y_i^2 = 3{,}764{,}318$$

$$\Sigma\, x_i y_i = 786{,}668$$

Using the methods of Section 2.35, we can compute

$$\bar{x} = 40.54 \qquad \bar{y} = 191.04$$

The variances and covariances, s_x^2, s_y^2, and s_{xy}, can be computed if desired, the variances as in Section 2.35 and the covariance from equation (*c*) of Section 2.52; however, their actual values are of no direct concern in this case. From equation (*a*) we compute

$$b_{y \cdot x} = \frac{100(786{,}668) - (4054)(19{,}104)}{100(169{,}430) - (4054)^2}$$

$$= \frac{1{,}219{,}184}{508{,}084}$$

$$= 2.400$$

This means that if we assume that the relationship is linear, we can expect a unit increase in the percentage of pig to be accompanied on the average by a 2.400-cwt increase in lime consumption. The linear relationship between lime consumption and percentage of pig is, using (*b*),

$$y - 191.04 = 2.400(x - 40.54)$$

The line corresponding to this equation is shown on the scatter diagram of Figure 2.6.

By squaring equation (*c*), we can compute r_{xy}^2 directly as

$$r_{xy}^2 = \frac{(1{,}219{,}184)^2}{(508{,}084)(11{,}468{,}984)} = 0.2551$$

and hence $r_{xy} = 0.505$. This means that approximately 25% of the variations in lime consumption can be explained by a linear dependence on the percentage of pig used; the remaining 75% must be due to other factors affecting lime consumption.

2.54. Limits for the Slope

The open-hearth technician who finds that he may expect the lime consumption to increase 2.400 cwt for each percent increase in pig, will probably wish to know how well this value reflects the true situation. If we are willing to make certain assumptions concerning the nature of the true situation, we can obtain confidence limits for the true slope in much the same fashion as we obtained limits for μ in Section 2.43.

Suppose that we had determined the lime consumption in a very large number of heats with a fixed percentage of pig. The lime consumption would vary, due to the many factors other than the percentage of pig which affect it, and we should obtain a distribution of the amounts of lime consumed. From these values we could compute an average value μ and a variance σ^2 as in the case of the large number of similar analyses which the chemist could have performed. In a similar fashion we could obtain an average value μ and a variance σ^2 for any fixed percentage of pig. We now assume that: (1) the variance σ^2 is the same for all percentages of pig, (2) the values of μ, when plotted against the percentage of pig, fall exactly on a line of slope β. It is this unknown β for which we wish to obtain limits.

The first step, as before, is to determine how much $b_{y\cdot x}$ can be expected to vary from β. It will be shown in Section 6.23 that, if, corresponding to the given set of x values, we repeatedly choose new sets of values for y from the observations available at each fixed x, then the values $b_{y\cdot x}$ computed from these new series of y values will scatter about β with a variance $\dfrac{\sigma^2}{\Sigma(x_i - \bar{x})^2}$. Next we need to replace σ^2 by an estimate computed from the available observations. Since σ^2 represents the variation from μ for any fixed x (by virtue of the first assumption), and since all values of μ fall on a line, it would be desirable to compute this estimate, using deviations of the y observations from this line. As we do not know the position of this line, we do the next best thing and compute the variance, using deviations of the y observations from the regression line given by (*b*) of the preceding section. Note the similarity between this operation and the replacement of μ by $\bar{x}$ in Section 2.42. In this way we obtain an estimate of σ^2 which is given by

$$(a) \qquad s_{y\cdot x}^2 = \frac{1}{n-2}\left[\Sigma y_i^2 - a\,\Sigma y_i - b\,\Sigma x_i y_i\right]$$

where $a = \bar{y} - b\bar{x}$ is the intercept of the computed regression line. This estimate has $n - 2$ degrees of freedom; i.e., it is as reliable as one obtained using $n - 2$ deviations from the unknown exact line.

The limits which have a given chance of containing β are now given by

$$(b) \qquad b_{y\cdot x} \pm t\,\frac{s_{y\cdot x}}{\sqrt{\Sigma(x_i - \bar{x})^2}}$$

where t is determined in exactly the same fashion as in Section 2.43 except that we now use $n - 2$ instead of $n - 1$ degrees of freedom. Also, as in that section, we are assuming the variations of the y observations from the unknown line to be approximately normal.

EXAMPLE. Referring to the example of the preceding section, where we had $b_{y\cdot x} = 2.400$, we obtain, using (*a*),

$$a = 191.04 - (2.400)(40.54)$$
$$= 93.744$$

$$s_{y\cdot x}^2 = \frac{3{,}764{,}318 - (93.744)(19{,}104) - (2.400)(786{,}668)}{98}$$
$$= \frac{85{,}429}{98}$$
$$= 871.72$$

Using the identity

$$\Sigma(x_i - \bar{x})^2 = \frac{n\,\Sigma x_i^2 - (\Sigma x)^2}{n}$$

we obtain from the computations of the previous example

$$\Sigma(x_i - \bar{x})^2 = 5080.84$$

and hence

$$\frac{s_{y\cdot x}}{\sqrt{\Sigma(x_i - \bar{x})^2}} = \sqrt{\frac{871.72}{5080.84}}$$
$$= \sqrt{0.171570}$$
$$= 0.414$$

From Table III of the Appendix the value of t for 98 degrees of freedom and a 5% risk is found to be 1.99. The 95% confidence limits for β are then

$$2.400 \pm (1.99)(0.414) = 2.400 \pm 0.824$$

i.e., we have a 95% confidence that, if our assumptions concerning the true situation are reasonably correct, the true change in lime consumption for a one percent increase in pig is between 1.58 and 3.22 cwt.

CHAPTER 3

Probability and Samples

3.1. Introduction

In the past chapter we have considered some of the methods used in describing, or summarizing, the information contained in a given set of data, and several specific instances were considered in which this information was used to draw inferences concerning a larger number of similar data which might have been obtained.

In using our *sample* of observations or measurements to make inferences concerning a *population* of similar observations or measurements, we are faced, as in the examples in Chapter 2, with the necessity of knowing something about the expected behavior of a sample (or statistics computed from it) from a given population. One method of accomplishing this is to take repeated samples from a given population, or from a close approximation to it, and then study the behavior of the statistics computed from these samples. We might, for example, study the distribution of the number of aces obtained in dealing a sample of 13 cards from a bridge deck of 52 cards, by simply dealing a large number of such samples, shuffling repeatedly between each deal, or mixing the cards in some other manner, to ensure "randomness" in our sampling procedure. Similarly, in Chapter 2 we might have determined the frequency of occurrence of 5 heads or 5 tails simply by tossing 5 coins a sufficient number of times. The difficulties in using this method are the time and cost involved in making studies sufficiently large to enable us to determine sampling behavior accurately, and the problem of avoiding non-randomness of some sort in the sampling procedure. Nevertheless, with the present developments in the field of high-speed computational machinery, the method is being used to determine sampling distributions in situations where mathematical methods are impractical or even impossible.

The mathematical method of studying sampling behavior can be described briefly as follows: We assume that our population can be represented by a *distribution function* by means of which we can make certain probability statements concerning the frequency of occurrence of values chosen "at random" from the population. Then by the use of the theory of probability we can obtain mathematically similar probability statements concerning the values of statistics computed from "random samples" from such a population. For example, if we assume that a given sample of observations is from a normally distributed population, we can show (Section 4.51) that the means of samples of size n will also be normally distributed. This fact, along with a knowledge of their average value and variance, enables us to make probability statements concerning the means of such samples.

The above procedure is certainly not peculiar to statistics. We have all, in elementary physics, taken the mathematical model $d^2s/dt^2 = k$ to represent the motion of a body under constant acceleration, and then used the elementary rules of calculus to obtain the equation $s = \frac{1}{2}kt^2$ for the distance traveled by such a body in time t starting from rest. Most of us have also obtained essentially the same result by observing the time necessary for a ball starting from rest to roll certain distances down an inclined plane, and have then shown by graphical analysis that the distance traveled was very nearly proportional to the square of the time.

Most of the statistical procedures given in later chapters are based on sample behavior derived by mathematical methods. In order that we may use these procedures with discretion, it is essential that we have some understanding of the principles on which they are based. In this chapter we shall describe the elements of the theory of probability, and then consider some elementary sampling theory concerning means and variances which does not depend on the assumption of a particular form of distribution for the population. In Chapter 4 we shall introduce the more complicated mathematical machinery which will be needed, directly or indirectly, in later applications. For a more complete discussion of the theory of probability, the reader is referred to [76] (Chapters 1–5 and 9 are particularly applicable to our present discussion) and to [1]. A complete discussion of sampling theory is given in [77].

3.2. Probability and Random Variables

3.21. Population and Random Variables

The word population is used in statistics to describe any collection of objects of a given type. This collection may be finite or infinite, real or imagined. Thus the population may consist of the 52 cards in a bridge deck, or the items in a particular lot of manufactured articles, or all the samples that could have been taken from a shipment of mixed fertilizer, or all the tosses of a coin that might be made. For the data on coke yields we could consider the population to consist of all possible yields that might be obtained in the future under similar operating conditions, or we could consider each value as coming from the population of yields which might have been obtained that day if the operating conditions or the quality of coal used had been different.

We must assume that the reader has an intuitive feeling for the phrase "selecting an object at random" from a given population. For example, suppose we had a population consisting of a number of ball bearings, all of similar size and shape. If we stir and shake these ball bearings for a long time in a box, and then have someone select one from the box, we should feel that each of the ball bearings had an equal chance of being chosen. Such an experiment would constitute a very close approach to a "random choice" from a given population. The elaborate sampling procedures often used in the chemical industries represent attempts to perform just such a random choice; i.e., we try to make certain that any small sample which might be drawn from the "population" of such samples has an equal chance of occurring in a particular instance.

By a *random variable* we shall mean some observable or measurable characteristic associated with each random choice from a population. For example, the suit of a card drawn at random from a deck of cards, the nitrogen content of a random sample from a shipment of fertilizer, the defectiveness or non-defectiveness of an item drawn at random from a lot, or the occurrence of heads or tails on a single toss of a coin are all random variables. It is possible that more than one random variable may be associated with a given population; for example, in the case of a playing card, we might consider its playing value as well as its suit as a random variable; or in the case of the fertilizer sample, we might also measure the available phosphoric acid content. For mathematical purposes we shall generally wish to think of random variables as numbers; in the deck of cards, for example, we could associate the four suits with the numbers 1, 2, 3, 4 and the values of the cards with the numbers 1,2,3, $\cdots$, 13, counting aces as 1 and kings as 13. Similarly, we can indicate a defect by a 1 and the absence of a defect by a 0; or a head by a 1 and a tail by a 0. Measurements of any sort, such as nitrogen content, are usually recorded as numbers.

3.22. Probability

Suppose that we represent by S a group, or *set*, of numbers. Then by the *probability that a random variable x belongs to the set of numbers S*, written $Pr(x \text{ in } S)$, we shall mean the relative frequency with which the objects associated with these numbers occur in the given population. If the set S consists of the number x_0 only, we shall write $Pr(x = x_0)$;

similarly, if S consists of those numbers less than or equal to x_0, or greater than x_1 and less than or equal to x_2, we shall write $Pr(x \leq x_0)$, or $Pr(x_1 < x \leq x_2)$, respectively.

As an example, consider the population of 52 cards. In this case, if we assign numbers as in the preceding section, we shall have the probability of an ace, or $Pr(x = 1)$, equal to 4/52 or 1/13; similarly, we would have $Pr(x \leq 2)$, or the probability that we would obtain an ace or a deuce in a single random choice equal to 8/52 or 2/13. Suppose that a lot of manufactured articles contained N items of which an unknown number M were defective; then we would have $Pr(x = 1) = Pr$ (item defective) $= M/N$, i.e., the relative frequency of defectives in the lot.

To compute similar probabilities in populations such as the samples from a shipment of fertilizer, or all possible tosses of a coin, a simple evaluation procedure is either impractical or impossible. The desired probability may then be approximated by the following relationship between random choices and probabilities: *Suppose that the process of choosing an object at random from a given population is repeated many times. Then the relative frequency, obtained in this manner, with which the random variable x belongs to the set S will approximate very closely the probability that x belongs to the set S.* Thus, if we perform the experiment of tossing a coin many times, and all these tosses can be considered random choices from the population of possible tosses, then the relative frequency of occurrence of heads will be approximately the probability of obtaining heads on a single toss. Similarly, if a card is chosen at random a great many times from a bridge deck of 52 cards, the relative frequency of aces and deuces should be very close to 2/13. In fact, any considerable deviation of the relative frequency from this value would be considered proof of the non-randomness of the individual choices.

There are certain similarities between the method described above and that of taking the limit of a sequence in ordinary calculus, but it should be noted that in the former case we cannot guarantee that for a sufficiently large number of random choices the relative frequency of values belonging to S will differ from $Pr(x \text{ in } S)$ by less than a given small number. We can only say that it is very unlikely that it will do so; i.e., denoting by n_S/n the relative frequency of values belonging to S in n random choices, we can write

$$\lim_{n \to \infty} Pr\left[\left|\frac{n_S}{n} - Pr(x \text{ in } S)\right| < \epsilon\right] = 1$$

Convergence of this type is known as *stochastic* convergence.

It should be noted carefully that probabilities are defined only with respect to a given population, and hence it is necessary to specify exactly the population with which we are dealing. We might, for example, consider the population consisting of 2 objects, a "head" and a "tail," and thus have a population for which $Pr(x = 0) = 1/2$. We would then expect a series of random choices of one of the two objects to give a relative frequency of heads (1's) of approximately 1/2. In this case we would consider the toss of a coin not as an object drawn at random from a population, but as an experimental means (and often a biased one) of making a random choice between 2 objects.

3.23. Laws of Probability, Independence

Let S_1 and S_2 be any 2 sets of numbers. We shall denote by $S_1 + S_2$ the set of all numbers belonging to either of the 2 sets. Then from the above definition of probability we can state the following: *If two sets of numbers S_1 and S_2 of possible values of x have no numbers in common, then the probability that x belongs to $S_1 + S_2$ is equal to the probability that x belongs to S_1 plus the probability that x belongs to S_2*, i.e.,

$$(a) \qquad Pr(x \text{ in } S_1 + S_2) = Pr(x \text{ in } S_1) + Pr(x \text{ in } S_2)$$

Using the example of the playing card, suppose that we have the $Pr(x = 1) = 1/13$ and $Pr(x = 13) = 1/13$; it follows immediately that $Pr(x = 1 \text{ or } 13) = 2/13$. Similarly with the population of samples from a shipment of fertilizer where x represents the nitrogen content of any random sample, suppose that $Pr(x \leq 7.90) = 0.13$ and $Pr(x \geq 8.15) = 0.05$; then $Pr(x \leq 7.90 \text{ or } x \geq 8.15) = 0.18$.

It will be seen that the above rule can be extended to a group of sets $S_1, S_2, S_3, \cdots$, none of which have any elements in common. In this case

$$(b) \qquad Pr(x \text{ in } S_1 + S_2 + S_3 + \cdots) = Pr(x \text{ in } S_1) + Pr(x \text{ in } S_2) + Pr(x \text{ in } S_3) + \cdots$$

In particular, if the sets $S_1, S_2, S_3, \cdots$ contain all possible values of the random variable x, then

$$(c) \qquad Pr(x \text{ in } S_1 + S_2 + S_3 + \cdots) = 1$$

We now consider 2 random variables x and y associated with some population. By the conditional probability that y belongs to a set of numbers S for a particular value of x, we shall mean the frequency with which the random variable y belongs to the set S in the subpopulation consisting of those objects for which x has the given value. We shall say that y is *independent* of x if the conditional probability that y belongs to a set S is the same as the original probability for the entire population for every possible value of x. For example, let us again consider our deck of cards and let the 2 random variables x and y associated with the population of 52 cards be the suit of cards (we take spades $= 1$, hearts $= 2$, diamonds $= 3$, and clubs $= 4$) and the value of the cards. Now if we consider the subpopulation for which $x = 1$, i.e., spades only, we find that the probability of drawing an ace in a random choice from these 13 cards is still 1/13, and similarly for $x = 2$, 3, and 4. Hence in this case y is independent of x. It can be shown that, if y is independent of x, x is independent of y, so that we usually speak of x and y as being independent. What we really mean by independence is that a knowledge of x does not give us any information concerning the possible value of y, and vice versa.

If 2 random variables x and y are independent, we may state the following: *Let S_x be a set of possible values of x, and S_y a set of possible values of y. Then, if an object is chosen at random from the original population, the probability that both x will belong to S_x, and y will belong to S_y, is equal to the product of the probability that x will belong to S_x and the probability that y will belong to S_y*, i.e.,

$$(d) \qquad Pr(x \text{ in } S_x, \quad y \text{ in } S_y) = Pr(x \text{ in } S_x) \cdot Pr(y \text{ in } S_y)$$

As illustration we return to our deck of cards. The probability that $x = 1$, i.e., that the card drawn is a spade, is 13/52 or 1/4. The probability that $y = 1$, i.e., that the card drawn is an ace, is 1/13. Then the probability of drawing the ace of spades, i.e., of obtaining a card which is simultaneously an ace and a spade, would be $1/4 \times 1/13 = 1/52$. This obviously agrees with the probability that we would have obtained for the ace of spades from our original definition.

As another example, suppose that we have established, by taking a large number of samples from our shipment of fertilizer, that the probability of obtaining a sample with nitrogen content greater than 8.15% is approximately 0.05, and that we also know that the probability of obtaining a ratio of available to total phosphoric acid greater than 0.965 is approximately 0.10. Then, if we make the reasonable assumption that there is no relationship between the nitrogen content and this ratio, we would obtain a probability of 0.005 that a given sample would show both a nitrogen content greater than 8.15% and a ratio greater than 0.965.

3.3. Average Values and Variances

3.31. One Random Variable

Given a population and an associated random variable x, let us designate by $x_1, x_2, \cdots, x_N$ the possible values of the random variable, and by (x_i) their probability of occurrence. Note that it follows from (c) of Section 3.23 that $\Sigma p(x_i) = 1$. We shall consider here only the case where the possible values of the random variable are discrete (although possibly infinite, i.e., we may have $N = \infty$); if the random variable can take on all values on some continuous scale of measurement, it will be shown in Chapter 4 that we need only replace the probabilities by a probability density, and the process of discrete summation by that of integration. In practice, measurements and observations are always discrete, since even if they are measured on a continuous scale they must be recorded to a certain number of decimal places, but for mathematical purposes it is frequently convenient to consider continuous variables.

For the given population the average value, or mean, is defined by

$$(a) \qquad \mu = \text{ave}\,(x) = \Sigma x_i p(x_i)$$

and the variance by

$$(b) \qquad \sigma^2 = \text{var}\,(x) = \text{ave}\,(x - \mu)^2 = \Sigma(x_i - \mu)^2 p(x_i)$$

Note the similarity between these definitions and the methods used in Chapter 2 to compute the mean and variance from a frequency distribution; we have now replaced relative frequencies by probabilities, and summed over all possible values rather than a selected group.

As an example, consider the value of the cards in a bridge deck. In this case each of the values 1, 2, 3, $\cdots$, 13 occurs with probability 1/13. Hence for this population

$$\begin{aligned} \text{ave}\,(x) &= 1 \cdot 1/13 + 2 \cdot 1/13 + \cdots + 13 \cdot 1/13 \\ &= 91/13 = 7 \end{aligned}$$

and

$$\begin{aligned} \sigma^2 &= \text{ave}\,(x - \mu)^2 \\ &= (-6)^2 \cdot 1/13 + (-5)^2 \cdot 1/13 + \cdots + 5^2 \cdot 1/13 + 6^2 \cdot 1/13 \\ &= 182/13 = 14 \end{aligned}$$

The standard deviation of the population is

$$\sigma = \sqrt{14} = 3.74$$

In the discrete case the average value of x does not necessarily have to be a value of x which can occur; for instance, if x can take on only the two values 0 and 1, each with probability 1/2, then the average value will be $0 \cdot 1/2 + 1 \cdot 1/2 = 1/2$. Also the average value need not be the most probable, or most "likely" value.

From (b) we have

$$\begin{aligned} \sigma^2 &= \Sigma(x_i - \mu)^2 p(x_i) \\ &= \Sigma(x_i^2 - 2\mu x_i + \mu^2)\, p(x_i) \\ &= \Sigma x_i^2 p(x_i) - 2\mu\, \Sigma x_i\, p(x_i) + \mu^2\, \Sigma p(x_i) \\ &= \text{ave}\,(x^2) - 2\mu^2 + \mu^2 = \text{ave}\,(x^2) - \mu^2 \end{aligned}$$

since $\mu = \Sigma x_i p(x_i)$ and $\Sigma p(x_i) = 1$. From this we obtain

(c) $$\text{ave}\,(x^2) = \sigma^2 + \mu^2$$

Also it can easily be seen that if c is any constant

(d) $$\text{ave}\,(c) = c$$

(e) $$\text{ave}\,(cx) = c\ \text{ave}\,(x)$$

3.32. Two Random Variables

We now consider two random variables x and y, and designate by x_i, $i = 1, 2, \cdots, N_x$, the possible values of x; and by y_j, $j = 1, 2, \cdots, N_y$, the possible values of y. Let $p(x_i, y_j)$ be the probability of occurrence of any pair (x_i, y_j); as before we must have $\Sigma_{ij} p(x_i, y_j) = 1$. Some of these probabilities may be zero, since there is no reason why all possible pairs should occur in the population considered. As in the previous section, we can define the probabilities $p(x_i)$ and $p(y_j)$; by (a) of Section 3.23 we have

(a) $$p(x_i) = \Sigma_j p(x_i, y_j)$$
$$p(y_j) = \Sigma_i p(x_i, y_j)$$
$$\Sigma_i p(x_i) = \Sigma_j p(y_j) = \Sigma_{ij} p(x_i, y_j) = 1$$

If x and y are independent, we have from (d) of Section 3.23

(b) $$p(x_i, y_j) = p(x_i) \cdot p(y_j)$$

We can now define average values and variances for x and y as in the previous section. Since it follows from (a) above that, for example,

(c) $$\begin{aligned} \text{ave}\,(x) &= \Sigma_{ij} x_i p(x_i, y_j) \\ &= \Sigma_i x_i\, \Sigma_j p(x_i, y_j) \\ &= \Sigma_i x_i p(x_i) \end{aligned}$$

it does not matter in this case whether we define these values in terms of the probabilities $p(x_i, y_j)$ of the pairs (x_i, y_j) or in terms of the probabilities $p(x_i)$ and $p(y_j)$. Thus we may safely speak of the average value and variance of x for a given population, regardless of whether x is considered in connection with other random variables defined for the same population.

We next consider the average value of the sum, $x + y$, of two random variables. By definition

(d) $$\begin{aligned} \text{ave}\,(x + y) &= \Sigma_{ij}(x_i + y_j)\, p(x_i, y_j) \\ &= \Sigma_{ij} x_i p(x_i, y_j) + \Sigma_{ij} y_j p(x_i, y_j) \\ &= \text{ave}\,(x) + \text{ave}\,(y) \end{aligned}$$

the last step following exactly as in (*c*) above. In words, this means that the average value of the sum of two random variables is the sum of their average values. This result does not depend on the independence or dependence of x and y.

The average value of the product of two random variables x and y is given by

$$\text{ave}\,(xy) = \Sigma_{ij} x_i y_j\, p(x_i, y_j)$$

If x and y are independent, we have

$$\begin{aligned} \text{ave}\,(xy) &= \Sigma_{ij} x_i y_j p(x_i)\, p(y_j) \\ &= \Sigma_i x_i\, p(x_i) \cdot \Sigma_j y_j\, p(y_j) \qquad (e) \\ &= \text{ave}\,(x) \cdot \text{ave}\,(y) \end{aligned}$$

or, in words, the average value of the product of two independent random variables is the product of their average values.

The covariance of x and y is defined by

$$\sigma_{xy} = \text{cov}\,(x, y) = \text{ave}\,(x - \mu_x)\,(y - \mu_y) \qquad (f)$$

where μ_x and μ_y are again used to denote ave (x) and ave (y), respectively. By a procedure similar to that used to obtain (*c*) of the preceding section we can show that

$$\text{cov}\,(x, y) = \text{ave}\,(xy) - \mu_x \mu_y \qquad (g)$$

from which, together with (*e*), it follows that cov $(x, y) = 0$ when x and y are independent.

3.33. Average Value and Variance of a Linear Combination

In many cases the measurements or observations which are of interest to the scientist are functions of other measurements or observations and he will frequently wish to know how the mean and variance of these functions are related to the means, variances, and covariances of the original measurements. These relationships are also required in the study of the sampling behavior of sample statistics, which are always functions of the observations in the sample.

The simplest case to consider is a linear combination. Suppose that $w = \Sigma_i a_i x_i$, where the a_i are constants and the x_i random variables with ave $(x_i) = \mu_i$, var $(x_i) = \sigma_i^2$, and cov $(x_i, x_j) = \sigma_{ij}$. Then we have by definition

$$\mu = \text{ave}\,(w) = \text{ave}\,(a_1 x_1 + a_2 x_2 + \cdots + a_n x_n)$$

By a repeated application of (*d*) of Section 3.32 and (*e*) of Section 3.31, we obtain

$$\begin{aligned} \mu &= a_1\,\text{ave}\,(x_1) + a_2\,\text{ave}\,(x_2) + \cdots + a_n\,\text{ave}\,(x_n) \\ &= \Sigma_i a_i \mu_i \qquad (a) \end{aligned}$$

The variance of w is by definition

$$\begin{aligned}\text{var}(w) &= \text{ave}(w^2) - \mu^2\\ &= \text{ave}\{(\Sigma_i a_i x_i)^2\} - (\Sigma a_i \mu_i)^2\\ &= \text{ave}(\Sigma_i a_i^2 x_i^2 + 2\Sigma_{i<j} a_i a_j x_i x_j) - \Sigma_i a_i^2 \mu_i^2 - 2\Sigma_{i<j} a_i a_j \mu_i \mu_j\end{aligned}$$

Using (a) above, we obtain

$$\begin{aligned}\text{var}(w) &= \Sigma_i a_i^2 \,\text{ave}(x_i^2) - \Sigma_i a_i^2 \mu_i^2 + 2\Sigma_{i<j} a_i a_j \,\text{ave}(x_i x_j)\\ &\qquad - 2\Sigma_{i<j} a_i a_j \mu_i \mu_j\\ &= \Sigma_i a_i^2 [\text{ave}(x_i^2) - \mu_i^2] + 2\Sigma_{i<j} a_i a_j [\text{ave}(x_i x_j) - \mu_i \mu_j]\end{aligned}$$

and, using (c) of Section 3.31 and (g) of Section 3.32, this becomes

$$\begin{aligned}\text{var}(w) &= \Sigma_i a_i^2 \,\text{var}(x_i) + 2\Sigma_{i<j} a_i a_j \,\text{cov}(x_i, x_j)\\ (b) \qquad &= \Sigma_i a_i^2 \sigma_i^2 + 2\Sigma_{i<j} a_i a_j \sigma_{ij}\end{aligned}$$

In the particular case where the x_i are mutually independent, all the covariances σ_{ij} are zero, and (b) reduces to

$$(c) \qquad \text{var}(w) = \Sigma_i a_i^2 \sigma_i^2$$

EXAMPLE 1. In many chemical procedures, quantities are determined by difference, i.e., as a gross weight minus a net weight, or as the optical density of a sample minus the optical density of a blank. In this case $w = x_2 - x_1$, and ave $(x) =$ ave $(x_2) -$ ave (x_1). If we consider the two measurements to be independent, then var $(x) =$ var $(x_2) +$ var (x_1); notice that these variances are added, not subtracted, as the coefficients a_i ($+1$ and -1 in this case) are squared in (c). If var $(x_2) =$ var $(x_1) = \sigma^2$, then var $(x) = 2\sigma^2$; and hence the standard deviation of the differences is $\sqrt{2}\sigma$, i.e., the variation in the difference measurements, as indicated by their standard deviation, will on the average be greater than those in the direct measurements by a factor of $\sqrt{2}$.

EXAMPLE 2. Suppose that we consider the single toss of an unbiased coin, which is a random variable with 2 possible values, 1 for heads, and 0 for tails, each having probability 1/2. Then for an individual toss, we have ave $(x) = \frac{1}{2}(1) + \frac{1}{2}(0) = \frac{1}{2}$, and var $(x) = \frac{1}{2}(1 - \frac{1}{2})^2 + \frac{1}{2}(0 - \frac{1}{2})^2 = \frac{1}{4}$. If we now make n independent tosses, and designate the results by $x_1, x_2, \cdots, x_n$, the total number of heads in the n tosses will be given by $x = x_1 + x_2 + \cdots + x_n$. Each of the x_i has $\mu_i = \frac{1}{2}$ and $\sigma_i^2 = \frac{1}{4}$, and we have for the average value and variance of the number of heads in n tosses

$$\begin{aligned}\text{ave}(x) &= \Sigma_i a_i \mu_i \\ &= \Sigma_i (1)\left(\tfrac{1}{2}\right) \\ &= n/2 \\ \text{var}(x) &= \Sigma_i a_i^2 \sigma_i^2 \\ &= \Sigma_i (1)^2 \left(\tfrac{1}{4}\right) \\ &= n/4\end{aligned}$$

the latter being a consequence of the independence of the tosses. In this case $\sigma_x = \sqrt{n}/2$, and hence the relative standard deviation (frequently called the coefficient of variation) is given by

$$\frac{\sigma_x}{\mu_x} = \frac{\sqrt{n}/2}{n/2} = \frac{1}{\sqrt{n}}$$

i.e., the relative, or percentage, scatter of the number of heads in n tosses about the average value of $n/2$ will become smaller as the number of tosses is increased, although the magnitude of the absolute deviation $\left|x - \frac{n}{2}\right|$ can be expected to increase as n is increased.

3.34. Approximate Average Value and Variance of an Arbitrary Function

To find the exact mean and variance of an arbitrary function of several random variables is, even for comparatively simple functions, quite a difficult procedure. However, if the function varies comparatively slowly so that in the region where values of the independent variables are likely to occur, say within one or two standard deviations of their mean, it can be adequately represented by the linear terms of its Taylor series expansion (i.e., by its tangent plane), then the results of the previous section can be used to obtain an approximate average value and variance. We shall consider the case of a function of two variables; the extension of the results to more than two variables is immediate.

Suppose we have $z = f(x, y)$, where x and y are random variables with average values μ_x and μ_y, variances σ_x^2 and σ_y^2 and covariance σ_{xy}. Expanding z about the point μ_x and μ_y, and neglecting terms of degree higher than the first in the deviations $x - \mu_x$ and $y - \mu_y$, we have

$$z \sim f(\mu_x, \mu_y) + f_x(\mu_x, \mu_y)(x - \mu_x) + f_y(\mu_x, \mu_y)(y - \mu_y)$$

where $f_x(\mu_x, \mu_y)$ and $f_y(\mu_x, \mu_y)$ are the partial derivatives of f with respect to x and y evaluated at the point (μ_x, μ_y). It then follows directly from previous results that

(a) $$\text{ave}(z) \sim f(\mu_x, \mu_y)$$

since ave $(x - \mu_x) =$ ave $(y - \mu_y) = 0$, and

(b) $$\text{var}(z) \sim f_x^2(\mu_x, \mu_y)\,\sigma_x^2 + f_y^2(\mu_x, \mu_y)\,\sigma_y^2 + 2f_x(\mu_x, \mu_y)f_y(\mu_x, \mu_y)\,\sigma_{xy}$$

since the variance of the constant $f(\mu_x, \mu_y)$ is zero. As before, the last term of (b) disappears when x and y are independent.

EXAMPLE 1. Let us consider $z = xy$, for which $f_x = y, f_y = x$. From (a) we obtain

$$\text{ave}(z) \sim \mu_x\mu_y$$

$$\text{var}(z) \sim \mu_y^2\sigma_x^2 + \mu_x^2\sigma_y^2 + 2\mu_x\mu_y\sigma_{xy}$$

If x and y are independent, we have

$$\text{var}(z) \sim \mu_y^2\sigma_x^2 + \mu_x^2\sigma_y^2$$

Designating ave (z) and var (z) by μ_z and σ_z^2, respectively, and dividing the left-hand side of the above expression by μ_z^2 and the right by its approximate value $\mu_y^2\mu_x^2$, we have

$$\frac{\sigma_z^2}{\mu_z^2} \sim \frac{\sigma_x^2}{\mu_x^2} + \frac{\sigma_y^2}{\mu_y^2}$$

or

$$\frac{\sigma_z}{\mu_z} \sim \sqrt{\left(\frac{\sigma_x}{\mu_x}\right)^2 + \left(\frac{\sigma_y}{\mu_y}\right)^2}$$

Thus under the conditions for which the approximations (a) and (b) are valid, we can say that, exclusive of any bias which may be present, the relative error (as measured by the standard deviation) in the product of two independent measurements will be approximately the square root of the sum of squares of their relative errors.

EXAMPLE 2. Let us consider the determination of the percentage of moisture in a sample of coal by weighing the original sample, heating in a toluene-jacketed oven using a nitrogen atmosphere, weighing the dried sample, and computing the percentage of moisture as

$$\%\text{ moisture} = \frac{\text{original weight} - \text{final weight}}{\text{original weight}} \times 100$$

Here the percentage of moisture z is related to the original weight x and the dry weight y by the relation

$$z = \frac{x - y}{x} = 1 - \frac{y}{x}$$

and assuming x and y to be independent we have approximately

$$\mu_z \sim 1 - \frac{\mu_y}{\mu_x}$$

$$\sigma_z^2 \sim \left(\frac{\mu_y}{\mu_x^2}\right)^2 \sigma_x^2 + \left(-\frac{1}{\mu_x}\right)^2 \sigma_y^2$$

$$\sim \frac{\mu_y^2}{\mu_x^2}\left[\frac{\sigma_x^2}{\mu_x^2} + \frac{\sigma_y^2}{\mu_y^2}\right]$$

or, since $\mu_z^2 \sim \dfrac{(\mu_x - \mu_y)^2}{\mu_x^2}$

$$\frac{\sigma_z^2}{\mu_z^2} \sim \frac{\mu_y^2}{(\mu_x - \mu_y)^2}\left[\frac{\sigma_x^2}{\mu_x^2} + \frac{\sigma_y^2}{\mu_y^2}\right]$$

In this case we may reasonably expect the standard deviations σ_x and σ_y of both initial weight and final weight to be constant over the range of sample sizes (and consequent final weights) which could be used in practice. However, we would expect $\sigma_y > \sigma_x$, since for a given sample σ_x would reflect only the error in weighing the sample, whereas σ_y would include any variations due to the actual heating operation in addition to variation due to errors in weighing. Now let us suppose that, for a fixed average percentage of moisture, the average sample size μ_x (and consequently the average final weight μ_y) are increased by a factor k; designating the new result by z', we see that

$$\mu_{z'} = \mu_z$$

$$\sigma_{z'}^2 = \frac{\sigma_z^2}{k^2}$$

$$\sigma_{z'} = \frac{\sigma_z}{k}$$

i.e., the standard deviation of the determination will be reduced by a factor of $1/k$.

It can also be seen from the above expressions that for constant σ_x and σ_y, and a fixed average sample size μ_x, both the absolute standard deviation

σ_z and the relative standard deviation σ_z/μ_z will decrease as the average percentage of moisture increases, although not proportionally.

3.35. Population and Sample

We have repeatedly stated that the problem of statistical inference is that of drawing conclusions concerning a population of possible events from a given random sample of these events. Frequently the information which we wish to obtain concerning the behavior of the population can be described or summarized by the same methods which we used to obtain the information in a sample; for example, in Sections 2.32 and 2.33 we computed the mean and variance of a sample and in Section 3.31 the average value and variance of a population by almost identical methods. We must distinguish carefully between population and sample values in these cases; in Section 2.4 it was necessary to differentiate between the average value μ of the population of all possible analyses and the mean $\bar{x}$ of the given analyses. The former represents (at least statistically) the information desired in this case, whereas the latter is just one of the statistics which can be computed from the sample, and used for purposes of inference. We could have obtained limits for μ by using the midrange and range of the sample, rather than the mean and variance; similarly, as shown in Section 2.44, we can obtain limits for the population median without any reference to the sample median. This distinction is also important when we describe sample behavior in terms of population parameters. We have stated, for example, that the variance of the sample means computed from random samples of n is σ^2/n, where σ^2 is the population variance. The quantity s^2/n is an estimate of this variance which will fluctuate from sample to sample, and these fluctuations must be considered when the estimate is used to set limits.

It should be emphasized that the inferences to be made from a given sample depend on the population from which it is assumed to be drawn. It is possible to consider a given sample in terms of several different populations from which it could have been obtained. For example, we could consider 5 analyses for carbon content by 5 different laboratory chemists as representative of the population of possible carbon determinations by these 5 chemists, or of the population of all possible carbon determinations by a chemist in the laboratory concerned, or even of the population of all possible carbon determinations by the method used. The conclusions drawn, or the inferences made, would depend on the population considered. It is also possible to consider a given set of data as the combination of two or more effects drawn at random from different populations, and we may be able to use the same set of data to make inferences concerning some or all of these populations.

3.4. Samples

3.41. Random Sampling

If we repeat the process of choosing an object at random from a population n times, the values $x_1, \cdots, x_n$ of a random variable x so obtained will be called a *random sample*. Since each choice is at random from the same population, these values can also be thought of as single observations associated with a series of independent and identical random variables. It will also be convenient at times to think in terms of a "superpopulation" consisting of all such series of n choices. This population will always be considered infinite, even when the original population is finite; it is important not to confuse the finite number of different samples which can occur with the infinite number of samples which could be drawn.

The importance of randomness in obtaining a sample cannot be overstressed. All the methods of inference which we shall use depend on the assumption that we have randomness in the sample obtained; even in the complex experimental designs which are planned in such a manner as to eliminate certain known effects, we must wherever feasible randomize the experiments to guard against unknown effects. It should also be emphasized that the process of randomization should not be left to an individual, for few people are capable of making a random selection. By using such mechanical aids as beads or chips, which can be thoroughly mixed up before each object is chosen, an almost random choice can be obtained, but it is usually more convenient to use tables of "random numbers" which have been prepared for this purpose. References to such tables, and a discussion of their use, are given in Section 8.39.

3.42. Average Value and Variance of a Sample Mean

Some facts concerning the sampling behavior of the mean of a random sample can be deduced from the results in Section 3.3. Since the observations $x_1, \cdots, x_n$ in a given random sample are all from the same population, they all have the same average value μ and the same variance σ^2. Thus we have

$$\begin{aligned} \text{ave}\,(\bar{x}) &= \text{ave}\left\{\frac{1}{n}\,\Sigma_i x_i\right\} \\ &= \frac{1}{n}\,\Sigma_i\,\text{ave}\,(x_i) \\ &= \frac{1}{n}\,\Sigma_i \mu \\ &= \frac{1}{n}\cdot n\mu = \mu \end{aligned} \tag{a}$$

This important result says that the average value of the mean of a sample of n observations from the same population is the same as the average value of each of the individual observations.

Now let us consider the variance of the mean of a sample of n observations. From Section 3.31 we have by definition

$$\text{var}(\bar{x}) = \text{ave}(\bar{x} - \mu)^2$$

and since we have shown above that ave $(\bar{x}) = \mu$, then

$$\sigma_{\bar{x}}^2 = \text{ave}(\bar{x}^2) - \mu^2 \qquad (b)$$

To evaluate the quantity ave $(\bar{x}^2)$, we expand and obtain

$$\text{ave}(\bar{x}^2) = \text{ave}\left[\left(\frac{x_1 + x_2 + \cdots + x_n}{n}\right)^2\right]$$

$$= \frac{1}{n^2}\,\text{ave}\left[\sum_{i=1}^{n} x_i^2 + \sum_{i \neq j} x_i x_j\right] \qquad (c)$$

where the quantity $\sum_{i \neq j} x_i x_j$ represents all the cross products of two different values of x. Now since x_i and x_j are independent for $i \neq j$, we have

$$\text{ave}(x_i x_j) = \text{ave}(x_i)\,\text{ave}(x_j) = \mu \cdot \mu = \mu^2$$

also we have ave $(x_i^2) = \sigma^2 + \mu^2$. Substituting these in (c) and remembering that there are n terms of the form x_i^2 and $n(n - 1)$ cross products of the form $x_i x_j$, we have

$$\text{ave}(\bar{x}^2) = \frac{1}{n^2}[n(\sigma^2 + \mu^2) + n(n - 1)\,\mu^2]$$

$$= \frac{1}{n^2}[n\sigma^2 + n\mu^2 + n(n - 1)\,\mu^2]$$

$$= \frac{1}{n^2}[n\sigma^2 + n^2\mu^2]$$

$$= \frac{\sigma^2}{n} + \mu^2$$

Substituting this in (b) we obtain

$$\sigma_{\bar{x}}^2 = \frac{\sigma^2}{n} + \mu^2 - \mu^2 = \frac{\sigma^2}{n} \qquad (d)$$

We have now shown that the variance of the mean of a sample of n is equal to the variance of an individual observation divided by the number of observations, a result which we stated without proof in 2.41. It follows that $\sigma_{\bar{x}}$, the standard deviation of the mean, is equal to $\sigma/\sqrt{n}$, i.e., the standard deviation of the mean of a sample of n is less by a factor $1/\sqrt{n}$ than that of an individual observation.

3.43. Average Value of the Sample Variance

We now consider the average value of s^2, the sample variance. We have

$$\text{ave}\,(s^2) = \text{ave}\left(\frac{\Sigma(x-\bar{x})^2}{n-1}\right) = \frac{1}{n-1}\,\text{ave}\,(\Sigma x^2 - n\bar{x}^2)$$

$$= \frac{1}{n-1}\,[\text{ave}\,(\Sigma x^2) - n\,\text{ave}\,(\bar{x}^2)]$$

$$= \frac{1}{n-1}\,[\Sigma\,\text{ave}\,(x^2) - n\,\text{ave}\,(\bar{x}^2)]$$

In the previous section we noted that

$$\text{ave}\,(x^2) = \sigma^2 + \mu^2$$

$$\text{ave}\,(\bar{x}^2) = \frac{\sigma^2}{n} + \mu^2$$

Substituting and remembering that the summation of the constant, ave (x^2), is simply n times that quantity, we have

$$\text{ave}\,(s^2) = \frac{1}{n-1}\left[n(\sigma^2+\mu^2) - n\left(\frac{\sigma^2}{n}+\mu^2\right)\right]$$

$$= \frac{1}{n-1}\,[n\sigma^2 + n\mu^2 - \sigma^2 - n\mu^2]$$

$$= \frac{1}{n-1}\,(n\sigma^2 - \sigma^2) = \sigma^2$$

We stated in Section 2.42 and have now proved that the average value of the sample variance s^2, when defined using a divisor of $n - 1$ and not n, is the population variance σ^2.

3.44. Sampling from Finite Populations

There will be many instances, both in practical problems and in later development of methods, where we shall wish to consider the problem of drawing at random a sample of size n from a finite population of N

objects. Since in this case all of the N values x_i are equally likely, we can define directly the population values

(a) $$\mu = \text{ave}\,(x) = \frac{\sum_{i=1}^{N} x_i}{N}$$

(b) $$\sigma^2 = \text{var}\,(x) = \frac{\sum_{i=1}^{N} (x_i - \mu)^2}{N - 1} = \frac{N}{N-1}\,\text{ave}\,(x_i - \mu)^2$$

It has been customary to divide by N rather than $N - 1$ in defining the population variance for finite populations, but, since the above definition will be much more convenient, and causes no conflict with the usual definition in the infinite case, we shall adopt it here and throughout this book. Some additional justification for this definition has been found [75].

It is also necessary to consider the covariance between any pair of elements drawn at random from the population *without replacement*, i.e., such that x_i and x_j must be two different elements of the population. To do this let us consider the new population of all possible ordered pairs x_i, x_j such that $x_i \neq x_j$. If the sampling is random all the $N(N - 1)$ such pairs are equally likely, and we have

(c) $$\text{ave}\,(x_i - \mu)(x_j - \mu) = \frac{\sum_{i \neq j}^{N} (x_i - \mu)(x_j - \mu)}{N(N-1)}$$

To evaluate the numerator, we note that by definition

$$\sum_{i=1}^{N} (x_i - \mu) = (x_1 - \mu) + (x_2 - \mu) + \cdots + (x_N - \mu) = 0$$

Squaring both sides of this equality, we have

(d) $$\sum_{i=1}^{N} (x_i - \mu)^2 + \sum_{i \neq j}^{N} (x_i - \mu)(x_j - \mu) = 0$$

and hence from (b)

(e) $$\sum_{i \neq j}^{N} (x_i - \mu)(x_j - \mu) = -(N - 1)\sigma^2$$

Substituting this result in (c), we have

(f) $$\text{ave}\,(x_i - \mu)(x_j - \mu) = \frac{-(N-1)\,\sigma^2}{N(N-1)} = -\frac{\sigma^2}{N}$$

We can now consider the mean $\bar{x}$ of a sample of n drawn from the above population without replacement. Since the value of $\bar{x}$ depends only on the particular combination of sample values obtained, and not on their order, it can be considered as a random choice from the population of $\binom{N}{n}$ equally likely combinations.* Hence the average value of $\bar{x}$ would be

$$\text{ave}\,(\bar{x}) = \frac{\sum_{\text{a.c.}} \bar{x}}{\binom{N}{n}} = \frac{1}{n\binom{N}{n}} \sum_{\text{a.c.}} (x_1 + x_2 + \cdots + x_n) \tag{g}$$

where the summation in this instance is over *all* the $\binom{N}{n}$ possible *combinations* of n values from the original population of N. To evaluate the sum on the right we note that each value x_i from the original population must occur with each possible combination of $n - 1$ of the remaining $N - 1$ values, or a total of $\binom{N-1}{n-1}$ times; hence

$$\sum_{\text{a.c.}} (x_1 + x_2 + \cdots + x_n) = \binom{N-1}{n-1} \sum_{i=1}^{N} x_i \tag{h}$$

and

$$\text{ave}\,(\bar{x}) = \frac{\binom{N-1}{n-1}}{n\binom{N}{n}} \sum_{i=1}^{N} x_i$$

$$= \frac{1}{n} \frac{(N-1)!}{(n-1)!\,(N-n)!} \cdot \frac{n!\,(N-n)!}{N!} \cdot \sum_{i=1}^{N} x_i$$

$$= \frac{\sum_{i=1}^{N} x_i}{N} = \mu \tag{i}$$

Similarly we have

$$\sigma_{\bar{x}}^2 = \text{ave}\,(\bar{x} - \mu)^2 = \frac{\sum_{\text{a.c.}} (\bar{x} - \mu)^2}{\binom{N}{n}}$$

$$= \frac{1}{n^2\binom{N}{n}} \sum_{\text{a.c.}} (x_1 + x_2 + \cdots + x_n - n\mu)^2$$

* For those readers not familiar with permutations and combinations, their definition and properties are summarized briefly in Appendix 3*A*.

(*j*)
$$= \frac{1}{n^2\binom{N}{n}} \sum_{\text{a.c.}} [(x_1 - \mu) + (x_2 - \mu) + \cdots + (x_n - \mu)]^2$$

$$= \frac{1}{n^2\binom{N}{n}} \sum_{\text{a.c.}} \left[\sum_{i=1}^{n} (x_i - \mu)^2 + \sum_{i \neq j}^{n} (x_i - \mu)(x_j - \mu) \right]$$

$$= \frac{1}{n^2\binom{N}{n}} \left[\sum_{\text{a.c.}} \sum_{i=1}^{n} (x_i - \mu)^2 + \sum_{\text{a.c.}} \sum_{i \neq j}^{n} (x_i - \mu)(x_j - \mu) \right]$$

To evaluate the first sum inside the brackets, we again note that each quantity $(x_i - \mu)^2$ will appear $\binom{N-1}{n-1}$ times in the summation; hence, using (*b*)

(*k*)
$$\sum_{\text{a.c.}} \sum_{i=1}^{n} (x_i - \mu)^2 = \binom{N-1}{n-1} \sum_{i=1}^{N} (x_i - \mu)^2$$

$$= (N-1)\binom{N-1}{n-1} \sigma^2$$

To evaluate the second sum we note that each pair x_i, x_j must appear with each possible combination of $n - 2$ of the remaining $N - 2$ population values, or a total of $\binom{N-2}{n-2}$ times; hence, using (*e*)

(*l*)
$$\sum_{\text{a.c.}} \sum_{i \neq j}^{n} (x_i - \mu)(x_j - \mu) = \binom{N-2}{n-2} \sum_{i \neq j}^{N} (x_i - \mu)(x_j - \mu)$$

$$= -(N-1)\binom{N-2}{n-2} \sigma^2$$

Substituting (*k*) and (*l*) in (*j*) we obtain

$$\sigma_{\bar{x}}^2 = \frac{(N-1)\sigma^2}{n^2} \left[\frac{\binom{N-1}{n-1}}{\binom{N}{n}} - \frac{\binom{N-2}{n-2}}{\binom{N}{n}} \right]$$

(*m*)
$$= \frac{(N-1)\sigma^2}{n^2} \left[\frac{n}{N} - \frac{n(n-1)}{N(N-1)} \right]$$

$$= \frac{\sigma^2}{n} \left(1 - \frac{n}{N} \right)$$

Note that the factor in brackets is less than 1 for $n \geq 1$, hence $\sigma_{\bar{x}}^2$ in the case of the finite sampling is in general less than the value σ^2/n previously obtained for the infinite case, approaching this value as N becomes large.

The consideration of the average value or variance of the sample variance or other sample statistics rapidly becomes complex, and we shall not consider this problem here. The interested reader is referred to [2, Vol. I], where a discussion of this topic and a bibliography of the more important contributions is available, or to [75], where the whole problem of finite sampling is considered from a new viewpoint.

3.5. Stratified Sampling

3.51. Estimating the Mean

On the basis of our knowledge concerning a population we can often divide the population into several subpopulations which are known to be essentially different. The question then arises as to whether we can obtain a sample mean with a smaller variance by substituting random sampling of the subpopulations for an overall random sampling procedure. The subpopulations are generally referred to in this instance as strata, and the sampling procedure used in this connection as stratified sampling. An example of the use of this procedure will be given in Chapter 8.

Suppose that we have a population of N objects which can be divided into k subpopulations, or strata, of size $N_1, N_2, \cdots, N_k, \sum_{i=1}^{k} N_i = N$, and let μ_i and σ_i^2 represent the average values and variances, respectively, of these strata. Then we have for the average value of the entire population

$$(a) \qquad \mu = \frac{1}{N} \Sigma_i N_i \mu_i$$

Now suppose that we draw a sample of n_i elements $x_{ij}, j = 1, \cdots, n_i$, at random from each of the k strata, and wish to obtain from these $n = \sum_{i=1}^{k} n_i$ elements an estimate m of the population mean μ. Our first problem is to determine weights w_{ij} such that

$$(b) \qquad m = \Sigma_{ij}\, w_{ij}\, x_{ij}$$

will not be systematically greater or smaller than the population mean to be determined, i.e., such that

$$(c) \qquad \text{ave}\,(m) = \mu$$

Substituting (a) and (b) into (c), we have

$$(d) \qquad \Sigma_{ij} w_{ij} \text{ ave}\,(x_{ij}) = \frac{1}{N} \Sigma_i N_i \mu_i$$

Now since x_{ij} is drawn at random from the ith population, ave $(x_{ij}) = \mu_i$. Hence

$$(e) \qquad \Sigma_{ij} w_{ij} \mu_i = \frac{1}{N} \Sigma_i N_i \mu_i$$

$$\Sigma_i \mu_i \left(\Sigma_j w_{ij} - \frac{N_i}{N} \right) = 0$$

and for this to hold for all values of μ_i it is sufficient that we have

$$(f) \qquad \Sigma_j w_{ij} = \frac{N_i}{N}$$

i.e., that the sum of the weights applied to the sample from the ith stratum be equal to the fraction of the population it contains.

3.52. Minimizing the Variance

The condition (f) of the previous section assures us that any estimate m based on the stratified sample x_{ij}, $i = 1, \cdots, k$, $j = 1, \cdots, n_i$ will have as its average value the population mean. We shall now consider what further restrictions are necessary on the weights w_{ij} and the size n_i of the random samples drawn from the various strata.

If we consider the sample from the ith stratum, we have

$$(a) \qquad \begin{aligned} \text{ave } (\Sigma_j w_{ij} x_{ij}) &= \Sigma_j w_{ij} \text{ ave } (x_{ij}) \\ &= \Sigma_j w_{ij} \mu_i \end{aligned}$$

and hence

$$\begin{aligned} \text{var } (\Sigma_j w_{ij} x_{ij}) &= \text{ave } (\Sigma_j w_{ij} x_{ij} - \Sigma_j w_{ij} \mu_i)^2 \\ &= \text{ave } (\Sigma_j w_{ij} (x_{ij} - \mu_i))^2 \\ &= \Sigma_j w_{ij}^2 \text{ ave } (x_{ij} - \mu_i)^2 + \Sigma_{j \neq k} w_{ij} w_{ik} \text{ ave } (x_{ij} - \mu_i)(x_{ik} - \mu_i) \end{aligned}$$

and hence, using (b) and (f) of Section 3.44,

$$(b) \qquad \begin{aligned} \text{var } (\Sigma_j w_{ij} x_{ij}) &= \Sigma_j w_{ij}^2 \sigma_i^2 \frac{N_i - 1}{N_i} - \Sigma_{j \neq k} w_{ij} w_{ik} \frac{\sigma_i^2}{N_i} \\ &= \sigma_i^2 \, \Sigma_j w_{ij}^2 - \frac{\sigma_i^2}{N_i} (\Sigma_j w_{ij})^2 \end{aligned}$$

Letting $\bar{w}_{i\cdot} = \dfrac{1}{n_i} \Sigma_j w_{ij}$, we have

$$\begin{aligned} \text{var } (\Sigma_j w_{ij} x_{ij}) &= \sigma_i^2 \left(\Sigma_j w_{ij}^2 - \frac{n_i^2}{N_i} \bar{w}_{i\cdot}^2 \right) \\ &= \sigma_i^2 \left[(\Sigma_j w_{ij}^2 - n_i \bar{w}_{i\cdot}^2) - \left(\frac{n_i^2}{N_i} - n_i \right) \bar{w}_{i\cdot}^2 \right] \\ &= \sigma_i^2 \left[\Sigma_j (w_{ij} - \bar{w}_{i\cdot})^2 + \frac{n_i (N_i - n_i)}{N_i} \bar{w}_{i\cdot}^2 \right] \end{aligned}$$

and hence the quantity in brackets is minimized when $\Sigma_j(w_{ij} - \bar{w}_{i\cdot})^2 = 0$, i.e., when $w_{ij} = \bar{w}_{i\cdot}$. Thus the variance of the contribution of the ith subsample to the overall estimate m of μ is minimized when all the sample values are equally weighted. Using the condition (f) of the previous section, we have

$$\bar{w}_{i\cdot} = \frac{1}{n_i}\,\Sigma_i w_{ij} = \frac{N_i}{n_i N} \tag{c}$$

hence

$$\begin{aligned} m &= \Sigma_{ij} w_{ij} x_{ij} \\ &= \Sigma_i \bar{w}_{i\cdot}\ \Sigma_j x_{ij} \\ &= \Sigma_i \frac{N_i}{N}\,\frac{\Sigma_j x_{ij}}{n_i} \qquad (d) \\ &= \Sigma_i \frac{N_i}{N}\,\bar{x}_{i\cdot} \end{aligned}$$

where $\bar{x}_{i\cdot}$ is the mean of the ith subsample. Since from 3.44 we have $\operatorname{var}(\bar{x}_{i\cdot}) = \dfrac{1}{n_i}\left(1 - \dfrac{n_i}{N_i}\right)\sigma_i^2$, we obtain immediately

$$\begin{aligned} \operatorname{var}(m) &= \Sigma_i \frac{N_i^2}{N^2}\operatorname{var}(\bar{x}_{i\cdot}) \\ &= \frac{1}{N^2}\,\Sigma_i \frac{N_i^2}{n_i}\left(1 - \frac{n_i}{N_i}\right)\sigma_i^2 \qquad (e) \end{aligned}$$

Up to this point we have considered only the determination of the weights w_{ij} based on a knowledge of the number N_i in the ith subpopulation. Now let us suppose that the variances σ_i^2 are also known, and consider the determination of the optimum values of the numbers of elements n_i to be drawn from each subpopulation. To do this we minimize the variance of m subject to the condition $\Sigma_i n_i = n$. Using the method of Lagrange multipliers, we obtain the equations

$$\frac{\partial}{\partial n_i}\left(\frac{1}{N^2}\,\Sigma_i \frac{N_i^2}{n_i}\,\sigma_i^2 - \frac{1}{N^2}\,\Sigma_i N_i \sigma_i^2 \lambda\, \Sigma_i n_i\right) = 0$$

$$-\frac{1}{N^2}\cdot\frac{N_i^2\sigma_i^2}{n_i^2} + \lambda = 0 \qquad i = 1, \cdots, k$$

which leads to the conditions

$$n_i^2 \propto \frac{N_i^2\sigma_i^2}{N^2} \tag{f}$$

$$n_i \propto \frac{N_i\sigma_i}{N}$$

which, along with the restriction $\Sigma_i n_i = n$, determines the values of the n_i. For these values we have

(g) $$\text{var}(m) = \frac{1}{N^2}\left[\frac{(\Sigma N_i\sigma_i)^2}{n} - \Sigma N_i\sigma_i^2\right]$$

3.53. The Case of Infinite Subpopulations

It frequently happens that in practical problems the N_i are sufficiently large to be considered infinite. As N becomes infinite the proportions N_i/N of values in the ith subpopulation approach the probabilities p_i of obtaining a value from the ith subpopulation. Assuming these to be known, we have from (d) of the previous section

(a) $$m = \Sigma_i p_i \bar{x}_{i\cdot}$$

and assuming the variances σ_i^2 are also known, we have from (f) and (g) of the previous section

(b) $$n_i \propto \sigma_i p_i$$

(c) $$\text{var}(m) = \frac{(\Sigma p_i\sigma_i)^2}{n}$$

since the second term in (g) of the previous section disappears as N becomes infinite.

In this case it is comparatively easy to demonstrate the essential superiority of the method of stratified sampling over that of random sampling from the entire population. Let us consider the special case where $p_1 = p_2 = \cdots = p_k = \frac{1}{k}$ and $\sigma_1^2 = \sigma_2^2 = \cdots = \sigma_k^2 = \sigma^2$. Then in stratified sampling we have from (c)

(d) $$\sigma_m^2 = \frac{\sigma^2}{n}$$

Now let us consider the variance of the mean of a random sample from the entire population. By definition, the variance of the population is

(e) $$\begin{aligned}\sigma_p^2 &= \Sigma_i p_i \text{ ave } (x_i - \mu)^2 \\ &= \Sigma_i p_i \text{ ave } [(x_i - \mu_i) + (\mu_i - \mu)]^2 \\ &= \Sigma_i p_i \text{ ave } (x_i - \mu_i)^2 + \Sigma_i p_i(\mu_i - \mu)^2 \\ &= \Sigma_i p_i\sigma_i^2 + \Sigma_i p_i(\mu_i - \mu)^2\end{aligned}$$

and, since $p_i = 1/k$ and $\sigma_i = \sigma$,

(f) $$\sigma_p^2 = \sigma^2 + \frac{\Sigma_i(\mu_i - \mu)^2}{k}$$

Hence the variance of the mean $\bar{x}$ of a random sample of n would be

$$(g) \qquad \sigma_{\bar{x}}^2 = \frac{\sigma_p^2}{n} = \frac{\sigma^2}{n} + \frac{\Sigma(\mu_i - \mu)^2}{nk}$$

and hence, since $\Sigma(\mu_i - \mu)^2$ is always positive or zero, we have

$$(h) \qquad \sigma_m^2 \leq \sigma_{\bar{x}}^2$$

the two variances being equal only when the means μ_i of the strata are all equal. This inequality can be shown to hold for any infinite population.

The methods of stratified sampling are unaffected by the nature of the stratification, whether it be a simple division into k classes or a cross classification into pq cells. It should be noted that the general reduction of variance in stratified sampling is due to the assumption of a knowledge of the N_i (or p_i) and the σ_i^2 which is not necessary to random sampling. Thus by making use of additional information concerning the population we have reduced the variance of the estimate of the mean. On the other hand, if this information is incorrect we may obtain a biased estimate, which would not occur if a completely random sample were employed.

APPENDIX 3A

Permutations and Combinations

3A.1. Permutations

The number of possible arrangements of x items which can be made using n distinct items is called the "number of permutations of n things taken x at a time" and will be designated by $P(n, x)$. For example, given the letters a, b, c, there are six distinct arrangements containing two letters: ab, ac, bc, ba, ca, cb. In general, given n items, we can have n arrangements containing one item and $n(n - 1)$ arrangements containing two items, since the first could be any one of the n items and the second any one of the remaining $n - 1$. Following this same argument, it is easily seen that we can have $P(n, x) = n(n - 1) \cdots (n - x + 1)$ arrangements containing x items. In the above example, $n = 3$ and $x = 2$, hence $P(3, 2) = 3 \cdot 2 = 6$. By multiplying and dividing $P(n, x)$ by the additional terms $(n - x)(n - x - 1) \cdots 1$, we obtain

$$P(n, x) = \frac{n!}{(n - x)!}$$

When $x = n$, we have $(n - x)! = 0! = 1$, hence $P(n, n) = n!$, i.e., there are $n!$ possible arrangements of all n objects.

3A.2. Combinations

By the "number of combinations of n things taken x at a time" designated by $C(n, x)$, we shall mean the number of ways in which x distinct objects can be chosen from n distinct objects *irrespective of the order of arrangement of the x objects*. Thus in the example above, there would be only 3 possible *combinations* of the letters a, b, c: ab, ac, and bc. The remaining 3 permutations simply represent the second possible arrangement of each of these combinations. In general, the number of permutations of n things taken x at a time will be the number of combinations of n things taken x at a time, times the number of permutations of the x things chosen; i.e., the total number of arrangements of n distinct items containing x items will be the number of ways in which x items can be chosen from n, times the number of ways in which they can be rearranged. Hence

$$P(n, x) = C(n, x)\, P(x, x)$$

and

$$C(n, x) = \frac{P(n, x)}{P(x, x)} = \frac{n!}{(n - x)!\, x!}$$

Notice that, since $0! = 1$, $C(n, 0) = 1$ and $C(n, n) = 1$. Also we have immediately $C(n, x) = C(n, n - x)$.

In statistical problems we are frequently concerned with the number of possible arrangements of n items of which x are of one distinct type and $n - x$ of another, but the two types are indistinguishable among themselves. This can be reduced to a problem in combinations by considering as distinct the n possible positions which one of the n items will fill in a given arrangement and then determining the number of ways in which x distinct positions can be chosen from the x items of the first type. Thus the number of possible arrangements, usually designated by $\binom{n}{x}$, is $\binom{n}{x} = C(n, x) = \dfrac{n!}{x!\,(n - x)!}$. Since $C(n, x) = C(n, n - x)$, we could have obtained the same result by choosing $n - x$ distinct positions for the $n - x$ items of the second type. This is exactly the situation which arises with the binomial expansion or binomial distribution, where we are considering the number of possible orders in which we might have obtained x successes in n trials, and the quantities $\binom{n}{x}$ are frequently called binomial coefficients. Tables of $C(n, x) = \binom{n}{x}$ for n from 2 to 200 and for all possible values of x are given in [6, p. 439 ff.].

The above argument can be extended to the case where there are a total of n items of which x_1 are of one type, x_2 of another, x_3 of another, etc., and $n = x_1 + x_2 + x_3 + \cdots$. In this case, on choosing x_1 of the n positions for the first type, x_2 of the remaining $n - x_1$ positions for the second type, and so on, we obtain

$$\binom{n}{x_1\, x_2\, x_3 \cdots} = \frac{n!}{(n - x_1)!\, x_1!} \frac{(n - x_1)!}{(n - x_1 - x_2)!\, x_2!} \frac{(n - x_1 - x_2)!}{(n - x_1 - x_2 - x_3)!\, x_3!} \cdots$$

$$= \frac{n!}{x_1!\, x_2!\, x_3! \cdots}$$

Since these are the quantities which arise in connection with the multinomial expansion or distribution, they are frequently called multinomial coefficients.

CHAPTER 4

Mathematical Machinery

4.1. Introduction

This chapter is devoted to the development of some mathematical concepts which underlie the methods to be used in later chapters. The primary new concept will be that of a distribution function, which can best be described as a mathematical function representing the behavior of a random variable associated with a given population. The Maxwellian distribution, which gives the distribution of the velocities of a large number of molecules, is one example of such a distribution; the binomial distribution, which as a special case gives the chances of obtaining various combinations of heads or tails in repeated tosses of an unbiased coin, is another. We have actually used a distribution function in the presentation of Section 3.3, where the probabilities $p(x_i)$ give the distribution of the possible values x_i in the population considered.

In this chapter we consider briefly the general theory of distribution functions and their associated moments, semi-invariants, and generating functions. This is followed by a more detailed consideration of the normal distribution and the distributions of some sample statistics for random samples of a normally distributed variable. The last section is devoted to a brief discussion of some of the more important discrete distributions. Although the methods employed are mathematical we shall indicate a number of points at which empirical sampling studies have shown the way to subsequent mathematical development or have confirmed the value of results obtained by such development.

4.2. Distribution Functions

4.21. Cumulative Distribution Function

With every random variable x there is associated a function $F(x)$, called the cumulative distribution function of x, which has the property $Pr(x \leq x_0) = F(x_0)$. We consider a population consisting of a deck of playing cards, and associate the values 1–13 of a random variable x with the ranks of the cards. Some values of $F(x)$ for this population are given in Table 4.1.

TABLE 4.1

x	$\frac{1}{2}$	1	$\frac{3}{2}$	2	3	4	5	6	7	8	9	10	11	12	13	14
$F(x)$	0	$\frac{1}{13}$	$\frac{1}{13}$	$\frac{2}{13}$	$\frac{3}{13}$	$\frac{4}{13}$	$\frac{5}{13}$	$\frac{6}{13}$	$\frac{7}{13}$	$\frac{8}{13}$	$\frac{9}{13}$	$\frac{10}{13}$	$\frac{11}{13}$	$\frac{12}{13}$	1	1

Notice that $F(x)$ is 0 for $x < 1$, since it is impossible to obtain a number less than 1; similarly, $F(x) = 1$ for $x \geq 13$, since x is certain to be ≤ 13. A graph of the function $F(x)$ is shown in Figure 4.1. In this example

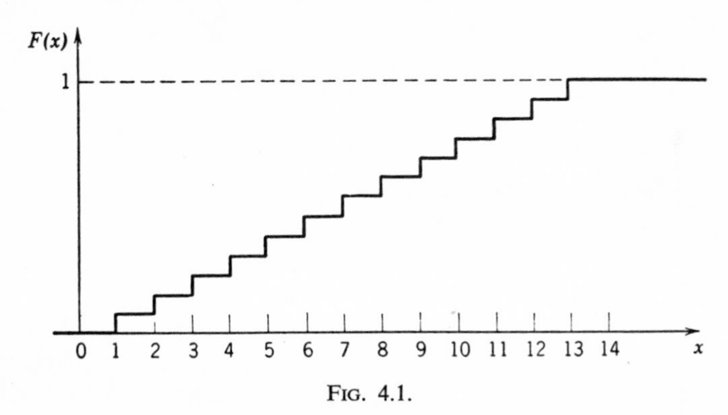

FIG. 4.1.

$F(x)$ can change values only at whole numbers, since x can take only integral values, and therefore the probability of obtaining any intermediary value is 0.

If we were to consider a random variable x which could take on all the values on a continuous scale of measurement, say, $0 \leq x \leq A$, then $F(x)$ might be represented by a continuous curve such as that of Figure 4.2. It is difficult to give an example of the occurrence of such a distribution in practice, since most observations must be recorded to a finite number of decimal places, and, even if this difficulty were overcome by some means, most of the physical measuring processes which we can envisage are ultimately restricted by the discrete nature of matter. However, it is frequently convenient, for both mathematical and practical purposes, to consider continuous scales of measurement.

The relationship between this definition of the cumulative distribution function associated with a given random variable and the cumulative frequency distribution which was prepared in Chapter 2 should be noted.

In that case we prepared a table giving the relative frequency or proportion of values in the sample less than or equal to a given value of x. In this case we have defined a function which gives the relative frequency or probability with which a given value drawn at random from the population will be less than or equal to x. In fact, if the size of the sample is increased, and we are dealing with a stable system, the cumulative graph prepared in Chapter 2 will become a closer approximation to the graph of the cumulative distribution function $F(x)$, in the same sense as that defined in Section 3.22 for the approach of the relative frequency in a large number of random choices to the probability of occurrence of a given set of values of x.

The cumulative distribution function $F(x)$ must either remain the same or increase, for as x_0 is increased more and more values of x will

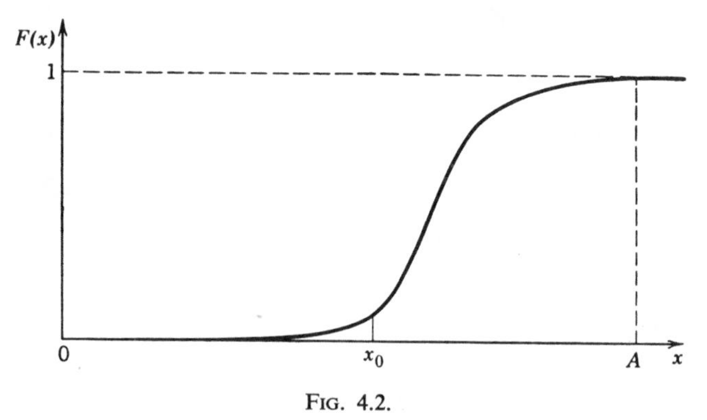

FIG. 4.2.

be included in the set $x \leq x_0$, and as the number of values of x increases the relative frequency of their occurrence in the population, and hence $Pr(x \leq x_0)$, must also increase. This is also true of the cumulative frequency distribution of Chapter 2.

4.22. Relationships between Cumulative Distribution Function and Probabilities

The cumulative frequency function defined in the preceding section specifies only the probabilities of sets of values of x such that $x \leq x_0$. We shall often be interested in other sets of possible values of x, particularly those of the form $x = x_0$, $x_1 < x \leq x_2$, or $x > x_0$. These can be computed by using the formulae of Section 3.23.

For instance, let us compute $Pr(x_1 < x \leq x_2)$. The two sets of values $x \leq x_1$ and $x_1 < x \leq x_2$ have no values in common; their sum is the set $x \leq x_2$. Hence by (*a*) of Section 3.23 we have

$$Pr(x \leq x_2) = Pr(x \leq x_1) + Pr(x_1 < x \leq x_2)$$

Hence

(a) $$Pr(x_1 < x \leq x_2) = Pr(x \leq x_2) - Pr(x \leq x_1)$$
$$= F(x_2) - F(x_1)$$

In a similar manner we obtain $Pr(x > x_0) = 1 - Pr(x \leq x_0)$ since x must obviously be either $> x_0$ or $\leq x_0$. We can define

(b) $$Pr(x = x_0) = Pr(x \leq x_0) - Pr(x < x_0)$$

This last probability can be computed from the relation

$$Pr(x < x_0) = F(x_0 - 0)$$

where $F(x_0 - 0)$ is the limit of $F(x)$ as $x \to x_0$ from below, and hence we have

(c) $$Pr(x = x_0) = F(x_0) - F(x_0 - 0)$$

For example, from Table 4.1 and Figure 4.1 it can be seen that, for any non-integral value x_0, $F(x_0)$ and $F(x_0 - 0)$ are equal; but for integral values, say 6, we have $F(x_0) = 6/13$ and $F(x_0 - 0) = 5/13$, and thus $Pr(x = 6) = 1/13$. If $F(x)$ is continuous, then $F(x_0 - 0) = F(x_0)$ for all values of x_0, and hence the probability of the random variable x assuming any particular value x is always 0. However, in this case there is always a probability of the random variable x falling in any interval over which $F'(x)$ is not identically zero, whereas in the first case the probability of the random variable x falling in any interval not containing one of the isolated points at which a jump occurs is 0.

Many types of cumulative distribution functions more complicated than those given above are possible. In fact, any function $F(x)$ defined for all real numbers x which satisfies the conditions

(d) $$F(-\infty) = 0$$
$$F(+\infty) = 1$$
$$F(x_1) \leq F(x_2) \qquad \text{for } x_1 < x_2$$

can be considered the cumulative frequency distribution of some random variable x. In this book we shall be interested only in the two types discussed above.

4.23. Discrete Probability Distribution. Probability Density Function

In the first case considered in the previous section the random variable x can take on only certain isolated values, and the jumps in the cumulative distribution function at these values indicate the probabilities with which these values occur. Such a distribution of a total probability of one among a discrete group of events will be called a *discrete probability distribution.* For the foregoing example, this distribution is given in Table 4.2, which indicates the association of the probability 1/13 with

TABLE 4.2

x	1	2	3	4	5	6	7	8	9	10	11	12	13
$p(x)$	$\frac{1}{13}$	$\frac{1}{13}$	$\frac{1}{13}$	$\frac{1}{13}$	$\frac{1}{13}$	$\frac{1}{13}$	$\frac{1}{13}$	$\frac{1}{13}$	$\frac{1}{13}$	$\frac{1}{13}$	$\frac{1}{13}$	$\frac{1}{13}$	$\frac{1}{13}$

each of the possible values of x. In tossing a coin, where we do not assume heads and tails to be equally likely, the probability distribution will consist of some undetermined probability $1 - p$ associated with the value 0, and a probability p associated with the value 1. In more general terminology a discrete probability distribution will consist of

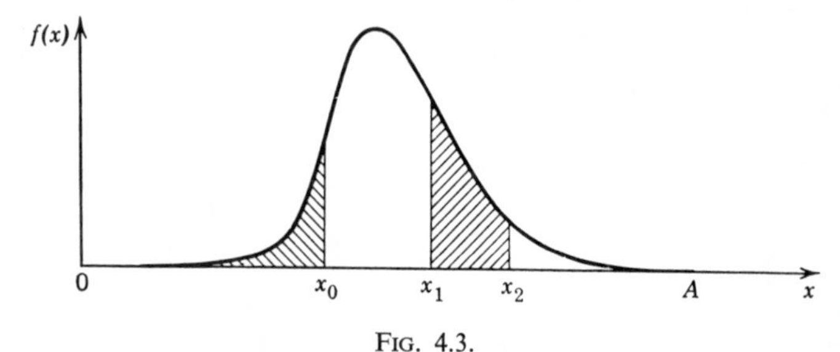

FIG. 4.3.

a series of probabilities $p(x_1) = Pr(x = x_1)$, $p(x_2) = Pr(x = x_2), \cdots$, associated with the possible discrete values $x_1, x_2, \cdots$, of the random variable x, such that $\Sigma_i p(x_i) = 1$. This is exactly the type of distribution which was considered in the last chapter in studying the properties of average values and variances.

In the second class of probability distributions that we wish to consider, where the function $F(x)$ increases continuously from 0 to 1, we can always find a function $f(x)$ such that

$$(a) \qquad F(x_0) = \int_{-\infty}^{x_0} f(x)\, dx$$

The function $f(x)$ is called a *probability density function,* and, since $F(+\infty) = 1$, we must have

$$(b) \qquad \int_{-\infty}^{+\infty} f(x)\, dx = 1$$

It follows from (a) of the preceding section that in this case

$$Pr(x_1 \leq x \leq x_2) = \int_{x_1}^{x_2} f(x)\, dx$$

$f(x)\,dx$ represents the probability that the random variable x will fall in the interval between x and $x + dx$, and will be called the probability element of x. As an example, corresponding to the cumulative distribution function given in Figure 4.2 we should have a probability density function like that of Figure 4.3. For any point x_0, the proportion of the area under the curve to the left of this point gives the ordinate at the point x_0 in Figure 4.2, or the $Pr(x \leq x_0)$. For any two values x_1 and x_2, the area underneath the curve between these points represents the $Pr(x_1 < x < x_2)$.

In later chapters we shall frequently speak of the "distribution function" or simply the "distribution" of a variable x, meaning either function, whichever is applicable. Similarly we shall speak of the random variable x as being "distributed" in a certain manner, implying that the values of x have a certain distribution or distribution function. We shall also speak of the "population distribution" or of a "population distributed in a certain manner" when it is clear which characteristic of the population we are considering.

4.24. Joint Cumulative Distribution Function of Two Variables

As an extension of the above discussion, we can consider a joint cumulative distribution function $F(x, y)$ of 2 random variables x and y, such that $Pr(x \leq x_0, y \leq y_0) = F(x_0, y_0)$, i.e., a function giving the probability of the joint occurrence of the two events $x \leq x_0$ and $y \leq y_0$. Such joint occurrences may be thought of as represented by points in the xy plane, in which case $F(x_0, y_0)$ is the probability of obtaining a point in the region falling to the left of the line $x = x_0$ and below the line $y = y_0$.

Suppose now we wish to compute the probability that a given point (x, y) will fall in any square region of the xy plane defined by the inequalities $x_1 < x \leq x_2$ and $y_1 < y \leq y_2$. A diagram of this region is shown in Figure 4.4. From this diagram, using (*a*) of Section 3.23 for sets of points in the plane rather than sets of numbers (which we could have regarded as sets of points on a line), we see that

$$Pr(x_1 < x \leq x_2,\ y_1 < y \leq y_2) = Pr(x \leq x_2,\ y \leq y_2) - Pr(x \leq x_1,\ y \leq y_2) - Pr(x \leq x_2,\ y \leq y_1) + Pr(x \leq x_1,\ y \leq y_1)$$

or

$$(a) \qquad Pr(x_1 < x \leq x_2,\ y_1 < y \leq y_2) = F(x_2, y_2) - F(x_1, y_2) - F(x_2, y_1) + F(x_1, y_1)$$

The cumulative distribution function $F(x, \infty) = G(x)$ will be called the *marginal distribution* of x and the distribution $F(\infty, y) = H(y)$ the marginal distribution of y. $G(x)$ and $H(y)$ are cumulative distribution functions of the single random variables x and y, respectively, and can be thought of as the distributions that would apply to one of the variables if the other variable were entirely disregarded and only the first or the second of the pair of observed values were used.

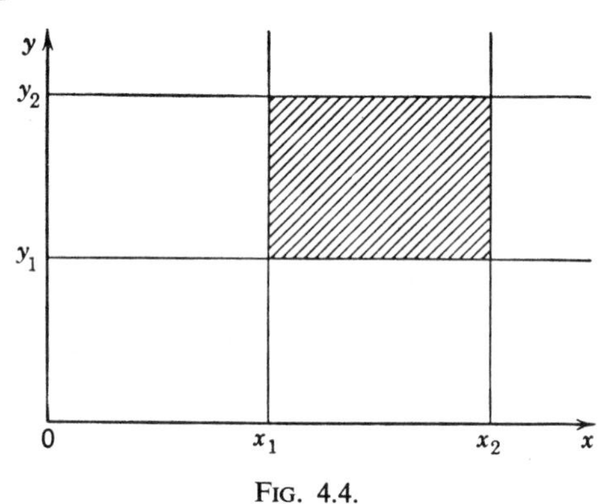

FIG. 4.4.

If x and y are independent, we have from (*d*) of Section 3.23,

$$Pr(x \leq x_0,\ y \leq y_0) = Pr(x \leq x_0) \cdot Pr(y \leq y_0)$$

and hence under this condition we must have

(*b*) $$F(x,\ y) = F(x,\ \infty) \cdot F(\infty,\ y) = G(x) \cdot H(y)$$

4.25. Discrete Probability Distributions and Probability Density Function of Two Variables

We shall again consider only those cumulative distribution functions of 2 random variables in which the marginal distributions of x and y are either both discrete or both continuous. If both marginal distributions are discrete, then, as in the case of a single variable, we can only have probabilities other than 0 at a group of isolated points given by all combinations of the possible values $x_1, x_2, \cdots$ of x and $y_1, y_2, \cdots$ of y. These probabilities can then be denoted by the quantities

(*a*) $$p(x_i,\ y_j) = Pr(x = x_i,\ y = y_j) \qquad i = 1, 2, \cdots ; j = 1, 2, \cdots$$

where $\Sigma_{ij} p(x_i,\ y_j) = 1$. If x and y are independent, we have

(*b*) $$p(x_i,\ y_j) = p(x_i) \cdot p(y_j)$$

where the $p(x_i)$ are the probabilities associated with the values of x, and the $p(y_j)$ are the probabilities associated with the values of y. This type of distribution has also been considered in Chapter 3.

If both marginal distributions are continuous there always exists a function $f(x, y)$ such that $F(x_0, y_0) = \int_{-\infty}^{x_0} \int_{-\infty}^{y_0} f(x, y)\, dy\, dx$ for all points (x_0, y_0) and $\int_{-\infty}^{\infty} \int_{-\infty}^{\infty} f(x, y)\, dy\, dx = 1$. This function will be called the *joint probability density function* of the variables x and y, and it follows from (*a*) of the previous section that the probability of the point (x, y) falling in any square area of the xy plane is given by

$$(c) \qquad Pr(x_1 \leq x \leq x_2, y_1 \leq y \leq y_2) = \int_{x_1}^{x_2} \int_{y_1}^{y_2} f(x, y)\, dy\, dx$$

The probability density function corresponding to the marginal distribution $F(x, \infty)$ will be

$$(d) \qquad g(x) = \int_{-\infty}^{+\infty} f(x, y)\, dy$$

since we have

$$F(x, \infty) = \int_{-\infty}^{x} \int_{-\infty}^{\infty} f(x, y)\, dy\, dx$$
$$= \int_{-\infty}^{x} \left[\int_{-\infty}^{\infty} f(x, y)\, dy \right] dx$$
$$= \int_{-\infty}^{x} g(x)\, dx$$

Similarly the probability density function of the marginal distribution $F(\infty, y)$ will be

$$(e) \qquad h(y) = \int_{-\infty}^{+\infty} f(x, y)\, dx$$

If x and y are independent, we have from (*b*) of the previous section

$$F(x, y) = F(x, \infty) \cdot F(\infty, y)$$

$$\int_{-\infty}^{x} \int_{-\infty}^{y} f(x, y)\, dy\, dx = \int_{-\infty}^{x} g(x)\, dx \cdot \int_{-\infty}^{y} h(y)\, dy$$
$$= \int_{-\infty}^{x} \int_{-\infty}^{y} g(x) \cdot h(y)\, dy\, dx$$

Hence in the case of independence, we must have

$$(f) \qquad f(x, y) = g(x) \cdot h(y)$$

i.e., the joint probability density function is the product of the probability density functions of the individual variables. The factoring of the joint probability density function into a function of x and a function of y is also sufficient to ensure the independence of x and y.

Although the probability associated with any particular point is in this case zero, it will again be convenient to think of the *probability element* $f(x, y)\,dx\,dy$ as defining the probability that x will fall in the interval from x to $x + dx$, and y in the interval y to $y + dy$. In the case of independence, it follows from (f) that

$$f(x, y)\,dx\,dy = g(x)\,dx \cdot h(y)\,dy \tag{g}$$

4.26. Distributions in More than Two Variables

In a manner completely analogous with that of the preceding section, we could define the joint cumulative distribution function $F(x, y, z, \cdots)$ of a series of random variables $x, y, z, \cdots$ and the associated discrete probability distribution $p(x_i, y_j, z_k, \cdots)$ or joint probability density function $f(x, y, z, \cdots)$. For the purposes of this book it will be sufficient to note a few extensions of the formulae already obtained.

The marginal distribution (discrete probability distribution or probability density function) of any variable can be obtained by integrating or summing over the remaining variables, i.e.,

$$p(x_i) = \Sigma_j \Sigma_k \cdots p(x_i, y_j, z_k, \cdots)$$

$$g(x) = \int_{-\infty}^{\infty} \int_{-\infty}^{\infty} \cdots f(x, y, z, \cdots)\,dy\,dz \cdots$$

If the random variables $x, y, z, \cdots$, are mutually independent, we have

$$p(x_i, y_j, z_k, \cdots) = p(x_i) \cdot p(y_j) \cdot p(z_k) \cdots$$

or

$$f(x, y, z, \cdots) = g(x) \cdot h(y) \cdot l(z) \cdots$$

i.e., the joint distribution of a series of independent random variables is simply the product of their individual distributions.

4.3. Moments, Semi-invariants, and Generating Functions

4.31. Moments and Moment Generating Functions

In Chapter 3 we defined the average value and variance of a random variable as a means of characterizing its behavior. We now see that these definitions were directly dependent on the distribution function of the random variable. In this and the following section we shall introduce two more extensive sets of parameters, both of which include the mean and variance, which can be used for this purpose. We shall frequently carry out derivations only for a variable x which is continuous and has a probability density function $f(x)$; completely analogous developments are possible for a discrete distribution, and we shall give the more important formulae for both cases. Mathematically, this duplicity can be avoided by using Stieltje's integration with respect to the cumulative distribution function $F(x)$.

The value

$$\mu_r' = \text{ave}\,(x^r) = \Sigma_i\, x_i^r p(x_i) \tag{a}$$

$$= \int_{-\infty}^{\infty} x^r f(x)\, dx$$

is called the rth moment of the variable x. In particular, we have $\mu_1' = \text{ave}\,(x) = \mu$. With certain general restrictions, which need not concern us, the distribution of the random variable x is completely specified by the moments μ_r'.

In many cases it is more convenient to consider the particular set of moments obtained by choosing the mean μ_1' as the origin. These *moments about the mean* are defined by

$$\mu_r = \text{ave}\,(x - \mu_1')^r = \Sigma_i\,(x_i - \mu_1')^r p(x_i) \tag{b}$$

$$= \int_{-\infty}^{\infty} (x - \mu_1')^r f(x)\, dx$$

By expanding the integrand using the binomial theorem, and integrating term by term, we can obtain expressions for the μ_r in terms of the μ_r' previously defined. The expressions for the first four are

$$\mu_1 = \mu_1' - \mu_1' = 0$$

$$\mu_2 = \mu_2' - \mu_1'^2 \tag{c}$$

$$\mu_3 = \mu_3' - 3\mu_2'\mu_1' + 2\mu_1'^2$$

$$\mu_4 = \mu_4' - 4\mu_3'\mu_1' + 6\mu_2'\mu_1'^2 - 3\mu_1'^4$$

In particular, since $\mu_2' = \text{ave}\,(x^2)$ and $\mu_1' = \mu$, we have

$$\mu_2 = \text{ave}\,(x^2) - \mu^2 = \sigma^2 \tag{d}$$

The function

$$M_x(\theta) = \text{ave}\,(e^{\theta x}) = \Sigma_i\, e^{\theta x_i} p(x_i) \tag{e}$$

$$= \int_{-\infty}^{\infty} e^{\theta x} f(x)\, dx$$

is called the moment generating function of the variable x. If such a function exists, we can expand $e^{x\theta}$ and integrate term by term, obtaining

$$
\begin{aligned}
M_x(\theta) &= \int_{-\infty}^{\infty} e^{\theta x} f(x)\, dx \\
&= \int_{-\infty}^{\infty} \left[1 + \theta x + \frac{\theta^2 x^2}{2!} + \cdots\right] f(x)\, dx \\
&= \int_{-\infty}^{\infty} f(x)\, dx + \theta \int_{-\infty}^{\infty} x f(x)\, dx + \frac{\theta^2}{2!} \int_{-\infty}^{\infty} x^2 f(x)\, dx + \cdots \\
&= 1 + \mu_1'\theta + \mu_2' \frac{\theta^2}{2!} + \cdots
\end{aligned} \qquad (f)
$$

i.e., if $M_x(\theta)$ is expanded in powers of θ the coefficient of $\theta^r/r!$ is μ_r'. From the MacLaurin expansion of the function $M_x(\theta)$ we obtain immediately

$$
\mu_r' = \left.\frac{d^r M_x(\theta)}{d\theta^r}\right]_{\theta=0} \qquad (g)
$$

In an analogous fashion the function

$$
\begin{aligned}
M_{x-\mu}(\theta) = \text{ave } [e^{\theta(x-\mu)}] &= \int_{-\infty}^{\infty} e^{\theta(x-\mu)} f(x)\, dx \\
&= e^{-\mu\theta} \int_{-\infty}^{\infty} e^{\theta x} f(x)\, dx \\
&= e^{-\mu\theta} M_x(\theta)
\end{aligned} \qquad (h)
$$

will generate moments about the mean.

Although these functions present a convenient method of determining the moments of a probability distribution, they are most useful because of two properties which make them extremely important in the derivation of the distributions of sampling statistics. These are:

(1) The moment generating function of the sum of two independent variables is the product of their individual moment generating functions, i.e., if x and y are 2 independent random variables,

$$
M_{x+y}(\theta) = M_x(\theta) \cdot M_y(\theta) \qquad (i)
$$

This, of course, is true only when the moment generating functions for x and y exist.

(2) Under certain general conditions which for our purposes will always be satisfied, the moment generating function for a particular distribution, if it exists, is unique, i.e., if we obtain a particular moment generating function for a random variable x we can conclude that it has the probability distribution associated with this moment generating function.

The first is easily seen to be true, since, if x and y are independent, $e^{\theta x}$ and $e^{\theta y}$ are independent, and hence

$$\text{ave}\,[e^{\theta(x+y)}] = \text{ave}\,(e^{\theta x} \cdot e^{\theta y}) = \text{ave}\,(e^{\theta x}) \cdot \text{ave}\,(e^{\theta y})$$

from which (i) follows immediately. This proof can be extended to any number of independent variables. The proof of the second property is beyond the scope of this book.

Moments and moment generating functions can be similarly defined for joint distributions of two or more random variables. Since they will not be required for later developments in this book, we shall not deal with them here. The interested reader is referred to [2, Vol. I] or [3]. For theoretical purposes the moment generating function, which frequently fails to exist because of the non-existence of higher-order moments, may be replaced by the characteristic function

$$\phi_x(\theta) = \text{ave}\,(e^{ix\theta})$$

which has essentially the same properties and exists under much more general circumstances.

4.32. Semi-invariants

Another set of parameters which can be used to characterize a distribution are the semi-invariants, or cumulants, which are usually designated by the Greek letter κ. They can be defined formally by the identity in θ

$$(a) \qquad e^{\left(\kappa_1\theta + \kappa_2 \frac{\theta^2}{2!} + \cdots\right)} = 1 + \mu_1'\theta + \mu_2' \frac{\theta^2}{2!} + \cdots$$

If $M_x(\theta)$ exists it follows immediately that

$$(b) \qquad \kappa_1\theta + \kappa_2 \frac{\theta^2}{2!} + \cdots = \ln M_x(\theta)$$

i.e., $\ln M_x(\theta)$ is a generating function for the semi-invariants, and is frequently called the semi-invariant generating function. From (h) of the preceding section we have, on taking logarithms of both sides

$$(c) \qquad \ln M_x(\theta) = \ln M_{x-\mu}(\theta) - \mu\theta$$

Hence defining the semi-invariants in terms of moments about the mean instead of moments about the origin only changes the value of κ_1. The same would be true for a shift of origin to any arbitrary point. The first four semi-invariants are

$$(d)\qquad \begin{aligned} \kappa_1 &= \mu_1' \\ \kappa_2 &= \mu_2 \\ \kappa_3 &= \mu_3 \\ \kappa_4 &= \mu_4 - 3\mu_2^2 \end{aligned}$$

An extensive tabulation of the relationships between the semi-invariants and moments, both about the mean and about an arbitrary origin, is given in [2, Vol. I, pp. 62–64].

The most important property of the semi-invariants follows directly from (*i*) of the preceding section. Taking the logarithms of both sides, we have, for x and y independent,

$$(e)\qquad \ln M_{x+y}(\theta) = \ln M_x(\theta) + \ln M_y(\theta)$$

and, designating the semi-invariants of $x + y$ by κ_r and those of x and y by κ_r' and κ_r'', respectively, it follows immediately from the fact that (*e*) is an identity in θ that we must have

$$(f)\qquad \kappa_r = \kappa_r' + \kappa_r''$$

i.e., the semi-invariants of the sum of two independent random variables are the sums of their respective semi-invariants of the same order. This property, which can easily be extended to more than two mutually independent variables, has made the use of semi-invariants preferable to the use of moments in the study of sampling.

Tukey [75] has defined the equivalent of a system of semi-invariants for finite populations. In this case additional quantities must be introduced to take care of the products of unlike elements which, like the cov (x_i, x_j), $i \neq j$, in Section 3.44, do not disappear. These quantities have been useful in the study of variance components, to be considered in Chapter 7, where the assumption of an infinite population of effects is often completely unrealistic.

4.33. Calculation and Interpretation of Semi-invariants

When either a large sample (say ≥ 50) or a large number of small samples is available, it is frequently desirable to compute the third and fourth semi-invariants of the sample or samples as a reflection of the nature of the underlying distribution. Except in the case of very large samples, semi-invariants of order higher than 4 are not usually computed, since their variation from sample to sample is too high for the values obtained to be useful.

The first 4 sample semi-invariants (called by R. A. Fisher k-statistics) are given by

(a)
$$k_1 = \frac{S_1}{n}$$

$$k_2 = \frac{nS_2 - S_1^2}{n(n-1)}$$

$$k_3 = \frac{n^2S_3 - 3nS_2S_1 + 2S_1^3}{n(n-1)(n-2)}$$

$$k_4 = \frac{(n^3 + n^2)S_4 - 4(n^2 + n)S_3S_1 - 3(n^2 - n)S_2^2 + 12nS_2S_1^2 - 6S_1^4}{n(n-1)(n-2)(n-3)}$$

where n is the number in the sample and $S_r = \Sigma_i x_i^r$, or, in the case of a frequency distribution, $S_r = \Sigma_i x_i^r f_i$. The first two of these are simply the previously defined sample mean and variance. These statistics have the important property that

(b)
$$\text{ave}\,(k_r) = \kappa_r$$

i.e., they are unbiased estimates of the semi-invariants of the population. The variances of the k-statistics (except for var (k_1) = var $(\bar{x})$, which is given in Section 3.42) will be considered only for the special case of the normal distribution, to be discussed later in this chapter. The interested reader is again referred to [2, Vol. I].

Since $\kappa_3 = \mu_3$ depends on the cube of the deviations from the average value μ, it will be zero only if the distribution is symmetric. For asymmetric, or skewed, distributions it will be positive or negative, depending on whether the larger deviations from the mean are in the positive or negative direction. The statistic k_3 depends on the skewness of the sample in a similar fashion, and reflects the asymmetry, or skewness, of the population. Figure 4.5 illustrates the type of distribution that might be expected if either positive or negative skewness were present. Instead of the value of k_3, which is in units which are the cube of those of the original measurements, the value $g_1 = \dfrac{k_3}{k_2^{3/2}} = \dfrac{k_3}{\sigma^3}$, which is independent of the original units, is usually used as a measure of skewness. Since k_2 must be positive, similar signs on g_1 and k_3 indicate the same type of skewness.

Since $\kappa_4 = \mu_4 - 3\mu_2^2$, and both μ_4 and μ_2 are based on even powers of the deviations, κ_4 will not reflect the direction of the deviations. However, because of the presence of the fourth power of the deviations, κ_4 will be affected to a greater degree than $\kappa_2 = \mu_2$ by the presence of large deviations. For the particular case of the normal distribution, it will be shown in Section 4.41 that $\mu_4 = 3\mu_2^2$, or $\kappa_4 = 0$; hence, using this as a base point, positive values of k_4 would reflect the presence of a greater number of large deviations than would be expected if the distribution were normal, and negative values of k_4 would reflect the presence of fewer large deviations than expected. k_4 is particularly sensitive as a measure of "contamination" in a distribution, i.e., the presence of a

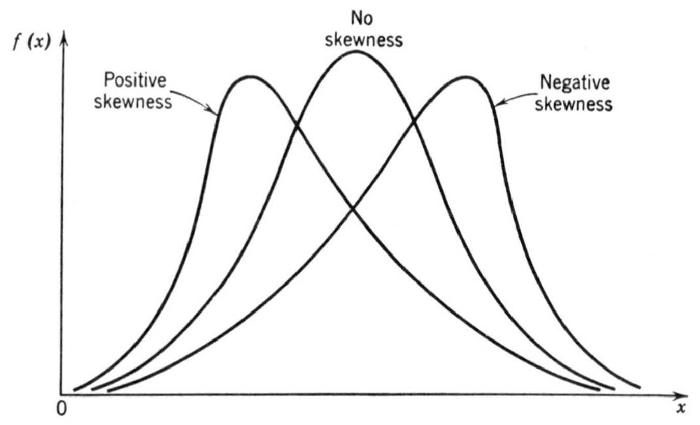

FIG. 4.5. Skewed distributions.

small percentage of widely scattered observations in an otherwise normal distribution. Contaminated distributions frequently arise in the study of chemical measurements, owing to the presence of occasional gross errors, a fact noted by "Student" [78] as early as 1927. The dimensionless value $g_2 = k^4/k_2^2$ is used as a measure more frequently than k_4, and the characteristic of the distribution which it measures is called kurtosis. Distributions with positive kurtosis ($\kappa_4 > 0$) are called leptokurtic, and with negative kurtosis ($\kappa_4 < 0$) platykurtic.

EXAMPLE 1. The probability distribution defined by

$$f(x) = 0 \qquad x < -a/2$$

$$f(x) = 1/a \qquad -a/2 < x < a/2$$

$$f(x) = 0 \qquad x > a/2$$

is known as the rectangular distribution because of the shape of its graph. For this distribution we have

$$\mu_r = \int_{-a/2}^{a/2} x^r \left(\frac{1}{a}\right) dx$$
$$= \frac{x^{r+1}}{a(r+1)}\Bigg]_{-a/2}^{+a/2}$$
$$= \frac{1}{(r+1)2^{r+1}}[a^r + (-a)^r]$$

Hence all odd moments are zero, and for the even moments

$$\mu_r = \frac{a^r}{(r+1)2^r} \qquad r = 2, 4, 6, \cdots$$

In particular

$$\mu_2 = a^2/12 \qquad \mu_4 = a^4/80$$

and hence

$$\kappa_3 = 0 \qquad \kappa_4 = \mu_4 - 3\mu_2^2 = -a^4/120$$

These values reflect the symmetry of the distribution, and the fact that the uniform scatter is abruptly terminated, with no possibility of the larger deviations which are present in the case of the normal distribution.

EXAMPLE 2. By extending Table 2.4 in Section 2.35 to include columns for $z_i^3 f_i$ and $z_i^4 f_i$, we obtain for the data of Table 2.1 the approximate power sums $S_0 = 364$, $S_1 = 141$, $S_2 = 2703$, $S_3 = 2331$, $S_4 = 64{,}827$. Substituting these in (*a*), we obtain as before, in units of z,

$$\bar{z} = k_1 = 0.3874 \qquad s_z^2 = k_2 = 7.2958$$

and, in addition,

$$k_3 = -2.1269 \quad k_4 = 17.1792$$

From the latter we compute

$$g_1 = \frac{k_3}{k_2^{3/2}} = -0.108 \qquad g_2 = \frac{k_4}{k_2^2} = 0.323$$

In Section 4.45 we shall consider methods of determining whether these values reflect real tendencies towards negative skewness and towards an excessive number of large deviations in comparison with the normal distribution or whether they could be due to sampling fluctuations.

4.4. The Normal Distribution

4.41. The Normal Distribution Function

One of the most frequent assumptions made in problems of statistical inference is that the observations are drawn from a population which could be described by a normal or Gaussian distribution function. Several reasons for this are:

(1) Distributions which are approximately normal are frequently encountered.

(2) The normal distribution is important as a "limiting distribution."

(3) The mathematical study of sampling from a normal population is comparatively simple and has been highly developed.

Because of the last point in particular, there is sometimes a tendency to assume a normal distribution without sufficient investigation as to whether this assumption is justified. On the other hand, it has also been shown that moderate departures from normality do not seriously affect the validity of many of the procedures which are based on the normal distribution.

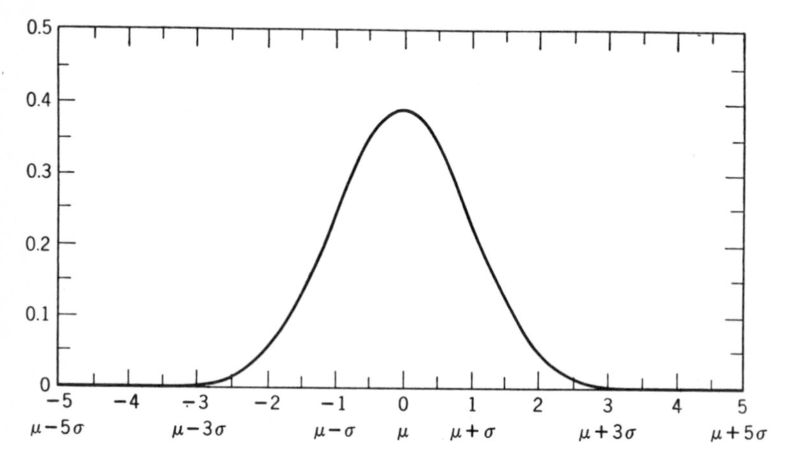

FIG. 4.6. Standard normal distribution.

The normal or Gaussian distribution function (or probability density function) is given by

$$(a) \qquad f(x) = \frac{1}{\sigma\sqrt{2\pi}} e^{-(x-\mu)^2/2\sigma^2}$$

where the factor $1/\sigma\sqrt{2\pi}$ is chosen so that $\int_{-\infty}^{\infty} f(x)\,dx = 1$. A graph of this distribution for $\mu = 0$ and $\sigma^2 = 1$ is shown in Figure 4.6. Notice that it is possible for x to have any value from $+\infty$ to $-\infty$; i.e., the function $f(x)$ is never 0 except for the values $+\infty$ and $-\infty$. However, about 99.99% of the area under the curve lies between $\mu - 4\sigma$ and $\mu + 4\sigma$, and about 99.74% between $\mu - 3\sigma$ and $\mu + 3\sigma$. Hence the apparent conflict between the infinite range of the normal distribution and the finite range of possible values which is generally the case in practice need be of very little concern if the range of possible values is greater than three or four standard deviations from the mean.

For the moment generating function of the normal distribution we have from (*e*) of Section 4.31

$$M_x(\theta) = \frac{1}{\sigma\sqrt{2\pi}} \int_{-\infty}^{\infty} e^{x\theta} e^{-(x-\mu)^2/2\sigma^2}\, dx$$

Integrating the right-hand side (see Appendix 4*A*), we obtain

$$(b) \qquad M_x(\theta) = e^{\mu\theta + (\sigma^2\theta^2/2)}$$

The first two derivatives of this function are

$$\frac{dM}{d\theta} = (\mu + \sigma^2\theta)\, e^{\mu\theta + (\sigma^2\theta^2/2)}$$

$$\frac{d^2M}{d\theta^2} = (\mu + \sigma^2\theta)^2\, e^{\mu\theta + (\sigma^2\theta^2/2)} + \sigma^2\, e^{\mu\theta + (\sigma^2\theta^2/2)}$$

Putting $\theta = 0$ in these expressions, we have from (*g*) of Section 4.31,

$$\mu_1' = \mu$$

$$\mu_2' = \mu^2 + \sigma^2$$

and hence

$$(c) \qquad \begin{aligned} \text{ave}\,(x) &= \kappa_1 = \mu_1' = \mu \\ \text{var}\,(x) &= \kappa_2 = \mu_2 = \mu_2' - \mu_1'^2 = \mu^2 + \sigma^2 - \mu^2 = \sigma^2 \end{aligned}$$

justifying the use of μ and σ as parameters in the original distribution. These results could have been obtained directly from (*a*) and (*b*) of Section 3.31 by integrating

$$\text{ave}\,(x) = \frac{1}{\sigma\sqrt{2\pi}} \int_{-\infty}^{\infty} x e^{-(x-\mu)^2/2\sigma^2}\, dx$$

$$\text{var}\,(x) = \frac{1}{\sigma\sqrt{2\pi}} \int_{-\infty}^{\infty} (x - \mu)^2 e^{-(x-\mu)^2/2\sigma^2}\, dx$$

We have used the moment generating function both to illustrate its use and to have it available for future reference.

For the generating function of moments about the mean we obtain

$$(d) \qquad \begin{aligned} M_{x-\mu}(\theta) &= e^{-\mu\theta} M_x(\theta) \\ &= e^{-\mu\theta} \cdot e^{\mu\theta + (\sigma^2\theta^2/2)} \\ &= e^{(\sigma^2\theta^2/2)} \end{aligned}$$

and for the semi-invariant generating function

(e) $$\ln M_{x-\mu}(\theta) = \sigma^2\theta^2/2$$

It follows that, except for $\kappa_1 = \mu$ and $\kappa_2 = \sigma^2$, all the semi-invariants of the normal distribution are zero. In particular, as we noted in Section 4.33, κ_3 and κ_4 are both zero. The first simply reflects the symmetry of the distribution, and the second establishes a basis for comparison with other distributions. Distributions for which κ_4 is positive will have longer "tails" than the normal distribution and those for which κ_4 is negative shorter ones.

If we transform the normal frequency function from the variable x to the variable $t = (x - \mu)/\sigma$, we obtain the distribution function

$$f(t) = (1/\sqrt{2\pi})e^{-t^2/2}$$

since in making the transformation we must transform the differential element $dx = \sigma\, dt$. Notice that the normal distribution so obtained has a mean of zero and a variance of 1, and that t, like x, can have all values between $-\infty$ and $+\infty$. A particular value of t is frequently called a standard normal deviate, and the distribution the unit normal distribution. Values of $f(t)$,

(f) $$F(t) = \frac{1}{\sqrt{2\pi}} \int_{-\infty}^{t} e^{-z^2/2}\, dz$$

and $1 - F(t)$ are given in Table I of the Appendix for t running from 0 to 4.00. Since the normal distribution is symmetric, $F(-t) = 1 - F(t)$. In order to find the area in the interval $x_1 \leq x \leq x_2$, we compute the values $t_1 = (x_1 - \mu)/\sigma$ and $t_2 = (x_2 - \mu)/\sigma$ and then determine from the tables the $Pr(t_1 \leq t \leq t_2) = F(t_2) - F(t_1)$. For example, if a given characteristic were normally distributed with $\mu = 100$ and $\sigma = 5$, the probability of obtaining a value between 96 and 108 would be:

$$\begin{aligned} F(t_2) - F(t_1) &= F\left(\frac{108 - 100}{5}\right) - F\left(\frac{96 - 100}{5}\right) \\ &= F(1.6) - F(-0.8) \\ &= 0.9452 - 0.2119 \\ &= 0.7333 \end{aligned}$$

Normal tables are frequently found which give values other than that of $F(t)$. Tables in the *Handbook of Chemistry and Physics* [4] give values of

$$\frac{1}{\sqrt{2\pi}} \int_{0}^{t} e^{-z^2/2}\, dz = F(t) - 0.5$$

for t ranging from 0 to 4.00. A particularly extensive table to 15 decimal places prepared by the New York Tables Project [5] gives values of

$$\frac{1}{\sqrt{2\pi}}\int_{-t}^{+t} e^{-z^2/2}\,dz = F(t) - F(-t)$$
$$= 2F(t) - 1$$
$$= 1 - 2F(-t)$$

Since we defined $t = (x - \mu)/\sigma$, we have $x - \mu = \sigma t$, and hence for a given t_0

$$Pr(|x - \mu| \leq \sigma t_0) = Pr(|t| \leq t_0)$$
$$Pr(|x - \mu| > \sigma t_0) = Pr(|t| > t_0)$$

Now suppose that we choose a value t_α of t such that

$$Pr(|t| > t_\alpha) = 2F(-t_\alpha) = \alpha$$

and hence

$$Pr(|t| \leq t_\alpha) = 1 - \alpha$$

Then a given value x of a normally distributed variable will have a probability α of deviating from its mean in absolute value by more than $t_\alpha\sigma$, and a probability $1 - \alpha$ of being within the interval $\mu - t_\alpha\sigma$ to $\mu + t_\alpha\sigma$. Table 4.3 lists the values of α, $1 - \alpha$, and t_α for some values of α and t_α which will be used frequently in later chapters.

TABLE 4.3

α	$1 - \alpha$	t_α
0.50	0.50	0.6745
0.3174	0.6826	1.00
0.10	0.90	1.6448
0.05	0.95	1.9600
0.0454	0.9546	2.00
0.02	0.98	2.3263
0.01	0.99	2.5758
0.0026	0.9974	3.00

Note that for $t_\alpha = 1$, 2, and 3 the values of $1 - \alpha$ are the relative frequencies which, as we stated in Section 2.36, should be approximated by the relative frequency of sample values in the intervals $\bar{x} \pm s$, $\bar{x} \pm 2s$, and $\bar{x} \pm 3s$. These observed frequencies will be approximate not only because they are determined from a sample but also because we must use $\bar{x}$ and s to replace the unknown population values μ and σ in determining the intervals to be considered.

4.42. Distribution of a Linear Combination of Normally Distributed Variables

One important property of the normal distribution which will frequently be used in later chapters is that any linear combination of independent normally distributed variables is itself normally distributed. We have seen (Section 3.33) that, if $x_1, x_2, \cdots x_n$, are independent variables with means $\mu_1, \mu_2, \cdots, \mu_n$, and variances $\sigma_1^2, \sigma_2^2, \cdots, \sigma_n^2$, then for any set of constants $a_1, \cdots, a_n$ the linear combination

$$(a) \qquad w = \Sigma a_i x_i$$

has mean and variance

$$(b) \qquad \begin{aligned} \text{ave}\,(w) &= \mu_w = \Sigma a_i \mu_i \\ \text{var}\,(w) &= \sigma_w^2 = \Sigma a_i^2 \sigma_i^2 \end{aligned}$$

By repeated use of the first property of moment generating functions, given by (*i*) of Section 4.31, we have

$$(c) \qquad M_w(\theta) = M_{a_1x_1}(\theta) \cdot M_{a_2x_2}(\theta) \cdots \cdot M_{a_nx_n}(\theta)$$

Since

$$\text{ave}\,(a_i x_i) = a_i\ \text{ave}\,(x_i) = a_i \mu_i$$

and

$$\text{var}\,(a_i x_i) = a_i^2\ \text{var}\,(x_i) = a_i^2 \sigma_i^2$$

we obtain by substituting these values in (*b*) of Section 4.41

$$(d) \qquad M_{a_ix_i}(\theta) = e^{a_i\mu_i\theta + (a_i^2\sigma_i^2\theta^2/2)}$$

Hence (*c*) becomes

$$(e) \qquad \begin{aligned} M_w(\theta) &= e^{a_1\mu_1\theta + (a_1^2\sigma_1^2\theta^2/2)} \cdot e^{a_2\mu_2\theta + (a_2^2\sigma_2^2\theta^2/2)} \cdot \cdots \cdot e^{a_n\mu_n\theta + (a_n^2\sigma_n^2\theta^2/2)} \\ &= e^{\theta \Sigma a_i\mu_i + \frac{\theta^2}{2}\Sigma a_i^2\sigma_i^2} \\ &= e^{\mu_w\theta + (\sigma_w^2\theta^2/2)} \end{aligned}$$

Since this is identical with the moment generating function of a normal distribution, we can conclude (Section 4.31) that w is normally distributed with mean μ_w and variance σ_w^2 given by (*b*).

This property of the normal distribution could also have been deduced immediately from the nature of its semi-invariants. We showed in Section 4.32 that the semi-invariants of a sum of independent variables will be the sum of their semi-invariants. Since a normally distributed variable is characterized by having all semi-invariants beyond the second equal to zero, then the corresponding semi-invariants of any linear combination of such variables will also be zero, which characterizes the linear combination as normally distributed.

Several particular cases of linear combinations will frequently occur in later chapters. One of these is the case

$$w = x \pm y$$

where x and y are independent and normally distributed with means μ_x and μ_y and variances σ_x^2 and σ_y^2. It follows immediately that w is normally distributed with

$$\text{ave}(w) = \text{ave}(x \pm y) = \mu_x \pm \mu_y$$

(*f*)

$$\text{var}(w) = \sigma_x^2 + \sigma_y^2$$

Notice that the variances are added for either the sum or the difference. A second case is that in which

$$w = \Sigma x_i$$

the x_i, $i = 1, \cdots, n$ being assumed independent and normally distributed with identical means μ and variances σ^2. In this instance w is normally distributed with

$$\text{ave}(w) = n\mu$$

(*g*)

$$\text{var}(w) = n\sigma^2$$

since all the a_i are equal to 1.

4.43. The Central Limit Theorem

Let us now consider the sum

$$w = \sum_{i=1}^{n} x_i = x_1 + x_2 + \cdots + x_n$$

where $x_1, x_2, \cdots, x_n$ are identically distributed independent variables, each having mean μ and finite variance σ^2. Then, under very minor restrictions which can always be assumed to hold in practice, the distribution of

$$t = \frac{w - \mu_w}{\sigma_w}$$

where μ_w and σ_w are given by (*g*) of the preceding section, approaches the unit normal distribution as n becomes infinite. The theorem is also true when the variables are not identically distributed if their variances are not too different. Essentially, this theorem states that the sum of a sufficient number of random variables will be approximately normally distributed, regardless of the distribution of the individual variables.

What constitutes a sufficient number of variables depends to a great extent on the shape of the distribution of the original variables. If the distributions are at least bell-shaped, with a single central point about which the values seem to cluster, then the sum of as few as 4 or 5 independent observations can be considered normally distributed for almost all practical purposes. This fact has been confirmed by experimental sampling, notably that done by W. A. Shewhart [26], who labeled chips in such a fashion as to duplicate closely a normal distribution, a rectangular distribution (all values between two limits equally probable), and a right triangular distribution (values from lowest to highest have linearly decreasing probabilities). By placing the chips in a bowl, and stirring between each drawing, he made 4000 drawings (replacing the chips drawn) from each population. He then showed, among other things, that the means of the 1000 samples of 4 so obtained are very nearly normally distributed with the expected mean and variance for each of the three distributions.

EXAMPLE. Table 4.4 gives the frequency distribution of sums obtained by summing the point values (assigned as in Section 3.21) of 50 groups of 5 cards each drawn as randomly as possible from a deck of 52 cards. The distribution of these sums is already assuming a normal shape, as can be seen from the histogram in Figure 4.7, even though each point

TABLE 4.4

FREQUENCY DISTRIBUTION OF SUMS OF SAMPLES OF FIVE FROM A RECTANGULAR POPULATION

Sum of Point Values	Frequency
12.5–17.5	2
17.5–22.5	1
22.5–27.5	6
27.5–32.5	12
32.5–37.5	11
37.5–42.5	8
42.5–47.5	8
47.5–52.5	1
52.5–57.5	1
	50

value is equally probable for any individual drawing. Note that, since, from Section 3.31, the expected value and variance of an individual drawing are 7 and 14, respectively, the expected value and variance of the sum of 5 drawings should be 5(7) = 35 and 5(14) = 70. The sample mean and variance for the ungrouped sums were 34.9 and 70.46, respectively.

4.44. Transformations to Approximate Normality

Very frequently a distribution which is apparently not normal can be made approximately normal by a transformation of the data, i.e., by using $\log x$, $\sqrt{x}$, or some other similar function of the observations rather than the observations themselves. It is frequently difficult to obtain the correct transformation in any particular instance, although in many cases

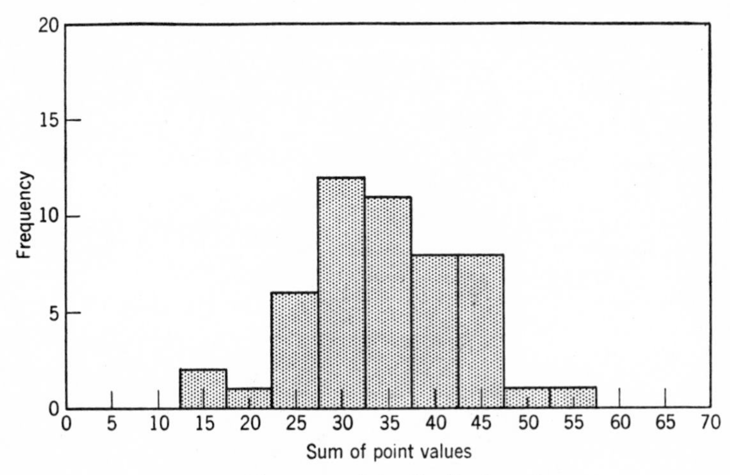

FIG. 4.7. Histogram of the distribution of sums.

an investigation of the sources of the variation in the data will provide some guidance. For example, in studying the distribution of a series of spectrographic analyses on a single standard the logarithmic transformation is the natural one to use, since the basis of most spectrographic analyses is the measurement of the density of a given spectral line on a photographic plate, and this density is proportional to the logarithm of the amount of the given material present. Another method of looking for the correct transformation is to plot the data on probability paper, using various functions of the original scale of measurement as the abscissa. Because the logarithmic transformation occurs so frequently, special probability paper (frequently called log probability paper) with the linear scale replaced by a logarithmic one, has been prepared for this purpose. The range of observed values may give some clue as to the type of scale which should be used. A variable like the radioactivity of air samples may vary over 3 or 4 orders of magnitude, making a logarithmic transformation almost mandatory as a preliminary step in examining the data. Similarly, in analyzing for small amounts of impurity, observations with an average of, say, 1 ppm, but with occasional

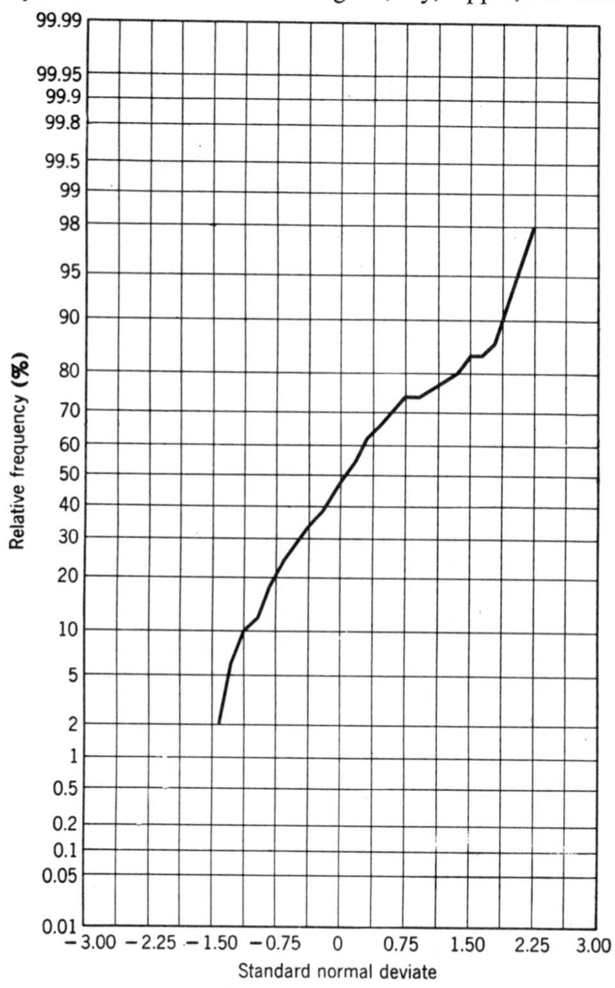

FIG. 4.8. Cumulative frequency polygon for samples of 50 from unit normal distribution.

values of 10 or 15 ppm, should certainly be transformed in some manner, in view of the fact that values below 0 are, at least theoretically, impossible.

4.45. Tests for Non-normality

If we are to have some means of determining when a transformation to normality is necessary, or when a transformation has given approximate normality, we must have some means of testing for non-normality. This is very difficult to do in the case of small samples. With moderately

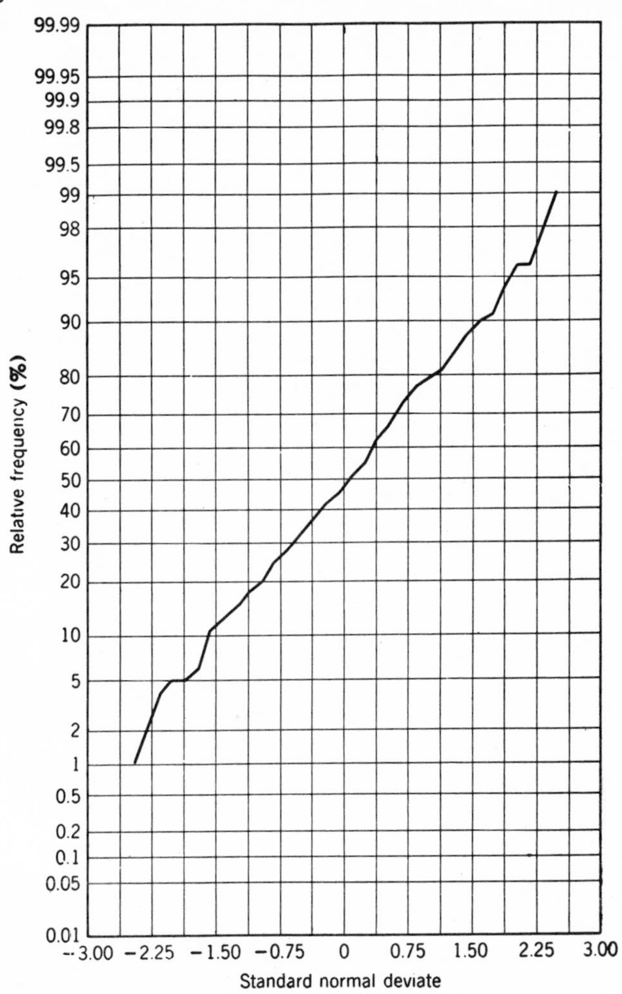

FIG. 4.9. Cumulative frequency polygon for samples of 100 from unit normal distribution.

large samples (say $n \geq 50$), or with a number of smaller samples, there are several means of checking the reasonableness of our assumption. One of the simplest ways of getting a rough check is to plot the data on probability paper as suggested in Chapter 2 and in the previous section. If such a plot seems to be fairly close to a straight line, it is reasonable to assume that we have approximately a normal distribution. In order to give a feeling for the way normal samples should look, Figures 4.8, 4.9, and 4.10 show the plots of random samples of size 50, 100, and 200 drawn

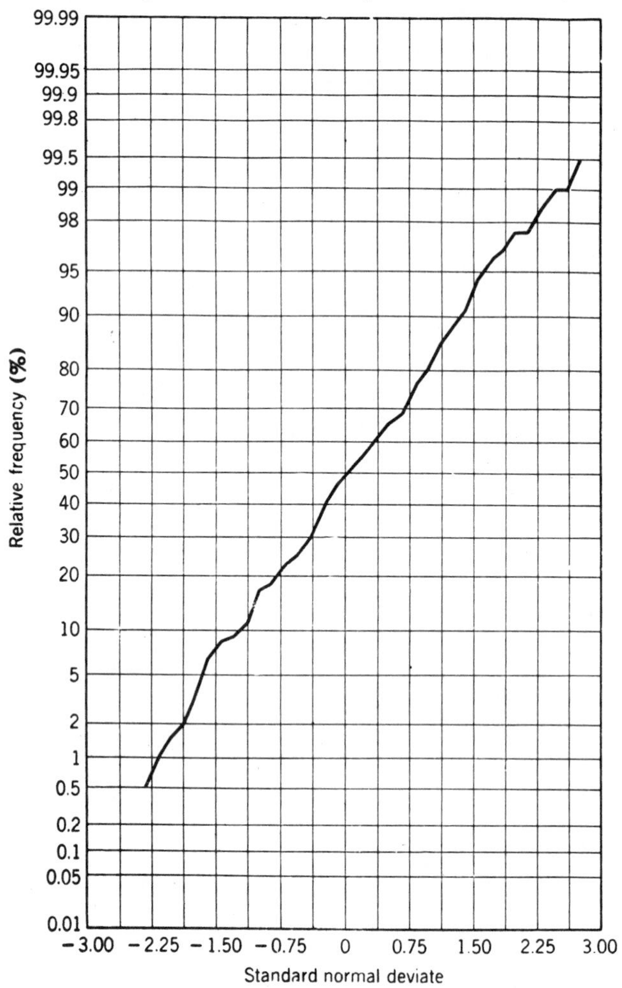

FIG. 4.10. Cumulative frequency polygon for samples of 200 from unit normal distribution.

from an artificially constructed normal population with mean 0 and standard deviation 1.

A second method is to compute k_3 and k_4 (see Section 4.33) and compare them with their average value 0 for samples from a normal distribution. To do this we must have a means of computing their expected variation from sample to sample. For random samples of any size from a normal distribution we have

$$\text{var}(k_3) = \frac{6n}{(n-1)(n-2)}\kappa_2^3$$

$$\text{var}(k_4) = \frac{24n(n-1)}{(n-1)(n-2)(n-3)}\kappa_2^4$$

However, since they depend on the unknown κ_2, they are not directly

TABLE 4.5. 5% AND 1% POINTS FOR g_1 AND g_2*

Size of Sample	g_1		g_2			
	Lower and Upper		Lower		Upper	
	5%	1%	1%	5%	5%	1%
50	0.550	0.812	—	—	—	—
75	0.454	0.664	—	—	—	—
100	0.395	0.576	−0.80	−0.62	0.87	1.53
125	0.354	0.514	−0.74	−0.57	0.78	1.34
150	0.324	0.469	−0.69	−0.53	0.71	1.22
175	0.301	0.434	−0.66	−0.50	0.66	1.11
200	0.282	0.406	−0.62	−0.47	0.62	1.04
250	0.253	0.362	−0.57	−0.44	0.55	0.91
300	0.231	0.331	−0.53	−0.40	0.50	0.82
350	0.214	0.306	−0.49	−0.37	0.47	0.75
400	0.201	0.286	−0.48	−0.35	0.43	0.69
450	0.189	0.270	−0.44	−0.33	0.40	0.65
500	0.180	0.256	−0.42	−0.32	0.38	0.62
550	0.171	0.244	−0.41	−0.30	0.37	0.59
600	0.163	0.234	−0.39	−0.29	0.35	0.55
650	0.157	0.225	−0.38	−0.28	0.35	0.53
700	0.151	0.215	−0.37	−0.27	0.32	0.51
750	0.146	0.208	−0.35	−0.26	0.31	0.49
800	0.142	0.202	−0.34	−0.25	0.30	0.47
850	0.138	0.196	−0.33	−0.25	0.29	0.46
900	0.134	0.190	−0.33	−0.24	0.29	0.44
950	0.130	0.185	−0.32	−0.23	0.28	0.43
1000	0.127	0.180	−0.31	−0.23	0.27	0.42

* Derived, by permission of Professor E. S. Pearson, from: Tables IV and V of R. C. Geary and E. S. Pearson, "Tests of Normality," *Biometrika* Office, University College, London, 1938.

useful except when n is sufficiently large to assume $k_2 = \kappa_2$. For very large samples (say $n > 500$) we can consider g_1 and g_2 (see Section 4.33) as being approximately normally distributed with average value 0 and variances $\sim 6/n$ and $\sim 24/n$. Hence, $t = g_1\sqrt{n/6}$ and $t = g_2\sqrt{n/24}$ are approximately standard normal deviates, and we can determine the probability α of a deviation as large or larger in absolute value (Section 4.41). If this probability is small, we suspect that the distribution from which the sample was taken is not normal. For smaller samples it is preferable to use Table 4.5, which was obtained from curves more closely approximating those of g_1 and g_2 than the normal when n is only moderately large [118]. Table 4.5 gives values which should be exceeded by $|g_1|$ only 5% and 1% of the time; if these values are exceeded, the normality of the underlying distribution would be questioned. Similarly, this table gives upper and lower 5% and 1% limits (not symmetric in this case) for g_2.

EXAMPLE. For the data of Table 2.1 in Chapter 2, we found in Section 4.33 that $g_1 = -0.108$ and $g_2 = 0.323$. Since neither of these values is excessive for $n = 364$, there is no evidence of non-normality from these tests. The cumulative distribution plotted on probability paper, which was given in Figure 2.5, also shows no evidences of non-normality.

4.46. The χ^2 Distribution

An important distribution which is closely related to the normal distribution is that of the sum of squares of independent standard normal deviates. In particular, if $t_1, t_2, \cdots, t_\nu$ are standardized normal variables, and we define

$$(a) \qquad \chi^2 = t_1^2 + t_2^2 + \cdots + t_\nu^2 = \sum_{i=1}^{\nu} t_i^2$$

then χ^2 has the distribution

$$(b) \qquad f(\chi^2) = \frac{1}{2^{\nu/2}\Gamma(\nu/2)} e^{-\chi^2/2} (\chi^2)^{(\nu/2)-1} \qquad 0 < \chi^2 < \infty$$

which is known as the *χ^2-distribution with ν degrees of freedom.*

This distribution is directly related to the familiar Maxwell distribution of molecular velocities. If we denote the components of the thermal velocity of a molecule of mass m by v_1, v_2, v_3, then it is known that for a gas in a steady state these components are independently distributed with the probability elements

$$(c) \qquad f(v_i)\, dv_i = \left(\frac{hm}{\pi}\right)^{1/2} e^{-hmv_i^2}\, dv_i \qquad i = 1, 2, 3$$

where $h = 1/2RT$. If we let $t_i = v_i\sqrt{2hm}$, $dv_i = dt_i/\sqrt{2hm}$, then the probability elements (*c*) become

$$(d) \qquad f(t_i)\,dt_i = \frac{1}{\sqrt{2\pi}}\,e^{-t_i^2/2}\,dt_i \qquad i = 1, 2, 3$$

i.e., each t_i has the unit normal distribution. Hence

$$\chi^2 = t_1^2 + t_2^2 + t_3^2 = 2hm(v_1^2 + v_2^2 + v_3^2) = 2hmv^2$$

where v is the total molecular velocity, will have the distribution (*b*) with $\nu = 3$. By direct substitution we obtain

$$f(\chi^2)\,d\chi^2 = \frac{1}{2^{3/2}\Gamma(3/2)}\,e^{-hmv^2}\,(2hmv^2)^{1/2}\,d(2hmv^2)$$

or

$$(e) \qquad f(v)\,dv = 4\pi\left(\frac{hm}{\pi}\right)^{3/2} e^{-hmv^2}\,v^2\,dv$$

as the probability element for the total molecular velocity of a molecule. As with any such probability distribution, integration from 0 to c_0 or between c_1 and c_2 will give the fraction of the molecules having velocities below c_0, or between c_1 and c_2.

The distribution (*b*) can be obtained directly, using an extension of the transformation to spherical coordinates usually used to obtain (*e*) from the joint distribution of v_1, v_2, and v_3. However, it is easier to show that this distribution is that of the sum of squares of ν independent standard normal deviates by the use of moment generating functions. If t has a unit normal distribution, then

$$\begin{aligned}
M_{t^2}(\theta) &= \frac{1}{\sqrt{2\pi}}\int_{-\infty}^{\infty} e^{t^2\theta}e^{-t^2/2}\,dt \\
&= \frac{1}{\sqrt{2\pi}}\int_{-\infty}^{\infty} e^{-1/2(1-2\theta)t^2}\,dt \\
&= \frac{(1-2\theta)^{-1/2}}{\sqrt{2\pi}}\int_{-\infty}^{\infty} e^{-1/2(1-2\theta)t^2}\,d(t\sqrt{1-2\theta}) \\
&= (1-2\theta)^{-1/2}
\end{aligned}$$

since (see Appendix 4*A*)

$$\int_{-\infty}^{\infty} e^{-z^2/2}\,dz = \sqrt{2\pi}$$

Hence the sum of ν independent values of t^2 would have the moment generating function

$$M_{\chi^2}(\theta) = (1 - 2\theta)^{-\nu/2} \tag{g}$$

It can be shown by direct integration (see Appendix 4*A*) that this is the moment generating function of the distribution (*b*).

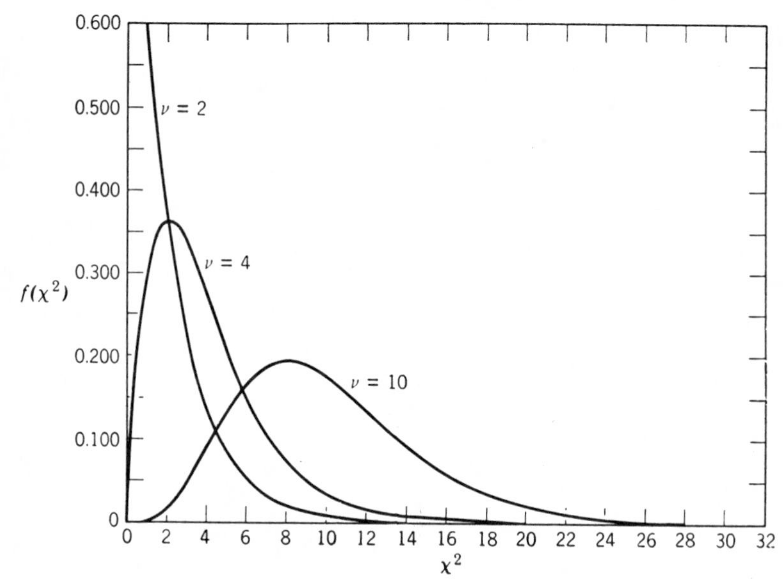

FIG. 4.11. χ^2-distribution for $\nu = 2$, 4, and 10.

The first three terms of the expansion of (*g*) in powers of θ are

$$M_{\chi^2}(\theta) = 1 + \nu\theta + \frac{\nu(\nu + 2)\theta^2}{2} + \cdots$$

and hence for χ^2 we have $\mu_1' = \nu$ and $\mu_2' = \nu^2 + 2\nu$. Thus

$$\begin{aligned} \text{ave}\,(\chi^2) &= \nu \\ \text{var}\,(\chi^2) &= \nu^2 + 2\nu - \nu^2 = 2\nu \end{aligned} \tag{h}$$

The range of the χ^2-distribution is $0 < \chi^2 < \infty$, since negative values of χ^2 are impossible. The distribution is highly skewed for small values of ν, becomes more symmetric as ν increases, and approaches normality for large ν. The last statement follows from the central limit theorem, since χ^2 is the sum of ν independent and identically distributed variables.

Figure 4.11 shows the χ^2-distribution for $\nu = 2$, 4, and 10. Because of the importance of this distribution in later applications, values of $\chi^2_{\nu,\alpha}$ such that $Pr(\chi_\nu^2 \geq \chi^2_{\nu,\alpha}) = \alpha$, where χ_ν^2 designates a variable with the χ^2-distribution with ν degrees of freedom, are given for various values of α and ν in Table II in the Appendix.

It follows from the nature of the moment generating function (g) that if two variables x and y are independent, and each has a χ^2-distribution with ν_1 and ν_2 degrees of freedom, respectively, their sum has a χ^2-distribution with $\nu_1 + \nu_2$ degrees of freedom. For we have

$$\begin{aligned} M_{x+y}(\theta) &= M_x(\theta) \cdot M_y(\theta) \\ &= (1 - 2\theta)^{-\nu_1/2} (1 - 2\theta)^{-\nu_2/2} \\ &= (1 - 2\theta)^{-(\nu_1+\nu_2)/2} \end{aligned}$$

which is the moment generating function of a χ^2-distribution with $\nu_1 + \nu_2$ degrees of freedom.

4.5. Sampling from the Normal Distribution

4.51. Distribution of Sample Mean

We now proceed to study the distribution of sample statistics computed from samples from a normal population. These distributions are basic to many of the methods of inference we shall use later.

For each observation $x_1, \cdots, x_n$ in a sample from a normal distribution with average value μ and variance σ^2, we have ave $(x_i) = \mu$ and var $(x_i) = \sigma^2$. It follows from Sections 4.42 and 3.33, since the sample values are normally distributed and independent, that any linear combination $w = \Sigma a_i x_i$ of the sample values will also be normally distributed with

$$\text{(a)} \qquad \begin{aligned} \text{ave}\,(w) &= \mu \Sigma a_i \\ \text{var}\,(w) &= \sigma^2\, \Sigma a_i^2 \end{aligned}$$

In particular, the sample mean is a linear combination of the sample values with $a_i = 1/n$; thus for samples from a normal population, $\bar{x}$ is also normally distributed. Since $\Sigma_i a_i = \Sigma_i(1/n) = 1$, and $\Sigma_i a_i^2 = \Sigma_i(1/n)^2 = 1/n$, the average value and variance of $\bar{x}$ are μ and σ^2/n, respectively, as was found in Section 3.42.

In simple terms this means that, if we draw samples of n at random from a normal distribution, the means of these samples will also form a normal distribution with the same mean, but smaller variance. This is illustrated in Figure 4.12, which shows the relative shape of the distribution of means for samples of $n = 4$ and $n = 9$ compared with the distribution sampled ($n = 1$). As another example, Figure 4.13 shows the histogram of the distribution of weekly means for the data of Table 2.1 superimposed on the histogram of daily values given in Figure 2.3. The roughly normal shape and the decreased standard deviation are what might be expected of the means of samples of 7. The standard deviation

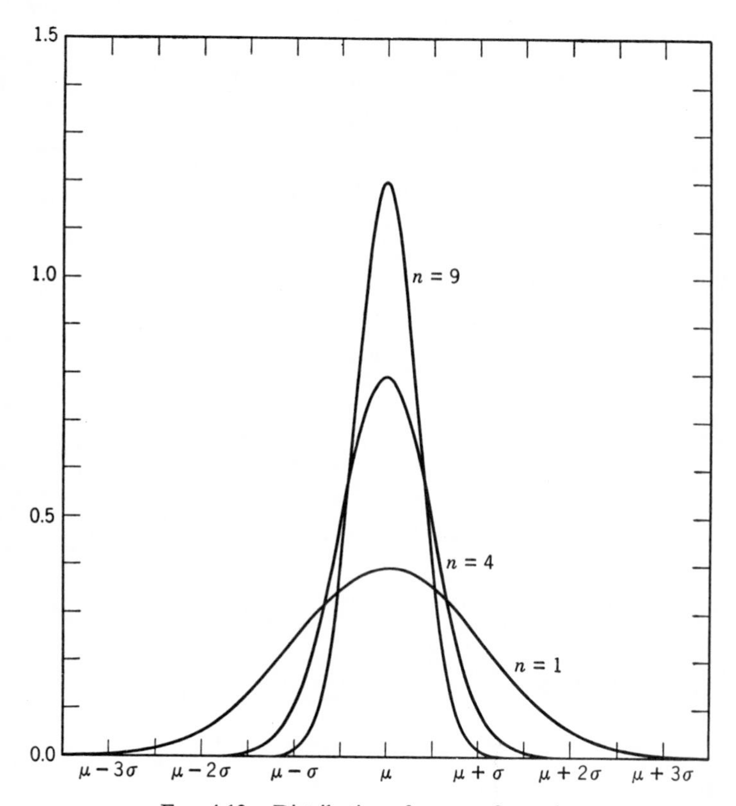

FIG. 4.12. Distribution of means of samples.

of the weekly means is 0.516 compared with the $1.351/\sqrt{n} = 0.511$ which would be expected. Although this example illustrates these considerations, it is not strictly correct to apply them here, since we are drawing from a restricted population, i.e., the 364 daily yields.

It follows immediately from the above considerations and Section 4.42 that, if $\bar{x}_1$ is the mean of a sample of n_1 observations from a normal distribution with mean μ_1 and variance σ_1^2, and $\bar{x}_2$ is the mean of an independent sample of n_2 observations from a normal distribution with mean μ_2 and variance σ_2^2, then $\bar{x}_1 + \bar{x}_2$ and $\bar{x}_1 - \bar{x}_2$ are normally distributed with

(b) $$\text{ave}\ (\bar{x}_1 \pm \bar{x}_2) = \mu_1 \pm \mu_2$$

and

(c) $$\text{var}\ (\bar{x}_1 \pm \bar{x}_2) = \sigma_1^2/n_1 + \sigma_2^2/n_2$$

4.52. Independence of Linear Combinations

We now consider two different linear combinations $w_1 = \Sigma a_i x_i$ and

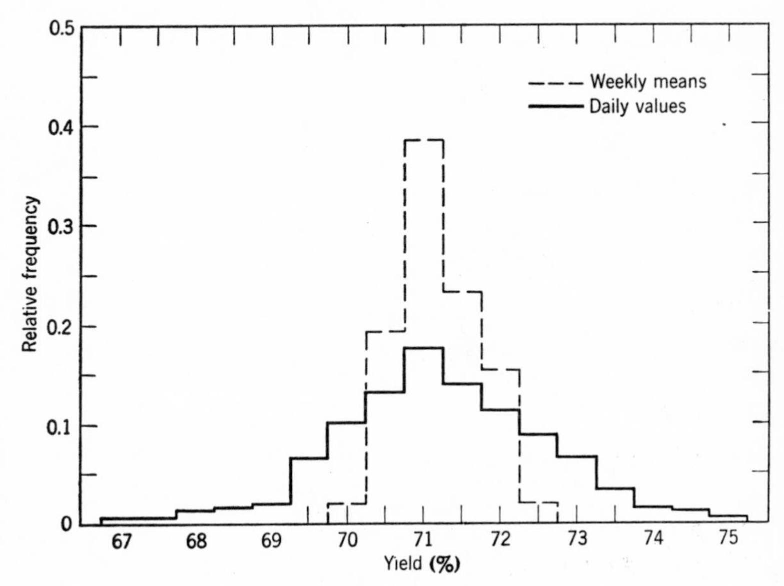

FIG. 4.13. Distribution of daily and weekly mean yields for data of Table 2.1.

$w_2 = \Sigma b_i x_i$ of the sample values, with means $\mu_1 = \mu \Sigma a_i$ and $\mu_2 = \mu \Sigma b_i$ and variances $\sigma_1^2 = \sigma^2 \Sigma a_i^2$ and $\sigma_2^2 = \sigma^2 \Sigma b_i^2$, respectively. Then since both are normally distributed, the condition $\text{cov}\ (w_1, w_2) = 0$ is sufficient to ensure their independence. We have

$$\begin{aligned}\text{cov}\ (w_1, w_2) &= \text{ave}\ (w_1 - \mu_1)\ (w_2 - \mu_2) \\ &= \text{ave}\ [\Sigma a_i(x_i - \mu)]\ [\Sigma b_i(x_i - \mu)] \\ &= \Sigma a_i b_i\ \text{ave}\ (x_i - \mu)^2 + \Sigma_{i \neq j} a_i b_j\ \text{ave}\ (x_i - \mu)\ (x_j - \mu)\end{aligned}$$

Since all the x_i are independent, $\text{ave}\ (x_i - \mu)\ (x_j - \mu) = 0$ for $i \neq j$; also $\text{ave}\ (x_i - \mu)^2 = \sigma^2$ for all i. Hence

$$\text{cov}\ (w_1, w_2) = \sigma^2\ \Sigma a_i b_i$$

and since σ^2 must be positive, cov $(w_1, w_2) = 0$ if and only if $\Sigma a_i b_i = 0$. Hence this condition is sufficient to ensure the independence of w_1 and w_2, and in this case w_1 and w_2 are said to be orthogonal. We could consider a third combination $w_3 = \Sigma c_i x_i$ orthogonal to both w_1 and w_2; in this case w_1, w_2, and w_3 would be mutually independent. In fact, exactly n such mutually orthogonal combinations of the n sample values can be chosen, and there are an infinite number of ways in which this can be done. Such a set of linear combinations is called an orthogonal transformation from the variables $x_1, \cdots, x_n$ to the variables $w_1, \cdots, w_n$.

4.53. Distribution of the Sample Variance

Now let us consider the sum of squared deviations

$$(a) \qquad (n-1)s^2 = \sum_{i=1}^{n} (x_i - \bar{x})^2$$

For the particular case $n = 2$, we have

$$(x_1 - \bar{x})^2 + (x_2 - \bar{x})^2 = \left(\frac{x_1 - x_2}{\sqrt{2}}\right)^2$$

i.e., the sum of squared deviations may be expressed as the square of the linear combination $x_1/\sqrt{2} - x_2/\sqrt{2}$, which, since $\Sigma a_i = 1/\sqrt{2} - 1/\sqrt{2} = 0$, and $\Sigma a_i^2 = \frac{1}{2} + \frac{1}{2} = 1$, has average value 0 and variance σ^2. Similarly, if $n = 3$ we have

$$(x_1 - \bar{x})^2 + (x_2 - \bar{x})^2 + (x_3 - \bar{x})^2 = \left(\frac{x_1 - x_2}{\sqrt{2}}\right)^2 + \left(\frac{x_1 + x_2 - 2x_3}{\sqrt{6}}\right)^2$$

It is easily seen that the linear combination of x_1, x_2, and x_3 in the second bracket on the right has average value 0 and variance σ^2; also we have

$$\sum_{i=1}^{3} a_i b_i = \frac{1}{\sqrt{2}} \cdot \frac{1}{\sqrt{6}} + \frac{-1}{\sqrt{2}} \cdot \frac{1}{\sqrt{6}} + 0 \cdot \frac{2}{\sqrt{6}} = 0$$

and hence the combinations are independent. Extending this argument to the case of n sample values, we can write

$$(b) \qquad \sum_{i=1}^{n} (x_i - \bar{x})^2 = \sum_{i=1}^{n-1} z_i^2$$

where

$$z_1 = \frac{1}{\sqrt{2}} x_1 - \frac{1}{\sqrt{2}} x_2$$

$$z_2 = \frac{1}{\sqrt{6}} x_1 + \frac{1}{\sqrt{6}} x_2 - \frac{2}{\sqrt{6}} x_3$$

(c) . . . .

$$z_r = \frac{1}{\sqrt{r(r+1)}} x_1 + \frac{1}{\sqrt{r(r+1)}} x_2 + \cdots + \frac{1}{\sqrt{r(r+1)}} x_r - \frac{r}{\sqrt{r(r+1)}} x_{r+1}$$

. . . .

$$z_{n-1} = \frac{1}{\sqrt{n(n-1)}} x_1 + \frac{1}{\sqrt{n(n-1)}} x_2 + \cdots + \frac{1}{\sqrt{n(n-1)}} x_{n-1} - \frac{n-1}{\sqrt{n(n-1)}} x_n$$

Each of the z_i has average value 0 and variance σ^2, and all are mutually orthogonal, and hence independent. It follows that z_i/σ is a standard normal deviate, and $\frac{1}{\sigma^2}\sum_{i=1}^{n-1} z_i^2$ the sum of squares of $n - 1$ independent standard normal deviates; therefore $\frac{\Sigma z_i^2}{\sigma^2} = \frac{\Sigma(x_i - \bar{x})^2}{\sigma^2} = \frac{(n-1)s^2}{\sigma^2}$ has the χ^2-distribution with $\nu = n - 1$ degrees of freedom. It can be seen immediately that each of the linear combinations in (c) above is orthogonal to the combination $\bar{x} = \Sigma_i \frac{x_i}{n}$; it follows that the sample mean is independent of all the z_i, and hence of the sample variance.

We can obtain the distribution of s^2 from the probability element of the distribution (b) of Section 4.46. Letting $\chi^2 = \frac{\nu s^2}{\sigma^2}$, we obtain

$$f(\chi^2)\, d(\chi^2) = \frac{1}{2^{\nu/2}\Gamma(\nu/2)} e^{-\nu s^2/2\sigma^2} \left(\frac{\nu s^2}{\sigma^2}\right)^{(\nu/2)-1} d\left(\frac{\nu s^2}{\sigma^2}\right)$$

and hence for the probability element of the distribution of s^2

$$\text{(d)} \qquad f(s^2)\, d(s^2) = \frac{1}{\Gamma(\nu/2)} \left(\frac{\nu}{2\sigma^2}\right)^{\nu/2} e^{-\nu s^2/2\sigma^2} (s^2)^{(\nu/2)-1}\, d(s^2)$$

where $\nu = n - 1$ is the number of degrees of freedom on which s^2 is based. Since from Section 4.46

$$\text{ave}\,(\chi^2) = \nu$$

and

$$\text{var}\,(\chi^2) = 2\nu$$

we have immediately

$$(e) \qquad \text{ave}\,(s^2) = \frac{\sigma^2}{\nu}\,\text{ave}\,(\chi^2) = \sigma^2$$

as we found in Section 3.43 and

$$(f) \qquad \text{var}\,(s^2) = \left(\frac{\sigma^2}{\nu}\right)^2 \text{var}\,(\chi^2) = \frac{2\sigma^4}{\nu} = \frac{2\sigma^4}{n-1}$$

The latter is true only for distributions for which $\kappa_4 = 0$, which includes the normal.

From (d) we can write for the probability element of the distribution of the sample standard deviation s

$$f(s)\,ds = \frac{2}{\Gamma(\nu/2)}\left(\frac{\nu}{2\sigma^2}\right)^{\nu/2} e^{-\nu s^2/2\sigma^2} s^{\nu-1}\,ds$$

and obtain

$$\text{ave}\,(s) = \frac{2}{\Gamma(\nu/2)}\left(\frac{\nu}{2\sigma^2}\right)^{\nu/2}\int_0^\infty s\cdot e^{-\nu s^2/2\sigma^2} s^{\nu-1}\,ds$$

$$= \left(\frac{2\sigma^2}{\nu}\right)^{1/2}\cdot\frac{\Gamma\left(\dfrac{\nu+1}{2}\right)}{\Gamma(\nu/2)}$$

or, putting $\nu = n - 1$,

$$\text{ave}\,(s) = \left(\frac{2\sigma^2}{n-1}\right)^{1/2}\cdot\frac{\Gamma(n/2)}{\Gamma\left(\dfrac{n-1}{2}\right)}$$

$$\sim \sigma\sqrt{\frac{n-3/2}{n-1}}$$

Since $\text{ave}\,(s^2) = \sigma^2$, we have

$$\text{var}\,(s) = \text{ave}\,(s^2) - [\text{ave}\,(s)]^2$$

$$= \sigma^2 - \frac{2\sigma^2}{n-1}\left[\frac{\Gamma(n/2)}{\Gamma\left(\dfrac{n-1}{2}\right)}\right]^2$$

$$\sim \sigma^2/2n$$

For most practical problems we shall be interested in the distribution of s^2.

Figure 4.14 shows the histogram of the sample variances for each week for the data of Table 2.1. In this case $n = 7$, $\nu = 6$. The dotted lines show the expected frequencies, using $\sigma^2 = 1.8252$, the overall variance for the data of Table 2.1 computed in Section 2.35. We can show, using the methods of Chapter 9, that the variations between the observed and expected frequencies are no more than might be expected for a random sample of 52 sample variances.

4.54. The "Student" t-Distribution

It was noted in Section 2.43, that even when we know the average value, variance, and distribution of the sample mean $\bar{x}$, this does not completely solve our problem. We must still develop some means of

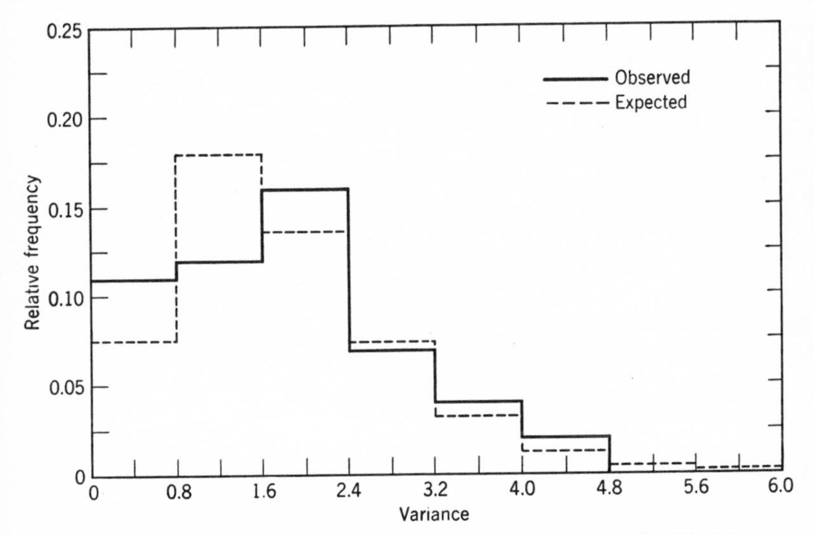

FIG. 4.14. Histogram of weekly sample variances for data of Table 2.1.

allowing for the fact that we have to use the sample variance s^2 rather than the population variance σ^2. In large samples no appreciable error will be caused by considering $\bar{x}$ as normally distributed with variance s^2/n; in small samples, which are the rule rather than the exception in physical, chemical, and industrial work, we cannot afford to ignore the appreciable variation in s^2 noted in the previous section.

This problem was first studied by W. S. Gosset, a chemist employed by Guinness Breweries, who wrote under the pseudonym of "Student." His papers on the subject, which began in 1907 and 1908 and have been collected [79], started the trend toward the development of exact statistical tests, i.e., tests independent of knowledge or assumptions concerning population parameters. His first step was to conduct a large-scale sampling experiment from which he deduced the independence of the sample mean and sample variance. Later he deduced the exact distribution of $z = (\bar{x} - \mu)/s$, after having obtained the distribution of s^2

by fitting a curve using the first 4 moments. The work of "Student" in this direction was later developed more completely, and its many applications were studied by R. A. Fisher, although it was some years before such exact tests were commonly used in statistical applications.

We shall consider the problem here somewhat more generally. Suppose that two variables u and v are independently distributed, u normally with zero mean and unit variance, and v as χ^2 with ν degrees of freedom. Then the variable

$$(a) \qquad t_\nu = \frac{u}{\sqrt{v/\nu}}$$

has a distribution with probability element

$$(b) \qquad f(t)\,dt = \frac{1}{\sqrt{\nu\pi}} \frac{\Gamma\left(\frac{\nu+1}{2}\right)}{\Gamma(\nu/2)} \left(1 + \frac{t^2}{\nu}\right)^{-(\nu+1)/2} dt$$

known as the "Student" t-distribution, or simply the t-distribution, with ν degrees of freedom.

Since u and v are independent, their joint probability element is given by

$$f(u, v)\,du\,dv = \frac{1}{\sqrt{2\pi}} e^{-u^2/2}\,du \cdot \frac{1}{2^{\nu/2}\Gamma(\nu/2)} e^{-v/2} v^{(\nu/2)-1}\,dv$$

$$= \frac{1}{\sqrt{2\pi}\,2^{\nu/2}\Gamma(\nu/2)} e^{-\frac{v}{2}\left(1+\frac{u^2}{v}\right)} v^{(\nu/2)-1}\,du\,dv$$

where $-\infty < u < \infty$, $0 < v < \infty$. Now, if we make the substitution

$$t = \frac{u}{\sqrt{v/\nu}}$$

$$z = \frac{v}{2}\left(1 + \frac{u^2}{v}\right) = \frac{v}{2}\left(1 + \frac{t^2}{\nu}\right)$$

we have $-\infty < t < +\infty$, $0 < z < \infty$, and

$$du\,dv = 2\sqrt{\frac{2}{\nu}}\sqrt{z}\left(1 + \frac{t^2}{\nu}\right)^{-3/2} dz\,dt$$

$$v = 2z\left(1 + \frac{t^2}{\nu}\right)^{-1}$$

Hence the probability element of the joint distribution of t and z becomes

$$f(t, z)\,dt\,dz = \frac{1}{\sqrt{2\pi}\,2^{\nu/2}\Gamma(\nu/2)}\,e^{-z}\left[2z\left(1+\frac{t^2}{\nu}\right)^{-1}\right]^{(\nu/2)-1} \cdot 2\sqrt{2/\nu}\,\sqrt{z}\left(1+\frac{t^2}{\nu}\right)^{-3/2} dt\,dz$$

$$= \frac{1}{\sqrt{\nu\pi}\,\Gamma(\nu/2)}\,e^{-z}\,z^{(\nu-1)/2}\left(1+\frac{t^2}{\nu}\right)^{-(\nu+1)/2} dt\,dz$$

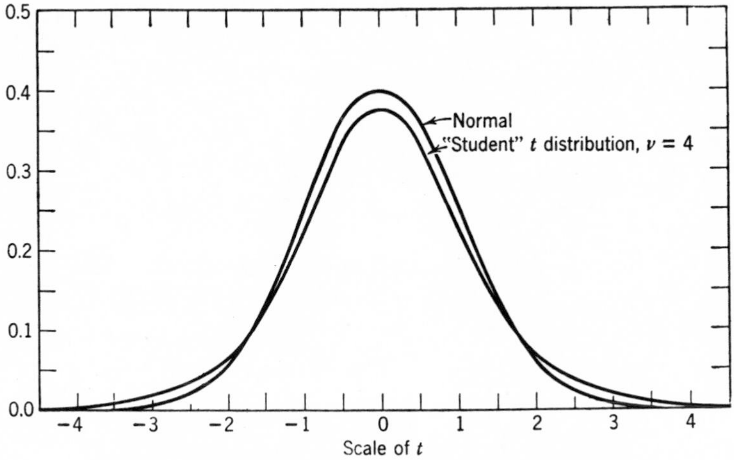

FIG. 4.15. Comparison of normal distribution and "Student" t-distribution.

To obtain the marginal distribution of t, we integrate with respect to z from 0 to ∞; hence

$$f(t)\,dt = \frac{1}{\sqrt{\nu\pi}\,\Gamma(\nu/2)}\left(1+\frac{t^2}{\nu}\right)^{-(\nu+1)/2} dt \cdot \int_0^\infty e^{-z}\,z^{(\nu-1)/2}\,dz$$

$$= \frac{\Gamma\left(\dfrac{\nu+1}{2}\right)}{\sqrt{\nu\pi}\,\Gamma(\nu/2)}\left(1+\frac{t^2}{\nu}\right)^{-(\nu+1)/2} dt$$

Note that this distribution depends only on the number of degrees of freedom ν. When ν becomes large the distribution of t approaches the normal distribution, as we might have expected; for small ν it is symmetric about $t = 0$, but with less probability near the center and more probability in the tails than the normal distribution. Because of the symmetry, ave $(t_\nu) = 0$, and it can be shown by direct integration, putting the integral in beta function form (see Appendix 4A), that the variance of t_ν is $\nu/(\nu - 2)$. A comparison of the normal distribution with the t-distribution for $\nu = 4$ is shown in Figure 4.15. Values of $t_{\nu,\alpha}$ such that

$$Pr(|t_\nu| > t_{\nu,\alpha}) = 1 - \int_{-t_{\nu,\alpha}}^{t_{\nu,\alpha}} f(t)\,dt = \alpha$$

are given for various values of ν and α in Table III of the Appendix. For $\nu = \infty$, these values are identical with the values of t_α given in Table 4.3.

For a sample from a normal distribution $\bar{x}$ is normally distributed with mean μ and variance σ^2/n, and therefore $t = \dfrac{\bar{x} - \mu}{\sigma}\sqrt{n}$ will be a standard normal deviate. Since $\dfrac{(n-1)s^2}{\sigma^2}$ has the χ^2-distribution with $n - 1$ degrees of freedom,

$$t_{n-1} = \frac{\dfrac{\bar{x} - \mu}{\sigma}\sqrt{n}}{\sqrt{\dfrac{(n-1)s^2}{\sigma^2} \Big/ n - 1}} = \frac{\bar{x} - \mu}{s}\sqrt{n}$$

will have the t-distribution with $n - 1$ degrees of freedom. This is the expression from which the limits in Section 2.43 were obtained. The general form developed in this section will be used to derive other similar expressions in later chapters.

4.55. The Distribution of the Ratio of Two Variances

Another distribution which will be of importance in later applications is that of the ratio $F = s_1^2/s_2^2$ of 2 variances s_1^2 and s_2^2 which are independent estimates based on ν_1 and ν_2 degrees of freedom, respectively, of the variances σ_1^2 and σ_2^2 of 2 normal populations. The exact distribution of $z = \frac{1}{2}\ln F = \frac{1}{2}(\ln s_1^2 - \ln s_2^2)$ was determined by R. A. Fisher, but Snedecor first studied the distribution of the variance ratio F, and published tables for its use.

We shall again consider this distribution somewhat more generally by supposing that u is a variable having the χ^2-distribution with ν_1 degrees of freedom, and v is a variable independent of u having the χ^2-distribution with ν_2 degrees of freedom. Then

$$(a) \qquad F_{\nu_1,\nu_2} = \frac{u/\nu_1}{v/\nu_2} = \frac{\nu_2 u}{\nu_1 v}$$

has the distribution

$$(b) \qquad f(F) = \frac{(\nu_1/\nu_2)^{\nu_1/2}}{\mathrm{B}(\nu_1/2,\, \nu_2/2)} F^{(\nu_2/2)-1} \left(1 + \frac{\nu_1}{\nu_2} F\right)^{-(\nu_1+\nu_2)/2}$$

Since u and v are independent, their joint distribution is given by

$$f(u, v)\, du\, dv = \frac{1}{2^{\nu_1/2}\Gamma(\nu_1/2)} e^{-u/2} u^{(\nu_1/2)-1}\, du \cdot \frac{1}{2^{\nu_2/2}\Gamma(\nu_2/2)} e^{-v/2} v^{(\nu_2/2)-1}\, dv$$

where $0 < u < \infty$, $0 < v < \infty$. This can be written as

$$f(u, v)\, du\, dv = \frac{1}{2^{(\nu_1+\nu_2)/2}\Gamma(\nu_1/2)\Gamma(\nu_2/2)} e^{-\frac{v}{2}\left(1+\frac{u}{v}\right)} u^{(\nu_1/2)-1} v^{(\nu_2/2)-1}\, du\, dv$$

Making the substitutions

$$F = \frac{\nu_2 u}{\nu_1 v}$$

$$z = \frac{v}{2}\left(1 + \frac{u}{v}\right)$$

and, integrating the resulting joint distribution of z and F over the range $0 < z < \infty$, we obtain the distribution given by (*b*). Note that this distribution depends only on the degrees of freedom ν_1 and ν_2 of the two variables distributed as χ^2; also that we could just as easily have considered the distribution of $F' = 1/F = \nu_1 v/\nu_2 u$ which would have an identical distribution except that ν_1 and ν_2 would be interchanged.

Table IV of the Appendix gives values $F_{\nu_1, \nu_2, \alpha}$ such that

$$Pr(F \geq F_{\nu_1, \nu_2, \alpha}) = \alpha$$

for various values of ν_1 and ν_2 and $\alpha = 0.50, 0.25, 0.10, 0.05, 0.025, 0.01$, and 0.005. To obtain values $F_{\nu_1, \nu_2, 1-\alpha}$ such that

$$Pr(F < F_{\nu_1, \nu_2, 1-\alpha}) = 1 - Pr(F \geq F_{\nu_1, \nu_2, 1-\alpha}) = 1 - (1 - \alpha) = \alpha,$$

we make use of the fact that $Pr(F \leq F_0) = Pr(F' \geq F_0')$; hence

$$F_{\nu_1, \nu_2, 1-\alpha} = \frac{1}{F_{\nu_2, \nu_1, \alpha}}$$

i.e., the lower critical values of F can be found from the upper critical values given in the table by taking the reciprocal of the value obtained for the degrees of freedom in the inverse order. In most applications only the upper critical values are needed.

In the particular case where s_1^2 and s_2^2 are independent estimates based on ν_1 and ν_2 degrees of freedom, respectively, of the same variance σ^2

of a normal population, $\nu_1 s_1^2/\sigma^2$ and $\nu_2 s_2^2/\sigma^2$ have the χ^2-distribution with ν_1 and ν_2 degrees of freedom, respectively, and hence

$$F = \frac{\dfrac{\nu_1 s_1^2}{\sigma^2} \cdot \dfrac{1}{\nu_1}}{\dfrac{\nu_2 s_2^2}{\sigma^2} \cdot \dfrac{1}{\nu_2}} = \frac{s_1^2}{s_2^2}$$

has the F-distribution with ν_1 and ν_2 degrees of freedom.

4.56. Distribution of Other Sample Statistics

The distributions of many other sample statistics for samples from a normal distribution have been studied, and some of these will be referred to later in connection with their applications. In many cases the exact distribution is not obtainable in a useful form, and it has been necessary to use numerical quadratures in order to obtain the distribution, or even critical values, for specific sample sizes.

The use of order statistics has undergone rapid development since World War II, and much work has been done on their distribution in random samples. A considerable portion of this work does not require the assumption of a normal distribution; it is sufficient for many purposes to assume only that the distribution function is continuous. It is this lack of dependence on the exact form of the distribution, in addition to ease of computation, which makes methods based on order statistics, and their ranks, so appealing. A complete review of this field is given in [21]. Of the order statistics mentioned in Chapter 2, the median and extreme values have been most extensively studied. For large samples the median of a sample from a normal distribution is approximately normally distributed with average value the population median (which is also the population mean, since the distribution is symmetric) and with variance $\sim \pi\sigma^2/2n$. This fact is rarely used, since the median is least useful in large samples, and for small samples it is preferable to obtain limits by the method of Section 2.44, which requires no assumptions concerning the underlying distribution. The distribution of extreme values has been studied both with and without the assumption of a normal population; a description of this work is given in the above reference.

Among the many possible systematic statistics, some of which will be mentioned in Chapter 5 in connection with problems of estimation, the range $R = x_{(n)} - x_{(1)}$, where $x_{(n)}$ and $x_{(1)}$ represent the largest and smallest values of a sample of n, has been most completely investigated. A complete tabulation of the distribution of R/σ for samples of size $2 \leq n \leq 20$ from a normal distribution with variance σ^2 is given in [16].

Equally important in applications is the "studentized" range R/s [80], [123], where the population standard deviation has been replaced by an independent estimate of the sample standard deviation. Lower and upper 5% and 1% points of this ratio for samples of $2 \leq n \leq 20$ from a normal distribution, where s^2 is an *independent* estimate of σ^2 based on ν degrees of freedom, are given for various values of ν from 1 to ∞ in [123], and are reproduced in Table 5.8, Section 5.53.

4.6. Discrete Probability Distributions

4.61. Binomial Distribution

Suppose that we have a random variable which has only two values, one of which occurs with probability p and the other with probability $1 - p$. Such a situation might arise when the result of a given trial of an experiment could be classed as a success or a failure, a given item could be classified as defective or non-defective, or a given characteristic might or might not be observed in a given individual. Now let us suppose that we make a series of n observations on this random variable, assuming that for each observation the probability of success is p, and the probability of failure is $1 - p$. The number of successes x in n such trials will again be a random variable, and its distribution is given by

$$(a) \qquad p(x) = \binom{n}{x} p^x(1-p)^{n-x} \qquad x = 0, 1, 2, \cdots, n$$

so that $\sum_{x=0}^{n} p(x) = [p + (1-p)]^n = 1$. Because the successive values of $p(x)$ are the terms of a binomial expansion, this distribution is usually referred to as the binomial distribution. The distribution depends only on the fixed number n of *independent* trials and the fixed probability p of the event occurring or not occurring on each trial.

From our considerations in the previous chapter it is easy to see how this distribution can be obtained. Since the trials are independent, the probability of obtaining any particular sequence of x successes and $n - x$ failures will be the product of the individual probability of a success in x cases and the probability of a failure in $n - x$ cases, i.e., $p^x(1-p)^{n-x}$. Then the probability of x heads, regardless of the order in which they occur, would simply be this probability multiplied by the number of possible sequences of n things containing exactly x successes and $n - x$ failures. This number is given by the quantity $\binom{n}{x}$ discussed in Appendix 3*A*, and multiplying this number by the probability of any particular sequence gives us the probability of x heads, regardless of the order in which they are obtained. For small values of n the actual probabilities associated with the various values of x can be readily computed directly; for larger values of n the computations can be simplified by using the table of $\binom{n}{x}$ referred to in Appendix 3*A*, or by the use of

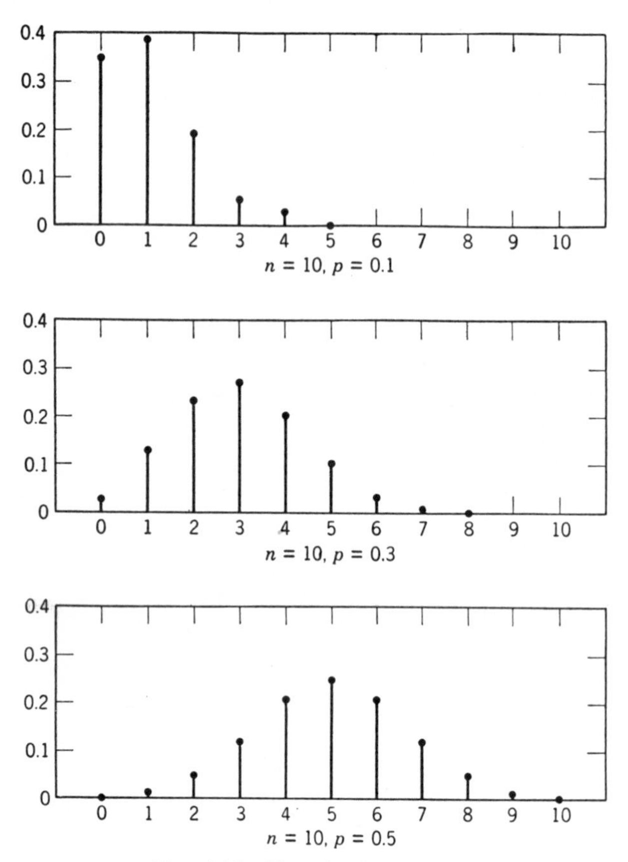

FIG. 4.16. Binomial distribution.

the incomplete beta function tables (see Appendix 4*A*). An extensive tabulation of the binomial distribution which gives both individual values and partial sums for $n = 2(1)49$ and $p = 0.01(0.01)0.50$ has been published [81].

As an example of the nature of this distribution, Table 4.6 gives the probabilities of getting 0, 1, 2, $\cdots$, 10 successes in a sample of size 10 when p is 0.1, 0.2, 0.3, 0.4, or 0.5. Graphs for $p = 0.1$. 0.3 and 0.5 are shown in Figure 4.16.

TABLE 4.6

x	Binomial: $n = 10$					Poisson
	$p = 0.5$	$p = 0.4$	$p = 0.3$	$p = 0.2$	$p = 0.1$	$\lambda = 1$
0	0.0010	0.0061	0.0283	0.1074	0.3487	0.3679
1	0.0098	0.0403	0.1211	0.2684	0.3874	0.3679
2	0.0439	0.1209	0.2335	0.3020	0.1937	0.1839
3	0.1172	0.2150	0.2668	0.2013	0.0574	0.0613
4	0.2051	0.2508	0.2001	0.0881	0.0112	0.0153
5	0.2461	0.2007	0.1029	0.0264	0.0015	0.0031
6	0.2051	0.1115	0.0368	0.0055	0.0001	0.0005
7	0.1172	0.0424	0.0090	0.0008	0.0000	0.0001
8	0.0439	0.0106	0.0014	0.0001		0.0000
9	0.0097	0.0016	0.0001	0.0000		
10	0.0010	0.0001	0.0000			

The moment generating function for the binomial distribution is given by

$$M_x(\theta) = \sum_{x=0}^{n} e^{x\theta} \cdot \binom{n}{x} p^x(1-p)^{n-x}$$

$$= \sum_{x=0}^{n} \binom{n}{x} (pe^\theta)^x(1-p)^{n-x}$$

$$= (pe^\theta + 1 - p)^n$$

Taking the first two derivatives

$$\frac{dM}{d\theta} = n(pe^\theta + 1 - p)^{n-1}pe^\theta$$

$$\frac{d^2M}{d\theta^2} = n(n-1)\,(pe^\theta + 1 - p)^{n-2}(pe^\theta)^2 + n(pe^\theta + 1 - p)^{n-1}pe^\theta$$

and, setting $\theta = 0$, or $e^\theta = 1$,

$$\mu_1' = np$$

(b)

$$\mu_2' = n(n-1)p^2 + np$$

and hence

$$\text{ave}\,(x) = \mu_1' = np$$

$$\text{var}\,(x) = \mu_2' - \mu_1'^2$$

(c)
$$= n(n-1)p^2 + np - n^2p^2$$
$$= n^2p^2 - np^2 + np - n^2p^2$$
$$= np(1-p)$$

From these we have for the proportion of successes x/n

(d)
$$\text{ave}\left(\frac{x}{n}\right) = \frac{1}{n}\text{ave}(x) = p$$
$$\text{var}\left(\frac{x}{n}\right) = \frac{1}{n^2}\text{var}(x) = \frac{p(1-p)}{n}$$

EXAMPLE. Suppose that it has been determined that the probability of a gross error in a particular analytical procedure is 0.05, i.e., about 1 time in 20, an analyst will make some sort of mistake leading to a large discrepancy in the result obtained. Then, if a single analytical result is to be obtained from a series of 4 such determinations, the respective probabilities of 0, 1, 2, 3, 4 gross errors are given in the accompanying table. The probability of at least 1 gross error per analysis is 0.1855, i.e., about $\frac{1}{5}$ of the series of determinations will contain at least 1 gross error.

x	$p(x)$
0	$p(0) = (0.95)^4 = 0.8145$
1	$p(1) = 4(0.05)(0.95)^3 = 0.1715$
2	$p(2) = 6(0.05)^2(0.95)^2 = 0.0134$
3	$p(3) = 4(0.05)^3(0.95) = 0.0006$
4	$p(4) = (0.05)^4 = 0.0000$

Conversely, if we observed that approximately 1 in 10 of the series of 4 determinations contained a gross error, then the solution of the equation

$$(1-p)^4 = 0.90$$

gives a value $p = 0.026$ for the probability of a gross error in a single determination; i.e., the relative frequency of gross errors is approximately 1 in every 40 determinations.

4.62. Multinomial Distribution

We now consider a random variable which can take on one of k different values, each with probability p_k, $\sum_{i=1}^{k} p_k = 1$. For example, the observed characteristic might be the hair color of a person chosen at random and classified as either blond, brown, or black, or we might classify the amount of precipitation obtained in a given experiment as heavy, moderate, slight, or no apparent precipitation. If a sample of n observations is taken from such a population, then the joint distribution of the new variables $x_1, \cdots, x_k$, representing the number of observations falling into each of the k categories, is

$$p(x_1, x_2, \cdots, x_k) = \frac{n!}{x_1!\, x_2! \cdots x_k!}(p_1)^{x_1}(p_2)^{x_2} \cdots (p_k)^{x_k} \tag{a}$$

This defines the probability of any possible set of values $x_1, \cdots, x_k$ such that $\sum_{i=1}^{k} x_i = n$. The sum of the probabilities over all such sets is

$$(p_1 + p_2 + \cdots + p_k)^n = 1^n = 1 \tag{b}$$

Because the given probability distribution represents the individual terms of the multinomial expansion of the left-hand side of (*b*), this distribution is called the *multinomial distribution.*

For any particular variable x_i it can be shown that, as in the binomial distribution,

$$\text{ave }(x_i) = np_i$$

$$\text{var }(x_i) = np_i(1 - p_i)$$

The variables x_i are not independent, and we have, for $i \neq j$,

$$\text{ave }(x_i x_j) = n(n-1)p_i p_j$$

$$\text{cov }(x_i, x_j) = -np_i p_j$$

As in the case of the binomial distribution similar formulae can be obtained for the proportions x_i/n. When $k = 2$, $p_1 = p$, and $p_2 = 1 - p$, $x_1 = x$, $x_2 = n - x$, the multinomial distribution reduces to the binomial distribution.

4.63. Poisson Distribution

The Poisson distribution arises when we have discrete events occurring randomly over a long period of time or over a large area and we consider as a random variable the frequency with which these events will occur in any small time interval or small area chosen at random. The number of possible events should be large (theoretically, infinite), but the probability of occurrence of any individual event in the time interval or area considered should be small. Examples of the kind of data which we would expect to find distributed in the Poisson form are the number of counts recorded by a Geiger counter in 1-minute intervals, the number of automobile collisions reported in a given section of a large city in time intervals of a given length, or the number of white blood cells counted in small equal areas of a prepared slide. The Poisson distribution is given by

$$p(x) = \frac{e^{-\lambda}\lambda^x}{x!} \qquad x = 0, 1, 2, \cdots$$

where λ is the expected number of events occurring on any given observation, i.e., ave $(x) = \lambda$. A derivation of this distribution from fundamental axioms is given in [76]; its derivation as a limiting form of the binomial distribution can be found in many places (e.g., [2, Vol. 1]).

The moment generating function for the Poisson distribution is given by

$$\begin{aligned} M_x(\theta) &= \sum_{x=0}^{\infty} e^{x\theta} \cdot \frac{e^{-\lambda}\lambda^x}{x!} \\ &= e^{-\lambda} \sum_{x=0}^{\infty} \frac{(\lambda e^{\theta})^x}{x!} \\ &= e^{-\lambda} \cdot e^{\lambda e^{\theta}} \\ &= e^{\lambda(e^{\theta}-1)} \end{aligned}$$

The first two derivatives are

$$\frac{dM}{d\theta} = e^{\lambda(e^{\theta}-1)}\lambda e^{\theta}$$

$$\frac{d^2M}{d\theta^2} = e^{\lambda(e^{\theta}-1)} \cdot (\lambda e^{\theta})^2 + e^{\lambda(e^{\theta}-1)} \cdot \lambda e^{\theta}$$

and for $\theta = 0$ we obtain

$$\begin{aligned} \mu_1' &= \lambda \\ \mu_2' &= \lambda^2 + \lambda \end{aligned}$$

and hence

$$\begin{aligned} \text{ave}\,(x) &= \mu_1' = \lambda \\ \text{var}\,(x) &= \mu_2' - \mu_1'^2 \\ &= \lambda^2 + \lambda - \lambda^2 \\ &= \lambda \end{aligned}$$

Hence both the mean and the variance of the Poisson distribution are equal to the expected number of events λ. This fact is extremely useful in the practical applications of the Poisson distribution, particularly in the counting of fundamental particles.

Note that in the definition of the Poisson distribution we assume that it is *possible* for the number of observed events to be infinite, even though the probability of occurrence of any individual event is small. This is never true in the examples that we have considered, but we need only ensure that the number of possible events which could be observed is sufficiently large, and the probability of occurrence of an individual event sufficiently small, if we are to be justified in applying the Poisson distribution. It is also important that the observations be made on equal

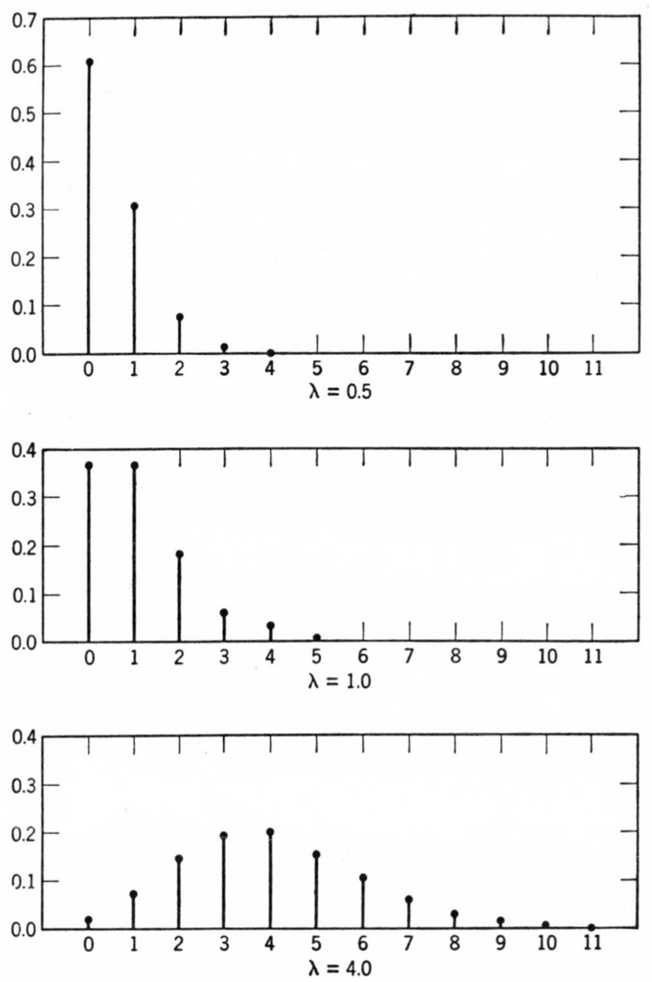

FIG. 4.17. Poisson distribution.

areas or on equal time intervals and that the expected value be the same for each observation. In our first example, that of the Geiger counter, these conditions would hold only if the number of particles available for disintegration was so large that those few particles which disintegrated and were counted did not appreciably affect the total number available for disintegration.

Excellent tables of the Poisson distribution which give values to 6 places of both $p(x)$ and $Pr(x \geq c) = \sum_{x=c}^{\infty} \frac{e^{-\lambda}\lambda^x}{x!}$ have been computed by E. C. Molina [7] for values of λ ranging from 0.001 to 100 ($\lambda = a$ in these tables). Figure 4.17 shows the graph of the Poisson distribution for $\lambda = 0.5$, $\lambda = 1$, and $\lambda = 4$. When the number of trials n is large and the probability p is small, the Poisson distribution with $\lambda = np$ may be used for ease in computation as an approximation to the binomial distribution. Thus, if we were considering the number of defective screws found in boxes of one thousand screws, where the average number of defectives was small (say < 50), we could use the Poisson distribution in place of the binomial distribution to compute the probability of finding different numbers of defective screws. As an illustration which probably represents the extreme limit of applicability of the Poisson distribution as an approximation to the binomial, the first 11 terms of the Poisson distribution for $\lambda = np = 1$ are given in Table 4.6 (Section 4.61) for comparison with the binomial distribution with $n = 10$, $p = 0.1$.

EXAMPLE. In a given chemical laboratory, the minor injury rate over a long period of time has been 3.42 minor injuries per 10,000 manhours worked. If the laboratory employs 210 people on a 40-hour week, then the total manhours worked per week would be 8400, hence the expected number of minor injuries per week would be $8400/10{,}000 \times 3.42 = 2.87$. If we assume that the number of minor injuries in a given week is a random variable having the Poisson distribution, then the probability that in a given week no injuries will occur is $p(0) = e^{-2.87} = 0.0567$, that 1 injury will occur is $p(1) = 2.87e^{-2.87} = 0.1627$, that 2 injuries will occur is $p(2) = \frac{(2.87)^2e^{-2.87}}{2} = 0.2335$, etc. Similarly the probability of 6 or more minor injuries in 1 week is $p(x \geq 6) = \sum_{x=6}^{\infty} \frac{e^{-2.87}(2.87)^x}{x!} = 0.0714$; i.e., there are only about 7 chances in 100 of 6 or more injuries occurring in 1 week unless there has been some increase in the long-term frequency.

4.64. Limiting Forms of the Binomial and Poisson Distributions

If we consider the number of successes x distributed in the binomial form as the sum of n random variables, each of which can take on the values 1 and 0 with probability p and $1 - p$, respectively, then it is easily seen that x is the sum of n independent and identically distributed random variables. Hence the central limit theorem is applicable, and, if we transform the distribution from the variable x to the variable

$$t = \frac{x - np}{\sqrt{np(1 - p)}} \qquad (a)$$

and then allow n to approach infinity, the probabilities of the values of t approach very closely those given by the normal frequency function with zero mean and unit variance. Hence we can compute approximately the probability of any given set of values of x by obtaining the probability of the corresponding set of values of t from a normal distribution. Since the values of x are discrete, the values of t will also be discrete, although as n becomes large the distances between the values of t will become smaller and smaller. To correct for this we compute the probability, using the normal distribution, of obtaining a value of t between the t_1 corresponding to $x_1 - \frac{1}{2}$, and the t_2 corresponding to $x_2 + \frac{1}{2}$, where x_1 and x_2 are the smallest and largest values of x, respectively, of the set of values of x for which the probability is desired. Hence we have

$$Pr(x_1 \leq x \leq x_2) \sim Pr(t_1 \leq t \leq t_2)$$

where $t_1 = \dfrac{x_1 - \frac{1}{2} - np}{\sqrt{np(1-p)}}$, and $t_2 = \dfrac{x_2 + \frac{1}{2} - np}{\sqrt{np(1-p)}}$. For example, if we wish to know the probability that in tossing an unbiased coin 1600 times we would obtain more than 830 heads, we could approximate it as follows:

$$\begin{aligned} Pr(x > 830) &= 1 - Pr(x \leq 830) \\ &\sim 1 - Pr\left(t < \frac{830 + \frac{1}{2} - 800}{20}\right) \\ &\sim 1 - Pr(t < 1.525) \\ &\sim 1 - 0.936 = 0.064 \end{aligned}$$

It is also true for the Poisson distribution that, if we transform from the variable x to the variable

$$t = \frac{x - \lambda}{\sqrt{\lambda}} \tag{b}$$

and then allow λ to approach infinity, the probability of the various values of t is very closely approximated by the normal frequency function with zero mean and unit variance. Hence we can again approximate the probability of occurrence of one of a given set of values of x by computing the quantities

$$t_1 = \frac{x_1 - \frac{1}{2} - \lambda}{\sqrt{\lambda}} \qquad t_2 = \frac{x_2 + \frac{1}{2} - \lambda}{\sqrt{\lambda}}$$

and then determine from the tables of the normal distribution the $Pr(t_1 < t < t_2)$. For example, if a given sample of a long half-lived material was known to be emitting alpha particles at the rate of 20,000 per minute, and if the effective geometry of a given alpha counter was 50%, the expected value of a 1-minute count would be 10,000. If we wish to know the probability of obtaining a single 1-minute count $\leq$ 10,100, we could approximate this probability by computing

$$t_2 = \frac{10{,}100 + \frac{1}{2} - 10{,}000}{100} = 1.005$$

and determining $Pr(t < 1.005)$ from tables of the normal distribution.

The above transformations will enable us to compute probabilities satisfactorily when, say, $\lambda = np > 50$, $p < 0.5$.

Other transformations which are more satisfactory when λ or $np < 50$, or for the purpose of making the variance independent of the unknown parameters p and λ, will be discussed in Chapter 9.

4.65. Hypergeometric Distribution

Let us now consider a finite population of N objects which can be classified according to some characteristic into X of one type and $N - X$ of another. For example, we might classify manufactured items as "defective" and "non-defective," possibly on the basis of several different tests, or we might classify a particular lot of returnable chemical containers as "failed after first use" or "did not fail after first use." For convenience we shall speak of "defectives" and "non-defectives," since it is in this sense that such a population most frequently arises.

If we could draw a sample of n objects from such a population one at a time, and each time replace the object drawn and allow it to have an equal chance of being redrawn on the next random selection, we should have a situation in which the probability $p = \frac{X}{N}$ of a defective was constant from drawing to drawing, and the number of defectives x in a sample of n would have the binomial distribution. This is equivalent to considering the infinite population of samples of one that can be drawn from such a population. On the other hand, we can consider the problem of drawing a sample of n objects from such a population without replacement, which is the same as considering a single choice from the population of possible samples containing n different objects. Since in this case each choice is dependent on the previous choices, the simplest way of obtaining this distribution is to compute directly the probability of obtaining exactly x defectives in a sample of n. To do this we determine the relative frequency with which samples of n will contain exactly x defectives. The total number of possible samples of size n is given by $\binom{N}{n}$. The number of samples which will contain exactly x defectives is

$$\binom{X}{x}\binom{N-X}{n-x} \tag{a}$$

which is the product of the number of ways x defectives can be chosen from the X defectives present and the number of ways in which the remaining $n - x$ non-defectives can be chosen from the $N - X$ non-defective items. Hence

$$p(x) = \frac{\binom{X}{x}\binom{N-X}{n-x}}{\binom{N}{n}} \tag{b}$$

This probability is defined for all values of x between 0 and n as in the case of the binomial. However, if the number of defectives x in the original population is less than n, $p(x)$ will be 0 for values of x greater than X. This distribution is frequently called the *hypergeometric distribution*, because the values $p(x)$ can be expressed as successive terms of a so-called hypergeometric series. It can be shown that

$$\text{ave}\,(x) = np$$

where $p = X/N$, as in the binomial distribution. However,

$$\text{var}\,(x) = np(1-p)\frac{N-n}{N-1}$$

which is less than that of the binomial distribution. If in the hypergeometric distribution we allow the quantities N and X to approach infinity in such a manner that the ratio $p = X/N$ becomes fixed, the hypergeometric distribution reduces to the binomial distribution.

In general, the binomial distribution will closely approximate the hypergeometric distribution if $p < 0.1$. If we also have $n/N < 0.1$, and N fairly large, the Poisson distribution is a satisfactory approximation.

The primary use of the hypergeometric distribution has been in connection with acceptance sampling, which is discussed in Chapter 9.

APPENDIX 4A

Gamma and Beta Functions

4A.1. Gamma Functions

The definite integral

$$(a) \qquad \Gamma(l) = \int_0^\infty e^{-z} z^{l-1}\, dz \qquad l > 0$$

is called the gamma function of the argument l. In particular

$$\Gamma(1) = \int_0^\infty e^{-z}\, dz = 1$$

A single integration by parts gives

$$(b) \qquad \Gamma(l + 1) = l\Gamma(l)$$

Repeated application of this property gives, for any positive integral value of k,

$$(c) \qquad \Gamma(l) = (l - 1) \cdots (l - k)\, \Gamma\,(l - k)$$

For $l = n + 1$, n an integer, and $k = n$, we have

$$(d) \qquad \Gamma(n + 1) = n(n - 1) \cdots (1)\Gamma\,(1) = n!$$

In particular, for $n = 0$, $\Gamma(1) = 0! = 1$. The gamma function can be considered as an extension of the definition of $n!$ to non-integral values of n.

From (c) we have

$$\Gamma(l + k) = (l + k - 1)\,(l + k - 2) \cdots (l + 1)l\Gamma(l)$$

This can be rewritten

$$(e) \qquad \Gamma(l) = \frac{\Gamma(l + k)}{(l + k - 1)\,(l + k - 2) \cdots (l + 1)l}$$

This form defines the value of $\Gamma(l)$ for negative non-integral values of l such that $l + k > 1$. If l has one of the integral values $0, -1, -2, \cdots, -k + 1$, then one of the factors $l(l + 1) \cdots (l + k - 1)$ will be 0, and hence $\Gamma(l)$ is infinite for l a negative integer or 0, since the above will hold for all positive integral k.

By means of (c) or (e) we can express the gamma function of any value for which it is defined in terms of the gamma function of a value between two given integers. Most tables of the gamma function give the values of $\Gamma(l)$ for $1 \le l \le 2$, since $\Gamma(1) = 1$ and $\Gamma(2) = 1! = 1$, and all the intermediary values are less than 1, and therefore most conveniently tabulated. Since for even moderately large integers $\Gamma(n + 1)$ is quite large, the calculation of its value by the above method becomes impractical, and $\log \Gamma(n + 1) = \log n!$ has been extensively tabulated.

For very large n, the approximation

$$(f) \qquad n! = \Gamma(n + 1) = \sqrt{2\pi}\, n^{n+1/2} e^{-n}$$

known as Stirling's formula, can be used. It is frequently written

$$\ln n! = \ln \Gamma(n + 1) = \tfrac{1}{2} \ln 2\pi n + n \ln n - n$$

The relative accuracy of this approximation increases with n. Its primary value is that it avoids dealing with the more cumbersome $n!$ or $\Gamma(n + 1)$ in the mathematical study of limiting situations involving these quantities.

It can be shown that

$$(g) \qquad \Gamma(l)\Gamma(1 - l) = \frac{\pi}{\sin l\pi}, \qquad 0 < l < 1$$

For $l = \frac{1}{2}$, we obtain $[\Gamma(\frac{1}{2})]^2 = \pi$, $\Gamma(\frac{1}{2}) = \sqrt{\pi}$, a particular result important in statistical applications. This result can also be proved by direct integration.

To illustrate the use of gamma functions, let us consider the problem of showing that

$$\frac{1}{\sigma\sqrt{2\pi}} \int_{-\infty}^{\infty} e^{-1/2\left(\frac{x-\mu}{\sigma}\right)^2} dx = 1$$

i.e., that the area under the normal distribution function as defined in Section 4.41 is unity. Denoting the integral above by I, and making the transformation $t = \dfrac{x - \mu}{\sigma}$, $dx = \sigma\, dt$, we have

$$I = \frac{1}{\sqrt{2\pi}} \int_{-\infty}^{\infty} e^{-t^2/2}\, dt$$

$$= \frac{2}{\sqrt{2\pi}} \int_{0}^{\infty} e^{-t^2/2}\, dt$$

since the function $e^{-t^2/2}$ is even. Now if we make the transformation $z = t^2/2$, $2z = t^2$, $dz = t\,dt$, $dt = dz/\sqrt{2z}$, noting that for $t = 0$, $z = 0$, and for $t = \infty$, $z = \infty$, we have

$$I = \frac{2}{\sqrt{2\pi}} \int_0^\infty e^{-z} \frac{dz}{\sqrt{2z}}$$

$$= \frac{2}{\sqrt{2}\,\sqrt{2\pi}} \int_0^\infty e^{-z} z^{-1/2}\,dz$$

The last integral is identical with that defining $\Gamma(\frac{1}{2})$, hence

$$I = \frac{1}{\sqrt{\pi}}\,\Gamma(\tfrac{1}{2}) = \frac{\sqrt{\pi}}{\sqrt{\pi}} = 1$$

This justifies the choice of the factor $\dfrac{1}{\sigma\sqrt{2\pi}}$ as the coefficient of the normal distribution function.

As a second example, let us consider the determination of the moment generating function of the χ^2 distribution. In this case we are required to integrate

$$M_{\chi^2}(\theta) = \frac{1}{2^{\nu/2}\Gamma(\nu/2)} \int_0^\infty e^{\chi^2\theta} e^{-\chi^2/2} (\chi^2)^{(\nu/2)-1} d(\chi^2)$$

$$= \frac{1}{\Gamma(\nu/2)} \int_0^\infty e^{-(\chi^2/2)(1-2\theta)} (\chi^2/2)^{(\nu/2)-1} d(\chi^2/2)$$

Making the substitution

$$z = \frac{\chi^2}{2}(1 - 2\theta)$$

from which

$$\frac{\chi^2}{2} = \frac{z}{1 - 2\theta}$$

$$d(\chi^2/2) = \frac{dz}{1 - 2\theta}$$

we obtain

$$M_{\chi^2}(\theta) = \frac{1}{\Gamma(\nu/2)} (1 - 2\theta)^{-\nu/2} \int_0^\infty e^{-z} z^{(\nu/2)-1}\,dz$$

The integral is by definition $\Gamma(\nu/2)$, and hence

$$M_{\chi^2}(\theta) = (1 - 2\theta)^{-\nu/2}$$

as was stated in Section 4.46.

4A.2. Beta Functions

The definite integral

$$B(l, m) = \int_0^1 z^{l-1}(1 - z)^{m-1}\, dz \tag{a}$$

is called the beta function of the arguments l and m. Obviously we have $B(1, 1) = 1$. If we make the transformation $z' = 1 - z$, $z = 1 - z'$, $dz = -dz'$, noting that, when $z = 0$, $z' = 1$, and when $z = 1$, $z' = 0$, we have

$$\begin{aligned} B(l, m) &= \int_1^0 (1 - z')^{l-1}(z')^{m-1}(-\, dz') \\ &= \int_0^1 (z')^{m-1}(1 - z')^{l-1}\, dz' \qquad (b) \\ &= B(m, l) \end{aligned}$$

It can be shown that

$$B(l, m) = \frac{\Gamma(l)\Gamma(m)}{\Gamma(l + m)} \tag{c}$$

Thus the value of $B(l, m)$ can be obtained from tables of the gamma function.

4A.3. Incomplete Gamma Function

The function

$$\Gamma_x(l) = \int_0^x e^{-z} z^{l-1}\, dz \qquad l > 0,\ 0 < x < \infty \tag{a}$$

is called the incomplete gamma function. Extensive tables of the ratio $I(x, l) = \Gamma_x(l)/\Gamma(l)$ are given in *Tables of the Incomplete Gamma Function*, [8]. (The tabulations are in terms of $u = x/\sqrt{l}$ and $p = l - 1$.)

The ratio $I(x, l)$ is itself the cumulative distribution function of the gamma distribution

$$f(x) = \frac{1}{\Gamma(l)}\, e^{-x} x^{l-1} \qquad l > 0,\ 0 < x < \infty \tag{b}$$

By an integration similar to that in the second example in Section 1 of this appendix, we obtain for the moment generating function of this distribution

$$M_x(\theta) = (1 - \theta)^{-l} \tag{c}$$

from which we can obtain in the usual manner

$$\mu_1' = l$$

$$\mu_2' = l(l + 1)$$

These can also be obtained directly. It follows that for this distribution

(*d*) $$\text{ave}(x) = \text{var}(x) = l$$

Since the χ^2-distribution for ν degrees of freedom is a gamma distribution with $x = \chi_\nu^2/2$ and $l = \nu/2$, we have

(*e*) $$\begin{aligned} Pr(\chi_\nu^2 \geq c) &= Pr(x \geq c/2) \\ &= 1 - I(c/2, \nu/2) \end{aligned}$$

which, when the above tables are available, makes it possible to determine $Pr(\chi_\nu^2 \geq c)$ more closely than from the tables giving $\chi^2_{\nu,\alpha}$ for specific values of α.

Tables of the incomplete gamma function can also be used to evaluate partial sums of the Poisson distribution. If x has a Poisson distribution with ave $(x) = \lambda$, then

$$Pr(x \geq c) = \sum_{x=c}^{\infty} \frac{e^{-\lambda}\lambda^x}{x!}$$

Differentiating both sides of this identity with respect to λ, we obtain

$$\begin{aligned} \frac{d}{d\lambda} Pr(x \geq c) &= \sum_{x=c}^{\infty} \frac{1}{x!}(x\lambda^{x-1}e^{-\lambda} - \lambda^x e^{-\lambda}) \\ &= \sum_{x=c}^{\infty} \frac{e^{-\lambda}\lambda^{x-1}}{(x-1)!} - \sum_{x=c}^{\infty} \frac{e^{-\lambda}\lambda^x}{x!} \\ &= \frac{e^{-\lambda}\lambda^{c-1}}{(c-1)!} \end{aligned}$$

and hence

$$\begin{aligned} Pr(x \geq c) &= \frac{1}{\Gamma(c)} \int_0^\lambda e^{-\lambda}\lambda^{c-1}\, d\lambda \\ &= I(\lambda, c) \end{aligned}$$

4A.4. Incomplete Beta Function

The incomplete beta function is defined as

(*a*) $$I_x(l, m) = \frac{1}{B(l, m)} \int_0^x z^{l-1}(1 - z)^{m-1}\, dz \qquad l, m > 0,\ 0 \leq x \leq 1$$

This function has also been extensively tabulated in *Tables of the Incomplete Beta Function* [9]. Values are given only for $l \geq m$, since if $l < m$ we can use the relationship $I_x(l, m) = 1 - I_{1-x}(m, l)$ and look up $I_{1-x}(m, l)$, $m > l$, in the tables.

As in the case of the incomplete gamma function, $I_x(l, m)$ is the cumulative distribution function of the beta distribution

$$(b) \qquad f(x) = \frac{1}{\mathrm{B}(l, m)} x^{l-1}(1 - x)^{m-1} \qquad l, m > 0, 0 < x < 1$$

Both the "Student" t-distribution and the F-distribution can, by appropriate transformations, be expressed in this form. By direct integration we obtain

$$\mu_1' = \frac{l}{l + m}$$

$$\mu_2' = \frac{l(l + 1)}{(l + m)(l + m + 1)}$$

and hence for this distribution

$$(c) \qquad \text{ave}(x) = \frac{l}{l + m}$$

$$\text{var}(x) = \frac{lm}{(l + m)^2 (l + m + 1)}$$

For $l = m$ the distribution is symmetric about ave $(x) = \frac{1}{2}$.

By a method identical with that used in the preceding section, we obtain

$$(d) \qquad I_{1-p}(n - c, c + 1) = \sum_{x=0}^{c} \binom{n}{x} p^x(1 - p)^{n-x}$$

The right-hand side is the $Pr(x \leq c)$ when x has the binomial distribution, and hence the incomplete beta function on the left gives the cumulative binomial distribution directly. The probability of a specific value of x can be obtained from the relationship

$$\begin{aligned} Pr(x = c) &= Pr(x \leq c) - Pr(x \leq c - 1) \\ &= I_{1-p}(n - c, c + 1) - I_{1-p}(n - c + 1, c) \end{aligned}$$

APPENDIX 4B

The Normal Bivariate Distribution

Although, in most cases where sets of two or more associated measurements arise in physical, chemical, or industrial problems, we shall want to consider only one of these as a random variable and the rest as fixed numbers, there are occasions when it is desirable to consider a set of associated measurements as drawn from a population which can be represented by a joint probability density function of the several variables being considered. The distribution used almost exclusively for this purpose is the multivariate normal distribution, on which the methods of multivariate analysis, which have proved so successful in many fields, are based. We shall give here a brief description of this distribution for the case of 2 variables; the case of more than 2 variables is discussed in detail in most texts on mathematical statistics (e.g., [2], [3]).

The normal bivariate distribution, or probability density function, is given by

$$(a)\quad f(x, y) = \frac{1}{2\pi\sigma_x\sigma_y\sqrt{1-\rho^2}}\, e^{-\frac{1}{2(1-\rho^2)}\left[\left(\frac{x-\mu_x}{\sigma_x}\right)^2 - 2\rho\left(\frac{x-\mu_x}{\sigma_x}\right)\left(\frac{y-\mu_y}{\sigma_y}\right) + \left(\frac{y-\mu_y}{\sigma_y}\right)^2\right]}$$

where the identity of the parameters μ_x, μ_y, σ_x^2, and σ_y^2, with the average values and variances of x and y, can be established as in the case of the univariate normal distribution from the definitions in Section 3.32. We can also obtain by direct integration $\text{cov}(x, y) = \rho\sigma_x\sigma_y$. It follows that $\rho = \dfrac{\text{cov}(x, y)}{\sigma_x\sigma_y}$ is the population correlation coefficient, corresponding to the sample value r defined in Section 2.53. In terms of the standardized variables $t_x = \dfrac{x-\mu_x}{\sigma_x}$ and $t_y = \dfrac{y-\mu_y}{\sigma_y}$, (a) becomes

$$(b)\qquad f(t_x, t_y) = \frac{1}{2\pi\sqrt{1-\rho^2}}\, e^{-\frac{1}{2(1-\rho^2)}(t_x^2 - 2\rho t_x t_y + t_y^2)}$$

Since ave $(t_x) = $ ave $(t_y) = 0$ and var $(t_x) = $ var $(t_y) = 1$, it follows immediately that cov $(t_x, t_y) = \rho$. Tables of $F(h, k) = Pr(t_x \leq h, t_y \leq k)$ are given for various values of h, k, and ρ in [14].

By integrating $f(x, y)$ with respect to y and x respectively the marginal distributions $g(x)$ and $h(y)$ are found to be normal distributions with means μ_x and μ_y and variances σ_x^2 and σ_y^2. If $\rho = 0$, it can easily be seen that

$$(c) \qquad f(x, y) = \frac{1}{\sigma_x \sqrt{2\pi}} e^{-\frac{1}{2}\left(\frac{x-\mu_x}{\sigma_x}\right)^2} \cdot \frac{1}{\sigma_y \sqrt{2\pi}} e^{-\frac{1}{2}\left(\frac{y-\mu_y}{\sigma_y}\right)^2}$$

$$= g(x) \cdot h(y)$$

i.e., in this case $\rho = 0$, or cov $(x, y) = 0$, implies the independence of x and y. Since both marginal distributions can be normal only if the joint distribution is multivariate normal, this means that for any two normally distributed variables x and y, cov $(x, y) = 0$ is sufficient to ensure their independence.

Note that for a constant value of the exponent, i.e., for

$$(d) \quad \left(\frac{x - \mu_x}{\sigma_x}\right)^2 - 2\rho\left(\frac{x - \mu_x}{\sigma_x}\right)\left(\frac{y - \mu_y}{\sigma_y}\right) + \left(\frac{y - \mu_y}{\sigma_y}\right)^2 = \lambda^2(1 - \rho^2)$$

the function $f(x, y)$, and hence the probability density, is constant. The points (x, y) satisfying this condition form an ellipse with its center at the point (μ_x, μ_y) known as a contour ellipse. The major axis of this ellipse has a positive slope when ρ is positive, and a negative slope when ρ is negative; for the special case $\sigma_x = \sigma_y$, the slope is either $+1$ or -1. When $\rho = 0$ the ellipse reduces to

$$(e) \qquad \left(\frac{x - \mu_x}{\sigma_x}\right)^2 + \left(\frac{y - \mu_y}{\sigma_y}\right)^2 = \lambda^2$$

which is an ellipse with its major axis parallel to the x axis if $\sigma_x > \sigma_y$, or parallel to the y axis if $\sigma_x < \sigma_y$. If $\sigma_x = \sigma_y$ the ellipse reduces to the circle

$$(f) \qquad (x - \mu_x)^2 + (y - \mu_y)^2 = \lambda^2\sigma^2$$

Figure 4.18 shows, for a given value of λ^2, the shape of these ellipses, which indicates the type of scatter to be expected of the points (x, y) for various combinations of σ_x, σ_y, and ρ.

In general, the left-hand side of (d) can by proper rotation of axes be transformed into the sum of squares of 2 independent quantities, each having a normal distribution with zero mean and unit variance. This is already true for (e) and (f), where the axes of the contour ellipse are parallel to the coordinate axes. Thus λ^2, considered as a function of

x and y, will have the χ^2-distribution with 2 degrees of freedom. If we now choose values of λ_α^2 such that $Pr(\chi_2^2 \geq \lambda_\alpha^2) = \alpha$, the probability of a point (x, y) falling outside the contour ellipse defined by λ_α^2 is exactly α. The contour ellipses and the values λ_α^2 for the two-dimensional normal distribution are analogous to the limits $\mu \pm t\sigma$ and the values t_α for

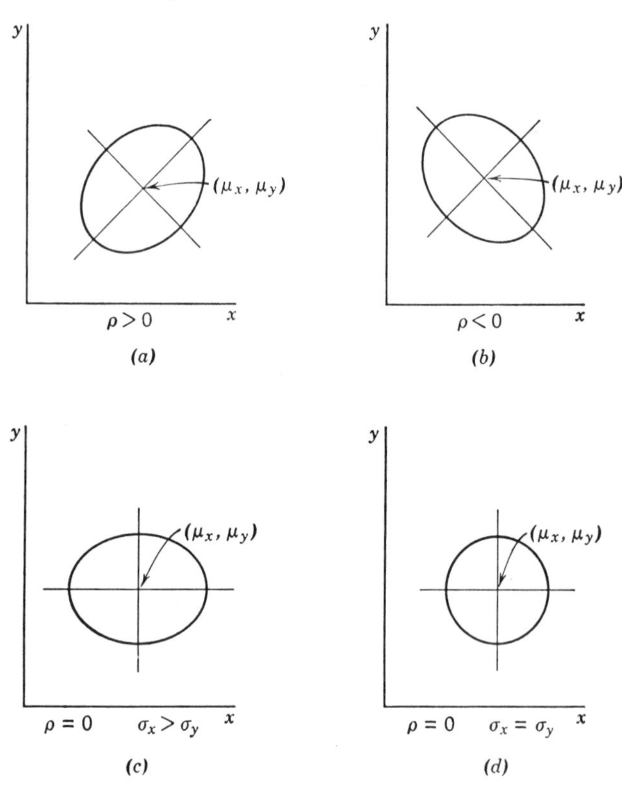

FIG. 4.18. Contour ellipses.

the one-dimensional normal distribution. Figure 4.19 shows the contour ellipses for $\rho = 0.5$, $\sigma_x = \sigma_y$, and $\alpha = 0.50$, 0.05, and 0.01.

The conditional distributions $f(y \mid x)$ and $f(x \mid y)$ of y for a given x and x for a given y are obtained by dividing $f(x, y)$ by the marginal distributions $g(x)$ and $h(y)$, respectively. In the case of the bivariate normal distribution it can be shown by simple algebraic rearrangement of the ratios $\dfrac{f(x, y)}{g(x)}$ and $\dfrac{f(x, y)}{h(y)}$ that these distributions are also normal. In particular, it can be determined from these distributions that the average value of y for a given x will fall on the line

(g) $$y = \mu_y + \rho \frac{\sigma_y}{\sigma_x}(x - \mu_x)$$

and the average value of x for a given y will fall on the line

(h) $$x = \mu_x + \rho \frac{\sigma_x}{\sigma_y}(y - \mu_y)$$

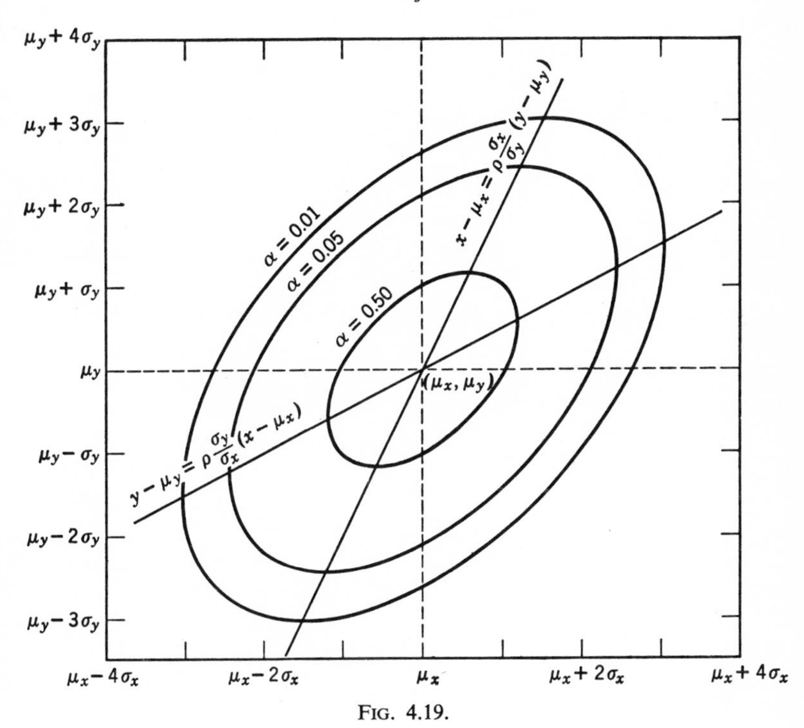

FIG. 4.19.

These lines are called the *regression lines* of y on x and x on y, respectively, and their slopes $\beta_{y \cdot x} = \rho \frac{\sigma_y}{\sigma_x}$ and $\beta_{x \cdot y} = \rho \frac{\sigma_x}{\sigma_y}$ are called the *regression coefficients* of y on x and x on y. The variance of y for a given x is

(i) $$\sigma^2_{y \cdot x} = \sigma_y^2(1 - \rho^2)$$

which is independent of x, and similarly the variance of x for a given y is

(j) $$\sigma^2_{x \cdot y} = \sigma_x^2(1 - \rho^2)$$

which is independent of y. These regression lines are always different, except in the trivial case when x and y are completely dependent, and $\rho = \pm 1$. The two regression lines, and their relationship to the contour ellipses, are shown in Figure 4.19. Both lines pass through the point μ_x, μ_y, and they represent the diameters of the ellipses obtained by taking the midpoints of lines parallel to the y and x axes, respectively. In terms of the standardized variables previously defined (g) and (h) reduce to

$$(k) \qquad \begin{aligned} t_y &= \rho t_x \\ t_x &= \rho t_y \end{aligned}$$

The product of the slope $\beta_{y \cdot x}$ of (g) and the slope $\beta_{x \cdot y}$ of (h) is ρ^2, i.e.,

$$(l) \qquad \rho = \sqrt{\beta_{y \cdot x}\, \beta_{x \cdot y}}$$

This is also obviously true for equations (k).

For samples of size n from a normal bivariate population, the sample means and variances for x and y individually must have the distribution previously obtained in Sections 4.51 and 4.53, since the marginal distributions are normal. The joint distribution of the variances and covariances has been developed by Wishart for the multivariate case, and a generalized t-distribution by Hotelling; both of these are immediately applicable to samples from a bivariate distribution. Most of the interest in the past has centered upon the distribution of the sample correlation coefficient r. For $\rho = 0$ the exact distribution was found by R. A. Fisher, who pointed out that

$$(m) \qquad t = \frac{r}{\sqrt{1 - r^2}} \sqrt{n - 2}$$

has the "Student" t-distribution with $\nu = n - 2$ degrees of freedom. However, for $\rho \neq 0$ the exact distribution cannot be evaluated in closed form, and for ρ appreciably different from zero the distribution approaches normality very slowly. F. N. David [15] has tabulated the exact distribution for various values of ρ and a large number of values of n. For moderately large values of n and with ρ not too large, the quantity

$$(n) \qquad z = \tfrac{1}{2} \ln \frac{1 + r}{1 - r} = \tanh^{-1} r$$

is very nearly normally distributed with

$$(o) \qquad \begin{aligned} \text{ave}\,(z) &= \tfrac{1}{2} \ln \frac{1 + \rho}{1 - \rho} + \frac{\rho}{2(n - 1)} \\ \text{var}\,(z) &= \frac{1}{n - 3} \end{aligned}$$

the latter being independent of ρ. This transformation, also due to R. A. Fisher, is of great practical importance.

CHAPTER 5

Statistical Inference

5.1. Introduction

Most statistical problems fall into one of two broad classes: (1) the estimation of one or more of the unknown parameters of the assumed population distribution, and (2) the testing of hypotheses concerning either the parameters of the model assumed or the validity of the model itself. The problems of group (1) can be further subdivided into those of point estimation, in which we try to obtain from the sample one single value which is in some sense the "best" estimate of an unknown population parameter; and the determination of confidence intervals, or the specification of a range of values within which we have a certain "confidence" that an unknown parameter will be included.

R. A. Fisher [12] has divided the process used to make statistical inferences into three distinct problems. The first problem is the *problem of specification*, i.e., the decision as to what type of mathematical model we shall use to represent the actual situation. Since statistical inferences are always made with respect to such a model, even though the model may specify only that the population can be represented by a continuous cumulative distribution function, the validity of our inferences will always depend to a greater or lesser degree on the validity of the model. The specification of a model usually consists of two parts: (1) the assumption of a distribution function or functions which will adequately represent the population of possible observations; (2) the identification of the parameters of this distribution, or these distributions, or possibly some non-parametric aspect of the distribution itself, with what we wish to know concerning the actual situation. Of these the latter is the more important, for, although we can develop methods of inference which depend either not at all, or very little, on the form of the distribution, there is no statistical way of avoiding the possibility of making a perfectly valid inference concerning a parameter of the model which is not the actual quantity desired. For example, the best statistical analysis of a set of 10 chemical analyses of an unknown substance cannot avoid giving the incorrect answer if the method is producing answers which are on the average 2% low.

The second problem is the *problem of estimation*, i.e., the problem of deciding what statistic or statistics computed from the sample will be used for the purpose of making the required inferences. Several general methods of solving this problem have been advanced, depending, of course, on a specific definition of what is meant by a "best estimate" or a "best test," but these are frequently difficult to apply, or are practically undesirable, in specific situations. The third problem is the *problem of distribution*; after we have chosen the proper statistic, we must know something about its distribution in random samples from the assumed population if we are to make probability statements concerning the validity of the inferences made. This problem has been discussed briefly in Chapter 4 for several much-used sample statistics, assuming a normal distribution for the original population. The solution of this problem for other cases has resulted in the tables of critical values of which some will be given or referred to in this chapter.

The last problem and that portion of the second dealing with general methods of inference and the properties of estimates and tests are primarily the problems of mathematical statistics. The first problem and that part of the second problem which deals with the statistics to be used in specific situations are those which face the applied statistician or experimenter. In this chapter we shall consider briefly some of the general principles used in statistical inferences, and then proceed to give the solution, or solutions, to the problems of inference in a number of specific situations which commonly occur.

5.2. General Principles of Statistical Inference

5.21. Point Estimation and Efficiency

One of the most frequent reasons for obtaining experimental data is to determine some unknown value. A chemist wishes to determine such things as the yield of a new process, the impurity in a certain type of raw material, the value of some physical characteristic of a new product, the precision of a new analytical method, or the frequency with which a certain process will fail. In the case of measurements we are always attempting to find the unknown "true value" of the quantity or characteristic being measured. A single value which is in some sense the best single "guess" as to the value of the unknown quantity being estimated is called a *point estimate*. The statistical problem of point estimation can be summarized as follows: We assume that our sample came from a population which can be represented by a particular distribution function, and that the quantity desired is represented by one of the parameters of this distribution function; under these circumstances what value calculated from the sample gives the best single estimate of the unknown quantity?

In order to determine the best estimate in any particular instance, we must define what is meant by the term "best." Typical qualities which we might like our estimate to have are:

(1) In the long run it should tend to give the "true value" of the parameter being estimated.

(2) In the long run it should be above or below this "true value" about equally frequently.

(3) In any particular case it should not deviate too far from the unknown "true value."

The first condition means that the average value of the statistic used should be the population parameter desired and the second that the median of the distribution of the statistic should be the desired parameter. The third condition can be expressed in several ways, depending on the measure of the scatter of the estimates about the desired parameter which we wish to use. The most common requirement is that the variance of the statistic used be the least among those statistics satisfying the first condition; alternatively we can require that the mean deviation, or even the range be minimized.

Designating a statistic to be used as an estimate of a parameter θ by $\hat{\theta}$, this estimate will be called *unbiased* if it has the property that

$$\text{ave}\,(\hat{\theta}) = \theta \tag{a}$$

Note that this gives a particular meaning to the concept of bias; if we think of bias as meaning "more frequently above than below," or vice versa, rather than "long-term average systematically in error," then unbiased means that the median of $\hat{\theta}$ be equal to θ. Fortunately, these two concepts of bias frequently coincide. If, in addition to satisfying (*a*), an estimate $\hat{\theta}$ has the property that among all possible estimates satisfying (*a*)

$$\text{var}\,(\hat{\theta}) = \text{a minimum} \tag{b}$$

then $\hat{\theta}$ will be called the *best unbiased* estimate of the parameter θ. Again, we have given a particular meaning to the word "best"; other criteria, as indicated above, could be used.

In many situations more than one unbiased estimate of a given parameter are available, one or none of which may be the best estimate in the above sense. If $\hat{\theta}_1$ and $\hat{\theta}_2$ are two such estimates of θ, then we shall say $\hat{\theta}_1$ is more efficient than $\hat{\theta}_2$, if

$$\text{(c)} \qquad \operatorname{var}(\hat{\theta}_1) < \operatorname{var}(\hat{\theta}_2)$$

If a best estimate $\hat{\theta}_0$ of θ does exist, then it is possible to define the *efficiency* of any estimate $\hat{\theta}$ by the ratio

$$E = \frac{\operatorname{var}(\hat{\theta}_0)}{\operatorname{var}(\hat{\theta})}$$

usually stated as 100 $E\%$. The concept of efficiency frequently enables us to decide in practical situations between several available statistics, particularly where we may be willing to sacrifice efficiency for ease of computation (see Section 2.34).

Several general methods of obtaining point estimates are available. One of them, known as the *method of moments*, is to obtain estimates of the population parameter or parameters, representing the unknown value or values desired, by equating a sufficient number of sample moments to the population moments to determine the desired parameters. A second method, which ensures under fairly general conditions that the estimates obtained will, for large samples, be approximately the best unbiased estimates and also approximately normally distributed, is the *method of maximum likelihood*. This method of obtaining point estimates, due to R. A. Fisher, is discussed in detail in Appendix 5A.

5.22. Confidence Intervals

If instead of determining from the sample one single value as an estimate of the unknown parameter θ, we determine two values A and B such that there is a given probability $1 - \alpha$ that $A < \theta < B$, then A and B will be called $100(1 - \alpha)\%$ confidence limits for the given parameter, and the interval between them a $100(1 - \alpha)\%$ confidence interval. Since the probability is α that this interval will not include θ, we are taking a $100\alpha\%$ risk of being wrong if we make the statement that A and B include the unknown value θ. The limits we determined in Sections 2.43, 2.44, and 2.54 were estimates of this type. It must be emphasized that we do not make the statement that θ has a probability $1 - \alpha$ of falling between the given limits; the value of θ is simply an unknown constant, and therefore we cannot make probability statements of this kind concerning it. What we do state is a probability that the limits which we select will include the fixed but unknown value θ. Since probability statements are involved, the determination of confidence limits will require some knowledge of the sampling distribution of the statistics involved in addition to the expected values and variances required to form a point estimate.

Just as we can make a point estimate by a number of methods, so we can choose many confidence intervals which have a probability $1 - \alpha$ of including the unknown value of the parameter θ. For example, if we have a sample of size n from a normal distribution, for which the population mean and median are both equal to the average value μ, then the methods of both Sections 2.43 and 2.44 can be used to determine $100(1 - \alpha)\%$ confidence limits for μ. We need some basis for deciding which interval will be preferred. In this case the criterion most frequently used is the *length* of the interval obtained. Since A and B are sample statistics, the length $B - A$ will vary from sample to sample, so that no one method will give the shortest interval in every case; however, assuming a given form for the distribution of the population, some methods of computing limits will on the average give shorter confidence intervals than others. We may sometimes choose to use intervals which are on the average longer than the best available, but possess other practical advantages which make their use desirable.

For a given method, the size of a confidence interval will depend on the probability α. We saw in Section 2.43 that the factor used to obtain 99% limits in the second example was larger than the factor used to obtain 95% limits in the first. The choice that must be made here is between a more general statement which has a high probability of being correct and a more specific statement which has a smaller probability of being correct. The size of the confidence interval will also depend on the sample size; for a fixed probability α of being correct, we can make a more specific statement concerning the unknown value of θ when n is large than when n is small.

5.23. Example of the Determination of Confidence Limits

We shall illustrate the general problem of finding confidence limits by considering the special case of computing limits from a sample of size n for the mean μ of a normal distribution with known variance σ^2. In this case the sample mean $\bar{x}$ is also normally distributed with mean μ and variance σ^2/n. Hence the quantity $t = \dfrac{\bar{x} - \mu}{\sigma}\sqrt{n}$ is a standardized normal variate. If we obtain from the tables of the areas of the normal distribution the value t_α of t such that $Pr(|t| > t_\alpha) = \alpha$, then the $Pr(-t_\alpha < t < t_\alpha) = Pr\left(-t_\alpha < \dfrac{\bar{x} - \mu}{\sigma}\sqrt{n} < t_\alpha\right) = 1 - \alpha$, regardless of the value of the unknown mean μ. However, the statement $\dfrac{\bar{x} - \mu}{\sigma}\sqrt{n} < t_\alpha$

is equivalent to the statement $\bar{x} - \frac{t_\alpha \sigma}{\sqrt{n}} < \mu$, since we have

$$\frac{\bar{x} - \mu}{\sigma}\sqrt{n} < t_\alpha$$

$$\bar{x} - \mu < \frac{t_\alpha \sigma}{\sqrt{n}}$$

since $\sigma \geq 0$

$$\bar{x} - \frac{t_\alpha \sigma}{\sqrt{n}} < \mu$$

Similarly, the statement $-t_\alpha < \frac{\bar{x} - \mu}{\sigma}\sqrt{n}$ is equivalent to the statement $\mu < \bar{x} + \frac{t_\alpha \sigma}{\sqrt{n}}$. Hence the statement that $\frac{\bar{x} - \mu}{\sigma}\sqrt{n}$ will fall between $-t_\alpha$ and t_α is equivalent to the statement that we shall have both $\bar{x} - \frac{t_\alpha \sigma}{\sqrt{n}} < \mu$ and $\bar{x} + \frac{t_\alpha \sigma}{\sqrt{n}} > \mu$. Combining these into one statement, we have

$$Pr\left(\bar{x} - \frac{t_\alpha \sigma}{\sqrt{n}} < \mu < \bar{x} + \frac{t_\alpha \sigma}{\sqrt{n}}\right) = 1 - \alpha$$

i.e., the interval between $\bar{x} - \frac{t_\alpha \sigma}{\sqrt{n}}$ and $\bar{x} + \frac{t_\alpha \sigma}{\sqrt{n}}$ is a $100(1 - \alpha)\%$ confidence interval for the unknown mean μ if the population variance σ^2 is known.

This situation is shown graphically in Figure 5.1. For any particular value of μ, say μ_0, the probability is $1 - \alpha$ that the mean $\bar{x}$ of a sample of n observations will fall between the points A and B. Hence, regardless of the value of μ, the probability is $1 - \alpha$ that the point $(\mu, \bar{x})$, where $\bar{x}$ is the mean of a sample of n observations from a normal distribution with mean μ and known variance σ^2, will fall in the region between the two lines shown. Hence for a given $\bar{x}$, say $\bar{x}_0$, the probability is $1 - \alpha$ that the unknown mean μ of the population from which the sample was drawn is included between C and D. Notice that, since in this case the variation in $\bar{x}$ depends only on the known variance σ^2 and not on the unknown mean μ, the lines defining the region which will include the point $(\mu, \bar{x})$ with a probability $1 - \alpha$ are parallel.

The preceding confidence interval gives 2 values computed from the sample which have a certain probability of including the unknown mean μ. In this instance these values are equally spaced from the sample mean, since it can be shown that such an interval is the smallest giving both upper and lower limits having a probability of $1 - \alpha$ of including the unknown mean. However, we are sometimes concerned only with the problem of determining a value from the sample which has probability $1 - \alpha$ of exceeding the unknown value of μ, or, conversely, of determining a value which has probability $1 - \alpha$ of being less than μ. Such intervals will be called one-sided confidence intervals, as opposed to the two-sided confidence interval given above. With reference to the general case we have simply chosen $B = \infty$ or $A = -\infty$.

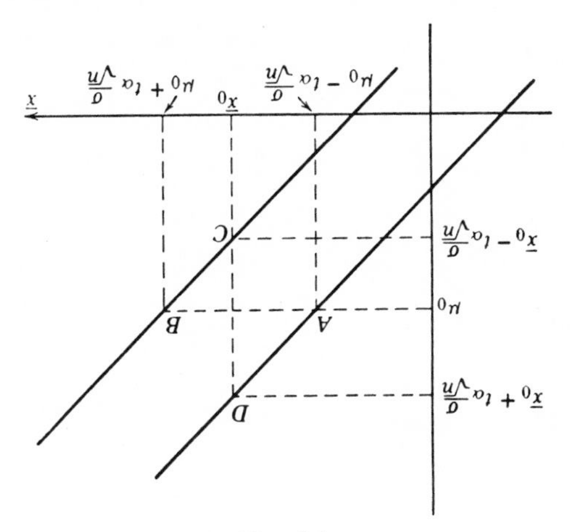

FIG. 5.1.

Suppose that we choose a value $t_{2\alpha}$ such that $Pr(|t| > t_{2\alpha}) = 2\alpha$. Then since the normal distribution is symmetric, we shall have $Pr(t < t_{2\alpha}) = 1 - \alpha$ and $Pr(t > -t_{2\alpha}) = 1 - \alpha$, or $Pr(\frac{\bar{x} - \mu}{\sigma}\sqrt{n} < t_{2\alpha})$ $= 1 - \alpha$ and $Pr(\frac{\bar{x} - \mu}{\sigma}\sqrt{n} > t_{2\alpha}) = 1 - \alpha$. By arguments similar to those used above, we find that these statements imply

$$Pr(\bar{x} - \frac{\sigma t_{2\alpha}}{\sqrt{n}} < \mu) = 1 - \alpha$$

$$Pr(\bar{x} + \frac{\sigma t_{2\alpha}}{\sqrt{n}} > \mu) = 1 - \alpha$$

Hence the intervals above $\bar{x} - \frac{\sigma t_{2\alpha}}{\sqrt{n}}$ and below $\bar{x} + \frac{\sigma t_{2\alpha}}{\sqrt{n}}$ are the required one-sided $100(1 - \alpha)\%$ confidence intervals for the unknown mean μ.

The above confidence intervals are applicable to any normally distributed variable with known standard deviation. For example, if we had 2 samples of n_1 and n_2 from normal populations with means μ_1 and μ_2

and variances σ_1^2 and σ_2^2, respectively, then $\bar{d} = \bar{x}_1 - \bar{x}_2$ would be normally distributed with mean $\delta = \mu_1 - \mu_2$ and variance $\sigma_\delta^2 = \sigma_1^2/n_1 + \sigma_2^2/n_2$. Hence, if σ_1^2 and σ_2^2 were known, we should have as a two-sided confidence interval for δ

$$\bar{d} - t_\alpha \left(\frac{\sigma_1^2}{n_1} + \frac{\sigma_2^2}{n_2}\right) < \delta < \bar{d} + t_\alpha \left(\frac{\sigma_1^2}{n_1} + \frac{\sigma_2^2}{n_2}\right)$$

and similar one-sided intervals. Equivalent expressions could also be written for the sum $\bar{x}_1 + \bar{x}_2$, or even for general linear combinations of the means.

EXAMPLE. As a concrete example of the use of these limits, suppose that data taken over a long period of time have established the standard deviation of an iron analysis as 0.12% iron. If 6 replicate determinations on a given ore sample give an average iron content of 32.56%, then a 95% confidence interval for the true iron content of the sample, assuming that a large number of replicate results of a similar nature would be approximately normally distributed with the given standard deviation, is given by

$$32.56 - 1.96 \frac{0.12}{\sqrt{6}} < \mu < 32.56 + 1.96 \frac{0.12}{\sqrt{6}}$$

$$32.46 < \mu < 32.66$$

since from Table 4.3, $t_{0.05} = 1.96$. We are, of course, also assuming the method to be free from systematic error. If we were interested only in establishing a lower limit for the possible iron content of the sample, we should obtain, since $t_{0.10} = 1.645$,

$$32.56 - 1.645 \frac{0.12}{\sqrt{6}} < \mu$$

$$\mu > 32.48$$

5.24. Tests of Significance

Another common reason for obtaining experimental data is to test some preconceived notion that we may have concerning the population from which the data are a sample. Thus we may wish to determine whether a new process gives a greater yield than an old one, whether a given analytical method gives correct results when used to analyze a standard solution of known concentration, or whether a given shipment of raw material contains less than a certain amount of a given impurity. The use of experimental data to answer questions of this sort is called *hypothesis testing*. The simplest case of testing a statistical hypothesis can be summarized briefly as follows: Suppose that we assume that the population in question can be described by a certain distribution function depending on a single unknown parameter θ. Then, on the basis of a sample $x_1, \cdots, x_n$ drawn from this population, we wish either to accept or reject the hypothesis that θ has some particular value θ_0. More generally, any statement concerning the statistical model of the population from which the sample is drawn can be considered an hypothesis to be tested. Thus we could consider the hypothesis that θ was greater than a given value θ_0, or the hypothesis that the population has a given functional form (as in Section 4.45), or even that the given sample actually represented a random sample from some continuous distribution.

The test of a statistical hypothesis generally takes the following form:

(1) From the observations $x_1, x_2, \cdots, x_n$ we compute some appropriate sample statistic.

(2) On the assumption that the hypothesis is true, we determine the probability of the sample statistic deviating from its expected value as far as or further than the value observed.

(3) If the probability so determined is less than some small value α, we reject the hypothesis.

Thus we reject an hypothesis when it is unlikely that the given sample would have been obtained if it were true. For example, if, on the basis of a sample $x_1, x_2, \cdots, x_n$ assumed to be from a normal distribution with known variance σ^2, we wished to test the hypothesis $\mu = \mu_0$, we should determine the probability of a sample mean deviating from the hypothetical mean by as much as $\bar{x} - \mu_0$, and reject the hypothesis if this probability were small. Tests such as this are frequently referred to as *tests of significance*, and if the hypothesis is rejected the true value of θ is said to be significantly different from the hypothetical value θ_0 at the significance level α. The hypothesis $\theta = \theta_0$ in this case is referred to as the *null hypothesis*, since we never conclude that $\theta = \theta_0$, but only that it does not differ significantly from θ_0. It should be noted that in every case the confidence limits for a parameter θ determine a test of the hypothesis $\theta = \theta_0$, since if θ_0 is not included between these limits we can conclude with $100\alpha\%$ risk, or at the significance level α, that $\theta \neq \theta_0$.

It must be recognized that in testing an hypothesis we can make two types of errors. The first, frequently called a type I error, is to reject the hypothesis when it is true; we have already fixed the probability of this type of error by choosing a significance level α. The second, or type II, error is that of failing to reject the hypothesis when it is false. To show the distinction between these two types of error, let us suppose that the hypothesis to be tested is that a certain process change did not increase the yield. Then making the statement that the yield did increase when in fact it did not would be an error of type I, whereas the statement that the yield did not increase when in fact it did would be an error of type II. These two errors could have entirely different consequences. A plot of the probability of an error of type II as a function of the possible values of θ is called the *operating characteristic curve* of a test. Such curves are very important in determining the degree to which a test will reject false hypotheses, and in determining the sample size necessary to obtain the sensitivity desired.

5.25. Example of a Test of Significance

As a specific example of the above considerations, let us examine in greater detail the problem of testing the hypothesis that the unknown mean μ is equal to some particular value μ_0 based on a sample $x_1, \cdots, x_n$ of observations on a normally distributed variable with mean μ and known variance σ^2. As before, the quantity $t = \frac{\bar{x} - \mu}{\sigma}\sqrt{n}$ is a standardized normal variable. Now suppose that we choose a value of the standardized normal variable t such that $Pr(|t| > t_\alpha) = \alpha$. Then, if the absolute value of $t_0 = \frac{\bar{x} - \mu_0}{\sigma}\sqrt{n}$ exceeds t_α, we shall reject the hypothesis that $\mu = \mu_0$, and say that the difference $\mu - \mu_0$ is significant at the $100\alpha\%$ level.

If the unknown mean μ actually equals μ_0, then $Pr(|t_0| > t_\alpha) = \alpha$; i.e., α is the probability of rejecting the hypothesis when it is true, or of saying the difference $\bar{x} - \mu_0$ is significant when it is not. Now suppose that the hypothesis is false, and that μ is actually equal to some other value μ_1. Then we want to investigate the probability that t_0 will have absolute value less than t_α, or that the hypothesis will not be rejected when it is false, an error of type II. The quantity $t_1 = \frac{\bar{x} - \mu_1}{\sigma}\sqrt{n}$ will in this case be a standardized normal variable. This can be written

$$t_1 = \frac{\bar{x} - \mu_0}{\sigma}\sqrt{n} - \frac{\mu_1 - \mu_0}{\sigma}\sqrt{n} = t_0 - d\sqrt{n}$$

where $d = \frac{\mu_1 - \mu_0}{\sigma}$. Thus $t_0 = t_1 + d\sqrt{n}$, and hence in this case

$$\begin{aligned} Pr(|t_0| < t_\alpha) &= Pr(-t_\alpha < t_0 < t_\alpha) \\ &= Pr(-t_\alpha < t_1 + d\sqrt{n} < t_\alpha) \\ &= Pr(-t_\alpha - d\sqrt{n} < t_1 < t_\alpha - d\sqrt{n}) \end{aligned}$$

Since in this case t_1 is a standardized normal variable, this probability can be determined from tables of the normal distribution. Note that because of the symmetry of the normal distribution, deviations $+d$ and $-d$ of the same magnitude but in opposite directions would give rise to the same probability of not rejecting the hypothesis. Note also that as n becomes large and d remains fixed the quantities $-t_\alpha - d\sqrt{n}$ and $t_\alpha - d\sqrt{n}$ will both become small (or both large if d is negative), and hence the probability that the standardized normal variable t_1 falls between them will approach zero. This means that for large n the probability of not rejecting a false hypothesis, i.e., a type II error, will be small.

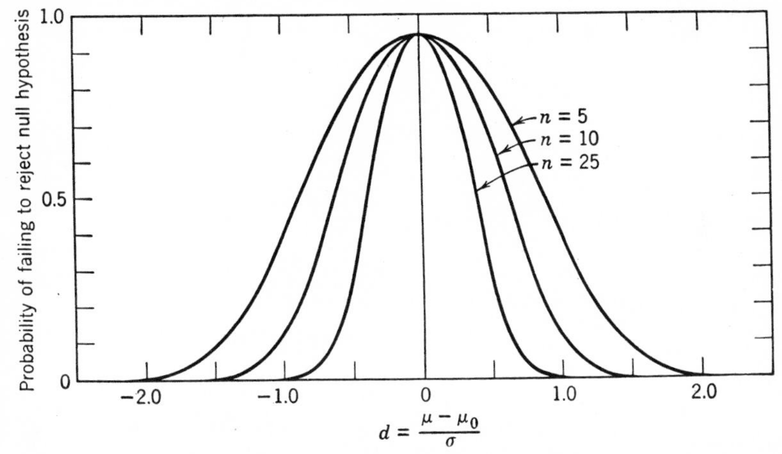

FIG. 5.2. Operating characteristic curves for test of $\mu = \mu_0$ against $\mu \neq \mu_0$, $\alpha = 0.05$, σ known.

Figures 5.2 and 5.3 show graphs, for $\alpha = 0.05$ and $\alpha = 0.01$, respectively, and several values of n, of the $Pr(|t_0| < t_\alpha)$, i.e., the probability of not rejecting the hypothesis $\mu = \mu_0$, plotted against d. These curves are the operating characteristic curves for this test. If $d = 0$, i.e., $\mu = \mu_0$, then the probability of not rejecting the hypothesis is $1 - \alpha = 0.95$ or 0.99. Sometimes instead of the $Pr(|t_0| < t_\alpha)$ the quantity $1 - Pr(|t_0| < t_\alpha) = Pr(|t_0| > t_\alpha)$, the probability of rejecting the hypothesis, is plotted against d. Such a curve is called the power curve of the test.

Notice that for values of d close to zero, i.e., for values of μ close to μ_0, the probability of not rejecting the hypothesis is almost as great as if the hypothesis were true. Thus our test is relatively insensitive to small deviations of the true mean from the hypothetical value being tested. However, as the difference between the true mean and the hypothetical mean becomes large the probability of not rejecting the hypothesis decreases rapidly, so that for large discrepancies the difference is almost certain to be adjudged significant.

EXAMPLE 1. In a chemical process it is very important that a certain solution to be used as a reactant have a pH of 8.30. A method for determining pH is available which for solutions of this type is known to give measurements which are approximately normally distributed (about the actual pH of the solution) with a known standard deviation σ of 0.02. Six determinations on a particular batch of reactant give the values 8.34, 8.29, 8.30, 8.31, 8.30, and 8.32. We wish on the basis of these measurements to test the hypothesis that this particular batch has a pH of 8.30.

Suppose we choose $\alpha = 0.05$. Then $t_\alpha = 1.96$; i.e., 1.96 is the standard normal deviate which will be exceeded in absolute value with a probability 0.05. For our sample $n = 6$, $\bar{x} = 8.31$; we are given $\sigma = 0.02$ and we wish to test the hypothesis $\mu_0 = 8.3$. Hence we have

$$t = \frac{\bar{x} - \mu_0}{\sigma}\sqrt{n} = \frac{8.31 - 8.3}{0.02}\sqrt{6} = 1.22$$

Thus we should not reject at the 5% level the hypothesis that the true pH was 8.3 on the basis of this sample.

Let us consider a different question. Suppose that it is very undesirable that the hypothesis $p\text{H} = 8.30$ be accepted if the true pH is greater than 8.33 or less than 8.27. Then, if we again choose $\alpha = 0.05$, how many pH measurements of the type described will be necessary so that the probability of failing to reject the hypothesis if either $\mu \geq 8.33$ or $\mu \leq 8.27$ is also to be less than 0.05?

Since the probability of failing to reject the hypothesis decreases as the difference $\mu - \mu_0$ becomes larger in absolute value, we need only consider the case $\mu_1 = 8.33$ (or alternatively $\mu_1 = 8.27$). In this case $d = \dfrac{8.33 - 8.30}{0.02} = 1.5$. Hence we wish to determine the smallest value of n such that

$$\begin{aligned} &Pr(-t_\alpha - d\sqrt{n} < t < t_\alpha - d\sqrt{n}) \\ &= Pr(-1.96 - 1.5\sqrt{n} < t < 1.96 - 1.5\sqrt{n}) < 0.05 \end{aligned}$$

where t is a standardized normal variable. This can be most easily done by choosing values of n and then determining the required probability from the normal tables. For $n = 5$, the probability is 0.0823; for $n = 6$, 0.0431. Thus 6 is the smallest number of measurements that will give the protection required.

As d is decreased the size of the sample required increases rapidly. For example, if we had required the same protection against the possibility that the true pH was greater than 8.31 or less than 8.29, the necessary number of measurements would have been 52. In most practical cases this number of measurements would be out of the question, so that we would either have to accept less protection against false hypotheses or develop a method of measurement with a smaller standard deviation σ.

This might lead us to ask the following question: How precise a method for determining *p*H is required if we are to have a probability less than 0.05 of failing to reject on the basis of 6 measurements the hypothesis that the *p*H is 8.30, when it is actually greater than 8.31 or less than 8.29? Again, using $\alpha = 0.05$, the smallest value of d for which

$$Pr(-1.96 - d\sqrt{6} < t < 1.96 - d\sqrt{6}) < 0.05$$

is 1.47; since $d = \dfrac{\mu_1 - \mu_0}{\sigma} = 0.01/\sigma$, we must have $0.01/\sigma > 1.47$, or

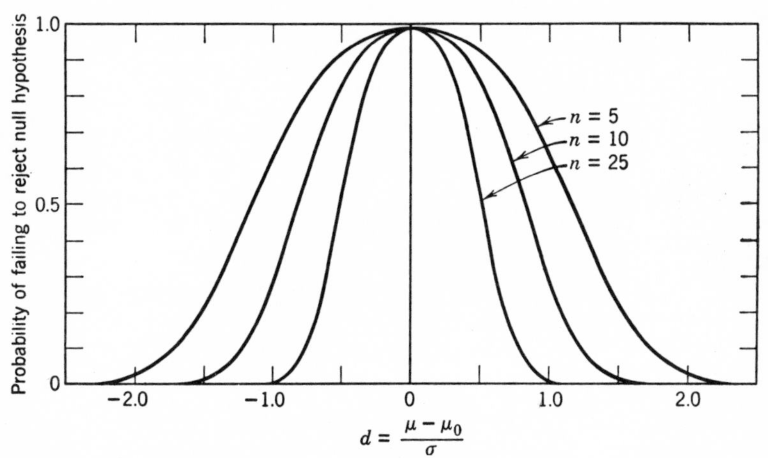

FIG. 5.3. Operating characteristics curves for test $\mu = \mu_0$ against $\mu \neq \mu_0$, $\alpha = 0.01$, σ known.

$\sigma < 0.0068$. Thus a method with a standard deviation less than 0.0068 would be required.

In the preceding test the hypothesis $\mu = \mu_0$ is just as likely to be rejected in the case $\mu < \mu_0$ as in the case $\mu > \mu_0$. In many cases we are interested in rejecting the null hypothesis as often as possible if $\mu > \mu_0$, but do not particularly care whether it is rejected if $\mu < \mu_0$; in fact, it might be even more desirable to fail to reject the hypothesis in this case than in the case $\mu = \mu_0$. For example, in testing the effect of a process modification designed to increase the product yield we might examine the hypothesis that the yield was unchanged, but we should only wish to reject this hypothesis if the yield were increased. For practical purposes a reduction in yield would be worse than an unchanged yield, and the real null hypothesis to be tested is that the product yield is not increased.

The natural procedure in this situation is to reject the null hypothesis

$\mu = \mu_0$ only when the actual value (not the absolute value) of t_0 is greater than $t_{2\alpha}$, where $t_{2\alpha}$ is chosen so that $Pr(|t| > t_{2\alpha}) = 2\alpha$, i.e., such that the area of the normal curve above $t_{2\alpha}$ is exactly the chosen significance level α. Then the probability of a type I error will again be α, since the $Pr(t_0 > t_{2\alpha})$ when $\mu = \mu_0$ will be α. However, the operating characteristic curve will be quite different, since in this case, if $\mu = \mu_1$ and we again let $d = \dfrac{\mu_1 - \mu_0}{\sigma}$, we shall have for the probability of not rejecting the hypothesis $\mu = \mu_0$ when $\mu = \mu_1$.

$$Pr(t_0 < t_{2\alpha}) = Pr(t_1 < t_{2\alpha} - d\sqrt{n})$$

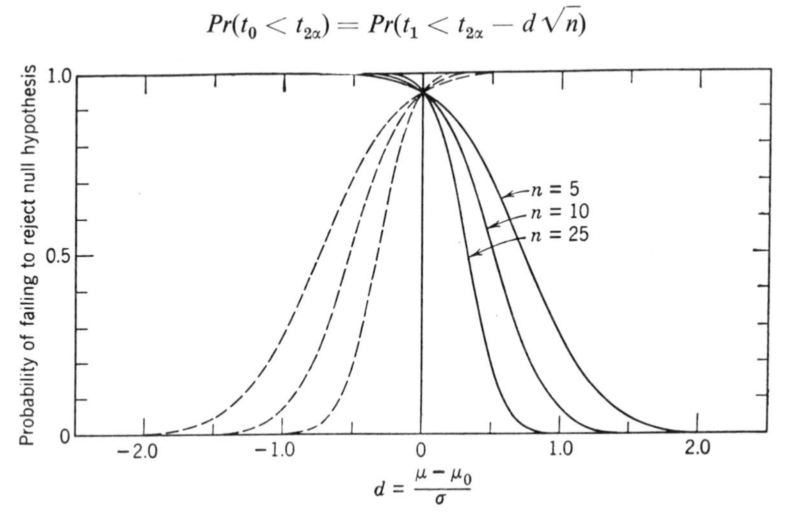

FIG. 5.4. Operating characteristic curves for test of $\mu \leq \mu_0$ against $\mu > \mu_0$, $\alpha = 0.05$, σ known.

As μ_1 becomes much larger than μ_0 this probability becomes quite small; however, when μ_1 is smaller than μ_0, we shall fail to reject the hypothesis even more frequently than if $\mu = \mu_0$. The operating characteristic for this test is shown in Figures 5.4 and 5.5 for $\alpha = 0.05$ and 0.01 and several values of n.

In a similar fashion the procedure of rejecting the hypothesis $\mu = \mu_0$ only if the value t_0 were less than $-t_{2\alpha}$ would be used if we wished to consider only the alternative values of μ less than μ_0. In this case the operating characteristic curves would simply be the reverse of that obtained in the previous instance. They are shown by the dotted lines in Figures 5.4 and 5.5. It should be emphasized that *the decision to make a one-sided test should be made a priori*, i.e., before the data are obtained and examined, if the given significance levels are to apply.

EXAMPLE 2. Suppose that in Example 1 it was important only that the pH should not be appreciably greater than 8.30. Then we should test the hypothesis $\mu = 8.30$ against the alternatives $\mu > 8.30$. Since $t = 1.22$ is less than $t_{2(0.05)} = t_{0.10} = 1.645$, we should still not reject the hypothesis.

Now suppose that we wish the probability of failing to reject the hypothesis when $\mu \geq 8.33$ to be less than or equal to 0.05. Then we must have n sufficiently large so that

$$Pr(t_1 < t_{2\alpha} - d\sqrt{n}) = Pr(t_1 < 1.645 - 1.5\sqrt{n}) \leq 0.05$$

Again, by substitution we find that for $n = 4$, the probability is 0.0877, and for $n = 5$, the probability is 0.0437. Hence 5 determinations will be sufficient to give this one-sided protection.

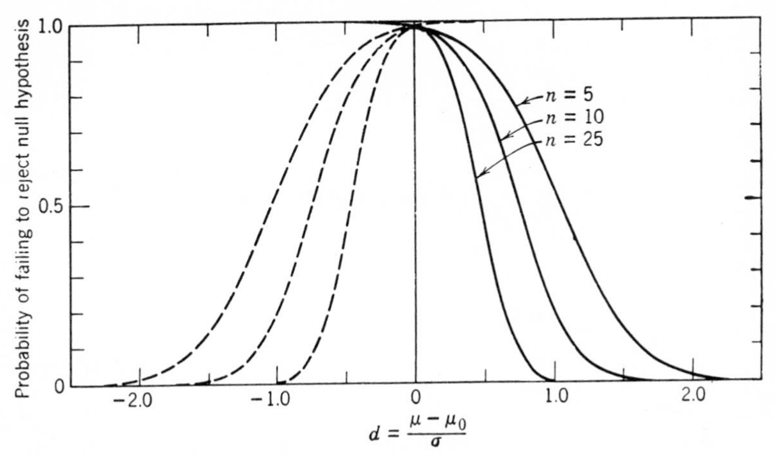

FIG. 5.5. Operating characteristic curves for test of $\mu \leq \mu_0$ against $\mu > \mu_0$, $\alpha = 0.01$, σ known.

The above tests are applicable to any normally distributed variable. For example, given 2 samples from normal distributions with known standard deviations, we could test the hypothesis that $\delta = 0$, or $\mu_1 = \mu_2$, by computing

$$t = \frac{\bar{x}_1 - \bar{x}_2}{\sigma_\delta} = \frac{\bar{x}_1 - \bar{x}_2}{\sqrt{\sigma_1^2/n_1 + \sigma_2^2/n_2}}$$

and proceeding in the above manner.

EXAMPLE 3. Using the method of analysis for iron mentioned in the Example of Section 5.23 which has a known standard deviation of 0.12%, 4 determinations were made on each of 2 samples. The average iron content for the first sample was 35.45%, that for the second 36.82%. To test the hypothesis that these results reflect no real difference between the iron content of the two samples we compute

$$t = \frac{\bar{x}_1 - \bar{x}_2}{\sqrt{\sigma_1^2/n_1 + \sigma_2^2/n_2}}$$

$$= \frac{16.82 - 16.45}{\sqrt{0.0144/4 + 0.0144/4}}$$

$$= \frac{0.37}{0.12\sqrt{\frac{1}{2}}} = \frac{0.37}{0.12}\sqrt{2}$$

$$= 4.36$$

Since $t_{0.01} = 2.576$, we reject the hypothesis of equality and conclude that there is a real difference in the iron content of the two samples.

EXAMPLE 4. Using the same analysis, in order to have a probability of 0.05 of detecting as significant at the 5% level a 1% difference between 2 samples containing approximately 30% iron, i.e., in order to detect a concentration difference of 0.3% iron, the number of analyses necessary on each sample would be obtained by solving for the smallest value of $n = n_1 = n_2$ such that

$$Pr(-t_\alpha - d\sqrt{n/2} < t < t_\alpha - d\sqrt{n/2}) < 0.05$$

where $d = \delta_1/\sigma = -0.3/0.12 = -2.5$. For $n = 2$

$$Pr(-1.96 + 2.5 < t < 1.96 + 2.5) = 0.2946$$

and similarly for $n = 3$ the probability is 0.1357, for $n = 4$, 0.0570, and for $n = 5$, 0.0233. Hence 5 determinations on each would be required.

5.26. Confidence and Significance Levels

In the preceding sections we have repeatedly spoken of a significance level α, or a confidence level $1 - \alpha$, or of a $100(1 - \alpha)\%$ confidence interval, or of a $100\alpha\%$ risk. In any practical problem it is necessary to choose a value of α with which to work. Our choice is frequently limited, since for many statistics only certain percentage points (i.e., values which are exceeded, either in actual or absolute value, with a given probability) of their sampling distributions have been computed and tabulated. Since it is customary to choose $\alpha = 0.05$ or $\alpha = 0.01$, i.e., to use a 5% or 1% significance level, to take a 5% or 1% risk, or to compute 95% or 99% confidence limits, as we have done in our examples up to this time, these levels of α, or possibly these and the 10% and 2% points required to obtain one-sided limits or tests at the 5% or 1% level, are most frequently given.

The choice of a significance or confidence level, within this limitation, is entirely up to the experimenter, and depends on his own feeling as to the risk of error he is willing to take in view of the possible consequences. However, the customary use of the 5% and 1% levels is not completely a matter of chance; these levels probably represent a comparatively good balance between the desire to be correct and the desire to be specific. Also in most people's minds the borderline between something which could possibly happen and something which is not likely to happen is somewhere in this region. Certainly 5% or 1% confidence limits are to be preferred to limits such as the "probable error," which has only a 50/50 chance of including the correct value. In this book we shall compute either 95% or 99% confidence limits, and in testing significance a value below the 5% level will be considered not significant, a value between the 5% and 1% level of questionable significance, and a value above 1% level significant. In the last case we shall occasionally designate a value above the 0.1% level as highly significant, and also in some instances we shall evaluate or bracket the actual probability of the event. In significance testing asterisks are frequently used to denote the significance of computed values, a single asterisk for questionably significant, a double asterisk for significant, and a triple asterisk for highly significant.

5.3. Inferences about Averages

5.31. Point Estimates

Since we have shown that ave $(\bar{x}) = \mu$, the sample mean is an unbiased estimate of the population mean μ. We can also show for a sample from a normal distribution that var $(\bar{x}) = \sigma^2/n$ is smaller than that of any other unbiased estimate of μ, so that the sample mean is the best estimate in this restricted case. However, the presence of any appreciable contamination (see Section 4.33) greatly decreases the efficiency of the mean as an estimate, and there are examples of theoretical distributions with long tails for which the mean of n observations is no better estimate of the average value than a single observation.

The median is an unbiased estimate of the population median, which coincides with the population mean when the distribution is symmetric. For samples from a normal distribution the median is less efficient than the mean; the efficiencies for several small sample sizes, and for the case

TABLE 5.1

EFFICIENCY OF MEDIAN FOR SAMPLES FROM A NORMAL DISTRIBUTION

Sample size	2	3	4	5	6	7	8	9	10	∞
Efficiency	1.00	0.74	0.84	0.70	0.78	0.68	0.74	0.67	0.72	0.64

of very large samples, are given in Table 5.1. The higher efficiencies for even sample sizes are due to the use of the two central values rather than a single value; for $n = 2$ the median and mean are identical. As was stated in Section 2.34, the median is much less affected by contamination than the mean, and is more efficient than the mean even when the amount of such contamination is quite small.

The midrange $\frac{x_{(n)} + x_{(1)}}{2}$, where $x_{(n)}$ and $x_{(1)}$ are the largest and smallest values in the sample, is quite efficient in small samples from normal populations, but rapidly loses efficiency as the sample size is increased. Mosteller [82] has shown that for large samples highly efficient estimates of the population mean can be obtained by averaging small numbers of order statistics. This method is particularly useful for large masses of data which are on punched cards or in some other form where they are easily ordered. The proper choices (given as percentiles) are tabulated most conveniently in a text by Dixon and Massey [20].

EXAMPLE 1. A series of 6 pilot runs on a denuder produced caustic concentrations of 68.5%, 69.2%, 68.2%, 68.6%, 68.8%, and 68.9%. *Assuming* that these results are a random sample of 6 concentrations from the large number of possible concentrations that might have been obtained and that if this large number of possible concentrations were available their frequency distribution could be represented closely by a normal frequency function, *then* we can say that the sample mean, 68.7%, of these 6 values represents the best estimate of the expected long-term yield under similar conditions.

In this case where only 6 values are available there is little chance of checking statistically the assumed randomness of the sample or normality of the population. The tests for non-normality given in the previous chapter would detect only the grossest departures from normality, and the same can be said of the tests for randomness to be discussed later. We might obtain the requisite experimental data to check these assumptions, but more often it will be necessary to base our decision on the available results.

This does not imply that we cannot proceed to use the estimate 68.7%; it only means that it should be used in the light of what we know concerning the conditions under which it was obtained. Thus our efforts to see that each test run was made under identical conditions may make us feel that the assumption of randomness is justified; similarly, we may have reason to believe from previous experience that yields of this type are approximately normally distributed, and have no reason to suspect that these results are any different in this respect. Given a certain statistical model, we deduce that a certain estimate is best; hence in a case where we feel justified in using the model, we can feel justified in expecting certain properties of the estimate obtained.

If it was felt that a certain proportion of the runs might have more variable yields, or in any event if we were willing to sacrifice efficiency in the ideal case for protection against this possibility, we might wish to consider the median, which for these runs is also 68.7%, as our estimate of the long-term yield.

EXAMPLE 2. The sample mean of the data given in Table 2.1 of Chapter 2 was 71.194. This value then represents our best estimate of the expected daily yield. In this case the assumption of normality has been tested in Section 4.45; the assumption of randomness is also subject to test, and will be discussed for this example in Chapter 11.

From reference [20] we find that in this case the average of the 10th, 30th, 50th, 70th, and 90th percentile values is an estimate of the mean with an efficiency of 93% for normal populations. Choosing the 36th, 109th, 182nd, 255th, and 328th values from the ordered data of Table 2.2, we obtain the estimate.

$$\frac{69.4 + 70.5 + 71.1 + 71.8 + 73.0}{5} = 71.16$$

of the population mean. There is no way of determining whether this is better or worse than the above estimate in this case; all that we know is that in the long run, and for certain populations, estimates obtained in this fashion vary more, but not much more, than the best estimate.

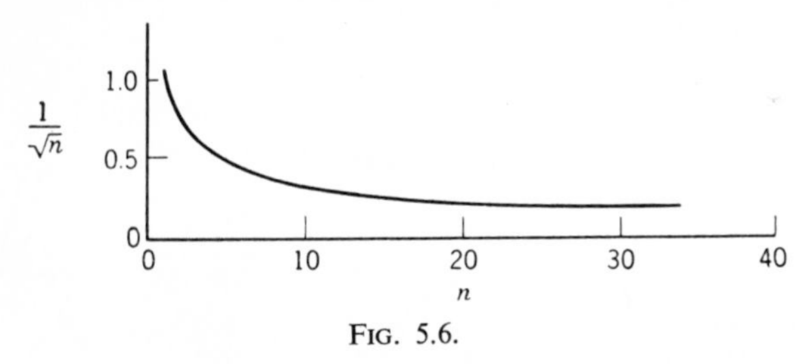

FIG. 5.6.

If the variance σ^2 is known, we can determine the number of observations required to make the variance σ^2/n of the estimate $\bar{x}$ as small as desired. However, it is the standard deviation $\sigma/\sqrt{n}$ which is used in interpreting the actual spread of the values of $\bar{x}$ about μ. Because of this fact the practical benefit of additional observations decreases as the number of observations becomes large, as can be seen in Figure 5.6 which shows, plotted against the sample size n, the factor $1/\sqrt{n}$ by which the standard deviation of an individual observation is decreased to obtain the standard deviation of the mean of a sample of n.

EXAMPLE 3. The density of replicate samples from carload lots of a given chemical can be shown to vary approximately normally about the true average density of a given lot with standard deviation 0.005g/cc. In order to obtain an estimate based on the mean of n samples which is within 0.002g/cc of the true average density for the lot in 90% of the cases, it is necessary that we have

$$1.645 \frac{\sigma}{\sqrt{n}} < 0.002$$

where 1.645 is the value t_α such that $Pr\left(|\bar{x} - \mu| \leq t_\alpha \frac{\sigma}{\sqrt{n}}\right) = 0.90$ (see Table 4.3). Putting $\sigma = 0.005$, we find that a minimum of 17 replicate samples is necessary to achieve the desired result.

5.32. Point Estimates for Linear Combinations

We now consider a series of samples of $n_1, n_2, \cdots, n_k$ from populations with means $\mu_1, \mu_2, \cdots, \mu_k$ and variances $\sigma_1^2, \sigma_2^2, \cdots, \sigma_k^2$. Then for each sample we have ave $(\bar{x}_i) = \mu_i$, and from Section 3.33.

$$\text{(a)} \qquad \text{ave } (\Sigma a_i \bar{x}_i) = \Sigma a_i \mu_i$$

Thus a linear combination of sample means is an unbiased estimate of the corresponding linear combination of population means. If the samples are independent, then the variance of this estimate is given by

$$\text{(b)} \qquad \text{var } (\Sigma a_i \bar{x}_i) = \sum \frac{a_i^2 \sigma_i^2}{n_i}$$

Now let us suppose that the variances σ_i^2 are known. Then by a procedure similar to that used in Section 3.52 we find that for a given linear combination and a fixed total number of observations $n = \Sigma n_i$, (b) is minimized when we choose (as nearly as possible)

$$\text{(c)} \qquad n_i = \frac{a_i \sigma_i}{\Sigma a_i \sigma_i} n \qquad i = 1, \cdots, k$$

When the n_i are so chosen the variance is given by

$$\text{(b)} \qquad \text{var } (\Sigma a_i \bar{x}_i) = \frac{(\Sigma a_i \sigma_i)^2}{n}$$

A slightly different situation arises when $\mu_1 = \mu_2 = \cdots = \mu_k = \mu$, i.e., all the populations have the same mean, and for fixed n_i we wish to choose the a_i so as to obtain the best estimate of μ. From (a) it follows that to have an unbiased estimate we must have $\Sigma a_i = 1$, since in this case ave $(\Sigma a_i x_i) = \mu \Sigma a_i$. Minimizing (*b*) with respect to the a_i subject to this condition, we find that the proper choice is given by

$$(e) \qquad a_i = \frac{w_i}{\Sigma w_i} \qquad i = 1, \cdots, k$$

where $w_i = n_i/\sigma_i^2$. For this choice we have

$$(f) \qquad \text{var}\,(\Sigma a_i \bar{x}_i) = \frac{1}{\Sigma w_i}$$

Notice that to use either (*c*) or (*e*) it is necessary to know the variances σ_i^2. In the first case we are almost invariably planning the estimation of some value which is a linear (or nearly linear) combination of different measurements, and wish to apportion the number of measurements of each type on the basis of a priori information concerning the variances of the measurements obtained from previous experience. However, in the second case we are frequently faced with the problem of estimating μ from several samples when the only information concerning the variances σ_i^2 is contained in the samples themselves. For example, we may have available for estimating the concentration of a compound in a solution several series of measurements which can be assumed free of systematic error, but which have different and unknown variances. The quantities $w_i' = n_i/s_i^2$, where the s_i^2 are the estimates of the variances σ_i^2 computed from the samples themselves, can be used to compute, using (*e*), weights a_i' which will give an unbiased estimate of μ. However, the variance of the estimate so obtained will be inflated above that given by (*f*), and is given approximately by

$$(g) \qquad \text{var}\,(\Sigma a_i' \bar{x}_i) \sim \frac{1}{\Sigma w_i'} \left[1 + 4 \sum \frac{a_i'(1 - a_i')}{n_i} \right]$$

which is correct to terms of order $1/n_i$ [83].

5.33. Confidence Intervals and Tests of Significance Using the t-Distribution

In Sections 5.23 and 5.25 we considered the problems of determining confidence intervals for, and testing hypotheses concerning, the unknown population mean. There we assumed the variance to be known, but ordinarily we have available only an estimate of this variance computed from the sample. It is at this point that the importance of the "Student" t-distribution becomes apparent, since for a sample of n from a normal distribution the ratio

$$(a) \qquad t_{n-1} = \frac{\bar{x} - \mu}{s} \sqrt{n}$$

obtained in Section 4.54 does not involve the unknown population variance. We can therefore proceed as in Sections 5.23 and 5.25, except that in place of a standardized normal deviate we shall use a variable having the t-distribution with the appropriate degrees of freedom.

If from Table III we obtain a value $t_{n-1,\alpha}$, then, since these values were chosen so that $Pr(|t_{n-1}| > t_{n-1,\alpha}) = \alpha$, it follows from (*a*) that

$$(b) \qquad Pr\left(-t_{n-1,\alpha} < \frac{\bar{x} - \mu}{s}\sqrt{n} < t_{n-1,\alpha}\right) = 1 - \alpha$$

As in Section 5.23 it can be shown that this is algebraically equivalent to

$$(c) \qquad Pr\left(\bar{x} - \frac{t_{n-1,\alpha}s}{\sqrt{n}} < \mu < \bar{x} + \frac{t_{n-1,\alpha}s}{\sqrt{n}}\right) = 1 - \alpha$$

and hence the limits

$$(d) \qquad \bar{x} \pm \frac{t_{n-1,\alpha}s}{\sqrt{n}}$$

are $100(1 - \alpha)\%$ confidence limits for μ. These are the limits which were computed in Section 2.43. If one-sided limits are desired, they can be obtained from (*d*) simply by replacing $t_{n-1,\alpha}$ by $t_{n-1,2\alpha}$, and choosing the appropriate sign for the limit desired.

Note that, where σ was known, we could choose the sample size n so as to make the length of the confidence interval as small as desired, since we could make the standard deviation $\sigma/\sqrt{n}$ of the estimate $\bar{x}$ of μ as small as desired. In this case, however, the quantity s is determined from each particular sample, and hence we cannot predict the length of the confidence interval in any particular instance. However, on the average the size decreases with increasing n. As n increases $t_{n-1,\alpha}$ approaches t_α, s approaches σ, and these confidence limits become almost identical with those based on the normal distribution with σ replaced by s. The customary differentiation between large and small samples appears somewhat artificial in this case since the limits (*d*) will be correct regardless of the sample size.

In testing the hypothesis $\mu = \mu_0$ we can again replace t by t_{n-1} which does not depend on the unknown σ. Hence the procedure will be to reject the hypothesis $\mu = \mu_0$ if the value of t_{n-1} computed from the sample exceeds in absolute value a given $t_{n-1,\alpha}$ chosen so that

$$Pr(|t_{n-1}| \geq t_{n-1,\alpha}) = \alpha$$

This procedure will ensure that the probability of rejecting the hypothesis when it is true will be equal to the given significance level α, regardless of the value of σ. Notice that this is equivalent to rejecting the hypothesis if μ_0 is not within the limits given by (*d*).

As in Section 5.25 we can evaluate the operating characteristic curve of this test in terms of the quantity $d = \dfrac{\mu - \mu_0}{\sigma}$. The curves for $\alpha = 0.05$ and several values of n are given in Figure 5.7. The procedure in this case is not so straightforward, and the given curves are not so useful as in the previous situation, since for a given difference $\mu_1 - \mu_0$ we cannot determine d without knowing the value of σ. We can determine for a given sample size the probability of failing to reject the null hypothesis

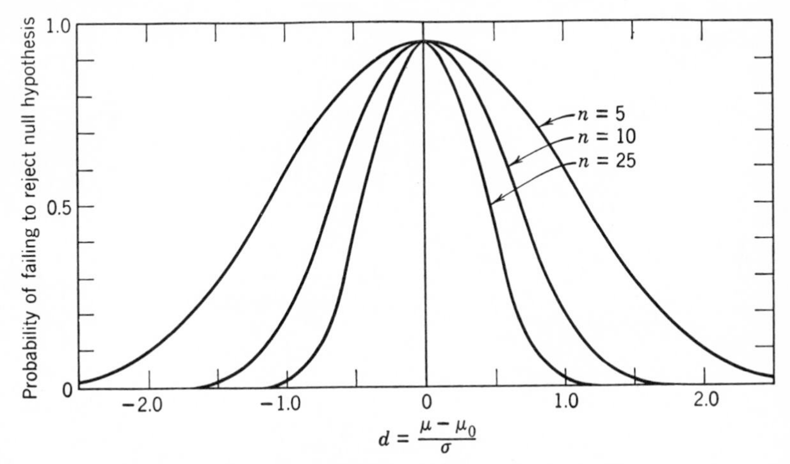

FIG. 5.7. Operating characteristic curves for test of $\mu = \mu_0$ against $\mu \neq \mu_0$, using "Student" t-distribution, $\alpha = 0.05$.

when the true mean is some value μ_1 which differs from μ_0 by a given multiple of the unknown standard deviation.

We can again define tests which are the best to use if we wish to consider the rejection of values on only one side of the hypothetical value μ_0. These would involve rejecting the hypothesis if the actual value of t_{n-1} were greater than $t_{n-1,2\alpha}$ in the event that it was important to reject the hypothesis only for values of $\mu > \mu_0$, or in rejecting the hypothesis if the actual value of t_{n-1} were less than $-t_{n-1,2\alpha}$ in the event that it was important that the hypothesis be rejected for values of $\mu < \mu_0$. These tests will have operating characteristic curves similar to those for the one-sided tests with σ^2 known; and again will depend on the quantity $d = \dfrac{\mu_1 - \mu_0}{\sigma}$ which contains the unknown σ. The curves for several values of n and $\alpha = 0.05$ are given in Figure 5.8.

It should be noted carefully that in the above tests we are testing the hypothesis that the sample came from a normal distribution with mean

μ and variance σ^2, against the alternative hypothesis that the sample came from a normal distribution with some other mean μ_1 but with the same variance σ^2; i.e., our hypothesis concerns only the mean, and the standard deviation whether known or unknown is assumed to be fixed, regardless of whether the hypothesis concerning the mean is correct. It should also be noted that the above tests are most frequently used in cases where the sample size is quite small and where the assumption of normality is very difficult to justify on the basis of the sample itself.

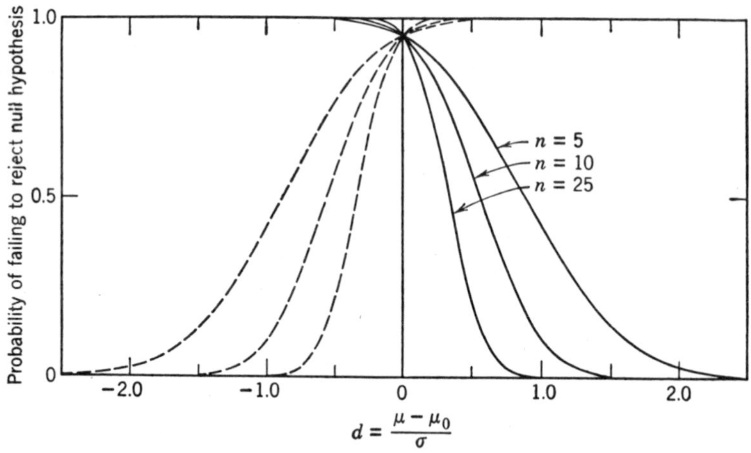

FIG. 5.8. Operating characteristic curves for test of $\mu \leq \mu_0$ against $\mu > \mu_0$, using "Student" t-distribution, $\alpha = 0.05$.

However, investigations which have been made concerning the validity of these tests indicate that they are not seriously affected by moderate departures from normality [95]. It will be apparent that failure to reject an hypothesis does not necessarily mean that the hypothesis is true. It could simply mean that the true value of the mean was not sufficiently different from the hypothetical value μ_0 to be detected on the basis of the sample given. Conversely, it should be mentioned that in rejecting the hypothesis statistically we do not imply that the difference between the true mean and the hypothetical value μ_0 is significant from a practical viewpoint.

EXAMPLE 1. Let us suppose that in Example 2 of Section 2.43, the methanol solution had actually contained 0.50% water. To test the hypothesis that the given sample came from a normal population with a mean of 0.50, we compute $\bar{x} = 0.552$, $s^2 = 0.001396$, $s = 0.037$, and hence

$$t_9 = \frac{0.552 - 0.500}{0.037} \sqrt{10} = 4.44$$

Since $t_{9, 0.01} = 3.25$, this value is significant at the 1% level, and we reject the hypothesis that the expected result is the actual water content of the solution. This is equivalent to saying that the 99% confidence levels do not include the value 0.50%.

EXAMPLE 2. For the 6 caustic concentrations of 68.5%, 69.2%, 68.6%, 68.2%, 68.8%, and 68.9% given in Example 1 of Section 5.31, $\bar{x} = 68.70$ and $s = 0.35$. Hence, since $t_{5, 0.05} = 2.57$, 95% confidence limits are given by

$$68.70 \pm \frac{(2.57)(0.35)}{\sqrt{6}} = 68.70 \pm 0.37$$

and we shall take a 5% risk of being wrong if we state that the unknown average concentration is between 68.33 and 69.07%. If we were concerned with determining only a lower limit for the unknown average, we should obtain, since $t_{5, 0.10} = 2.015$,

$$68.70 - \frac{2.015(0.35)}{\sqrt{6}} = 68.41$$

and hence we can also state with 5% risk that the unknown average yield is greater than 68.41.

5.34. Confidence Intervals and Tests of Significance Based on Order Statistics

Order statistics are frequently used to provide more simply computed confidence limits and hypothesis tests for population parameters. These methods in general have the advantage of requiring no assumptions other than the continuity of the distribution sampled. On the other hand, the confidence intervals obtained will be larger, or the probability of failing to reject false hypotheses greater than would be true if the assumption of normality, or near-normality, were justified and we used the methods of Section 5.33.

One of the simplest applications of order statistics arises in obtaining confidence limits for population percentiles. For, if μ_p represents the value of the random variable x such that $Pr(x \leq \mu_p) = p$, then it can be shown [21] that the probability of μ_p lying between the kth and lth observations in order of magnitude is

$$(a) \qquad Pr(x_{(k)} < \mu_p < x_{(l)}) = I_p(k, n - k + 1) - I_p(l, n - l + 1)$$

where $I_x(l, m)$ is the incomplete beta function. In particular, if we choose $p = 0.5$ and $l = n - k + 1$, we obtain for the probability that the symmetric limits $x_{(k)}$ and $x_{(n-k+1)}$ will contain the median of the population

(*b*) $$Pr(x_{(k)} < \mu_{0.5} < x_{(n-k+1)}) = I_{0.5}(k, n-k+1) - I_{0.5}(n-k+1, k)$$
$$= 1 - 2I_{0.5}(n-k+1, k)$$

since from Appendix 4*A* $I_x(l, m) = 1 - I_{1-x}(m, l)$.

These limits are identical with those obtained in Section 2.44. In that case we considered the probability that the highest and lowest values in the sample would contain the median between them, which is equal to 1 minus the probability that all the sample values will be either above or below the median. Hence for a sample of n

(*c*) $$Pr(x_{(1)} < \mu_{0.5} < x_{(n)}) = 1 - Pr(y = 0, n)$$

where y has the binomial distribution with $p = 0.5$ and the number of events equal to the sample size n. Since, for $p = 0.5$, $Pr(y = 0) = Pr(y = n)$, this becomes

(*d*) $$Pr(x_{(1)} < \mu_{0.5} < x_{(n)}) = 1 - 2Pr(y = 0)$$

By an identical argument, the probability that $\mu_{0.5}$ will lie between the kth value from the bottom and the kth value from the top is

(*e*) $$Pr(x_{(k)} < \mu_{0.5} < x_{(n-k+1)}) = 1 - 2Pr(y \leq k - 1)$$

The probability on the right is the sum of the first k terms of the binomial distribution for n events and $p = 0.5$; from (*d*) of Section 4 of Appendix 4*A*, this is given exactly by $I_{0.5}(n - k + 1, k)$, and hence (*e*) and (*b*) are equivalent. Obviously (*a*) can also be obtained by a slight generalization of the above argument.

Since k must be an integer, we can never choose k so that these intervals will contain $\mu_{0.5}$ with a probability of exactly 0.95 or 0.99. This is equivalent to saying that, since the binomial distribution is discrete, it is impossible to obtain exact percentage points for a given value of α. We can, however, choose from tables of the binomial distribution for $p = 0.5$ and various n the maximum value of k such that $Pr(y \leq k - 1) = p_1 < \alpha/2$. For these values of k it follows from (*e*) that

$$Pr(x_{(k)} < \mu_{0.5} < x_{(n-k+1)}) > 1 - \alpha$$

i.e., the values $x_{(k)}$ and $x_{(n-k+1)}$ are at least a $100(1 - \alpha)\%$ confidence interval for the median. A tabulation of such values for k is given in Table 5.2. One-sided confidence intervals can be obtained simply by choosing either the upper limit or the lower limit of the two-sided confidence interval for *twice* the significance level desired.

EXAMPLE 1. For the 6 caustic concentrations used in Example 2 of the previous section, a 95% confidence interval for the population median would be given immediately by the first and sixth values, i.e.,

$$Pr(68.2 < \text{median} < 69.2) \geq 0.95$$

No 99% confidence interval can be obtained in this case.

The significance test corresponding to the above intervals consists merely in counting the number of values above and below the hypothetical value μ_0 of the population median $\mu_{0.5}$. If for a sample of n the smaller of these two numbers is less than the value of k for the desired significance level obtained from Table 5.2, then we reject the hypothesis

TABLE 5.2

PERCENTAGE POINTS OF THE SYMMETRIC BINOMIAL DISTRIBUTION

(Values of k such that $Pr(y < k) = Pr(y \leq k - 1) < \frac{\alpha}{2}$ where y has the binomial distribution with $p = 0.5$)

Sample Size	Significance Level			
n	$\alpha = 0.10$	$\alpha = 0.05$	$\alpha = 0.02$	$\alpha = 0.01$
5	1	—	—	—
6	1	1	—	—
7	1	1	1	—
8	2	1	1	1
9	2	2	1	1
10	2	2	1	1
11	3	2	2	1
12	3	3	2	2
13	4	3	2	2
14	4	3	3	2
15	4	4	3	3
16	5	4	4	3
17	5	5	4	3
18	6	5	4	4
19	6	5	5	4
20	6	6	5	4
21	7	6	5	5
22	7	6	6	5
23	8	7	6	5
24	8	7	6	6
25	8	8	7	6
30	11	10	9	8
35	13	12	11	10
40	15	14	13	12
45	17	16	15	14
50	19	18	17	16

$\mu_{0.5} = \mu_0$. One-sided tests can again be made by counting the values either above or below μ_0, and comparing this number with the value of k for twice the desired significance level. Because this test depends only on the sign of the deviations of the sample values from the hypothetical value, and not on their magnitude, it has been called the sign test.

Note that either confidence intervals or significance tests for the population median are equivalent to those for the population mean if the distribution is symmetric.

The range has been used in the development of substitute t-ratios which are quick and easy to calculate and have relatively high efficiencies (as compared with the "Student" t) for small samples [119, 120]. For example, if we replace the sample standard deviation s by the range $R = x_{(n)} - x_{(1)}$, we can consider the ratio

$$t' = \frac{\bar{x} - \mu_0}{R} \tag{f}$$

as a test of the hypothesis $\mu = \mu_0$. Percentage points of this ratio, i.e., values $t'_{n,\alpha}$ such that for a sample of n from a normal distribution

$$Pr(|t'| > t'_{n,\alpha}) = \alpha$$

are given in Table 5.3 for $\alpha = 0.05$ and 0.01 and $n = 2$ to 10. The factor $\sqrt{n}$ and the correction for the bias of R which might have been expected to appear in (f) are implicit in these values. The use of this ratio above $n = 10$ is not recommended. Values of t' greater in absolute value than the tabulated figures are considered significant in the usual manner.

TABLE 5.3*

Sample Size	Significance Levels	
	$\alpha = 0.05$	$\alpha = 0.01$
2	6.353	31.828
3	1.304	3.008
4	0.717	1.316
5	0.507	0.843
6	0.399	0.628
7	0.333	0.507
8	0.288	0.429
9	0.255	0.374
10	0.230	0.333

* Reproduced by permission of Professor E. S. Pearson from Table 9 of "The Use of the Range in Place of the Standard Deviation in the t-Test," *Biometrika*, *34* (1947), 41–67, by E. Lord.

As an additional simplification we could replace the sample mean $\bar{x}$ by the midrange $\frac{x_{(1)} + x_{(n)}}{2}$, or $2\bar{x}$ by $x_{(1)} + x_{(n)}$. Neglecting the constant divisor 2, which can be absorbed in the percentage points $t''_{n,\alpha}$, this leads to the ratio

$$t'' = \frac{x_1 + x_n - 2\mu_0}{R}$$

as a test of the hypothesis $\mu = \mu_0$. Values of $t''_{n,\alpha}$ for $\alpha = 0.05$ and 0.01 and $n = 2$ to 10 are given in Table 5.4 for samples from a normal distribution.

TABLE 5.4*

Sample Size	Significance Levels	
	$\alpha = 0.05$	$\alpha = 0.01$
2	12.71	63.66
3	2.60	6.04
4	1.48	2.74
5	1.02	1.70
6	0.86	1.32
7	0.75	1.10
8	0.66	0.95
9	0.60	0.85
10	0.55	0.78

EXAMPLE 2. For the now familiar caustic concentrations of Example 2 of the previous section, we have, considering the hypothesis $\mu = 68.0\%$,

$$t' = \frac{68.7 - 68.0}{1.0} = 0.7$$

$$t'' = \frac{68.2 + 69.2 - 2(68.0)}{1.0} = 1.4$$

For $n = 6$, both of these ratios are significant at the 1% level; hence we should have concluded in either case that the mean yield was significantly different from 68.0%. Note that these tests are two sided.

* Reproduced by permission from Table 5 of "On the Range-Midrange Test and Some Tests with Bounded Significance Levels," *Annals Math. Stat.*, XX (1949), 257–67, by J. E. Walsh.

EXAMPLE 3. For the 10 results for the percentage of water used in Example 2 of Section 2.43, we have, considering the hypothesis $\mu = 0.50$,

$$t' = \frac{0.552 - 0.500}{0.14} = 0.37$$

$$t'' = \frac{0.50 + 0.64 - 2(0.50)}{0.14} = 1.00$$

Again both of these ratios are significant at the 1% level, as is the usual t-test.

5.35. Distribution-Free Tolerance Limits

The limits based on order statistics in the preceding section are examples of the simplest cases of non-parametric estimation. Non-parametric inferences are those which can be made from a sample with respect to the distribution as a whole, rather than with respect to one or more parameters of an assumed form of the distribution. For example, the conclusion that 2 independent samples came from populations with different distributions would be a non-parametric inference; the conclusion that they came from 2 normally distributed populations with equal variances but different means would be a parametric inference.

Another problem for which a non-parametric solution has been found is that of determining from a sample, limits within which a given fraction P of the population may be expected to fall. Such limits are called tolerance limits. Assuming only that the population sampled has a continuous distribution function, Wilks [3, 21] showed that the probability of including a fraction P of the population between the kth and $(n - k + 1)$st values in order of magnitude from a sample of n is

$$(a) \qquad 1 - I_p(n - 2k + 1, 2k) = I_{1-p}(2k, n - 2k + 1)$$

Such limits are called $100P\%$ distribution-free tolerance limits; they are directly related to the confidence intervals for μ_p given by (a) of the previous section.

In practice, the problem with which we are generally confronted is that of determining the sample size n necessary to ensure a given probability $1 - \alpha$ of including $100P\%$ of the population values between $x_{(k)}$ and $x_{(n-k+1)}$, where k is usually chosen small, say $k = 1$ or 2. For the case $k = 1$, i.e., where we are considering the probability that $100P\%$ of the population will fall between the largest and the smallest values in the sample, the problem reduces to determining the smallest value of n such that

$$(b) \qquad I_{1-p}(2, n - 1) \geq 1 - \alpha$$

Since

$$I_{1-p}(2, n-1) = \frac{1}{B(2, n-1)} \int_0^{1-P} z(1-z)^{n-2}\, dz$$

$$= \frac{\Gamma(n+1)}{\Gamma(2)\,\Gamma(n-1)} \left[-\int_0^{1-P} \cdot (1-z)^{n-1}\, dz + \int_0^{1-P} (1-z)^{n-2}\, dz \right]$$

$$= \frac{n!}{1!\,(n-2)!} \left[\frac{(1-z)^n}{n} - \frac{(1-z)^{n-1}}{n-1} \right]_0^{1-P}$$

$$= n(n-1) \left[\frac{(1-z)^n}{n} - \frac{(1-z)^{n-1}}{n-1} \right]_0^{1-P}$$

$$= (n-1)\,(P)^n - n(P)^{n-1} + 1$$

we need to determine, for a given α and P, the smallest value of n such that

$$(n-1)\,(P)^n - n(P)^{n-1} + 1 \geq 1 - \alpha$$

or

$$(c) \qquad n(P)^{n-1} - (n-1)\,(P)^n \leq \alpha$$

Similarly, it can be determined that the sample size n necessary to ensure that the probability will be $1 - \alpha$ of $100P\%$ of the population falling either below the largest value or above the smallest value in the sample is given by the smallest value of n such that

$$I_{1-p}(1, n) \geq 1 - \alpha$$

An integration similar to that performed above gives the equivalent result

$$(d) \qquad (P)^n \leq \alpha$$

which can be solved for the desired value of n.

Scheffé and Tukey [84] have developed an approximate solution for the minimum sample size n in the two-sided case which, unlike the exact solution given by (c), is not restricted to $k = 1$. The values of n are given by

$$(e) \qquad \begin{aligned} n &\sim 1.03k + 4.74\chi^2_{4k,\,\alpha} - 1 \qquad P = 0.90 \\ n &\sim 1.01k + 9.75\chi^2_{4k,\,\alpha} - 1 \qquad P = 0.95 \\ n &\sim 1.00k + 49.75\chi^2_{4k,\,\alpha} - 1 \qquad P = 0.99 \end{aligned}$$

where $\chi^2_{4k,\,\alpha}$ is the previously defined percentage point of the χ^2-distribution for $4k$ degrees of freedom, and can be obtained from Table III in the Appendix.

Methods of computing tolerance limits have also been developed for the case in which a normal distribution can be assumed. A description of these methods and extensive tables of the necessary factors are given in [19], Chapter 2.

EXAMPLE. We wish to obtain a sample such that the probability is 0.90 that the largest value is greater than 95% of the population. Since the minimum value of n such that

$$(P)^n < \alpha$$

$$(0.95)^n < 0.10$$

or

$$n \log 0.95 < \log 0.10$$

is $n = 45$, this is the size of the random sample required. Similarly, if we wished the probability to be 0.90 that 95% of the population was included between the largest and smallest sample values, we should have to obtain the smallest value of n for which

$$n(0.95)^{n-1} - (n-1)(0.95)^n < 0.10$$

The solution is best accomplished by trial and error, and we obtain $n = 77$ as the required sample size. In this case we obtain from the approximation for $P = 0.95$, since $k = 1$ and $\chi_4{}^2{}_{,\ 0.10} = 7.78$,

$$n \sim 1.01 + (9.75)(7.78) - 1$$
$$\sim 75.9$$

which is certainly close enough for most purposes.

5.4. Inferences about the Variance

5.41. Point Estimates

Let us now consider the problem of estimating from a sample of n observations the population variance σ^2, or the population standard deviation σ. Since we have shown in Section 3.43 that ave $(s^2) = \sigma^2$, the sample variance s^2 is an unbiased estimate of σ^2. For samples from a normally distributed population, the variance $2\sigma^4/n - 1$ of s^2 can be shown to be the smallest of any unbiased estimate not depending on the unknown mean μ of the population. Hence s^2 is in this case the best estimate of σ^2 available. If the population mean μ is known, as is very seldom true in practice, then the actual mean square of the deviations from μ is an unbiased estimate of the population variance σ^2, and, for a normally distributed population is the best estimate.

The sample standard deviation, the mean deviation, and the range can all be used as estimates of σ. All are biased, and the last two are not even consistent, i.e., the bias does not disappear for large sample sizes. Hence in each case it is necessary to correct for this bias by dividing the value obtained from the sample by its average value expressed in units of the population standard deviation. For example, the average value of the range of a sample of 5 is 2.326σ; hence to obtain an estimate of σ we should divide the range of a sample of 5 by 2.326. Similarly, the average value of the mean deviation about the median in samples of 10 is 0.74σ; thus we could obtain an estimate of σ by dividing the mean deviation for a sample of 10 by 0.74. These average values for samples of size 2 to 10 from a normal population are given in Table 5.5. For $n > 10$ the use of the range is not recommended, and the bias of the sample standard deviation can in most cases be neglected, although these factors, designated by c_2', are given for values up to $n = 25$ in Table 11.1.

The mean deviation corrected for bias has a high efficiency for small samples which decreases to 0.88 as the number in the sample becomes large. The range is easily computed and quite efficient in small samples, but its efficiency decreases rapidly as the sample size increases, and for large samples its efficiency is 0. The efficiencies of these estimates

TABLE 5.5

	Mean Deviation		Range		Sample Standard Deviation
n	Average Value*	Efficiency†	Average Value*	Efficiency†	Average Value*
2	0.56	1.00	1.128	1.00	0.798
3	0.56	0.99	1.693	0.99	0.886
4	0.66	0.91	2.059	0.98	0.921
5	0.66	0.94	2.326	0.96	0.940
6	0.70	0.90	2.534	0.93	0.951
7	0.70	0.92	2.704	0.91	0.960
8	0.73	0.90	2.847	0.89	0.965
9	0.73	0.91	2.970	0.87	0.969
10	0.74	0.89	3.078	0.85	0.973
∞	0.798	0.88	—	0	1.000

* In units of the population standard deviation.
† Relative to the sample standard deviation.

relative to the sample standard deviation are also given in Table 5.5 for samples from a normal population.

In cases where small samples are analyzed repeatedly, especially in quality control work, the range is used because of its simplicity of calculation. As mentioned in Section 2.34, in cases where the distribution of the observations is "contaminated" by a relatively small number of more variable observations, which is frequently true of chemical measurements, the mean deviation is less affected by such contamination than the standard deviation, and consequently may be a more efficient estimate of σ.

For large samples which are ordered it is possible, as with the mean, to estimate σ with a high efficiency from a selected group of observations. The details of the proper choices, and the factors required to make the estimates unbiased, are again given in [20] both for estimates of σ alone and for joint estimates of μ and σ from the same selected group.

EXAMPLE 1. For the data of Example 2 of Section 2.43 we have $n = 10$, $s = 0.037$, $R = 0.14$, and, for the mean deviation about the median, 0.024. From these three values we obtain, after correcting for bias, the estimates

$$\hat{\sigma} = \frac{0.037}{0.973} = 0.038 \tag{1}$$

$$\hat{\sigma} = \frac{0.024}{0.74} = 0.032 \tag{2}$$

$$\hat{\sigma} = \frac{0.14}{3.08} = 0.045 \tag{3}$$

We have no way of determining which of these is the best in this particular instance, but if the distribution sampled were normal we know that in the long run the first will vary least from the true value. Notice that the single large observation has had the greatest influence on the estimate based on the range, and the least on the estimate based on the mean deviation.

EXAMPLE 2. From the yields for the first week of Table 2.1 we found in Section 2.35 that $s = 1.28$, $R = 3.9$, and the mean deviation about the median was 0.86. Correcting for bias, we have as estimates of σ the values 1.33, 1.44, and 1.23, respectively. In this case we might compare these values with the estimate $s = 1.351$ obtained for the entire data of Table 2.1.

5.42. Joint Estimates from Several Samples

It may happen that we wish to estimate the value of σ^2 from several different samples, each obtained from a population with the same variance

σ^2 but having different means (if the means were equal, we could consider the entire sample as having been drawn from a single population and estimate in the usual manner). Suppose that we represent a particular sample value from one of these samples by x_{ij}, where $i = 1, \cdots, k$ indicates the particular sample to which x_{ij} belongs, and $j = 1, \cdots, n_i$ indicates the particular one of the n_i values in the ith sample. For the ith sample the sample variance is

$$(a) \qquad s_i^2 = \frac{\sum_{j=1}^{n_i} (x_{ij} - \bar{x}_i)^2}{n_i - 1}$$

and an unbiased joint estimate of σ^2 from all k samples is given by

$$(b) \qquad s^2 = \frac{(n_1 - 1)s_1^2 + (n_2 - 1)s_2^2 + \cdots + (n_k - 1)s_k^2}{n_1 + n_2 + \cdots + n_k - k}$$

Since $(n_i - 1)s_i^2 = \sum_{j=1}^{n_i} (x_{ij} - \bar{x}_i)^2$, it can be seen that this estimate is obtained by dividing the overall total of the sum of squares of the deviations of each sample value from its own sample mean by the total number of degrees of freedom.

For convenience let $N = n_1 + n_2 + \cdots + n_k$. Then we can write

$$(c) \qquad \frac{(N - k)s^2}{\sigma^2} = \frac{(n_1 - 1)s_1^2}{\sigma^2} + \frac{(n_2 - 1)s_2^2}{\sigma^2} + \cdots + \frac{(n_k - 1)s_k^2}{\sigma^2}$$

If we now assume that each of the populations from which the k samples were drawn was normally distributed, then each of the quantities $\frac{(n_i - 1)s_i^2}{\sigma^2}$ has the χ^2-distribution with $n_i - 1$ degrees of freedom. Since the sum of a series of quantities having the χ^2-distribution also has the χ^2-distribution with degrees of freedom equal to the total number of degrees of freedom in the quantities added, it follows that the quantity $\frac{(N - k)s^2}{\sigma^2}$ has the χ^2-distribution with $(n_1 - 1) + (n_2 - 1) + \cdots + (n_k - 1) = (N - k)$ degrees of freedom. Hence, except for the fact that we must subtract k degrees of freedom rather than 1 degree of freedom, the estimate s^2 obtained from k samples behaves in exactly the same way as a sample variance obtained from a single sample.

This method of estimating the variance σ^2 from a series of samples is of fundamental importance in the analysis of variance to be discussed in Chapter 7. For the special case $k = 2$, which occurs frequently, the above estimate reduces to

$$s^2 = \frac{(n_1 - 1)s_1^2 + (n_2 - 1)s_2^2}{n_1 + n_2 - 2}$$

(d)

$$= \frac{\sum_{j=1}^{n_1} (x_{1j} - \bar{x}_1)^2 + \sum_{j=1}^{n_2} (x_{2j} - \bar{x}_2)^2}{n_1 + n_2 - 2}$$

and the quantity

(e)

$$\frac{(n_1 + n_2 - 2)s^2}{\sigma^2}$$

has the χ^2-distribution with $n_1 + n_2 - 2$ degrees of freedom.

The above method of estimating σ^2 is very useful in the common situation where the sampling variation or the analytical variation in a particular measurement must be estimated from a small number of determinations on each of a large number of batches or samples of different materials having various unknown means. In particular, if $n_1 = n_2 = \cdots = n_k = 2$, as in the case when k pairs of duplicate determinations or duplicate samples are available, we can write

(f)

$$s_i^2 = \frac{(x_{i1} - x_{i2})^2}{2} = \frac{d_i^2}{2}$$

where d_i is the difference between the duplicate values obtained. Thus we have

$$s^2 = \frac{\dfrac{(2-1)d_1^2}{2} + \dfrac{(2-1)d_2^2}{2} + \cdots + \dfrac{(2-1)d_k^2}{2}}{2k - k}$$

(g)

$$= \frac{\frac{1}{2}\sum_{i=1}^{k} d_i^2}{k} = \frac{\sum_{i=1}^{k} d_i^2}{2k}$$

i.e., the estimate of σ^2 is obtained by taking the sums of squares of the differences between the duplicate observations and dividing by twice the number of pairs, or, what is equivalent in this case, by the total number of observations. This estimate will have k degrees of freedom, one from each pair of measurements.

It should also be noted that in this case we can use the average range

(h)

$$\bar{R} = \frac{\sum_{i=1}^{k} R_i}{k} = \frac{\sum_{i=1}^{k} |d_i|}{k}$$

to estimate σ. Since the expected value of the range for a sample of two is 1.128σ, the expected value of $\bar{R}$ will also be 1.128σ, and an unbiased estimate of σ will be $\bar{R}/1.128$. This is equivalent to using the mean deviation rather than the standard deviation of the differences d_i, and hence the relative efficiency of $\bar{R}$ is about 88% for large k.

In Section 5.62 a method will be given for testing the validity of the assumption that each of the k samples has the same variance σ^2 prior to the use of any of the above methods of estimation.

EXAMPLE 1. The following data [94] are the results of 16 determinations of the ratio of reacting weights of silver and iodine, classified according to the 2 different preparations of iodine and 5 different preparations of silver used.

RATIO OF IODINE TO SILVER

Iodine	Silver	Ratio
I	*A*	1.176422
	A	1.176425
	B	1.176441
	B	1.176441
	C	1.176429
	C	1.176420
	C	1.176437
	D	1.176449
	D	1.176450
	E	1.176455
II	*A*	1.176399
	A	1.176440
	A	1.176418
	B	1.176423
	B	1.176413
	D	1.176461
Average		1.1764327

These data, coded as differences from 1.176400 multiplied by 10^6, and classified according to silver-iodine combinations, are given below.

	Silver				
Iodine	*A*	*B*	*C*	*D*	*E*
I	22 25	41 41	29 20 37	49 50	55
II	−1 40 18	23 13		61	

For all 16 values, the mean is 32.7, or 1.1764327 in terms of the original units, and the standard deviation 16.8, or 0.0000168 in original units. However, this latter estimate includes the effects of any variation between silver-iodine combinations. To obtain a joint estimate of the variation in determinations within the particular silver-iodine combinations we compute the following table of estimate s_i^2 of the variances within each class, omitting those classifications containing 0 or 1 determinations, since these provide no estimate of the variance.

Iodine	Silver	$\sum_{j=1}^{n_i} (x_{ij} - \bar{x}_i)^2$	$\nu_i = n_i - 1$	s_i^2
I	A	4.5	1	4.5
	B	0	1	0.0
	C	144.7	2	72.3
	D	0.5	1	0.5
II	A	842.0	2	421.0
	B	50.0	1	50.0

Each of the quantities $\sum_{j=1}^{n_i} (x_{ij} - \bar{x}_i)^2$ is computed in the usual manner from the n_i results within a particular classification. Note that in the case of $n_i = 2$, $\sum_{j=1}^{n_i} (x_{ij} - \bar{x}_i)^2 = \frac{(x_{i1} - x_{i2})^2}{2}$; for example, for combination IA, we have

$$\frac{(25 - 22)^2}{2} = 4.5$$

For $n_i > 2$, as in IC, we use the form $\Sigma x^2 - \frac{(\Sigma x)^2}{n} = 2610 - 2465.33 = 144.67$. There is a wide disparity between the estimates s_i^2; this is to be expected in view of the small numbers of degrees of freedom involved.

The joint estimate s^2 of the variance within classifications is given by the total of the sums of squares of deviations divided by the total number of degrees of freedom, i.e.,

$$s^2 = \frac{4.5 + 0.0 + \cdots + 50.0}{1 + 1 + \cdots + 1} = \frac{1041.7}{8} = 129.54$$

from which $s = 11.4$, or 0.0000114 in original units. This represents an estimate of the variation due to experimental technique only, exclusive of any variation between preparation combinations. In Chapter 7 we shall consider whether the decrease in this variation below that for the data as a whole is indicative of any significant variation due to preparations.

EXAMPLE 2. Duplicate determinations of the percentage of solids in wet brewers yeast on each of 10 samples by each of 2 methods gave the results shown in the accompanying table.

Sample	Method I			Method II		
	Det. 1	Det. 2	Diff.	Det. 1	Det. 2	Diff.
1	13.13	13.12	0.01	13.06	13.17	−0.11
2	13.36	13.48	−0.12	13.44	13.43	0.01
3	19.97	19.89	0.08	20.18	20.21	−0.03
4	14.71	14.76	−0.05	14.78	14.83	−0.05
5	13.12	13.19	−0.07	13.07	13.15	−0.08
6	13.09	13.17	−0.08	13.12	13.06	0.06
7	17.88	17.92	−0.04	18.16	18.24	−0.08
8	18.17	18.15	0.02	18.23	18.18	0.05
9	17.60	17.69	−0.09	17.70	17.70	0.00
10	16.89	16.92	−0.03	16.86	16.90	−0.04

From (g) we obtain as an estimate of the variance of Method I

$$s^2 = \frac{\Sigma d_i^2}{2k} = \frac{0.0457}{20} = 0.002285 \qquad s = 0.048$$

and for Method II

$$s^2 = \frac{\Sigma d_i^2}{2k} = \frac{0.0361}{20} = 0.001805 \qquad s = 0.042$$

Each estimate is based on $k = 10$ degrees of freedom.

Similarly, from (h), we obtain for Method I

$$\bar{R} = \frac{0.59}{10} = 0.059$$

and for Method II

$$\bar{R} = \frac{0.51}{10} = 0.051$$

Dividing these values by 1.128 to correct for bias, we obtain 0.052 and 0.045, respectively, as estimates of σ for the two methods, which compare favorably with the standard deviations obtained above.

5.43. Confidence Intervals and Tests of Significance for the Variance

We can obtain confidence intervals for the unknown variance σ^2 of a normally distributed variable or variables from any estimate s^2 based on ν degrees of freedom, by using the fact that the quantity

$$\chi_\nu^2 = \frac{\nu s^2}{\sigma^2} \tag{a}$$

has the χ^2-distribution with ν degrees of freedom. However, it should be pointed out before we begin that in this case the lack of normality in applications will have much more effect on the results obtained than was true with confidence intervals for the mean based on the "Student" t-distribution.

To obtain an upper confidence limit for σ^2 at the confidence level $1 - \alpha$, we first obtain from Table II the value $\chi^2_{\nu, 1-\alpha}$. Since this value was chosen so that

$$Pr(\chi_\nu^2 \geq \chi^2_{\nu, 1-\alpha}) = 1 - \alpha \tag{b}$$

it follows from (*a*) that the statement

$$\frac{\nu s^2}{\sigma^2} > \chi^2_{\nu, 1-\alpha} \tag{c}$$

will have a probability $1 - \alpha$ of being correct. By simple algebra it can be seen that this statement is equivalent to the statement

$$\frac{\nu s^2}{\chi^2_{\nu, 1-\alpha}} > \sigma^2 \tag{d}$$

and hence the left-hand side of the inequality (*d*) represents a $100(1 - \alpha)\%$ upper confidence limit for σ^2.

Similarly, we can choose a value $\chi^2_{\nu, \alpha}$ such that

$$Pr(\chi_\nu^2 \geq \chi^2_{\nu, \alpha}) = \alpha \tag{e}$$

Then the statement

$$\frac{\nu s^2}{\sigma^2} < \chi^2_{\nu, \alpha} \tag{f}$$

will have a probability $1 - \alpha$ of being correct. This statement is equivalent to

$$\frac{\nu s^2}{\chi^2_{\nu, \alpha}} < \sigma^2 \tag{g}$$

and hence the left-hand side of (*g*) also represents a $100(1 - \alpha)\%$ lower confidence limit for σ^2.

The above one-sided intervals are appropriate when we are interested only in either how large or how small the unknown value σ^2 may be. We can also obtain a two-sided confidence interval for the unknown variance σ^2 by choosing two values χ^2_{ν,α_1} and $\chi^2_{\nu,1-\alpha_2}$ such that

$$Pr(\chi_\nu^2 \geq \chi^2_{\nu,\alpha_1}) = \alpha_1 \tag{h}$$

$$Pr(\chi_\nu^2 \geq \chi^2_{\nu,1-\alpha_2}) = 1 - \alpha_2$$

where $\alpha_1 + \alpha_2 = \alpha$. In this case the choice $\alpha_1 = \alpha_2 = \alpha/2$, i.e., the choice of the two values of χ_ν^2 which cut off equal areas of each tail of the χ^2-distribution, does not lead to the shortest confidence interval, as it did in the case of the mean. However, the optimum choice of α_1 and α_2 is not simple, and, since for moderately large ν, say ≥ 20, the choice of equal tails is a very close approximation to the best choice, we shall consider here the two-sided confidence interval based on the choice $\alpha_1 = \alpha_2 = \alpha/2$. Then the statement

$$\chi^2_{\nu,1-\alpha/2} < \frac{\nu s^2}{\sigma^2} < \chi^2_{\nu,\alpha/2} \tag{i}$$

will have a probability of $1 - \alpha$ of being correct. It can be shown as before that this statement is equivalent to saying that we shall have

$$\frac{\nu s^2}{\chi^2_{\nu,\alpha/2}} < \sigma^2 < \frac{\nu s^2}{\chi^2_{\nu,1-\alpha/2}} \tag{j}$$

and hence the set of values between the 2 sample functions given represents a $100(1 - \alpha)\%$ confidence interval for σ^2.

EXAMPLE 1. In Example 2 of the previous section we obtained as estimates of the variance of the two methods of analysis $s^2 = 0.002285$ and $s^2 = 0.001805$, respectively, each based on 10 degrees of freedom. Since $\chi^2_{10,0.95} = 3.940$, a 95% upper confidence limit for σ^2, the true variance for Method I, is given by

$$\sigma^2 < \frac{10(0.002285)}{3.940} < 0.005799$$

$$\sigma < 0.076$$

and similarly for Method II

$$\sigma^2 < \frac{10(0.001805)}{3.940} < 0.004581$$

$$\sigma < 0.068$$

We now consider a test of the simple hypothesis that the unknown variance σ^2 has some particular value σ_0^2. We shall consider in detail the case occurring most frequently in practice where we do not particularly mind if we fail to reject the hypothesis when the true value of σ^2 is less than σ_0^2, but we should like to reject the hypothesis as frequently as possible if the true value of σ^2 is greater than σ_0^2. The best test in this case consists of rejecting the hypothesis, i.e., concluding $\sigma^2 > \sigma_0^2$, if the variance estimate s^2 is sufficiently large that the quantity χ_ν^2 defined by (*a*) is greater than the value $\chi^2_{\nu,\alpha}$ defined by (*e*).

Now let us suppose that the true value of σ^2 is some value $\sigma_1^2 > \sigma_0^2$. The error of type II, or that of failing to reject the hypothesis when it is false, will be given in this case by

$$(k) \qquad Pr\left(\frac{\nu s^2}{\sigma_0^2} < \chi^2_{\nu,\alpha}\right) = \beta$$

when the true value of σ^2 is given by σ_1^2. The statement inside the parentheses is identical with the statement

$$(l) \qquad \frac{\nu s^2}{\sigma_1^2} < \frac{\sigma_0^2}{\sigma_1^2}\chi^2_{\nu,\alpha}$$

But, if σ_1^2 is the true value of σ^2, then the quantity $\nu s^2/\sigma_1^2$ has the χ^2-distribution with ν degrees of freedom. Hence, when $\sigma^2 = \sigma_1^2$, the probability (*k*) is given by

$$(m) \qquad \beta = Pr\left(\chi_\nu^2 < \frac{\sigma_0^2}{\sigma_1^2}\chi^2_{\nu,\alpha}\right)$$

The operating characteristic curves for the above test, plotted against the quantity $\rho = \sigma_1^2/\sigma_0^2$ for various values of ν and $\alpha = 0.05$ and 0.01, and extensive tables of the factors $\rho(\alpha, \beta, \nu)$, by which the true variance σ_1^2 must exceed the hypothetical variance σ_0^2 in order that the probability of rejecting the hypothesis $\sigma^2 = \sigma_0^2$ will be β for a test with significance level α based on an estimate with ν degrees of freedom, are given in [19], Chapter 8.

Completely analogous tests can be derived for testing the hypothesis $\sigma^2 = \sigma_0^2$ against the alternatives $\sigma^2 < \sigma_0^2$, or against the alternatives $\sigma^2 \neq \sigma_0^2$. In each case the test is equivalent to determining whether the hypothetical value σ_0^2 falls in the corresponding one-sided or two-sided confidence intervals, given above. The tables given in [19] can also be used to compute the errors of type II for the one-sided test against the alternatives $\sigma^2 < \sigma_0^2$.

Note that all the confidence intervals and hypothesis tests in the last two sections are based on an estimate s^2 of σ^2 with ν degrees of freedom.

They are applicable whether this estimate is from a single sample, or from a joint estimate from several samples, or is obtained in some other fashion. For a single sample we should have $\nu = n - 1$, for a joint estimate $\nu = N - k$, etc.

EXAMPLE 2. A method of analyzing for $Al(NO_3)_3$ which will have a standard deviation less than or equal to 1g/l at a concentration of approximately 200g/l is desired. Ten analyses on the same sample using a proposed method gave a variance based on 9 degrees of freedom of 3.02 $(g/l)^2$ and a mean value of 204.25. We wish to determine if this observed variance is consistent with the hypothesis that the true variance σ^2 is less than or equal to $\sigma_0^2 = 1$. Since

$$\chi_9^2 = \frac{9(3.02)}{1.00} = 27.2$$

is much greater than $\chi^2_{9,\,0.01} = 21.67$, and very close to $\chi^2_{9,\,0.001} = 27.88$, we conclude that the observed result is highly significant, and that the proposed method is not capable of the precision desired.

EXAMPLE 3. Suppose that in the above example we wished to determine the number of replicate results that would be required to assure that a proposed method with a true standard deviation greater than 1.5 would fail to be rejected with a relative frequency of 0.10, or 1 time in 10, if the above test were conducted at a significance level of 0.05. Here we have $\rho = \dfrac{\sigma_1^2}{\sigma_0^2} = \dfrac{(1.5)^2}{(1.0)^2} = 2.25$, $\alpha = 0.05$, and $\beta = 0.10$, and hence we wish to determine the smallest value of $\nu = n - 1$ such that

$$Pr\left(\chi_\nu^2 < \frac{\chi^2_{\nu,\,0.05}}{2.25}\right) \leq 0.10$$

or such that

$$\chi^2_{\nu,\,0.90} > \frac{\chi^2_{\nu,\,0.05}}{2.25}$$

From Table II we obtain the required value of ν to be between 25 and 26, or, directly from the tables referred to above, we find that $\rho(0.05, 0.10, 25) = 2.276$, $\rho(0.05, 0.10, 26) = 2.249$, which means that the required number to make $\rho < 2.25$ is very close to 26. Hence the smallest number of analyses necessary would be $n = 26 + 1 = 27$. Thus this comparatively large number of analyses is necessary to be certain that methods with a true standard deviation less than or equal to 1.0 g/l will be rejected only about 1 time in 20, and methods with a true standard deviation greater than 1.5 g/l will be accepted only about 1 time in 10. We have no control over the decision which will be made if the true standard deviation is between these two values.

5.5. Comparison of Means

5.51. Comparison of Two Means

One of the problems that arises most frequently in practice is that of comparing 2 population means μ_1 and μ_2 on the basis of samples from these populations. If we are willing to assume a priori that $\sigma_1^2 = \sigma_2^2 = \sigma^2$, the t-distribution can be applied to the problem of testing the significance of, and obtaining confidence intervals for, the difference $\delta = \mu_1 - \mu_2$ of the means of 2 populations assumed to be normally distributed. In this case, given independent samples of n_1 and n_2 from the 2 distributions, we found in Section 5.42 that the quantity

$$(a) \qquad \frac{(n_1 + n_2 - 2)s^2}{\sigma^2}$$

has the χ^2-distribution with $n_1 + n_2 - 2$ degrees of freedom when s^2 is the joint estimate of σ^2 given by (d) of that section. Also from Section 4.51 it follows that

$$(b) \qquad t = \frac{\bar{d} - \delta}{\sqrt{\dfrac{\sigma^2}{n_1} + \dfrac{\sigma^2}{n_2}}} = \frac{\bar{d} - \delta}{\sigma\sqrt{\dfrac{1}{n_1} + \dfrac{1}{n_2}}}$$

where $\bar{d} = \bar{x}_1 - \bar{x}_2$, is a standardized normal variate. Forming the ratio given by (a) of Section 4.54, we find that

$$(c) \qquad t = \frac{\bar{d} - \delta}{s\sqrt{\dfrac{1}{n_1} + \dfrac{1}{n_2}}}$$

has the "Student" t-distribution with $\nu = n_1 + n_2 - 2$ degrees of freedom. Hence in this case we can obtain confidence intervals for the quantity $\delta = \mu_1 - \mu_2$ exactly as for a single mean with unknown variance, replacing $\bar{x}$ by $\bar{d}$, μ by δ, the degrees of freedom $n - 1$ by $n_1 + n_2 - 2$, $1/\sqrt{n}$ by $\sqrt{\dfrac{1}{n_1} + \dfrac{1}{n_2}}$, and using the joint estimate s. It follows directly from (d) of Section 5.33 that the two-sided $100(1 - \alpha)\%$ confidence limits for δ are given by

$$(d) \qquad \bar{d} \pm t_{n_1+n_2-2,\,\alpha}\, s\sqrt{\frac{1}{n_1} + \frac{1}{n_2}}$$

and the one-sided $100(1 - \alpha)\%$ confidence limits are given by replacing α by 2α and choosing the appropriate sign. Similarly, we can test the hypothesis $\delta = \delta_0$ against the alternatives $\delta \neq \delta_0$ by computing the value of t given by (c) with $\delta = \delta_0$ and rejecting the hypothesis if this value exceeds $t_{n_1+n_2-2,\,\alpha}$ in absolute value for the chosen significance level. For the one-sided test against the alternatives $\delta > \delta_0$ we should reject the hypothesis $\delta = \delta_0$, if the computed value exceeded $t_{n_1+n_2-2,\,2\alpha}$ in actual value. In particular, if δ_0 is 0 we have a test of the hypothesis $\mu_1 = \mu_2$, i.e., a test of the null hypothesis that the two means are identical. This test is one of the simplest analyses of variance, a topic to which we return in Chapter 7.

The problem of testing the hypothesis $\delta = \delta_0$, or obtaining confidence intervals for δ, if we cannot assume a priori that $\sigma_1^2 = \sigma_2^2$, has received much attention. A solution of this problem, known as the Behrens-Fisher problem, which seems to be most satisfactory for practical purposes has been given by B. L. Welch [22] and Alice A. Aspin [85, 86]. Except where there is a sound basis for the assumption that $\sigma_1^2 = \sigma_2^2$, this method should be preferred to the preceding one. To apply this solution we first compute the estimated variances $s_{\bar{x}_1}{}^2 = s_1^2/n_1$ and $s_{\bar{x}_2}{}^2 = s_2^2/n_2$ of the means $\bar{x}_1$ and $\bar{x}_2$. Then

$$(e) \qquad t = \frac{\bar{d} - \delta}{\sqrt{s_{\bar{x}_1}{}^2 + s_{\bar{x}_2}{}^2}}$$

has approximately the t-distribution with ν degrees of freedom, where ν is given by

$$(f) \qquad \frac{1}{\nu} = \frac{1}{\nu_1}\left(\frac{s_{\bar{x}_1}{}^2}{s_{\bar{x}_1}{}^2 + s_{\bar{x}_2}{}^2}\right)^2 + \frac{1}{\nu_2}\left(\frac{s_{\bar{x}_2}{}^2}{s_{\bar{x}_1}{}^2 + s_{\bar{x}_2}{}^2}\right)^2$$

and $\nu_1 = n_1 - 1$ and $\nu_2 = n_2 - 1$ are the degrees of freedom for s_1^2 and s_2^2, respectively. This ratio can therefore be used as before to determine confidence limits and test hypotheses; for example, two-sided $100(1 - \alpha)\%$ confidence limits are given by

$$(g) \qquad \bar{d} \pm t_{\nu,\,\alpha}\sqrt{s_{\bar{x}_1}{}^2 + s_{\bar{x}_2}{}^2}$$

where $t_{\nu,\,\alpha}$ is the value of t obtained from Table III for the degrees of freedom ν given by (f), and the hypothesis $\delta = \delta_0$ can be tested by rejecting the hypothesis if the value given by (e) with $\delta = \delta_0$ exceeds $t_{\nu,\,\alpha}$ in absolute value. Exact values of the 5% and 1% points of the ratio given by (e) have been determined [86], but the approximation given is satisfactory for most practical purposes.

In testing the hypothesis $\delta = 0$ for small samples of equal size $n_1 = n_2 = n$, we may use the substitute t ratio

$$t''' = \frac{\bar{x}_1 - \bar{x}_2}{R_1 + R_2}$$

where R_1 and R_2 are the ranges of the two samples [119]. Table 5.6

gives percentage points of t''' for $\alpha = 0.05$ and 0.01 and n from 2 to 10, based on the assumptions that $\sigma_1^2 = \sigma_2^2$ and that the distributions are normal.

TABLE 5.6*

Sample Size	Significance Levels	
	$\alpha = 0.05$	$\alpha = 0.01$
2	1.714	3.958
3	0.636	1.046
4	0.406	0.618
5	0.306	0.448
6	0.250	0.357
7	0.213	0.300
8	0.186	0.260
9	0.167	0.232
10	0.152	0.210

* Reproduced by permission of Professor E. S. Pearson from Table 10 of "The Use of the Range in Place of the Standard Deviation in the t-Test," *Biometrika*, *34* (1947), 41–67, by E. Lord.

EXAMPLE 1. The following series of measurements represent 10 determinations of the percentage of chlorine in each of 2 batches of a polymer. In this case, since the same analytical method was used and

Determinations	I	II
1	58.59	55.71
2	58.45	56.65
3	59.64	56.72
4	58.64	57.56
5	58.00	58.27
6	57.03	56.58
7	57.33	57.08
8	57.80	57.13
9	58.04	57.92
10	58.41	56.21

the sampling methods and batches tested were similar, there seems to be no a priori reason to doubt that the assumption of equal variance for the two sets of measurements is valid. Hence we proceed to compare the means of the two sets, using the t-test based on this assumption. From the observations, subtracting the factor 55 for computational convenience, we have

	I	II
Σx	31.93	19.83
Σx^2	106.8293	44.7817
$\Sigma(x - \bar{x})^2$	4.8768	5.4588
$\bar{x}$	3.193	1.983

and hence we have as an estimate of δ

$$\bar{d} = 3.193 - 1.983 = 1.210$$

and as a joint estimate of σ^2

$$s^2 = \frac{4.8768 + 5.4588}{18} = 0.5742$$

from which we obtain $s = 0.758$. Since $t_{18,\,0.05} = 2.10$, 95% confidence limits for δ are given by

$$1.21 \pm 2.10(0.758)\sqrt{1/5} = 1.21 \pm 0.71$$

and, since these limits do not include 0, the difference is significant at the 5% level. Since

$$t = \frac{1.210}{0.758}\sqrt{5} = 3.57$$

which exceeds $t_{18,\,0.01} = 2.88$, the difference is also significant at the 1% level. For the substitute t-ratio we have

$$t''' = \frac{1.210}{2.61 + 2.56} = 0.234$$

which is also significant at the 1% level.

EXAMPLE 2. The following data give the results of 4 analyses for nickel by each of two methods:

	% Nickel	
Determination	Aqueous	Alcoholic
1	3.28	3.25
2	3.28	3.27
3	3.29	3.26
4	3.29	3.25

In this case there seems to be no a priori reason for assuming equal variance for the two methods, and we use the test developed by Aspin and Welch. From the data we obtain

	Aqueous	Alcoholic
$\bar{x}$	3.285	3.2575
s^2	0.0000333	0.0000917
$s_{\bar{x}}^2$	0.0000083	0.0000229

and hence
$$\bar{d} = 0.0275$$
$$s_{\bar{d}}^2 = 0.0000312$$
$$s_{\bar{d}} = 0.0056$$

and, moving the decimal place 6 places to the right for convenience,

$$\frac{1}{\nu} = \frac{1}{3}\left[\left(\frac{8.3}{31.2}\right)^2 + \left(\frac{22.9}{31.2}\right)^2\right]$$
$$= \tfrac{1}{3}(0.071 + 0.539) = 0.203$$

and hence we have approximately $\nu = 4.93$. By linear interpolation in Table III the value of t for the 5% significance level corresponding to 4.93 degrees of freedom is ~2.58 and for the 1% level ~4.07; hence 95% confidence limits for the difference between the results obtained by the two methods are given by

$$0.028 \pm (2.58)\,(0.0056) = 0.028 \pm 0.014$$

and 99% confidence limits by

$$0.028 \pm (4.07)\,(0.0056) = 0.028 \pm 0.023$$

Since neither set of limits includes zero, the difference is significant at both the 5% and the 1% levels.

5.52. Paired Comparisons

A special case of the difference between 2 means arises when we have k pairs of values x_{1j} and $x_{2j} (j = 1, \cdots, k)$, which are independent observations from populations with means μ_{1j} and μ_{2j} and variances σ_1^2 and σ_2^2, respectively. Then, if the difference $\delta = \mu_{1j} - \mu_{2j}$ is constant for all k pairs, the quantities $d_j = x_{1j} - x_{2j}$, $j = 1, \cdots, k$ will be a sample of k values from a population with mean δ and variance $\sigma_1^2 + \sigma_2^2$, and hence estimates of δ and hypothesis tests concerning δ can be made exactly as in the case of a single sample considered in Section 5.3. For example, if we assume all the populations concerned to be normally distributed, then the distribution of the differences is normal, and, as in Section 5.31,

(*a*)
$$t = \frac{(\bar{d} - \delta_0)\sqrt{k}}{s_d}$$

where
$$\bar{d} = \frac{\Sigma_j d_j}{k} = \frac{\Sigma_j (x_{1j} - x_{2j})}{k} = \bar{x}_1 - \bar{x}_2$$

(b)
$$s_d^2 = \frac{\Sigma_j(d_j - \bar{d})^2}{k - 1}$$

has the t-distribution with $k - 1$ degrees of freedom. A two-sided test of the hypothesis $\delta = \delta_0$ is given by rejecting the hypothesis when the value of t given by (a) with $\delta = \delta_0$ is greater in absolute value than $t_{k-1,\alpha}$, and one-sided tests are given by rejecting the hypotheses $\delta \leq \delta_0$ and $\delta \geq \delta_0$ when $t > t_{k-1,2\alpha}$ or $t < -t_{k-1,2\alpha}$, respectively. The operating characteristic curves for these tests are the same as those for the corresponding tests given in Figures 5.7 and 5.8, if we take $\nu = k - 1$ and $d = \dfrac{\delta_1 - \delta_0}{\sqrt{\sigma_1^2 + \sigma_2^2}}$. Similarly $100(1 - \alpha)\%$ confidence limits for δ are given by

(c)
$$\bar{d} \pm t_{k-1,\alpha}\, s_d$$

and the corresponding one-sided limits are obtained in the usual fashion.

Example 1. The accompanying data give the results obtained by the standard dichromate titrimetric method and a new spectrophotometric method based on the use of thioglycolic acid for the iron content of 21 different samples of iron ore, together with the differences reported as (new method — standard method):

Sample	Dichromate Method	Thioglycolic Acid Method	Difference
1	28.22	28.27	+0.05
2	33.95	33.99	+0.04
3	38.25	38.20	−0.05
4	42.52	42.42	−0.10
5	37.62	37.64	+0.02
6	36.84	36.85	+0.01
7	36.12	36.21	+0.09
8	35.11	35.20	+0.09
9	34.45	34.40	−0.05
10	52.83	52.86	+0.03
11	57.90	57.88	−0.02
12	51.52	51.52	0.00
13	49.59	49.52	−0.07
14	52.20	52.19	−0.01
15	54.04	53.99	−0.05
16	56.00	56.04	+0.04
17	57.62	57.65	+0.03
18	34.30	34.39	+0.09
19	41.73	41.78	+0.05
20	44.44	44.44	0.00
21	46.48	46.47	−0.01

From the differences we obtain $\bar{d} = +0.0086$, $s_d^2 = 0.002813$, $s_d = 0.0530$. To test the hypothesis $\delta = 0$, we compute

$$t_{20} = \frac{0.0086}{0.0530}\sqrt{21} = 0.74$$

Since $t_{20,\,0.05} = 2.086$, we do not reject the hypothesis, and conclude that these data give no evidence of a systematic difference between the methods.

The most important advantage of such paired comparisons is that the limits for the average difference, and the test of significance for the observed difference, are based only on the variation in the differences; other variations which have equal effects on 2 paired observations do not influence the accuracy of the comparison. For example, in determining the difference between the alpha counting rate of 2 samples, the average difference between pairs of short counts is frequently more desirable, owing to long-term fluctuations in voltage and counter efficiency, than the difference of 2 longer counts. Such a procedure represents one of the simplest examples of an experiment in which the data are collected in a manner designed to allow more powerful methods of statistical analysis to be employed. More complicated examples of the same type are considered in Chapters 7 and 8. Note also that in this case it is immaterial whether $\sigma_1^2 = \sigma_2^2$, as long as $\sigma_1^2 + \sigma_2^2$ is constant.

The methods of Section 5.34 are also applicable to paired comparisons; in fact, the sign test, which for the hypothesis $\delta = 0$ requires only that we know the sign of the difference, was developed primarily for this purpose. Confidence intervals for the median difference $\delta_{0.5}$, which are confidence intervals for the mean difference δ if the distribution of differences is symmetric, are also obtainable. These methods are less efficient than those based on the t-distribution when the distributions are normal, but they do not require this assumption to be valid. In fact, the sign test of the hypothesis $\delta = 0$ does not even require that the variance $\sigma_1^2 + \sigma_2^2$ remain constant from pair to pair.

Another simple test of the significance of the average difference between paired values which does not require the assumption of normality, but does take cognizance of the magnitude of the differences, is the simplest of a series of tests called *signed rank tests*. In order to perform this test we calculate the differences between the pairs of observations and then rank them in order of magnitude without regard to sign. The next step is to sum the ranks of all the differences having negative signs and all the differences having positive signs. Let us call the smaller of these two sums (usually known as the Wilcoxon sum, from the name of the person first to design this test) W. If W is less than or equal to the percentage point W_α, given in Table 5.7, we shall reject the hypothesis that the means are equal at the significance level α. Note that, as for the sign test, it is necessary for the number of pairs to be at least 6 to reject the hypothesis at the 5% level, and at least 8 to reject the hypothesis at the 1% level.

TABLE 5.7

PERCENTAGE POINTS FOR SIGNED RANKS*

The probability is $\sim\alpha$ that the Wilcoxon sum W obtained from k ranks, assigned positive or negative signs at random with equal probability, will be less than the values tabulated.

Number of Pairs k	Significance Levels			
	$\alpha = 0.10$	$\alpha = 0.05$	$\alpha = 0.02$	$\alpha = 0.01$
5	0.6	—	—	—
6	2.1	0.6	—	—
7	3.7	2.1	0.3	—
8	5.8	3.7	1.6	0.3
9	8.1	5.7	3.1	1.6
10	10.8	8.1	5.1	3.1
11	13.9	10.8	7.2	5.1
12	17.5	13.8	9.8	7.2
13	21.4	17.2	12.7	9.8
14	25.7	21.1	15.9	12.7
15	30.4	25.3	19.6	15.9
16	35.6	29.9	23.6	19.5
17	41.2	34.9	28.0	23.4
18	47.2	40.3	32.7	27.7
19	53.6	46.1	37.8	32.4
20	60.4	52.3	43.4	37.5
21	67.6	58.9	49.3	42.9
22	75.3	66.0	55.6	48.7
23	83.9	73.4	62.3	54.9
24	91.9	81.3	69.4	61.5
25	100.9	89.5	76.9	68.5

* The values in this table were given by J. W. Tukey in *Memorandum Report 17*, "The Simplest Signed Rank Test," Statistical Research Group, Princeton University, 1949. The decimal places, obtained by interpolation, were used for setting limits; they are retained for the purpose of judging significance in the case of tied ranks, which is the only way non-integral values of W can arise from the significance tests described in the text.

If we wish to consider only the alternative hypotheses $\mu_1 > \mu_2$, i.e., the possibility of a positive difference, we should choose the sum of the ranks of the negative differences, regardless of whether it was the smaller of the two sums, and compare it with the percentage point $W_{2\alpha}$ for twice the desired significance level.

This test may be used to test the hypothesis $\mu = \mu_0$ by replacing the first or the second of the paired observations by the hypothetical mean μ_0 in each instance, i.e., by considering the signed rank of the deviations from μ_0 rather than the differences between pairs. A description of other similar tests can be found in [87], and the use of a similar technique to obtain confidence limits is discussed in [121].

EXAMPLE 2. The following data give the saponification found and the equivalent theoretical value for a series of glycol esters of maleic, succinic, phthalic, adipic, sebacic, and oxalic acids. The first results are the average of duplicate determinations. Differences are given as (experimental result − theoretical result).

Compound	Saponification Found	Equivalent Theory	Difference	Rank of Difference (absolute) value)
1	164.65	165.2	−0.55	6
2	178.6	179.2	−0.60	7
3	188.2	189.2	−1.00	12
4	200.6	201.3	−0.70	8.5
5	188.5	189.2	−0.70	8.5
6	145.6	145.2	+0.40	5
7	130.25	131.15	−0.90	11
8	199.95	199.2	+0.75	10
9	160.9	161.2	−0.30	3
10	155.1	155.2	−0.10	2
11	131.15	131.15	0.00	—
12	174.15	174.2	−0.05	1
13	129.75	130.1	−0.35	4

Although theoretically zero differences cannot occur, and would seldom occur in practice if sufficient decimal places were carried, in situations such as the above they do occur, and are usually counted half positive and half negative. Adopting this convention for the single zero difference in the above table we have 2.5 positive and 10.5 negative differences. From Table 5.2 we have, for $n = 13$ and $\alpha = 0.05$, $k = 3$; thus on the basis of the sign test we would conclude that the experimental results were systematically different from the theoretical results at the 5% level. Note that for $\alpha = 0.02$, $k = 2$, and hence the difference is not significant at the 2% level.

To use the signed rank test, we rank the differences in absolute value, ignoring the zero difference, and assigning the average rank 8.5 to the tied differences of 0.70 which rank 8th and 9th. Obviously the smaller sum of ranks corresponds to the positive differences, and hence in this case $W = 15$. Since for $n = 12$, $W_{0.05} = 13.8$ and $W_{0.10} = 17.5$, W is below the 5% level but exceeds the 10% level of significance.

To make the usual t-test in this case, we obtain $\bar{d} = -0.315$, $s_d^2 = 0.2589$, $s_d = 0.509$, and

$$t_{12} = \frac{-0.315}{0.509}\sqrt{13}$$
$$= -2.23$$

Since $t_{12,\,0.05} = 2.181$, we would reject the hypothesis at almost exactly the 5% level on the basis of this test.

5.53. Comparison of More than Two Means

The comparison of groups of means will be considered in detail as a part of the analysis of variance. The remarks in this section serve only as an introduction to the more complete discussion to follow.

Let us suppose that we have a group of k independent samples, each containing n_i observations, from which we obtain estimates $\bar{x}_1, \bar{x}_2, \cdots, \bar{x}_k$ of the means $\mu_1, \mu_2, \cdots, \mu_k$ of the populations sampled. Typical questions which we might like to ask are

(1) Can all these samples be considered as having been drawn from populations with the same mean μ?

(2) Can these samples be divided into several groups drawn from populations with means $\mu_1, \cdots, \mu_r$, $r < k$?

(3) If the samples are from populations which are themselves a random sample of possible populations, what is the variance σ_μ^2 of the population means in this superpopulation?

(4) How do any two given means μ_a and μ_b, or any two linear combinations of means, compare?

(5) What confidence limits can be assigned to an individual mean on the basis of information available from all k samples?

Question (1) is the basic question asked in the analysis of variance, and question (2) is a logical consequence of a negative answer to (1). Question (3) is the simplest case whose answer is provided by the calculation of variance components, also to be discussed in Chapter 7. Question

(4) can be answered by the methods of Section 5.51, except that here a question arises of the a priori or a posteriori choice of the comparisons to be made; this is discussed in Section 7.35. Question (5), and a simpler, although possibly less efficient, method of answering questions (1) and (2) than that to be discussed later, will be considered here. The answers to all these questions, except (4), where the methods of Aspin and Welch are available, require that we assume all the samples to be from populations with equal variance σ^2, so that the joint estimate s^2 from all k samples (see Section 5.42) is a valid estimate of the variance of a single sample. Also in many cases the use of equal sample sizes, i.e., $n_i = n$ for all i, is required for simplicity.

Assuming both of these conditions to be satisfied, then under the null hypothesis $\mu_1 = \mu_2 = \cdots = \mu_k = \mu$ indicated by question (1), all the sample means $\bar{x}_i$ have average value μ and variance σ^2/n. In this case a quick way of checking the null hypothesis is to see if the range of these means is consistent with the estimate $s_{\bar{x}} = \sqrt{s^2/n}$ of their standard deviation by using the studentized range discussed in Section 4.56. The use of this ratio is valid here, since the joint estimate s^2 depends only on the sample variances s_i^2 of the k samples, which are independent of the sample means. The tables of percentage points referred to in Section 4.56 are reproduced in Table 5.8; note that they are based on the assumption of normally distributed populations. It should be remembered that in this case the sample size is k, not n, since we are considering the range of the k means, and that the independent estimate $s_{\bar{x}}$ is based on $N - k = k(n - 1)$ degrees of freedom.

EXAMPLE 1. Table 7.7 of Section 7.31 lists the means of sets of 5 determinations by 10 different analysts on a standard sample of iron. Arranged in order of magnitude, they are (in percentage of iron): 2.937, 2.938, 2.950, 2.951, 2.951, 2.955, 2.955, 2.958, 2.977, 2.984.

In Table 7.8 of Section 7.32, an independent estimate of σ^2 based on 40 degrees of freedom is found to be $s^2 = 0.0000934$. Hence the estimated variance of a mean of 5 determinations is

$$s_{\bar{x}}^2 = \frac{0.0000934}{5} = 0.00001868$$

and the estimated standard deviation

$$s_{\bar{x}} = 0.00432$$

The studentized range of these means is thus

$$w = \frac{0.047}{0.00432} = 10.9$$

which is beyond the 1% level for 10 means and 40 degrees of freedom. Hence we conclude there must be differences among the analysts.

We can attempt to determine which analysts differ from the group by dropping off the extreme means and retesting the remainder of the group. In this case the values 2.977 and 2.984 seem to be most discrepant; if these two values are dropped, we have for the remaining sample of 8

$$w = \frac{0.021}{0.00432} = 4.86$$

which is significant at the 5% level but not at the 1% level. If in addition we drop the two lowest values, we have for the remaining 6 values

$$w = \frac{0.008}{0.00432} = 1.85$$

which is far below the 5% point. Thus there is definite evidence that analysts 1 and 8 are obtaining higher results on the average than the remaining analysts, and some evidence to the effect that analysts 6 and 7 are obtaining low results, although it is not at all conclusive in view of our arbitrary method of choice of comparisons.

Question (5) is easily answered for the slightly more general case where the n_i are arbitrary. If we assume the populations sampled to be normally distributed, then the quantity

$$(a) \qquad t = \frac{\bar{x}_i - \mu_i}{\sigma}\sqrt{n_i}$$

is a standardized normal deviate. Also under this assumption $(N - k)s^2/\sigma^2$ has, as shown in Section 5.42, the χ^2-distribution with $N - k$ degrees of freedom. By forming the ratio given in Section 4.54, we find that the quantity

$$(b) \qquad t_{N-k} = \frac{\bar{x}_i - \mu_i}{s}\sqrt{n_i}$$

has the "Student" t-distribution with $N - k$ degrees of freedom. Using this fact, we obtain

$$(c) \qquad \bar{x} \pm t_{N-k,\,\alpha}\, s/\sqrt{n_i}$$

as the $100(1 - \alpha)\%$ confidence limits for the mean μ_i. The limits given by (c) will be narrower than if they had been based only on a single series of measurements, since in that case the value $t_{N-k,\,\alpha}$ would be replaced by the larger value $t_{n_i-1,\,\alpha}$.

TABLE 5.8

PERCENTAGE POINTS OF STUDENTIZED RANGE*

The range of a sample of size n, from a normal population, divided by an independent estimate of the standard deviation of the population based on ν degrees of freedom, will exceed the values tabulated with probability α.

$\alpha = 0.05$

ν \ n	2	3	4	5	6	7	8	9	10	11	12	13	14	15	16	17	18	19	20
1	18.0	26.7	32.8	37.2	40.5	43.1	45.4	47.3	49.1	50.6	51.9	53.2	54.3	55.4	56.3	57.2	58.0	58.8	59.6
2	6.09	8.28	9.80	10.89	11.73	12.43	13.03	13.54	13.99	14.39	14.75	15.08	15.38	15.65	15.91	16.14	16.36	16.57	16.77
3	4.50	5.88	6.83	7.51	8.04	8.47	8.85	9.18	9.46	9.72	9.95	10.16	10.35	10.52	10.69	10.84	10.98	11.12	11.24
4	3.93	5.00	5.76	6.31	6.73	7.06	7.35	7.60	7.83	8.03	8.21	8.37	8.52	8.67	8.80	8.92	9.03	9.14	9.24
5	3.61	4.54	5.18	5.64	5.99	6.28	6.52	6.74	6.93	7.10	7.25	7.39	7.52	7.64	7.75	7.86	7.95	8.04	8.13
6	3.46	4.34	4.90	5.31	5.63	5.89	6.12	6.32	6.49	6.65	6.79	6.92	7.04	7.14	7.24	7.34	7.43	7.51	7.59
7	3.34	4.16	4.68	5.06	5.35	5.59	5.80	5.99	6.15	6.29	6.42	6.54	6.65	6.75	6.84	6.93	7.01	7.08	7.16
8	3.26	4.04	4.53	4.89	5.17	5.40	5.60	5.77	5.92	6.05	6.18	6.29	6.39	6.48	6.57	6.65	6.73	6.80	6.87
9	3.20	3.95	4.42	4.76	5.02	5.24	5.43	5.60	5.74	5.87	5.98	6.09	6.19	6.28	6.36	6.44	6.51	6.58	6.65
10	3.15	3.88	4.33	4.66	4.91	5.12	5.30	5.46	5.60	5.72	5.83	5.93	6.03	6.12	6.20	6.27	6.34	6.41	6.47
11	3.11	3.82	4.26	4.58	4.82	5.03	5.20	5.35	5.49	5.61	5.71	5.81	5.90	5.98	6.06	6.14	6.20	6.27	6.33
12	3.08	3.77	4.20	4.51	4.75	4.95	5.12	5.27	5.40	5.51	5.61	5.71	5.80	5.88	5.95	6.02	6.09	6.15	6.21
13	3.06	3.73	4.15	4.46	4.69	4.88	5.05	5.19	5.32	5.43	5.53	5.63	5.71	5.79	5.86	5.93	6.00	6.06	6.11
14	3.03	3.70	4.11	4.41	4.64	4.83	4.99	5.13	5.25	5.36	5.46	5.56	5.64	5.72	5.79	5.86	5.92	5.98	6.03
15	3.01	3.67	4.08	4.37	4.59	4.78	4.94	5.08	5.20	5.31	5.40	5.49	5.57	5.65	5.72	5.79	5.85	5.91	5.96
16	3.00	3.65	4.05	4.34	4.56	4.74	4.90	5.03	5.15	5.26	5.35	5.44	5.52	5.59	5.66	5.73	5.79	5.84	5.90
17	2.98	3.62	4.02	4.31	4.52	4.70	4.86	4.99	5.11	5.21	5.31	5.39	5.47	5.55	5.61	5.68	5.74	5.79	5.84
18	2.97	3.61	4.00	4.28	4.49	4.67	4.83	4.96	5.07	5.17	5.27	5.35	5.43	5.50	5.57	5.63	5.69	5.74	5.79
19	2.96	3.59	3.98	4.26	4.47	4.64	4.79	4.92	5.04	5.14	5.23	5.32	5.39	5.46	5.53	5.59	5.65	5.70	5.75
20	2.95	3.58	3.96	4.24	4.45	4.62	4.77	4.90	5.01	5.11	5.20	5.28	5.36	5.43	5.50	5.56	5.61	5.66	5.71
24	2.92	3.53	3.90	4.17	4.37	4.54	4.68	4.81	4.92	5.01	5.10	5.18	5.25	5.32	5.38	5.44	5.50	5.55	5.59
30	2.89	3.48	3.84	4.11	4.30	4.46	4.60	4.72	4.83	4.92	5.00	5.08	5.15	5.21	5.27	5.33	5.38	5.43	5.48
40	2.86	3.44	3.79	4.04	4.23	4.39	4.52	4.63	4.74	4.82	4.90	4.98	5.05	5.11	5.17	5.22	5.27	5.32	5.36
60	2.83	3.40	3.74	3.98	4.16	4.31	4.44	4.55	4.65	4.73	4.81	4.88	4.94	5.00	5.06	5.11	5.15	5.20	5.24
120	2.80	3.36	3.69	3.92	4.10	4.24	4.36	4.47	4.56	4.64	4.71	4.78	4.84	4.90	4.95	5.00	5.04	5.09	5.13
∞	2.77	3.32	3.63	3.86	4.03	4.17	4.29	4.39	4.47	4.55	4.62	4.68	4.74	4.80	4.84	4.89	4.93	4.97	5.01

TABLE 5.8 (*continued*)

$\alpha = 0.01$

ν \ n	2	3	4	5	6	7	8	9	10	11	12	13	14	15	16	17	18	19	20
1	90.0	134	164	186	202	216	227	237	246	253	260	266	272	277	282	286	291	295	298
2	14.0	18.9	22.3	24.7	26.6	28.2	29.5	30.7	31.7	32.6	33.4	34.2	34.8	35.5	36.0	36.5	37.0	37.5	38.0
3	8.26	10.56	12.17	13.34	14.25	15.00	15.65	16.20	16.69	17.13	17.53	17.89	18.23	18.54	18.83	19.09	19.33	19.56	19.79
4	6.51	8.08	9.17	9.97	10.58	11.10	11.55	11.93	12.26	12.56	12.84	13.09	13.32	13.53	13.73	13.92	14.09	14.25	14.40
5	5.62	6.83	7.65	8.26	8.73	9.12	9.46	9.76	10.02	10.25	10.46	10.65	10.83	10.99	11.14	11.29	11.42	11.54	11.66
6	5.24	6.32	7.03	7.56	7.97	8.31	8.61	8.87	9.10	9.30	9.49	9.65	9.81	9.95	10.08	10.21	10.32	10.43	10.54
7	4.94	5.89	6.52	6.98	7.35	7.65	7.91	8.14	8.34	8.52	8.68	8.82	8.96	9.08	9.20	9.31	9.42	9.52	9.61
8	4.74	5.63	6.20	6.63	6.96	7.24	7.47	7.68	7.86	8.03	8.18	8.31	8.44	8.55	8.66	8.76	8.86	8.95	9.03
9	4.60	5.42	5.96	6.35	6.66	6.91	7.13	7.33	7.50	7.65	7.79	7.91	8.02	8.13	8.23	8.33	8.41	8.50	8.58
10	4.48	5.26	5.77	6.14	6.43	6.67	6.88	7.06	7.22	7.36	7.49	7.60	7.71	7.81	7.91	7.99	8.08	8.15	8.23
11	4.39	5.14	5.62	5.98	6.25	6.47	6.67	6.84	6.99	7.13	7.25	7.36	7.46	7.56	7.65	7.73	7.81	7.88	7.95
12	4.32	5.04	5.50	5.84	6.10	6.32	6.51	6.67	6.81	6.94	7.06	7.17	7.26	7.36	7.44	7.52	7.60	7.67	7.73
13	4.26	4.96	5.40	5.73	5.98	6.19	6.37	6.53	6.67	6.79	6.90	7.01	7.10	7.19	7.27	7.35	7.42	7.49	7.55
14	4.21	4.89	5.32	5.64	5.88	6.08	6.26	6.41	6.54	6.66	6.77	6.87	6.96	7.05	7.13	7.20	7.27	7.34	7.40
15	4.17	4.83	5.25	5.56	5.80	5.99	6.16	6.31	6.44	6.55	6.66	6.76	6.85	6.93	7.00	7.07	7.14	7.20	7.26
16	4.13	4.78	5.19	5.49	5.72	5.91	6.08	6.22	6.35	6.46	6.56	6.66	6.74	6.82	6.90	6.97	7.03	7.09	7.15
17	4.10	4.73	5.14	5.43	5.66	5.85	6.01	6.15	6.27	6.38	6.48	6.57	6.66	6.73	6.81	6.87	6.94	7.00	7.05
18	4.07	4.70	5.09	5.38	5.60	5.79	5.95	6.08	6.20	6.31	6.41	6.50	6.58	6.65	6.73	6.79	6.85	6.91	6.97
19	4.05	4.66	5.05	5.34	5.55	5.73	5.89	6.02	6.14	6.25	6.34	6.43	6.51	6.58	6.65	6.72	6.78	6.84	6.89
20	4.02	4.63	5.02	5.30	5.51	5.69	5.84	5.97	6.09	6.19	6.28	6.37	6.45	6.52	6.59	6.66	6.71	6.77	6.82
24	3.96	4.54	4.91	5.17	5.37	5.54	5.69	5.81	5.92	6.02	6.11	6.19	6.26	6.33	6.39	6.45	6.51	6.57	6.61
30	3.89	4.45	4.80	5.05	5.24	5.40	5.53	5.65	5.76	5.85	5.93	6.01	6.08	6.14	6.20	6.26	6.31	6.36	6.41
40	3.82	4.36	4.70	4.93	5.11	5.26	5.39	5.50	5.60	5.69	5.77	5.84	5.90	5.96	6.02	6.07	6.12	6.17	6.21
60	3.76	4.28	4.60	4.82	4.99	5.13	5.25	5.36	5.45	5.53	5.60	5.67	5.73	5.78	5.83	5.88	5.93	5.98	6.01
120	3.70	4.20	4.50	4.71	4.87	5.00	5.12	5.21	5.30	5.38	5.44	5.50	5.56	5.61	5.66	5.71	5.75	5.79	5.83
∞	3.64	4.12	4.40	4.60	4.76	4.88	4.99	5.08	5.16	5.23	5.29	5.35	5.40	5.45	5.49	5.53	5.57	5.61	5.64

* Reproduced by permission of Professor E. S. Pearson from: "Extended and Corrected Tables of the Upper Percentage Points of the 'Studentized' Range," computed by Joyce M. May, *Biometrika*, 39 (1952), 192–3.

EXAMPLE 2. In Example 2 of Section 5.42 we obtained from the 10 pairs of results by Method I an estimate $s = 0.048$ of the standard deviation based on 10 degrees of freedom. Assuming this to be an estimate of the common variance of the method for all samples, 95% confidence limits for a result on an individual sample, say sample 7, would be

$$17.90 \pm t_{10,\,0.05}\frac{0.048}{\sqrt{2}} = 17.90 \pm \frac{(2.228)\,(0.048)}{1.414}$$

$$= 17.90 \pm 0.075$$

A confidence interval can be obtained using only the two determinations carried out on sample 7. In this case $s^2 = \dfrac{(0.04)^2}{2} = 0.0008$, $s = 0.028$, and the confidence interval is

$$17.90 \pm t_{1,\,0.05}\frac{0.028}{\sqrt{2}} = 17.90 \pm \frac{(12.706)\,(0.028)}{1.414}$$

$$= 17.90 \pm 0.25$$

which, in spite of the smaller value of s, really reflects our lack of confidence in any estimate of variation based on 1 degree of freedom.

Note that, in placing confidence limits on a series of means by the above method, we are taking $100\alpha\%$ risk of not including a given mean μ_i within the limits. We may occasionally wish to think in terms of the risk we take of having one or more incorrect statements among the limits assigned to a group, or batch, of means in the above fashion. Designating this risk by α', it is easily seen that if the k sets of limits were independent, the probability $1 - \alpha'$ of being correct in each case would be $(1 - \alpha)^k$, where α is the risk corresponding to a given set of limits. Although the individual means are independent, the limits are not, since we have used a common joint estimate s in their determination. This means that the cases in which the individual limits are incorrect will not be scattered randomly throughout the batches of k, but will be to some extent clustered in those batches where the joint estimate s chances to be small. As a result there will be more instances than would be expected of no individual limits in error, and we shall have

$$1 - \alpha' > (1 - \alpha)^k$$

$$(d) \qquad \alpha' < 1 - (1 - \alpha)^k$$

equality being approached as the number of degrees of freedom on which s is based becomes large. In practice we can use (d) to determine an upper limit for the risk in assigning limits to a batch of means. This will not differ too greatly from the true risk if s is based on a fairly large number of degrees of freedom compared with the number of means in the batch.

Thus, for example, if we wish to take less than a 5% risk of assigning incorrect limits to one or more of a batch of 10 means, we should use $\alpha = 0.9948$ in obtaining the limits (c) for individual means. Since most tables (including Table III) do not give the values $t_{N-k,\alpha}$ for such odd values of α, "batching factors" are given in Table 5.9, depending on the number of means k and the degrees of freedom $\nu = N - k$, which will approximately convert values of $t_{N-k,\alpha'}$ obtained from Table III to values of $t_{N-k,\alpha}$. These factors can, of course, be used wherever the change in significance level given by (d) is applicable to limits based on the t-distribution.

EXAMPLE 3. In Example 2, if we wish to take only 1 chance in 20 of having 1 or more of the 10 means fall outside of the limits given, we include the batching factor $1.43 + \dfrac{1.7}{10} = 1.60$, and obtain, again using sample 7 as an example,

$$17.90 \pm (1.60)(0.075) = 17.90 \pm 0.120$$

TABLE 5.9

BATCHING FACTORS FOR "STUDENT'S" t*

Factors by which percentage points of the "Student" t-distribution for ν degrees of freedom should be multiplied if the limits obtained are to have $100\alpha\%$ risk of 1 or more errors per batch of k rather than $100\alpha\%$ risk per individual limit.

k	$\alpha = 0.05$	$\alpha = 0.01$
2	$1.14 + 0.4/\nu$	$1.09 + 0.4/\nu$
3	$1.22 + 0.7/\nu$	$1.14 + 0.7/\nu$
4	$1.27 + 0.9/\nu$	$1.17 + 0.9/\nu$
5	$1.31 + 1.0/\nu$	$1.20 + 1.0/\nu$
6	$1.34 + 1.2/\nu$	$1.22 + 1.2/\nu$
7	$1.37 + 1.3/\nu$	$1.24 + 1.3/\nu$
8	$1.39 + 1.5/\nu$	$1.25 + 1.5/\nu$
9	$1.41 + 1.6/\nu$	$1.27 + 1.6/\nu$
10	$1.43 + 1.7/\nu$	$1.28 + 1.7/\nu$
12	$1.47 + 1.8/\nu$	$1.31 + 1.8/\nu$
15	$1.49 + 2.1/\nu$	$1.32 + 2.2/\nu$
20	$1.54 + 2.4/\nu$	$1.35 + 2.5/\nu$
25	$1.57 + 2.7/\nu$	$1.38 + 2.8/\nu$
30	$1.60 + 3.0/\nu$	$1.40 + 3.1/\nu$
40	$1.63 + 3.4/\nu$	$1.42 + 3.5/\nu$
50	$1.68 + 3.8/\nu$	$1.44 + 3.8/\nu$

* Computed by, and reproduced by permission of, Professor J. W. Tukey, Princeton University.

Such limits for all 10 samples will include all the unknown sample means with less than a 5% risk of 1 or more errors. With the original limits, the chance of 1 or more errors in the batch of 10 would have been less than

$$\alpha' = 1 - (0.95)^{10} \sim 0.40$$

5.6. Comparison of Variances

5.61. Comparison of Two Variances

The F-distribution developed in Section 4.55 can be applied to the problem of comparing 2 variances, or obtaining confidence intervals for the ratio $\phi = \sigma_1^2/\sigma_2^2$. We found there that, for 2 estimates s_1^2 and s_2^2 of the variances σ_1^2 and σ_2^2 of normally distributed variables, based on ν_1 and ν_2 degrees of freedom, respectively, the quantity

$$F = \frac{1}{\phi}\frac{s_1^2}{s_2^2} \tag{a}$$

has the F-distribution with ν_1 and ν_2 degrees of freedom. In the special case $\sigma_1^2 = \sigma_2^2$, or $\phi = 1$, where both s_1^2 and s_2^2 are estimates of the same variance, (*a*) reduces to the ratio

$$F = s_1^2/s_2^2 \tag{b}$$

of the sample variances. It should be pointed out before starting to develop the tests of this and the following section that departures from normality have a much greater effect on the above ratios, and hence on these tests, than upon the corresponding tests for means; it is safe to say that these tests are for the equality of the variances and/or the normality of the underlying distributions. Fortunately, in the extensive use of similar variance ratios in Chapter 7 in connection with the analysis of variance, moderate departures from normality are not so important.

In testing hypotheses concerning variances, we are usually interested in the hypothesis $\sigma_1^2 \leq \sigma_2^2$ as opposed to $\sigma_1^2 > \sigma_2^2$, i.e., in concluding that one variance is either greater or less than another. In practice two cases arise:

(1) Where there is some reason to prefer the conclusion $\sigma_1^2 \leq \sigma_2^2$ unless $\sigma_1^2 > \sigma_2^2$, implying that we wish to accept σ_1^2 unless it is proved definitely greater than σ_2^2.

(2) Where if $\sigma_1^2 = \sigma_2^2$ it is a matter of complete indifference whether we conclude $\sigma_1^2 < \sigma_2^2$ or $\sigma_1^2 > \sigma_2^2$.

For example, in the first case we might be choosing between two methods on the basis of their precision, and one might be preferable on the basis of time required and adaptability for routine use unless it were definitely proved to be less precise. The second case might arise when the difference between the methods consisted only in the choice of 2 equally available reagents for use in a given step.

In either case, the best procedure is to reject the hypothesis $\sigma_1^2 \leq \sigma_2^2$ and conclude $\sigma_1^2 > \sigma_2^2$ when

$$s_1^2/s_2^2 \geq F_{\nu_1, \nu_2, \alpha}$$

where $F_{\nu_1, \nu_2, \alpha}$ is chosen from Table IV. If $\sigma_1^2 = \sigma_2^2$, s_1^2/s_2^2 has the F-distribution with ν_1 and ν_2 degrees of freedom, and, since $F_{\nu_1, \nu_2, \alpha}$ was chosen so that

$$(c) \qquad Pr(F \geq F_{\nu_1, \nu_2, \alpha}) = \alpha$$

the probability of concluding $\sigma_1^2 > \sigma_2^2$ will be exactly α. If $\sigma_1^2 \neq \sigma_2^2$, or $\phi \neq 1$, then $s_1^2/s_2^2\phi$ has the F-distribution with ν_1 and ν_2 degrees of freedom, and

$$(d) \qquad \begin{aligned} Pr(s_1^2/s_2^2 \geq F_{\nu_1, \nu_2, \alpha}) &= Pr(s_1^2/s_2^2\phi \geq F_{\nu_1, \nu_2, \alpha}/\phi) \\ &= Pr(F \geq F_{\nu_1, \nu_2, \alpha}/\phi) \end{aligned}$$

If $\phi \leq 1$, we have

$$F_{\nu_1, \nu_2, \alpha}/\phi \geq F_{\nu_1, \nu_2, \alpha}$$

$$Pr(F \geq F_{\nu_1, \nu_2, \alpha}/\phi) \leq Pr(F \geq F_{\nu_1, \nu_2, \alpha}) = \alpha$$

and hence for $\sigma_1^2 \leq \sigma_2^2$ we are even less likely to conclude $\sigma_1^2 > \sigma_2^2$. Thus in using this procedure the probability of concluding $\sigma_1^2 > \sigma_2^2$ when actually $\sigma_1^2 \leq \sigma_2^2$ is at most α.

We must now consider the probability β of concluding $\sigma_1^2 \leq \sigma_2^2$. Since

$$\beta = Pr(s_1^2/s_2^2 < F_{\nu_1, \nu_2, \alpha})$$

it follows from (c) that when $\sigma_1^2 = \sigma_2^2$ we have $\beta = 1 - \alpha$. When $\phi \neq 1$ we obtain, using (d),

$$(e) \qquad \begin{aligned} \beta &= 1 - Pr(F \geq F_{\nu_1, \nu_2, \alpha}/\phi) \\ &= Pr(F < F_{\nu_1, \nu_2, \alpha}/\phi) \end{aligned}$$

and by an argument similar to the above we find that when $\phi > 1$, or $\sigma_1^2 > \sigma_2^2$, the probability of concluding $\sigma_1^2 \leq \sigma_2^2$ is at most $1 - \alpha$.

The distinction between the two cases above lies in the choice of α. In the first case we wish to conclude $\sigma_1^2 \leq \sigma_2^2$ unless we are relatively certain that $\sigma_1^2 > \sigma_2^2$; hence we choose α small, say 0.05 or 0.01, so that the probability of concluding $\sigma_1^2 > \sigma_2^2$ when $\sigma_1^2 \leq \sigma_2^2$ will be small. As a consequence the probability of concluding $\sigma_1^2 \leq \sigma_2^2$ when actually

$\sigma_1^2 > \sigma_2^2$, which was shown above to be at most $1 - \alpha$, could be quite large if σ_1^2 were not much greater than σ_2^2. In the second case we express our indifference by choosing $\alpha = 1 - \alpha = 0.50$, so that the probability of an incorrect decision is in any event at most 0.50, with neither decision preferred if $\sigma_1^2 = \sigma_2^2$.

In the first case, it is frequently desirable to determine how large the ratio $\phi = \sigma_1^2/\sigma_2^2$ must be to make the probability β of concluding $\sigma_1^2 < \sigma_2^2$ less than a given value. Since it follows from (*e*) that

$$1 - \beta = Pr(F \geq F_{\nu_1, \nu_2, \alpha}/\phi)$$

we must have by definition

$$F_{\nu_1, \nu_2, 1-\beta} = F_{\nu_1, \nu_2, \alpha}/\phi$$

and, since we have seen in Section 4.55 that

$$F_{\nu_1, \nu_2, 1-\beta} = \frac{1}{F_{\nu_2, \nu_1, \beta}}$$

we obtain

$$(f) \qquad \phi = F_{\nu_1, \nu_2, \alpha} \cdot F_{\nu_2, \nu_1, \beta}$$

as the smallest value of ϕ for which this probability will be less than β.

Values of $\phi(\alpha, \beta, \nu_1, \nu_2)$ which are solutions of (*f*) are given for $\alpha = 0.05$ and 0.01, various values of β, and a wide range of values of ν_1 and ν_2 in [19, Chapter 8]. This reference also contains operating characteristic curves for the above tests.

EXAMPLE 1. Suppose that in Example 2 of Section 5.42 we wish to test whether the variance of Method I is less than that of Method II, and that because of other considerations we wish to accept Method I unless we have definite evidence that $\sigma_1^2 > \sigma_2^2$. To do this we compute

$$F_{10, 10} = \frac{s_1^2}{s_2^2} = \frac{0.002285}{0.001805} = 1.27$$

Since this is well below the value $F_{10, 10, 0.05} = 2.97$, we would not reject the hypothesis $\sigma_1^2 \leq \sigma_2^2$ on the basis of these results.

EXAMPLE 2. In the above example, suppose that it were desired to determine the number of pairs of duplicate determinations necessary by each method in order to make the probability of failing to reject Method I when $\sigma_1^2 = 2\sigma_2^2$ less than or equal to 0.05, the test being conducted at the 5% significance level. Since each pair of duplicates corresponds to 1 degree of freedom in the estimate, we want to determine ν such that

$$\phi > F_{\nu, \nu, \alpha} \cdot F_{\nu, \nu, \beta}$$

where $\phi = 2$ and $\alpha = \beta = 0.05$. This condition reduces to

$$F^2_{\nu, \nu, 0.05} < 2$$

$$F_{\nu, \nu, 0.05} < 1.414$$

From Table IV it is easily determined that ν, the number of pairs by each method, must be approximately 100.

Since this would undoubtedly be impractical, we might determine how our conditions could be relaxed. Let us first consider taking $\beta = 0.50$ rather than 0.05, i.e., taking 1 chance in 2 rather than 1 chance in 20 of failing to reject the hypothesis when $\sigma_1^2 = 2\sigma_2^2$. We then wish to determine the smallest value of ν such that

$$F_{\nu, \nu, 0.05} \cdot F_{\nu, \nu, 0.50} < 2$$

From the tables of F, or directly from the tables of $\phi(\alpha, \beta, \nu_1, \nu_2)$ referred to above, we find that the required number of pairs is now 24. We can also consider retaining the condition $\beta = 0.05$, but requiring the probability of failing to reject the hypothesis to have this value if $\sigma_1^2 < 4\sigma_2^2$, i.e., if $\sigma_1 < 2\sigma_2$. Then, solving the equation

$$F^2_{\nu, \nu, 0.05} < 4$$

$$F_{\nu, \nu, 0.05} < 2$$

we find the required number of pairs again to be approximately 24. Even with these relaxations, the number of pairs of determinations required is still quite large.

Since $F_{\nu_1, \nu_2, 1-\alpha}$ is defined so that $Pr(F \geq F_{\nu_1, \nu_2, 1-\alpha}) = 1 - \alpha$, we have from (a)

$$(g) \qquad Pr\left(\frac{1}{\phi}\frac{s_1^2}{s_2^2} > F_{\nu_1, \nu_2, 1-\alpha}\right) = 1 - \alpha$$

and hence

$$(h) \qquad Pr\left(\phi < \frac{s_1^2}{s_2^2 F_{\nu_1, \nu_2, 1-\alpha}}\right) = 1 - \alpha$$

Thus

$$(i) \qquad \frac{s_1^2}{s_2^2 F_{\nu_1, \nu_2, 1-\alpha}} = \frac{s_1^2}{s_2^2} F_{\nu_2, \nu_1, \alpha}$$

is a $100(1 - \alpha)\%$ upper confidence limit for the ratio ϕ. Similarly $\dfrac{s_1^2}{s_2^2} F_{\nu_2, \nu_1, 1-\alpha}$ would be a $100(1 - \alpha)\%$ lower confidence limit, and two-sided limits could be obtained by choosing both upper and lower limits for half the desired value of α.

In making a quick comparison of 2 variances, it is frequently possible to make use of the fact that

$$z = \tfrac{1}{2}(\ln s_1^2 - \ln s_2^2) = \tfrac{1}{2}\ln \frac{s_1^2}{s_2^2} \tag{j}$$

is approximately normally distributed with

$$(k) \qquad \text{ave}(z) = -\frac{1}{2}\left(\frac{1}{\nu_1} - \frac{1}{\nu_2}\right)$$

$$\text{var}(z) = \tfrac{1}{2}\left(\frac{1}{\nu_1} + \frac{1}{\nu_2}\right)$$

5.62. Tests for the Homogeneity of Several Variances

We shall now consider the problem of testing the equality of a set of variances $\sigma_1^2, \sigma_2^2, \cdots, \sigma_k^2$, i.e., testing the hypothesis $\sigma_1^2 = \sigma_2^2 = \cdots = \sigma_k^2$ on the basis of the sample variances $s_1^2, s_2^2, \cdots, s_k^2$, based on $\nu_1, \nu_2, \cdots, \nu_k$ degrees of freedom, respectively.

As examples of circumstances in which this problem arises, we might wish to test whether a series of results obtained by several analysts using a given analytical procedure were equally precise, or to sample the output of several different production lines in order to determine whether variation in product was equal on the respective lines, or even to test whether an analytical method gave equal precision on a series of samples having varying amounts of different impurities present. A test of this type is also necessary to check the frequent assumption that a joint estimate of the variance can be compounded from a number of different groups of data.

Several tests are available for testing this hypothesis, and each can be constructed so that the significance level, or the probability of a type I error, is α. However, the tests will differ in the probability of an error of type II not only because of their inherent construction but also because of the many possible ways in which the above hypothesis could be false. For example, a test which would have a small error of type II when the variances, instead of being equal, were more or less randomly scattered might have a large error of type II if the variances, instead of being equal, actually consisted of $k - 1$ which were equal and one which differed from the other $k - 1$ by a considerable amount. In other words, the probability of an error of type II depends not only on the construction of the significance test itself but also on the possible alternatives against which we are trying to guard, or the ways in which we think the hypothesis may be wrong.

The test most frequently used in the situation where instead of equality of the variances we have a more or less random non-homogeneity is called *Bartlett's test*. In using this test we calculate from the k sample variances the joint estimate $s^2 = \frac{\Sigma \nu_i s_i^2}{\nu}$, where $\nu = \nu_1 + \nu_2 + \cdots + \nu_k$, described in Section 5.42. Using this, we compute the quantity

$$B = \frac{1}{C}(\nu \ln s^2 - \Sigma \nu_i \ln s_i^2)$$

$$= \frac{2.30259}{C}(\nu \log s^2 - \Sigma \nu_i \log s_i^2)$$

where

$$C = 1 + \frac{\Sigma(1/\nu_i) - 1/\nu}{3(k - 1)}$$

For values of ν_i of 5 or more the distribution of B is satisfactorily approximated by the χ^2-distribution with $k - 1$ degrees of freedom. Hence we would reject the hypothesis $\sigma_1^2 = \sigma_2^2 = \cdots = \sigma_k^2$, if the value of B were greater than $\chi^2_{k-1,\alpha}$ where α is the chosen significance level. Notice that the quantity $1/C$ must be less than 1, since C is greater than 1; hence, if the value of B is not significant when $C = 1$, it will be unnecessary to calculate the value of C.

When the ν_i are smaller than 5 the χ^2-approximation may introduce an appreciable error. To overcome this difficulty Merrington and Thompson [122] have computed tables from which the exact percentage points of

$$M = \nu \ln s^2 - \Sigma \nu_i \ln s_i^2$$

$$= 2.30259(\nu \log s^2 - \Sigma \nu_i \log s_i^2)$$

$$= BC$$

can be approximated for $\alpha = 0.05$ and $\alpha = 0.01$. Table 5.10 gives approximate maximum values of these percentage points for $k = 3$–15 and various values of

$$c_1 = \Sigma_i \left(\frac{1}{\nu_i}\right) - \frac{1}{\nu} = 3(k - 1)(C - 1)$$

These maximum values differ little from the exact values when the ν_i are nearly equal, and their use in place of the exact points will ensure at least the stated significance level.

Another test which is more appropriate if we suspect a single one of the variances of being appreciably greater than the remaining $k - 1$

TABLE 5.10

PERCENTAGE POINTS OF M*

Top value $\alpha = 0.05$; bottom value $\alpha = 0.01$

k \ c_1	0.0	0.5	1.0	1.5	2.0	2.5	3.0	3.5	4.0	4.5	5.0	6.0	7.0	8.0	9.0	10.0
3	5.99	6.47	6.89	7.20	7.38	7.39	7.22	—	—	—	—	—	—	—	—	—
	9.21	9.92	10.47	10.78	10.81	10.50	9.83	—	—	—	—	—	—	—	—	—
4	7.81	8.24	8.63	8.96	9.21	9.38	9.43	9.37	9.18	—	—	—	—	—	—	—
	11.34	11.95	12.46	12.86	13.11	13.18	13.03	12.65	12.03	—	—	—	—	—	—	—
5	9.49	9.88	10.24	10.57	10.86	11.08	11.24	11.32	11.31	11.21	11.02	—	—	—	—	—
	13.28	13.81	14.30	14.71	15.03	15.25	15.34	15.28	15.06	14.66	14.07	—	—	—	—	—
6	11.07	11.43	11.78	12.11	12.40	12.65	12.86	13.01	13.11	13.14	13.10	12.78	—	—	—	—
	15.09	15.58	16.03	16.44	16.79	17.07	17.27	17.37	17.37	17.24	16.98	16.03	—	—	—	—
7	12.59	12.94	13.27	13.59	13.88	14.15	14.38	14.58	14.73	14.83	14.88	14.81	14.49	—	—	—
	16.81	17.27	17.70	18.10	18.46	18.77	19.02	19.21	19.32	19.35	19.28	18.84	17.92	—	—	—
8	14.07	14.40	14.72	15.03	15.32	15.60	15.84	16.06	16.25	16.40	16.51	16.60	16.49	16.16	—	—
	18.48	18.91	19.32	19.71	20.07	20.39	20.67	20.90	21.08	21.20	21.25	21.13	20.64	19.76	—	—
9	15.51	15.83	16.14	16.44	16.73	17.01	17.26	17.49	17.70	17.88	18.03	18.22	18.26	18.12	17.79	—
	20.09	20.50	20.90	21.28	21.64	21.97	22.26	22.52	22.74	22.91	23.03	23.10	22.91	22.41	21.56	—
10	16.92	17.23	17.54	17.83	18.12	18.39	18.65	18.89	19.11	19.31	19.48	19.75	19.89	19.89	19.73	19.40
	21.67	22.06	22.45	22.82	23.17	23.50	23.80	24.08	24.32	24.52	24.69	24.90	24.90	24.66	24.15	23.33
11	18.31	18.61	18.91	19.20	19.48	19.76	20.02	20.26	20.49	20.70	20.89	21.21	21.42	21.52	21.49	21.32
	23.21	23.59	23.97	24.33	24.67	25.00	25.31	25.59	25.85	26.08	26.28	26.57	26.70	26.65	26.38	25.86
12	19.68	19.97	20.26	20.55	20.83	21.10	21.36	21.61	21.84	22.06	22.27	22.62	22.88	23.06	23.12	23.07
	24.72	25.10	25.46	25.81	26.15	26.48	26.79	27.08	27.35	27.59	27.81	28.16	28.39	28.46	28.37	28.07
13	21.03	21.32	21.60	21.89	22.16	22.43	22.69	22.94	23.18	23.40	23.62	23.99	24.30	24.53	24.66	24.70
	26.22	26.58	26.93	27.28	27.62	27.94	28.25	28.54	28.81	29.07	29.30	29.70	29.99	30.16	30.19	30.06
14	22.36	22.65	22.93	23.21	23.48	23.75	24.01	24.26	24.50	24.73	24.95	25.34	25.68	25.95	26.14	26.25
	27.69	28.04	28.39	28.73	29.06	29.38	29.69	29.98	30.26	30.52	30.77	31.19	31.53	31.77	31.89	31.88
15	23.68	23.97	24.24	24.52	24.79	25.05	25.31	25.56	25.80	26.04	26.26	26.67	27.03	27.33	27.56	27.73
	29.14	29.49	29.83	30.16	30.49	30.80	30.11	31.40	31.68	31.95	32.20	32.66	33.03	33.32	33.51	33.59

* Reproduced by permission of Professor E. S. Pearson from: "Tables for Testing the Homogeneity of a Set of Estimated Variances," *Biometrika*, 33 (1946), 296–304, by Maxine Merrington and Catharine M. Thompson.

has been developed by Cochran for the case $\nu_1 = \nu_2 = \cdots = \nu_k = \nu$. The procedure in this case is to compute the value

$$g = \frac{\text{largest of the } s_i^2}{\sum_{i=1}^{k} s_i^2}$$

i.e., the ratio of the largest of the sample variances to their total. Tables of values $g_{k,\nu,\alpha}$ for which $Pr(g \geq g_{k,\nu,\alpha}) \sim \alpha$ are given in [19, Chapter 15, Tables 15.1 and 15.2].

Frequently it is possible to examine a group of variance estimates quickly (possibly graphically) by making use of the fact that $\ln s^2$ has approximate variance $2/\nu$. Note that this is essentially the approximation from which we obtain the fact that $2z = \ln s_1^2 - \ln s_2^2$ is approximately normally distributed with variance $2\left(\frac{1}{\nu_1} + \frac{1}{\nu_2}\right)$, or z with variance $\frac{1}{2}\left(\frac{1}{\nu_1} + \frac{1}{\nu_2}\right)$.

EXAMPLE 1. In determining the consistency of tests of tensile strength as a measure of the variation in iron production, 6 bars were poured under as nearly as possible controlled conditions at each of 7 foundries. For each of the foundries the accompanying table gives the type of meehanite metal cast and the type of testing machine used, together with the original results in tons per square inch and the derived mean, variance, and standard deviation.

CONSISTENCY TESTS FOR TENSILE TESTING

Foundry	1	2	3	4	5	6	7
Type meehanite	GD	GC	GA	GD	GC	GD	GB
Testing machine	Avery	Buckton	Avery	Buckton	Buckton (hand operated)	Buckton	Avery
Tensile obtained, tons/sq in.,							
Bar 1	17.70	18.40	25.25	17.00	19.86	17.55	23.00
2	18.00	19.20	26.47	15.75	20.00	17.68	22.70
3	17.93	19.84	25.35	18.90	19.29	17.80	21.80
4	16.63	19.16	23.26	17.50	18.11	17.26	22.60
5	17.06	19.04	24.85	20.00	19.11	17.43	22.00
6	17.46	19.72	22.20	17.70	18.42	17.40	21.70
Average	17.46	19.23	24.56	17.81	19.13	17.52	22.30
$\Sigma_i(x_i - \bar{x})^2$	1.4189	1.3429	12.0919	10.9920	2.8602	0.1950	1.4400
s_i^2	0.2838	0.2686	2.4184	2.1984	0.5720	0.0390	0.2880
s_i	0.53	0.52	1.56	1.48	0.76	0.20	0.54

For these 7 variance estimates, we have

$$s^2 = \frac{1.4189 + 1.3429 + \cdots + 1.4400}{5(7)}$$
$$= 0.8669$$
$$s = 0.93$$

and hence

$$M = 2.30259(35 \log s^2 - 5\Sigma \log s_i^2)$$
$$= 2.30259[35(9.93797 - 10) - 5(67.41561 - 70)]$$
$$= 2.30259(10.75090)$$
$$= 24.75$$
$$c_1 = \frac{7}{5} - \frac{1}{35} = \frac{48}{35} = 1.37$$

Since for $k = 7$ and $c_1 = 1.5$, the critical values of M for $\alpha = 0.05$ and $\alpha = 0.01$ are 13.59 and 18.10, respectively, there is definite indication that the testing variation differs for the different foundries, and the joint estimate $s^2 = 0.8669$ has little meaning for an individual foundry. These differences do not seem to be associated either with the type of testing machine used or the average tensile strength, since the two unusually high variances were obtained on the two different types of testing machine, and are associated with the highest and one of the lower mean tensile strengths

5.7. Sequential Tests

In the previous sections of this chapter we have discussed tests of hypotheses and the calculation of limits based on a sample of n, where the sample size n has been fixed. Beginning with the work of Wald during World War II there has been rapid development of sequential tests, in which the observations are made one by one, and at each stage one of three decisions is made: to accept the hypothesis, reject the hypothesis, or take another observation. The primary advantage of such a procedure, reduction in the average sample size, is due largely to the fact that the decision to accept or reject may become obvious after the first few observations and no additional sampling need be done. The greatest disadvantage is frequently associated with the practical aspects of taking a sequential sample, rather than a sample of n at a given time. This difficulty can sometimes be partially overcome, and the advantages of sequential testing to a large extent retained, by examining successive samples of some fixed size. The present section is intended to be only a brief introduction to the possibilities of sequential testing; for a more complete account the reader is referred to [92].

In order to fix our ideas concerning sequential tests, let us consider the simple case of choosing between the hypotheses $\theta = \theta_0$ and $\theta = \theta_1$, where the distribution of the variable x to be observed is given by $p(x, \theta)$. Then a sequential test will consist of a procedure whereby after any observation, say the mth, we decide on the basis of the sample $x_1, \cdots, x_m$ whether to conclude $\theta = \theta_0$, conclude $\theta = \theta_1$, or take another observation. To protect ourselves against incorrect decisions, we should like to have the probability of concluding $\theta = \theta_1$ when $\theta = \theta_0$ less than some small value α, and similarly the probability of concluding $\theta = \theta_0$ when $\theta = \theta_1$ less than some small value β. α and β are the errors of the first and second kind for the hypothesis $\theta = \theta_0$, and together they define the strength of the sequential procedure used. For procedures with a given strength, that requiring the lowest average sample size is to be preferred.

Most sequential test procedures are based on the sequential probability ratio test. To use this test we compute the probability (or probability density) p_{1m} that the sample $x_1, \cdots, x_m$ came from the distribution $p(x, \theta_1)$ and the probability p_{0m} that it came from $p(x, \theta_0)$. Then the test procedure consists of accepting $\theta = \theta_0$ if

$$\frac{p_{1m}}{p_{0m}} \leq B \tag{a}$$

accepting $\theta = \theta_1$ if

$$\frac{p_{1m}}{p_{0m}} \geq A \tag{b}$$

and taking another observation if

$$B \leq \frac{p_{1m}}{p_{0m}} \leq A \tag{c}$$

Since for $x_1, x_2, \cdots, x_m$ independent we have

$$\begin{aligned} p_{1m} &= p(x_1, \theta_1) \cdot \cdots \cdot p(x_m, \theta_1) \\ p_{0m} &= p(x_1, \theta_0) \cdot \cdots \cdot p(x_m, \theta_0) \end{aligned} \tag{d}$$

it follows that

$$\ln \frac{p_{1m}}{p_{0m}} = \sum_{i=1}^{m} z_i \tag{e}$$

where

$$z_i = \ln \frac{p(x_i, \theta_1)}{p(x_i, \theta_0)} \tag{f}$$

and from (*a*), (*b*), and (*c*) it follows that in terms of the z_i, which are additive and often of simpler form than the corresponding probability ratios, the test becomes

$$\text{accept } \theta = \theta_0 \quad \text{if} \quad \sum_{i=1}^{m} z_i \leq \log B$$

$$(g) \qquad \text{accept } \theta = \theta_1 \quad \text{if} \quad \sum_{i=1}^{m} z_i \geq \log A$$

$$\text{take another observation} \quad \text{if} \quad \log B \leq \sum_{i=1}^{m} z_i \leq \log A$$

This process is repeated at each stage of the test until a decision to accept $\theta = \theta_0$ or $\theta = \theta_1$ is reached; the process is certain to terminate if the observations are independent. For practical purposes, a test of strength approximately (α, β) is obtained by choosing

$$(h) \qquad A = \frac{1 - \beta}{\alpha} \qquad B = \frac{\beta}{1 - \alpha}$$

Note that these limits do not depend on the distribution $p(x, \theta)$.

As an example of the above method let us consider a test that the mean μ of a normal distribution with known variance is less than some hypothetical value μ_0 against the alternative that $\mu \geq \mu_1$. We have seen in Section 5.25 that it suffices to consider the probability of concluding $\mu \geq \mu_1$ when $\mu = \mu_0$, since for $\mu < \mu_0$ this probability can only decrease; similarly, we can define β with respect to the hypothesis $\mu = \mu_1$. Hence the above considerations apply, and we can formulate the proper sequential probability ratio test based on μ_0, μ_1, α, β and the known variance σ^2. In this case we have

$$(i) \qquad p(x, \mu) = \frac{1}{\sigma\sqrt{2\pi}} e^{-\frac{1}{2\sigma^2}(x-\mu)^2}$$

where we are assuming σ^2 known. Since

$$(j) \qquad \ln p(x, \mu) = \ln \frac{1}{\sigma\sqrt{2\pi}} - \frac{1}{2\sigma^2}(x - \mu)^2$$

we have from (f)

$$(k) \qquad \begin{aligned} z_i &= \ln p(x_i, \mu_1) - \ln p(x_i, \mu_0) \\ &= \frac{1}{2\sigma^2}\left[(x_i - \mu_0)^2 - (x_i - \mu_1)^2\right] \\ &= \frac{d}{\sigma}(x_i - \bar{\mu}) \end{aligned}$$

where $d = \dfrac{\mu_1 - \mu_0}{\sigma}$ and $\bar{\mu} = \dfrac{\mu_1 + \mu_0}{2}$. Hence the proper procedure after m observations is, from (g) and (h),

$$\text{conclude } \mu \leq \mu_0 \quad \text{if } \frac{d}{\sigma}\sum_{i=1}^{m}(x_i - \bar{\mu}) \leq \ln\frac{\beta}{1-\alpha}$$

$$(l)\ \text{conclude } \mu \geq \mu_0 \quad \text{if } \frac{d}{\sigma}\sum_{i=1}^{m}(x_i - \bar{\mu}) \geq \ln\frac{1-\beta}{\alpha}$$

$$\text{take another observation} \quad \text{if } \ln\frac{\beta}{1-\alpha} \leq \frac{d}{\sigma}\sum_{i=1}^{m}(x_i - \bar{\mu}) \leq \ln\frac{1-\beta}{\alpha}$$

The inequalities on the right can be reduced to

$$T_m \leq \frac{\sigma}{d}\ln\frac{\beta}{1-\alpha} + m\bar{\mu}$$

$$(m) \qquad T_m \geq \frac{\sigma}{d}\ln\frac{1-\beta}{\alpha} + m\bar{\mu}$$

and

$$\frac{\sigma}{d}\ln\frac{\beta}{1-\alpha} + m\bar{\mu} \leq T_m \leq \frac{\sigma}{d}\ln\frac{1-\beta}{\alpha} + m\bar{\mu},$$

respectively, where $T_m = \sum_{i=1}^{m} x_i$ is the sum of the first m observations.

If for each successive observation we plot T_m against m, then the parallel lines

$$L_0 : T_m = a_0 + bm$$

$$(n)$$

$$L_1 : T_m = a_1 + bm$$

where

$$a_0 = \frac{\sigma}{d}\ln\frac{\beta}{1-\alpha}$$

$$(o) \qquad a_1 = \frac{\sigma}{d}\ln\frac{1-\beta}{\alpha}$$

$$b = \bar{\mu}$$

define regions such that if (m, T_m) falls above L_1 we conclude $\mu \geq \mu_1$; if it falls below L_0 we conclude $\mu \leq \mu_0$; if it falls between L_1 and L_0 we continue making observations. This type of graph, shown in Figure 5.9, is convenient to use in practice. Alternatively we can compute for each value of m from L_0 and L_1 acceptance numbers a_m and rejection numbers r_m for the hypothesis $\mu \leq \mu_0$ with which each successive value of T_m can be compared, the test continuing as long as $a_m < T_m < r_m$.

Note that we have no control over what will happen when $\mu_0 < \mu < \mu_1$, although the probability of acceptance of the hypothesis $\mu \leq \mu_0$ for all values of μ can be computed and together constitute the operating characteristic curve of the test. A simple method of determining this curve roughly is obtained from the fact that the three points $(\mu_0, 1 - \alpha)$, (μ_1, β), and $(\bar{\mu}, \gamma)$, where

$$\gamma = \frac{\ln \dfrac{1-\beta}{\alpha}}{\ln \dfrac{1-\beta}{\alpha} - \ln \dfrac{\beta}{1-\alpha}}$$

lie on the curve, and that the curve must approach 1 and 0 as μ becomes very small or very large, respectively.

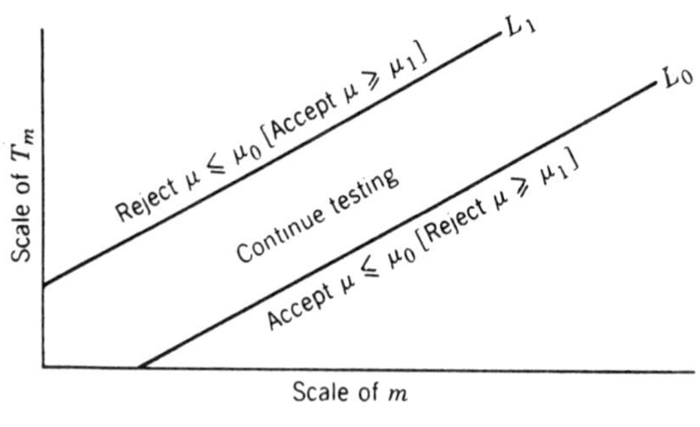

FIG. 5.9.

The above test is typical of the procedures used in sequential testing, although in most of the more practical situations additional approximations are required to overcome the difficulties which arise when the hypotheses to be tested become more complicated and the distribution of the observations less easily handled. Reference will be made in other chapters to places where sequential procedures can be used, and the application of sequential procedures in other situations is discussed in [92] and [93].

EXAMPLE. Let us consider a sequential test for the situation described in Example 1 of Section 5.25. Let us suppose that we are interested in the case where it is required that the pH of the reactant be less than 8.30, and it is important that we reject this hypothesis when the pH is greater than 8.33. Since, using a pH meter, determinations might reasonably be made on a sequence of independent small samples, we can consider the use of a sequential test. Assuming the observations to be normally distributed, with known standard deviation $\sigma = 0.02$, and, choosing $\mu_0 = 8.30$, $\mu_1 = 8.33$, $\alpha = \beta = 0.05$, we have

$$d = \frac{8.33 - 8.30}{0.02} = 1.5 \qquad \bar{\mu} = \frac{8.33 + 8.30}{2} = 8.315$$

$$a_0 = \frac{0.02}{1.5} \ln \frac{0.05}{0.95} = (0.0133)(-2.94) = -0.039$$

$$a_1 = \frac{0.02}{1.5} \ln \frac{0.95}{0.05} = (0.0133)(2.94) = +0.039$$

$$b = 8.315$$

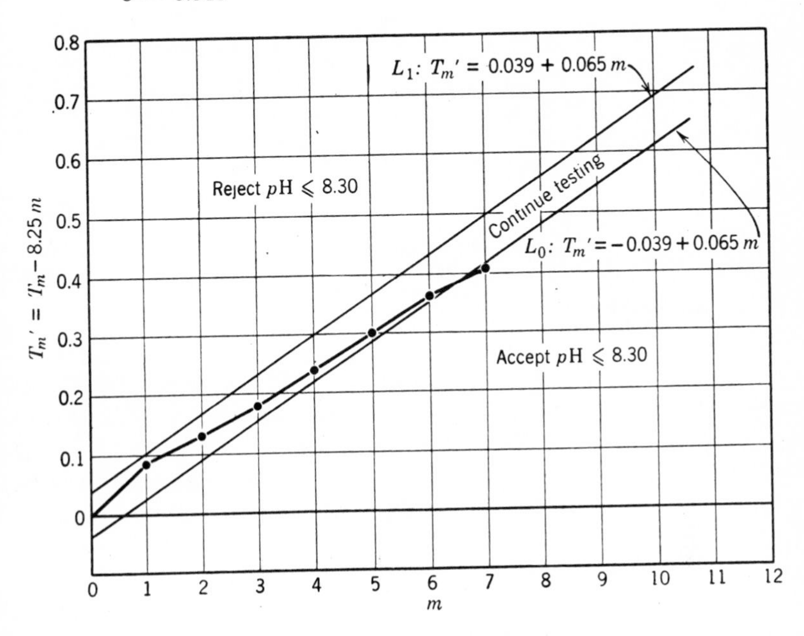

FIG. 5.10. Example of sequential test.

and hence

$$L_0 : T_m = -0.039 + 8.315m$$

$$L_1 : T_m = 0.039 + 8.315m$$

From these equations we can prepare Table 5.11 which gives the acceptance and rejection numbers and also provides space for the observations and their totals, or we can prepare a graph such as that shown in Figure 5.10. For graphical purposes we have subtracted 8.25 from the value of μ, and hence 8.25 should be subtracted from each observation, and $8.25m$ from their total. The 6 observations given in the previous example are shown both in the table and on the chart; no decision has yet been reached on the basis of these 6 observations. However, if we add an additional observation of 8.30, the total falls below the acceptance number, and we accept the material as satisfactory.

TABLE 5.11

m	a_m	x_m	T_m	r_m
1	8.276	8.34	8.34	8.354
2	16.591	8.29	16.63	16.669
3	24.906	8.30	24.93	24.984
4	33.221	8.31	33.24	33.299
5	41.536	8.30	41.54	41.614
6	49.851	8.32	49.86	49.929
7	58.166	8.30	58.16	58.244
8	66.481			66.559
9	74.796			74.874
10	83.111			83.189

5.8. Measurement Errors

Measurement errors are always present in experimental data, although in some cases they may be so small as to be unimportant. In field experimentation the errors in measuring the yield are usually negligible in comparison with the variations introduced by differences in natural soil fertility. In technological applications weighing processes are usually relatively error-free, so that in a performance test on a blast furnace, for example, the errors introduced by the weighing processes could be made very small in comparison with those introduced by variation in performance and by uncertainties in moisture and ash contents. However, in laboratory scale or pilot plant research, and in much physical and chemical research, the situation may be just the opposite; we may have difficulty in developing methods of measurement which are adequate for the purposes at hand. For example, in a pilot plant experiment to determine the effect of small changes in the feed composition on the extraction efficiency, an appreciable portion of our experimental error might be due to the analytical methods used, and, in determining the degree of preferred orientation in a metal by X-ray crystallography, instrumental errors may make an appreciable contribution to the error in the experimental results.

Probably the first requirement that we make of a method of measurement is that it be reproducible, i.e., repeated measurements of the same quantity should not differ too much from one another. The ability of a method of measurement to give reproducible results is referred to as the *precision* of the method; a method of measurement which is precise will repeatedly give results that are close together; a less precise method will give results which are farther apart. The precision of a method of measurement is determined by the scatter, or dispersion, of the results it produces, and hence the estimated variance or standard deviation of a series of repeated measurements of the same quantity is a convenient indication of the precision of the method. It should be noted that this relationship is inverse; high precision is reflected by small variances, low precision by high variances.

The second characteristic of a method of measurement with which we may be concerned is its bias. Bias refers to the systematic errors of measurement which are present to the same extent in each measurement. Typical of this type of error would be those due to loss of precipitate during washing in a gravimetric method of analysis, or incorrect calibration of an instrument. Bias can only be measured with respect to some standard; we could not detect the bias in the instrument except by comparison with some other method of measurement, or the bias in the analysis except by comparison with determinations by a second method, or by an analysis of the wash water. Thus in obtaining the bias of a method we are simply comparing it with another method which we have some reason (physical, chemical, or engineering, but not statistical) to believe is less biased. Fortunately methods of measurement are most often used in experiments involving comparisons, where we are interested in knowing only that the measurements are unbiased relative to one another and sufficiently precise to detect the experimental differences involved. To use an extreme example, if we use one yardstick to measure the kitchen, and a hardware dealer uses a second to measure the linoleum, the linoleum will fit the kitchen, provided that each set of measurements is sufficiently precise and the yardsticks are of the *same* length. The fit will not be affected if each yardstick is 35 in. long. Thus in most experimental work we are interested in methods which are relatively unbiased and are sufficiently precise so that the contribution of the variance in the measurements to the variance of the observations is not excessive. In other cases it is sufficient to calibrate the methods or instruments used against some standard; thus scales calibrated with weights certified by the National Bureau of Standards (again by comparison with standard weights) enable us to make meaningful comparisons of weights obtained at different places and times.

Occasionally it is desirable to obtain a value for some physical constant, which, although it will be expressed in terms of standard fundamental units, is to be determined by an absolute method. Such a procedure is necessary to establish the standards referred to in the previous paragraph if these standards are not fixed by definition, or to determine physical constants such as the velocity of light, the electron charge, or Planck's constant which cannot be fixed by definition. Where these constants are sufficiently important in the structure of present physical theory it is often convenient to suppose that some "true" value of them exists, so as to simplify the processes of reasoning and deduction. Such a concept does not imply that this value can be determined by measurement since, as we have seen, the measuring process which gave the true value as the average value of a large series of measurements would be one which was free from bias, and we can only define bias quantitatively in terms of a known standard. Without the standard we are unable from statistical considerations to identify an unbiased procedure.

The reader may choose, as is common in technological applications, to reject the concept of a true measurable value and substitute a measurable value defined only in terms of the standardized technique used in determining the constant. In this way an abrasion index for rubber or a volatile matter content of a coal is defined. Alternatively it may be argued that, since the true value would be determined by a method which was free from bias, and since sources of bias have in the past been identified from non-statistical considerations and then eliminated, it is not too much to hope that this process will continue to a goal which is not defined by the sequence of developments.

However we choose to think about this matter, the fact remains that our only confidence in present measurements must come from our ability to eliminate known sources of error and to use techniques which as far as we know will lead to an unbiased result, or it must come from the knowledge that our measurements are carried out by a technique which when properly executed leads by definition to the standard result. In the first case the method may be said to be the most accurate now available, and in the second the method is accurate by definition. "Accurate" used in this sense is synonymous with "unbiased," and implies nothing about the precision of the method.

APPENDIX 5A

Maximum Likelihood Estimation

5A.1. Method of Maximum Likelihood

The method of maximum likelihood, due to R. A. Fisher, is a means of obtaining from a sample, point estimates of population parameters which are, under fairly general conditions:

(1) *Consistent*, i.e., as the size of the sample is increased the estimate converges stochastically to the desired population value as a limit.

(2) *Asymptotically efficient*, i.e., as the size of the sample is increased the variance of the estimate approaches the minimum variance for any consistent estimate of the given parameter.

(3) *Approximately normally distributed*, i.e., the distribution of the estimate approaches normality as n becomes large.

The first condition does not imply that these estimates are unbiased; they may be seriously biased in small samples. Also the convergence to normality of point (3) may be quite slow.

The method of finding maximum likelihood estimates may be stated as follows: suppose we have a sample $x_1, \cdots, x_n$ from a distribution $f(x, \theta)$, depending on the single parameter θ. The joint distribution of the n independent sample values would then be

$$(a) \qquad \prod_{i=1}^{n} f(x_i, \theta)$$

This joint probability distribution, considered as a function of θ, is called the likelihood function of θ. Now, if there exists a value of θ, say $\theta = \hat{\theta}$, which maximizes this function, then $\hat{\theta}$ is consistent, asymptotically efficient, and for large samples approximately normally distributed with

$$(b) \qquad \text{ave}\,(\hat{\theta}) = \theta$$

$$\text{var}\,(\hat{\theta}) = \frac{1}{n\ \text{ave}\left(\dfrac{\partial \log f(x, \theta)}{\partial \theta}\right)^2}$$

and is called a maximum likelihood estimate of θ. In particular, var $(\hat{\theta})$ is the variance required of an efficient estimate. Note that this procedure is essentially that of choosing as our estimate that value of the parameter θ for which the given sample is most likely to occur.

The proof of the above statements is beyond the scope of this book. It requires the assumptions that the first, second, and third derivatives of $f(x, \theta)$ with respect to θ exist, and that

$$\frac{\partial \ln f(x, \theta)}{\partial \theta}$$

has mean zero and finite second and third moments. Frequently the solution for $\hat{\theta}$, which can be carried out by the usual methods of differentiation, considering the x_i, $i = 1, \cdots, n$, as constants, is more conveniently obtained by maximizing

$$L(\theta) = \ln \prod_{i=1}^{n} f(x_i, \theta) = \sum_{i=1}^{n} \ln f(x_i, \theta)$$

which attains its maximum at the same value as $\prod_{i=1}^{n} f(x_i, \theta)$.

The method of maximum likelihood can be extended to 2 or more parameters. Let us consider the case of 2 parameters θ_1 and θ_2. Then

$$\begin{aligned} L(\theta_1, \theta_2) &= \ln \prod_{i=1}^{n} f(x_i, \theta_1, \theta_2) \\ &= \sum_{i=1}^{n} \ln f(x_i, \theta_1, \theta_2) \\ &= \sum_{i=1}^{n} l(x_i, \theta_1, \theta_2) \end{aligned}$$

where $l(x, \theta_1, \theta_2) = \log f(x, \theta_1, \theta_2)$. Under conditions which require that

$$\text{ave}\left(\frac{\partial l}{\partial \theta_1}\right) = \text{ave}\left(\frac{\partial l}{\partial \theta_2}\right) = 0$$

and that any moments of the second or third order of these derivatives be finite, the values $\theta_1 = \hat{\theta}_1$ and $\theta_2 = \hat{\theta}_2$ which maximize $L(\theta_1, \theta_2)$ are *joint maximum likelihood estimates* which have, for sufficiently large samples, a joint bivariate normal distribution with

$$\begin{aligned} \text{ave}(\hat{\theta}_1) &= \theta_1 \\ \text{ave}(\hat{\theta}_2) &= \theta_2 \\ \text{var}(\hat{\theta}_1) &= \frac{1}{n} \frac{\text{ave}\left(\dfrac{\partial l}{\partial \theta_2}\right)^2}{D} \end{aligned}$$

$$\text{var}\,(\hat{\theta}_2) = \frac{1}{n}\,\frac{\text{ave}\left(\dfrac{\partial l}{\partial \theta_1}\right)^2}{D}$$

$$\text{cov}\,(\hat{\theta}_1, \hat{\theta}_2) = -\frac{1}{n}\,\frac{\text{ave}\left(\dfrac{\partial l}{\partial \theta_1}\cdot\dfrac{\partial l}{\partial \theta_2}\right)}{D}$$

where

$$D = \begin{vmatrix} \text{ave}\left(\dfrac{\partial l}{\partial \theta_1}\right)^2 & \text{ave}\left(\dfrac{\partial l}{\partial \theta_1}\cdot\dfrac{\partial l}{\partial \theta_2}\right) \\ \text{ave}\left(\dfrac{\partial l}{\partial \theta_1}\cdot\dfrac{\partial l}{\partial \theta_2}\right) & \text{ave}\left(\dfrac{\partial l}{\partial \theta_2}\right)^2 \end{vmatrix}$$

5A.2. Examples of Determination of Maximum Likelihood Estimates

Let us first consider the maximum likelihood estimate for the single parameter λ of the Poisson distribution

$$f(x, \lambda) = \frac{e^{-\lambda}\lambda^x}{x!}$$

We have for the joint distribution of a sample $x_1, \cdots, x_n$

$$f(x_1, \cdots, x_n, \lambda) = \prod_{i=1}^{n} \frac{e^{-\lambda}\lambda^{x_i}}{x_i!}$$

and hence

$$L(\lambda) = \ln \prod_{i=1}^{n} \frac{e^{-\lambda}\lambda^{x_i}}{x_i!}$$

$$= \sum_{i=1}^{n} (-\lambda + x_i \ln \lambda - \ln x_i!)$$

Alternatively, we could have noted that

$$l(x, \lambda) = \ln f(x, \lambda) = -\lambda + x \ln \lambda - \ln x!$$

and

$$L(\lambda) = \sum_{i=1}^{n} \ln l(x_i, \lambda) = \sum_{i=1}^{n} (-\lambda + x_i \ln \lambda - \ln x_i!)$$

Taking the derivative of $L(\lambda)$ with respect to λ, and equating to 0, we have

$$\Sigma_i \left(-1 + \frac{x_i}{\hat{\lambda}}\right) = 0$$

or

$$\Sigma_i (x_i - \hat{\lambda}) = 0$$

i.e.

$$\Sigma_i x_i - n\hat{\lambda} = 0$$

hence

$$\hat{\lambda} = \frac{\Sigma_i x_i}{n} = \bar{x}$$

i.e., the sample mean is the maximum likelihood estimate of λ in this case.

Also we have

$$\frac{\partial \ln f(x, \lambda)}{\partial \lambda} = \frac{\partial l}{\partial \lambda} = -1 + \frac{x}{\lambda}$$

and hence, since, from Section 4.63, ave $(x) = \lambda$ and ave $(x^2) = \lambda^2 + \lambda$, we have

$$\text{ave}\left(\frac{\partial l}{\partial \lambda}\right) = \text{ave}\left(-1 + \frac{x}{\lambda}\right) = -1 + \frac{1}{\lambda}\text{ ave }(x) = 0$$

which satisfies the condition ave $\left(\frac{\partial l}{\partial \lambda}\right) = 0$, and

$$\text{ave}\left(\frac{\partial l}{\partial \lambda}\right)^2 = \text{ave}\left(-1 + \frac{x}{\lambda}\right)^2$$

$$= 1 - 2 \cdot \frac{1}{\lambda}\text{ ave }(x) + \frac{1}{\lambda^2}\text{ ave }(x^2)$$

$$= 1 - 2 + \frac{1}{\lambda^2}(\lambda^2 + \lambda)$$

$$= \frac{1}{\lambda}$$

Hence, for large n, $\bar{x}$ is normally distributed with mean λ and variance $\dfrac{1}{n \cdot \dfrac{1}{\lambda}} = \dfrac{\lambda}{n}$. Since for large n we have seen (Section 4.64) that the Poisson distribution approaches a normal distribution with mean and variance both λ, this result is to be expected.

As a second example, we consider the estimation of the two parameters μ and σ^2 of the normal distribution from a random sample $x_1, \cdots, x_n$. In this case we have from (*a*) of Section 4.41

$$l(x, \mu, \sigma^2) = \ln f(x, \mu, \sigma^2)$$

$$= -\frac{1}{2}\ln 2\pi - \frac{1}{2}\ln \sigma^2 - \frac{1}{2\sigma^2}(x - \mu)^2$$

and
$$L(\mu, \sigma^2) = \Sigma_i l(x_i, \mu, \sigma^2)$$
$$= -\frac{n}{2} \ln 2\pi - \frac{n}{2} \ln \sigma^2 - \frac{1}{2\sigma^2} \Sigma_i (x_i - \mu)^2$$

Taking the partial derivatives of $L(\mu, \sigma^2)$ with respect to μ and σ^2 and equating to 0, we obtain the simultaneous equations

$$-\frac{1}{\sigma^2} \Sigma_i \cdot (x_i - \mu) = 0$$
$$-\frac{n}{2\sigma^2} + \frac{1}{2(\sigma^2)^2} \Sigma_i (x_i - \mu)^2 = 0$$

from which we obtain

$$\hat{\mu} = \bar{x}$$
$$\hat{\sigma}^2 = \frac{1}{n} \Sigma_i (x_i - \bar{x})^2$$

as the maximum likelihood estimates of μ and σ^2, respectively. Note that $\hat{\sigma}^2$ is not an unbiased estimate, since, as we have seen, ave $(\hat{\sigma}^2) = \frac{n}{n-1}$ ave $(s^2) = \frac{n}{n-1} \sigma^2$; however, this bias disappears as n becomes large.

Since
$$\frac{\partial l}{\partial \mu} = -\frac{1}{\sigma^2} (x - \mu)$$
$$\frac{\partial l}{\partial \sigma^2} = -\frac{1}{2\sigma^2} + \frac{1}{2(\sigma^2)^2} (x - \mu)^2$$

we have
$$\text{ave}\left(\frac{\partial l}{\partial \mu}\right) = -\frac{1}{\sigma^2} \text{ ave } (x - \mu) = 0$$
$$\text{ave}\left(\frac{\partial l}{\partial \sigma^2}\right) = -\frac{1}{2\sigma^2} + \frac{1}{2(\sigma^2)^2} \text{ ave } (x - \mu)^2$$
$$= -\frac{1}{2\sigma^2} + \frac{1}{2(\sigma^2)^2} \cdot \sigma^2$$
$$= 0$$

which shows that the stated necessary conditions are satisfied to at least this extent. Also, using the formula of Section 4.41,

$$\text{ave}\left(\frac{\partial l}{\partial \mu}\right)^2 = \frac{1}{(\sigma^2)^2} \text{ ave } (x - \mu)^2$$
$$= \frac{\sigma^2}{(\sigma^2)^2} = \frac{1}{\sigma^2}$$

$$\text{ave}\left(\frac{\partial l}{\partial \mu}\cdot\frac{\partial l}{\partial^2 \sigma}\right) = \frac{1}{2(\sigma^2)^2}\text{ ave }(x-\mu) - \frac{1}{2(\sigma^2)^3}\text{ ave }(x-\mu)^3$$

$$= 0$$

$$\text{ave}\left(\frac{\partial l}{\partial \sigma^2}\right)^2 = \frac{1}{4(\sigma^2)^2} - \frac{1}{2(\sigma^2)^3}\text{ ave }(x-\mu)^2 + \frac{1}{4(\sigma^2)^4}\text{ ave }(x-\mu)^4$$

$$= \frac{1}{4\sigma^4} - \frac{\sigma^2}{2\sigma^6} + \frac{3\sigma^4}{4\sigma^8}$$

$$= \frac{1}{2\sigma^4}$$

and therefore $D = 1/2\sigma^6$. Hence

$$\text{var }(\hat{\mu}) = \frac{1}{n}\cdot\frac{1/2\sigma^4}{1/2\sigma^6} = \frac{\sigma^2}{n}$$

$$\text{var }(\hat{\sigma}^2) = \frac{1}{n}\cdot\frac{1/\sigma^2}{1/2\sigma^6} = \frac{2\sigma^4}{n}$$

$$\text{cov }(\hat{\mu}, \hat{\sigma}^2) = 0$$

Thus in large samples the maximum likelihood estimates $\bar{x}$ and $\hat{\sigma}^2$ of μ and σ^2 will be independent and normally distributed with these variances. As we have seen, the property of independence also holds for any value of n, as does the normality of the distribution of $\bar{x}$ with mean μ and variance σ^2/n. However, not only is the estimate $\hat{\sigma}^2$ biased for small n, but it is approximately normally distributed only for quite large n.

5A.3. Relation of the Principle of Least Squares to Maximum Likelihood

Given a series of sample values x_i whose average value is a function $\mu(\theta_1, \cdots, \theta_m)$ of m unknown parameters, we frequently choose as estimates of these parameters those values which minimize the sum of squares

$$Q = \Sigma_i\{x_i - \mu(\theta_1, \cdots, \theta_m)\}^2$$

of the deviations of the sample values from their average value. For example, $\bar{x}$ is a least squares estimate of ave $(x) = \mu$, since, as shown in Section 2.33, the sum of squares of the deviations about the arithmetic mean $\bar{x}$ is a minimum.

If we include the additional restrictions that the x_i be normally distributed with average value $\mu(\theta_1, \cdots, \theta_m)$ and variance σ^2, then the least squares estimates of $\theta_1, \cdots, \theta_m$ are also maximum likelihood estimates. For in this case we have

$$L(\theta_1, \cdots, \theta_m) = -\frac{n}{2} \ln 2\pi - \frac{n}{2} \ln (\sigma^2) - \frac{1}{2\sigma^2} \Sigma_i \{x_i - \mu(\theta_1, \cdots, \theta_m)\}^2$$

$$= -\frac{n}{2} \ln 2\pi - \frac{n}{2} \ln (\sigma^2) - \frac{Q}{2\sigma^2}$$

and

$$\frac{\partial L}{\partial \theta_1} = \frac{1}{2\sigma^2} \cdot \frac{\partial Q}{\partial \theta_1}$$

$$\frac{\partial L}{\partial \theta_m} = \frac{1}{2\sigma^2} \cdot \frac{\partial Q}{\partial \theta_m}$$

$$\frac{\partial L}{\partial \sigma^2} = -\frac{n}{2\sigma^2} + \frac{Q}{2(\sigma^2)^2}$$

When these derivatives are equated to 0, the first m equations are identical with those for minimizing Q, i.e., determining $\hat{\theta}_1, \cdots, \hat{\theta}_m$ by least squares. Also in this case the maximum likelihood estimate of σ^2 is

$$\hat{\sigma}^2 = \frac{Q}{n}$$

As before, this estimate is in general biased.

Throughout the discussion of regression in Chapter 6, the least squares estimates obtained are also maximum likelihood estimates.

CHAPTER 6

Relationships between Variables

6.1. Introduction

6.11. Pairs of Measurements or Observations

The problems discussed in the preceding chapter were mostly concerned with series of measurements or observations of a single characteristic of objects drawn at random from one or more populations. In this chapter we shall concern ourselves with some of the problems that arise when two (or more) observations or measurements are made on a given object and as a consequence we obtain a sample consisting of pairs (or triplets, or greater numbers) of values. Correlation problems are concerned with the degree of dependence between these pairs of values, whereas regression problems are concerned with the nature of the relationship between them if any exists.

As we have seen in Chapter 2, the first step in the analysis of pairs of measurements is to plot them. If the relationship between the measurements is very nearly exact, we may be able to draw or sketch the desired relationship very easily. For example, Table 6.1 gives the data obtained

TABLE 6.1

CONCENTRATION OF STANDARD SOLUTION: 1.052 mg CH_2O/ml
(Data courtesy of K. H. Roberts, Princeton University)

Run	ml CH_2O Sol.	Optical Density
1	0.1	0.086
2	0.3	0.269
3	0.5	0.445
4	0.6	0.538
5	0.7	0.626
6*	0.9	0.782

* This point not used in determining standard curve.

in preparing a standard curve for the determination of formaldehyde by the addition of chromatropic acid and concentrated sulphuric acid and the reading of the consequent purple color on a Beckman Model DU Spectrophotometer at 570 mμ. These points are plotted in Figure 6.1, and a straight line has been drawn through them as closely as possible by eye, run 6 being omitted as indicated. Similarly, Table 6.2 gives data

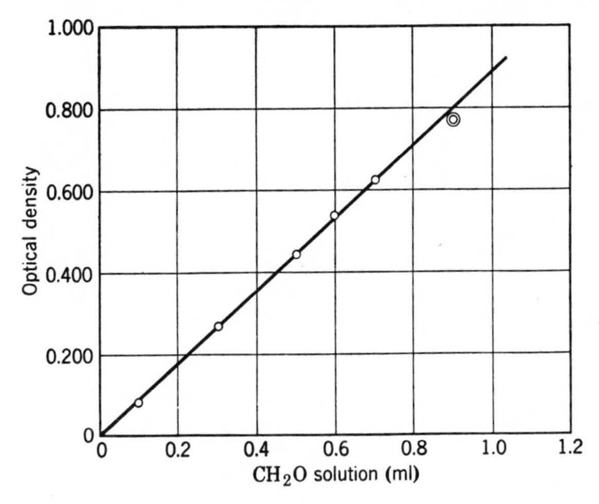

FIG. 6.1. Concentration of CH_2O solution: 1.052 mg/ml.

TABLE 6.2

(Data from Wsislawski, *The Chemical Analyst*, *35*, No. 4)

Concentration (mg/ml)	Colorimeter Reading
40	69
50	175
60	272
70	335
80	390
90	415

obtained from an experiment to study the relationship between the amount of β-erythroidine in an aqueous solution and the colorimeter reading of the turbidity. In this case the relationship is obviously not linear; however, it is still fairly easy to sketch a smooth curve through these points. Such a curve is shown in Figure 6.2.

In the above cases a graphical representation of the relationship of the observed y's to the observed x's is easily obtained. Actually in these instances the lines or curves drawn by different people would probably

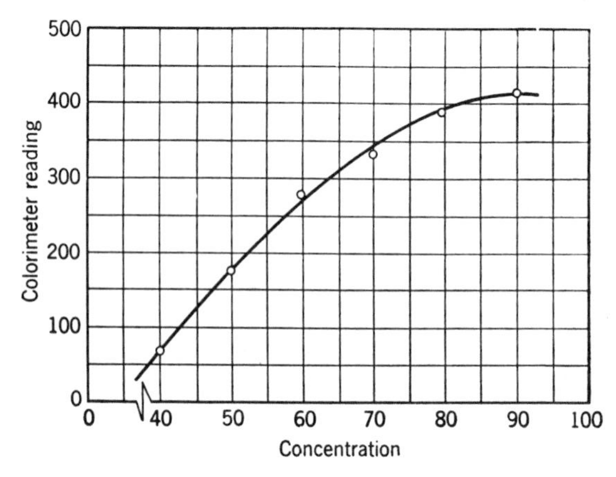

FIG. 6.2.

TABLE 6.3

CARBON CONTENT OF 36 SAMPLES OF BALL CLAY*

x = carbon determined by combustion.

y = Carboniferous material from rational analysis.

y	x	y	x	y	x	y	x	y	x	y	x
2.46	1.53	0.64	0.25	−0.01	0.14	6.14	4.18	9.68	6.55	0.35	0.16
1.54	0.87	0.78	0.29	4.53	2.98	0.52	0.22	4.08	2.54	7.49	5.06
0.70	0.28	0.12	0.12	9.94	6.84	0.40	0.38	2.80	1.43	1.41	0.86
−0.40	0.27	2.36	1.50	3.68	2.15	0.46	0.24	3.93	2.74	−0.50	0.16
4.82	3.07	2.14	1.31	1.84	1.35	2.80	1.79	8.22	6.08	15.89	11.43
0.30	0.25	0.08	0.31	0.97	0.40	2.09	0.58	0.28	0.75	0.18	0.19

* Private communication from the Director of the British Ceramic Research Association.

be very nearly the same, and all would be fairly close to the true relationship. The situation is not always so simple. Consider the data given in Table 6.3 which comprise the results obtained from measurements of the carbon content of 36 samples of ball clays from South Devon.

The carbon content was determined by the classical combustion method (x) and by a "difference" method from the ultimate analysis of the clays (y). These data, plotted in Figure 6.3, indicate a clear linear relationship between the two series of measurements, but the exact slope of the line and its intercept on the y axis are uncertain. Since the difference estimate is at best a measure of the carboniferous material in the clay, there is no

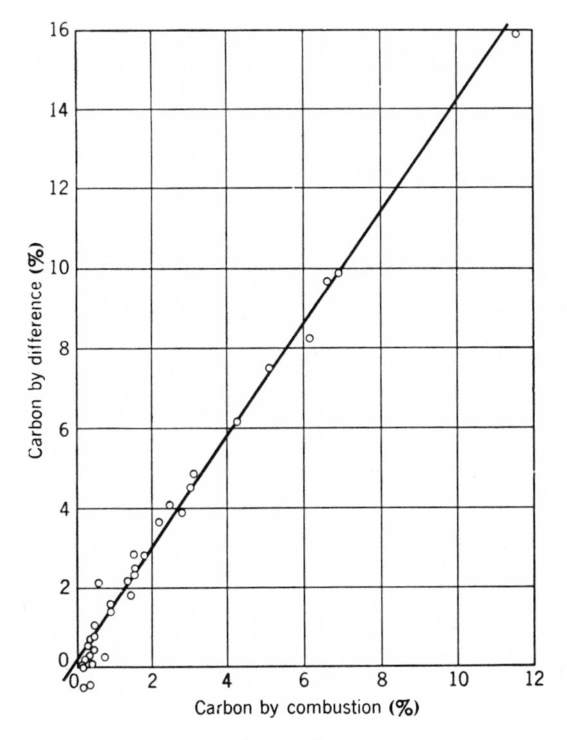

FIG. 6.3.

theoretical basis for anticipating a unit slope in this case. The closeness of the relationship is nevertheless of interest since it indicates that within this region the carbon content of the carboniferous material is approximately constant, and that a good approximation to the carbon content of such a clay may be obtained from the rational analysis.

As a final example we consider a less certain relationship. Table 6.4 gives the result of a certain physical measurement on a series of steel bars, along with their carbon content. The graph of these data is shown in Figure 6.4. In this case we should not only hesitate to draw any curve representing the relationship between the test result and the carbon content but might well ask whether any relationship exists at all.

TABLE 6.4

(Data given by H. S. O'Connor, *Industrial Quality Control,* Vol. IV, No. 2)

Percent Carbon	Test Result
0.07	62.8
0.09	73.5
0.09	68.7
0.10	73.5
0.10	72.7
0.10	69.9
0.11	71.6
0.12	75.9
0.13	71.6

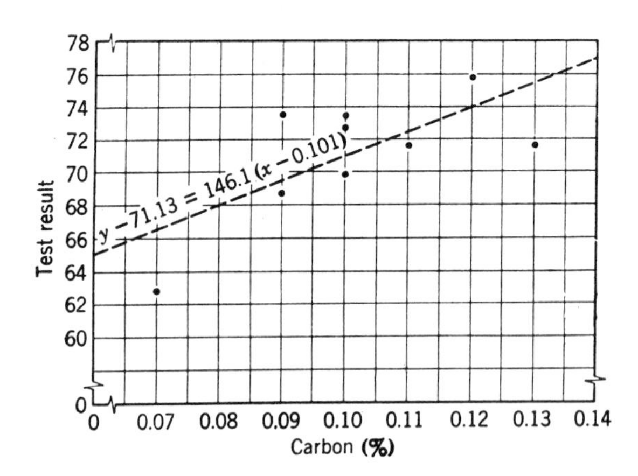

FIG. 6.4.

In the last two examples we need a more systematic method of determining the relationship, if any, that exists between the values of y and the values of x. As in previous chapters we shall proceed by constructing a model of the manner in which these pairs of measurements arose, and use this model to develop quantitative methods.

6.12. Ways in which Pairs of Measurements Can Occur

We shall wish to distinguish carefully between two ways in which pairs of values (x_i, y_i) can arise. The first way is described theoretically by considering one value, usually y_i, as an observation on a random variable, and the second, x_i, as some known constant associated with this random variable. Technical interest will usually center in the nature of the relationship between the average value of the random variable y_i and the associated value x_i, as for instance in the following examples:

(1) The calibration of an inferential measurement against a series of accepted standards. This situation arises frequently not only in the calibration of instruments but also in the comparison of analytical procedures, where in order to give meaning to the results obtained from a given test method we obtain a series of observations y_i on samples of which the compositions x_i are known accurately. In this instance the accuracy of the values x_i may be due to the fact that the sample was deliberately constituted from pure materials by a gravimetric method of high accuracy.

(2) Investigational work in which time is one of the variables. In this instance the time variable can frequently be measured to a sufficient degree of accuracy so that it may be regarded, relatively at least, as a known constant. Thus in investigating the effect of arsenical impurities in the gas stream upon a contact acid platinum catalyst we might carry out an extended test on a sample of catalyst under standardized conditions. Analyses of gas samples y_i, withdrawn at regular intervals would provide a measure of the conversion efficiency of the catalyst after various fixed periods of exposure x_i.

(3) Circumstances in which one measurement is repeated many times for a single associated observation of the second variable. These circumstances frequently occur in industrial plant investigations when a large number of measurements of a carefully controlled temperature, or reflux ratio, or feed material composition, may be available for each associated observation of the product quality. The mean value of the first factor may be regarded as the known constant x_i; the associated observation of the product quality is represented by the random variable y_i.

This model involving a fixed and certain value x_i will never represent the true state of affairs in practice, where no observation is completely devoid of uncertainty. Nevertheless, in many circumstances it is a good approximation to the conditions of observation.

The second situation in which we shall be interested is that in which the observations x_i and y_i each represent measurements of random variables associated with different characteristics of the same object. In this instance we may be interested in the nature of the relationship between x_i and y_i or in the extent of this interrelationship. The distinction between this case and the previous one, though clear enough from the theoretical viewpoint, is much less distinct in the technical data to which the methods are applied. A number of examples of instances in which the problems are clearly those of the latter type may, however, be quoted:

(1) Two laboratory tests are carried out on each of a series of samples

of a single material. If the material is a technical substance such as rubber or coal the laboratory tests may be empirical in the sense of determining some criterion defined in terms of the test result, the abrasion resistance of the rubber, the coking power of the coal. We usually want to estimate from our data the relationship between the estimates of this property given by the two tests. We might also wish to know the extent of the relationship between x_i and y_i in order to determine to what extent the tests were measuring the same fundamental property.

(2) In each of a series of pilot plant runs under controlled conditions the proportion of two impurities in the product is determined by analysis. In this case we should probably be interested in the association between the proportion of the two impurities, since technically this might be taken to indicate whether a common cause for these fluctuations should be sought.

(3) In a unit in which the properties of the raw material varied randomly, a series of tests on this material would normally be made before passing it to the process. We might well be interested in the extent of the relationship between individual laboratory tests and those aspects of the product quality which were of economic importance, since any test which showed no significant relationship could be discontinued, or carried out less frequently, with economic advantage.

Although the mechanical methods of analysis are similar for these two types of problem, the underlying assumptions are completely different, and we shall consider them separately. In practice the similarity of the method of analysis frequently leads to the use of either method without a critical examination of the relationship between the observations and the assumptions made in employing the theory.

6.2. Linear Regression with One Dependent Variable

6.21. Statistical Model for Linear Regression

The simplest statistical model we can use in the first case is obtained by assuming that each y_i is an observation on a random variable y which is normally distributed with constant variance σ^2 and mean $\alpha + \beta x_i$. This situation is illustrated in Figure 6.5 which indicates that for each of the fixed values x_i there is a normally distributed population of observations y_i. It is assumed that the variance of each of these distributions is the same, and that the average values of the distributions satisfy the relationship

$$y = \alpha + \beta x$$

Using the available experimental data, we wish to test this model of the population and obtain estimates of α and β. For example, for the data of Table 6.3 we would assume that the possible observations of the percentage of carboniferous material calculated from the ultimate analysis, which were associated with a fixed carbon content x_i determined by the

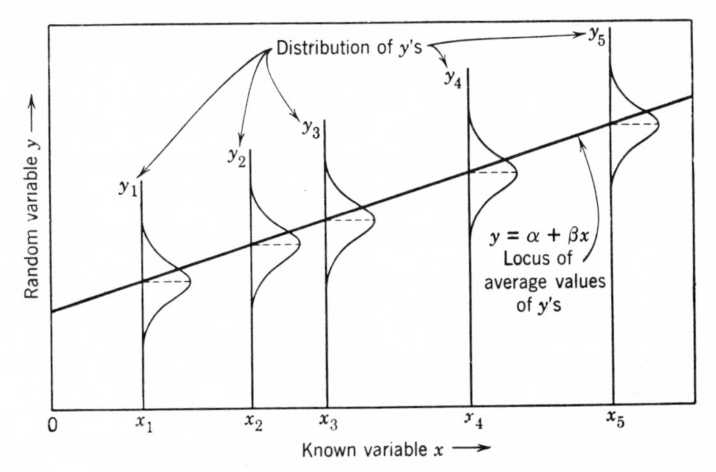

FIG. 6.5. Graphical representation of model for simple linear regression.

classical method, were distributed about a mean value $\alpha + \beta x_i$ with a variance σ^2 independent of x_i.

6.22. Estimation of α and β

In this instance maximum likelihood estimates of α and β (see Appendix 5A) can be obtained by the method of least squares, since the observations y_i are assumed to be normally distributed about $\mu(\alpha, \beta) = \alpha + \beta x_i$ with constant variance σ^2. Hence we wish to minimize

$$(a) \qquad Q = \Sigma(y_i - \alpha - \beta x_i)^2$$

i.e., the sum of squares of the deviations of the observations y_i from their mean values $\alpha + \beta x_i$. To obtain the estimates a and b which minimize the above expression we determine $\partial Q/\partial\alpha$ and $\partial Q/\partial\beta$, set the resulting expressions equal to zero, and solve simultaneously for a and b. Since the derivative of a sum is the sum of the derivatives, we obtain

$$\frac{\partial Q}{\partial \alpha} = -2\Sigma(y_i - \alpha - \beta x_i)$$

and

$$\frac{\partial Q}{\partial \beta} = -2\Sigma x_i(y_i - \alpha - \beta x_i)$$

Setting these expressions equal to zero, at the same time replacing the true values α and β by the estimates a and b, performing the indicated summations, and rearranging we obtain the equations:

$$na + b\Sigma x_i = \Sigma y_i \tag{b}$$
$$a\Sigma x_i + b\Sigma x_i^2 = \Sigma x_i y_i$$

These are frequently known as the normal regression equations.

Since the quantities n, Σx_i, Σx_i^2, $\Sigma x_i y_i$, and Σy_i can be computed from the sample values, these equations can be solved for the quantities a and b. Using the method of determinants, we have immediately

$$b = \frac{\begin{vmatrix} n & \Sigma y_i \\ \Sigma x_i & \Sigma x_i y_i \end{vmatrix}}{\begin{vmatrix} n & \Sigma x_i \\ \Sigma x_i & \Sigma x_i^2 \end{vmatrix}} = \frac{n\Sigma x_i y_i - \Sigma x_i \Sigma y_i}{n\Sigma x_i^2 - (\Sigma x_i)^2}$$

$$= \frac{\Sigma(x_i - \bar{x})(y_i - \bar{y})}{\Sigma(x_i - \bar{x})^2} \tag{c}$$

and from the first equation

$$a = \frac{\Sigma y_i - b\Sigma x_i}{n} = \bar{y} - b\bar{x} \tag{d}$$

The quantity $\tilde{y}_0 = a + bx_0$ is the best estimate of the average value $\alpha + \beta x_0$ of the random variable y associated with a given x_0. Note that, since $a = \bar{y} - b\bar{x}$, we can write

$$\tilde{y}_0 = \bar{y} - b\bar{x} + b\bar{x}_0 = \bar{y} + b(x_0 - \bar{x}) \quad \text{or} \quad \tilde{y}_0 - \bar{y} = b(x_0 - \bar{x}) \tag{e}$$

EXAMPLE 1. As an illustration we consider the data of Table 6.3. We take the carbon contents determined by the combustion method to be the known observations, since this method is of high precision, whereas the carboniferous material is computed from the results of a number of less precise analytical determinations by a method involving assumptions which are known to be only approximately correct. In fact the relationship for these data is probably better than would have been anticipated from technical considerations.

The following totals are first computed:

$$n = 36 \quad \Sigma y = 102.71 \quad \Sigma x = 69.25$$
$$\Sigma y^2 = 826.6842 \quad \Sigma xy = 510.8425$$
$$\Sigma x^2 = 354.0245$$

We then evaluate b determinantally as

$$b = \frac{\begin{vmatrix} 36 & 102.71 \\ 69.25 & 510.8425 \end{vmatrix}}{\begin{vmatrix} 36 & 69.25 \\ 69.25 & 354.0245 \end{vmatrix}} = \frac{11{,}277.6625}{7{,}949.3195} = 1.418695$$

and a as

$$a = \frac{102.71 - (1.418695)(69.25)}{36} = 0.124038$$

Alternatively we can transform our data to observations about the sample means $\bar{y}$ and $\bar{x}$, writing

$$S(x) = \Sigma_i(x_i - \bar{x}) \qquad = 0$$

$$S(y) = \Sigma_i(y_i - \bar{y}) \qquad = 0$$

$$S(y^2) = \Sigma_i(y_i - \bar{y})^2 \qquad = \Sigma_i(y_i)^2 - \frac{(\Sigma_i y_i)^2}{n}$$

$$= 826.6842 - 293.0373 = 533.6469$$

$$S(x^2) = \Sigma_i(x_i - \bar{x})^2 \qquad = \Sigma_i(x_i)^2 - \frac{(\Sigma_i x_i)^2}{n}$$

$$= 354.0245 - 133.2101 = 220.8144$$

$$S(xy) = \Sigma_i(x_i - \bar{x})(y_i - \bar{y}) = \Sigma_i(x_i y_i) - \frac{\Sigma x_i \Sigma y_i}{n}$$

$$= 510.8425 - 197.5741 = 313.2684$$

Then $$b = \frac{S(xy)}{S(x^2)} = \frac{313.2684}{220.8144} = 1.418695$$

and $$a = \frac{102.71 - (1.418695)(69.25)}{36} = 0.124038$$

as before. In either case the final equation becomes

$$y = 0.1240 + 1.4187x$$

for x values ranging approximately from 0 to 10% carbon content.

An estimate of the variance σ^2 of the observations y_i is given by

$$s^2_{y \cdot x} = \frac{\Sigma(y_i - \tilde{y}_i)^2}{n - 2} \tag{f}$$

where $\Sigma(y_i - \tilde{y}_i)^2 = \Sigma(y_i - a - bx_i)^2$ is the actual minimum sum of squares of the deviations corresponding to the estimates a and b. The division by $n - 2$ in this instance is due to the necessity of correcting for bias in estimating σ^2. Since both of the estimates a and b depend on the observations y_i, 2 linear restrictions are placed on these observations, and $\Sigma(y_i - \tilde{y}_i)^2$ can be expressed as the sum of squares of $n - 2$ independent linear combinations with zero mean and variance σ^2. From this it follows that $\dfrac{(n-2)s^2_{y\cdot x}}{\sigma^2}$ has the χ^2-distribution with $n - 2$ degrees of freedom.

The sum of squares of the deviations can be written

$$\begin{aligned}\Sigma(y_i - \tilde{y}_i)^2 &= \Sigma(y_i - a - bx_i)^2 \\ &= \Sigma(y_i - a - bx_i)(y_i - a - bx_i) \\ &= \Sigma y_i(y_i - a - bx_i) - a\Sigma(y_i - a - bx_i) \\ &\qquad - b\Sigma x_i(y_i - a - bx_i)\end{aligned}$$

The summations in the last two of the three right-hand forms are identical with those which were placed equal to zero in order to solve for the estimates a and b. Hence we have

$$(g)\qquad \begin{aligned}\Sigma(y_i - a - bx_i)^2 &= \Sigma y_i(y_i - a - bx_i) \\ &= \Sigma y_i^2 - a\Sigma y_i - b\Sigma x_i y_i\end{aligned}$$

This represents the most convenient form for computing the sum of squares of the deviations $y_i - \tilde{y}_i$ directly from the original data.

It can also be shown, using (e), that

$$(h)\qquad \Sigma(y_i - \tilde{y}_i)^2 = \Sigma(y_i - \bar{y})^2 - b^2\Sigma(x_i - \bar{x})^2$$

an identity which will be of importance in later considerations.

EXAMPLE 2. Continuing the analysis of Example 1, we have

$$\begin{aligned}\Sigma_i(y_i - \tilde{y}_i)^2 &= \Sigma y_i^2 - a\Sigma y_i - b\Sigma x_i y_i \\ &= 826.684 - 12.741 - 724.732 \\ &= 89.211\end{aligned}$$

$$s^2_{y\cdot x} = \frac{89.211}{34} = 2.624 \qquad s_{y\cdot x} = 1.620$$

or, if the second method were employed for the original calculation,

$$\begin{aligned}\Sigma_i(y_i - \tilde{y}_i)^2 &= S(y^2) - b^2 S(x^2) \\ &= 533.647 - 444.432 \\ &= 89.215\end{aligned}$$

$$s^2_{y\cdot x} = \frac{89.215}{34} = 2.624 \qquad s_{y\cdot x} = 1.620$$

6.23. Confidence Limits and Tests of Significance for β

Since the estimates a and b are sample statistics, in the sense of the preceding chapter, based on the observations y_i and the known values x_i, we can expect them to vary from sample to sample according to some distribution. The quantity

$$b = \frac{\Sigma(x_i - \bar{x})(y_i - \bar{y})}{\Sigma(x_i - \bar{x})^2}$$

can be written in the form

$$b = \frac{\Sigma(x_i - \bar{x})y_i}{\Sigma(x_i - \bar{x})^2} \tag{a}$$

since we have

$$\begin{aligned} \Sigma(x_i - \bar{x})(y_i - \bar{y}) &= \Sigma(x_i - \bar{x})y_i - \Sigma(x_i - \bar{x}\bar{y}) \\ &= \Sigma(x_i - \bar{x})y_i - \bar{y}\Sigma(x_i - \bar{x}) \\ &= \Sigma(x_i - \bar{x})y_i \end{aligned} \tag{b}$$

since $\Sigma(x_i - \bar{x}) = 0$. Because the x_i are known constants and not random variables, the other samples we might have obtained would consist of a different set of observations y_i, corresponding to the identical values of x_i. Thus if (a) is written in the form

$$b = \frac{x_1 - \bar{x}}{\Sigma(x_i - \bar{x})^2} y_1 + \frac{x_2 - \bar{x}}{\Sigma(x_i - \bar{x})^2} y_2 + \cdots + \frac{x_n - \bar{x}}{\Sigma(x_i - \bar{x})^2} y_n \tag{c}$$

we can see that b is a linear function of the normally distributed random variables $y_1, \cdots, y_n$ with coefficients $a_i = \dfrac{x_i - \bar{x}}{\Sigma(x_i - \bar{x})^2}$ and hence by Section 4.42 is also normally distributed. In particular

$$\text{ave}(b) = \Sigma_i \left[\frac{x_i - \bar{x}}{\Sigma(x_i - \bar{x})^2}(\alpha + \beta x_i)\right] = \beta \tag{d}$$

$$\text{var}(b) = \Sigma_i \left\{\frac{(x_i - \bar{x})^2}{[\Sigma(x_i - \bar{x})^2]^2}\right\}\sigma^2 = \frac{\sigma^2}{\Sigma(x_i - \bar{x})^2} \tag{e}$$

We have now shown that the estimate b is normally distributed with mean β and variance σ_b^2 given by (e). If σ^2 were known, we could use the estimate b and variance σ_b^2 to form confidence intervals for β or test hypotheses concerning this unknown constant just as the sample mean $\bar{x}$ and its variance σ^2/n are used in the case of the mean μ of a normal distribution with known variance σ^2. Since σ^2 is usually unknown, and the ratio $\dfrac{s^2_{y\cdot x}(n-2)}{\sigma^2}$ has the χ^2-distribution with $n-2$ degrees of freedom, the ratio

$$(f) \qquad t_{n-2} = \frac{\dfrac{b-\beta}{\sigma/\sqrt{\Sigma(x_i-\bar{x})^2}}}{\sqrt{\dfrac{s^2_{y\cdot x}(n-2)}{\sigma^2}\Big/ n-2}} = \frac{(b-\beta)\sqrt{\Sigma(x_i-\bar{x})^2}}{s_{y\cdot x}}$$

has the "Student" t-distribution with $n-2$ degrees of freedom. This quantity can then be used to test the hypothesis $\beta = \beta_0$ in a manner completely analogous to the use of the t-distribution to test hypotheses concerning an unknown mean. Similarly

$$(g) \qquad b - \frac{t_{n-2,\,\alpha}\, s_{y\cdot x}}{\sqrt{\Sigma(x_i-\bar{x})^2}} < \beta < b + \frac{t_{n-2,\,\alpha}\, s_{y\cdot x}}{\sqrt{\Sigma(x_i-\bar{x})^2}}$$

is a $100(1-\alpha)\%$ confidence interval for β.

EXAMPLE. For the data of Table 6.3, using the results of the examples of the preceding section, we have for the estimated value of $\sigma_b{}^2$

$$s_b{}^2 = \frac{s^2_{y\cdot x}}{S(x^2)} = \frac{2.624}{220.8144} = 0.01188$$

A proposed approximate value of β for these data is 1.5. We can test this approximation by using the estimates b and $s_b{}^2$, obtaining

$$t_{34} = \frac{b-\beta}{s_b} = \frac{-0.0813}{0.1090} = -0.746$$

Entering the table of "t" for 34 degrees of freedom, we obtain a probability level of about 0.46, so that our estimate of β is consistent with the hypothesis that the true value is 1.5.

A 95% confidence interval for β may also be written by noting that $t_{34,\,0.05} = 2.03$, and thus at this confidence level

$$1.4187 - 2.03(0.109) < \beta < 1.4187 + 2.03(0.109)$$

$$1.20 < \beta < 1.64$$

6.24. Confidence Limits and Tests of Significance for Predicted Values

Since, from Section 6.22 (*e*)

$$\tilde{y}_0 = \bar{y} + b(x_0 - \bar{x})$$

is a linear combination of the normally distributed variables $\bar{y}$ and b, the estimate $\tilde{y}_0$ of the average value of y for a given x_0 is normally distributed with

$$\text{ave}\,(\tilde{y}_0) = \alpha + \beta x_0$$

$$\text{var}\,(\tilde{y}_0) = \frac{\sigma^2}{n} + \sigma^2 \frac{(x_0 - \bar{x})^2}{\Sigma(x_i - \bar{x})^2} = \sigma^2 \left[\frac{1}{n} + \frac{(x_0 - \bar{x})^2}{\Sigma(x_i - \bar{x})^2}\right]$$

Again, if σ^2 is unknown, we find that

$$(a) \qquad t_{n-2} = \frac{\tilde{y}_0 - (\alpha + \beta x_0)}{s_{y\cdot x}\sqrt{\dfrac{1}{n} + \dfrac{(x_0 - \bar{x})^2}{\Sigma(x_i - \bar{x})^2}}}$$

has the "Student" t-distribution with $n - 2$ degrees of freedom and can be used to test hypotheses concerning the mean $\alpha + \beta x_0$ of y for a given x_0.

The $100(1 - \alpha)\%$ confidence interval for $\alpha + \beta x_0$ is given immediately by

$$(b) \qquad \tilde{y}_0 - t_{n-2,\,\alpha} s_{y\cdot x}\sqrt{\frac{1}{n} + \frac{(x_0 - \bar{x})^2}{\Sigma(x_i - \bar{x})^2}} < \alpha + \beta x_0 < \tilde{y}_0 + t_{n-2,\,\alpha} s_{y\cdot x}\sqrt{\frac{1}{n} + \frac{(x_0 - \bar{x})^2}{\Sigma(x_i - \bar{x})^2}}$$

EXAMPLE. A 95% confidence interval for the expected value $\tilde{y}_0 = a + bx_0$ in the previous example is

$$0.1240 + 1.4187x_0 - (2.03)\,(1.62)\sqrt{\frac{1}{36} + \frac{(x_0 - 1.924)^2}{220.814}} < \alpha + \beta x_0 < 0.1240 + 1.4187x_0 + (2.03)\,(1.62)\sqrt{\frac{1}{36} + \frac{(x_0 - 1.924)^2}{220.814}}$$

The limits of this interval are tabulated below for a number of values of x_0. For $x_0 = 0$ the confidence interval includes the origin, so that, if we accept the 0.05 level as a criterion, the data are not inconsistent with

x_0	Upper Limit	y_0	Lower Limit	x_0	Upper Limit	y_0	Lower Limit
0	0.335	0.124	−0.087	6	8.957	8.636	8.315
1	1.721	1.543	1.356	7	10.435	10.055	9.675
2	3.128	2.961	2.794	8	11.916	11.474	11.032
3	4.562	4.380	4.198	9	13.397	12.892	12.388
4	6.016	5.799	5.581	10	14.879	14.311	13.743
5	7.484	7.218	6.952				

the hypothesis that the true regression line has a zero intercept. If these were the only data available, we should naturally employ the relationship $y = 0.1240 + 1.4187x$ to predict y for a given x, or x for a given determination of y. However, if the relationship $y = 1.50x$ were based on extensive previous experience, we should probably employ the latter relationship since the results from this set of observations provide no basis for rejecting the hypothesis that this relationship is correct.

It should be noted that the variance of the estimate b of β decreases as the value of $\Sigma(x_i - \bar{x})^2$ increases, so that this estimate becomes less variable as the number of values x_i for which an observation y_i is made increases, or as the scatter of the x_i about $\bar{x}$ increases. This means that in practical situations we shall get the best estimate of the unknown slope β by choosing the points x_i as widely scattered as possible. In fact, in the above situation where we have assumed a priori that the relationship is linear, the best estimate of β would be obtained by concentrating all the values of x_i equally at the two extreme points of the possible range of values of x_i. This should only be done when convincing evidence is available that the relationship must be linear, since under these conditions we would have no opportunity to check the possible non-linearity of the regression. Methods of testing for non-linearity and the selection of the most effective arrangements in the case of linear relationships are considered in later sections. Also note that the variance of the estimate $\tilde{y}_0 = a + bx_0$ of $\alpha + \beta x_0$ is a minimum for $x_0 = \bar{x}$, in which case $\tilde{y}_0 = \bar{y}$ has variance σ^2/n. In this special case (*b*) reduces to

$$(c) \qquad \bar{y} - t_{n-2,\,\alpha}\frac{s_{y \cdot x}}{\sqrt{n}} < \alpha + \beta\bar{x} < \bar{y} + t_{n-2,\,\alpha}\frac{s_{y \cdot x}}{\sqrt{n}}$$

which is identical with the confidence interval for a mean given in Section 5.33 for the case of σ^2 unknown except that the variance estimate s_y^2 based on $n - 1$ degrees of freedom has been replaced by the estimate $s^2_{y \cdot x}$ based on $n - 2$ degrees of freedom. The latter estimate will be smaller than the first whenever the variations in the y_i can be partially explained by the measurements x_i; this represents the increase in accuracy of the estimate of the mean $\bar{y}$ of measurements y_i at $x_0 = \bar{x}$ due to the consideration of their dependence on the x_i. The variance of $\tilde{y}_0$ increases as $x_0 - \bar{x}$ becomes large, i.e., as the predicted value $\tilde{y}_0$ is computed for values of x_0 farther and farther from the mean of the x_i for which observations on y were made. If $x_0 = 0$, then $\tilde{y}_0 = a$ is an estimate of α. For this case we have

$$\text{var}\,(\tilde{y}_0) = \sigma_a^{\,2} = \sigma^2\left[\frac{1}{n} + \frac{\bar{x}^2}{\Sigma(x_i - \bar{x})^2}\right]$$

and, by putting $x_0 = 0$ in (*a*) and (*b*),

$$(d) \qquad t_{n-2} = \frac{a - \alpha}{s_{y \cdot x}\sqrt{\dfrac{1}{n} + \dfrac{\bar{x}^2}{\Sigma(x_i - \bar{x})^2}}}$$

$$(e) \qquad a - t_{n-2,\,\alpha} s_{y \cdot x}\sqrt{\frac{1}{n} + \frac{\bar{x}^2}{\Sigma(x_i - \bar{x})^2}} < \alpha < a + t_{n-2,\,\alpha} s_{y \cdot x}\sqrt{\frac{1}{n} + \frac{\bar{x}^2}{\Sigma(x_i - \bar{x})^2}}$$

It happens very frequently in chemical applications that we wish to predict the value of the independent variable x associated with an observation, or series of observations, on the dependent variable y. Thus, for example, in using a straight line fitted to the data of Table 6.1, we should ordinarily wish to predict the concentration of formaldehyde corresponding to a given reading, or series of readings, of the optical density, even though in fitting the line we should consider the optical density reading as the random variable and the concentrations of the standard solutions used as the known constants.

From 6.22 (*e*), the estimated value of x corresponding to an observation y_0 would be

$$(f) \qquad x_0 = \bar{x} + \frac{y_0 - \bar{y}}{b}$$

Here x_0 is a function (not linear) of 3 normally distributed variables: the new observation y_0, the mean $\bar{y}$ of the observations from which the line was determined, and the slope b. Although x_0 will not in general be normally distributed, we can say from Sections 3.34 and 4.42 that x_0 will be approximately normally distributed about the true value of x for which the observation y_0 was made, with

$$\text{var}\,(x_0) \sim \left(\frac{\partial x_0}{\partial y_0}\right)^2 \sigma_{y_0}{}^2 + \left(\frac{\partial x_0}{\partial \bar{y}}\right)^2 \sigma_{\bar{y}}{}^2 + \left(\frac{\partial x_0}{\partial b}\right)^2 \sigma_b{}^2$$

under the conditions given there, the most important in this instance being that $b/\sigma_b{}^2$ be sufficiently large. Since

$$\frac{\partial x_0}{\partial y_0} = \frac{1}{b}$$

$$\frac{\partial x_0}{\partial \bar{y}} = -\frac{1}{b}$$

$$\frac{\partial x_0}{\partial b} = \frac{(y_0 - \bar{y})}{b^2}$$

we have

$$\text{var}(x_0) \sim \frac{\sigma_{y_0}{}^2 + \sigma_{\bar{y}}{}^2}{b^2} + \frac{\sigma_b{}^2}{b^4}(y_0 - \bar{y})^2$$

If y_0 is the mean of m observations, all at the same unknown value of x, then $\sigma_{y_0}{}^2 = \sigma^2/m$; also, from the above $\sigma_{\bar{y}}{}^2 = \sigma^2/n$, and from (*e*) of the previous section, $\sigma_b{}^2 = \dfrac{\sigma^2}{\Sigma(x_i - \bar{x})^2}$. Substituting these values, we have

$$(g) \qquad \text{var}(x_0) \sim \frac{\sigma^2}{b^2}\left[\left(\frac{1}{m} + \frac{1}{n}\right) + \frac{(y_0 - \bar{y})^2}{b^2\Sigma(x_i - \bar{x})^2}\right]$$

6.25. Estimation of β when $\alpha = 0$

Situations frequently arise in which it is possible to assume a priori that the intercept α is 0, and that we have ave $(y) = \beta x$. In this case the best estimate of the single parameter β is obtained by minimizing the sum of squares $Q = \Sigma_i(y_i - \beta x_i)^2$. By the previous method we obtain

$$(a) \qquad \frac{\partial Q}{\partial \beta} = 2\Sigma x_i(y_i - \beta x_i)$$

and hence the estimate b of β is obtained from the equation $\Sigma x_i(y_i - bx_i) = 0$. Solving this, we have

$$(b) \qquad b = \frac{\Sigma x_i y_i}{\Sigma x_i{}^2}$$

Notice that this estimate is very similar to that for the case $\alpha \neq 0$, except that the actual values of x and y are used instead of the deviations from their means. The best estimate of the variance σ^2 is given in this case by

$$(c) \qquad s^2_{y \cdot x} = \frac{\Sigma(y_i - bx_i)^2}{n - 1}$$

The quantity $(n - 1)\dfrac{s^2_{y \cdot x}}{\sigma^2}$ has the χ^2-distribution with $n - 1$ degrees of freedom, since in this case only the value b has been computed from the observations. As before, the actual sum of squares of the deviations is most easily computed by using the form

$$(d) \qquad \begin{aligned} \Sigma(y_i - bx_i)^2 &= \Sigma y_i{}^2 - b\Sigma x_i y_i \\ &= \Sigma y_i{}^2 - \frac{(\Sigma x_i y_i)^2}{\Sigma x_i{}^2} \end{aligned}$$

It can be shown by a method completely analogous to that of Section 6.23 that the quantity

$$(e) \qquad t_{n-1} = \frac{(b - \beta)\sqrt{\Sigma x_i{}^2}}{s_{y \cdot x}}$$

has the t-distribution with $n - 1$ degrees of freedom and that a $100(1 - \alpha)\%$ confidence interval for β is given by

$$(f) \qquad b - t_{n-1,\alpha} \frac{s_{y \cdot x}}{\sqrt{\Sigma x_i^2}} < \beta < b + t_{n-1,\alpha} \frac{s_{y \cdot x}}{\sqrt{\Sigma x_i^2}}$$

Confidence limits and tests of significance for predicted values follow exactly as in the preceding section.

EXAMPLE. Let us consider the data from Table 6.1. Using only the first 5 runs, we have $n = 5$, $\Sigma x_i^2 = 1.20$, $\Sigma x_i y_i = 1.0728$, $\Sigma y_i^2 = 0.959102$. Hence

$$b = \frac{1.0728}{1.20} = 0.894,$$

$$\Sigma(y_i - bx_i)^2 = 0.959102 - \frac{(1.0728)^2}{1.20} = 0.000019$$

$$s^2_{y \cdot x} = 0.00000475$$

$$s_{y \cdot x} = 0.0022$$

$$\frac{t_{4,\,0.05}\, s_{y \cdot x}}{\sqrt{\Sigma x_i^2}} = \frac{(2.776)\,(0.0022)}{\sqrt{1.20}} = 0.006$$

and a 95% confidence interval for β would be

$$0.894 - 0.006 < \beta < 0.894 + 0.006$$

$$0.888 < \beta < 0.900$$

Table 6.5 gives both the observed values and the estimated values $\tilde{y}_i$ for each of the 6 runs based on the slope computed from the first 5 runs only.

TABLE 6.5

Run	y_i (Observed)	$\tilde{y}_i$ (Predicted)
1	0.086	0.089
2	0.269	0.268
3	0.445	0.447
4	0.538	0.536
5	0.626	0.626
6*	0.782	0.804

* Not included in computation of b.

Notice the comparatively large discrepancy (which was apparent graphically) between the last observed value and the predicted value. This observation can be shown to be significantly different from the value predicted *on the basis of the first* 5 *runs.* However, if we include the sixth run in our computations, we obtain

$$b = 0.884$$

$$s^2_{y \cdot x} = 0.0000648$$

$$s_{y \cdot x} = 0.0080$$

$$0.869 < \beta < 0.899$$

and *based on these figures the discrepancy of the sixth point is no longer significant at the* 5% *limit.* This point was dropped because it is *known* that Beer's law, which states that the optical density is directly proportional to the concentration, fails to hold at high concentrations (low transmissions) and a tailing off from a straight line at high concentrations is commonly observed in standardizations of this type. It is important to realize that this point was dropped on the basis of past chemical experience; on the basis of these 6 runs, there is no reason to suspect statistically an unusual discrepancy from a straight line relationship on the part of any one point.

In actually using this standard curve, we should determine the optical density and wish to obtain the milligrams of formaldehyde present. To determine the factor for making this conversion, we divide b, which represents units of optical density/milliliter of standard, into the concentration of the standard, which is in milligrams per milliliter. We obtain

Best estimate of factor: 1.177 mg CH_2O/unit of optical density
95% Confidence limits for factor: $1.169 <$ true factor < 1.185

Note that, under the assumption that most of the variation is in the determination of the optical density and very little in the preparation and aliquoting of the standard solution, we should always consider the optical density the dependent, or random variable, and the size of the aliquot the independent, or known variable, in estimating the slope, regardless of the way in which it was to be used later. The above factor, expressed in milligrams, also depends on the sample size.

6.26. Transformation of Variables to Obtain Linear Regressions

Situations frequently arise in which the measurements x_i and y_i are not linearly related, but one or both of them can easily be transformed in such a manner that a linear relationship results. One of the most frequent examples is the case where the relationship between the expected value of the y's and x's is of the form $y = ce^{\beta x}$. In this case we shall have ln $y = \ln c + \beta x$, and we can obtain estimates of $\alpha = \ln c$ and β by fitting a straight line to the values ln y_i and x_i exactly as in Section 6.22. If c is known a priori to be unity, then $\alpha = \ln c = 0$, and the method of Section 6.25 can be used to estimate β. It should be noted that the above situation is applicable where the points show a linear trend on semi-log graph paper. If logarithms to the base 10 are used in the calculation, the line will become $\log y = \log c + 0.4343\beta x$ and hence the estimate obtained for the slope must be divided by 0.4343 to obtain β.

EXAMPLE 1. Let us consider the data in the first two columns of 6.6, which represent a typical run in the determination of the rate constant for the decomposition of benzazide in various solvents. The solvent used in this case was dioxane. The data give the volume of N_2 liberated up to the specified time by the reaction

$$RCON_3 \rightarrow RNCO + N_2$$

TABLE 6.6

Data from Newman, Lee, and Garett, *J.A.C.S.*, *69* (1947), 113

Time of Reaction (hours)	Vol of N_2 at S.T.P. (ml)	$\frac{V_\infty - V}{V_\infty}$	$\log \frac{(V_\infty - V)}{V_\infty}$	
			Observed	Estimated
0.20	12.62	0.8674	−0.0618	−0.0586
0.57	30.72	0.6773	−0.1692	−0.1670
0.92	44.59	0.5316	−0.2744	−0.2695
1.22	52.82	0.4452	−0.3514	−0.3573
1.55	60.66	0.3628	−0.4403	−0.4540
1.90	68.20	0.2836	−0.5473	−0.5565
2.25	73.86	0.2242	−0.6494	−0.6590
2.63	78.59	0.1745	−0.7582	−0.7703
3.05	82.02	0.1384	−0.8589	−0.8933
3.60	86.29	0.0936	−1.0287	−1.0544
4.77	91.30	0.0410	−1.3872	−1.3971
5.85	93.59	0.0169	−1.7721	−1.7135

The first-order kinetic equation, which is applicable here, is of the form

$$\log (V_\infty - V) = \log V_\infty - 0.4343kt$$

where k is the rate constant desired. Rewriting in the form

$$\log \frac{(V_\infty - V)}{V_\infty} = -0.4343kt$$

and assuming V_∞ to be known, an assumption which we shall later reconsider, we have an equation of the form $y = \beta x$, where $x = t$, $y = \log \dfrac{V_\infty - V}{V_\infty}$, and $\beta = -0.4343k$. Here $\alpha = 0$ a priori, since, for $t = 0$, $V = 0$; hence $\dfrac{V_\infty - V}{V_\infty} = 1$ and $\log \dfrac{V_\infty - V}{V_\infty} = 0$.

The third and fourth columns of Table 6.6 give the values of $\dfrac{V_\infty - V}{V_\infty}$ and $\log \dfrac{V_\infty - V}{V_\infty}$ computed from the observed volumes of N_2, using $V_\infty = 95.20$. From the first and fourth columns, representing x and y, respectively, we obtain

$$\Sigma x_i^2 = 99.9295 \qquad \Sigma x_i y_i = -29.274205 \qquad \Sigma y_i^2 = 8.58180933$$

from which we obtain, using Section 6.25,

$$b = \frac{\Sigma x_i y_i}{\Sigma x_i^2} = \frac{-29.274205}{99.9295} = -0.2929$$

and hence the best estimate of k based on these data is

$$k = -\frac{b}{0.4343} = 0.6744 \text{ hr}^{-1}$$

Also we have

$$\Sigma(y_i - bx_i)^2 = \Sigma y_i^2 - \frac{(\Sigma x_i y_i)^2}{\Sigma x_i^2}$$

$$= 8.58180933 - 8.57583675$$

$$= 0.00597258$$

$$s^2_{y \cdot x} = \frac{\Sigma(y_i - bx_i)^2}{n - 1} = \frac{0.00597258}{11}$$

$$= 0.00054296$$

$$s_{y \cdot x} = 0.0233$$

$$\frac{t_{11,\,0.05} s_{y \cdot x}}{\sqrt{\Sigma x_i^2}} = \frac{(2.201)\,(0.0233)}{\sqrt{99.9295}} = 0.0051$$

from which we obtain for the 95% confidence interval for β

$$-0.2929 - 0.0051 < \beta < -0.2929 + 0.0051$$

$$-0.2980 < \beta < -0.2878$$

and, converting into terms of k by dividing by -0.4343,

$$0.6626 < k < 0.6862$$

The values of $\log \dfrac{V_\infty - V}{V_\infty}$ estimated from the equation are given in Table 6.6, and a graph of the values $\dfrac{V_\infty - V}{V_\infty}$ on a logarithmic scale against time is shown in Figure 6.6. It will be noted that the last point seems to be particularly discrepant from the computed line, and actually this point was not used in determining the slope graphically in the paper

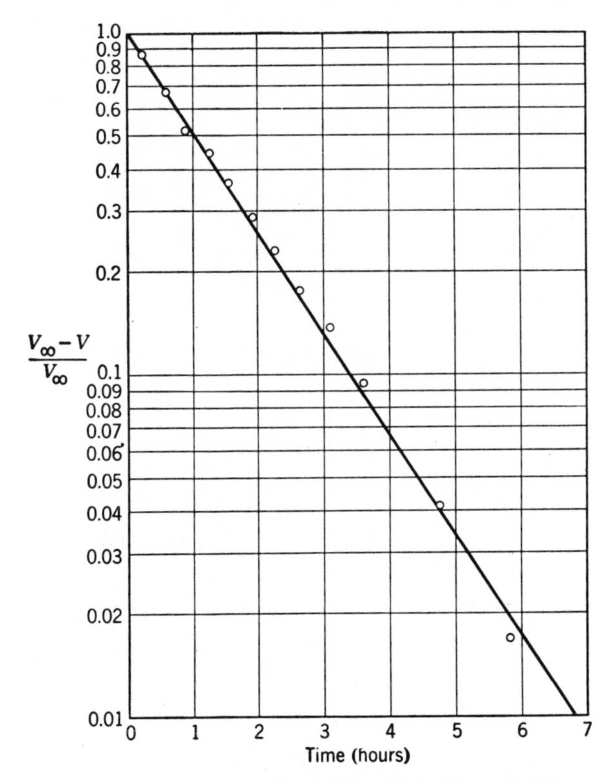

FIG. 6.6. Decomposition of benzazide ($V_\infty = 95.20$).

from which these data were taken. Note also that the deviations are somewhat regular, the first 3 being low, the next 8 high, and the last point low. This feature indicates non-randomness on the part of the deviations (see Chapter 11, Section 11.32) and consequently makes the validity of the above estimates somewhat doubtful.

One feature which might be questioned is the assumption of the value $V_\infty = 95.20$. Let us suppose that, instead of considering V_∞ known, we estimate it from the data by minimizing the sum of squares with respect to both V_∞ and β. This could be done by the usual method of differentiating the sum of squares of residuals with respect to both of these parameters, and solving the resulting equations, but the equations thus obtained are quite complicated. A more direct method is to compute,

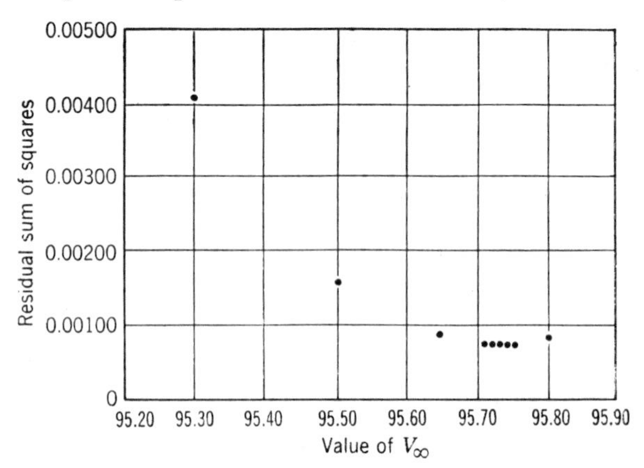

FIG. 6.7.

TABLE 6.7

$\frac{V_\infty - V}{V_\infty}$	Log $\frac{V_\infty - V}{V_\infty}$	
	Observed	Calculated
0.8682	−0.0614	−0.0563
0.6791	−0.1681	−0.1604
0.5342	−0.2723	−0.2590
0.4482	−0.3485	−0.3434
0.3663	−0.4362	−0.4363
0.2875	−0.5414	−0.5348
0.2284	−0.6413	−0.6333
0.1790	−0.7472	−0.7403
0.1431	−0.8444	−0.8585
0.0985	−1.0066	−1.0133
0.0462	−1.3354	−1.3426
0.0223	−1.6517	−1.6466

by the procedure used above, the residual sum of squares for various assumed values of V_∞, and to determine graphically that value which gives the minimum. Such a graph for the above problem is given in Figure 6.7. The value of V_∞ which gave the minimum residual sum of squares was 95.72, although the residuals for 95.73 and 95.71 were little different. Table 6.7 gives the values of $\frac{V_\infty - V}{V_\infty}$ and both the observed and estimated values of $\log \frac{V_\infty - V}{V_\infty}$, for $V_\infty = 95.72$. For this table, using the x values from Table 6.6, we have

$$\Sigma x_i^2 = 99.9295 \qquad \Sigma x_i y_i = -28.128097 \qquad \Sigma y_i^2 = 7.91824601$$

and hence

$$b = \frac{-28.128097}{99.9295} = -0.2815$$

$$\Sigma(y_i - bx_i)^2 = 7.91824601 - 7.91748023$$

$$= 0.00076578$$

$$s^2_{y \cdot x} = \frac{0.00076578}{10} = 0.00007658$$

$$s_{y \cdot x} = 0.00835$$

$$\frac{t_{11,\,0.05} s_{y \cdot x}}{\sqrt{\Sigma x_i^2}} = \frac{(2.201)\,(0.00875)}{\sqrt{99.9295}} = 0.0019$$

$$-0.2834 < \beta < -0.2796$$

Thus the use of $V_\infty = 95.72$ has reduced the residual sum of squares to about $^1/_8$ its former value, and has appreciably changed our estimate of b. In fact, the new value of b is not within the original confidence limits for b, which is not surprising, since there is every reason to believe that the model on which these limits were based was incorrect.

From this analysis we obtain for our estimate of k

$$k = \frac{b}{-0.4343} = 0.648$$

and as confidence limits for k

$$0.644 < k < 0.652$$

This procedure is not exact since we have assumed that the process of estimating V_∞ is the equivalent of an additional linear constraint upon the observations. It is precisely in those cases in which the estimating equation is not linear in the observations that the graphical technique is advantageous, so that when this technique is employed the confidence interval obtained will usually be approximate.

A plot of the data and the line obtained with $V_\infty = 95.72$ is shown in

Figure 6.8. There is no longer any significant systematic behavior of the differences between observed and estimated values.

Another situation in which the data can be transformed to the linear case by taking the logarithms of both the pairs of measurements is that in which the relationship is of the form $y = cx^{\beta}$, and hence $\log y = \log c + \beta \log x$. We can proceed to obtain estimates of $\alpha = \log c$ and β by

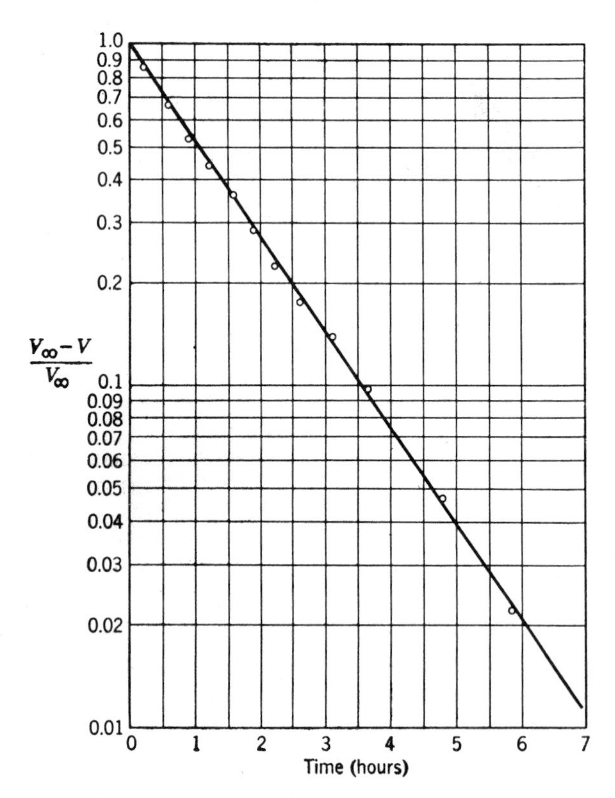

FIG. 6.8. Decomposition of benzazide ($V_{\infty} = 95.72$).

methods identical with those in Section 6.21 with the x_i and y_i replaced by $\log x_i$ and $\log y_i$. Again, if c is known a priori to be unity, we can assume $\alpha = \log c = 0$, and use the methods of Section 6.25.

EXAMPLE 2. A general example of this type is to be found in heat, mass, or momentum transfer where an attempt is made to systematize the results by dimensional analysis. The resulting equation in the case of heat transfer is frequently of the form $(\text{Nu}) = C(\text{Re})^a(\text{Pr})^b(\text{Gr})^c$ where Nu, Re, Pr, and Gr are dimensionless groups. If, on taking logarithms, we can assume the dependent variable to be normally distributed, we have a linear equation log (Nu) = log $C + a$ log (Re) + b log (Pr) + c log (Gr). The data of Table 6.8 represent measurement

TABLE 6.8

(Data from: C. Y. Chen, G. A. Hawkins, and H. L. Solberg, "Heat Transfer in Annuli," *A.S.M.E. Trans.*, *68* (1946), 99–106)

Log Reynolds Number	Log Nusselt Number (corrected)
2.305	0.254
2.455	0.285
2.480	0.306
2.574	0.329
2.683	0.419
2.696	0.363
2.789	0.454
2.825	0.488
2.841	0.464
2.936	0.519
3.076	0.652
3.092	0.656
3.173	0.683
3.218	0.722
3.250	0.736
3.270	0.727
3.303	0.740
3.318	0.755

of the Nusselt and Reynolds numbers obtained during streamline flow of water through an annulus. These data are plotted in Figure 6.9, and,

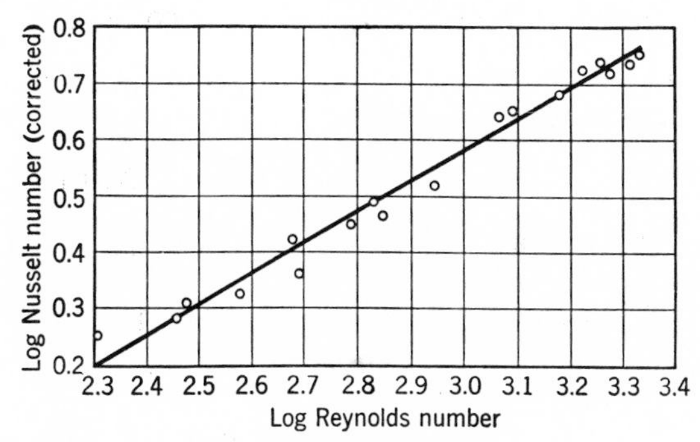

FIG. 6.9.

although a linear relationship appears reasonable, there is sufficient scatter to render the gradient of the line uncertain. We are interested to determine whether the gradient of our fitted line differs significantly from 0.45 since this exponent of the Reynolds number has been suggested for design purposes.

From Table 6.8 we have

$$N = 18 \qquad \Sigma y = 9.552 \qquad \Sigma x = 52.284$$

$$\Sigma y^2 = 5.614928 \quad \Sigma xy = 28.723569$$

$$\Sigma x^2 = 153.653380$$

$$b_{y \cdot x} = \frac{\begin{vmatrix} 18 & 9.552 \\ 52.284 & 28.723569 \end{vmatrix}}{\begin{vmatrix} 18 & 52.284 \\ 52.284 & 153.653380 \end{vmatrix}} = 0.54777$$

$$\Sigma(y - \tilde{y})^2 = \Sigma(y - \bar{y})^2 - b\Sigma(x - \bar{x})(y - \bar{y}) = 0.010175$$

$$s^2_{y \cdot x} = \frac{\Sigma(y - \tilde{y})^2}{(18 - 2)} = 0.0006359$$

$$s_b = \frac{s_{y \cdot x}}{\sqrt{\Sigma(x - \bar{x})^2}} = 0.01887$$

Testing the estimate of $\beta = 0.548$ against the proposed value of 0.45, we have

$$t = \frac{0.548 - 0.45}{0.01887} = 5.18^{***}$$

which for 16 degrees of freedom is beyond the 0.001 point.

If we wish to form confidence limits for β at the 5% level, we note that $t_{116,\, 0.05} = 2.12$, and thus

$$0.548 - 2.12(0.0189) < \beta < 0.548 + 2.12(0.0189)$$

$$0.508 < \beta < 0.588$$

The line of best fit to the points may be evaluated as

$$y - \bar{y} = 0.548(x - \bar{x})$$

$$y = 1.061 + 0.548x$$

This line has been superimposed on Figure 6.9.

It should be noted that in the above models we are assuming that log y_i is normally distributed with constant variance. In practice it may happen that the logarithmic transformation is required to obtain a linear relationship, but that the untransformed observations y_i and not their logarithms are distributed normally and with constant variance. In this case the above method can be used but it is not necessarily the best procedure in the maximum likelihood sense.

It is also possible that in some instances y will have a linear regression with some function of the known x's. For example, the expected values of y may be given by the relationships $y = \alpha + \beta \log x$, or $y = \beta \sin x$. In this case we simply use $x_i' = \log x_i$, or $x_i' = \sin x_i$, and proceed to estimate the constants α and β, or β, as before. Similarly, if the relationship is $y = ce^{\beta x}$, or $y = cx^{\beta}$, as considered above, and β is known a priori, the best procedure for estimating c would be to use $x_i' = e^{\beta x_i}$ or $x_i' = x_i^{\beta}$, and then determine the constant c by the methods of Section 6.25. This avoids the difficulty of the preceding paragraph, since the deviations $y_i - cx_i'$ are still measured in terms of the original variable y. For example, given a series of measurements y_i of the amount of a material of known half-life at times t_i, and if the measurements y_i are assumed to have normally distributed errors of measurement of the same magnitude (variance) for all values of t, then the best estimate of the amount present at time $t = 0$ would be obtained by determining the *slope* of the line $y = cx$ best fitting the points y_i, x_i where $x_i = e^{-\lambda t_i}$, λ being the decay constant of the given material.

6.27. Unequal Residual Variances in Regression Analysis

The discussion of Section 6.2 has been restricted to the case in which the residual components in the model were distributed about a mean value $\alpha + \beta x$ with variance σ^2 independent of x. If the variance is not independent of x, we can distinguish cases in which the variance can be expressed as a function of x from those in which the magnitudes of the variances are more or less randomly distributed with respect to x. An example of the first type might be afforded by the experiment to determine the proportion of radioactive material of known half-life present at $t = 0$. In this case, if the total time required for the experiment is appreciable compared with the half-life, the variance of the observations will decrease sufficiently as the experiment proceeds to render the previous model unsatisfactory. As an illustration of the second type we consider an experiment in which the observations at the different values of the x variate must be made with different instruments which are unbiased but vary in precision. If the instruments were allocated in a random fashion to the experiments, we should not anticipate a systematic variation of σ^2 with the value of the x variate.

In either event it would be logical, if the variances of the observations were known, to minimize the sum of squares $Q = \Sigma_i(y_i - a - bx_i)^2/\sigma_i^2$ To do this we set

$$\frac{\partial Q}{\partial a} = -2a\Sigma_i(y_i - a - bx_i)/\sigma_i^2 = 0$$

$$\frac{\partial Q}{\partial b} = -2b\Sigma_i(y_i x_i - ax_i - bx_i^2)/\sigma_i^2 = 0$$

If we write w_i for $1/\sigma_i^2$, the estimating equations can be written

$$\Sigma_i w_i y_i = a\Sigma_i w_i + b\Sigma_i w_i x_i$$

$$\Sigma_i w_i x_i y_i = a\Sigma_i w_i x_i + b\Sigma_i w_i x_i^2$$

Furthermore, if we divide throughout by $\Sigma_i w_i/n$ and write $a_i = \dfrac{nw_i}{\Sigma_i w_i}$, we obtain

$$\Sigma_i a_i y_i = na + b\Sigma_i a_i x_i$$

$$\Sigma_i a_i x_i y_i = a\Sigma_i a_i x_i + b\Sigma_i a_i x_i^2$$

These are exactly the equations obtained by weighting each of the original observations in proportion to the reciprocal of its variance, i.e., by regarding an observation of variance σ^2 as the equivalent of n independent observations of variance $n\sigma^2$. The estimating equations may be solved in the usual manner for a and b, and exact confidence intervals for these estimates may be computed.

Where the functional relationship between x and σ^2 is known, the weighting of the observations may be chosen in accordance with this relationship. If the form of the relationship is not known, or if no relationship exists, it is possible to use approximate weightings when a number of *independent y* measurements are available for each value of the x variate. In the first case the variance estimate for each value of x may be determined and plotted against the corresponding value of x in order to investigate the type of relationship present. If some relationship is apparent, it may be used as a semi-empirical basis for weighting. In the second case the variance estimates for the individual x variates may be used in place of the true variances in the weighting procedure. This would lead to the use of weighted observations in all regression analyses in which more than one y observation was made at each value of the x variate. The decision to use weights or to refrain from using them is essentially non-statistical unless the individual variance estimates can be shown to be heterogeneous by the use of the Bartlett or Cochran test. If it seems reasonable on technical grounds to assume that the observations have equal residual variance and if the analysis does not disprove this assumption at a chosen significance level, an unweighted procedure will be used. If technical considerations suggest that the hypothesis of equal variances is unreasonable, the observations will be weighted in all cases according to the variance estimates, provided that the answer is of sufficient importance to justify the additional effort.

6.3. Multiple Regression

6.31. Linear Regression with More than One Independent Variable

In many instances when control data are being analyzed it is not reasonable to assume that the observed values of y are influenced by only one known quantity. In laboratory technique it is sometimes possible, although not necessarily desirable, to complete a series of tests at a number of levels of one factor, maintaining other factors constant, but this is rarely the situation in industrial work. A technique is therefore required which will enable observations to be fitted to a model in which a number of factors are considered to influence the observed value of the dependent variable. The concepts of regression can easily be extended to the case where the expected value of the random variable y depends linearly on more than one known quantity x_i. We shall consider the case of two independent variables, and the extension to more than two independent variables will be obvious.

In this case we assume each observation y_i to be from a normal distribution with mean $\alpha + \beta_1 x_{1i} + \beta_2 x_{2i}$ and variance σ^2. Under this assumption the maximum likelihood estimates of α, β_1, and β_2 will be obtained by minimizing the sum of squares:

$$(a) \qquad Q = \Sigma(y_i - \alpha - \beta_1 x_{1i} - \beta_2 x_{2i})^2$$

Taking the derivatives of Q with respect to α, β_1, and β_2, equating them to zero, and replacing the parameters α, β_1, and β_2, by the estimates a, b_1, and b_2, respectively, we obtain the equations

$$(b) \qquad \begin{aligned} na + b_1\Sigma x_{1i} \quad + b_2\Sigma x_{2i} &= \Sigma y_i \\ a\Sigma x_{1i} + b_1\Sigma x_{1i}^2 \quad + b_2\Sigma x_{1i}x_{2i} &= \Sigma x_{1i}y_i \\ a\Sigma x_{2i} + b_1\Sigma x_{1i}x_{2i} + b_2\Sigma x_{2i}^2 &= \Sigma x_{2i}y_i \end{aligned}$$

The solution of these equations for $a, b_1,$ and b_2 by any convenient method gives the desired estimates. As in the previous section the best unbiased estimate of the variance σ^2 is given by

$$(c) \qquad s^2_{y \cdot x_1 x_2} = \frac{\Sigma(y_i - \tilde{y}_i)^2}{n - 3}$$

where $\tilde{y}_i = a + b_1x_{1i} + b_2x_{2i}$. Also in a completely analogous fashion we have

$$(d)\qquad \begin{aligned}\Sigma(y_i - \tilde{y}_i)^2 &= \Sigma(y_i - a - b_1x_{1i} - b_2x_{2i})^2 \\ &= \Sigma y_i^2 - a\Sigma y_i - b_1\Sigma x_{1i}y_i - b_2\Sigma x_{2i}y_i\end{aligned}$$

In this case the estimate $s^2_{y\cdot x_1x_2}$ involves the three quantities a, b_1, and b_2 depending linearly on the observations y_i and hence is based on $n - 3$ degrees of freedom.

For purposes of generalization to more than two independent variables, and for the purposes of the next section, it is sometimes more convenient to consider the regression equations (*b*) in a slightly different form. From the first equation in (*b*) we obtain, dividing through by n and solving for a,

$$(e)\qquad a = \bar{y} - b_1\bar{x}_1 - b_2\bar{x}_2$$

If a is replaced by this value in the remaining equations, we obtain

$$(\bar{y} - b_1\bar{x}_1 - b_2\bar{x}_2)\Sigma x_{1i} + b_1\Sigma x_{1i}^2 + b_2\Sigma x_{1i}x_{2i} = \Sigma x_{1i}y_i$$

$$(\bar{y} - b_1\bar{x}_1 - b_2\bar{x}_2)\Sigma x_{2i} + b_1\Sigma x_{1i}x_{2i} + b_2\Sigma x_{2i}^2 = \Sigma x_{2i}y_i$$

These equations can be rearranged in the form

$$b_1\Sigma x_{1i}(x_{1i} - \bar{x}_1) + b_2\Sigma x_{1i}(x_{2i} - \bar{x}_2) = \Sigma x_{1i}(y_i - \bar{y})$$

$$b_1\Sigma x_{2i}(x_{1i} - \bar{x}_1) + b_2\Sigma x_{2i}(x_{2i} - \bar{x}_2) = \Sigma x_{2i}(y_i - \bar{y})$$

By a procedure identical with that used in (*b*) of Section 6.23 it can be shown that

$$\Sigma x_{1i}(x_{1i} - \bar{x}_1) = \Sigma(x_{1i} - \bar{x}_1)^2$$

$$\Sigma x_{1i}(x_{2i} - \bar{x}_2) = \Sigma(x_{1i} - \bar{x}_1)(x_{2i} - \bar{x}_2)$$

$$\Sigma x_{1i}(y_i - \bar{y}) = \Sigma(x_{1i} - \bar{x}_1)(y_i - \bar{y})$$

and similarly for the three sums in the last equation. Hence these equations become

$$(f)\qquad \begin{aligned}b_1\Sigma(x_{1i} - \bar{x}_1)^2 + b_2\Sigma(x_{1i} - \bar{x}_1)(x_{2i} - \bar{x}_2) &= \Sigma(x_{1i} - \bar{x}_1)(y_i - \bar{y}) \\ b_1\Sigma(x_{1i} - \bar{x}_1)(x_{2i} - \bar{x}_2) + b_2\Sigma(x_{2i} - \bar{x}_2)^2 &= \Sigma(x_{2i} - \bar{x}_2)(y_i - \bar{y})\end{aligned}$$

If we introduce the notation

$$S(x_1^2) = \Sigma(x_{1i} - \bar{x}_1)^2 = \Sigma x_{1i}^2 - \frac{(\Sigma x_{1i})^2}{n}$$

$$S(x_1x_2) = \Sigma(x_{1i} - \bar{x}_1)(x_{2i} - \bar{x}_2) = \Sigma x_{1i}x_{2i} - \frac{\Sigma x_{1i}\,\Sigma x_{2i}}{n}$$

$$S(x_1y) = \Sigma(x_{1i} - \bar{x}_1)(y_i - \bar{y}) = \Sigma x_{1i}y_i - \frac{\Sigma x_{1i}\Sigma y_i}{n}$$

etc., the equation (*f*) can be written

$$(g)\qquad \begin{aligned} b_1S(x_1^2) + b_2S(x_1x_2) &= S(x_1y) \\ b_1S(x_1x_2) + b_2S(x_2^2) &= S(x_2y) \end{aligned}$$

which indicates their similarity to the original regression equations except for the fact that we are now dealing with deviations from the mean instead of the original measurements. Note that these equations are identical with those that would have been obtained if we had assumed originally that the expected value of y was given by an equation of the form

$$y - \bar{y} = \beta_1(x_1 - \bar{x}_1) + \beta_2(x_2 - \bar{x}_2)$$

and had determined the estimates b_1 and b_2 by minimizing

$$Q = \Sigma\{(y_i - \bar{y}) - \beta_1(x_{1i} - \bar{x}_1) - \beta_2(x_{2i} - \bar{x}_2)\}^2$$

In this case by a procedure similar to that used in obtaining (*d*) we can write

$$\begin{aligned} \Sigma(y_i - \tilde{y}_i)^2 &= \Sigma(y_i - \bar{y})^2 - b_1\Sigma(x_{1i} - \bar{x}_1)(y_i - \bar{y}) - b_2\Sigma(x_{2i} - \bar{x}_2)(y_i - \bar{y}) \\ &= S(y^2) - b_1S(x_1y) - b_2S(x_2y) \end{aligned}$$

where $\tilde{y}_i = \bar{y} + b_1(x_{1i} - \bar{x}_1) + b_2(x_{2i} - \bar{x}_2)$, which represents the simplest manner of computing the sum of the squares of the residuals.

Using the regression equations in the form (*g*), we shall carry through a determinantal solution originally advanced by R. A. Fisher. The actual computations can be carried out more compactly by the methods of Appendix 6A, but we are interested here in illustrating the principles involved in computing and testing the regression coefficients. The method is general, but for convenience in description we shall consider the case of 3 exact measurements x_{1i}, x_{2i}, x_{3i}, each influencing the observed value of a dependent variable y_i, it being understood that x_1, x_2, and x_3 may have any form and may be mutually dependent. For this case the normal equations (*g*) become

$$(h)\qquad \begin{aligned} b_1S(x_1^2) + b_2S(x_1x_2) + b_3S(x_1x_3) &= S(x_1y) \\ b_1S(x_2x_1) + b_2S(x_2^2) + b_3S(x_2x_3) &= S(x_2y) \\ b_1S(x_3x_1) + b_2S(x_3x_2) + b_3S(x_3^2) &= S(x_3y) \end{aligned}$$

The solution for the regression coefficients may be written in determinantal form as

$$b_1 = \frac{\begin{vmatrix} S(x_1y) & S(x_1x_2) & S(x_1x_3) \\ S(x_2y) & S(x_2^2) & S(x_2x_3) \\ S(x_3y) & S(x_3x_2) & S(x_3^2) \end{vmatrix}}{\begin{vmatrix} S(x_1^2) & S(x_1x_2) & S(x_1x_3) \\ S(x_2x_1) & S(x_2^2) & S(x_2x_3) \\ S(x_3x_1) & S(x_3x_2) & S(x_3^2) \end{vmatrix}}$$

with similar forms for b_2 and b_3. Designating the lower determinant by Δ and expanding the upper determinant by the first column, we obtain

$$b_1 = \frac{1}{\Delta}\left(S(x_1y)\begin{vmatrix} S(x_2^2) & S(x_2x_3) \\ S(x_3x_2) & S(x_3^2) \end{vmatrix} - S(x_2y)\begin{vmatrix} S(x_1x_2) & S(x_1x_3) \\ S(x_3x_1) & S(x_3^2) \end{vmatrix} + S(x_3y)\begin{vmatrix} S(x_1x_2) & S(x_1x_3) \\ S(x_2^2) & S(x_2x_3) \end{vmatrix} \right)$$

Making similar expansions of the solution for b_2 and b_3, we can write more briefly

$$(i) \qquad \begin{aligned} b_1 &= c_{11}S(x_1y) + c_{12}S(x_2y) + c_{13}S(x_3y) \\ b_2 &= c_{21}S(x_1y) + c_{22}S(x_2y) + c_{23}S(x_3y) \\ b_3 &= c_{31}S(x_1y) + c_{32}S(x_2y) + c_{33}S(x_3y) \end{aligned}$$

where, for example,

$$c_{32} = \frac{1}{\Delta}\begin{vmatrix} S(x_1^2) & S(x_1x_2) \\ S(x_3x_1) & S(x_3x_2) \end{vmatrix} = c_{23}$$

the latter equality following from the fact that $S(x_i, x_j) = S(x_j, x_i)$. The elements

$$\begin{matrix} c_{11} & c_{21} & c_{31} \\ c_{12} & c_{22} & c_{32} \\ c_{13} & c_{23} & c_{33} \end{matrix}$$

constitute the inverse of the matrix corresponding to the determinant Δ. It follows that c_{11}, c_{12}, and c_{13} can be obtained by solving the equations

$$(j) \qquad \begin{aligned} c_{11}S(x_1^2) + c_{12}S(x_1x_2) + c_{13}S(x_1x_3) &= 1 \\ c_{11}S(x_2x_1) + c_{12}S(x_2^2) + c_{13}S(x_2x_3) &= 0 \\ c_{11}S(x_3x_1) + c_{12}S(x_3x_2) + c_{13}S(x_3^2) &= 0 \end{aligned}$$

The elements c_{21}, c_{22}, c_{23} and c_{31}, c_{32}, c_{33} are obtained by solving similar equations with the right-hand side replaced by 0, 1, 0 and 0, 0, 1. The three regression coefficients may then be evaluated by the use of equations (*i*).

After having obtained estimates of the regression coefficients, it will usually be necessary to test for significance, and often to decide whether a particular β is greater or less than some fixed value B prescribed by economic considerations. For example, in a process in which the price obtained for a product was scaled according to its performance in a given test, we should be interested in determining the relationship between throughput and test performance. If the regression coefficient so obtained exceeded a value B which we could compute from works costing data, it would be economic to increase the throughput of the plant at the expense of the quality of the product. Since B itself will usually be a function of the throughput, the increase could be continued until β no longer differed from B at the chosen significance level.

In order to make these tests it is necessary to know the form of the distribution of the estimates b_i. Equations (i) indicate that, for fixed values of the x's, the b_i are linear combinations of independent normal deviates and are therefore normally distributed. It may also be shown that the average value of b_i is β_i. In order to form confidence estimates for β we therefore need to know, or estimate from our data, $\sigma_b{}^2$. Rewriting the first of equations (i) in the form

$$b_1 = \Sigma_i(y_i - \bar{y})\,[c_{11}(x_{1i} - \bar{x}_1) + c_{12}(x_{2i} - \bar{x}_2) + c_{13}(x_{3i} - \bar{x}_3)]$$

we see that

$$\begin{aligned}\text{var }(b_1) &= \sigma^2\,\Sigma_i[c_{11}(x_{1i} - \bar{x}_1) + c_{12}(x_{2i} - \bar{x}_2) + c_{13}(x_{3i} - \bar{x}_3)]^2 \\ &= \sigma^2\,\{c_{11}[c_{11}S(x_1{}^2) + c_{12}S(x_1x_2) + c_{13}S(x_1x_3)] \\ &\quad + c_{12}[c_{11}S(x_2x_1) + c_{12}S(x_2{}^2) + c_{13}S(x_2x_3)] \\ &\quad + c_{13}[c_{11}S(x_3x_1) + c_{12}S(x_3x_2) + c_{13}S(x_3{}^2)]\}\end{aligned}$$

where σ^2 is the constant variance of the observations y_i. Since $c_{12} = c_{21}$, and $c_{13} = c_{31}$, the first bracket is simply the sum of the products of the elements of the first row of the determinant Δ and their respective cofactors, divided by Δ, and therefore equal to unity. The second and third brackets correspond to the products of the elements of the second and third rows of the determinant Δ and the cofactors of the first row, divided by Δ, and are therefore zero. These conclusions correspond exactly to the equations (j). Thus we obtain

$$\text{var }(b_1) = \sigma^2 \cdot c_{11}$$

and by similar procedures

$$\text{var }(b_2) = \sigma^2 \cdot c_{22}$$

$$\text{var }(b_3) = \sigma^2 \cdot c_{33}$$

The coefficient b_i is therefore distributed normally about β_i with variance $\sigma^2 c_{ii}$. By a similar procedure we can show that

$$\text{cov}\,(b_1, b_2) = c_{12}\sigma^2$$

$$\text{cov}\,(b_1, b_3) = c_{13}\sigma^2$$

$$\text{cov}\,(b_2, b_3) = c_{23}\sigma^2$$

If the estimate from our data of σ^2 is $s^2 = \dfrac{\Sigma(y_i - \tilde{y}_i)^2}{n - 3 - 1}$, then $\dfrac{b_i - \beta_i}{s\sqrt{c_{ii}}}$ has the t-distribution with $n - 4$ degrees of freedom, since under these conditions $\dfrac{(n-4)s^2}{\sigma^2}$ has the χ^2-distribution with $n - 4$ degrees of freedom, owing to the 4 linear conditions imposed on the y_i in minimizing the sum of squares of residuals with respect to α, β_1, β_2, and β_3. The confidence interval for β_i at a probability level α may thus be written as

$$b_i - t_{n-4,\,\alpha} s\sqrt{c_{ii}} < \beta_i < b_i + t_{n-4,\,\alpha} s\sqrt{c_{ii}}$$

and our estimate b_i may be tested against any prescribed value B_i.

The variance of the predicted value of y associated with a particular $X_1, X_2, \cdots, X_k$ can be determined since the variances and covariances of the b's are known. We have

$$\tilde{y} = a + b_1(X_1 - \bar{x}_1) + b_2(X_2 - \bar{x}_2) + \cdots + b_k(X_k - \bar{x}_k)$$

$$\text{var}\,(\tilde{y}) = \sigma^2 \left[\frac{1}{n} + \Sigma_j(X_j - \bar{x}_j)^2 c_{jj} + \Sigma_{j \neq k}(X_j - \bar{x}_j)\,({}_kX - \bar{x}_k)c_{jk}\right]$$

and this can be estimated by substituting s^2 for σ^2. Since y is normally distributed and the estimate s^2 is distributed as $\dfrac{\sigma^2\chi^2}{n - k - 1}$ we can form confidence limits for y based on the "Student" t-distribution.

In addition to providing a method of estimating the standard error of the regression coefficients, this technique of solution is useful in cases where, for each fixed set of measurements $x_{1i}, x_{2i}, \cdots, x_{ki}$, measurements are made of more than one dependent variable. This situation may occur in large-scale work, where having once instrumented the unit for measurement purposes we desire to collect a wide range of data without prolonging the test. Under these circumstances, if measurements $y_{1i}, y_{2i}, \cdots, y_{mi}$ are made at each of a set of values of the fixed measurements $x_{1i}, x_{2i}, \cdots, x_{ki}$, the inverse matrix need be computed only once. The regression coefficients of the variable y_h are then computed as

$$b_{hj} = c_{j1}S(y_h x_1) + c_{j2}S(y_h x_2) + \cdots + c_{jk}S(y_h x_k)$$

Estimates of the variance of the regression coefficients may be computed in the usual way, the diagonal elements of the inverse matrix and the appropriate estimate s_h^2 of the variance σ_h^2 of y_h being used.

As in the case of a single independent variable, the above procedures place no restriction on the form of the independent variables $x_1, \cdots, x_k$. Hence they can be applied to situations where the expected value of y is given by relationships of the type $y = \beta_1 \sin t + \beta_2 \cos t$, or $y = \alpha + \beta_1 e^{-\lambda_1 t} + \beta_2 e^{-\lambda_2 t}$ (assuming in the latter case that λ_1 and λ_2 are known) by letting $x_1 = \sin t$, $x_2 = \cos t$, or $x_1 = e^{-\lambda_1 t}$, $x_2 = e^{-\lambda_2 t}$.

A situation that is of frequent occurrence in chemistry and physics is the case where the random variable y is assumed to be a polynomial function of a single known variable x. Here we assume that for each pair of observed values (x_i, y_i) we have ave $(y_i) = \alpha + \beta_1 x_i + \beta_2 x_i^2 + \cdots + \beta_k x_i^k$, where k is the degree of the polynomial to be fitted. It can be seen that this is simply a special case of the preceding discussion with $x_{1i} = x_i$, $x_{2i} = x_i^2$, $x_{3i} = x_i^3$, etc. Note that not only the sums of squares but also the cross products will be sums of powers; for example, $\Sigma x_{1i} x_{2i} = \Sigma x_i^3$, $\Sigma x_{2i} x_{3i} = \Sigma x_i^5$, etc. In general fitting a kth degree polynomial will require the sums of powers of the x_i to $2k$, so that as the degree of the polynomial increases the numerical labor involved both in obtaining the necessary power sums and in solving the equations for the estimates $a, b_1, b_2, \cdots, b_k$ increases rapidly.

EXAMPLE. Tests carried out to determine the influence of gas inlet temperature and rotor speed on the tar content of a gas stream are reported by Badger, *J. Soc. Chem. Ind.*, *167* (1946), 65. The tests were carried out on a high-speed 3-stage turbo exhauster at rotor speeds from 2400 to 3900 rpm and at inlet temperatures ranging from 43° to 69° F, using a stream of carburetted water gas containing tar fog. The results are recorded in Table 6.9, and on plotting Figure 6.10 there is an indication that:

(1) Increased rotor speed reduces the tar content of the exit gas stream, and the relationship is only approximately linear.
(2) At lower inlet temperatures the tar content of the exit gases is less.

It would therefore appear reasonable to attempt to represent the data by an equation in the form

$$\text{tar content} = a + b_1\,(\text{speed}) + b_2\,(\text{speed})^2 + b_3\,(\text{temp})$$

For convenience in computation we code the results as follows:

$$x_1 = \text{rotor speed (rpm)} - 2{,}400$$

$$x_2 = x_1^2$$

$$x_3 = \text{inlet temperature (°F)} - 40$$

$$y = \text{tar content (grains/100 cu ft)}$$

The coded data are also given in Table 6.9.

Computing the sums, sums of squares, and sums of products we obtain

$N = 31$

$\Sigma y = 1{,}269.5$ $\Sigma x_1 = 18{,}325$ $\Sigma x_2 = 15{,}344{,}375$ $\Sigma x_3 = 572.5$

$\Sigma y^2 = 58{,}510.75$ $\Sigma yx_1 = 642{,}750$ $\Sigma yx_2 = 499{,}783{,}125$ $\Sigma yx_3 = 24{,}374.25$

$\Sigma x_1^2 = 14{,}344{,}375$ $\Sigma x_1x_2 = 14{,}882{,}171{,}875$ $\Sigma x_1x_3 = 361{,}787.5$

$\Sigma x_2^2 = 15{,}915{,}763{,}671{,}875$ $\Sigma x_2x_3 = 308{,}076{,}562.5$

$\Sigma x_3^2 = 12{,}019.25$

TABLE 6.9

Rotor Speed (rpm)	Temperature (°F)	Tar Content (grains per 100 cu ft)	x_1	x_2	x_3	y
2400	54.5	60.0	0	0	14.5	60.0
2450	56.0	61.0	50	2,500	16.0	61.0
2450	58.5	65.0	50	2,500	18.5	65.0
2500	43.0	30.5	100	10,000	3.0	30.5
2500	58.0	63.5	100	10,000	18.0	63.5
2500	59.0	65.0	100	10,000	19.0	65.0
2700	62.5	44.0	300	90,000	12.5	44.0
2700	65.5	52.0	300	90,000	25.5	52.0
2700	68.0	54.5	300	90,000	28.0	54.5
2750	45.0	30.0	350	122,500	5.0	30.0
2775	45.5	26.0	375	140,625	5.5	25.0
2800	48.0	23.0	400	160,000	8.0	23.0
2800	63.0	54.0	400	160,000	23.0	54.0
2900	58.5	36.0	500	250,000	18.5	36.0
2900	64.5	53.5	500	250,000	24.5	53.5
3000	66.0	57.0	600	360,000	26.0	57.0
3075	57.0	33.5	675	455,625	17.0	33.5
3100	57.5	34.0	700	490,000	17.5	34.0
3150	64.0	44.0	750	562,500	24.0	44.0
3200	57.0	33.0	800	640,000	17.0	33.0
3200	64.0	39.0	800	640,000	24.0	39.0
3200	69.0	53.0	800	640,000	29.0	53.0
3225	68.0	38.5	825	680,625	28.0	38.5
3250	62.0	39.5	850	722,500	22.0	39.5
3250	64.5	36.0	850	722,500	24.5	36.0
3250	48.0	8.5	850	722,500	8.0	8.5
3500	60.0	30.0	1100	1,210,000	20.0	30.0
3500	59.0	29.0	1100	1,210,000	19.0	29.0
3500	58.0	26.5	1100	1,210,000	18.0	26.5
3600	58.0	24.5	1200	1,440,000	18.0	24.5
3900	61.0	26.5	1500	2,250,000	21.0	26.5

In order to reduce computing errors when a calculator is employed, it is customary to select the coding system so that changes of sign do not occur within a column. This usually means that a large number of

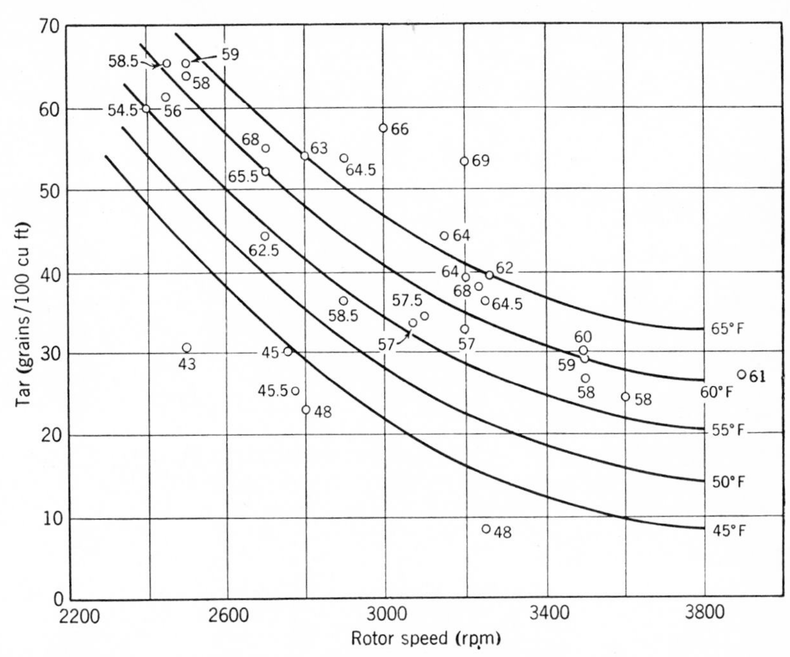

FIG. 6.10. Note: Figures given with plotted points are inlet temperatures in °F.

figures will occur in the sums of squares and cross products. These should not usually be rounded off until corrected to the sample mean by use of the expressions

$$S(x_jx_k) = \Sigma_i(x_{ij} - x_j)(x_{ik} - x_k) = \Sigma_i x_{ij}x_{ik} - \frac{\Sigma_i x_{ij}\Sigma_i x_{ik}}{N}$$

Carrying out this process we obtain

$S(y^2) = 6.522677(10^3)$ $\quad S(yx_1) = 1.076883(10^5)$ $\quad S(yx_2) = 1.285938(10^8)$ $\quad S(yx_3) = 9.294516(10^2)$

$S(x_1^2) = 4.511936(10^6)$ $\quad S(x_1x_2) = 5.811666(10^9)$ $\quad S(x_1x_3) = 2.336613(10^4)$

$S(x_2^2) = 8.320607(10^{12})$ $\quad S(x_2x_3) = 2.470061(10^7$

$S(x_3^2) = 1.446468(10^3$

We can then write out the determinant

$$\Delta = \begin{vmatrix} 4.511936(10^6) & 5.811666(10^9) & 2.336613(10^4) \\ 5.811666(10^9) & 8.320607(10^{12}) & 2.470061(10^7) \\ 2.336613(10^4) & 2.470061(10^7) & 1.446468(10^3) \end{vmatrix} = 4.861067(10^{21})$$

The inverse matrix may be computed by solving 3 sets of equations of the type (j), but unless special methods are used this is most easily done by replacing each term in the above determinant by its cofactor and dividing each element in the resulting matrix by Δ. Thus, for example, we have by definition

$$c_{21} = c_{12} = \frac{-5.811666(10^9) \cdot 1.446468(10^3) - 2.336613(10^4) \cdot 2.470061(10^7)}{4.861067(10^{21})} = -1.610600(10^{-9})$$

The complete inverse matrix is

$$c_{11} = +2.350385(10^{-6}) \quad c_{12} = -1.610600(10^{-9}) \quad c_{13} = -1.046452(10^{-5})$$

$$c_{21} = -1.610600(10^{-9}) \quad c_{22} = +1.230265(10^{-12}) \quad c_{23} = +5.008897(10^{-9})$$

$$c_{31} = -1.046452(10^{-5}) \quad c_{32} = +5.008897(10^{-9}) \quad c_{33} = +7.748477(10^{-4})$$

The regression coefficients can be evaluated immediately from equations (j) as

$$b_1 = -2.350385(10^{-6}) \cdot 1.076883(10^5) + 1.610600(10^{-9}) \cdot 1.285938(10^8)$$
$$-1.046452(10^{-5}) \cdot 9.294516(10^2) = -0.055721$$

Similarly

$$b_2 = +0.000019893 \qquad b_3 = +1.20297$$

In order to form confidence intervals for the regression coefficients we proceed to determine s^2 using the relationship

$$\Sigma_i(y_i - \tilde{y}_i)^2 = S(y^2) - b_1 S(yx_1) - b_2 S(yx_2) - b_3 S(yx_3)$$
$$= 6522.677 - 6000.483 + 2558.168 - 1118.102$$
$$= 1962.260$$

$$s^2 = \frac{\Sigma_i(y_i - \tilde{y}_i)^2}{31 - 4} = 72.6763$$

$$s = 8.525$$

Then

$$s_{b1} = s\sqrt{c_{11}} = 0.01307$$
$$s_{b2} = s\sqrt{c_{22}} = 9.456(10^{-6})$$
$$s_{b3} = s\sqrt{c_{33}} = 0.2373$$

and from the tables $t_{27,\,0.05} = 2.05$, so that the confidence intervals at the 5% level are

$$0.05572 - 2.05(0.01307) < \beta_1 < 0.05572 + 2.05(0.01307)$$

$$0.0289 < \beta_1 < 0.0825$$

and, similarly for β_2 and β_3

$$0.0000005 < \beta_2 < 0.0000393$$

$$0.717 \qquad < \beta_3 < 1.689$$

It is apparent that both the linear effect of speed and the effect of temperature are highly significant, whereas the apparent non-linearity as measured by the quadratic term is just significant at the 5% level.

The fitted equation may be written

$$y - \bar{y} = -0.05572(x_1 - \bar{x}_1) + 0.00001989(x_2 - \bar{x}_2) + 1.203(x_3 - \bar{x}_3)$$

or

$$y = 41.8274 - 0.05572x_1 + 0.00001989x_2 + 1.203x_3$$

Returning to the original variables of speed (rpm) and temperature (°F), we have tar content = 12.869 − 0.1512 (speed) + 0.00001989 (speed²) + 1.203 (temp). Using this equation, lines of constant temperature have been superimposed on Figure 6.10, and it is seen that the fitted curves pass through a minimum value at a speed of about 3900 rpm. This conflicts with technical experience, and indicates that, although our model is a reasonably satisfactory description of the data within the region in which our measurements have been made, it does not provide a sound basis for extrapolation. Extrapolation should only be used in cases where some physical law, i.e., previous experience, leads us to suppose that the form of model which we choose is applicable to a much wider range than that from which our data were taken.

6.32. Polynomial Regression on Equally Spaced Observations

If the values of x_i are equally spaced, as is frequently the case when the choice of the values x_i at which y_i is to be observed is under the control of the experimenter, it is always possible to reduce these values to the form $0, \pm 1, \pm 2, \cdots, \pm \frac{n-1}{2}$ if n is odd, or $\pm\frac{1}{2}, \pm 3/2, \cdots, \pm \frac{n-1}{2}$ if n is even, by the transformation $z_i = \frac{x_i - x_0}{w}$, where $x_0 = \bar{x}$ is the middle value of the x_i if n is odd, and the midpoint of the two middle values if n is even, and w is the common distance between the x_i. With these values the determination of the estimates $a, b_1, b_2, \cdots, b_k$ for the polynomial regression of y on z is somewhat simplified, since all the sums of odd powers of z vanish, owing to the symmetry of the z_i about 0, and all the remaining sums, except those involving the observations y_i, depend only on n. None of these advantages are lost if for even n the values $\pm 1, \pm 3, \cdots, \pm(n-1)$ are used to avoid fractions. Since the y_i have not been changed, the residual sum of squares is not affected. If the regression equation were desired in terms of the original variable x, it would be necessary to perform a back substitution for z in the equation determined, but for most purposes this is unnecessary, since it is much simpler to obtain the computed values $\tilde{y}_i$ directly from the equation in z, or, if a predicted value $\tilde{y}_0$ for a particular x_0 is desired, to first obtain z_0 and then use this value in the equation in z directly.

The computations can be further simplified, and additional advantages secured, by assuming the regression in the form

$$\text{ave}\,(y_i) = \alpha' + \beta_1'\xi_{1i} + \beta_2'\xi_{2i} + \cdots + \beta_k'\xi_{ki}$$

where $\xi_1, \xi_2, \xi_3, \cdots, \xi_k$ are polynomials of degree $1, 2, 3, \cdots, k$, respectively, in x, which have the property that $\Sigma_i \xi_{ri}\xi_{si} = 0$, $r \neq s$. Functions having this property are called orthogonal. Note that an equation involving ξ_i, which is of degree 1, will be linear; an equation involving ξ_1 and ξ_2 will be quadratic, since ξ_2 is quadratic in x; and in general an equation involving $\xi_1, \xi_2, \cdots, \xi_k$ will be of degree k in x. Hence fitting a linear equation involving $\xi_1, \cdots, \xi_k$ is equivalent to fitting a polynomial of degree k. By definition all the cross-product terms in the normal equations for determining the estimates $a', b_1', b_2' \cdots,$ b_k', of the regression coefficients of y on $\xi_1, \xi_2, \cdots, \xi_k$ vanish, and for equally spaced x's the sums $\Sigma\xi_{1i}, \Sigma\xi_{2i}, \cdots, \Sigma\xi_{ki}$ are also zero (this may be thought of as defining the existence of an orthogonal polynomial $\xi_0 = 1$ of degree zero). Hence the normal equations reduce to

$$\text{(a)}\quad \begin{array}{llll} na' + 0 & + 0 & + \cdots + 0 & = \Sigma y_i \\ 0 + b_1'\,\Sigma_i\,\xi_{1i}^2 + 0 & & + \cdots + 0 & = \Sigma\xi_{1i}y_i \\ 0 + 0 & + b_2'\,\Sigma_i\,\xi_{2i}^2 & + \cdots + 0 & = \Sigma\xi_{2i}y_i \\ \vdots & \vdots & \ddots & \vdots \\ 0 + 0 & + 0 & + \cdots + b_k'\Sigma_i\xi_{ki}^2 & = \Sigma\xi_{ki}y_i \end{array}$$

and we obtain immediately

$$a' = \frac{\Sigma_i y_i}{n} = \bar{y}$$

$$b_1' = \frac{\Sigma_i \xi_{1i} y_i}{\Sigma_i \xi_{1i}^2}$$

(*b*)
$$b_2' = \frac{\sum_i \xi_{2i} y_i}{\sum_i \xi_{2i}^2}$$
$$b_r' = \frac{\sum_i \xi_{ri} y_i}{\sum_i \xi_{ri}^2}$$
$$b_k' = \frac{\sum_i \xi_{ki} y_i}{\sum_i \xi_{ki}^2}$$

For the case of equally spaced x's, the first three orthogonal polynomials, in terms of the z's, are

(*c*)
$$\xi_{1i} = z_i$$
$$\xi_{2i} = z_i^2 - \frac{n^2 - 1}{12}$$
$$\xi_{3i} = z_i^3 - \frac{3n^2 - 7}{20} z_i$$

and in general we can obtain additional values by using the recursion formula

(*d*)
$$\xi_{r+1} = \xi_r \xi_1 - \frac{r^2(n^2 - r^2)}{4(4r^2 - 1)} \xi_{r-1}$$

Alternate ξ's contain only even or only odd powers of z, owing to the symmetry of these values about zero. Since the values of z_i depend only on n, the corresponding values of ξ_{ri} depend only on n, and hence can be determined and tabulated for various n.

Unfortunately, the values ξ_{ri} for a given n are frequently fractional, and at other times contain a common factor. To simplify the computation as much as possible, values ξ'_{ri} which represent the smallest set of integers such that $\xi'_{ri} = \lambda_r \xi_{ri}$ have been determined. The values λ_r are common to all ξ_{ri} for a particular n, and represent the coefficient of the highest power of z in the expression for ξ_r' in terms of z, since the coefficient of the highest power of z in the expression for ξ_r is always unity. In terms of the ξ'_{ri} we have

(*e*)
$$a'' = \frac{\sum_i y_i}{n} = \bar{y} = a'$$
$$b_1'' = \frac{\sum_i \xi'_{1i} y_i}{\sum_i \xi'^2_{1i}} = \frac{b_1'}{\lambda_1}$$
$$b_2'' = \frac{\sum_i \xi'_{2i} y_i}{\sum_i \xi'^2_{2i}} = \frac{b_2'}{\lambda_2}$$
$$b_r'' = \frac{\sum_i \xi_{ri} y_i}{\sum_i \xi'^2_{ri}} = \frac{b_r'}{\lambda_r}$$
$$b_k'' = \frac{\sum_i \xi'_{ki} y_i}{\sum_i \xi'^2_{ki}} = \frac{b_k'}{\lambda_k}$$

Values of ξ'_{ri} for $n = 3$ to 75 and $r = 1$ to 5 are given in [13]. For values of $n > 8$ only the values corresponding to positive z_i are given. The values for negative z_i are identical for r even, and of opposite sign for r odd. Values of the divisor $D_r = \Sigma\xi'_{ri}{}^2$ and the constant λ_r are also given. Thus the coefficients b_r'' for the regression of y on the ξ''s are obtained simply by multiplying the values of y_i by the appropriate values ξ'_{ri} and dividing by D_r. If the equation is desired in terms of the ξ's, the coefficients b'_r can be determined from the b''_r by multiplying by the given values of λ_r.

This method has the particular advantage that each successive coefficient is independent of the preceding one, so that going from a quadratic to a cubic equation, for example, requires only the additional computation of b''_3. In the usual method the entire solution for the coefficients a, b_1, b_2, b_3, would be required, since in general the coefficients of x and x^2 and the constant term will be different for the quadratic and cubic equations. Also, because of the orthogonality of the ξ_r's, the reductions of the sums of squares of the residuals due to each successive power of x are independent, and we have

$$\begin{aligned}
\Sigma_i(y_i - \tilde{y}_i)^2 &= \Sigma_i(y_i - \bar{y})^2 - b'_1{}^2\Sigma_i\xi_{1i}{}^2 - b'_2{}^2\Sigma_i\xi_{2i}{}^2 - \cdots - b'_k{}^2\Sigma_i\xi_{ki}{}^2 \\
(f) \qquad &= \Sigma_i(y_i - \bar{y})^2 - b''_1{}^2\Sigma_i\xi'_{1i}{}^2 - b''_2{}^2\Sigma_i\xi'_{2i}{}^2 - \cdots - b''_k{}^2\Sigma_i\xi'_{ki}{}^2 \\
&= \Sigma_i(y_i - \bar{y})^2 - b''_1{}^2 D_1 - b''_2{}^2 D_2 - \cdots - b''_k{}^2 D_k
\end{aligned}$$

The quantity $\Sigma_i(y_i - \bar{y})^2$ could have been expressed as $\Sigma_i y_i^2 - na'^2 = \Sigma_i y_i^2 - na''^2 = \Sigma_i y_i^2 - n\bar{y}^2$. Each successive reduction in the residual sum of squares is independent and we can immediately assess the importance of each additional power of x added. Since every b_r'' (or b_r') is a linear function of the original independent observations y_i (the ξ'_{ri} depend only on n), they are all normally distributed with ave $(b''_r) = \beta'_r/\lambda_r$ and var $(b''_r) = \sigma^2/D_r$. Hence, using the estimate $s^2 = \dfrac{\Sigma(y_i - \tilde{y}_i)^2}{n - k - 1}$ of σ^2, we can test whether a given $b_r''$is significantly different from zero by computing

$$t_{n-k-1} = \frac{b_r''\sqrt{D_r}}{s}$$

which has the t-distribution with $n - k - 1$ degrees of freedom, and comparing it with the value of $t_{n-k-1,\,\alpha}$ for the desired significance level. In Chapter 7 a completely equivalent method for testing the significance of the reduction in the sum of squares due to a particular power of x will be discussed.

When using this method, we can easily determine the expected variance of a computed value $\tilde{y}_i$. We have

$$\tilde{y}_i = a'' + b_1''\xi'_{1i} + b_2''\xi'_{2i} + b_3''\xi'_{3i} + \cdots + b_k''\xi'_{ki}$$

hence $\tilde{y}_i$ is a linear function of the independently and normally distributed sample statistics b_r'' and the constant $a'' = \bar{y}$, which is also normally distributed with ave $(a'') = \alpha'$ and var $(a'') = \sigma^2/n$. Hence by the theorem of Section 4.42, $\tilde{y}_i$ is also normally distributed with

$$\begin{aligned}
\text{ave}\,(\tilde{y}_i) &= \text{ave}\,(\alpha'') + \xi_{1i}'\,\text{ave}\,(b_1'') + \cdots + \xi_{ki}'\,\text{ave}\,(b_k'') \\
&= \alpha' + \frac{\beta_1'\xi_{1i}'}{\lambda_1} + \cdots + \frac{\beta_k'\xi_{ki}'}{\lambda_k} \\
&= \alpha' + \beta_1'\xi_{1i} + \cdots + \beta_k'\xi_{ki}
\end{aligned}$$

as we originally assumed, and

$$\begin{aligned}
\text{var}\,(\tilde{y}_i) &= \text{var}\,(a'') + \xi_{1i}'^2\,\text{var}\,(b_{1i}'') + \cdots + \xi_{ki}'^2\,\text{var}\,(b_{ki}'') \\
&= \sigma^2\left(\frac{1}{n} + \frac{\xi_{1i}'^2}{D_1} + \cdots + \frac{\xi_{ki}'^2}{D_k}\right) \\
&= \sigma^2\left(\frac{1}{n} + \sum_{r=1}^{k} \frac{\xi_{ri}'^2}{D_r}\right)
\end{aligned}$$

A confidence interval for the average value of y corresponding to a particular x_i is given by

$$\tilde{y}_i - t_{n-k-1,\,\alpha}\, s\sqrt{\frac{1}{n} + \sum_{r=1}^{k}\frac{\xi_{ri}'^2}{D_r}} < \text{ave}\,(y_i) < \tilde{y}_i + t_{n-k-1,\,\alpha}\, s\sqrt{\frac{1}{n} + \sum_{r=1}^{k}\frac{\xi_{ri}'^2}{D_r}}$$

One disadvantage of the above method is that the fitted equation is obtained in terms of $\xi_1', \cdots, \xi_k'$, or $\xi_1, \cdots, \xi_k$, rather than in terms of the original x's. For the purpose of obtaining computed values, or for merely determining the nature of the dependence by examining the residual sum of squares after the addition of successive terms, this does not matter, since no return to the original x's is necessary. However, if extrapolated values are desired, the ξ_{ri} for corresponding values of z_i (assuming the extrapolated value also to be equally spaced) must be computed from the equations (c) since values of ξ_{ri} are tabulated only for the n points observed. If the original coefficients $\alpha, \beta_1, \cdots, \beta_k$ have physical importance, so that it is desirable to have the original estimates $a, b_1, \cdots, b_k$ of these coefficients, then it is necessary to transform the equation in ξ_r' to ξ_r, using the λ_r, in ξ_k to z using equations (c), and in z to x, using the transformation $z = \dfrac{x - x_0}{w}$. Factors which simplify this procedure are given in [11].

EXAMPLE. The data recorded in Table 6.10 and illustrated in Figure 6.11 represent measurements made during an investigation of the influence

TABLE 6.10

Annealing Temp. °C	Density − 2.2350
450	0.00144
475	0.00034
500	0.00016
525	0.00074
550	0.00013
575	0.00174
600	0.00248
625	0.00402
650	0.00485
675	0.00505
700	0.00585
725	0.00620
750	0.00618

of annealing temperature upon the density of a high silica borosilicate glass. The annealing treatment was carried out at temperatures from 450 to 750°C by 25°C intervals and was prolonged until constant density was reached. Tests were carried out on a number of specimens cut from a sample plate. Although there is an obvious systematic variation of density with temperature, this appears non-linear. It is instructive to fit a regression line by the orthogonal polynomial method and determine which of the variations from linearity can be regarded as statistically significant on the basis of these data.

In order to simplify the arithmetic the results should be coded, the following transformations being convenient:

$$z_i = \frac{T_i - 600}{25} \qquad y_i = (\text{Density} - 2.23500)\,(10^5)$$

This transformation gives Table 6.11.

TABLE 6.11

z	−6	−5	−4	−3	−2	−1	0	1	2	3	4	5	6
y	144	34	16	74	13	174	248	402	485	505	585	620	618

As an illustration, the first 5 orthogonal polynomials for 13 observations will be computed and listed in Table 6.12, together with the reduced values ξ_{ri}', the factors λ_r, and the sums of squares $(\Sigma\xi'^2_{ri}) = D_r$ of the reduced values. Ordinarily these values would be obtained directly from the tables referred to above.

The first row of the table may be written down immediately since $\xi_{1i} = z_i$. As typical terms for the next row, for which the general form is $\xi_{2i} = z_i^2 - \dfrac{n^2 - 1}{12}$, we consider $z = -4$ and $z = +5$, the remaining

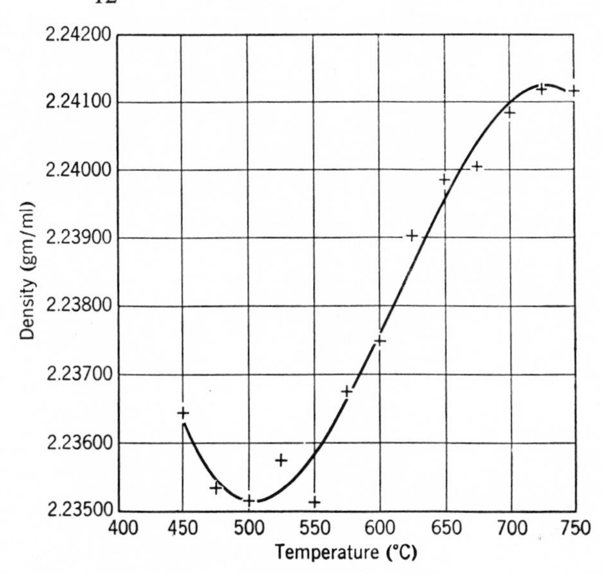

FIG. 6.11.

terms being computed in a similar fashion.

$$\xi_{2,\,-4} = 16 - \frac{12(14)}{12} = +2$$

$$\xi_{2,\,+5} = 25 - \frac{12(14)}{12} = +11$$

For the third-order polynomial $\xi_{3i} = z_i^3 - \dfrac{3n^2 - 7}{20}\, z_i$ we again compute the terms $z = -4$ and $z = +5$.

$$\xi_{3,\,-4} = -64 - \frac{(507 - 7)(-4)}{20} = 36$$

$$\xi_{3,\,+5} = 125 - \frac{(507 - 7)(+5)}{20} = 0$$

TABLE 6.12

z	−6	−5	−4	−3	−2	−1	0	+1	+2	+3	+4	+5	+6	λ	D
ξ_1	−6	−5	−4	−3	−2	−1	0	+1	+2	+3	+4	+5	+6	1	182
ξ_2	+22	+11	+2	−5	−10	−13	−14	−13	−10	−5	+2	+11	+22	1	2,002
ξ_3	−66	0	+36	+48	+42	+24	0	−24	−42	−48	−36	0	+66	1	20,592
ξ_4	$\frac{+1188}{7}$	$\frac{-792}{7}$	$\frac{-1152}{7}$	$\frac{-648}{7}$	$\frac{+132}{7}$	$\frac{+768}{7}$	$\frac{+1008}{7}$	$\frac{+768}{7}$	$\frac{+132}{7}$	$\frac{-648}{7}$	$\frac{-1152}{7}$	$\frac{-792}{7}$	$\frac{+1188}{7}$	1	$\frac{9,801,792}{49}$
ξ_5	$\frac{-2640}{7}$	$\frac{+3960}{7}$	$\frac{+2160}{7}$	$\frac{-1320}{7}$	$\frac{-3120}{7}$	$\frac{-2400}{7}$	0	$\frac{+2400}{7}$	$\frac{+3120}{7}$	$\frac{+1320}{7}$	$\frac{-2160}{7}$	$\frac{-3960}{7}$	$\frac{+2640}{7}$	1	$\frac{89,107,200}{49}$
ξ_1'	−6	−5	−4	−3	−2	−1	0	+1	+2	+3	+4	+5	+6	1	182
ξ_2'	+22	+11	+2	−5	−10	−13	−14	−13	−10	−5	+2	+11	+22	1	2,002
ξ_3'	−11	0	+6	+8	+7	+4	0	−4	−7	−8	−6	0	+11	1/6	572
ξ_4'	+99	−66	−96	−54	+11	+64	+84	+64	+11	−54	−96	−66	+99	7/12	68,068
ξ_5'	−22	+33	+18	−11	−26	−20	0	+20	+26	+11	−18	−33	+22	7/120	6,188

The fourth- and fifth-order terms may be computed by the recursion formula:

$$r+1=4, \quad z_i=-4, \quad \xi_{4,\,-4}=(+36)(-4)-\frac{9(160)(2)}{4(35)}=\frac{-1152}{7}$$

$$z_i=+5, \quad \xi_{4,\,+5}=0-\frac{9(160)(11)}{4(35)}=\frac{-792}{7}$$

$$r+1=5, \quad z_i=-4, \quad \xi_{5,\,-4}=\frac{-1152}{7}(-4)-\frac{16(153)(36)}{4(63)}=\frac{2160}{7}$$

$$z_i=+5, \quad \xi_{5,\,+5}=\frac{-792}{7}(5)-0=\frac{-3960}{7}$$

Examination of the table indicates that the factors 6, 12/7 and 120/7 may be removed from ξ_{3i}, ξ_{4i}, and ξ_{5i} to obtain the tabulated values ξ_{ri}'. The values of λ_r are the reciprocals of these factors.

We now proceed to compute

$$\frac{\Sigma\xi'_{1i}y_i}{\Sigma\xi'^{\,2}_{1i}}=b''_1=57.775 \qquad b'_1=\lambda_1 b''_1=57.775$$

$$\frac{\Sigma\xi'_{2i}y_i}{\Sigma\xi'^{\,2}_{2i}}=b''_2=3.15934 \qquad b'_2=\lambda_2 b''_2=3.15934$$

$$\frac{\Sigma\xi'_{3i}y_i}{\Sigma\xi'^{\,2}_{3i}}=b''_3=-10.2517 \qquad b'_3=\lambda_3 b''_3=-1.70862$$

$$\frac{\Sigma\xi'_{4i}y_i}{\Sigma\xi'^{\,2}_{4i}}=b''_4=+0.095286 \qquad b'_4=\lambda_4 b''_4=0.055584$$

$$\frac{\Sigma\xi'_{5i}y_i}{\Sigma\xi'^{\,2}_{5i}}=b''_5=-0.391241 \qquad b'_5=\lambda_5 b''_5=-0.022822$$

$$\Sigma y_i=3918 \qquad \Sigma y_i^2=1{,}879{,}976 \qquad \Sigma(y_i-\bar{y})^2=\Sigma y_i^2-\frac{(\Sigma y_i)^2}{13}$$
$$=699{,}151$$

$$\Sigma(y_i-\tilde{y}_i)^2=\Sigma(y_i-\bar{y})^2-(b''_1)^2\Sigma\xi'^{\,2}_{1i}-(b''_2)^2\Sigma\xi'^{\,2}_{2i}-\cdots-(b''_5)^2\Sigma\xi'^{\,2}_{5i}$$
$$=699{,}151-607{,}504-19{,}983-60{,}116-618-947=9983$$

$$s^2=\frac{\Sigma(y_i-\tilde{y}_i)^2}{n-k-1}=\frac{9983}{7}=1426$$

Testing the b_5'' term, we obtain

$$t_7=\frac{b_5''\sqrt{D_5}}{s}=\frac{\sqrt{947}}{\sqrt{1426}}=-0.81$$

which is not significantly different from zero. We may therefore include its contribution to the sum of squares to give an extra degree of freedom in the residual before testing b_4'', so that

$$s^2 = \frac{9983 + 947}{8} = 1366.2$$

Testing the b_4'' term on the basis of this new estimate

$$t_8 = \frac{\sqrt{618}}{\sqrt{1366.2}} = 0.67$$

which again is not significant. To test the b_3'' term we form

$$s^2 = \frac{9983 + 947 + 618}{9} = 1283.1$$

$$t_9 = \frac{\sqrt{60{,}033}}{\sqrt{1283.1}} = 6.83^{***}$$

and, similarly for b_2'' and b_1'', we have

$$b_2'' : t_9 = \sqrt{\frac{19{,}983}{1283.1}} = 3.95^{***}$$

$$b_1'' : t_9 = \sqrt{\frac{607{,}506}{1283.1}} = 21.8^{***}$$

The linear, quadratic, and cubic components are all significant and the equation becomes:

$$y = \bar{y} + 57.775\xi_1' + 3.1593\xi_2' + 10.2517\xi_3'$$

The values of y may then be computed for the various z values (Table 6.13) and plotted as shown in the figure.

TABLE 6.13

z	-6	-5	-4	-3	-2	-1	0	$+1$	$+2$	$+3$	$+4$	$+5$	$+6$
y	137.0	47.3	15.1	30.3	82.5	161.6	257.2	359.1	457.1	540.9	600.3	625.0	604.8

The reader interested in more complete discussions of these and other methods of calculation is referred to [11] and [12]. The notation used here is similar to that used in [12]. In [11] ξ_r is designated by T_t, ξ_r' by V_t, D_r by N_t, and λ_r by S_{tt}, to mention the more important changes.

6.33. Orthogonal Combinations in Multiple Regression

The use of orthogonal combinations, such as the polynomials introduced in the last section in connection with polynomial regression on a single equally spaced independent variable, can be extended to the general case. There is little computational advantage in this extension, since the proper combinations must be computed from the data for individual examples, but the methods involved frequently afford additional insight into the nature of the data considered. These methods, like those of Section 6.31, include the special case of polynomial regression on a single unequally spaced variate, in which case the combinations used will be polynomials as in the previous section, but they will be dependent on the spacing of the observations in the example under consideration.

We shall represent the observations by a model

$$(a)\quad \text{ave}\,(y) = b_{y0}C_0 + b_{y1.0}C_1 + b_{y2.01}C_2 + \cdots + b_{yk.012\cdots(k-1)}\,C_k$$

where $C_0, C_1, \cdots, C_k$ are linear combinations of the independent variables $x_1, x_2, \cdots, x_k$ such that C_j involves only $x_1, x_2, \cdots, x_j$. The notation used in writing the regression coefficients will be justified by later developments. By defining a variable $x_0 \equiv 1$, we can write

$$(b)\qquad C_j = \sum_{s=0}^{j} c_{js}x_s$$

where the coefficients c_{js} are to be chosen so that $\Sigma C_iC_j = 0$ for $i \neq j$. Since these conditions result in $\frac{1}{2}k(k+1)$ independent equations, and since the combinations C_j involve a total of $\frac{1}{2}(k+1)(k+2)$ coefficients c_{sj}, we may arbitrarily choose $k+1$ coefficients, and it is customary to choose $c_{00} = c_{11} = c_{22} = \cdots = c_{kk} = 1$. Since $C_0 = c_{00} = 1$ we have $\Sigma C_0C_j = \Sigma c_{00}C_j = \Sigma C_j$, and hence the condition $\Sigma C_0C_j = 0$ implies $\Sigma C_j = 0$. Similarly, since $C_1 = c_{01} + c_{11}x_1 = c_{01} + x_1$, we have

$$\begin{aligned}\Sigma C_1C_j &= \Sigma c_{01}C_j + \Sigma x_1C_j \\ &= c_{01}\Sigma C_j + \Sigma x_1C_j\end{aligned}$$

and the two conditions $\Sigma C_0C_j = 0$ and $\Sigma C_1C_j = 0$ imply that $\Sigma x_1C_j = 0$. By continuing this process we see that in general the conditions $\Sigma C_iC_j = 0$, $i < j$, imply that $\Sigma x_iC_j = 0$, $i < j$, and that $\Sigma C_j^2 = \Sigma x_jC_j$. Thus for each j we have the equations

$$(c)\qquad \begin{aligned}
&nc_{j0} + c_{j1}\Sigma x_1 + c_{j2}\Sigma x_2 + \cdots + c_{j,\,j-1}\Sigma x_{j-1} + \Sigma x_j = 0\\
&c_{j0}\Sigma x_1 + c_{j1}\Sigma x_1^2 + c_{j2}\Sigma x_1x_2 + \cdots + c_{j,\,j-1}\Sigma x_1x_{j-1} + \Sigma x_1x_j = 0\\
&\qquad\cdot \qquad\qquad \cdot \qquad\qquad \cdot\\
&\qquad\cdot \qquad\qquad \cdot \qquad\qquad \cdot\\
&\qquad\cdot \qquad\qquad \cdot \qquad\qquad \cdot\\
&c_{j0}\Sigma x_{j-1} + c_{j1}\Sigma x_{j-1}x_1 + c_{j2}\Sigma x_{j-1}x_2 + \cdots + c_{j,\,j-1}\Sigma x^2_{j-1} + \Sigma x_{j-1}x_j = 0
\end{aligned}$$

which we can solve for the coefficients $c_{j0}, c_{j1}, \cdots, c_{j,j-1}$ of C_j. Reference to (*b*) of Section 6.31 indicates that these coefficients are identical, except for a change of sign, with the regression coefficients of the variable x_j on the variables $x_1, \cdots, x_{j-1}$, since the above equations are the normal equations for this regression. It follows that the linear combination C_j reflects that portion of the variation in x_j which is linearly independent of the variation in $x_1, \cdots, x_{j-1}$.

Owing to the orthogonality of the linear combinations C_j determined as above, we have for the regression coefficient of y on C_j

$$(d) \qquad b_{yj.012\cdots(j-1)} = \frac{\Sigma y C_j}{\Sigma C_j^2}$$

If we define

$$(e) \qquad \Delta^{(j)} = \begin{vmatrix} n & \Sigma x_1 & \Sigma x_2 & \cdots & \Sigma x_j \\ \Sigma x_1 & \Sigma x_1^2 & \Sigma x_1 x_2 & \cdots & \Sigma x_1 x_j \\ \Sigma x_2 & \Sigma x_2 x_1 & \Sigma x_2^2 & \cdots & \Sigma x_2 x_j \\ \cdot & \cdot & \cdot & & \cdot \\ \cdot & \cdot & \cdot & & \cdot \\ \cdot & \cdot & \cdot & & \cdot \\ \Sigma x_j & \Sigma x_j x_1 & \Sigma x_j x_2 & \cdots & \Sigma x_j^2 \end{vmatrix}$$

the solutions of equations (*c*) are given by

$$c_{js} = (-1)^{j-s+1} \frac{\Delta_{js}^{(j)}}{\Delta^{(j-1)}}$$

where $\Delta_{js}^{(j)}$ indicates the determinant obtained by deleting the *j*th row and *s*th column of $\Delta^{(j)}$ and is the cofactor of the element $\Sigma x_j x_s$ of $\Delta^{(j)}$. Hence

$$C_j = \frac{1}{\Delta^{(j-1)}} \sum_{s=0}^{j} (-1)^{j-s+1} \Delta_{js}^{(j)} x_s$$

and

$$\Sigma C_j^2 = \Sigma x_j C_j = \frac{1}{\Delta^{(j-1)}} \sum_{s=0}^{j} (-1)^{j-s+1} \Delta_{js}^{(j)} \Sigma x_j x_s$$

$$= \frac{\Delta^{(j)}}{\Delta^{(j-1)}}$$

Similarly, we have

$$\Sigma y C_j = \frac{1}{\Delta^{(j-1)}} \sum_{j=0}^{s} (-1)^{j-s+1} \Delta_{js}^{(j)} \Sigma y x_s$$

where the summation is the expansion of the determinant

$$(f) \qquad \Delta_y^{(j)} = \begin{vmatrix} n & \Sigma x_1 & \Sigma x_2 & \cdots \Sigma y \\ \Sigma x_1 & \Sigma x_1^2 & \Sigma x_1 x_2 & \cdots \Sigma x_1 y \\ \Sigma x_2 & \Sigma x_2 x_1 & \Sigma x_2^2 & \cdots \Sigma x_2 y \\ \cdot & \cdot & \cdot & \cdot \\ \cdot & \cdot & \cdot & \cdot \\ \cdot & \cdot & \cdot & \cdot \\ \Sigma x_j & \Sigma x_j x_1 & \Sigma x_j x_2 & \Sigma x_j y \end{vmatrix}$$

which is identical with $\Delta^{(j)}$ except that y has replaced x_j in the last column. Thus we have

$$(g) \qquad b_{yj.012\cdots(j-1)} = \frac{\Delta_y^{(j)}}{\Delta^{(j)}}$$

and, as for the orthogonal polynomials in the previous section, we have from (d)

$$(h) \qquad \text{var}\,(b_{yj.012\cdots(j-1)}) = \frac{\sigma^2}{\Sigma C_j^2} = \sigma^2 \frac{\Delta^{(j-1)}}{\Delta^{(j)}}$$

where as usual σ^2 is the variance of the y observations, assumed to be constant for all values of the independent variables. As before, an unbiased estimate of σ^2 is given by

$$s^2 = \frac{\Sigma(y - \tilde{y})^2}{n - k - 1}$$

where

$$(i) \qquad \begin{aligned} \Sigma(y - \tilde{y})^2 &= \Sigma y^2 - \Sigma_j b^2_{yj.012\cdots(j-1)} \cdot \Sigma C_j^2 \\ &= \Sigma y^2 - \Sigma_j b_{yj.012\cdots(j-1)} \cdot \Sigma y C_j \end{aligned}$$

Using this estimate, we can test the significance of a particular coefficient by forming the ratio

$$(j) \qquad t_{n-k-1} = \frac{b_{yj.012\cdots(j-1)}\sqrt{\Sigma C_j^2}}{s}$$

which has the t-distribution with $n - k - 1$ degrees of freedom. Because of the orthogonality of the C_j the estimates of the regression coefficients are independent, and we have for an estimated value $\tilde{y}$ of the mean y associated with a particular set of values of the independent variables

$$(k) \qquad \text{var}\,(\tilde{y}) = \sigma^2\left[\frac{C_0^2}{\Sigma C_0^2} + \frac{C_1^2}{\Sigma C_1^2} + \cdots + \frac{C_k^2}{\Sigma C_k^2}\right]$$

where the orthogonal combinations in the numerator are evaluated, using the given values of $x_1, x_2, \cdots, x_k$. Using this variance and the estimate s^2, we can form confidence intervals for the predicted value $\tilde{y}$.

We noted above that the linear combinations C_j corresponded to that portion of the variation in x_j remaining after the linear regression on $x_1, \cdots, x_{j-1}$ had been eliminated. This regression can also be expressed in terms of the linear combinations $C_0 = 1, C_1, \cdots, C_{j-1}$, so that

$$C_j = x_j - b_{j0} - b_{j1.0}C_1 - b_{j2.01}C_2 - \cdots - b_{j(j-1).012\cdots(j-2)}\, C_{j-1}$$

A slight extension of this concept enables us to define for $r = j, j+1, \cdots, k$ the new variables

$$(l) \qquad x_{r.012\cdots(j-1)} = x_r - b_{r0} - b_{r1.0}x_{1.0} - b_{r2.01}x_{2.01} - \cdots - b_{r(j-1).012\cdots(j-2)}x_{(j-1).012\cdots(j-2)}$$

which represent the variation remaining in the variables $x_j, x_{j+1}, \cdots, x_k$ after their dependence on $x_1, \cdots, x_{j-1}$ has been removed, or, stated in an equivalent fashion, the components of $x_j, x_{j+1}, \cdots, x_k$ which are independent of $x_1, \cdots, x_{j-1}$. In particular, $C_j = x_{j.012\cdots(j-1)}$, and the original regression equation could have been written

$$(m) \qquad \text{ave}\,(y) = b_{y0} + b_{y1.0}x_{1.0} + b_{y2.01}x_{2.01} + \cdots + b_{yk.012\cdots(k-1)}x_{k.012\cdots(k-1)}$$

from which it is easily seen that the coefficient $b_{yj.012\cdots(j-1)}$ can be interpreted as a measure of the regression of y on that part of x_j which is linearly independent of $x_1, \cdots, x_{j-1}$. The $b_{yj.012\cdots(j-1)}$ are known as partial regression coefficients.

It follows from (l) and the orthogonality properties of the $C_j = x_{j.012\cdots(j-1)}$ that

$$b_{rj.012\cdots(j-1)} = \frac{\Sigma x_{r.012\cdots(j-1)}x_{j.012\cdots(j-1)}}{\Sigma x^2_{j.012\cdots(j-1)}} = \frac{\Sigma x_r x_{j.012\cdots(j-1)}}{\Sigma x^2_{j.012\cdots(j-1)}}$$

In particular, $b_{jj} = 1$, and

$$b_{yj.012\cdots(j-1)} = \frac{\Sigma y x_{j.012\cdots(j-1)}}{\Sigma x^2_{j.012\cdots(j-1)}}$$

as previously defined by (d). Also, since $x_0 = 1$, we have $b_{j0} = \bar{x}_j$, $b_{y0} = \bar{y}$. Thus we have $x_{j.0} = x_j - \bar{x}_j$, so that the first stage of the orthogonalization procedure requires only that we change from the original observations to deviations about their means. This step is frequently divorced from the remainder of the procedure, and the 0 subscript is omitted from the above formulation, so that, for example, we have

$$(n) \qquad x_{r.12\cdots(j-1)} = (x_r - \bar{x}_r) - b_{r1}(x_1 - \bar{x}_1) - b_{r2.1}x_{2.1} - \cdots - b_{r(j-1).12\cdots(j-2)}x_{(j-1).12\cdots(j-2)}$$

$$\text{ave}\,(y) = \bar{y} + b_{y1}(x_1 - \bar{x}_1) + b_{y2.1}x_{2.1} + \cdots + b_{yk.12\cdots(k-1)}x_{k.12\cdots(k-1)}$$

It should be noted that, although the coefficients $b_{yj.012\cdots(j-1)}$ are independent, their value depends on the particular $j - 1$ variables previously eliminated, and the practical advantage of the above methods may depend to a large extent on the order of elimination used. If we have some technical basis for ordering the insertion of the factors because of known cause-effect relationships, then the partial regression coefficients and the significance tests on these coefficients will provide the answers to sound technical questions such as: "Given that a part of the variation in factor C is caused by variations in factors A and B, do the variations in C which are independent of A and B appear to affect the variable Y?"

If, on the other hand, the associated variations in A and B were caused by factor C, this might not be a sound technical question. The problem of cause-effect relationships is essentially technical so that the decision as to the form of question must also be technical. In cases where no prior knowledge is available, it is often advantageous to insert the independent variates in the model in a number of different orders and examine the resulting regression coefficients.

If a large number of data are to be examined, the calculational procedure is extensive and the methods considered in the appendix should be employed. With a relatively small number of observations it is possible to calculate the individual values of the variables $x_{r.012\cdots(j-1)}$, $r = j, j + 1, \cdots, k$, and undertake a regression analysis with these values for the independent variates. This procedure has the advantage that at each stage of the computation we can observe the effect on the individual variates of the variables eliminated. The procedure involves successive computations in the following stages:

(1) Compute the k regressions of $x_1, x_2, \cdots, x_k$ on $x_0 \equiv 1$, and correct the values of $x_1, x_2, \cdots, x_k$ for each observation on the basis of these regressions. This is equivalent, as noted previously, to replacing the individual observations by deviations about their means. In the following discussion we shall omit the 0 subscript.

(2) Compute the regressions of $x_2 - \bar{x}_2$, $x_3 - \bar{x}_3, \cdots, x_k - \bar{x}_k$ on $x_1 - \bar{x}_1$, and correct the values of these variables to obtain the values of $x_{2.1}, x_{3.1}, \cdots, x_{k.1}$.

(3) Compute the regressions of $x_{3.1}, x_{4.1}, \cdots, x_{k.1}$ on $x_{2.1}$, and obtain the values of $x_{3.12}, x_{4.12}, \cdots, x_{k.12}$.

(4) Repeat the process until finally $x_{k.12\cdots(k-2)}$ is corrected for its regression on $x_{(k-1).12\cdots(k-2)}$.

At each stage we can also compute the residuals

$$
\begin{aligned}
y_0 &= y - \bar{y} \\
y_{01} &= y_1 = y - \bar{y} - b_{y1}(x_1 - \bar{x}_1) \\
y_{12} &= y - \bar{y} - b_{y1}(x_1 - \bar{x}_1) - b_{y2.1}x_{2.1} \\
&\vdots \\
y_{12\cdots k} &= y - \bar{y} - b_{y1}(x_1 - \bar{x}_1) - b_{y2.1}x_{2.1} - \cdots \\
&\qquad - b_{yk.12\cdots(k-1)}x_{k.12\cdots(k-1)}
\end{aligned}
$$

so as to note the effect of the consideration of each successive variable in explaining the variation in the individual observations of y.

EXAMPLE. This procedure is illustrated for the case of a regression on 3 x-variates by the following hypothetical data:

Yield	865	861	844	859	870	861	820	843	859	864
Factor 1	17	14	15	10	4	13	9	6	14	2
Factor 2	131	137	142	139	151	134	131	145	130	150
Factor 3	77	69	72	80	81	85	72	75	80	79

The variates are to be considered in order of their numerals. The computation is given in Table 6.14 in the following stages:

(1) The means

$$
\bar{y} = 854.6 \qquad \bar{x}_2 = 139.0
$$
$$
\bar{x}_1 = 10.4 \qquad \bar{x}_3 = 77.0
$$

are computed from the original data and the deviations of the individual values from these means tabulated in the first four columns.

(2) The quantities $\Sigma(x_1 - \bar{x}_1)^2$, $\Sigma(x_1 - \bar{x}_1)(x_2 - \bar{x}_2)$, $\Sigma(x_1 - \bar{x}_1)(x_3 - \bar{x}_3)$, and $\Sigma(x_1 - \bar{x}_1)(y - \bar{y})$ are computed, and the regression coefficients b_{21}, b_{31}, and b_{y1} determined.

(3) The values of $x_{2.1} = (x_2 - \bar{x}_2) - b_{21}(x_1 - \bar{x}_1)$, $x_{3.1} = (x_3 - \bar{x}_3) - b_{31}(x_1 - \bar{x}_1)$, and $y_1 = (y - \bar{y}) - b_{y1}(x_1 - \bar{x}_1)$ are computed and recorded in the next three columns.

TABLE 6.14

Observation	$x_1 - \bar{x}_1$	$x_2 - \bar{x}_2$	$x_3 - \bar{x}_3$	$y - \bar{y}$	$x_{2.1}$	$x_{3.1}$	y_1	$x_{3.12}$	y_{12}	y_{123}
1	6.6	−8	0	10.4	−1.65	1.37	10.53	1.19	12.61	10.39
2	3.6	−2	−8	6.4	1.46	−7.25	6.47	−7.09	4.62	17.87
3	4.6	3	−5	−10.6	7.43	−4.05	−10.51	−3.23	−19.90	−13.86
4	−0.4	0	3	4.4	−0.38	2.92	4.39	2.88	4.87	−0.51
5	−6.4	12	4	15.4	5.84	2.67	15.28	3.32	7.90	1.69
6	2.6	−5	8	6.4	−2.50	8.54	6.45	8.26	9.61	−5.83
7	−1.4	−8	−5	−34.6	−9.35	−5.29	−34.63	−6.32	−22.81	−11.00
8	−4.4	6	−2	−11.6	1.76	−2.91	−11.68	−2.71	−13.90	−8.83
9	3.6	−9	3	4.4	−5.53	3.75	4.47	3.14	11.46	5.59
10	−8.4	11	2	9.4	2.91	0.26	9.24	0.58	5.56	4.47
Sum of squares	231.84				230.13			207.14		906.69
Sum of cross products		−223.2	−48.0	−4.40		−25.44	290.83		387.20	
Regression coefficients		−0.9627	−0.2070	−0.0190		−0.1106	1.2638		1.8693	

(4) The quantities $\Sigma x^2_{2.1}$, $\Sigma x_{2.1}x_{3.1}$, and $\Sigma y_1 x_{2.1}$ are computed, and the regression coefficients $b_{32.1}$ and $b_{y2.1}$ are determined.

(5) The values of $x_{3.12} = x_{3.1} - b_{32.1}x_{2.1}$ and $y_{12} = y_1 - b_{y2.1}x_{2.1}$ are computed and tabulated in the next two columns.

(6) The quantities $\Sigma x^2_{3.12}$ and $\Sigma y_{12}x_{3.12}$ are computed, and the coefficient $b_{y3.12}$ is determined.

(7) The residuals $y_{123} = y - \tilde{y} = y_{12} - b_{y3.12}x_{3.12}$ are computed and recorded in the last column, from which the sum of squares of residuals can be computed directly.

If there is no interest in the residual y's at each stage, the last stage can be omitted as well as the computation of residuals at each previous stage, since it is easily confirmed that

$$\Sigma(x_1 - \bar{x}_1)(y - \bar{y}) = \Sigma(x_1 - \bar{x}_1)y$$

$$\Sigma y_1 x_{2.1} = \Sigma(y - \bar{y})x_{2.1} = \Sigma y x_{2.1}$$

$$\Sigma y_{12}x_{3.12} = \Sigma(y - \bar{y})x_{3.12} = \Sigma y x_{3.12}$$

and the residual sum of squares can be computed from the relation

$$\begin{aligned}\Sigma(y - \tilde{y})^2 &= \Sigma(y - \bar{y})^2 - b_{y1}\Sigma(y - \bar{y})(x_1 - \bar{x}_1) - b_{y2.1}\Sigma(y - \bar{y})x_{2.1} \\ &\quad - b_{y3.12}\Sigma(y - \bar{y})x_{3.12} \\ &= \Sigma y^2 - n\bar{y}^2 - b_{y1}\Sigma y(x_1 - \bar{x}_1) - b_{y2.1}\Sigma y x_{2.1} - b_{y3.12}\Sigma y x_{3.12}\end{aligned}$$

In this example, we have

$$\begin{aligned}\Sigma(y - \tilde{y})^2 &= 1998.40 - (-0.0190)(-4.40) - (1.2638)(290.83) \\ &\quad - (1.8693)(387.20) \\ &= 906.97\end{aligned}$$

which agrees within errors of rounding with the sum of residuals obtained from the last column of Table 6.14.

The regression equation is given by

$$\begin{aligned}y &= \bar{y} + b_{y1}(x_1 - \bar{x}_1) + b_{y2.1}x_{2.1} + b_{y3.12}x_{3.12} \\ &= 854.6 - 0.0190(x_1 - \bar{x}_1) + 1.2638x_{2.1} + 1.8693x_{3.12}\end{aligned}$$

from which the computed values $\tilde{y}$ can be obtained if required. The residual variance estimate is

$$s^2 = \frac{\Sigma(y - \tilde{y})^2}{n - 4} = \frac{907}{6} = 151.2$$

and the variances of the regression coefficients are

$$s^2_{b_{y1}} = \frac{s^2}{\Sigma x_1{}^2} = 0.652 \qquad s^2_{b_{y2.1}} = \frac{s^2}{\Sigma x^2_{2.1}} = 0.657$$

$$s^2_{b_{y3.12}} = \frac{s^2}{\Sigma x^2_{3.12}} = 0.730$$

Since $t_{6,\,0.05} = 2.45$, the coefficient of $x_{3.12}$ is significant at the 5% level. Note that the effect of the factors, as measured by the level at which the estimates of the regression coefficients were significant, was in exactly the reverse order to that used in the calculation, and that these three factors together have explained only about 50% of the variation in the observed yields. We could now change the order of insertion of the variables and repeat the procedure, excepting, of course, the computation of $(y - \bar{y})$, $(x_1 - \bar{x}_1)$, $(x_2 - \bar{x}_2)$, and $(x_3 - \bar{x}_3)$, which is independent of the order of the x variates.

6.4. Correlation

6.41. Estimation of the Correlation Coefficient

When a situation arises where we are really sampling from a normal bivariate population (see Appendix 4*B*), we are frequently interested in the correlation coefficient ρ (or its square ρ^2) as a measure of the degree of dependence between the two population characteristics being considered. Since ave $(r) = \rho$, where r is the sample correlation coefficient defined in Section 2.53, this furnishes an unbiased estimate of ρ with

$$\text{(a)} \qquad \text{var}\,(r) = \frac{(1 - \rho^2)^2}{n}$$

As the sampling distribution of r is quite complicated, the direct calculation of confidence limits for ρ is difficult. From extensive calculations F. N. David [15] has obtained charts, two of which are reproduced in Figures 6.12 and 6.13, from which confidence intervals for ρ can be obtained directly. For a value of r computed from a sample of size n, we simply determine from these charts the values of ρ at which the curves for the given value of n intersect the line with abscissa r; these are the desired confidence limits with 5% or 1% risk. For values of n greater than, say, 50, and r not too large we can use the transformation given by (j) of Appendix 4*B*. We first compute

$$\text{(b)} \qquad z = \frac{1}{2} \ln \frac{1 + r}{1 - r} = \tanh^{-1} r$$

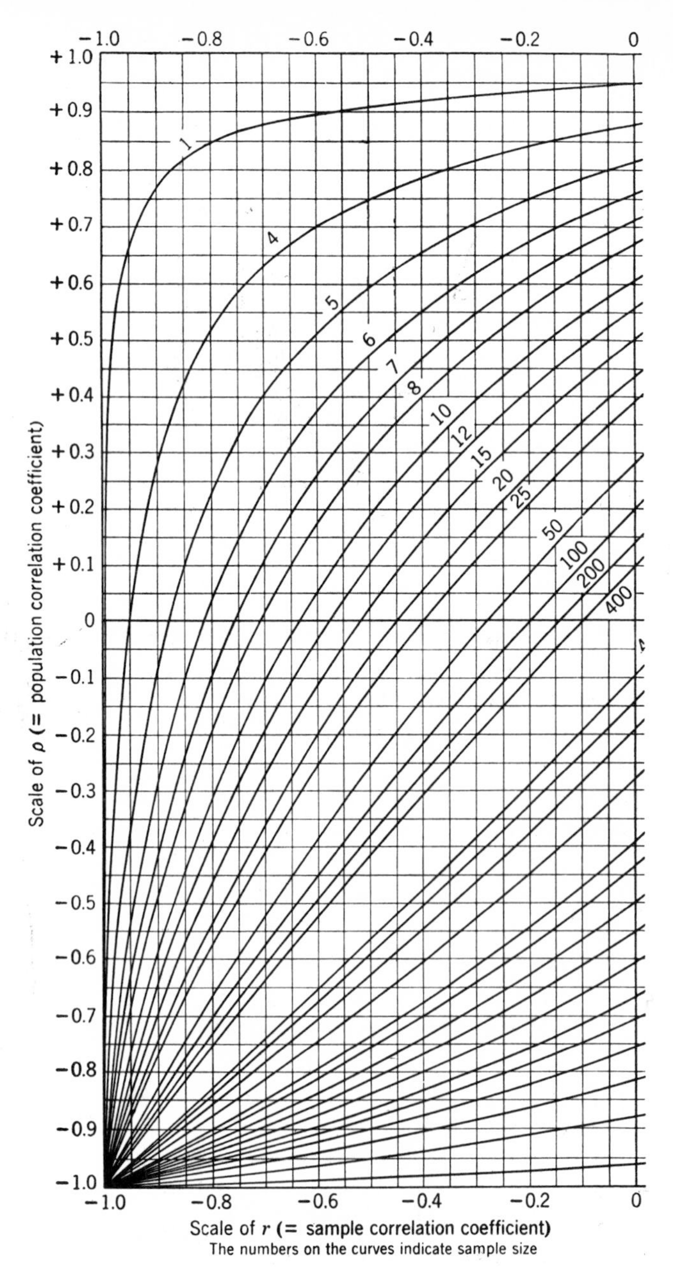

FIG. 6.12. Confidence belts. Chance of rejecting the hypothesis when true $= 0.025 + 0.025 = 0.05$. Reproduced by permission of Professor

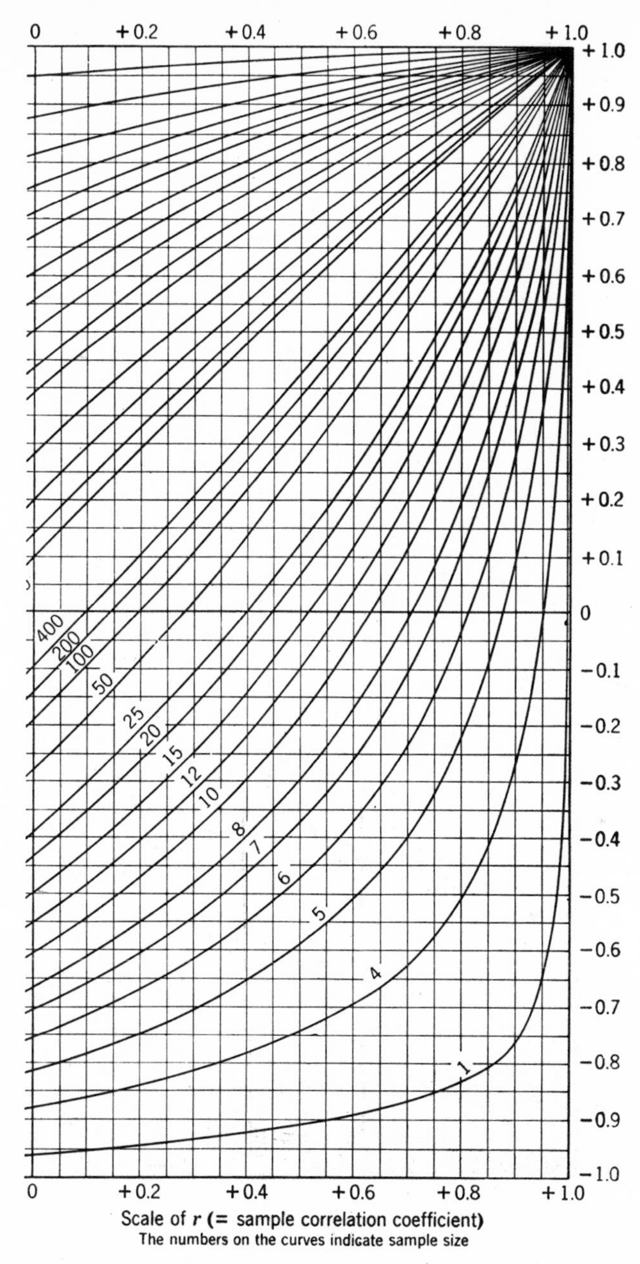

E. S. Pearson from *Tables of the Correlation Coefficient*, The Biometrika Office, University College, London, 1938, by F. N. David.

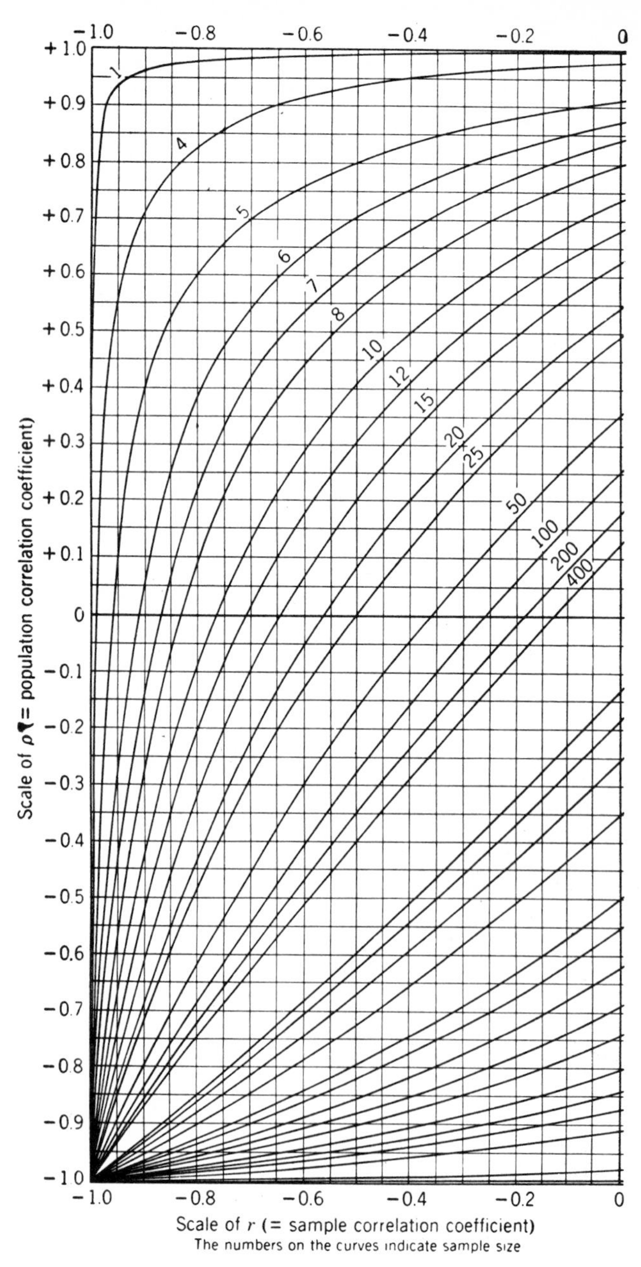

FIG. 6.13. Confidence belts. Chance of rejecting the hypothesis when true = 0.005 + 0.005 = 0.01. Reproduced by permission of Professor

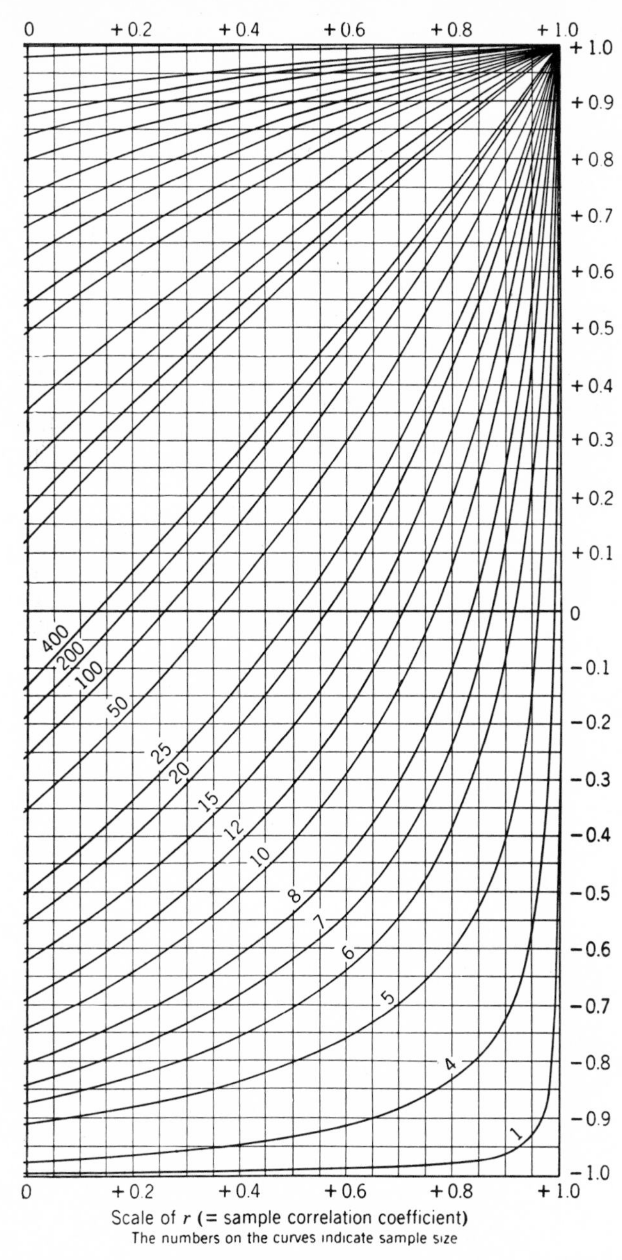

E. S. Pearson from *Table of the Correlation Coefficient*, The Biometrika Office, University College, London, 1938, by F. N. David.

and then determine the $100(1 - \alpha)\%$ confidence limits

$$z \pm \frac{t_\alpha}{\sqrt{n-3}} \tag{c}$$

for z, where t_α is the two-sided percentage point of the normal distribution. These limits can then be retransformed by the use of

$$r = \frac{e^{2z} - 1}{e^{2z} + 1} = \tanh z \tag{d}$$

and the two values r_2 and r_1 so obtained represent approximate $100(1 - \alpha)\%$ confidence limits for ρ.

To test the hypothesis $\rho = \rho_0 \neq 0$, we can proceed either by determining whether ρ_0 falls within the above confidence intervals or, if $n \geq 50$ and ρ_0 is not too large, by computing z from (*b*) and

$$\text{ave}\,(z) = \frac{1}{2} \ln \frac{1 + \rho_0}{1 - \rho_0} + \frac{\rho_0}{2(n-1)} \tag{e}$$

and comparing the ratio

$$t = \frac{z - \text{ave}\,(z)}{\sqrt{n-3}} \tag{f}$$

with the percentage points for a standard normal deviate. For the particular case $\rho = 0$ it can be shown that the quantity

$$t_{n-2} = \frac{r}{\sqrt{1 - r^2}} \sqrt{n - 2}$$

has the t-distribution with $n - 2$ degrees of freedom; thus in this case we can make an exact test of the hypothesis $\rho = 0$. It will be shown in Chapter 7 that an identical test can be made by the use of analysis of variance techniques. Table 6.15 gives for various values of $\nu = n - 2$ the 5%, 1%, and 0.1% points of the distribution of r when $\rho = 0$.

EXAMPLE 1. In Section 2.53 we obtained for the data of Table 2.5, $r = 0.5051$. To determine whether this value is significant, we compute

$$t_{98} = \frac{0.5051}{\sqrt{1 - (0.5051)^2}} \sqrt{98} = 5.794$$

Since for 98 degrees of freedom the t-distribution is very closely approximated by the normal distribution for which $t_{0.001} = 3.29$, we conclude that this value is highly significant, and definitely indicates the presence of positive correlation between percentage pig and lime consumption. A

TABLE 6.15

PERCENTAGE POINTS FOR r WHEN $\rho = 0$*

Degrees of Freedom	Percentage Points			Degrees of Freedom	Percentage Points		
$\nu = n - 2$	$\alpha = 0.05$	$\alpha = 0.01$	$\alpha = 0.001$	$\nu = n - 2$	$\alpha = 0.05$	$\alpha = 0.01$	$\alpha = 0.001$
1	0.99692	0.99988	—	16	0.4683	0.5897	0.7084
2	0.95000	0.99000	0.99900	17	0.4555	0.5751	0.6932
3	0.8783	0.95873	0.99116	18	0.4438	0.5614	0.6787
4	0.8114	0.91720	0.97406	19	0.4329	0.5487	0.6652
5	0.7545	0.8745	0.95074	20	0.4227	0.5368	0.6524
6	0.7067	0.8343	0.92493	25	0.3809	0.4869	0.5974
7	0.6664	0.7977	0.8982	30	0.3494	0.4487	0.5541
8	0.6319	0.7646	0.8721	35	0.3246	0.4182	0.5189
9	0.6021	0.7348	0.8471	40	0.3044	0.3932	0.4896
10	0.5760	0.7079	0.8233	45	0.2875	0.3721	0.4648
11	0.5529	0.6835	0.8010	50	0.2732	0.3541	0.4433
12	0.5324	0.6614	0.7800	60	0.2500	0.3248	0.4078
13	0.5139	0.6411	0.7603	70	0.2319	0.3017	0.3799
14	0.4973	0.6226	0.7420	80	0.2172	0.2830	0.3568
15	0.4821	0.6055	0.7246	90	0.2050	0.2673	0.3375
				100	0.1946	0.2540	0.3211

* Reproduced by permission from Table VI of "Statistical Tables for Biological, Agricultural and Medical Research," by R. A. Fisher and F. Yates, Oliver and Boyd, Ltd., Edinburgh, Second Edition, 1943.

direct comparison of r with one 0.1% point for $\nu = 98$ in Table 6.15 leads to the same conclusion.

To estimate confidence limits for this value, we obtain

$$z = \tanh^{-1} 0.5051 = 0.556$$

$$\sigma_z = \sqrt{\frac{1}{97}} = 0.102$$

Hence, since $t_{0.05} = 1.96$, 95% confidence limits in terms of z are

$$0.556 - 1.96\,(0.102) < z_\rho < 0.556 + 1.96(0.102)$$

$$0.356 < z_\rho < 0.756$$

Retransforming these values into values of r, we have finally

$$0.342 < \rho < 0.639$$

as a 95% confidence limit for the true correlation coefficient.

EXAMPLE 2. For the data of Table 6.4, we have $n = 9$, $r = 0.70$. From Table 6.15 we see that this value is above the 5% level, but below the 1% level, so that there is some indication of correlation, although additional data would be required to give definite confirmation. A confidence interval in this case is most easily obtained from Figure 6.12, from which we have, for $r = 0.70$, $n = 9$, and $\alpha = 0.05$, $0.05 < \rho < 0.92$. Since the observed correlation is based on so few points, the confidence interval is quite large. In this case the use of z gives a confidence interval $0.07 < \rho < 0.93$, which is a close approximation to the exact limits even when $n = 9$.

6.42. Estimation of the Regression Lines

If the correlation coefficient gives some indication of a dependence of x and y, we may be interested in determining the nature of this dependence. For a sample from a normal bivariate population this dependence is reflected by the regression lines (or regression coefficients) given by (g) and (h) of Appendix 4B. Depending on the nature of the cause and effect relationship which we believe to be indicated, we may be interested in either or both of these lines.

Estimates of the regression coefficients and regression lines are given by direct substitution of the corresponding sample values. Hence we have

$$b_{y \cdot x} = r \frac{s_y}{s_x}$$

(a)

$$b_{x \cdot y} = r \frac{s_x}{s_y}$$

and

$$(y - \bar{y}) = b_{y \cdot x}(x - \bar{x})$$

(b)

$$(x - \bar{x}) = b_{x \cdot y}(y - \bar{y})$$

To estimate the variances $\sigma^2_{y \cdot x}$ and $\sigma^2_{x \cdot y}$ an additional correction for bias is necessary, and we have

$$s^2_{y \cdot x} = \frac{n - 1}{n - 2} s_y^2(1 - r)^2 = (1 - r)^2 \frac{\Sigma(y_i - \bar{y})^2}{n - 2}$$

(c)

$$s^2_{x \cdot y} = \frac{n - 1}{n - 2} s_x^2(1 - r)^2 = (1 - r)^2 \frac{\Sigma(x_i - \bar{x})^2}{n - 2}$$

These estimated regression coefficients, regression lines, and estimated variances are the same as those which we obtained using the methods of Section 6.2. We have

$$b_{y \cdot x} = r \frac{s_y}{s_x}$$

$$= \frac{\Sigma(x_i - \bar{x})(y_i - \bar{y})}{(n-1)s_x s_y} \cdot \frac{s_y}{s_x}$$

$$= \frac{\Sigma(x_i - \bar{x})(y_i - \bar{y})}{(n-1)s_x^2}$$

$$= \frac{\Sigma(x_i - \bar{x})(y_i - \bar{y})}{\Sigma(x_i - \bar{x})^2}$$

since

$$s_x^2 = \frac{(x_i - \bar{x})^2}{n-1}$$

and this is the definition of b given in Section 6.22. Also the estimate $s^2_{y \cdot x}$ of $\sigma^2_{y \cdot x}$ is identical with the estimate $s^2_{y \cdot x}$ of σ^2 (thus justifying the notation) since

$$s^2_{y \cdot x} = \frac{n-1}{n-2} s_y^2(1 - r)^2$$

$$= \frac{n-1}{n-2}(s_y^2 - r^2 s_y^2)$$

$$= \frac{n-1}{n-2}(s_y^2 - b^2 s_x^2)$$

since

$$b = b_{y \cdot x} = r \frac{s_y}{s_x}$$

$$= \frac{n-1}{n-2} \frac{\Sigma(y_i - \bar{y})^2 - b^2 \Sigma(x_i - \bar{x})^2}{n-1}$$

and hence, using (h) of Section 6.22, we have

$$s^2_{y \cdot x} = \frac{\Sigma(y_i - \tilde{y}_i)^2}{n-2}$$

which is the result obtained in (f) of Section 6.22. This relationship might be expected, since, if we regard either the x or the y values in our sample as fixed, the distributions of the other variable corresponding to these fixed values are all normal with constant variance and average values which lie on the desired regression line. The difficulty lies in the fact that in this case a new sample would have different values of both variables, not just of the one variable, albeit the same regression line would obtain for the second sample.

Since

$$s^2_{y\cdot x} = \frac{n-1}{n-2} s_y^2(1 - r^2_{xy})$$

we have

$$1 - r^2_{xy} = \frac{(n-2)s^2_{y\cdot x}}{(n-1)s_y^2}$$

(*d*)
$$1 - r^2_{xy} = \frac{\Sigma(y_i - \tilde{y}_i)^2}{\Sigma(y_i - \bar{y})^2}$$

$$r^2_{xy} = \frac{\Sigma(y_i - \bar{y})^2 - \Sigma(y_i - \tilde{y}_i)^2}{\Sigma(y_i - \bar{y})^2}$$

Thus the square of the correlation coefficient indicates the relative portion of the variation in the y's explained by the dependence on x. In the case of simple linear regression the correlation coefficient has no meaning except as a convenient index of this reduction. It should also be noted that in the present case the estimated regression line of y on x would be used to predict y for a given x, and the regression line of x on y to predict x for a given y, whereas in the preceding case there is only one possible line which would be used to estimate y for a given x or x for a given y.

EXAMPLE. For the data of Table 6.4, letting $x =$ percent carbon and $y =$ test result, we have

$$\bar{x} = 0.101 \qquad \bar{y} = 71.13$$

$$s_x = 0.017 \qquad s_y = 3.55$$

and, from Example 2 of the previous section, $r = 0.70$. Hence for the regression of y on x we obtain

$$y - 71.13 = 0.70\frac{3.55}{0.017}(x - 0.101)$$

$$y - 71.13 = 146.1(x - 0.101)$$

and

$$s^2_{y\cdot x} = s_y^2(1 - r^2)$$

$$= 12.60(0.51) = 6.43$$

$$s_{y\cdot x} = 2.54$$

Although this line, which is shown in Figure 6.4, may be taken as indicative of the nature of any existing linear regression, it should be remembered that it is based on only 9 values, and may be appreciably different from the true regression line, even if the latter is linear.

6.43. Rank Correlation

In cases where a quick estimate of the correlation coefficient is desired, or where, for purposes of economy, or owing to the inadequacy of the methods of measurement, only the comparative sizes of the variables x and y can be observed, the *rank correlation* is frequently used. This is merely the correlation between the positions or ranks, of each pair of observations, usually measured from lowest to highest, when the observations on each variable are arranged in order of magnitude. We shall consider only the case where, owing to one of the above reasons, the rankings have replaced some continuous scale of measurement.

The ranks $x_1', \cdots, x_n'$ and $y_1', \cdots, y_n'$ corresponding to observed or potentially observable sets of values $x_1, \cdots, x_n$ and $y_1, \cdots, y_n$ will simply be arrangements of the numbers from 1 to n in some order. Hence we have immediately

$$\Sigma x_i' = \Sigma y_i' = 1 + 2 + 3 + \cdots + n = \frac{n(n+1)}{2}$$

$$\Sigma x'^2_i = \Sigma y'^2_i = 1^2 + 2^2 + 3^2 + \cdots + n^2 = \frac{n(n+1)(2n+1)}{6}$$

Hence

$$\Sigma(x_i' - \bar{x}')^2 = \Sigma(y_i' - \bar{y}')^2 = \frac{n(n+1)(2n+1)}{6} - \frac{1}{n} \cdot \left(\frac{n(n+1)}{2}\right)^2$$

$$= \frac{4n^3 + 6n^2 + 2n - 3n^3 - 6n^2 - 3n}{12}$$

$$= \frac{n^3 - n}{12} = \frac{n(n^2-1)}{12}$$

The cross product $\Sigma x_i'y_i'$ depends, of course, on the arrangement of the x_i' and y_i' in any particular case. However, its computation can be simplified by letting $d_i' = x_i' - y_i'$, and noting that

$$\Sigma d'^2_i = \Sigma(x_i' - y_i')^2 = \Sigma x'^2_i + \Sigma y'^2_i - 2\Sigma(x_i'y_i')$$

hence

$$2\Sigma x_i'y_i' = \frac{n(n+1)(2n+1)}{3} - \Sigma d'^2_i$$

and

$$\Sigma(x_i' - \bar{x}')(y_i' - \bar{y}') = \frac{n^3 - n}{12} - \frac{\Sigma d'^2_i}{2}$$

Substituting in (c) of Section 2.53, we obtain for the correlation between the ranks

$$r' = \frac{\dfrac{n(n^2-1)}{12} - \dfrac{\Sigma d'^2_i}{2}}{\dfrac{n(n^2-1)}{12}}$$

$$= 1 - \frac{6\Sigma d'^2_i}{n(n^2-1)}$$

When the ranks of each pair of values are identical $\Sigma d'^2_i = 0$ and $r' = +1$; and $r' = -1$ when $\Sigma d'^2_i = \dfrac{n(n^2-1)}{3}$, which is the true case when the ranks are exactly opposite, i.e., when we have the pairs

$$(1, n), \quad (2, n-1), \cdots, (n-1, 2), \quad (n, 1)$$

Thus $r' = +1$ represents complete agreement in rank, and $r' = -1$ the greatest possible discordance. For 2 arrangements chosen at random the expected value of r' is 0. If the pairs ranked are assumed to be from a bivariate normal distribution with correlation ρ, then $2 \sin \dfrac{\pi r'}{6}$ is an estimate of ρ with an efficiency in large samples of about 90%.

However, the greatest asset of rank correlation is that, like order statistics, its distribution can be studied, and used in making significance tests, if we assume only that the observed pairs are from 2 independent continuous populations. The exact distribution of $\Sigma d'^2_i$ (which is equivalent to r') based on possible permutations of the order of ranking has been computed for $n = 2$ through 7 and approximated by a type II distribution for $n = 8, 9, 10$. Certain percentage points have also been computed from a normal approximation for $11 \leq n \leq 30$ [23, 24]. Table 6.16 gives limits which will be equalled or exceeded by chance values of Σd_i^2 with probability $\leq \alpha$ for $\alpha = 0.10, 0.05, 0.02$, and 0.01. If we wish to test for either positive or negative association only, we should consider only values below the lower limit or above the upper limit, respectively, as significant, and halve the significance levels given.

EXAMPLE. During an investigation of factors responsible for the deterioration of large-scale units 14 tests, involving pilot plant trials and small-scale laboratory tests, were carried out. The results from the laboratory investigation were obtained in the form of a standard chart interpreted by examination of a number of characteristics. The pilot plant trials carried out under conditions more closely approaching those of the industrial unit provided quantitative results. In the latter case the variances in different tests were known to be unequal, and a rank correlation was employed. The data are given in Table 6.17.

TABLE 6.16

LIMITS FOR SUMS OF SQUARES OF RANK DIFFERENCES*

n	$\alpha = 0.10$		$\alpha = 0.05$		$\alpha = 0.02$		$\alpha = 0.01$	
	Lower	Upper	Lower	Upper	Lower	Upper	Lower	Upper
5	2	38	0	40	0	40	—	—
6	6	64	4	66	2	68	0	70
7	16	96	12	100	6	106	4	108
8	30	138	22	146	14	154	10	158
9	48	192	38	202	26	214	20	220
10	72	258	58	272	42	288	34	296
11	105	335	84	357	58	382	40	400
12	144	428	117	455	85	487	63	509
13	191	537	158	570	119	609	93	635
14	247	663	207	703	161	749	125	781
15	313	807	266	854	211	909	174	946
16	391	969	335	1025	271	1089	227	1133
17	480	1152	416	1216	341	1291	290	1342
18	582	1356	508	1430	422	1516	363	1575
19	698	1582	613	1667	514	1766	447	1833
20	828	1832	732	1928	620	2040	544	2116
21	973	2107	865	2215	738	2342	653	2427
22	1135	2407	1013	2529	872	2670	775	2767
23	1314	2734	1178	2870	1020	3028	912	3136
24	1511	3089	1360	3240	1184	3416	1064	3536
25	1727	3473	1559	3641	1365	3835	1233	3968
26	1962	3888	1778	4072	1564	4286	1418	4432
27	2219	4333	2016	4536	1781	4771	1621	4931
28	2497	4811	2275	5033	2018	5290	1842	5466
29	2797	5323	2556	5564	2275	5845	2083	6037
30	3122	5868	2859	6131	2553	6437	2345	6645

* Reproduced by permission from "Distributions of Sums of Squares of Rank Differences for Small Numbers of Individuals," *Annals Math. Stat.*, IX (1938), 133–48, by E. G. Olds; and "The 5% Significance Levels for Sums of Squares of Rank Differences and a Correction," *Annals Math. Stat.*, XX (1949), 117–18, by E. G. Olds.

TABLE 6.17

RANKING OF 14 SAMPLES BY PILOT SCALE AND LABORATORY TESTS

Ranking by Pilot Scale Test	Ranking by Lab. Test	Ranking by Pilot Scale Test	Ranking by Lab. Test
1	5	8	11
2	4	9	7
3	2	10	3
4	6	11	13
5	9	12	8
6	1	13	12
7	10	14	14

The differences in rank are given by −4, −2, +1, −2, −4, 5, −3, −3, 2, 7, −2, 4, 1, 0. Thus

$$\Sigma d'^2_i = 158$$

$$n(n^2 - 1) = 2730$$

$$r = 1 - \frac{(6)\,(158)}{2730} = 0.653$$

The value 158 for $\Sigma d'^2_i$ is just significant at the 0.01 level, and it is clear that there is a significant relationship between the results from the pilot scale test and the standard method of interpreting the laboratory scale test results. Examination of the difference in ranking also suggests that a large part of the residual divergence is due to the fact that rankings 1 and 3 by the laboratory test conflicted with the pilot plant rankings of these samples. An analysis of the reasons for this ranking indicated that a modification of the interpretation of the laboratory test data might be expected to result in closer agreement in future samples, which proved to be true. It should be noted that any modification suggested by this test series can be tested only by reference to the results of independent experimental work.

6.44. Correlations of Three or More Variables

The entire theory of correlation and the associated regression lines and planes can be extended to more than two variables if we consider our sample to have been drawn from a normal multivariate population. For such a population we can define correlation coefficients of various types which indicate the interdependence of these variables, and regression planes which indicate the nature of this dependence. For a complete summary of this subject the reader is referred to [2, Vol. I] or [91].

The methods of estimation in Section 6.3 are directly applicable to the estimation of any particular regression plane in which we may be interested, and of the associated regression coefficients. The correlation coefficient between any two variables can be defined in terms of their marginal distribution, and all the estimation methods of the preceding sections are applicable. In addition, we can in this case define the partial correlation coefficients (as opposed to the simple correlation coefficient) between two variables as a measure of the linear relationship between any two variables which remains after any dependence on the remaining variables has been removed. In terms of the multiple regression coefficients in Section 6.31 these can be defined by relationships of the form

(*a*) $$\rho_{12.34\cdots} = \sqrt{\beta_{12.34\cdots} \cdot \beta_{21.34\cdots}}$$

Note the analogy between this definition and (*l*) of Appendix 4*B*. The estimate of $\rho_{12.34\cdots}$ from a sample would be given by

(*b*) $$r_{12.34\cdots} = \sqrt{b_{12.34\cdots} \cdot b_{21.34\cdots}}$$

although other equivalent forms are frequently desirable for ease of computation. We can also consider a multiple correlation coefficient $\rho^2{}_{1.23\cdots}$ which reflects the total dependence of one variable on the remaining variables. This is most conveniently defined by considering the extent to which a particular regression plane will explain the variation in a given dependent variable, i.e., by

(*c*) $$\rho^2{}_{1.23\cdots} = 1 - \frac{\sigma^2{}_{1.23\cdots}}{\sigma_1{}^2}$$

The corresponding sample estimate, after correction for bias as in (*d*) of Section 6.42, is

(*d*) $$r^2{}_{1.23\cdots} = 1 - \frac{(n-k)s^2{}_{1.23\cdots}}{(n-1)s_1{}^2}$$

where k is the number of variables being considered. Partial correlation coefficients, like simple correlation coefficients, lie between -1 and $+1$, the sign indicating the nature of the dependence; multiple correlation coefficients are always less than 1, but are restricted to positive values, since sign in this case has no meaning. As in the case of simple correlation, it is more frequently the squares of these coefficients which have practical meaning.

We saw in Appendix 4*B* that for the bivariate normal distribution the correlation coefficient ρ could be expressed in terms of the variances

σ_x^2 and σ_y^2 and the covariances σ_{xy} of x and y. In the multivariate case, all the simple, partial, and multiple correlation coefficients can be expressed as functions of the variances and covariances

$$
(e) \qquad \begin{matrix} \sigma_1^2 & \sigma_{12} & \cdots\cdots & \sigma_{1k} \\ & \sigma_2^2 & \cdots\cdots & \sigma_{2k} \\ & & \ddots & \vdots \\ & & & \sigma_k^2 \end{matrix}
$$

of the variables $x_1, \cdots, x_k$, which, along with the k average values, define the multivariate normal distribution with which we are dealing.

The estimates of the various correlations are the same functions of the sample variance and covariances

$$
(f) \qquad \begin{matrix} s_1^2 & s_{12} & \cdots\cdots & s_{1k} \\ & s_2^2 & \cdots\cdots & s_{2k} \\ & & \ddots & \vdots \\ & & & s_k^2 \end{matrix}
$$

Since both the sample and population covariances are symmetrical in the sense that cov $(x_i, x_j) =$ cov (x_j, x_i) the elements above the diagonal in (e) and (f) may be inserted in the corresponding positions below the diagonal. The resulting square arrays are referred to as the population variance-covariance matrix and the sample variance-covariance matrix, respectively.

The sampling distributions of both partial and multiple correlation coefficients computed from samples from a normal multivariate population have been obtained under the assumption that the corresponding population parameters are zero, and can be used to test for the existence of such correlations in the population. Details of this type of test are given in previous references, and a convenient summary of computational techniques is given in [30].

6.45. Discriminant Functions

In the preceding sections on correlation we have been dealing with problems of estimation of parameters where a bivariate or multivariate normal population can be assumed to represent the population. Much work has also been done in developing multivariate analogues of univariate tests of significance such as the t-test and the χ^2-test. These problems are generally referred to as multivariate analysis. Essentially, we are extending tests of significance from single population characteristics to sets of population characteristics. One primary difficulty arises in determining what characteristics, or which combination of characteristics, best reflect the population differences we are trying to study. A complete discussion of multivariate analysis is beyond the scope of this book, but we shall examine one aspect of the above problem, that of discriminatory analysis.

The problem which we shall consider is: Suppose that we have n_1 individuals known to be from one population, and n_2 individuals known to be from another population. For each of the n_1 and n_2 individuals we observe a number of characteristics $x_1, x_2, \cdots, x_k$. Then what linear combination

$$(a) \qquad X = a_1x_1 + a_2x_2 + \cdots + a_kx_k$$

of these k characteristics will be best in assigning an unknown individual to one of the two populations; i.e., what single derived value X will, in some sense, best reflect the difference between the two populations? We shall assume that for each population the k characteristics have a multivariate normal distribution, with different means, but common variances and covariances. X is commonly called a discriminant function.

From the n_1 observations on the k characteristics for the first population we can compute the means $\bar{x}_{11}, \bar{x}_{12}, \cdots, \bar{x}_{1k}$ for each characteristic, and the sums of squares and cross products of deviations from the mean

$$\begin{matrix} S_1(x_1^2) & S_1(x_1x_2) \cdots & S_1(x_1x_k) \\ & S_1(x_2^2) \quad \cdots & S_1(x_2x_k) \\ & \cdot & \cdot \\ & \quad \cdot & \cdot \\ & \qquad \cdot & \cdot \\ & & S_1(x_k^2) \end{matrix}$$

exactly as in Section 6.31. Similarly, for the n_2 observations on the second variable, we obtain the means $\bar{x}_{21}, \bar{x}_{22}, \cdots, \bar{x}_{2k}$ and the sums of squares and cross products of deviations from the mean

$$\begin{matrix} S_2(x_1^2) & S_2(x_1x_2) \cdots & S_2(x_1x_k) \\ & S_2(x_2^2) \quad \cdots & S_2(x_2x_k) \\ & \cdot & \cdot \\ & \quad \cdot & \cdot \\ & \qquad \cdot & \cdot \\ & & S_2(x_k^2) \end{matrix}$$

From these we can obtain an estimate of the differences

(b) $$d_i = \bar{x}_{1i} - \bar{x}_{2i}$$

between the means of the k characteristics for the two populations and joint estimates

$$\begin{matrix} s_{11}(=s_1^2) & s_{12} & \cdots & s_{1k} \\ & s_{22}(=s_2^2) & \cdots & s_{2k} \\ & & \ddots & \vdots \\ & & & s_{kk}(=s_k^2) \end{matrix}$$

of their common variances and covariances, using the general formula

(d) $$s_{ij} = \frac{S_1(x_i x_j) + S_2(x_i x_j)}{n_1 + n_2 - 2} = \frac{S(x_i x_j)}{n_1 + n_2 - 2}$$

where $S(x_i x_j) = S_1(x_i x_j) + S_2(x_i x_j)$. Now, since X is a linear combination of $x_1, \cdots, x_k$, we can obtain estimates

(e) $$\begin{aligned} \bar{X}_1 &= a_1\bar{x}_{11} + a_2\bar{x}_{12} + \cdots + a_k\bar{x}_{1k} \\ \bar{X}_2 &= a_1\bar{x}_{21} + a_2\bar{x}_{22} + \cdots + a_k\bar{x}_{2k} \end{aligned}$$

of the mean of X in the two populations, and consequently an estimate

(f) $$D = \bar{X}_1 - \bar{X}_2 = a_1 d_1 + a_2 d_2 + \cdots + a_k d_k$$

of the difference between the two populations in terms of X. Also, using (b) of 3.33 we obtain a joint estimate

(g) $$s_X^2 = \Sigma_{ij} a_i a_j s_{ij}$$

of the variance of X based on variations within the two populations. Note that both of these involve the unknown coefficients $a_1, \cdots, a_k$.

In making the usual test between the means of 2 populations, we should base our test on the sample value

$$t = \frac{\bar{x}_1 - \bar{x}_2}{s\sqrt{\dfrac{1}{n_1} + \dfrac{1}{n_2}}}$$

and reject the hypothesis if this value were sufficiently large. Thus it seems reasonable to choose the $a_1, \cdots, a_k$ so that the ratio

(h) $$t = \frac{\bar{X}_1 - \bar{X}_2}{s_X\sqrt{\dfrac{1}{n_1} + \dfrac{1}{n_2}}} = \frac{D}{s_X\sqrt{\dfrac{1}{n_1} + \dfrac{1}{n_2}}}$$

has the maximum absolute value, which is the same as choosing X so that the population difference is most likely to be termed significant. Since the factor $\sqrt{\frac{1}{n_1} + \frac{1}{n_2}}$ does not involve the a_i, this is equivalent to maximizing the absolute value of D/s_X, or simply to maximizing the ratio D^2/s_X^2, with respect to the a_i.

To do this we differentiate D^2/s_X^2 with respect to each a_i and equate to 0, which gives in general

$$\frac{2D}{s_X^2} \cdot \frac{\partial D}{\partial a_i} - \frac{D^2}{(s_X^2)^2} \cdot \frac{\partial (s_X^2)}{\partial a_i} = 0$$

or

$$\frac{1}{2} \frac{\partial (s_X^2)}{\partial a_i} = \frac{s_X^2}{D} \cdot \frac{\partial D}{\partial a_i}$$

From (f) and (g) we have

$$\frac{\partial D}{\partial a_i} = d_i$$

$$\frac{\partial (s_X^2)}{\partial a_i} = 2\Sigma_j a_j s_{ij}$$

Substituting these in the above, we have finally k equations of the form

$$\Sigma_j a_j s_{ij} = \frac{s_X^2}{D} d_i$$

or, expanding,

$$
\begin{aligned}
a_1 s_{11} + a_2 s_{12} + \cdots + a_k s_{1k} &= \frac{s_X^2}{D} d_1 \\
a_1 s_{21} + a_2 s_{22} + \cdots + a_k s_{2k} &= \frac{s_X^2}{D} d_2 \\
\cdot \qquad \cdot \qquad & \\
\cdot \qquad \cdot \qquad & \\
\cdot \qquad \cdot \qquad & \\
a_1 s_{k1} + a_2 s_{k2} + \cdots + a_k s_{kk} &= \frac{s_X^2}{D} d_k
\end{aligned}
\qquad (i)
$$

The value of s_X^2/D occurring in each term on the right-hand side is arbitrary so far as the problem of maximizing the discrimination is concerned, since multiplying the right-hand side by a constant increases or decreases the value of every a_i without changing their relationship to each other. If we choose s_X^2/D equal to $\dfrac{1}{n_1 + n_2 - 2}$, and then multiply each equation by $n_1 + n_2 - 2$, we obtain, using (d)

$$
\begin{aligned}
a_1 S(x_1^2) + a_2 S(x_1 x_2) + \cdots + a_k S(x_1 x_k) &= d_1 \\
a_1 S(x_2 x_1) + a_2 S(x_2^2) + \cdots + a_k S(x_2 x_k) &= d_2 \\
\cdot \qquad\qquad \cdot \qquad\qquad\qquad\qquad & \quad \cdot \\
\cdot \qquad\qquad \cdot \qquad\qquad\qquad\qquad & \quad \cdot \\
\cdot \qquad\qquad \cdot \qquad\qquad\qquad\qquad & \quad \cdot \\
a_1 S(x_k x_1) + a_2 S(x_k x_2) + \cdots + a_k S(x_k^2) &= d_k
\end{aligned}
$$

The above equations are very similar to the regression equations obtained in Section 6.21, and they can be solved directly for $a_1, \cdots, a_k$ by any of the methods given for solving regression equations in 6.31 or Appendix 6*A*. In this case there seems to be little point in obtaining the inverse elements c_{ij} rather than solving directly for $a_1, \cdots, a_k$, since a second set of differences $d_1, \cdots, d_k$ would arise only in connection with different sums $S(x_i x_j)$. Note that the solutions obtained can still be multiplied by another arbitrary constant if this is convenient.

The solution of the above equations can become tedious if, as is frequently true in practice, the number of characteristics k is large. An alternative method of procedure, valid when it seems reasonable to assume that the intercorrelations ρ_{ij} between the characteristics are all equal or approximately equal to some value ρ, somewhat reduces the computation involved.

Let us reduce each set of measurements to standard units with the origin at the common mean of the two sets

$$
(k) \qquad \bar{x}_i = \frac{n_1 \bar{x}_{1i} + n_2 \bar{x}_{2i}}{n_1 + n_2}
$$

using the joint estimate $\sqrt{s_{ii}} = \sqrt{s_i^2} = s_i$ of the standard deviation. Then, denoting these standardized values by primes, we have

$$
(l) \qquad
\begin{aligned}
\bar{x}'_{1i} &= \frac{\bar{x}_{1i} - \bar{x}_i}{s_i} = \frac{n_2}{n_1 + n_2} \frac{\bar{x}_{1i} - \bar{x}_{2i}}{s_i} = \frac{n_2}{n_1 + n_2} \frac{d_i}{s_i} \\
\bar{x}'_{2i} &= \frac{\bar{x}_{2i} - \bar{x}_i}{s_i} = \frac{-n_1}{n_1 + n_2} \frac{\bar{x}_{1i} - \bar{x}_{2i}}{s_i} = \frac{-n_1}{n_1 + n_2} \frac{d_i}{s_i} \\
d_i' &= \bar{x}'_{1i} - \bar{x}'_{2i} = \frac{d_i}{s_i}
\end{aligned}
$$

and the matrix (c) of variances and covariances reduces to the correlation matrix

$$\begin{matrix} 1 & r_{12} & r_{13} \cdots r_{1k} \\ & 1 & r_{23} \cdots r_{2k} \\ & & 1 \;\cdots r_{3k} \\ & & \cdot \quad \cdot \\ & & \cdot \quad \cdot \\ & & \cdot \quad \cdot \\ & & \quad 1 \end{matrix}$$

where $r_{ij} = \dfrac{S(x_i x_j)}{\sqrt{S(x_i^2)S(x_j^2)}}$.

Now, for a given set of measurements $x_1, \cdots, x_k$, we define 2 new measurements

$$Q = x_1' + x_2' + \cdots + x_k'$$

and

$$P = c_1 x_1' + c_2 x_2' + \cdots + c_k x_k'$$

where the primes denote standardization by the transformation

$$x_i' = \frac{x_i - \bar{x}_i}{s_i}$$

and the c_i are weights determined by

$$c_i = \frac{kd_i}{\sum_{i=1}^{k} d_i} - 1 \qquad (n)$$

where $\Sigma_i c_i = 0$. Q and P are referred to as the size and shape, respectively, of the individual measured. As before, we have for the given sets of individuals

$$\begin{aligned} \bar{Q}_1 &= \bar{x}'_{11} + \bar{x}'_{12} + \cdots + \bar{x}'_{1k} \\ \bar{Q}_2 &= \bar{x}'_{21} + \bar{x}'_{22} + \cdots + \bar{x}'_{2k} \\ D_Q &= \bar{Q}_1 - \bar{Q}_2 = d_1' + d_2' + \cdots + d_k' \\ \bar{P}_1 &= c_1\bar{x}'_{11} + c_2\bar{x}'_{12} + \cdots + c_k\bar{x}'_{1k} \\ \bar{P}_2 &= c_1\bar{x}'_{21} + c_2\bar{x}'_{22} + \cdots + c_k\bar{x}'_{2k} \\ D_P &= \bar{P}_1 - \bar{P}_2 = c_1 d_1' + c_2 d_2' + \cdots + c_k d_k' \\ s_Q^2 &= \Sigma_{ij} r_{ij} \\ s_P^2 &= \Sigma_{ij} c_i c_j r_{ij} \\ s_{PQ} &= \Sigma_{ij} c_i r_{ij} \end{aligned} \qquad (o)$$

Under the conditions previously specified, $r_{PQ} = \dfrac{s_{PQ}}{s_P s_Q}$ should be negligible.

We now proceed to determine α and β such that

$$X = \alpha Q + \beta P$$

is the best discriminant function between the two populations based on the two measurements Q and P. From (i), choosing $s_X{}^2/D = 1$, we should have

$$\alpha s_Q{}^2 + \beta s_{PQ} = D_Q$$

$$\alpha s_{PQ} + \beta s_P{}^2 = D_P$$

from which

$$\alpha = \frac{D_Q s_P{}^2 - D_P s_{PQ}}{s_Q{}^2 s_P{}^2 - s^2{}_{PQ}}$$

$$\beta = \frac{D_P s_Q{}^2 - D_Q s_{PQ}}{s_Q{}^2 s_P{}^2 - s^2{}_{PQ}}$$

Since we can divide these values by any arbitrary constant, a process which corresponds to a different choice of $s_X{}^2/D$, we choose to make $\beta = 1$, and obtain

$$\alpha = \frac{D_Q s_P{}^2 - D_P s_{PQ}}{D_P s_Q{}^2 - D_Q s_{PQ}} \tag{p}$$

Using this value of α and $\beta = 1$

$$X = \alpha Q + P$$

$$= (\alpha + c_1)x_1' + (\alpha + c_2)x_2' + \cdots + (\alpha + c_k)x_k' \tag{q}$$

$$= a_1'x_1' + a_2'x_2' + \cdots + a_k'x_k'$$

is the approximate discriminant function based on the standardized observations $x_1', \cdots, x_k'$.

The procedures can be outlined as follows:

A. For the exact solution.

(1) Obtain the means and the sums and cross products of deviations for each population.

(2) Obtain the differences (b) and the joint sums $S(x_i x_j) = S_1(x_i x_j) + S_2(x_i x_j)$ [see (d)].

(3) Substitute these values in (j), and solve for the coefficients $a_1, \cdots, a_k$ of the discriminant function. Usually these are divided by one of the a_i to make one of the coefficients unity. Substitute these in (a) to obtain the discriminant function.

B. For the approximate solution.

(1) and (2) exactly as above.

(3) Obtain the d_i' from (*l*) and the correlation matrix (*m*).

(4) Compute the c_i from (*n*).

(5) Compute D_Q, D_P, s_Q^2, s_P^2, and s_{PQ} from (*o*).

(6) Compute α from (*p*).

(7) Compute the coefficients a_i' of X in (*q*).

Note that procedure *B* does not involve the solution of any simultaneous equations, although the computations are otherwise somewhat more complicated.

APPENDIX 6A

The Solution of Regression Equations

6A.1. Introduction

As we have seen in Chapter 6, the solution of regression and correlation problems involves calculations which soon become tedious as the number of independent variables is increased, particularly in the solution of the linear equations necessary, in the non-orthogonal case, to determine the estimates $b_1, \cdots, b_k$ of the regression coefficients, or the inverse elements c_{ij}, $i < j = 1, \cdots, k$, used to determine these coefficients and to test their significance. This appendix is devoted to a brief summary of the procedures available for the systematic solution of linear simultaneous equations, particularly that known as the abbreviated Doolittle method, and their application to regression and correlation problems. Although for small numbers of independent variables, say 2 or 3, or for an occasional problem of a more complex nature, the usual methods, as exemplified in the text of Chapter 6, may suffice, it is almost essential if any consistent use is to be made of regression or correlation techniques to have a method which reduces the necessary calculations to a minimum.

In regression problems the equations involved are symmetric, and the methods to be considered here will be designed for this case only. In some instances they may be of more general application, but the reader interested in a complete discussion of the methods available for solving simultaneous equations is referred to [25].

6A.2. Exact Solution of Linear Equations. Evaluation of Determinants

For simplicity we consider the case of 3 equations in 3 unknowns, and generalize where necessary. These equations can be written as

$$
(a)\qquad
\begin{aligned}
a_{11}x_1 + a_{12}x_2 + a_{13}x_3 &= a_{1y}\\
a_{21}x_1 + a_{22}x_2 + a_{23}x_3 &= a_{2y}\\
a_{31}x_1 + a_{32}x_2 + a_{33}x_3 &= a_{3y}
\end{aligned}
$$

or, symbolically,

$$(b)\qquad \begin{array}{ccc|c} a_{11} & a_{12} & a_{13} & a_{1y} \\ a_{21} & a_{22} & a_{23} & a_{2y} \\ a_{31} & a_{32} & a_{33} & a_{3y} \end{array}$$

where the unknowns x_1, x_2, and x_3 have been omitted and the point of equality between the known coefficients is indicated by a vertical bar.

The two usual methods of solving simultaneous equations involve successive elimination or the use of determinants. Thus in the above case we could proceed by multiplying the second equation of (*a*) by a_{11}, the first by a_{21}, and then subtracting, obtaining

$$(c)\qquad (a_{11}a_{22} - a_{21}a_{12})x_2 + (a_{11}a_{23} - a_{21}a_{13})x_3 = (a_{11}a_{2y} - a_{21}a_{1y})$$

Similarly, by eliminating x_1 from the first and third equations we obtain

$$(d)\qquad (a_{11}a_{32} - a_{31}a_{12})x_2 + (a_{11}a_{33} - a_{31}a_{13})x_3 = (a_{11}a_{3y} - a_{31}a_{1y})$$

By the same procedure we can then eliminate x_2 between (*c*) and (*d*) and obtain a single equation which can be solved for x_3. To evaluate x_2 we substitute this solution in either (*c*) or (*d*), and to evaluate x_1 we substitute for both x_2 and x_3 in one of the equations (*a*).

Using determinants, it is known that, for example,

$$(e)\qquad x_3 = \frac{\begin{vmatrix} a_{11} & a_{12} & a_{1y} \\ a_{21} & a_{22} & a_{2y} \\ a_{31} & a_{32} & a_{3y} \end{vmatrix}}{\begin{vmatrix} a_{11} & a_{12} & a_{13} \\ a_{21} & a_{22} & a_{23} \\ a_{31} & a_{32} & a_{33} \end{vmatrix}}$$

Similar expressions hold for x_1 and x_2, the coefficients of the unknown to be determined being replaced in the numerator determinant by the column of coefficients on the right-hand side of the equations. The problem of obtaining the solutions can thus be related to that of evaluating the determinants involved. We shall develop our methods from the point of view of elimination, and indicate the relationship of the terms obtained to the values of the determinants involved.

If we use the notation

$$A_{22\cdot 1} = a_{11}a_{22} - a_{21}a_{12} = \begin{vmatrix} a_{11} & a_{12} \\ a_{21} & a_{22} \end{vmatrix}$$

$$A_{23\cdot 1} = a_{11}a_{23} - a_{21}a_{13} = \begin{vmatrix} a_{11} & a_{13} \\ a_{21} & a_{23} \end{vmatrix}$$

$$\vdots$$

$$A_{ij\cdot 1} = a_{11}a_{ij} - a_{i1}a_{1j} = \begin{vmatrix} a_{11} & a_{1j} \\ a_{i1} & a_{ij} \end{vmatrix}$$

then equations (c) and (d) can be written

(c') $$A_{22\cdot 1}x_2 + A_{23\cdot 1}x_3 = A_{2y\cdot 1}$$

(d') $$A_{32\cdot 1}x_2 + A_{33\cdot 1}x_3 = A_{3y\cdot 1}$$

Performing a similar reduction to eliminate x_2 from these equations, we obtain

(f) $$A_{33\cdot 12}x_3 = A_{3y\cdot 12}$$

where

$$A_{33\cdot 12} = A_{22\cdot 1}A_{33\cdot 1} - A_{32\cdot 1}A_{23\cdot 1} = \begin{vmatrix} A_{22\cdot 1} & A_{23\cdot 1} \\ A_{32\cdot 1} & A_{33\cdot 1} \end{vmatrix}$$

$$A_{3y\cdot 12} = A_{22\cdot 1}A_{3y\cdot 1} - A_{32\cdot 1}A_{2y\cdot 1} = \begin{vmatrix} A_{22\cdot 1} & A_{2y\cdot 1} \\ A_{32\cdot 1} & A_{3y\cdot 1} \end{vmatrix}$$

and, in general,

$$A_{ij\cdot 12} = A_{22\cdot 1}A_{ij\cdot 1} - A_{i2\cdot 1}A_{2j\cdot 1} = \begin{vmatrix} A_{22\cdot 1} & A_{2j\cdot 1} \\ A_{i2\cdot 1} & A_{ij\cdot 1} \end{vmatrix}$$

Solving (f) for x_3, we have

(g) $$x_3 = \frac{A_{3y\cdot 12}}{A_{33\cdot 12}}$$

or, in general,

$$x_n = \frac{A_{ny \cdot 12 \cdots (n-1)}}{A_{nn \cdot 12 \cdots (n-1)}}$$

Substituting the value of x_3 obtained from (g) in either (c') or (d'), say (c'), and solving for x_2, we have

(h) $$x_2 = \frac{A_{2y \cdot 1} - A_{23 \cdot 1} x_3}{A_{22 \cdot 1}}$$

and, substituting x_2 and x_3 in the first of equations (a),

(i) $$x_1 = \frac{a_{1y} - a_{13} x_3 - a_{12} x_2}{a_{11}}$$

which completes the solution.

The solution can be outlined symbolically, starting with (b), as in Table 6A.1. Only those values essential to the solution are retained, and,

TABLE 6A.1

x_1	x_2	x_3	y	Sum
a_{11}	a_{12}	a_{13}	a_{1y}	a_{1s}
a_{21}	a_{22}	a_{23}	a_{2y}	a_{2s}
a_{31}	a_{32}	a_{33}	a_{3y}	a_{3s}
	$A_{22 \cdot 1}$	$A_{23 \cdot 1}$	$A_{2y \cdot 1}$	$A_{2s \cdot 1}$
	$A_{32 \cdot 1}$	$A_{33 \cdot 1}$	$A_{3y \cdot 1}$	$A_{3s \cdot 1}$
		$A_{33 \cdot 12}$	$A_{3y \cdot 12}$	$A_{3s \cdot 12}$
x_1	x_2	x_3		
$x_1 + 1$	$x_2 + 1$	$x_3 + 1$		

since combinations of the type $ab - cd$ and $\dfrac{a - bc}{d}$ can be evaluated on any modern calculator as one continuous operation, no intermediary steps are involved in obtaining the above table. Also note that up to the solution for x_3 by equation (f) no divisions are involved, so that the solution to this point is exact as far as errors introduced by calculation are concerned. The last column introduced is a check column. a_{1s}, a_{2s}, and a_{3s} are obtained by summing the corresponding rows, and are then carried throughout the reduction exactly like the column a_{1y}, a_{2y}, and a_{3y}. At each stage the rows should sum to the value obtained, and the solutions obtained by using this column in place of the y column should be exactly 1 unit greater than the true solutions.

The operations down to the heavy line are concerned with the systematic elimination of the unknowns, and this process is known as the forward solution. The process of substitution to find the values of x_1, x_2, and x_3 is known as the back solution. In the forward solution the elimination at each stage was accomplished by evaluating a series of 2×2 determinants. If we take the first reduction as an example, these determinants consist of the term a_{11} corresponding to the row and column to be eliminated, called the pivotal term, or pivot, the term a_{ij} corresponding to the element $A_{ij \cdot 1}$ to be evaluated, and the other two elements in the same rows and columns as these two. This general method of reducing determinants is known as *pivotal condensation.* It can be shown [35] that in general the value of the determinant $|a_{ij}|$; $i, j = 1, \cdots, n$ is given by

$$\frac{A_{nn \cdot 12 \cdots (n-1)}}{A_{(n-2)(n-2) \cdot 12 \cdots (n-3)} \cdot A^2_{(n-3)(n-3) \cdot 12 \cdots (n-4)} \cdots A_{22.1}^{\,n-3} \cdot a_{11}^{\,n-2}}$$

In Table 6*A*.1

$$\begin{vmatrix} a_{11} & a_{12} & a_{13} \\ a_{21} & a_{22} & a_{23} \\ a_{31} & a_{32} & a_{33} \end{vmatrix} = \frac{A_{33 \cdot 12}}{a_{11}}$$

and

$$\begin{vmatrix} a_{11} & a_{12} & a_{1y} \\ a_{21} & a_{22} & a_{2y} \\ a_{31} & a_{32} & a_{3y} \end{vmatrix} = \frac{A_{3y \cdot 12}}{a_{11}}$$

and hence from (e)

$$x_3 = \frac{\dfrac{A_{3y \cdot 12}}{a_{11}}}{\dfrac{A_{33 \cdot 12}}{a_{11}}} = \frac{A_{3y \cdot 12}}{A_{33 \cdot 12}}$$

as we found previously.

EXAMPLE 1. Consider the following set of linear equations in 3 unknowns:

$$2x_1 - x_2 + x_3 = 1$$

$$x_1 + x_2 + x_3 = 6$$

$$x_1 + x_2 - x_3 = 0$$

The solution of these equations by the method of Table 6A.1 is shown in Table 6A.2.

TABLE 6A.2

x_1	x_2	x_3	y	Sum
2	1	−1	1	3
1	1	1	6	9
1	1	−1	0	1
	1	3	11	15
	1	−1	−1	−1
		−4	−12	−16
1	2	3		
2 = 1 + 1	3 = 2 + 1	4 = 3 + 1		

The answers can be checked by substituting into the original equations as well as by carrying the check column. In the remaining discussion we shall omit this column, as it can be added at will to any of the methods discussed in exactly the same manner as above. Note that the above solution does not require symmetry.

EXAMPLE 2. Consider the evaluation of the determinant

$$\begin{vmatrix} 2 & 4 & 1 & 3 \\ -1 & 3 & 0 & -1 \\ 6 & 5 & 1 & 2 \\ -3 & 2 & -1 & 1 \end{vmatrix}$$

With the value 2 in the upper left-hand corner used as a pivot, the first reduction gives

$$\begin{vmatrix} 10 & 1 & 1 \\ -14 & -4 & -14 \\ 16 & 1 & 11 \end{vmatrix}$$

the second reduction

$$\begin{vmatrix} -26 & -126 \\ -6 & 94 \end{vmatrix}$$

and the third reduction $A_{44\cdot 123} = 3200$. Hence the value of the determinant is $\dfrac{3200}{2^2(10)} = 80$.

The above method (frequently called the method of multiplication and subtraction) has two disadvantages: (1) as the number of equations to be solved, or the order of the determinant, becomes large the numbers involved may become very large (or very small); (2) at each stage of the reduction process, we are required to evaluate a complete determinant of one order less than the previous one, which is quite tedious when the number of unknowns is large. The first situation can be eased somewhat, without sacrificing exactness, by a method (known as multiplication and subtraction with division or the method of determinants) which makes use of the fact that the terms of the above solution are exactly divisible by preceding pivotal terms, but in general this objection can be overcome by introducing division. The second disadvantage can be removed to a large extent by making use of the symmetry of the equations that we wish to consider.

6A.3. Method of Single Division. The Abbreviated Doolittle Method

Let us again consider the three equations of (*a*) of the previous section. An alternative method of eliminating x_1 is to divide the first equation by a_{11}, obtaining

(*a*)
$$x_1 + \frac{a_{12}}{a_{11}} x_2 + \frac{a_{13}}{a_{11}} x_3 = \frac{a_{1y}}{a_{11}}$$

or, letting $b_{1i} = \dfrac{a_{1i}}{a_{11}}$,

(*b*)
$$x_1 + b_{12}x_2 + b_{13}x_3 = b_{1y}$$

If this equation is multiplied by a_{21} and a_{31}, respectively, and then subtracted from the second and third equations, we obtain

(*c*)
$$\begin{aligned} (a_{22} - a_{21}b_{12})x_2 + (a_{23} - a_{21}b_{13})x_3 &= (a_{2y} - a_{21}b_{1y}) \\ (a_{32} - a_{31}b_{12})x_2 + (a_{33} - a_{31}b_{13})x_3 &= (a_{3y} - a_{31}b_{1y}) \end{aligned}$$

Using the notation

$$a_{22\cdot1} = a_{22} - a_{21}b_{12}$$

$$a_{23\cdot1} = a_{23} - a_{21}b_{13}$$

$$a_{32\cdot1} = a_{32} - a_{31}b_{12}$$

or, in general,

$$a_{ij\cdot1} = a_{ij} - a_{i1}b_{1j} \tag{d}$$

equations (*c*) can be written

$$\begin{aligned} a_{22\cdot1}x_2 + a_{23\cdot1}x_3 &= a_{2y\cdot1} \\ a_{32\cdot1}x_2 + a_{33\cdot1}x_3 &= a_{3y\cdot1} \end{aligned} \tag{e}$$

Note that, if the original coefficients a_{ij} are comparable in size, these new coefficients will be of the same order of magnitude, since $a_{i1}b_{1j} = \dfrac{a_{i1}a_{1j}}{a_{11}}$ is of the same order of magnitude as a_{ij}.

If we now divide the first equation of (*e*) by $a_{22\cdot1}$, we obtain

$$x_2 + b_{23\cdot1}x_3 = b_{2y\cdot1} \tag{f}$$

where $b_{23\cdot1} = \dfrac{a_{23\cdot1}}{a_{22\cdot1}}$, $b_{2y\cdot1} = \dfrac{a_{2y\cdot1}}{a_{22\cdot1}}$, and, in general, $b_{2i\cdot1} = \dfrac{a_{2i\cdot1}}{a_{22\cdot1}}$. Multiplying (*f*) by $a_{32\cdot1}$ and subtracting from the second equation of (*e*), we obtain

$$(a_{33\cdot1} - a_{32\cdot1}b_{23\cdot1})x_3 = (a_{3y\cdot1} - a_{32\cdot1}b_{2y\cdot1})$$

or

$$a_{33\cdot12}x_3 = a_{3y\cdot12} \tag{g}$$

where, in general,

$$a_{ij\cdot12} = a_{ij\cdot1} - a_{i2\cdot1}b_{2j\cdot1} \tag{h}$$

From (*g*) we obtain

$$x_3 = \frac{a_{3y\cdot12}}{a_{33\cdot12}} = b_{3y\cdot12} \tag{i}$$

Substituting this value in (*f*) gives

$$x_2 = b_{2y\cdot1} - b_{23\cdot1}x_3 \tag{j}$$

and these two values substituted in (*b*) give

$$x_1 = b_{1y} - b_{12}x_2 - b_{13}x_3 \tag{k}$$

which completes the solution.

This solution is shown schematically in Table 6*A*.3, starting with (*b*) of the previous section.

The general procedure is as follows:

I. Obtain (1) by dividing the first row by a_{11}.

II. Obtain section (2) by the use of (*d*), i.e., by subtracting from the corresponding original coefficient the product of the first element in the same row and the b_{1i} in the same column.

III. Obtain (3) by dividing the first row of section (2) by $a_{22\cdot1}$.

IV. Obtain (4) by the use of (*h*), i.e., by subtracting from the corresponding element of section (2) the product of the first element in the same row and the value of $b_{2i\cdot1}$ in the same column.

V. Obtain the solutions x_3, x_2, and x_1, in this order, using (*i*), (*j*), (*k*), and the values of $b_{2i\cdot1}$ and b_{1i} from (3) and (1).

VI. If a sum-check column has not been used, check the results by substitution in the original equations.

TABLE 6*A*.3

	x_1	x_2	x_3	y
	a_{11}	a_{12}	a_{13}	a_{1y}
	a_{21}	a_{22}	a_{23}	a_{2y}
	a_{31}	a_{32}	a_{33}	a_{3y}
(1)	1	b_{12}	b_{13}	b_{1y}
(2)		$a_{22\cdot1}$	$a_{23\cdot1}$	$a_{2y\cdot1}$
		$a_{32\cdot1}$	$a_{33\cdot1}$	$a_{3y\cdot1}$
(3)		1	$b_{23\cdot1}$	$b_{2y\cdot1}$
(4)			$a_{33\cdot12}$	$a_{3y\cdot12}$
(5)	x_1	x_2	x_3	

In the general case steps identical with I–II and III–IV are repeated until a single equation

$$a_{nn\cdot12\cdots(n-1)}\,x_n = a_{ny\cdot12\cdots(n-1)}$$

is obtained from which x_n is determined. This value is then substituted in the counterpart of (3) to obtain x_{n-1}, these two values are substituted in the counterpart of (1) to obtain x_{n-2}, and so on, until all the unknowns have been obtained.

EXAMPLE 1. Table 6*A*.4 gives the solution by the method of Table 6*A*.3 of the regression equations for the example of Section 6.31. In this case the unknowns x_1, x_2, and x_3 are the regression coefficients b_1, b_2, and b_3, and the sums $S(x_1^2)$, $S(x_1x_2)$, $S(x_{1y})$, $\cdots$, are the coefficients a_{11}, a_{12}, a_{1y}, $\cdots$. Since in a solution of this type, it is difficult to work with powers of 10, we have adopted the coding $x_1' = 10^{-3}x_1$, $x_2' = 10^{-6}x_2$, $x_3' = 10^{-1}x_3$, and $y' = 10^{-1}y$, from which $S(y'^2) = 10^{-2}S(y^2)$, $S(x_1'y') = 10^{-4}S(x_1y)$, $S(x_1'^2) = 10^{-6}S(x_1^2)$, $S(x_1'x_2') = 10^{-9}S(x_1x_2)$, etc. Note that the solutions b_1', b_2', and b_3' thus obtained must be multiplied by $\frac{10^{-3}}{10^{-1}} = 10^{-2}$, $\frac{10^{-6}}{10^{-1}} = 10^{-5}$, and $\frac{10^{-1}}{10^{-1}} = 1$, respectively, to obtain the b_1, b_2, and b_3 for the original data. This solution gives the regression coefficients directly, but not the values in the c_{ij} matrix necessary to test their significance.

TABLE 6*A*.4

1	2	3	4
4.511936	5.811666	2.336613	−10.76883
5.811666	8.320607	2.470061	−12.85938
2.336613	2.470061	14.464680	9.294516
1.000000	1.288065	0.517874	−2.386743
	0.834803	−0.539649	1.011573
	−0.539649	13.254609	14.871411
	1.000000	−0.646439	1.211751
		12.905759	15.525331

$b_1' = -5.572213$ $\quad b_2' = 1.989402$ $\quad b_3' = 1.202977$
$b_1 = -0.055722$ $\quad b_2 = 0.000019894$ $\quad b_3 = 1.20298$

While in the above solution the numbers involved tend to stay of the same order of magnitude, we must perform and record more operations than in the method of the previous section owing to the addition of the divisions to obtain (1) and (3) of Table 6*A*.3. However, by omitting intermediate calculations which need not be recorded and by using the symmetry properties of our equations, we can to a large extent overcome this difficulty.

Let us consider the first of these modifications. In obtaining $a_{33\cdot12}$ in Table 6*A*.3, we had from (*h*), for $i=j=3$,

$$a_{33\cdot12} = a_{33\cdot1} - a_{32\cdot1}b_{23\cdot1}$$

Now, to obtain $a_{33\cdot1}$, we had from (*d*)

$$a_{33\cdot1} = a_{33} - a_{31}b_{13}$$

Combining these steps by substitution, we have

$$a_{33\cdot12} = a_{33} - a_{31}\,b_{31} - a_{32\cdot1}b_{23\cdot1}$$

Hence $a_{33\cdot12}$ can be computed directly without first calculating $a_{33\cdot1}$. For example, in Table 6*A*.4 we could have obtained directly

$$\begin{aligned} a_{33\cdot12} &= 14.464680 - (2.336613)(0.517874) - (-0.539649)(-0.646439) \\ &= 12.905759 \end{aligned}$$

and

$$\begin{aligned} a_{3y\cdot12} &= 9.294516 - (2.336613)(-2.386743) - (-0.539649)(1.211751) \\ &= 15.525331 \end{aligned}$$

This means that at each stage of the reduction we need to compute only the first row and first column. This is illustrated schematically for the case of 4 variables in Table 6*A*.5.

TABLE 6*A*.5

	x_1	x_2	x_3	x_4	y
	a_{11}	a_{12}	a_{13}	a_{14}	a_{1y}
	a_{21}	a_{22}	a_{23}	a_{24}	a_{2y}
	a_{31}	a_{32}	a_{33}	a_{34}	a_{3y}
	a_{41}	a_{42}	a_{43}	a_{44}	a_{4y}
(1)	1	b_{12}	b_{13}	b_{14}	b_{1y}
(2)		$a_{22\cdot1}$	$a_{23\cdot1}$	$a_{24\cdot1}$	$a_{2y\cdot1}$
		$a_{32\cdot1}$	—	—	—
		$a_{42\cdot1}$	—	—	—
(3)		1	$b_{23\cdot1}$	$b_{24\cdot1}$	$b_{2y\cdot1}$
(4)			$a_{33\cdot12}$	$a_{34\cdot12}$	$a_{3y\cdot12}$
			$a_{43\cdot12}$	—	—
(5)			1	$b_{34\cdot12}$	$b_{3y\cdot12}$
(6)				$a_{44\cdot123}$	$a_{4y\cdot123}$
(7)	x_1	x_2	x_3	x_4	

Line (5) is obtained from the first row of section (4) in the usual manner, and the back solutions (7) are obtained just as before, since none of the b's have been omitted. To illustrate the calculation, several typical terms in (4) and (6) would be computed as follows:

$$a_{43\cdot12} = a_{43} - a_{41}b_{13} - a_{42\cdot1}b_{23\cdot1}$$

$$a_{34\cdot12} = a_{34} - a_{31}b_{14} - a_{32\cdot1}b_{24\cdot1}$$

$$a_{44\cdot123} = a_{44} - a_{41}b_{14} - a_{42\cdot1}b_{24\cdot1} - a_{43\cdot12}b_{34\cdot12}$$

$$a_{4y\cdot123} = a_{4y} - a_{41}b_{1y} - a_{42\cdot1}b_{2y\cdot1} - a_{43\cdot12}b_{3y\cdot12}$$

The saving in labor is quite small for this number of equations, but in the general case we must at each stage compute and record only $2k + 1$, rather than $k(k + 1)$ values. On the other hand, each computation has become more complicated.

We now consider the case where the coefficients of the unknowns $x_1, \cdots, x_n$ are symmetric, or $a_{ij} = a_{ji}$. The solution can be further simplified in this case, since the first row and first column, second row and second column, etc., are identical. The derived coefficients of the reduced equations at each stage are also symmetric; for example,

$$\begin{aligned} a_{ji\cdot1} &= a_{ji} - a_{j1}b_{1i} \\ &= a_{ij} - \frac{a_{j1}a_{1i}}{a_{11}} \\ &= a_{ij} - \frac{a_{1j}a_{i1}}{a_{11}} \\ &= a_{ij} - a_{i1}b_{1j} = a_{ij\cdot1} \end{aligned}$$

and similarly for the coefficients at successive stages. This can be seen in Table 6A.4, where we have for example $a_{23\cdot1} = a_{32\cdot1} = -0.539649$. Thus if the original array of coefficients is symmetric, the elements in the first column of (2) and (4) of Table 6A.5 are repetitions of those in the corresponding first rows. If these elements are omitted we obtain the scheme for computation shown in Table 6A.6. Of the original array of coefficients those in parentheses are also commonly omitted.

The procedure in this case is as follows:

I. Copy the first row. This step, although not necessary, is very convenient for later calculations.

TABLE 6A.6

	x_1	x_2	x_3	x_4	y
	a_{11}	a_{12}	a_{13}	a_{14}	a_{1y}
	(a_{21})	a_{22}	a_{23}	a_{24}	a_{2y}
	(a_{31})	(a_{32})	a_{33}	a_{34}	a_{3y}
	(a_{41})	(a_{42})	(a_{43})	a_{44}	a_{4y}
(1)	a_{11}	a_{12}	a_{13}	a_{14}	a_{1y}
(2)	1	b_{12}	b_{13}	b_{14}	b_{1y}
(3)		$a_{22\cdot 1}$	$a_{23\cdot 1}$	$a_{24\cdot 1}$	$a_{2y\cdot 1}$
(4)		1	$b_{23\cdot 1}$	$b_{24\cdot 1}$	$b_{2y\cdot 1}$
(5)			$a_{33\cdot 12}$	$a_{34\cdot 12}$	$a_{3y\cdot 12}$
(6)			1	$b_{34\cdot 12}$	$b_{3y\cdot 12}$
(7)				$a_{44\cdot 123}$	$a_{4y\cdot 123}$
(8)				1	$b_{4y\cdot 123} = x_4$
(9)	x_1	x_2	x_3	x_4	

II. Obtain row (2) by dividing row (1) by a_{11}.

III. Obtain row (3) by using

$$a_{2i\cdot 1} = a_{2i} - b_{12}a_{1i}$$

i.e.,

$$a_{22\cdot 1} = a_{22} - b_{12}a_{12}$$

$$a_{23\cdot 1} = a_{23} - b_{12}a_{13}$$

etc. Note that in each instance the multiplier b_{12} is used in connection with the proper element of (1).

IV. Obtain row (4) by dividing the elements of row (3) by $a_{22\cdot 1}$.

V. Obtain row (5) by using

$$a_{3i\cdot 12} = a_{3i} - b_{13}a_{1i} - b_{23\cdot 1}a_{2i\cdot 1}$$

i.e.,

$$a_{33\cdot 12} = a_{33} - b_{13}a_{13} - b_{23\cdot 1}a_{23\cdot 1}$$

$$a_{34\cdot 12} = a_{34} - b_{13}a_{14} - b_{23\cdot 1}a_{24\cdot 1}$$

etc. Note that in each instance b_{13} and $b_{23\cdot 1}$ are used as multipliers in connection with the proper elements of (1) and (3).

VI. Obtain rows (6) and (7) by repeating the process of steps II–III and IV–V.

VII. Obtain $x_4 = b_{4y\cdot 123} = \dfrac{a_{4y\cdot 123}}{a_{44\cdot 123}}$

VIII. The above completes the forward solution. Since all the b's of the previous solutions are retained, the back solution is identical with that already given.

The above procedure is easily extended to any number of equations, steps similar to II–III, IV–V, and VI–VII being repeated until a single equation

$$a_{nn\cdot 12\cdots(n-1)}x_n = a_{ny\cdot 12\cdots(n-1)} \tag{l}$$

and the solution

$$x_n = b_{ny\cdot 12\cdots(n-1)} \tag{m}$$

are obtained. Using this value, the preceding equation is solved for x_{n-1}, and so on until all the unknowns have been obtained.

EXAMPLE 2. Table 6*A*.7 gives the solution, following the form of Table 6*A*.6, for the example of Table 6*A*.4. Note that the line −0.539649, 13.254609, 14.871411 has disappeared completely from the calculation, as would all but the first of each succeeding set of equations. Some typical calculations are

$$\begin{aligned} a_{23\cdot 1} &= a_{23} - b_{12}a_{13} \\ &= 2.470061 - (1.288065)(2.336613) \\ &= 0.539649 \end{aligned}$$

$$\begin{aligned} a_{2y\cdot 1} &= a_{2y} - b_{12}a_{1y} \\ &= -12.85938 - (1.288065)(-10.76883) \\ &= 1.011573 \end{aligned}$$

$$\begin{aligned} a_{33\cdot 12} &= a_{33} - b_{13}a_{13} - b_{23\cdot 1}a_{23\cdot 1} \\ &= 14.464680 - (0.517874)(2.33613) - (-0.646439)(-0.539649) \\ &= 12.905759 \end{aligned}$$

$$\begin{aligned} a_{3y\cdot 12} &= a_{3y} - b_{13}a_{1y} - b_{23\cdot 1}a_{2y\cdot 1} \\ &= 9.294516 - (0.517874)(-10.76883) - (-0.646439)(1.011573) \\ &= 15.525331 \end{aligned}$$

$$\begin{aligned} b_2 = x_2 &= b_{2y\cdot 1} - b_{23\cdot 1}x_3 \\ &= 1.211751 - (-0.646439)(1.202977) \\ &= 1.989402 \end{aligned}$$

$$\begin{aligned} b_1 = x_1 &= b_{1y} - b_{13}x_3 - b_{12}x_2 \\ &= -2.386743 - (0.517874)(1.202977) - (1.288065)(1.989402) \\ &= -5.572213 \end{aligned}$$

As might be expected, the above method can also be used to evaluate determinants. It can be shown [35] that the value of the determinant $|a_{ij}|, i, j = 1, \cdots, n$, is

$$(n) \qquad |a_{ij}| = a_{11} \cdot a_{22\cdot 1} \cdot a_{33\cdot 12} \cdot \cdots \cdot a_{nn\cdot 12 \cdots (n-1)}$$

TABLE 6A.7

1	2	3	y
4.511936	5.811666	2.336613	−10.76883
(5.811666)	8.320607	2.470061	−12.85938
(2.336613)	(2.470061)	14.464680	9.294516
4.511936	5.811666	2.336613	−10.76883
1.000000	1.288065	0.517874	−2.386743
	0.834803	−0.539649	1.011573
	1.000000	−0.646439	1.211751
		12.905759	15.525331
		1.000000	1.202977
−5.572213	1.989402	1.202977	

i.e., the product of the consecutive diagonal, or pivotal, terms. For example, in Table 6A.6 we should have for the value of the determinant of coefficients of x_1, x_2, x_3, x_4

$$\begin{vmatrix} a_{11} & \cdots & a_{14} \\ \cdot & & \cdot \\ \cdot & & \cdot \\ \cdot & & \cdot \\ a_{41} & \cdots & a_{44} \end{vmatrix} = a_{11} \cdot a_{22\cdot 1} \cdot a_{33\cdot 12} \cdot a_{44\cdot 123}$$

or, for the determinant of coefficients of x_1, x_2, x_3 in the first equations

$$\begin{vmatrix} a_{11} & \cdots & a_{13} \\ \cdot & & \cdot \\ \cdot & & \cdot \\ \cdot & & \cdot \\ a_{31} & \cdots & a_{33} \end{vmatrix} = a_{11} \cdot a_{22\cdot 1} \cdot a_{33\cdot 12}$$

As a numerical example, the determinant of the coefficients of b_1, b_2, and b_3 for the data of Example 2 would be, from Table 6A.7 or Table 6A.4,

$$(4.511936)\,(0.834803)\,(12.905759) = 48.61053$$

which, when multiplied by 10^{20} due to coding, is the value obtained in the example of Section 6.31.

Note that it follows from the two equations above that

$$a_{44\cdot 123} = \frac{\begin{vmatrix} a_{11} & \cdots & a_{14} \\ \cdot & & \cdot \\ \cdot & & \cdot \\ \cdot & & \cdot \\ a_{41} & \cdots & a_{44} \end{vmatrix}}{\begin{vmatrix} a_{11} & \cdots & a_{13} \\ \cdot & & \cdot \\ \cdot & & \cdot \\ \cdot & & \cdot \\ a_{31} & \cdots & a_{33} \end{vmatrix}}$$

It is true in general that, if $\Delta = |a_{ij}|, i, j = 1, 2, \cdots, n$ and Δ_{nn} is the same determinant with the last row and column omitted, then

$$(o) \qquad a_{nn\cdot 12\cdots(n-1)} = \frac{\Delta}{\Delta_{nn}}$$

The identification of various determinants of coefficients with the product of terms in the solution makes it possible to relate this method to the determinantal definition of the partial regression coefficients defined in Section 6.33. For we had there

$$(p) \qquad b_{yj\cdot 12\cdots(j-1)} = \frac{\Delta_y^{(j)}}{\Delta^{(j)}}$$

where $\Delta^{(j)}$ is the determinant of the sum of squares and cross products of the variables $1, x_1, x_2, \cdots, x_j$, and $\Delta_y^{(j)}$ is the same determinant with x_j replaced by y in the last column. If we now let $a_{0j} = \Sigma x_i$, $a_{0y} = \Sigma y$, $a_{ij} = \Sigma x_i x_j$, and $a_{iy} = \Sigma x_i y$, then we have from (n)

$$(q) \qquad \begin{aligned} \Delta^{(j)} &= a_{jj\cdot 012\cdots(j-1)} \cdot a_{(j-1)(j-1)\cdot 012\cdots(j-2)} \cdots a_{22\cdot 01} \cdot a_{11\cdot 0} \cdot a_{00} \\ \Delta_y^{(j)} &= a_{jy\cdot 012\cdots(j-1)} \cdot a_{(j-1)(j-1)\cdot 012\cdots(j-2)} \cdots a_{22\cdot 01} \cdot a_{11\cdot 0} \cdot a_{00} \end{aligned}$$

and hence from (p)

$$(r) \qquad b_{yj\cdot 12\cdots(j-1)} = \frac{a_{jy\cdot 012\cdots(j-1)}}{a_{jj\cdot 012\cdots(j-1)}}$$

which is exactly the definition of $b_{yj\cdot 12\cdots(j-1)}$ given by (l) and (m). Since the first step in the orthogonalization procedure amounts merely to changing from the original x's to deviations from their mean and taking

$b_y = a - \bar{y}$, we can obtain all other values of the b's by starting with the squares and cross products of deviations as the a_{ii} and a_{ij}, respectively, as in previous examples. This has no effect on the b's, and the starting point used is merely a matter of taste or individual convenience. Note that, since $b_{ij \cdot 12 \ldots} = b_{ji \cdot 12 \ldots}$, the entries in Table 6*A*.6 are just sufficient to define the orthogonalized variables

$$
\begin{aligned}
x_{2\cdot 1} &= (x_2 - \bar{x}_2) - b_{21}(x_1 - \bar{x}_1) \\
x_{3\cdot 12} &= (x_3 - \bar{x}_3) - b_{31}(x_1 - \bar{x}_1) - b_{32\cdot 1}x_{2\cdot 1} \qquad (s) \\
x_{4\cdot 123} &= (x_4 - \bar{x}_4) - b_{41}(x_1 - \bar{x}_1) - b_{42\cdot 1}x_{2\cdot 1} - b_{43\cdot 12}x_{3\cdot 12}
\end{aligned}
$$

and the regression equation

$$(y - \bar{y}) = b_{y1}(x_1 - \bar{x}_1) + b_{y2\cdot 1}x_{2\cdot 1} + b_{y3\cdot 12}x_{3\cdot 12} + b_{y4\cdot 123}x_{4\cdot 123} \qquad (t)$$

The reader can easily verify that substituting the variables (*s*) into (*t*) to obtain the regression equation in terms of the deviations of the original variates is equivalent to the usual back solution.

It also follows immediately from (*o*) and (*q*) that

$$\frac{\Delta_y^{(j)}}{\Delta^{(j-1)}} = a_{jy\cdot 12\cdots(j-1)}$$

so that

$$
\begin{aligned}
b_{jy\cdot 12\cdots(j-1)}\Sigma y c_j &= b_{jy\cdot 12\cdots(j-1)}\frac{\Delta_y^{(j)}}{\Delta^{(j-1)}} \qquad (u) \\
&= b_{jy\cdot 12\cdots(j-1)}a_{jy\cdot 12\cdots(j-1)}
\end{aligned}
$$

is the reduction in the sum of squares of the y's due to the jth orthogonalized variable C_j. This permits the direct computation of the residual sum of squares, for, again using Table 6*A*.6 as an example, we can introduce into the y column the additional term $a_{yy} = S(y^2) = \Sigma_i(y_i - \bar{y})^2$, and proceed one step further to calculate

$$
\begin{aligned}
a_{yy\cdot 1234} &= a_{yy} - b_{1y}a_{1y} - b_{2y\cdot 1}a_{2y\cdot 1} - b_{3y\cdot 12}a_{3y\cdot 12} - b_{4y\cdot 123}a_{4y\cdot 123} \qquad (v) \\
&= \Sigma_i(y_i - \tilde{y}_i)^2
\end{aligned}
$$

Thus $a_{yy\cdot 1234}$ represents the sum of squares of residuals from the regression and

$$s^2_{y\cdot 1234} = \frac{a_{yy\cdot 1234}}{n - 4 - 1} \qquad (w)$$

or, in general,

$$(x) \qquad s^2_{y\cdot 12\cdots k} = \frac{a_{yy\cdot 12\cdots k}}{n-k-1}$$

We also obtain, using (h) of Section 6.33,

$$(y) \qquad \text{var}\,(b_{yj\cdot 12\cdots(j-1)}) = \frac{\sigma^2}{a_{jj\cdot 12\cdots(j-1)}}$$

A check on the back solution is afforded at this point, since, as we have seen,

$$(z) \qquad a_{yy\cdot 1234} = \Sigma_i(y_i - \tilde{y}_i)^2$$
$$= S(y^2) - b_1S(x_1y) - b_2S(x_2y) - b_3S(x_3y) - b_4S(x_4y)$$

which should, within error of routine calculation, be equal to the value obtained by using (v).

For example, using the data of Table 6.9, which we have already considered in Examples 1 and 2, and coding as before by multiplying by 10^{-2}, we obtain

$$a_{yy} = S(y^2) = \Sigma_i(y_i - \bar{y})^2 = 65.22677$$

and hence

$$\Sigma_i(y_i - \tilde{y}_i)^2 = a_{yy\cdot 123} = 65.22677 - (-10.76883)(-2.386743)$$
$$- (1.011573)(1.211751) - (15.525331)(1.202977) = 19.62195$$

which agrees, after decoding by multiplying by 10^2 and within errors of calculation, with the value obtained previously using (z).

EXAMPLE 3. In a research project on the development of swelling pressures in British coking coals, control measurements made in a moveable wall test oven were analysed to provide information on the effect of a number of variables on the time required to coke the charge. The variables included in the analysis were oven width (x_1), bulk density of charge (x_2), volatile matter of coal (x_3), flue temperature (x_4), and coking time (y); and 63 sets of observations were available. The sums and sums of squares obtained from the original observations, where

oven width $= 12.00 + x_1/100$ (in.)
bulk density $= 50 + x_2/10$ (lb/ft³)
volatile matter $= x_3/10$ (%)
flue temperature $= 1300 + x_4$ (°C)

are*

$N = 63$ $\Sigma x_1 = 2314$ $\Sigma x_2 = 1231$ $\Sigma x_3 = 16{,}552$ $\Sigma x_4 = 2999$ $\Sigma y = 2066$
$\Sigma x_1^2 = 133{,}492$ $\Sigma x_1x_2 = 53{,}437$ $\Sigma x_1x_3 = 608{,}274$ $\Sigma x_1x_4 = 97{,}151$ $\Sigma x_1y = 108{,}510$
$\Sigma x_2^2 = 35{,}957$ $\Sigma x_2x_3 = 308{,}821$ $\Sigma x_2x_4 = 55{,}220$ $\Sigma x_2y = 57{,}248$
$\Sigma x_3^2 = 4{,}502{,}418$ $\Sigma x_3x_4 = 801{,}137$ $\Sigma x_3y = 510{,}217$
$\Sigma x_4^2 = 157{,}411$ $\Sigma x_4y = 86{,}522$
$\Sigma y^2 = 173{,}858$

* Data by courtesy of the Director, British Coke Research Association.

TABLE 6A.8

$S(x_1^2) = 48{,}498.4127$ $S(x_1x_2) = 8222.1746$ $S(x_1x_3) = 316.4127$ $S(x_1x_4) = -13{,}002.7460$ $S(x_1y) = 32{,}625.4921$
$S(x_2^2) = 11{,}903.6508$ $S(x_2x_3) = 14{,}559.8254$ $S(x_2x_4) = 3379.5079$ $S(x_2y) = 16{,}879.0159$
$S(x_3^2) = 153{,}708.1427$ $S(x_3x_4) = 13{,}209.2540$ $S(x_3y) = -32{,}583.5079$
$S(x_4^2) = 14{,}649.0794$ $S(x_4y) = -11{,}826.1587$
$S(y^2) = 106{,}106.3175$

x_1	x_2	x_3	x_4	y
4.84984127	0.82221746	0.03164127	−1.30027460	3.26254921
	1.19036508	−1.45598254	−0.33795079	1.68790159
		15.37084127	1.32092540	−3.25835079
			1.46490794	−1.18261587
				10.61063175
4.84984127	0.82221746	0.03164127	−1.30027460	3.26254921
1	0.16953492	0.00652419	−0.26810663	0.67271258
	1.05097051	−1.46134684	−0.11750884	1.13478557
	1	−1.39047369	−0.11180984	1.07975015
		13.33867050	1.16601569	−1.70174680
		1	0.08741619	−0.12757994
			1.00122841	−0.03226438
			1	−0.03222480
				6.97244050

$b_1 = 0.51185322$ $b_2 = 0.90266747$ $b_3 = -0.12476297$ $b_4 = -0.03222480$

From these the sums of squares and cross products of deviations were computed, and are given in Table 6*A*.8, which also gives the solution of the resulting equations, all coefficients having been multiplied by 10^{-4} for convenience. This change of decimal point has no effect upon the regression coefficients b_1, b_2, b_3, and b_4, but it does affect the term $a_{yy \cdot 1234} = 6.97244050$, which must be multiplied by 10^4 to obtain the residual sum of squares

$$\Sigma(y_i - \tilde{y}_i)^2 = 69{,}724.4050$$

in terms of the y's as originally coded. Notice that all 4 variables together have explained less than half of the variation in coking time. The contributions of the 4 independent variables to this reduction were, in the order taken,

$$x_1 : (3.26255)(0.67271) = 2.19475$$

$$x_2 : (1.13479)(1.07975) = 1.22529$$

$$x_3 : (-1.70175)(-0.12758) = 0.21711$$

$$x_4 : (-0.03226)(-0.03222) = 0.00104$$

It should be emphasized again that these reductions depend on order, so that the reduction due to x_4 is the part which x_4 has contributed in addition to x_1, x_2, and x_3, and does not indicate the relative effect of x_4 alone. The check value for the residual sum of squares was obtained by the usual method, using the regression coefficients.

6A.4. Solution of Sets of Equations. Computation of Inverse Elements

In some cases a series of sets of equations are to be solved, and the coefficients a_{ij} of the x's are the same in each set but the a_{iy} vary from set to set. The back solution must then be repeated for each set, but in the forward solution only those operations carried out on the y column need be repeated. Hence, by adding several columns of y's and following the usual procedure, in both the forward and the back solutions, for each, we obtain the solution of several sets of equations of the above type simultaneously.

In particular, such a procedure is convenient in determining the inverse elements c_{ij}, considered in Section 6.22. To obtain these inverse elements in the case of 4 independent variables it is necessary to solve the 4 sets of equations obtained by replacing the constant terms of equations (*a*) of the previous section by the 4 sets of numbers 1, 0, 0, 0; 0, 1, 0, 0; 0, 0, 1, 0; and 0, 0, 0, 1, respectively. The scheme for this computation is shown in Table 6*A*.9 where columns 1, 2, 3, and 4, containing the above constants, replace the y column of Table 6*A*.6.

TABLE 6A.9

x_1	x_2	x_3	x_4	1	2	3	4
a_{11}	a_{12}	a_{13}	a_{14}	1	0	0	0
	a_{22}	a_{23}	a_{24}	0	1	0	0
		a_{33}	a_{34}	0	0	1	0
			a_{44}	0	0	0	1
a_{11}	a_{12}	a_{13}	a_{14}	1	0	0	0
1	b_{12}	b_{13}	b_{14}	b_{15}	0	0	0
	$a_{22\cdot 1}$	$a_{23\cdot 1}$	$a_{24\cdot 1}$	$a_{25\cdot 1}$	1	0	0
	1	$b_{23\cdot 1}$	$b_{24\cdot 1}$	$b_{25\cdot 1}$	$b_{25\cdot 2}$	0	0
		$a_{33\cdot 12}$	$a_{34\cdot 12}$	$a_{35\cdot 12}$	$a_{36\cdot 12}$	1	0
		1	$b_{34\cdot 12}$	$b_{35\cdot 12}$	$b_{36\cdot 12}$	$b_{37\cdot 12}$	0
			$a_{44\cdot 123}$	$a_{45\cdot 123}$	$a_{46\cdot 123}$	$a_{47\cdot 123}$	1
			1	$b_{45\cdot 123}$	$b_{46\cdot 123}$	$b_{47\cdot 123}$	$b_{48\cdot 123}$
c_{11}	c_{12}	c_{13}	c_{14}				
(c_{21})	c_{22}	c_{23}	c_{24}				
(c_{31})	(c_{32})	c_{33}	c_{34}				
(c_{41})	(c_{42})	(c_{43})	c_{44}				

Since the c's are symmetric, those below the diagonal (in parentheses) need not be computed unless desired as a check. Note that we have immediately

$$c_{14} = b_{45\cdot 123} \qquad c_{34} = b_{47\cdot 123}$$

$$c_{24} = b_{46\cdot 123} \qquad c_{44} = b_{48\cdot 123}$$

When the c's have been obtained, we can evaluate the regression coefficients and test their significance exactly as in Section 6.22.

Example. Table 6A.10 shows the calculation of the inverse elements c_{ij} for Example 3 of the preceding section, according to the method of Table 6A.9. Each coefficient of the original equations has been multiplied by 10^{-4}, and each element of the inverse should be multiplied by 10^{-4} to give the inverse in terms of the variables as coded originally. In the computation and testing of the regression coefficients these adjustments cancel automatically.

TABLE 6A.10

x_1	x_2	x_3	x_4	1	2	3	4
4.84984127	0.82221746	0.03164127	−1.30027460	1	0	0	0
	1.19036508	−1.45598254	−0.33795079	0	1	0	0
		15.37084127	1.32092540	0	0	1	0
			1.46490794	0	0	0	1
4.84984127	0.82221746	0.03164127	−1.30027460	1	0	0	0
1	0.16953492	0.00652419	−0.26810663	0.2619232	0	0	0
	1.05097051	−1.46134684	−0.11750884	−0.16953492	1	0	0
	1	−1.39047369	−0.11180984	−0.16131273	0.95150148	0	0
		13.33867050	1.16601569	−0.24225804	1.39047369	1	0
		1	0.08741619	−0.01816208	0.10424380	0.07496999	0
			1.00122841	0.27032823	−0.00974007	−0.08741619	1
			1	0.26999657	−0.00972812	−0.08730894	0.99877310
0.31092807	−0.18919641	−0.04176415	0.26999657				
	1.09654450	0.10509420	−0.00972812				
		0.08260220	0.08730894				
			0.99877310				

Using the elements c_{ij} from Table 6*A*.10, and the values of a_{1y}, a_{2y}, a_{3y}, and a_{4y} from Table 6*A*.8, we have, for example,

$$\begin{aligned} b_2 &= (-0.18919641)(3.26254291) + (1.09654450)(1.68790159) \\ &\quad + (0.10509420)(-3.25835079) + (-0.00972812)(-1.18261587) \\ &= 0.90266866 \end{aligned}$$

and similarly for b_1, b_3, and b_4. To test the significance of this and the other regression coefficients obtained in Table 6*A*.8, we obtain, using the residual sum of squares already computed in Table 6*A*.8 (or computed in an identical fashion in Table 6*A*.10),

$$s^2_{y\cdot 1234} = \frac{\Sigma_i(y_i - \tilde{y}_i)^2}{n-5} = \frac{6.97244}{58} = 0.120214$$

$$s_{y\cdot 1234} = 0.3467$$

Using the values of c_{11}, c_{22}, c_{33}, and c_{44} from Table 6*A*.10, we have

$$t_{b_1} = \frac{b_1}{s_{y\cdot 1234}\sqrt{c_{11}}} = \frac{0.51185}{0.3467\sqrt{0.310928}} = \frac{0.51185}{0.19332} = 2.65$$

and similarly

$$t_{b_2} = \frac{0.90267}{0.3467\sqrt{1.096544}} = \frac{0.90267}{0.36306} = 2.48$$

$$t_{b_3} = \frac{-0.12476}{0.3467\sqrt{0.082602}} = \frac{-0.12476}{0.09964} = -1.25$$

$$t_{b_4} = \frac{-0.03222}{0.3467\sqrt{0.998773}} = \frac{-0.03222}{0.34649} = -0.09$$

From Table III, using the values for 60 degrees of freedom as the closest to 58, we have $t_{60,\,0\cdot 05} = 2.00$ and $t_{60,\,0\cdot 01} = 2.66$. Hence the regression on oven width is significant at almost exactly the 0.01 level, the regression on bulk density is significant at between the 0.05 and 0.01 level, and the other regressions are not significant.

CHAPTER 7

Analysis of Variance

7.1. Introduction

The analysis of variance is a technique which is of great power and importance in the examination of experimental data. If a set of observations can be classified according to one or more criteria, then the total variation between the members of the set can be broken up into components which can be attributed to the different criteria of classification. By testing the significance of these components it is then possible to determine which of the criteria are associated with a significant proportion of the overall variation. To carry out the analysis it is necessary to assume for the data a model which involves a number of parameters and properties, and a part of the technique consists of estimating these parameters.

The advantages of the analysis of variance depend to a considerable extent upon 2 factors:

(1) The systematic arrangement of the observations and the resulting orthogonality of important comparisons.

(2) The accuracy of the assumptions concerning the properties of the model.

As a result it has become customary in designing an experimental procedure to adopt methods which are chosen with these factors in mind. In this chapter we consider only the analysis of the data; in Chapter 8 we consider the forms of experimental design which allow these methods of analysis to be employed with the greatest advantage.

7.2. Basic Analysis of a One-Way Classification

7.21. Example

The data of Table 7.1 taken from Davies [63] give the results obtained from dye trials on each of 5 preparations of Naphthalene Black 12B made from each of 6 samples of H acid intermediate.

TABLE 7.1

YIELDS OF NAPHTHALENE BLACK 12B

Sample of H acid	1	2	3	4	5	6
Individual yields in grams of standard color	1440	1490	1510	1440	1515	1445
	1440	1495	1550	1445	1595	1450
	1520	1540	1560	1465	1625	1455
	1545	1555	1595	1545	1630	1480
	1580	1560	1605	1595	1635	1520
Mean	1505	1528	1564	1498	1600	1470

If these 30 values are considered as a single sample, we obtain $\bar{x} = 1527.5$ and $s^2 = 3972.0$ as estimates of the mean and variance of the population consisting of all possible preparations from all possible intermediates. Since the data are classified as to which preparations were made on which intermediates, we can use this additional information concerning the way in which the data arose in attempting to analyze the causes of the variability in the results. In this instance it is natural to ask whether there has been an appreciable contribution to the overall variability of the data owing to the use of 6 different intermediates rather than a single intermediate in the preparation. One way of answering this question which leads to a convenient significance test is to obtain, instead of the single variance estimate given above, 2 independent variance estimates, one which includes the possible variability due to the choice of different intermediates, and one which does not, and then to compare these two estimates.

By the method of Section 5.42 we can obtain a joint estimate of the variation σ^2 between preparations on the same intermediate. The value of this estimate is $s^2 = 2451$, and it has $30 - 6 = 24$ degrees of freedom, or 4 for each sample of 5 preparations from the same intermediate. This estimate, as stated in Section 5.42, is not affected by possible variation in the average results for different intermediates, but the assumption that the variance σ^2 is the same for all intermediates, which is the basis for making this joint estimate, is subject to test. Such a test can be made by the methods of Section 5.62.

Now let us consider the variation between the means of the 5 preparations on each intermediate. Each of these means is independent of the sample variance of the sample from which it was computed (see Section 4.53) and is obviously independent of the other sample variances, since we assume all 30 results to be independent. Thus any estimate of the variance between these means will be independent of the joint estimate obtained above. The sample variance between the 6 mean values is $s^2 = 2254.3$, which has 5 degrees of freedom. Since each mean value is obtained from a sample of 5 preparations, the expected variation in these mean values due to the variation in preparations made on the same intermediates would be $\sigma^2/5$; in addition, any variation σ_ξ^2 in the average values for the 6 intermediates would be reflected directly in this estimate. If we assume that the variations due to preparation and choice of intermediates are independent, we obtain $\sigma^2/5 + \sigma_\xi^2$ as the average value of this estimate. Hence $5(2254.3) = 11{,}272$ is an estimate of $\sigma^2 + 5\sigma_\xi^2$.

Now, in place of our original estimate $s^2 = 3972$ with $29 = 30 - 1$ degrees of freedom, we have 2 independent estimates as follows:

D.F.	Estimate	Average Value
5	11,272	$\sigma^2 + 5\sigma_\xi^2$
24	2,451	σ^2

If there has been no contribution to the overall variability from differences between the average values for the various intermediates, then we shall have $\sigma_\xi^2 = 0$ and both of the above estimates will have average value σ^2. Their ratio will then be distributed as F with 5 and 24 degrees of freedom. Hence we can test the hypothesis $\sigma_\xi^2 = 0$ by computing this ratio and rejecting the hypothesis, i.e., concluding $\sigma_\xi^2 > 0$, if this ratio exceeds $F_{n_1, n_2, \alpha}$ for a chosen significance level α. In this case we have for the ratio of the two estimates $11{,}272/2451 = 4.60$; since $F_{5, 24, 0.05} = 2.62$ and $F_{5, 24, 0.01} = 3.90$, we reject the hypothesis that these are estimates of the same variance, and conclude that there is an increase in the overall variation due to the use of different intermediates. Since 11,272 is an estimate of $\sigma^2 + 5\sigma_\xi^2$, and 2451 is an estimate of σ^2, then $\dfrac{11{,}272 - 2451}{5} = 1764$ is an estimate of the variance σ_ξ^2 between the average values for the different intermediates.

7.22. General Model for One-Way Classification with Equal Numbers

Let us suppose that we are given $N = np$ observations $x_{i\alpha}$ which represent n replicate results, $\alpha = 1, \cdots, n$, in each of p classes, $i = 1, \cdots, p$. We further suppose that

$$\text{ave}\,(x_{i\alpha}) = \mu + \xi_i$$

so that the average value of each observation in a particular class is the combination of an overall mean plus a variation from this overall mean, or an effect due to the characteristic associated with that class. In addition, let us assume that all observations are normally distributed about these average values with common variance σ^2, so that each class is a sample of n from a normal distribution with mean $\mu + \xi_i$ and variance σ^2. An equivalent way of stating the above model which is frequently used is to assume that for each observation we have

$$x_{i\alpha} = \mu + \xi_i + \varepsilon_{i\alpha}$$

i.e., each observation is the sum of an overall mean μ, an effect ξ_i due to the class in which the observation occurs, and a random érror $\varepsilon_{i\alpha}$ which represents the variation of the particular value $x_{i\alpha}$ from the average value of the ith class. In the case of the example in Table 7.1 we are in fact assuming that the samples of H acid differ in such a fashion that the average yields to be anticipated differ from sample to sample by amounts $(\xi_i - \xi_j)$ but that the variation in yield in the 5 tests carried out on any particular sample is normally distributed with a variance σ^2, which is the same for all samples.

Now let us designate by $\bar{x}_i$ the mean of the n observations in the ith class. Then we can write

$$(a) \qquad x_{i\alpha} - \bar{x} = (x_{i\alpha} - \bar{x}_i) + (\bar{x}_i - \bar{x})$$

where $\bar{x}$ is the overall mean of the $N = np$ observations. Squaring both sides and summing over i and α, we have

$$(b) \qquad \begin{aligned} \Sigma_{i\alpha}(x_{i\alpha} - \bar{x})^2 &= \Sigma_{i\alpha}(x_{i\alpha} - \bar{x}_i)^2 + \Sigma_{i\alpha}(x_i - \bar{x}_i)(\bar{x}_i - \bar{x}) + \Sigma_{i\alpha}(\bar{x}_i - \bar{x})^2 \\ &= \Sigma_{i\alpha}(x_{i\alpha} - \bar{x}_i)^2 + \Sigma_i(\bar{x}_i - \bar{x})\,\Sigma_\alpha(x_{i\alpha} - \bar{x}_i) + n\Sigma_i(\bar{x}_i - \bar{x})^2 \end{aligned}$$

since the term $\bar{x}_i - \bar{x}$ does not involve α. But we know that the sum of the deviations of any group of observations from their mean is zero, so that $\Sigma_\alpha(x_{i\alpha} - \bar{x}_i) = 0$ for all classes $i = 1, \cdots, p$, and (b) reduces to

$$(c) \qquad \Sigma_{i\alpha}(x_{i\alpha} - \bar{x})^2 = \Sigma_{i\alpha}(x_{i\alpha} - \bar{x}_i)^2 + n\Sigma_i(x_i - \bar{x})^2$$

If we now designate the overall sum of squares on the left-hand side by S, the first sum of squares on the right by $S_{\alpha(i)}$, the subscript being read as "α within i," and the second by S_i, we can write (c) as

$$(d) \qquad S = S_{\alpha(i)} + S_i$$

$S_{\alpha(i)}$ is computed from the deviations of the $N = np$ observations from the p sample means, and hence has $N - p = np - p = p(n - 1)$ degrees of freedom. Similarly, S_i is computed from the deviations of the p independent class means from the overall mean, and hence has $p - 1$ degrees of freedom. These two sums of squares add up to the overall sum of squares S, and account for the $N - 1$ degrees of freedom on which it is based. $\dfrac{S_{\alpha(i)}}{N-p}$ is a joint estimate of σ^2, the variation within the p classes, and is identical with that described in Section 5.42 for a joint estimate of σ^2 from a series of samples, and $S_i/(p - 1)$ is an estimate of $\sigma^2 + n\sigma_\xi^2$, where σ_ξ^2 is the variance of the population of effects from which the ξ_i were drawn. These statements can be summarized in Table 7.2, usually called an analysis of variance table.

TABLE 7.2

ANALYSIS OF VARIANCE TABLE:

ONE CLASSIFICATION WITH REPLICATION

Source of Estimate	Sum of Squares	Degrees of Freedom	Mean Square	Average Mean Square
Between classes	$S_i = n\Sigma_i(\bar{x}_i - \bar{x})^2$	$p - 1$	$\dfrac{S_i}{p-1}$	$\sigma^2 + n\sigma_\xi^2$
Within classes	$S_{\alpha(i)} = \Sigma_{i\alpha}(x_{i\alpha} - \bar{x}_i)^2$	$N - p$	$\dfrac{S_{\alpha(i)}}{N-p}$	σ^2
Total	$S = \Sigma_{i\alpha}(x_{i\alpha} - \bar{x})^2$	$N - 1$	—	—

The estimate of the basic variance σ^2, the "within classes" estimate in this case, is frequently spoken of as the *error*, or the *error variance estimate*. Under the hypothesis $\sigma_\xi^2 = 0$ both estimates have average value σ^2, and we can test this hypothesis by computing the ratio of the between-classes estimate to the within-classes, or error, estimate, and by concluding $\sigma_\xi^2 > 0$ at the significance level α if this ratio is greater than $F_{p-1,\,N-p,\,\alpha}$. The component σ_ξ^2 may be estimated by subtracting the within-classes estimate from the between-classes estimate and dividing by n.

7.23. Computations

As in previous chapters, the definitive forms of the sums of squares given above do not represent the best forms for computation. Let us define $T_i = \Sigma_\alpha x_{i\alpha}$, the sum of the n observations in the ith class, and $T = \Sigma_{i\alpha} x_{i\alpha}$, the sum of all $N = np$ observations. Then $\bar{x}_i = \dfrac{T_i}{n}$, and $\bar{x} = \dfrac{T}{N}$, and we can show by the algebraic procedures used in Section 2.36

that

(a) $$S = \Sigma_{i\alpha}(x_{i\alpha} - \bar{x})^2 = \Sigma_{i\alpha}x_{i\alpha}^2 - T^2/N$$

(b) $$S_{\alpha(i)} = \Sigma_{i\alpha}(x_{i\alpha} - \bar{x}_i)^2 = \Sigma_{i\alpha}x_{i\alpha}^2 - \frac{\Sigma_i T_i^2}{n}$$

(c) $$S_i = n\Sigma_i(\bar{x}_i - \bar{x})^2 = \frac{\Sigma_i T_i^2}{n} - \frac{T^2}{N}$$

Hence the calculations are most easily performed as follows:

(1) Obtain the sums of the observations in each class, T_i.

(2) Obtain the grand total of all observations, T, by summing the class totals.

TABLE 7.3

OBSERVATIONS AND SUMS

Sample of H acid	1	2	3	4	5	6
	40	90	110	40	115	45
	40	95	150	45	195	50
Yield of dyestuff—1400	120	140	160	65	225	55
	145	155	195	145	230	80
	180	160	205	195	235	120
Sums	525	640	820	490	1000	350

$T = 3825$ $\qquad \Sigma_{i\alpha}x_{i\alpha}^2 = 602{,}875$

$T^2/N = 487{,}687.5$ $\qquad \Sigma_i T_i^2/n = 544{,}045$

(3) Obtain the sum of squares of the individual values $\Sigma_{i\alpha}x_{i\alpha}^2$. On most computing machines the sum of the individual observations are obtained simultaneously and should check the grand total previously obtained.

(4) Obtain the sum of squares of the class totals, $\Sigma_i T_i^2$, again using the sum as a check. Without removing from the machine, divide by n, obtaining $\Sigma_i T_i^2/n$ directly.

(5) Compute the quantity T^2/N, frequently called the *correction for the mean.*

(6) Obtain the between-classes sum of squares by using (*c*), i.e., by subtracting the correction for the mean obtained in (5) from the result obtained in (4), and enter in the analysis of variance table.

(7) Obtain the "total" sum of squares using (*a*), i.e., by subtracting the correction for the mean obtained in (5) from the result obtained in (3), and enter in the analysis of variance table.

(8) Obtain the within-classes sum of squares using (*b*), i.e., by subtracting the result obtained in (4) from the result obtained in (3). This sum of squares can also be obtained by difference.

(9) Fill in the proper degrees of freedom in the table, check that both the sums of squares and the degrees of freedom give the correct totals, and compute the variance estimates.

This completes the analysis of variance table, and we can then carry out the significance test or estimate σ_ξ^2 as described in Section 7.22. Tables 7.3 and 7.4 show all the necessary recorded calculations for the example of Section 7.21.

If we are to use this estimate of σ_ξ^2 as a basis for decisions about a process, it is desirable to have a confidence interval in addition to the

TABLE 7.4
ANALYSIS OF VARIANCE

Source of Estimate	Sum of Squares	D.F.	Mean Square
Between samples	56,357.5 (= 544,045 − 487,687.5)	5	11,272
Within samples	58,830.0 (= 602,875 − 544,045)	24	2,451
Total	115,187.5 (= 602,875 − 487,687.5)	29	

$$F_{5,\,24} = \frac{11{,}272}{2451} = 4.60^{**} \qquad F_{5,\,24,\,0\cdot05} = 2.62 \qquad F_{5,\,24,\,0\cdot01} = 3.90$$

$$\text{Estimate of } \sigma^2 : \; 2451$$

$$\text{Estimate of } \sigma_\xi^2 : \; \frac{11{,}272 - 2451}{5} = 1764$$

point estimate. This cannot be obtained in a direct manner since the estimate $\hat{\sigma}_\xi^2$, although derived from the two estimates which are distributed in the χ^2-form, is not itself distributed in this form. A number of methods of procedure have been developed for this situation, among them an approximation to the exact fiducial limits due to Bross [96]. It appears that, for this type of test, the fiducial and confidence limits are similar and we may regard the values which we derive as good approximations to the confidence limits.

The method depends upon the use of 3 values of the variance ratio to specify each limit:

F = sample variance ratio, based on n_1 and n_2 degrees of freedom.

$F_{\alpha/2}$ = critical variance ratio corresponding to the required confidence interval, based on n_1 and n_2 degrees of freedom.

$F'_{\alpha/2}$ = critical variance ratio corresponding to the required confidence interval, based on n_1 and ∞ degrees of freedom.

If the estimated variance component is $\hat{\sigma}_\xi^2$ then the lower confidence limit is given by

$$(e) \qquad \underline{L}_{\alpha/2}\,\hat{\sigma}_\xi^2 = \frac{(F/F_{\alpha/2}) - 1}{(F'_{\alpha/2}F/F_{\alpha/2}) - 1}\,\hat{\sigma}_\xi^2 = \frac{F - F_{\alpha/2}}{F'_{\alpha/2}F - F_{\alpha/2}}\,\hat{\sigma}_\xi^2$$

and the corresponding upper limit by

$$(f) \quad \bar{L}_{1-\alpha/2}\hat{\sigma}_\xi^2 = \frac{(F/F_{(1-\alpha/2)}) - 1}{(F'_{(1-\alpha/2)}F/F_{(1-\alpha/2)}) - 1}\,\hat{\sigma}_\xi^2 = \frac{F - F_{(1-\alpha/2)}}{F'_{(1-\alpha/2)}F - F_{(1-\alpha/2)}}\,\hat{\sigma}_\xi^2$$

where we obtain $F_{(1-\alpha/2)}$ and $F'_{(1-\alpha/2)}$ from the usual tables of the variance ratio (see Section 4.55) as

$$F_{n_1,\,n_2,\,1-\alpha/2} = \frac{1}{F_{n_2,\,n_1,\,\alpha/2}} \qquad F_{n,\,\infty,\,1-\alpha/2} = \frac{1}{F_{\infty,\,n,\,\alpha/2}}$$

For the estimated value of the variance component σ_ξ^2 obtained from Table 7.4 we have

$$F = 4.60 \qquad F_{5,\,24,\,0.05} = 2.62 \qquad F_{5,\,\infty,\,0.05} = 2.21$$

$$F_{24,\,5,\,0.05} = 4.53 \qquad F_{\infty,\,5,\,0.05} = 4.36$$

$$F_{5,\,24,\,0.95} = 0.221 \qquad F_{5,\,\infty,\,0.95} = 0.229$$

$$\underline{L}_{0.05} = \frac{4.60 - 2.62}{(2.21)(4.60) - 2.62} = 0.26$$

$$\bar{L}_{0.95} = \frac{4.60 - 0.221}{(0.229)(4.60) - 0.221} = 5.25$$

$$\text{Est}\,(\sigma_\xi^2) = 1764 = \hat{\sigma}_\xi^2$$

$$460 = \underline{L}_{0.05}\hat{\sigma}_\xi^2 < \sigma_\xi^2 < \bar{L}_{0.95}\hat{\sigma}_\xi^2 = 9260$$

so that the range 460–9260 constitutes a 90% confidence interval.

In this example the sample variance ratio was significant, and we obtained a positive lower confidence limit. If the sample variance ratio is not significant, the lower confidence limit is obviously zero, whereas the upper limit may be computed as [97]

$$(g) \qquad \bar{L}_{0.95} = (F - F_{0.95})\,\frac{\hat{\sigma}^2}{nF'_{0.95}}$$

In the analysis of this example Tables 7.3 and 7.4 were prepared, the first for convenience in computation and the second as a systematic expression of the results from the analysis of variance. In reporting the conclusions derived from the analysis, care should be taken to express them in terms of the original variables. Thus, for example, the mean yield for sample 1 is computed as $\frac{525}{5} + 1400 = 1505$ and is reported as a part of the original table. Since the coding procedure employed in obtaining Table 7.3 involved a change of origin (addition or subtraction of a constant) but no change of scale (multiplication or division by a constant) the variance components of Table 7.4 may be referred directly to the original data. If a change of scale had been introduced, for example, a division of the results in Table 7.3 by 5, then the variance components would be corrected by multiplication by 5^2 before they were referred to the original data.

7.24. Unequal Numbers of Observations in Each Class

Although it is usually desirable, in the interests of experimental efficiency and simplicity of analysis, to have the number of replicate observations, n, equal for every class, data will frequently be encountered where, owing to a lack of design or to the loss of part of the data, or perhaps to the deliberate placing of emphasis on certain effects, the number of observations in the various classes will be unequal. Let us now suppose that we have p classes and n_i, $i = 1, \cdots, p$, observations in each class, and let $N = \sum_{i=1}^{p} n_i$. Then the analysis of the preceding sections can be modified in the following manner:

(1) Identity (b) of Section 7.22 will be replaced by

$$\Sigma_{i\alpha}(x_{i\alpha} - \bar{x})^2 = \Sigma_{i\alpha}(x_{i\alpha} - \bar{x}_i)^2 + \Sigma_i n_i(\bar{x}_i - \bar{x})^2$$

Note that, since the number n_i is no longer constant for all classes, it must be kept inside the summation.

(2) The computational forms (b) and (c) of Section 7.23 will be replaced by

$$S_i = \Sigma_i n_i(\bar{x}_i - \bar{x})^2 = \Sigma_i \frac{T_i^2}{n_i} - \frac{T^2}{N}$$

and

$$S_{\alpha(i)} = \Sigma_{i\alpha}(x_i - \bar{x}_i)^2 = \Sigma_{i\alpha} x_{i\alpha}^2 - \Sigma_i \frac{T_i^2}{n_i}$$

Note that the major change here is that the square of each class total is divided by the number in the particular class and then summed, rather than summing the squares of the totals and then dividing by the common number n. This automatically weights the contribution of each class mean to the variation between class means, as indicated in (1).

(3) The above changes will enable the computation of an analysis of variance table similar to Table 7.2, except that the degrees of freedom for error will now be

$$\Sigma_i(n_i - 1) = \Sigma_i(n_i) - p$$

and the average value of the between-classes estimate of the variance will now be replaced by

$$\sigma^2 + \frac{(N^2 - \Sigma_i n_i^2)}{(p-1)N} \sigma_\xi^2$$

An estimate of the variance component σ_ξ^2 may be obtained from this expression by a method similar to that previously employed, but a convenient method for determining the confidence limits of this estimate is not available.

As an example let us consider the data of Table 7.5 (Brownlee [44]) which represent the throughputs obtained before failure due to corrosion of a series of acid pots, classified according to the foundry in which they were manufactured.

TABLE 7.5

Foundry	Throughputs Obtained	Mean
A	84, 60, 40, 47, 34	53.0
B	67, 92, 95, 40, 98, 60, 59, 108, 86	78.3
C	46, 93, 100	79.7

We should like to determine whether or not there are variations in the average throughput to be expected from the product of the different foundries, and, in particular, whether there is any evidence to support the conclusion that the product of foundry A gives a lower average throughput than that of foundries B and C. For these data we have

$$\Sigma_{i\alpha} x_{i\alpha}^2 = (84)^2 + (60)^2 + \cdots + (93)^2 + (100)^2 = 95{,}709$$

and, since the 3 class totals are $T_1 = 265$, $T_2 = 705$, and $T_3 = 239$, we obtain

$$\Sigma_i \frac{T_i^2}{n_i} = \frac{(265)^2}{5} + \frac{(705)^2}{9} + \frac{(239)^2}{3} = 88{,}310$$

From the overall total $T = 1209$ we obtain

$$\frac{T^2}{N} = \frac{(1209)^2}{17} = 85{,}981$$

The results of these calculations may be collected into the analysis of variance in Table 7.6.

TABLE 7.6

Source of Estimate	Sum of Squares	D.F.	Mean Square
Between foundries	2329 = (88,310 − 85,981)	2 = (3 − 1)	1164.5
Within foundries	7399 = (95,709 − 88,310)	14 = (17 − 3)	528.5
Total	9728 = (95,709 − 85,981)	16 = (17 − 1)	

To test the hypothesis that there is no appreciable variation between the average lives of the product of the different foundries, we form the variance ratio 1164.5/528.5 = 2.20. Since this is less than the 5% level for 2 and 14 degrees of freedom ($F_{2,\,14,\,0.05} = 3.74$) there is insufficient evidence to reject the hypothesis and conclude that there is a real difference between the products of the three foundries. Although it would probably be of little interest in this case, we shall compute the estimate of the variance σ_ξ^2 between the average values for the three foundries to illustrate the procedure. Since $\dfrac{N^2 - \Sigma n_i^2}{N(p-1)} = \dfrac{17 - (5^2 + 9^2 + 3^2)}{17(3-1)} = \dfrac{174}{34} = 5.118$, we have 1164.5 as an estimate of $\sigma^2 + 5.118\sigma_\xi^2$, and 528.5 as an estimate of σ^2. Hence $\dfrac{1164.5 - 528.5}{5.118} = 124.3$ is an estimate of σ_ξ^2.

7.25. Interpretation of Low F-Ratios

In a typical analysis of variance of the type of Table 7.4 it may happen that on testing the hypothesis $\sigma_\xi^2 = 0$ by forming the appropriate variance ratio we obtain a value less than one. We should anticipate this effect in at least 50% of similar tests carried out on data for which the hypothesis was true, and the negative value for the estimate of σ_ξ^2 which results in these cases may be attributed to sample fluctuations about an average value of zero.

If, proceeding by the method of Section 4.55, we find that the variance ratio is significantly low at our chosen probability level, we are led to suspect the model which we have assumed to describe the data, and in particular the assumed randomness of the residual effects. The effect of non-randomness may be anticipated by supposing that in a table such as Table 7.1 the tests were not carried out in a random order, but that the vertical sequence represents the chronological sequence of days in which the tests were undertaken. Any effects of chronological sequence would then increase the within-samples estimate of the variance while leaving the between-samples estimate unaffected, and as a consequence would reduce the variance ratio. In general, if we obtain a significantly low variance ratio, we suspect that the experimental design was not randomized with respect to some important factor. If this is true, the value of the experimental results is seriously impaired, and one of the fundamental techniques of experimental design discussed in Chapter 8 was developed to guard against such a possibility.

7.3. Other Considerations in One-Way Classification

7.31. Interpretation of the Effects ξ_i in the General Model

The effects ξ_i associated with each class of observations may be interpreted in a number of ways, and the model selected and the method of analysis used should be designed to suit the interpretation which seems appropriate. It is often advantageous to present a model as a preliminary to a practical analysis of variance, since the analyst is then required to define the effects which he estimates and the subsequent reader is informed of the interpretation which he proposes. We shall consider here 2 extreme cases which frequently arise. The first is that in which the effects ξ_i are a random sample from a large population of possible effects, and, since $\Sigma\xi_i$ over the whole population of ξ_i is defined to be zero, $\hat{\sigma}_\xi^2$ is an estimate of the population variance of these effects. In the above example, the samples of H acid used represent a sample of 6 from a comparatively large population of samples that might have been chosen, and hence the estimate of σ_ξ^2 obtained is an estimate of the variance to be expected in the yield of dyestuff due to variation in the intermediate.

The second case is that in which we are interested only in the values of the effects ξ_i themselves, and they may be considered the entire population of effects with which we are concerned. In this instance the variance σ_ξ^2 of this population is not of primary interest, and we are generally much more concerned with the individual effects and their possible relationships. For example, if an experiment were conducted to determine, on the basis of the yields of 6 trial runs using raw material from each of 4 available sources, which of these 4 materials should be used routinely in the process in question, we should be interested in the individual average values $\mu + \xi_i$ for the yield of each raw material and their relationship to each other; a knowledge of σ_ξ^2 alone would give us little useful information.

As a numerical example, suppose that the data in Table 7.7 represent the results of an experiment in which the concentration of iron in a standard solution was determined 5 times by each of 10 different analysts. One extreme would be the case in which we were interested only in the increase in variation of the results due to the use of different analysts.

This might be true if the applicability of the method for routine use by a large number of laboratories were being tested. On the other hand, an identical experiment might be made using the 10 analysts performing the analysis routinely in a given laboratory in order to evaluate the systematic differences between their results, and the interest would then be primarily in the average results for the different analysts.

TABLE 7.7

STANDARD SOLUTION 2.95% IRON

Analyst	Determination 1	2	3	4	5	Mean
1	2.963	2.996	2.979	2.970	2.979	2.977
2	2.958	2.964	2.955	2.932	2.941	2.950
3	2.956	2.945	2.963	2.950	2.975	2.958
4	2.948	2.960	2.953	2.944	2.950	2.951
5	2.953	2.961	2.961	2.953	2.949	2.955
6	2.941	2.940	2.931	2.942	2.930	2.937
7	2.963	2.928	2.925	2.940	2.934	2.938
8	2.987	2.989	2.988	2.983	2.974	2.984
9	2.946	2.950	2.955	2.969	2.954	2.955
10	2.956	2.947	2.947	2.960	2.954	2.951
						2.956

7.32. Estimation of the Effects ξ_i

Let us consider the above data as an example of this second situation. Our model assumes that $x_{i\alpha} = \mu + \xi_i + \varepsilon_{i\alpha}$, where $i = 1, \cdots, 10$, and $\alpha = 1, \cdots, 5$, and that the population of analysts includes only the 10 analysts from whom results have been obtained. This means that we are not to make any inferences about analysts in general, but only about the particular ones who have taken part in the test. The value μ then refers to the average value which would have been obtained by these analysts if the same conditions obtained while a much larger number of tests were carried out. The analysis of variance of the data, computed by coding as $x =$ determination $- 2.9\%$, is given in Table 7.8, and indicates that there is an appreciable variance σ_ξ^2 among the average results for each of the 10 analysts.

TABLE 7.8

ANALYSIS OF VARIANCE

Source of Estimate	Sums of Squares	D.F.	Mean Square	Average Value of Mean Square
Between analysts	0.0101517	9	0.0011280	$\sigma^2 + 5\sigma_\xi^2$
Between replicates	0.0037368	40	0.0000934	σ^2
Total	0.0138885	49		

If the hypothesis $\sigma_\xi^2 = 0$ were correct, then the two mean squares would each be an estimate of σ^2. We form the variance ratio $\dfrac{0.0011280}{0.0000934} = 12.07$, and, since $F_{9,\,40,\,0.01} = 2.88$ and $F_{9,\,40,\,0.001} = 4.0$, we see that the hypothesis $\sigma_\xi^2 = 0$ is untenable. In this case we should like to go a little further, and indicate which analysts differ, and to what extent.

Let us first examine the problem of estimating the overall mean μ and the effects ξ_i due to the individual analyst. If we consider the mean of the observations obtained by one of the analysts we have

$$\bar{x}_i = \mu + \xi_i + 1/5 \sum_{\alpha=1}^{5} \varepsilon_{i\alpha}$$

Since the $\varepsilon_{i\alpha}$ are randomly distributed with population mean zero, the average value of the final term is zero, and $\bar{x}_i$ is an unbiased estimate of $\mu + \xi_i$.

We now consider the mean value of the $\bar{x}_i$ which is given by

$$\bar{x} = \mu + \sum_{i=1}^{10} \xi_i + 1/50 \sum_{i=1}^{10} \sum_{\alpha=1}^{5} \varepsilon_{i\alpha}$$

Since the 10 analysts are the whole of the population, $\sum_{i=1}^{10} \xi_i = 0$. Furthermore, the average value of the final term is zero so that $\bar{x}$ is an unbiased estimate of μ. If the $\varepsilon_{i\alpha}$ are normally distributed, the estimates $\bar{x}_i$ and $\bar{x}$ are the best estimates of $\xi_i + \mu$ and μ in the maximum likelihood sense. Also

$$\begin{aligned} \text{var } \bar{x}_i &= \text{ave } (\bar{x}_i)^2 - (\text{ave } \bar{x}_i)^2 \\ &= (\mu + \xi_i)^2 + 2(\mu + \xi_i) \text{ ave } (\varepsilon_{i\alpha}) + 1/5 \text{ ave } (\varepsilon_{i\alpha})^2 - (\mu + \xi_i)^2 \\ &= 1/5 \text{ ave } (\varepsilon_{i\alpha})^2 = \sigma^2/5 \end{aligned}$$

$$\begin{aligned}\text{var } \bar{x} &= \text{ave } (\bar{x})^2 - (\text{ave } \bar{x})^2 \\ &= \mu^2 + 2\mu \text{ ave } (\varepsilon_{i\alpha}) + 1/50 \text{ ave } (\varepsilon_{i\alpha})^2 - \mu^2 \\ &= 1/50 \text{ ave } (\varepsilon_{i\alpha})^2 = \sigma^2/50\end{aligned}$$

For the general case of N observations, n from each of p classes which constitute all possible classes, we have

$$\begin{aligned}\text{ave } (\bar{x}_i) &= \text{ave } (T_i/n) = \mu + \xi_i \\ \text{ave } (\bar{x}) &= \text{ave } (T/N) = \mu \\ \text{var } \bar{x}_i &= \sigma^2/n \\ \text{var } \bar{x} &= \sigma^2/N\end{aligned}$$

If the numbers of observations in the classes differ and if n_i is the number in the ith class

$$\begin{aligned}\text{ave } (\bar{x}_i) &= \text{ave } (T_i/n_i) = \mu + \xi_i \\ \text{ave } \frac{1}{p} (\Sigma_i \bar{x}_i) &= \text{ave } \frac{1}{p} (\Sigma_i T_i/n_i) = \mu \\ \text{var } \bar{x}_i &= \sigma^2/n_i \\ \text{var } \frac{1}{p} (\Sigma_i \bar{x}_i) &= \frac{\sigma^2}{p} \Sigma_i \frac{1}{n_i}\end{aligned}$$

and $\bar{x}$ is not in this case an unbiased estimate of μ since $\Sigma_i n_i \xi_i$ is not in general zero.

The best estimates of the average values $\mu + \xi_i$ for the particular analysts are given by the class means $\bar{x}_i$, and the best estimates of the effects ξ_i themselves are given by $\bar{x}_i - \bar{x}$, although generally it is the estimate of $\mu + \xi_i$ which is of most interest. Thus in the example given above, the means of the 5 determinations by each analyst give the best estimate of the average result of an analysis by that analyst. If each class can be considered a sample of n from a normal distribution with variance σ^2, the class mean is also normally distributed with variance σ^2/n and standard deviation $\sigma/\sqrt{n}$.

7.33. Comparison of Class Effects

The above considerations make it clear that the difference $\bar{x}_a - \bar{x}_b$ between any two class means is the best estimate of the difference $(\mu + \xi_a) - (\mu + \xi_b) = \xi_a - \xi_b$ between the effects corresponding to the two classes. From the general theorem on linear combinations of independent normally distributed variables, this estimate is also normally distributed with variance $\sigma^2/n + \sigma^2/n = 2\sigma^2/n$ in the case of equal numbers in each class, or $\sigma^2/n_a + \sigma^2/n_b = \sigma^2(1/n_a + 1/n_b)$ in the case of unequal numbers. Hence, using the error estimate s^2 of σ^2, the ratio

$$t = \frac{\bar{x}_a - \bar{x}_b}{s\sqrt{\dfrac{2}{n}}}$$

has the "Student" t-distribution with $p(n - 1)$ degrees of freedom in the case of equal numbers, and the ratio

$$t = \frac{\bar{x}_a - \bar{x}_b}{s\left(\dfrac{1}{n_a} + \dfrac{1}{n_b}\right)^{1/2}}$$

has the same distribution with $\Sigma n_i - p$ degrees of freedom in the case of unequal numbers in each class, under the hypothesis $\xi_a - \xi_b = 0$. We can use this ratio to test the significance of the difference between any two class means. It differs from the usual test of the significance of the difference between the two means only in the use of the estimate s of σ which is based on all classes.

Let us suppose, for example, that, for the data of Table 7.7, we have some particular reason for suspecting a difference between analysts 1 and 2. We then compute

$$t = \frac{29.77 - 29.50}{\sqrt{0.00934}\sqrt{2/5}} = \frac{0.27}{0.061} = 4.43$$

and, since this exceeds the value $t_{40,\,0.05} = 2.70$, we conclude that our suspicion of a real difference between the two analysts is justified.

An equivalent test could be obtained by remembering that, under the hypothesis $\xi_a - \xi_b = 0$, ave $(\bar{x}_a - \bar{x}_b) = 0$, and hence the average value of the square of the difference is its variance. Thus, considering the case of equal numbers in each class, $(\bar{x}_a - \bar{x}_b)^2$ is an estimate of $2\sigma^2/n$ with 1 degree of freedom, and $\dfrac{n(\bar{x}_a - \bar{x}_b)^2}{2}$ is an estimate of σ^2 based on 1 degree of freedom. The ratio of the latter estimate to the error estimate s^2 would then have the F-distribution with 1 and $p(n - 1)$ degrees of freedom, and this ratio could be used to test the hypothesis $\xi_a = \xi_b$. Since this ratio is the square of the ratio defining t if there are equal numbers in each class, and, since it is known that the F-distribution with 1 and $p(n - 1)$ degrees of freedom is equivalent to the distribution of t^2 where t has $p(n - 1)$ degrees of freedom, this test is equivalent to the one previously given. For example, in the above test of the difference between analysts 1 and 2, we might have computed $\frac{5(29.77 - 29.50)^2}{2} = \frac{5(0.27)^2}{2} = 0.1822.$ On comparing this with the error variance of 0.00934, we obtain a variance ratio of 19.51, which is significant at the 1% level for 1 and 40 degrees of freedom. The square root of this ratio is 4.41, which, except for errors of calculation, is identical with the value of t obtained above.

7.34. General Linear Comparisons Based on Single Degrees of Freedom

A similar method of testing can be applied to any linear combination $\Sigma_i a_i \bar{x}_i$ of the class means such that $\Sigma_i a_i = 0$. Under the hypothesis $\Sigma_i a_i \xi_i = 0$, it follows that $\frac{n(\Sigma_i a_i \bar{x}_i)^2}{\Sigma_i a_i^2}$ is an estimate of σ^2 with 1 degree of freedom and that the ratio of this estimate to the error variance can be used to test the given hypothesis. It can be shown that $\frac{(\Sigma_i a_i \bar{x}_i)^2}{\Sigma_i a_i^2/n_i}$ is a similar estimate in the case of unequal numbers. As an example, suppose that, in a test to compare 5 methods of analysis, 4 analyses were performed by each method. Then the basic analysis of variance would be as shown in the accompanying table. Now let us suppose that 2 methods, say

Source of Estimate	Degrees of Freedom
Between methods	4
Within methods	15
Total	19

1 and 2, included agitation of the solution and the remaining 3 did not. Then, if we wished to compare the means of the agitated and non-agitated results, we could use the comparison

$$\frac{\bar{x}_1 + \bar{x}_2}{2} - \frac{\bar{x}_3 + \bar{x}_4 + \bar{x}_5}{3}$$

which would have average value zero if there were no real difference due to agitation. Multiplication by a constant does not affect the average value or the sum of the coefficients, since both are zero in this case, and we can avoid fractions by using the comparison,

$$3\bar{x}_1 + 3\bar{x}_2 - 2\bar{x}_3 - 2\bar{x}_4 - 2\bar{x}_5$$

If the value thus obtained is squared, multiplied by the number of analyses ($n = 4$) performed by each method, and divided by $3^2 + 3^2 + 2^2 + 2^2 + 2^2 = 30$, the resulting quantity will be an estimate of σ^2 with 1 degree of freedom, and it can be compared with the error estimate in the usual manner to test the hypothesis of no difference due to agitation.

The computation of the sum of squares for such comparisons can be made directly from the totals by replacing the means $\bar{x}_i$ by the corresponding totals T_i, and dividing by n. Thus in the above calculation we could have computed

$$\frac{(3T_1 + 3T_2 - 2T_3 - 2T_4 - 2T_5)^2}{30n}$$

The general formula for the sum of squares would be $\dfrac{(\Sigma_i a_i T_i)^2}{n\Sigma_i a_i^2}$, or in the case of unequal numbers, $\dfrac{(\Sigma_i a_i T_i)^2}{\Sigma_i n_i a_i^2}$.

The single degree of freedom for such a linear comparison represents one of the $p - 1$ degrees of freedom for variation between class means. Hence, if the sum of squares obtained is subtracted from the total sum of squares between classes, the remaining sum of squares has $p - 2$ degrees of freedom. In the above example the sum of squares between methods would have 4 degrees of freedom; after the removal of the sum of squares representing the comparison between methods including and not including agitation, the remaining sum of squares would have 3 degrees of freedom, 1 corresponding to the variation within the two methods which received agitation and 2 to the variation within the three which did not.

More than one *independent* comparison can be made simultaneously on the same set of class means, the second comparison representing 1 of the remaining degrees of freedom. A second linear comparison, $\Sigma_i b_i \bar{x}_i$, satisfying the condition $\Sigma_i b_i = 0$, is independent of the first if $\Sigma_i a_i b_i = 0$, i.e., if the cross product of the coefficients of the two comparisons add up to zero, which implies that the linear combinations are orthogonal (see Section 4.52). If in the above example we also wished to test separately the difference between the two methods using agitation, we could use the linear comparison $\bar{x}_1 - \bar{x}_2$, the coefficients 1, -1, 0, 0, 0 being orthogonal to the coefficients 3, 3, -2, -2, -2 of the previous comparison, and to the coefficients of the mean effect 1, 1, 1, 1, 1. Since $\Sigma_i b_i = 0$, the corresponding sum of squares would be

$$\frac{n(\bar{x}_1 - \bar{x}_2)^2}{2}$$

since $(-1)^2 + (1)^2 = 2$, or, using totals, $\dfrac{(T_1 - T_2)^2}{2n}$. The remaining sum of squares would now have 2 degrees of freedom, and would represent the variation within the methods not receiving agitation. We then have the accompanying breakdown of the sum of squares for the analysis of variance. If required, 2 more independent comparisons could be made between the last 3 means corresponding to the 2 remaining degrees of freedom.

Source of Estimate	Degrees of Freedom
Between averages for methods using agitation and those not using agitation	1
Between methods using agitation	1
Between methods not using agitation	2
Total between methods	4
Within methods	15
Total	19

In making comparisons of this kind it is desirable to tabulate them so that the relationships between the coefficients become apparent. Table 7.9 gives the coefficients corresponding to the breakdown of the 4 degrees of freedom between methods into 4 independent comparisons, the last 2 being an arbitrary subdivision of the 2 degrees of freedom for the variation between the methods not using agitation. For each of these comparisons

TABLE 7.9

Comparison	Coefficients					Divisor
	1	2	3	4	5	
Between averages of the first 2 methods and last 3 methods	3	3	−2	−2	−2	30
Between first 2 methods	1	−1	0	0	0	2
Between last 3 methods	0	0	2	−1	−1	6
	0	0	0	1	−1	2

the sum of the coefficients is zero, and for any pair of comparisons the sum of the cross products is zero. The divisors given are the sums of squares of the coefficients; they would be used in computing the sum of squares corresponding to the single degree of freedom for each comparison.

The breakdown into single degrees of freedom in any particular situation will be dictated by the characteristics of the experiment. Two cases frequently used are the differences between pairs of results in all possible independent combinations, illustrated in Table 7.10 for the case of 4 means, and the comparison of single means with the average of the preceding means, illustrated for the case of 5 means in Table 7.11.

TABLE 7.10

Comparison	Coefficients				Divisor
	1	2	3	4	
Between pair 1, 2 and pair 3, 4	1	1	−1	−1	4
Between pair 1, 3 and pair 2, 4	1	−1	1	−1	4
Between pair 1, 4 and pair 2, 3	1	−1	−1	1	4

TABLE 7.11

Comparison	Coefficients					Divisor
	1	2	3	4	5	
Between 1 and 2	1	−1	0	0	0	2
Between mean of 1, 2 and 3	1	1	−2	0	0	6
Between mean of 1, 2, 3 and 4	1	1	1	−3	0	12
Between mean of 1, 2, 3, 4 and 5	1	1	1	1	−4	20

The tests of significance used in this section are based on the assumption that the comparisons to be made are chosen before the data are analyzed. This situation is likely to occur only if the experiment was designed to test the results of theoretical work, or of a previous experiment. If the tests to be made are decided on the basis of a preliminary examination, or if the same data are analyzed in a number of ways, then the sense in which a confidence statement is to be interpreted will be affected. The statistician's solution to the problem may be to call for a confirmatory experiment, and, if the results obtained from the first series of tests indicate conclusions which are of sufficient economic importance when compared with the cost of the *independent* confirmatory work, this may be practicable. In many instances replication will be impracticable and it is desirable to estimate the true level of significance of a given inference. If the number of tests or comparisons which would have been made before rejecting the results as showing no significant effects is known and if each of these comparisons is independent, then the appropriate significance level may be computed. In cases where the comparisons are not independent some alternative approach is required.

7.35. Confidence Limits, Comparisons and Classifications of Means

In examining the means obtained from an analysis of variance table it seems reasonable to consider the types of statements which we might wish to make. The first relates to the confidence limits for the individual independent estimates of the means, and for any particular mean $\bar{x}_i$ based on n observations we can specify confidence limits as

$$\bar{x}_i \pm t_{n(p-1),\,\alpha}\, s/\sqrt{n}$$

implying that in a large number of similar tests the true mean would fall outside the limits in approximately 100 α% of the cases. Now in a single analysis of variance the number of means for which confidence limits were to be set might be p, and the events that the true means fall outside the confidence limits would not be independent since a common value of s is used in the t statistic. Nevertheless, the average number of such events per analysis over a large number of independent analyses would be simply $p\alpha$. If, however, we wish to consider the number of analyses in which $0, 1, 2, \cdots, p$ such events will occur, the dependence of the events in a single analysis is of importance. This problem can arise if, for example, we wish to make a statement on the proportion of analyses in which one or more events, i.e., one or more failures of the confidence intervals to include the true class means has occurred. If the joint dependence on the common variance estimate were absent, the probability of observing one or more events would be $\alpha' = 1 - (1 - \alpha)^p$. If the events have a positive correlation within analyses, this probability will be somewhat reduced. In any case, for the small values of α' and p which are usual in most analyses the use of $\alpha' = 1 - (1 - \alpha)^p$ would lead to a significance level which was not unduly conservative. Factors for converting $t_{n(p-1),\,\alpha'}$ to $t_{n(p-1),\,\alpha}$ have been given in Table 5.9, Section 5.53.

If the means are to be compared, it is apparent that $(p - 1)$ independent comparisons can be made. If these consist of the differences between $(p - 1)$ class means and the class mean for a standard class, the considerations of the previous section indicate that in fixing confidence limits for these differences we shall have an average of $(p - 1)\,\alpha$ events per analysis and that an analysis will include at least one event in somewhat less than $100(1 - (1 - \alpha)^{p-1})\%$ of all such analyses. If the comparisons are not predetermined, but are selected after an examination of the results, we have in fact $\frac{1}{2}p(p - 1)$ possible comparisons which are interdependent. On the hypothesis that the class mean estimates are normally distributed, these comparison estimates should be distributed as the possible differences between random samples of n observations from a normal population of variance σ^2/n. In particular the greatest comparison difference should be distributed as the range in such a sample. Since for the analysis in question we have available only an estimate s^2/n of σ^2/n, the ratio

$$\frac{\text{greatest comparison difference}}{s^2/n} = W_{p,\,N-p}$$

will be distributed as the studentized range. A confidence interval for any comparison will therefore be

$$\bar{x}_i - \bar{x}_j - W_{p,\,N-p,\,\alpha} s\sqrt{n} < \xi_i - \xi_j < \bar{x}_i - \bar{x}_j + W_{p,\,N-p,\,\alpha} s\sqrt{n}$$

and in a series of independent analyses the proportion in which one or more of these limits are incorrect will be α.

In the case of the data of Table 7.1 we have

$$s^2 = \text{within samples variance estimate} = 2451$$

$$N - p = 24 \quad p = 6 \quad n = 5 \quad s^2/n = 490.2$$

and from Table 5.8 we have $W_{6,\,20,\,0.05} = 4.45$, so that if a comparison difference is to be judged significant the difference in estimates must exceed $\sqrt{490.2} \times 4.45 = 98.6$. Ranking the mean values of Table 7.1, we have

Rank	1	2	3	4	5	6
Sample	5	3	2	1	4	6
Mean	1600	1564	1528	1505	1498	1470

and it appears that samples 4 and 6 are significantly lower than sample 5 and that the comparisons between $5 \sim 1$ and $3 \sim 6$ are sufficiently near the significance level to be considered probably significant.

If we wish to obtain a grouping of the means rather than a series of pair comparisons, an alternative method due to Tukey [42] may be employed. The examination in this case consists of 3 stages, in the first of which the means are ranked, and adjacent means are tested by using the t-test at an unadjusted significance level of say 5%. From results obtained by tests on tables of random normal deviates it is suggested that the significance level achieved in this way will be less than 5%. Any groups of results separated by this device are then treated separately in the second stage, where in any group which contains more than 2 observations the group mean $\bar{x}_m$ is calculated and the greatest difference $|\bar{x}_m - \bar{x}_i|$ from all means of the group is obtained. This value is then checked by an approximation to the distribution of $\dfrac{\bar{x}_m - \bar{x}_i}{s}$ developed by Nair [98].

If the extreme value of the group is separated by this technique, the process is repeated on the reduced group until no further means are separated. Finally, the individual groups formed by the two stages together are checked against the residual variance ratio estimate. The exact procedure is:

(1) Choose a significance level α.

(2) From the error estimate s^2 of σ^2 based on m degrees of freedom compute the difference between the means of 2 classes, $t_{m,\alpha} s_{\bar{x}_i} \sqrt{2}$, where $s_{\bar{x}_i} = \sqrt{s^2/n}$ is the estimate of the standard deviation of a single class mean due to error, which would be significant at the 100 α% level (see Section 5.51).

(3) Arrange the class means in order of magnitude, and divide them into subgroups, considering any gap between 2 means greater than the difference calculated in (2) as a division between 2 groups.

(4) If no group contains more than 2 means, the process terminates, and the groups are declared significantly different from each other; the two means in a group are considered as showing no difference at the level chosen.

(5) In any group of 3 or more class means, compute the grand mean $\bar{x}_m$ of the group. Then determine the largest difference d_L of the differences $\bar{x}_i - \bar{x}_m$ for all class means in the group. Then, if the group contains exactly 3 class means, compute

$$t = \frac{\dfrac{d_L}{s_{\bar{x}_i}} - \dfrac{1}{2}}{3\left(\dfrac{1}{4} + \dfrac{1}{m}\right)}$$

or, if the group contains more than 3 means, compute

$$t = \frac{\dfrac{d_L}{s_{\bar{x}_i}} - \dfrac{6}{5}\log k}{3\left(\dfrac{1}{4} + \dfrac{1}{m}\right)}$$

where k is the number of means in the subgroup and m the degrees of freedom in the estimate of the error variance. If the value obtained exceeds the critical value t_α of a standardized normal variable for the significance level α, the class mean used in obtaining d_L should be separated into a new group. This process should be repeated for all groups until no new groups are formed. If 2 or more adjacent class means are separated from the same group, they should be considered as 1 new group, and, if there are 3 or more, the above process is again applied to this new group.

(6) Test the homogeneity of each subgroup of 3 or more which remains by computing the sum of squares of the deviations of the class means in a given group from the group mean, obtaining the variance estimate, and comparing this estimate with the estimate $s^2_{\bar{x}_i}$ obtained earlier using the F-test with $s^2_{\bar{x}_i}$ in the denominator.

Now let us consider the example of the iron analyses, choosing $\alpha = 0.05$. From the analysis of variance we have the error estimate $s^2 = 0.00934$ of σ^2 based on 40 degrees of freedom. Hence $s^2_{\bar{x}_i} = \dfrac{0.00943}{5} = 0.001868$, and $s_{\bar{x}_i} = 0.043$. The difference between 2 means which would be significant at the 5% level is (2.021) (0.043) (1.414) = 0.123, where 2.021 is the 5% value of the "Student" t with 40 degrees of freedom. The class means arranged in order of magnitude are 29.37, 29.38, 29.50, 29.51, 29.51, 29.55, 29.55, 29.58, 29.77, and 29.84. Since the difference 29.77 − 29.58 = 0.19 is the only gap greater than 0.123, the means are at this point separated into 2 groups consisting of the first 8 and the last 2 (in order). Since the latter group contains only 2 means, we are finished with it, and proceed to apply the second test to the remaining group of $k = 8$ means.

The overall average for this group is 24.49, and the largest deviation from this average is $d_L = 29.49 - 29.37 = 0.12$. Hence we have

$$t = \frac{0.12/0.043 - 6/5(\log 8)}{3(1/4 + 1/40)}$$

$$= \frac{40}{33}(2.79 - 1.08) = 2.07$$

Since this is greater than $t_{0.05} = 1.96$, we separate the mean 29.37 from the group. Repeating the process for the $k = 7$ remaining means, we have for the largest difference from the overall mean $d_L = 29.51 - 29.38 = 0.13$ and

$$t = \frac{0.13/0.043 - 6/5(\log 7)}{3(1/4 + 1/40)}$$

$$= \frac{40}{33}(3.02 - 1.01) = 2.44$$

Since this is significant at the 5% level, we separate the class mean 29.38 from the group, and, since it is adjacent to the mean 29.37 previously separated, these two form a single group. Repeating the process for the $k = 6$ remaining means gives $d_L = 29.58 - 29.533 = 0.047$ and $t = 0.19$. Hence this process terminates.

At this point we have divided the class means into 3 groups.

Group	Analysts
29.37, 29.38	6, 7
29.50, 29.51, 29.51, 29.55, 29.55, 29.58	2, 3, 4, 5, 9, 10
29.77, 29.84	1, 8

Since the means in the 2 groups of 2 class means were not significantly different by the first test, there is no point in testing their internal variability again. For the larger group of 6, the variance between the means is 0.000996, an estimate based on 5 degrees of freedom. The ratio of this estimate to the estimate $s^2_{\bar{x}_i} = 0.001868$ based on 40 degrees of freedom obtained from the error variance is 0.53, which is obviously non-significant, and not unreasonably low.

Hence we should conclude from this analysis that the 10 analysts can be divided into 3 groups, and that, although the average results for the 3 groups are significantly different, there is no evidence of any unusual variability within the groups.

As a second example, suppose that in the example of Section 7.21 the batches of H acid represented samples from 6 different production processes between which it was necessary to choose. Then from the estimate $s^2 = 2451$ of the error variance based on 24 degrees of freedom we obtain $s^2_{\bar{x}_i} = \frac{2451}{5} = 490.2$, $s_{\bar{x}_i} = 22.1$. Hence the difference between 2 class means necessary for significance at the 5% level is $t_{24,\,0.05}\, s_{\bar{x}_i}\sqrt{2} = 2.064(22.1)(1.414) = 64.3$. The class means in order of magnitude are: 1470, 1498, 1505, 1528, 1564, 1600. Since none of the differences between successive means are significant, we consider them as a single group and pass to the next test. The group mean for the single group is the overall mean 1527.5. The largest difference d_L is 1600 — 1527.5 = 72.5. Since there are $k = 6$ class means in the group, and 24 degrees of freedom in the error estimate, we have

$$t = \frac{72.5/22.1 - 6/5(\log 6)}{3(1/4 + 1/24)} = 24/21(3.51 - 0.93) = 2.95$$

Since this value is greater than $t_{0.05} = 1.96$, we separate off the mean 1600. Repeating the process on the subgroup of $k = 5$, we have $d_L = 1564 - 1513 = 51$, and $t = 24/21[51/22.1 - 6/5(\log 5)] = 24/21(2.31 - 0.84) = 1.68$. Since this is less than 1.96, the process terminates.

Now let us test the variability of the remaining group of 5. For this group the variance of the means, based on 4 degrees of freedom, is 1241; the ratio of this to the estimate $s^2_{\bar{x}_i} = 490.2$ based on 24 degrees of freedom obtained from the error variance is 2.53. Since $F_{4,\,24,\,0.05} = 2.78$, these 5 samples of intermediate show no unusually large variability. We therefore conclude that of the 6 processes, the fifth (original order) gave a significantly greater yield than the remaining 5, between which there is no preference at the 5% level.

7.36. Testing Hypotheses Concerning the Overall Mean

In a one-way classification we might also wish to test the hypothesis that the overall mean μ was equal to some predetermined value μ_0. In the example of the preceding section we might wish to determine whether the average analysis obtained by the central group of 6 analysts differed from the standard value of the solution used for the experiment.

We have

$$\Sigma_{i\alpha}(x_{i\alpha} - \mu)^2 = \Sigma_{i\alpha}(x_{i\alpha} - \bar{x})^2 + N(\bar{x} - \mu)^2$$

where the sum of squares on the left has N degrees of freedom and those on the right $N - 1$ and 1 degree of freedom, respectively. Hence, using (*c*) of Section 7.22, we have

$$\Sigma_{i\alpha}(x_{i\alpha} - \mu)^2 = \Sigma_{i\alpha}(x_{i\alpha} - \bar{x}_i)^2 + n\Sigma_i(\bar{x}_i - \bar{x})^2 + N(\bar{x} - \mu)^2$$

Given a hypothetical value μ_0 of the mean μ, we can form the analysis of variance in Table 7.12. This analysis is identical with that of Table 7.2 except that an additional line has been added to represent the variation of the sample mean from the hypothetical mean.

The average value of the estimate obtained from the difference between the sample mean and the hypothetical mean is given in a form which contains both the cases considered in Section 7.31, P representing the number of possible classes (or effects) from which the p classes (or effects)

TABLE 7.12

GENERAL ANALYSIS OF VARIANCE TABLE:

ONE-WAY CLASSIFICATION, INCLUDING TEST FOR $\mu = \mu_0$

Source of Estimate	Sum of Squares	D.F.	Mean Square	Average Mean Square
Mean	$N(\bar{x} - \mu_0)^2$	1	$N(\bar{x} - \mu_0)^2$	$\sigma^2 + n\left(1 - \frac{p}{P}\right)\sigma_\xi^2 + N(\mu - \mu_0)^2$
Between classes	$n\,\Sigma_i(\bar{x}_i - \bar{x})^2$	$p - 1$	$\dfrac{n\,\Sigma_i(\bar{x}_i - \bar{x})^2}{p - 1}$	$\sigma^2 + n\sigma_\xi^2$
Within classes	$\Sigma_{i\alpha}(x_{i\alpha} - \bar{x}_i)^2$	$N - p = p(n - 1)$	$\dfrac{\Sigma_{i\alpha}(x_{i\alpha} - \bar{x}_i)^2}{N - p}$	σ^2
Total	$\Sigma_{i\alpha}(x_{i\alpha} - \mu_0)^2$	N		

in the experiment constitute a random sample. We have for the average value of the variance of the mean estimate

$$\text{var}(\bar{x}) = \text{ave}(\bar{x}^2) - (\text{ave}(\bar{x}))^2$$

If the p classes from which the observations are taken are a random sample from the population of P possible classes, and if the residual components in each class have average value zero and are randomly distributed

$$\text{ave}(\bar{x}) = \mu + \text{ave}\,\frac{1}{p}\left(\sum_i^p \xi_i\right) + \text{ave}\,\frac{1}{np}\left(\sum_{i\alpha}^{pn} \varepsilon_{i\alpha}\right) = \mu$$

$$\text{ave}(\bar{x}^2) = \mu^2 + \text{ave}\,\frac{1}{p^2}\left(\sum_i^p \xi_i\right)^2 + \text{ave}\,\frac{1}{n^2p^2}\left(\sum_{i\alpha}^{pn} \varepsilon_{i\alpha}\right)^2$$

$\text{ave}\,\frac{1}{p^2}\left(\sum_i^p \xi_i\right)^2$ is the variance of the mean of a random sample of p from a finite population P, and by Section 3.44 this is simply $\left(1 - \frac{p}{P}\right)\frac{\sigma_\xi^2}{p}$. Also $\text{ave}\,\frac{1}{n^2p^2}\left(\sum_{i\alpha}^{pn} \varepsilon_{i\alpha}\right)^2$ is the variance of a random sample of np from an infinite population and is equal to σ^2/np. Thus

$$\text{var}(\bar{x}) = \left(1 - \frac{p}{P}\right)\frac{\sigma_\xi^2}{p} + \frac{\sigma^2}{np}$$

$$\begin{aligned} \text{ave}\, N(\bar{x} - \mu_0)^2 &= N(\mu - \mu_0)^2 + N\,\text{var}(\bar{x}) \\ &= N(\mu - \mu_0)^2 + n\left(1 - \frac{p}{P}\right)\sigma_\xi^2 + \sigma^2 \end{aligned}$$

In the second case of Section 7.31, where the classes represent the total population of effects with which we are concerned, we have $p = P$, and the average value of the mean square based on the difference between sample and hypothetical means reduces to $\sigma^2 + N(\mu - \mu_0)^2$. Hence to test the hypothesis $\mu = \mu_0$ we should compare the mean estimate and the within-classes, or error, estimate of σ^2. For example, if we were testing to determine whether a particular group of analysts were obtaining, on the average, the correct result, we should test the deviation of the overall average from the true value against the error estimate. Note that since the F-test with 1 and $N - p$ degrees of freedom is equivalent to the t-test with $N - p$ degrees of freedom, this is equivalent to the t-test for the significance of the difference between the overall mean and the true value μ_0, where the error variance is used as an estimate of σ^2.

In the other case, where the classes used in the experiment can be considered as chosen at random from a large number of possible classes, so that P is large compared with p, we have $1 - p/P \sim 1$, and hence the average value of the variance estimate obtained from the mean is $\sigma^2 + n\sigma_\xi^2 + N(\mu - \mu_0)^2$. In this case the proper test of the hypothesis $\mu = \mu_0$ is made by comparing the mean estimate with the between-classes mean square, and not with the error estimate. If we were testing the possible accuracy of a new analytical method, and had chosen the analysts for this experiment as a random sample of the large number of analysts who might possibly be required to use the method, then the second significance test would be appropriate. In this instance, we wish to know whether results obtained by analysts in general differ significantly from the true value; the particular results obtained by the sample of analysts employed in the experiment are of interest only in so far as they represent the whole class of analysts. This test is equivalent to the t-test for the significance of the difference of the overall mean from the hypothetical value when the variance estimate is based on differences between class means.

EXAMPLE 1. Let us consider the subgroup of 6 analysts found to be homogeneous among themselves in Section 7.35. For this group, $N = 30$, since we are dealing with only 6 of the 10 analysts, and $N(\bar{x} - \mu_0)^2 = 30(29.533 - 29.50)^2 = 0.0327$, using the standard value of the solution as μ_0. Comparing this with the estimate $s^2 = 0.00943$ of the error variance, we obtain the variance ratio $F_{1,40} = 3.50$; since $F_{1,40,0.05} = 4.08$ and $F_{1,40,0.01} = 7.31$, we concluded that this group as a whole is not obtaining values significantly different from the standard.

EXAMPLE 2. Let us consider the data of Section 7.21 under the original premise that these samples are a random sample from a large number of possible samples, and test the hypothesis $\mu = 1500$, i.e., that a given preparation made from a given sample has an expected yield of 1500 grams. We then have for the estimate $N(\bar{x} - \mu_0)^2$, based on 1 degree of freedom, $30(1527.5 - 1500)^2 = 30(27.5)^2 = 22{,}687.5$. In this case we should compare this estimate with the estimate 11,272 based on 6 degrees of freedom, obtained from the variation between the class means. This gives a variance ratio of 2.01, which is not significant at the 5% level for 1 and 6 degrees of freedom. If we had tested this estimate against the error term, we would have obtained a variance ratio of 9.26, which for 1 and 24 degrees of freedom is significant at the 1% level. The distinction here is that, if we could make repeated preparations on each of these same 6 samples, the average yield would be expected to be greater than 1500 grams; however, if a new sample of intermediate is to be used for each series of preparations, we have no definite assurance that the long-term average yield will be greater than 1500 grams per preparation.

7.4. Models and Populations

7.41. Introduction

We have seen in the previous section that the method of procedure and the model assumed in the analysis of variance depend upon the type of population to which the method is to apply. In this section we consider a number of types of model and the assumptions which must be made in using these models as a basis for the analysis of variance procedure. The consequences of failure in the assumptions, and a number of methods for avoiding such failures, will also be considered.

7.42. One-Way Classifications

In Section 7.3 we specified a mathematical model to describe the data which we were to examine by variance analysis in the form

(*a*) $$x_{i\alpha} = \mu + \xi_i + \varepsilon_{i\alpha}$$

It is convenient to restate the assumptions which we built into such a model in order to simplify the analysis. The first is that the components $\varepsilon_{i\alpha}$ are random variables, a property of the data which we assume at the outset, and which in experimental work we endeavour to introduce by deliberately randomizing the effects of those factors which will enter into $\varepsilon_{i\alpha}$. If the fluctuations in the data are not random, then we should not employ statistical methods which are based on this assumption in the analysis. In addition we assume that the random components of the various observations are of equal variance and uncorrelated. These latter assumptions are not essential to the analysis, in the sense that we might still proceed if the correlation coefficients and the relative sizes of the variances were known, but most of the advantages of the techniques would then be lost.

The assumptions concerning the ξ_i which we make depend upon the type of experiment which the model is supposed to describe and upon the scope of the inferences which we make.

I. If the ξ_i represent the mean values of subsets of observations, and the inference from the experiment is to be restricted to the subsets which have been examined, there are no further assumptions concerning the ξ_i.

II. Alternatively, the subsets ξ_i may be regarded as a sample from an infinite population of possible subsets, in which case, if the sample were obtained in a random manner, we may form an estimate of the variance of this population.

III. The ξ_i may be regarded as a random sample from a finite population of possible effects in which case models I and II are limiting cases of III.

These assumptions are sufficient to allow a complete one-way analysis of variance to be made, and it is only in the hypothesis testing or in the fixing of confidence limits that further assumptions concerning the distribution of the $\varepsilon_{i\alpha}$ and the ξ_i are necessary. It is usual to assume that the $\varepsilon_{i\alpha}$ and, in the case of model II, the ξ_i are normally distributed. If the $\varepsilon_{i\alpha}$ or ξ_i have zero covariance and are normally distributed, they are also independent, although this property is not essential to the procedure.

7.43. Two-Way Classifications

It is often desirable to collect experimental data so that they may be classified according to more than one factor or grouping, and under these conditions more complicated models are required.

7.431. *Simple Crossed Classifications with No Interaction.* If the data consist of n observations from each cell of a $p \times q$ classification, so that there are in all npq observations, we may attempt to describe them by a model of the form

$$(a) \qquad x_{ij\alpha} = \mu + \xi_i + \eta_j + \varepsilon_{ij\alpha} \qquad i = 1, \cdots, p \quad j = 1, \cdots, q \quad \alpha = 1, \cdots, n$$

This is equivalent to the assumption that any of the nq observations from the ith class of the factor having p classes contain an effect ξ_i, that any of the np observations from the jth class of the factor having q classes contain the effect η_j, and that these effects are additive. The residual components $\varepsilon_{ij\alpha}$ are still present and are subject to the same assumptions as those put forward in the one-way classification. The class effects may be described by a model of type I in which the ξ_i and η_j represent the means of individual subsets in which we are interested, or by a model of type II or III in which the ξ_i and η_j represent random samples from a population of possible ξ_i and η_j such that cov $(\xi_i,\eta_j) = 0$. The important new assumption is that the effects ξ_i and η_j in any cell are additive.

If the class effects are not additive, the factors of the two classifications are said to interact, and the model must be modified in order to describe this situation. If we continued to employ the simple model, we should confuse the systematic interaction effects and the random residual components. We take

$$(b) \qquad x_{ij\alpha} = \mu + \xi_i + \eta_j + \lambda_{ij} + \varepsilon_{ij\alpha}$$

as a description of the observations, where λ_{ij} represents the systematic departure of the average value of all observations in the ijth cell from the sum of the first 3 terms of the model. We also impose on the λ_{ij} the condition that for any j the sum of the interaction effects for all possible i classes is zero and that for any i the sum of the interaction effects over all possible j classes is zero. In this case "all possible" implies all classes of that population to which any inferences from the analysis are to be applied.

7.432. *Nested Classifications.* A third important type of two-way classification, frequently described as nested, arises in practice. Let us suppose that the one-way classification of Table 7.1 refers to results obtained from 6 random samples of H acid produced from naphthalene supplied by a particular tar distiller. We further suppose that the experiment was carried out 4 times, the supplier of naphthalene being changed for each experiment. If we attempt to describe the data by model (a) we are in effect supposing that variations in the yield of intermediate are due to the supplier of the naphthalene and to the number of the sample, and that all samples having a particular number will have a common effect. This is illogical since the sample numbers only represent identification *within* the material from a particular supplier, and if the samples were randomly selected such an effect cannot exist. Since model (b) also supposes the existence of these effects, it is not suited to the data from this experiment. What is required is a model of the form

$$x_{ij\alpha} = \mu + \xi_i + \eta_{j(i)} + \varepsilon_{ij\alpha}$$

in which the ξ_i are the effect upon the yield of the suppliers from whom the raw material was obtained, and the $\eta_{j(i)}$—"η sub j within i"—represent the effects due to variation within samples from a given supplier. The $\eta_{j(i)}$ are subject to the condition that, for any given supplier s, $\Sigma_j \eta_{j(s)} = 0$ when the summation extends over all possible samples from this supplier, and to this extent they resemble interactions. They differ from the interactions in that it is not a condition on the $\eta_{j(i)}$ that $\Sigma_i \eta_{j(i)} = 0$, when the summation extends over the jth sample from each possible supplier, and it is sometimes convenient in the actual analysis to regard the $\eta_{j(i)}$ as the combination of interaction and main effects. Since the residual random components are nested within i and j, they might be written $\varepsilon_{\alpha(ij)}$, although this is not customary.

7.433. *Many-Way Classifications.* When a number of factors of classification are involved in an analysis it is possible to have more complicated models in which the factors are of any of the three types considered and many different interaction effects are present. We shall consider the analysis of examples of this type in Section 7.5, deriving the models which seem appropriate as we proceed. In general such models represent elaborations of the type of model discussed in this section.

7.44. The Effect of Failure in the Assumptions concerning the Model Parameters

If any of the assumptions which are made in specifying the mathematical model are incorrect, the variance analysis is likely to be prejudiced. The assumptions are not equally important in this respect, and it is convenient to consider them in order.

(1) *Non-Randomness of Residuals.* Failure in this assumption invalidates the statistical procedure completely, since our conclusions must be restricted to the particular set of measurements which were obtained and cannot be related to the model parameters. If collection of the data is undertaken with this point in mind, the difficulty should not arise (Section 8.3). In cases where the data are to be analyzed without a knowledge of the sampling procedure it is sometimes possible to detect evidence of gross non-randomness by the methods of Chapter 11.

(2) *Failure in Distribution Assumptions.* Since the analysis of variance depends on a series of algebraic identitites and the estimation procedures upon the assumption of randomness and absence of correlation, these procedures are not affected by departure of the actual distribution from the normal form which is usually assumed. It also appears that the F-test as used in the analysis of variance and "Student" t-test are little affected by the type of departure from normality which may be anticipated in most experimental work. The cases in which the incorrect assumption of a normal distribution may lead to substantial errors in practice appear to be limited to the estimation of confidence intervals for variance components, and, if the actual distribution is skew, to the specification of single-sided confidence limits for the mean [99].

(3) *Failure in Variance and Covariance Assumptions.* If the variance of the random components in an analysis differs from class to class, the normal method of analysis leads to a loss of efficiency in the estimation of effects and a distortion of the significance level of comparisons. This is because all observations are considered of equal value in deriving the estimates, which is untrue under these conditions, and because a common variance estimate is used in examining all comparisons when a series of different variance estimates would be more appropriate. In cases where a number of replicate determinations are available in each cell of the classification it may be possible to estimate the variance separately for each cell and to test for heterogeneity by means of Bartlett's or Cochran's test (Section 5.62). If the variances appear to change from cell to cell, a number of alternatives are available. At worst it may be necessary to proceed with the analysis in the normal way, interpreting the apparent

conclusions with more or less reserve, according to the degree of heterogeneity. In the simpler experimental designs, such as randomized blocks, certain Latin squares [100], and experiments resulting in one-way classifications, it is possible to employ in testing a particular comparison between means an estimate of the appropriate error term rather than the mean error estimate for the whole of the experiment. In Table 7.1 the variance estimates within each of the columns taken singly are 3975, 1107.5, 1442.5, 4720, 2500, and 9625. These estimates give no evidence of heterogeneity, so that we are justified in using a joint variance estimate in the analysis. If the Bartlett test had suggested heterogeneity, we might have employed the Aspin and Welch method (Section 5.51) in testing the significance of differences between individual means. The effective degrees of freedom for the t-test would then be reduced, so that we should in effect have avoided the assumption of equal variance within classes by sacrificing efficiency in the testing procedure.

An alternative procedure which may be desirable in cases where a sufficient number of replicates are available in each cell is to weight the cell means in proportion to the residual variance estimates. If these estimates represented the true values of the variances, this procedure would be exact. Since a considerable number of degrees of freedom are required to obtain a reliable estimate of a variance, and since the subsequent computational procedure is tedious when the weights are unequal, this method is rarely worth while unless the number of observations in each cell is 15 or more [99].

In a number of technical problems there is reason to anticipate that the variance in individual cells or classes will be related to the mean value of the measured property in that class. If the form of the relationship is known or can be approximately determined by plotting cell variance estimates against cell mean estimates, it is usually possible to find a transformation of the data which stabilizes the variance. The types of transformation which are commonly employed will be discussed in Section 7.45, but it should be noted that a transformation which stabilizes the variance may be of little value if the additivity of the effects is prejudiced.

The assumption of zero covariance between random components cannot usually be tested. The effect of the failure of the assumption on the analysis of variance may be understood by considering the case of a single cell of a two-way classification in which pairs of random components in n repeated determinations have a correlation coefficient ρ. The true variance of the cell total is estimated by

$$(\varepsilon_1 + \varepsilon_2 + \cdots + \varepsilon_n)^2 = \sum_i \varepsilon_i^2 + \sum_{\substack{i \\ i \neq j}} \sum_j \varepsilon_i \varepsilon_j$$

of which the average value is $n\sigma^2 + n(n-1)\rho\sigma^2$. The true variance of the cell mean is then $\sigma^2/n[1 + (n-1)\rho]$. In fact we estimate the variance of the cell mean as

$$\frac{1}{n(n-1)} \sum_n (\varepsilon_i - \bar{\varepsilon})^2 = \frac{1}{n(n-1)} (\Sigma \varepsilon_i^2 - n\bar{\varepsilon}^2)$$

The average value of the above expression is $\sigma^2(1 - \rho)/n$, so that, if there is a positive correlation between errors, the true variance is considerably underestimated. The correlation coefficient is usually unknown, and cannot be estimated from the data unless the factors involved are known to be additive in their effects, since an abnormally high average value for the mean value of the random components is indistinguishable from an interaction. The argument may be extended to the case where the errors in a whole class of a two-way classification are correlated, when the high average value of the class mean due to correlation of errors may be taken incorrectly as evidence of a class effect.

In experimental work correlation between errors usually results from the effect of some additional factor which has been ignored in the experimental procedure, but which has influenced a number of the results in a cell or class in the same way. Such factors cannot always be avoided or controlled, but if they are recognized as potentially important their effect upon the validity of the analysis may be minimized by ensuring that they will, in the long run, contribute equally to each observation. This process, known as randomization, together with a number of its more sophisticated modifications will be considered in the next chapter.

(4) *Non-Additivity of Effects.* Provided that the interaction effects are sufficiently large, the absence of additivity of effects in data which are subjected to variance analysis can be detected when replicate determinations are available to provide an estimate of the residual variance. The procedure involved will be discussed later in the chapter when the details of the analysis of two-way classifications are considered. In order to examine the effect of non-additivity it is necessary to recall the two limiting types of classification in Section 7.42, and to consider their application to the additive and non-additive models

(*a*) $$x_{ij\alpha} = \mu + \xi_i + \eta_j + \varepsilon_{ij\alpha} \qquad \begin{array}{l} i = 1, \cdots, p \\ j = 1, \cdots, q \\ \alpha = 1, \cdots, n \end{array}$$

(*b*) $$x_{ij\alpha} = \mu + \xi_i + \eta_j + \lambda_{ij} + \varepsilon_{ij\alpha}$$

where $\Sigma\xi_i = \Sigma\eta_j = \Sigma\lambda_{ij} = 0$, the summation taking place over all possible parameters in the population being considered.

If the classifications under each factor can be regarded as comprising all the possible classifications to which we wish our results to apply, the interaction effects λ_{ij} which exist in the non-additive case comprise all the possible interaction effects in the population. Any effect ξ_i will therefore be estimated from a sum which includes all the possible interaction effects $\lambda_{i1}, \lambda_{i2}, \cdots, \lambda_{iq}$ an equal number of times, and, since we define the λ_{ij} such that $\Sigma_j \lambda_{ij} = 0$, the mean estimate will be unaffected by the presence of the interactions. However, in order to distinguish between the interaction effects and the residual random component, the value of the former must be estimated for each cell. This reduces the number of degrees of freedom available within the cell for the estimation of the residual variance, and the efficiency of any comparisons which we may wish to make between mean effects is less than it would have been had the λ_{ij} parameters been absent.

If the classifications of one of the factors are to be regarded as a random sample from all the possible classifications of this factor, the second factor being of the type considered above, then the mean value in one of the random classes (i) will be estimated from the sum of all the observations falling in this class. This sum will include equal numbers from each possible class of the second factor and will therefore include equal numbers of all the possible interaction effects $\lambda_{i1}, \lambda_{i2}, \cdots, \lambda_{iq}$. Since by definition $\Sigma_j \lambda_{ij} = 0$, the mean estimate ξ_i will be unaffected by the interaction effects, but, as above, the degrees of freedom available for the residual will be reduced. In computing a class mean for the second factor η_j, we employ the sum of all the np observations in this class. This sum therefore includes each of the interaction effects $\lambda_{1j}, \lambda_{2j}, \cdots, \lambda_{pj}$ n times, but, since $1, \cdots, p$ represent only a sample of p classes from a larger population of classes, the interaction effects represent a sample of p from a larger population of possible interaction effects. Our estimate of η_j will therefore be affected by sampling fluctuations in the sum of the interaction effects $\lambda_{1j}, \cdots, \lambda_{pj}$ and by the residual fluctuations, and the variance estimate used in a comparison of two such class effects will be larger. In addition, the previous discussion on the effect of correlation of the random components $\varepsilon_{ij\alpha}$ is applicable in this case to the interaction effects λ_{ij} which we have regarded as random components in the model. If the p classes cannot be regarded as a random sample from the population of classes, our inferences concerning the class effects must be restricted to the classes actually examined and cannot be extended to the population. In the event that the classes of both factors of classification may be regarded as a random sample from a population of possible samples, the arguments of the previous paragraph indicate that the class mean estimates of both factors will be subject to increased variance, owing to the sampling procedure involved in selecting the interactions.

It is clear that the presence of interaction effects in a two-way classification involves a reduction in the accuracy of the main effect comparisons. Although, for the sake of simplicity, the problem has been considered in terms of a two-way classification only, we shall see in later examples that this conclusion applies equally well to n-way classifications where non-additivity may assume more complicated forms. Since this is true we should examine the sources of interaction effects in the analysis to see which are characteristic of the system from which the data are drawn, and which may be introduced by the methods of collection or analysis employed. In a large number of technical problems characteristic interaction effects are to be expected, and if our conclusions are to be of practical value we shall usually wish to include a sample of these effects in the analysis. On the other hand, the presence of an interaction effect is recognized by comparing the deviation from additivity in a particular cell with the residual variance estimate. If the appropriate residual variance is underestimated because of heterogeneity in the cell variances or the correlation of the residual components, or if some gross error has occurred in the measurement or recording of the observations in one cell, we may be led to suspect interaction effects where none exist. A second important way in which interaction effects may be introduced is in the choice of the scale of observation. The substitution of a scale of observation which is natural (i.e., which experience or theory suggests is additive), may suffice to eliminate these effects. Common cases of this type occur in the use of the reciprocal absolute temperature in thermodynamic problems and the logarithm of the transfer coefficients in problems concerned with the transfer of heat or mass.

7.45. Transformation of Data

Transformation of the scale of observation has been suggested as a means of obtaining homogeneity of variance and additivity of effects. It will not usually be possible to obtain both requirements by this means. If previous experience in an experimental field suggests that additivity is desirable, and may be obtained by the use of a suitable scale, it should be possible to arrange an experiment so that the variances of the cell means are approximately homogeneous by varying the number of observations in the individual cells. Alternatively, provided that a sufficient number of replicates per cell can be arranged, the cell means may be weighted in inverse proportion to the estimated variances and analyzed by the more complex methods appropriate to this case. In many instances the data are insufficient to justify the use of estimated weights for the cell means, and a transformation is suggested only when the data are examined. In this case it is probably best, since data are usually costly and calculations cheap, to conduct the analysis on the transformed and untransformed scales and examine both results. An intelligent combination of the conclusions is then required.

In a large class of data the effects are not additive under any transformation, so that the use of a transformation designed to stabilize the variance can do no harm in this respect. Such a transformation will usually be possible in cases where the magnitude of the variance is related to the average value m of the measurements in a particular cell. The nature of the relationship may be known in advance, or, if a sufficient number of cell variance estimates are available, may be determined from the data by graphical methods, e.g., plotting $\log s_m^2$ against $\log \bar{x}_{ij}$. In any event, if we write

$$\sigma_m^2 = f(m)$$

then for any transformation $\phi(m)$ we have approximately

$$\sigma_\phi^2 = \left(\frac{d\phi}{dm}\right)^2 f(m)$$

or, if σ_ϕ^2 is to have constant value k^2,

$$\phi = \int \frac{k\,dm}{\sqrt{f(m)}}$$

A number of the more common transformations follow.

(*a*) $$f(m) = m$$

A relationship of this kind occurs with data having a Poisson distribution, and may be approximately true in chemical analysis if the investigation involves a number of materials containing relatively small percentages of the component to be determined. The appropriate transformation is $\phi = \sqrt{x}$ if the data represent scale factors, or relatively large counts. If small counts are involved the transformation $\phi = \sqrt{x + \frac{1}{2}}$ has been proposed and used by Yates, whereas Tukey suggests $\phi = \sqrt{x} + \sqrt{x + 1}$ (see section 9.2). Either transformation gives a sufficiently stable variance for cells containing 3 or more counts.

(*b*) $$f(m) = m(1 - m)$$

This is the appropriate relationship for counted data distributed in the binomial form and leads to the transformation

$$\phi(m) = \sin^{-1}\sqrt{m}$$

In this case the correction for continuity may be made by substituting $(m + {}^1/_2)$ for m, but Bartlett suggests the use of m for all intermediate values and $^1/_4$ and $n - {^1/_4}$ for the extreme values 0 and n. Results distributed in this form may arise in exposure or corrosion tests where a number of samples of each of several types of material are subjected to test. It is not uncommon to find that the extent of corrosion, or of failure of the surface film, is difficult to measure quantitatively, since once local failure occurs on a part of a specimen failure spreads rapidly over the remainder of the surface of that specimen. The $\sqrt{x}$ or $\sin^{-1}\sqrt{x}$ transformation will then be used according to the number of replicates and the proportion of failures involved.

(c) $$f(m) = m^2$$

This corresponds to the case of a constant coefficient of variation, and the appropriate transformation is

$$\phi(m) = \log(m)$$

In many chemical analyses the relationship between the mean and variance for a given test method appears to be best described by

$$\phi(m) = a + bm + cm^2$$

so that when the range of variation of m in a particular experiment is limited to fairly large values the $\log(m)$ transformation may be appropriate. Alternatives such as $\log(m + 1)$ or $\log(m + c)$ have been proposed as better approximations for such cases.

A number of other transformations have been proposed, a simple summary and bibliography being given by Bartlett [101]. With all these transformations data in which the residual component was originally normally distributed will no longer have this property after the transformation. This is not usually very serious since

1. Systematic heterogeneity of variance is frequently accompanied by skewness in the distribution of the random components, which is partially corrected by the transformation.

2. Moderate departures from normality are relatively unimportant in variance analysis.

3. The common types of transformation do not result in substantial loss of efficiency in the estimates.

A more radical method of procedure is to rank the observations and carry out a variance analysis on the ranks. In this case the variance is automatically stabilized, but the analysis is largely qualitative. If the data are completely inconsistent with the assumptions in the analysis of variance, or if the cost of the experiment can be materially reduced by the use of ranks rather than measurements, this technique may be worth while.

7.5. Two-Way Classifications

7.51. Types of Two-Way Classifications

In many cases the collection of data in a form in which it can be analyzed according to a single criterion of classification is not a very efficient method of investigation. If we arrange matters so that the data can be classified according to 2 or more criteria, it is frequently possible to obtain information on a number of points from 1 set of experimental data. This is analogous in many ways to the use of multiple regression in place of simple regression methods. In each case we employ the same estimate of the residual variance to test a number of hypotheses. The number of observations required in such an experiment is slightly larger than the number required to test a single hypothesis with comparable rigor, but usually very much smaller than the number which would be required to test a series of hypotheses singly.

Before considering the methods which can be employed to achieve this increased efficiency in experimental design, it is necessary to consider extensions of the technique of analysis of variance which make the analysis of the results from such experiments possible. The two methods of classification which may be distinguished have been discussed in Section 7.4 where the appropriate models were developed. We now consider the application of these models.

7.52. Analysis of Two Nested Classifications

The data of Table 7.13 (Wernimont, [74]) provide an example of a nested classification. An experiment was designed to test the homogeneity of the copper content of a series of bronze castings from the same pour. Two samples were taken from each of 11 castings, and each sample was analyzed in duplicate.

The sample descriptions A and B differentiate between the two random samples taken from a particular casting; they do not imply that sample A from any casting will be any more closely related to the A-samples from the remaining castings than to the B-samples from these castings. Clearly we require a model which does not include a main effect due to sample letters, since the sample classification is "nested" with the main classification of castings. The appropriate model is

$$x_{ij\alpha} = \mu + \xi_i + \lambda_{j(i)} + \varepsilon_{ij\alpha}$$

where the ξ_i represent the effects associated with the castings, $i = (1, \cdots, p)$, and the castings represent a random sample of p from a population of P possible castings and μ is defined so that $\sum_i^P \xi_i = 0$. The $\lambda_{j(i)}$ represent

TABLE 7.13

COPPER CONTENT OF BRONZE CASTINGS

Casting	Sample	Percent Copper		
		X_1	X_2	Mean
1	*A*	85.54	85.56	85.55
	B	85.51	85.54	85.52
2	*A*	85.54	85.60	85.57
	B	85.25	85.25	85.25
3	*A*	85.72	85.77	85.74
	B	84.94	84.95	84.94
4	*A*	85.48	85.50	85.49
	B	84.98	85.02	85.00
5	*A*	85.54	85.57	85.56
	B	85.84	85.84	85.84
6	*A*	85.72	85.86	85.79
	B	85.81	85.91	85.86
7	*A*	85.72	85.76	85.74
	B	85.81	85.84	85.82
8	*A*	86.12	86.12	86.12
	B	86.12	86.20	86.16
9	*A*	85.47	85.49	85.48
	B	85.75	85.77	85.76
10	*A*	84.98	85.10	85.04
	B	85.90	85.90	85.90
11	*A*	85.12	85.17	85.14
	B	85.18	85.24	85.21
			Overall mean	85.57

the fluctuations between q random samples drawn from each of the castings. We imagine that the population of possible samples per casting is Q and define the effects ξ_i such that $\sum_j^Q \lambda_{j(i)} = 0$. Finally the $\varepsilon_{ij\alpha}$ are taken to be random variables normally distributed with variance σ^2, and the $\lambda_{j(i)}$ are defined so that, for any ij, ave $(\varepsilon_{ij\alpha}) = 0$. For the data of the table we have $p = 11$, $q = 2$, $n = 2$, $N = npq = 44$.

If we designate the mean value for a given class, or casting, by $\bar{x}_{i\cdot}$, and that for a given subclass, or sample, by $\bar{x}_{ij}$, we have over the whole population

$$\begin{aligned}
\text{ave}\,(\bar{x}) &= \mu + \text{ave}\,\frac{1}{p}\,\Sigma_i\xi_i + \text{ave}\,\frac{1}{pq}\,\Sigma_{ij}\lambda_{j(i)} + \text{ave}\,\frac{1}{pqn}\,\Sigma_{ij\alpha}\varepsilon_{ij\alpha} \\
&= \mu \\
\text{ave}\,(\bar{x}_{i\cdot}) &= \mu + \xi_i + \text{ave}\,\frac{1}{q}\,\Sigma_j\lambda_{j(i)} + \text{ave}\,\frac{1}{qn}\,\Sigma_{j\alpha}\varepsilon_{ij\alpha} \\
&= \mu + \xi_i \\
\text{ave}\,(\bar{x}_{ij}) &= \mu + \xi_i + \lambda_{j(i)} + \text{ave}\,\frac{1}{n}\,\Sigma_\alpha\varepsilon_{ij\alpha} \\
&= \mu + \xi_i + \lambda_{j(i)}
\end{aligned}$$

Thus

$$\text{ave}\,(\bar{x}_{i\cdot}) - \text{ave}\,(\bar{x}) = \xi_i$$

$$\text{ave}\,(\bar{x}_{ij}) - \text{ave}\,(\bar{x}_{i\cdot}) = \lambda_{j(i)}$$

and we can employ $\bar{x}$, $(\bar{x}_{i\cdot} - \bar{x})$ and $(\bar{x}_{ij} - x_{i\cdot})$ as unbiased estimates of μ, ξ_i, and $\lambda_{j(i)}$. If we denote our estimates of μ, ξ_i, $\lambda_{j(i)}$, and $\varepsilon_{ij\alpha}$ by $\hat{\mu}$, $\hat{\xi}_i$, $\hat{\lambda}_{j(i)}$, and $\hat{\varepsilon}_{ij\alpha}$, we see that for these estimates

$$\Sigma_i\hat{\xi}_i = \Sigma_i(\bar{x}_{i\cdot} - \bar{x}) = 0$$

$$\Sigma_j\hat{\lambda}_{j(i)} = \Sigma_j(\bar{x}_{ij} - \bar{x}_{i\cdot}) = 0 \quad \text{(for each } i\text{)}$$

$$\Sigma_\alpha\hat{\varepsilon}_{ij\alpha} = \Sigma_\alpha(x_{ij\alpha} - \bar{x}_{ij}) = 0 \quad \text{(for each } ij\text{)}$$

Since the sums of squares of the residual effect estimates will be minimized when the estimates have zero mean in each cell, an equivalent procedure for deriving the estimating equations would be to select the $\hat{\mu}$, $\hat{\xi}_i$, and $\hat{\lambda}_{j(i)}$ so as to minimize the sum of squares

$$Q = \Sigma_{ij\alpha}(x_{ij\alpha} - \mu - \xi_i - \lambda_{j(i)})^2$$

subject to the conditions $\Sigma_i\xi_i = 0$ and $\Sigma_j\lambda_{j(i)} = 0$ for each i. This is equivalent to minimizing the unrestricted sum

$$Q' = \Sigma_{ij\alpha}(x_{ij\alpha} - \mu - \xi_i - \lambda_{j(i)})^2 + m_1\Sigma\xi_i + m'_i\Sigma\lambda_{j(i)}$$

and we have

$$(a) \qquad \frac{\partial Q'}{\partial \mu} = 0 = -2\hat{\mu}\Sigma_{ij\alpha}(x_{ij\alpha} - \hat{\mu} - \hat{\xi}_i - \hat{\lambda}_{j(i)})$$

(b) $$\frac{\partial Q'}{\partial \xi_i} = 0 = -2\hat{\xi}_i \Sigma_{j\alpha}(x_{ij\alpha} - \hat{\mu} - \hat{\xi}_i - \hat{\lambda}_{j(i)}) + m_1\hat{\xi}_i \quad \text{(for each } i\text{)}$$

(c) $$\frac{\partial Q'}{\partial \lambda_{j(i)}} = 0 = -2\hat{\lambda}_{j(i)}\Sigma_\alpha(x_{ij\alpha} - \hat{\mu} - \hat{\xi}_i - \hat{\lambda}_{j(i)}) + m'_i\hat{\lambda}_{j(i)} \text{ (for each } ij\text{)}$$

Dividing each of the p equations of type (b) by $-2\hat{\xi}_i$ and summing, we obtain

$$0 = \Sigma_{ij\alpha}(x_{ij\alpha} - \hat{\mu} - \hat{\xi}_i - \hat{\lambda}_{j(i)}) - \frac{pm_1}{2}$$

but from (a)

$$\Sigma_{ij\alpha}(x_{ij\alpha} - \hat{\mu} - \hat{\xi}_i - \hat{\lambda}_{j(i)}) = 0$$

so that $m_1 = 0$, and we can show similarly that each $m'_i = 0$. From (a), since $\Sigma_i\hat{\xi}_i = 0 = \Sigma_j\hat{\lambda}_{j(i)}$, we have

$$\hat{\mu} = \frac{1}{pqn}\Sigma_{ij\alpha}x_{ij\alpha} = \bar{x}$$

from (b)

$$\hat{\xi}_i = \frac{1}{qn}\Sigma_{j\alpha}x_{ij\alpha} - \bar{x} = \bar{x}_{i\cdot} - \bar{x}$$

and from (c)

$$\hat{\lambda}_{j(i)} = \frac{1}{n}\Sigma_\alpha x_{ij\alpha} - \bar{x}_{i\cdot} = \bar{x}_{ij} - \bar{x}_{i\cdot}$$

Thus the least squares estimates and the unbiased estimates obtained by considering the average values of the sample, class, and subclass means are identical.

By a method similar to that given in Section 7.22 we can derive the algebraic identity:

(d) $$\Sigma_{ij\alpha}(x_{ij\alpha} - \bar{x})^2 = \Sigma_{ij\alpha}(x_{ij\alpha} - \bar{x}_{ij})^2 + n\Sigma_{ij}(\bar{x}_{ij} - \bar{x}_{i\cdot})^2 + nq\Sigma_i(\bar{x}_{i\cdot} - \bar{x})^2$$

which can be written, in the notation of Section 7.22,

$$S = S_{\alpha(ij)} + S_{j(i)} + S_i$$

where $S_{\alpha(ij)}$ is the sum of squares of the deviations of the $N = npq$ observations from the pq means $\bar{x}_{ij}$ of the subclasses, and hence has $N - pq$ degrees of freedom. From it we obtain a joint estimate of σ^2 based on variation in the replicate observations within the subclasses. $S_{j(i)}$ is the sum of squares of the deviations of the pq subclass means from the means of the p classes, and hence has $pq - p = p(q - 1)$ degrees of freedom. The variance $S_{j(i)}/p(q - 1)$ is an estimate of $\sigma^2 + n\sigma_\lambda^2$. S_i as before is based on the deviations of the p class means from the overall mean, and has $p - 1$ degrees of freedom. The variance $S_i/(p - 1)$ is an estimate of $\sigma^2 + n\left(1 - \frac{q}{Q}\right)\sigma_\lambda^2 + nq\sigma_\xi^2$. These results are summarized in the accompanying analysis of variance table.

TABLE 7.14

GENERAL ANALYSIS OF VARIANCE*:

TWO NESTED CLASSIFICATIONS WITH REPLICATION

Source of Estimate	Sum of Squares	D.F.	Average Value of Estimate
Between classes	$S_i = nq\,\Sigma_i(\bar{x}_{i\cdot} - \bar{x})^2$	$p - 1$	$\sigma^2 + n\left(1 - \frac{q}{Q}\right)\sigma_\lambda^2 + nq\sigma_\xi^2$
Between subclasses (within classes)	$S_{j(i)} = q\,\Sigma_{ij}(\bar{x}_{ij} - \bar{x}_{i\cdot})^2$	$p(q - 1)$	$\sigma^2 + n\sigma_\lambda^2$
Within subclasses	$S_{\alpha(ij)} = \Sigma_{ij\alpha}(x_{ij\alpha} - \bar{x}_{ij})^2$	$N - pq$	σ^2
Total	$S = \Sigma_{ij\alpha}(x_{ij\alpha} - \bar{x})^2$	$N - 1$	

* As the estimates of the variance are the sums of squares divided by the corresponding degrees of freedom, we shall omit the column of mean squares in the general tables.

In order to estimate σ_λ^2, we subtract the within-subclasses estimate from the between-subclasses (within-classes) estimate and divide by n. We estimate σ_ξ^2 as

$$\frac{1}{nq}\left[\left(\text{mean square between classes}\right) - \left(1 - \frac{q}{Q}\right)\left(\text{mean square between subclasses}\right) - \frac{q}{Q}\left(\text{mean square within subclasses}\right)\right]$$

To test for an appreciable between-subclasses variation σ_λ^2, we should use the ratio of the between-subclasses mean square to the within-subclasses mean square, since both of these estimate σ^2 under the hypothesis $\sigma_\lambda^2 = 0$. To test for the existence of an appreciable variation σ_ξ^2 between castings from the same pour, we should use the ratio of the between-classes mean square to the combination

$$\left[\left(1 - \frac{q}{Q}\right)\left(\text{mean square between subclasses}\right) + \frac{q}{Q}\left(\text{mean square within subclasses}\right)\right]$$

If σ_ξ^2 is negligible the ratio will not have the F-distribution unless either $q = Q$ or $q/Q \to 0$. In the first case $\left(1 - \frac{q}{Q}\right) = 0$, and the denominator is simply the mean square within subclasses, which is distributed as $\frac{\sigma^2\chi^2}{N - pq}$, and in the second the denominator is the mean square between subclasses, which is distributed as $\frac{(\sigma^2 + n\sigma_\lambda^2)\chi^2}{p(q-1)}$, provided that we assume the $\varepsilon_{ij\alpha}$ and $\lambda_{j(i)}$ to be normally distributed. The above example is a good illustration of this differentiation; if the two samples from each casting represent a small portion of the total casting, then the variation between samples would make an appreciable contribution to the variation between the sample means for the different castings. On the other hand, if the samples represented all possible samples from a given casting, that is, the whole casting, then there would obviously be no variation in the mean for a particular casting due to sampling, and any variation in these means would have to be due either to errors in determination or differences between castings. For this example we suppose the ratios p/P and q/Q to be very small.

The computations leading to the analysis of variance in Table 7.16 are shown in Table 7.15. As before, if we define T_{ij} to be the total of the observations in the jth subclass of the ith class, and $T_{i\cdot}$ to be the total of all observations in the ith class, we have the algebraic identities

$$(e) \qquad S_i = nq\,\Sigma_i(\bar{x}_{i\cdot} - \bar{x})^2 = \frac{\Sigma_i T_{i\cdot}^2}{nq} - \frac{T^2}{N}$$

$$(f) \qquad S_{j(i)} = n\Sigma_{ij}(\bar{x}_{ij} - \bar{x}_{i\cdot})^2 = \frac{\Sigma_{ij} T_{ij}^2}{n} - \frac{\Sigma_i T_{i\cdot}^2}{nq}$$

$$(g) \qquad S_{\alpha(ij)} = \Sigma_{ij\alpha}(x_{ij\alpha} - \bar{x}_{ij})^2 = \Sigma_{ij\alpha} x_{ij}^2 - \frac{\Sigma_{ij} T_{ij}^2}{n}$$

$$(h) \qquad S = \Sigma_{ij\alpha}(x_{ij\alpha} - \bar{x})^2 = \Sigma_{ij\alpha} x_{ij\alpha}^2 - \frac{T^2}{N}$$

which give the simplest computational forms for the sums of squares in terms of the 4 quantities T^2/N, $\Sigma_i T^2_{i\cdot}/nq$, $\Sigma_{ij} T^2_{ij}/n$, and $\Sigma_{ij\alpha} x^2_{ij\alpha}$.

TABLE 7.15
DATA AND SUMS
(84.00 subtracted from all entries)

Casting	Sample	Determination		Total for Sample T_{ij}	Total for Casting $T_{i\cdot}$
		1	2		
1	*A*	1.54	1.56	3.10	6.15
	B	1.51	1.54	3.05	
2	*A*	1.54	1.60	3.14	5.64
	B	1.25	1.25	2.50	
3	*A*	1.72	1.77	3.49	5.38
	B	0.94	0.95	1.89	
4	*A*	1.48	1.50	2.98	4.98
	B	0.98	1.02	2.00	
5	*A*	1.54	1.57	3.11	6.79
	B	1.84	1.84	3.68	
6	*A*	1.72	1.86	3.58	7.30
	B	1.81	1.91	3.72	
7	*A*	1.72	1.76	3.48	7.13
	B	1.81	1.84	3.65	
8	*A*	2.12	2.12	4.24	8.56
	B	2.12	2.20	4.32	
9	*A*	1.47	1.49	2.96	6.48
	B	1.75	1.77	3.52	
10	*A*	0.98	1.10	2.08	5.88
	B	1.90	1.90	3.80	
11	*A*	1.12	1.17	2.29	4.71
	B	1.18	1.24	2.42	
Sum		69.00		69.00	69.00

$$T^2/N = 108.2045$$
$$\Sigma_i T_{i\cdot}^2/nq = 445.6304/4 = 111.4076$$
$$\Sigma_{ij} T^2_{ij}/n = 226.6158/2 = 113.3079$$
$$\Sigma_{ij\alpha} x^2_{ij\alpha} = 113.3430$$

TABLE 7.16

ANALYSIS OF VARIANCE

Source of Estimate	Sum of Squares	D.F.	Mean Square
Between castings	3.2031	10	0.3202
Between samples (within castings)	1.9003	11	0.1728
Within samples	0.0351	22	0.0016
Total	5.1385	43	

Hypothesis Testing. $\sigma_\lambda^2 = 0$ (no appreciable variation among the true copper content of different samples from a given casting, as compared with the variation among determinations on the same sample), $F = 0.1728/0.0016 = 108.0$, highly significant.

$\sigma_\xi^2 = 0$ (no appreciable variation among the true average copper contents of the various castings as compared with the variations among samples from the same casting), $F = 0.3202/0.1728 = 1.853$, not significant at the 5% level ($F_{10,\,11,\,0.05} = 2.86$).

Estimation.

Estimate of σ^2 (variation between determinations): 0.0016 ($\sigma = 0.04$)

Estimate of σ_λ^2 (variation between samples):

$$\frac{0.1728 - 0.0016}{2} = 0.0856 \quad (\sigma_\lambda = 0.30)$$

Estimate of σ_ξ^2 (variation between castings):

$$\frac{0.3203 - 0.1728}{4} = 0.0369 \quad (\sigma_\xi = 0.19)$$

Confidence Intervals. Since s^2 is an estimate of σ^2 distributed as $\dfrac{\sigma^2\chi^2}{N - pq}$, we have for a 90% confidence interval

$$\frac{s^2\chi^2_{0.95}}{N - pq} < \sigma^2 < \frac{s^2\chi^2_{0.05}}{N - pq}$$

or since $\chi^2_{22,\,0.95} = 12.34 \quad \chi^2_{22,\,0.05} = 33.92$

$$0.0009 < \sigma^2 < 0.0025$$

For the 90% confidence interval for σ_λ^2 we use (*e*) and (*f*) of Section 7.23 with

$$F = 108 \quad F_{11,\,22,\,0.05} = 2.25 \quad F_{11,\,\infty,\,0.05} = 1.85$$

$$F_{22,\,11,\,0.05} = 2.6 \quad F_{\infty,\,11,\,0.05} = 2.4$$

$$0.0856\left[\frac{108 - 2.25}{(1.85)(108) - 2.25}\right] < \sigma_\lambda^2 < 0.0856\left[\frac{(108)(2.6) - 1}{\frac{(108)(2.6)}{2.4} - 1}\right]$$

$$0.046 < \sigma_\lambda^2 < 0.206$$

For the 90% confidence interval for σ_ξ^2 we use (*g*) of Section 7.23, since the variance ratio $F = 1.853$ is not significant at the 5% level. The table values required in this case are $F_{11,\,10,\,0.05} = 2.95$ and $F_{\infty,\,10,\,0.05} = 2.5$, and the confidence interval is

$$0 < \sigma_\xi^2 < \left(0.3202 - \frac{0.1728)}{2.95}\right)\frac{2.5}{4}$$

$$0 < \sigma_\xi^2 < 0.163$$

It is unlikely in this example that we should wish to place confidence limits on either the $\lambda_{j(i)}$ or the ξ_i since these effects represent random sample effects. It is possible that we should wish to place confidence limits on the mean μ of the population of castings. In the general case we have

$$\text{var}\,(\bar{x}) = \text{ave}\,(\bar{x}^2) - (\text{ave}\,(\bar{x}))^2$$

and, since

$$\text{ave}\,(\bar{x}) = \mu$$

$$\text{ave}\,(\bar{x}^2) = \text{ave}\left(\mu + \frac{1}{p}\Sigma_i\xi_i + \frac{1}{pq}\Sigma_{ij}\lambda_{j(i)} + \frac{1}{pqn}\Sigma_{ij\alpha}\varepsilon_{ij\alpha}\right)^2$$

$$= \mu^2 + \text{ave}\left[\frac{1}{p^2}(\Sigma_i\xi_i)^2 + \frac{1}{p^2q^2}(\Sigma_{ij}\lambda_{j(i)})^2 + \frac{1}{p^2q^2n^2}(\Sigma_{ij\alpha}\varepsilon_{ij\alpha})^2\right]$$

$$= \mu^2 + \frac{1}{p}\left(1 - \frac{p}{P}\right)\sigma_\xi^2 + \frac{1}{pq}\left(1 - \frac{q}{Q}\right)\sigma_\lambda^2 + \frac{1}{pqn}\sigma^2$$

where the last step follows from the considerations in Section 7*A*.1 of the appendix to this chapter, we have

$$\text{var}\,(\bar{x}) = \frac{1}{p}\left(1 - \frac{p}{P}\right)\sigma_\xi^2 + \frac{1}{pq}\left(1 - \frac{q}{Q}\right)\sigma_\lambda^2 + \frac{1}{pqn}\sigma^2$$

In our example we supposed p/P and q/Q to be very small so that $\left(1 - \frac{p}{P}\right) \sim 1$ and $\left(1 - \frac{q}{Q}\right) \sim 1$. Thus

$$\text{var}\,(\bar{x}) = \frac{1}{N}\,\text{(average mean square between castings)}$$

and an estimate of this variance is given by

$$\sigma_{\bar{x}}^2 = \frac{0.3202}{44} = 0.00728 \qquad \sigma_{\bar{x}} = 0.085$$

If we suppose the ξ_i, $\lambda_{j(i)}$, and $\varepsilon_{ij\alpha}$ to be normally distributed, we have a 95% confidence interval for the mean as

$$85.568 - 0.085\, t_{10,\,0.05} < \mu < 85.568 + 0.085\, t_{10,\,0.05}$$

or, since $t_{10,\,0.05} = 2.23$,

$$85.35 < \mu < 85.76$$

Conclusions. The F-test of the ratio of the between-samples estimate to the within-samples estimate is significant, indicating that there is appreciably more variation between the duplicate samples from the 11 castings than between the duplicate analyses performed on each sample. However, the F-test for the ratio of the between-castings variance to the within-castings (between-samples) variance is not significant, indicating that in this case the apparent variation between castings could (in the sense that the probability is greater than 1 in 20) be due to the variations found from sample to sample within a given casting. In other words, there is no definite evidence that σ_ξ^2 is significantly greater than zero.

In the estimation and significance testing procedure we have assumed a model of type II, and exact significance tests are available since the denominator required for any F-test is the preceding mean square estimate. When a model of type III is employed this is not true, since the denominator must be compounded from a number of mean square estimates. An approximate procedure for this case is to carry out an F-test using the unbiased denominator obtained by compounding and an equivalent number of degrees of freedom n'. If the unbiased denominator is obtained by compounding a number of mean square estimates as

$$\text{M.S.} = a_1(\text{M.S.})_1 + a_2(\text{M.S.})_2 + \cdots$$

where $(\text{M.S.})_i$ is based on n_i degrees of freedom, then n' is given by

$$\frac{[\text{M.S.}]^2}{n'} = \frac{a_1^2(\text{M.S.})_1^2}{n_1} + \frac{a_2^2(\text{M.S.})_2^2}{n_2} + \cdots$$

or

$$n' = \frac{(\text{M.S.})^2}{a_1^2 \dfrac{(\text{M.S.})_1^2}{n_1} + \dfrac{a_2^2(\text{M.S.})_2^2}{n_2} + \cdots}$$

This procedure, due to Satterthwaite [103], and previously to Welch [125], is the equivalent of replacing the exact distribution of a composite mean square estimate by a χ^2-distribution of equal variance.

7.53. Analysis of Two Crossed Classifications

The data in Table 7.17 give the results of duplicate determinations of the total solid content of wet brewers' yeast for 10 different samples using 3 drying periods. For convenience, the constant factor 10% has been subtracted from each result. Since each sample is analyzed in duplicate, using each drying time, the classifications "drying period" and "samples" are completely crossed, while the duplicate determinations give us 2 replications for each combination of sample and drying period. The classes in each classification correspond to the columns and rows of the table, and it is customary to speak of "rows" and "columns" to mean the corresponding classes of each separate classification, and of the "class," or "cell," corresponding to a particular combination of the two classifications. Thus in the above table we have 3 columns corresponding to drying periods, 10 rows corresponding to the 10 samples used, and 30 classes or cells, each containing 2 replicate observations.

In general, we shall have observations $x_{ij\alpha}$, $i = 1, \cdots, p$ rows, $j = 1, \cdots, q$ columns, $\alpha = 1, \cdots, n$ replicates in each cell, which we shall assume to be described by the model

$$x_{ij\alpha} = \mu + \xi_i + \eta_j + \lambda_{ij} + \varepsilon_{ij\alpha}$$

Here the ξ_i represent the row effects, i.e., in the example, the variation in the true total solid content of the various samples, and the η_j represent column effects, i.e., the variations in the average analytical results due to drying time. The "interaction" effects λ_{ij} represent any variations which may be peculiar to a particular combination of sample and drying time and the effects $\varepsilon_{ij\alpha}$ are normally distributed random components with average value zero for each ij.

The 10 samples of yeast will be considered as random samples from a large population of such samples. We shall wish to extend any conclusions

TABLE 7.17

Percent Total Solids Determined by Method *V* Using Different Drying Periods

Sample	Percent Total Solids		
	3 Hours	6 Hours	9 Hours
1*a*	3.24	3.16	2.96
1*b*	3.56	3.26	3.01
2*a*	3.92	3.81	3.76
2*b*	3.86	3.80	3.75
3*a*	9.13	8.86	8.70
3*b*	9.23	8.79	8.75
4*a*	8.35	8.11	7.94
4*b*	8.29	8.24	7.99
5*a*	5.51	5.06	4.84
5*b*	5.53	5.11	4.80
6*a*	6.63	6.61	6.60
6*b*	6.65	6.57	6.55
7*a*	9.29	8.96	8.84
7*b*	9.28	9.12	9.03
8*a*	8.76	8.39	8.23
8*b*	8.72	8.43	8.27
9*a*	8.03	7.86	7.72
9*b*	8.11	7.84	7.79
10*a*	6.61	6.32	6.21
10*b*	6.77	6.23	6.13

which we may make to this population. The scale classification 3 hours', 6 hours', or 9 hours' drying time cannot be treated in this way. These levels have been chosen on the basis of convenience and cannot be regarded as a random sample from any population except one which includes only these drying periods. The statistical inference concerning the effect of drying period will therefore be applicable only to these values. If the observations vary systematically with the drying period, we may be prepared to extend our conclusions to all drying periods from 3 to 9 hours, but this extension will be based on technical experience and not upon statistical evidence.

When the summations extend over the whole population we have

$$\Sigma_i \xi_i = \Sigma_j \eta_j = \Sigma_i \lambda_{ij} = \Sigma_j \lambda_{ij} = 0$$

so that, designating a row mean by $\bar{x}_{i}.$, a column mean by $\bar{x}._j$, and a class mean by $\bar{x}_{ij}$

$$\begin{aligned}
\text{ave}\,(\bar{x}) &= \mu + \frac{1}{10}\,\text{ave}\,(\Sigma_i \xi_i) + \frac{1}{3}\,\text{ave}\,(\Sigma_j \eta_j) + \frac{1}{30}\,\text{ave}\,(\Sigma_{ij} \lambda_{ij}) \\
&\qquad + \frac{1}{60}\,\text{ave}\,(\Sigma_{ij\alpha} \varepsilon_{ij\alpha}) \\
&= \mu \\
\text{ave}\,(\bar{x}_i.) &= \mu + \xi_j + \frac{1}{3}\,\text{ave}\,(\Sigma_j \eta_j) + \frac{1}{3}\,\text{ave}\,(\Sigma_j \lambda_{ij}) + \frac{1}{6}\,\text{ave}\,(\Sigma_{j\alpha} \varepsilon_{ij\alpha}) \\
&= \mu + \xi_i \\
\text{ave}\,(\bar{x}\,._j) &= \mu + \eta_j + \frac{1}{10}\,\text{ave}\,(\Sigma_i \xi_i) + \frac{1}{10}\,\text{ave}\,(\Sigma_i \lambda_{ij}) + \frac{1}{20}\,\text{ave}\,(\Sigma_{j\alpha} \varepsilon_{ij\alpha}) \\
&= \mu + \eta_j \\
\text{ave}\,(\bar{x}_{ij}) &= \mu + \xi_i + \eta_j + \lambda_{ij} + \frac{1}{2}\,\text{ave}\,(\Sigma_\alpha \varepsilon_{ij\alpha}) \\
&= \mu + \xi_j + \eta_i + \lambda_{ij}
\end{aligned}$$

Thus, in the example, the grand mean, row means, column means, and cell means lead to unbiased estimates of the model parameters, since we have

$$\begin{aligned}
\text{ave}\,(\bar{x}_{ij} - \bar{x}_i. - \bar{x}\,._j + \bar{x}) &= \text{ave}\,(\bar{x}_{ij}) - \text{ave}\,(\bar{x}_i.) - \text{ave}\,(\bar{x}\,._j) + \text{ave}\,(\bar{x}) \\
&= \lambda_{ij} \\
\text{ave}\,(\bar{x}_i. - \bar{x}) &= \text{ave}\,(\bar{x}_i.) - \text{ave}\,(\bar{x}) \\
&= \xi_i \\
\text{ave}\,(\bar{x}\,._j - \bar{x}) &= \text{ave}\,(\bar{x}\,._j) - \text{ave}\,(\bar{x}) \\
&= \eta_j \\
\text{ave}\,(\bar{x}) &= \mu
\end{aligned}$$

In using this method of estimation we are in fact imposing the condition that the sum of our estimates of row effects and column effects shall each be zero. Also the sum of the interaction effect estimates in any row or column and the sum of the residual estimates in any cell will be zero. This is equivalent to the least squares solution with

$$\Sigma_i\hat{\xi}_i = \Sigma_j\hat{\eta}_j = \Sigma_i\hat{\lambda}_{ij} = \Sigma_j\hat{\lambda}_{ij} = 0$$

The algebraic breakdown of the overall sum of squares in this instance can be written as

$$S = S_{\alpha(ij)} + S_{ij} + S_i + S_j$$

where S, $S_{\alpha(ij)}$, and S_i are identical with the corresponding definitions of Section 7.52, and

$$S_j = np\Sigma_j(\bar{x}_{.j} - \bar{x})^2$$
$$S_{ij} = n\Sigma_{ij}(\bar{x}_{ij} - \bar{x}_{i.} - \bar{x}_{.j} + \bar{x})^2$$

This corresponds to the further breakdown of the within-classes (between-subclasses) variation $S_{j(i)}$ of Section 7.52 into 2 parts with $q - 1$ and $(p - 1)(q - 1)$ degrees of freedom, respectively, and to the division of the subclass effects $\lambda_{j(i)}$ into the effects η_i common to all columns, and the effects λ_{ij} peculiar to a particular combination of row and column. The analysis of variance corresponding to this breakdown and the average values of the variance estimates are given in Table 7.18.

In place of the computational form (f) for $S_{j(i)}$ given in section 7.52, we have

(*a*) $$S_j = np\,\Sigma_j(\bar{x}_{.j} - \bar{x})^2 = \frac{\Sigma_j T_{.j}^2}{np} - \frac{T^2}{N}$$

and

(*b*) $$S_{ij} = n\,\Sigma_{ij}(\bar{x}_{ij} - \bar{x}_{i.} - \bar{x}_{.j} + \bar{x})^2$$
$$= \frac{\Sigma_{ij}T^2_{ij}}{n} - \frac{\Sigma_i T_{i.}^2}{nq} - \frac{\Sigma_j T_{.j}^2}{np} + \frac{T^2}{N}$$

where $T_{.j}$ is the sum of the np observations in the jth column.

The ratio of the estimate based on the interactions to the estimate based on variation between replicates is used to test the hypothesis $\sigma_\lambda^2 = 0$, and the difference between these estimates divided by n gives an estimate of σ_λ^2. To determine the proper test of the hypothesis $\sigma_\xi^2 = 0$ we must consider the average value of the between-rows estimate given in Table 7.18. In this case we have

ave [M.S. for rows]

$$= \text{ave}\left[\frac{nq}{p-1}\,\Sigma_i(\bar{x}_{i.} - \bar{x})^2\right]$$

$$= \frac{nq}{p-1}\,\text{ave}\left[\Sigma_i\left(\mu + \xi_i + \frac{1}{q}\,\Sigma_j\eta_j + \frac{1}{q}\Sigma_j\lambda_{ij} + \frac{1}{qn}\,\Sigma_{j\alpha}\varepsilon_{ij\alpha} - \mu\right.\right.$$
$$\left.\left. - \frac{1}{p}\,\Sigma_i\xi_i - \frac{1}{q}\,\Sigma_j\eta_j - \frac{1}{pq}\,\Sigma_{ij}\lambda_{ij} - \frac{1}{pqn}\,\Sigma_{ij\alpha}\varepsilon_{ij\alpha}\right)^2\right]$$

TABLE 7.18
GENERAL ANALYSIS OF VARIANCE:
TWO CROSSED CLASSIFICATIONS WITH REPLICATIONS

Source of Estimate	Sum of Squares	D.F.	Average Value of Estimate
Between rows	$S_i = nq\, \Sigma_i(\bar{x}_{i\cdot} - \bar{x})^2$	$p - 1$	$\sigma^2 + n\left(1 - \frac{q}{Q}\right)\sigma_\lambda^2 + nq\,\sigma_\xi^2$
Between columns	$S_j = np\, \Sigma_j(\bar{x}_{\cdot j} - \bar{x})^2$	$q - 1$	$\sigma^2 + n\left(1 - \frac{p}{P}\right)\sigma_\lambda^2 + np\,\sigma_\eta^2$
Interaction	$S_{ij} = n\, \Sigma_{ij}(\bar{x}_{ij} - \bar{x}_{i\cdot} - \bar{x}_{\cdot j} + \bar{x})^2$	$(p - 1)(q - 1)$	$\sigma^2 + n\sigma_\lambda^2$
Within cells (replicates)	$S_{\alpha(ij)} = \Sigma_{ij\alpha}(x_{ij\alpha} - \bar{x}_{ij})^2$	$N - pq$	σ^2
Total	$S = \Sigma_{ij\alpha}(x_{ij\alpha} - \bar{x})^2$	$N - 1$	

$$= \frac{nq}{p-1} \text{ ave} \left[\Sigma_i \left(\xi_i - \frac{1}{p} \Sigma_i \xi_i \right)^2 \right] + \frac{nq}{p-1} \text{ ave} \left[\Sigma_i \left(\frac{1}{q} \Sigma_j \lambda_{ij} - \frac{1}{pq} \Sigma_{ij} \lambda_{ij} \right)^2 \right]$$

$$+ \frac{nq}{p-1} \text{ ave} \left[\Sigma_i \left(\frac{1}{qn} \Sigma_{j\alpha} \varepsilon_{ij\alpha} - \frac{1}{pqn} \Sigma_{ij\alpha} \varepsilon_{ij\alpha} \right)^2 \right]$$

the last step following from the assumed independence of the effects, so that the average value of the cross-product terms is zero in each case. Again referring to the appendix to this chapter where necessary, we note that

$$\text{var}(\xi_i) = \sigma_\xi^2 \qquad \text{var}(\Sigma_j \lambda_{ij}) = q \left(1 - \frac{q}{Q} \right) \sigma_\lambda^2$$

$$\text{var}(\Sigma_{j\alpha} \varepsilon_{ij\alpha}) = qn\sigma^2$$

and hence

ave [M.S. for rows]

$$= \frac{nq}{p-1} \left[(p-1)\sigma_\xi^2 + \frac{(p-1)q}{q^2} \left(1 - \frac{q}{Q} \right) \sigma_\lambda^2 + \frac{(p-1)qn}{q^2 n^2} \sigma^2 \right]$$

$$= nq\sigma_\xi^2 + n \left(1 - \frac{q}{Q} \right) \sigma_\lambda^2 + \sigma^2$$

Thus, if $\frac{q}{Q} \sim 0$, this mean square will be tested against the interaction mean square, if $\frac{q}{Q} = 1$ against the residual mean square, and, if $0 < \frac{q}{Q} < 1$, against a term of average value $\left(1 - \frac{q}{Q} \right) \sigma_\lambda^2 + \sigma^2$ compounded from the interaction and residual mean squares. In the latter case the approximate method discussed in Section 7.32 should be employed to determine the equivalent number of degrees of freedom for the denominator.

It should be remarked that the term $N(\bar{x} - \mu_0)^2$ with 1 degree of freedom can be added to the general analysis of either this section or of Section 7.52 exactly as it was in the case of the one-way classification in Table 7.12. The average value of this estimate for the crossed classification will be

$$\text{ave}\,[npq\,(\bar{x} - \mu_0)^2] = npq \text{ ave}\,[(\mu - \mu_0)^2] + \frac{nq}{p} \text{ ave}\,[(\Sigma_i \xi_i)^2]$$

$$+ \frac{np}{q} \text{ ave}\,[(\Sigma_j \eta_j)^2] + \frac{n}{pq} \text{ ave}\,[(\Sigma_{ij} \lambda_{ij})^2]$$

$$+ \frac{1}{npq} \text{ ave}\,[(\Sigma_{ij\alpha} \varepsilon_{ij\alpha})^2]$$

$$= npq(\mu - \mu_0)^2 + nq \left(1 - \frac{p}{P} \right) \sigma_\xi^2 + np \left(1 - \frac{q}{Q} \right) \sigma_\eta^2$$

$$+ n \left(1 - \frac{p}{P} \right) \left(1 - \frac{q}{Q} \right) \sigma_\lambda^2 + \sigma^2$$

TABLE 7.19

Sample	Drying Time			Total
	3 Hours	6 Hours	9 Hours	
1	6.80	6.42	5.97	19.19
2	7.78	7.61	7.51	22.90
3	18.41	17.65	17.45	53.51
4	16.64	16.35	15.93	48.92
5	11.04	10.17	9.64	30.85
6	13.28	13.18	13.15	39.61
7	18.57	18.08	17.87	54.52
8	17.48	16.82	16.50	50.80
9	16.14	15.70	15.51	47.35
10	13.38	12.55	12.34	38.27
Total	139.52	134.53	131.87	405.92

The computations and the analysis of variance table for the above example have been made. Table 7.19 has been prepared by adding the duplicate results from Table 7.17, and hence the entries represent the sums T_{ij}. The row totals and column totals are the sums $T_{i\cdot}$ and $T_{\cdot j}$, respectively. From the original table we obtain by squaring each item (repeating the operation, or using the simultaneously obtained sum as a check) $\Sigma_{ij\alpha} x^2_{ij\alpha} = 2990.2668$. From the above table we obtain

$$\Sigma_{ij} T^2_{ij}/n = 5980.2346/2 = 2990.1173$$

$$\Sigma_i T^2_{i\cdot}/nq = 17{,}929.5130/6 = 2988.2522$$

$$\Sigma_j T^2_{\cdot j}/np = 54{,}953.8482/20 = 2747.6924$$

and

$$T^2/N = (405.92)^2/60 = 2746.1841$$

again using the sum as a check or repeating the operation in obtaining each sum of squares. Using the algebraic identities (*e*), (*g*) and (*h*) of Section 7.52 and (*a*) and (*b*) of the present section, we obtain the accompanying analysis of variance table.

Since the inferences concerning the effect of drying time are to be restricted to the 3 levels examined, $p = P$ and the average value of the mean square between samples has no component from the interaction

TABLE 7.20

ANALYSIS OF VARIANCE FOR DATA OF TABLE 7.17

Source of Estimate	Sum of Squares	D.F.	Mean Square	Average Value of Mean Square
Between samples	242.0681	9	26.8965	$\sigma^2 + 6\sigma_\xi^2$
Between drying times	1.5083	2	0.7542	$\sigma^2 + 2\sigma_\lambda^2 + 20\sigma_\eta^2$
Interaction	0.3568	18	0.0198	$\sigma^2 + 2\sigma_\lambda^2$
Between determinations within classes (error)	0.1495	30	0.0050	σ^2
Total	244.0827	59		

variance. In simple cases this final column of the analysis of variance table is frequently omitted, but this is not recommended since it is useful as an indication of the proper significance tests.

Significance Testing.

(1) $\sigma_\lambda^2 = 0$. We test this hypothesis by forming the variance ratio

$$F = \frac{0.0198}{0.0050} = 3.96^{**} \qquad F_{18,\,30,\,0.01} = 2.9$$

(2) $\sigma_\eta^2 = 0$. Hypothesis (1) is clearly rejected so that the proper test is given by the ratio

$$F = \frac{0.7542}{0.0198} = 38.1^{***} \qquad F_{2,\,18,\,0.001} = 10.4$$

(3) $\sigma_\xi^2 = 0$. In this case we use the residual mean square as a variance estimate in the ratio $\dfrac{26.8965}{0.0050} = 5379^{***}$.

Estimation. The unusually large variation between samples is of little concern in this experiment, as no attempt was made to control the level of the samples; indeed, the use of different samples is an integral part of the experimental design since it enables us to make comparisons between drying rates on these samples. The two important facts in the above analysis concerning sample variations are: (1) the systematic variations between samples can be eliminated from the study of other effects;

(2) the variation between samples can, if desired, be estimated. The estimate obtained from the present analysis would be

$$\sigma_\xi^2 = \frac{26.8965 - 0.0050}{6} = 4.48192, \qquad \sigma_\xi = 2.12$$

The experiment was primarily designed to detect differences due to drying time. The variation σ_η^2 between the drying times is of little concern, since the 3-hour, 6-hour, and 9-hour drying times chosen represent 3 possible choices on a continuous scale of drying times which might have been chosen, and σ_η^2 is certainly directly dependent on this choice.

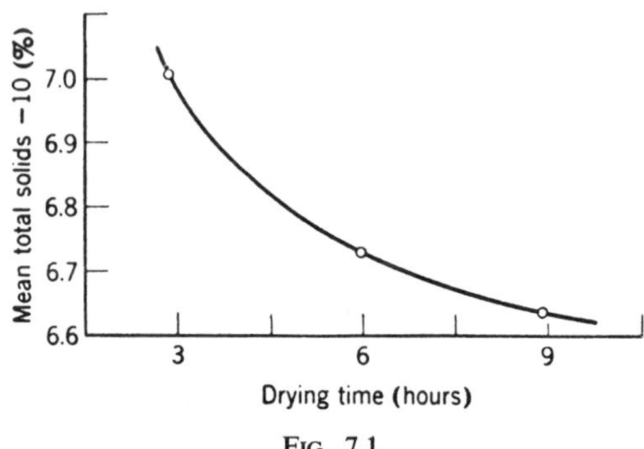

FIG. 7.1.

What we are really interested in here is the *dependence* of the analytical result on drying time. The significance of the variation between drying times indicates that the average analytical results obtained on all the types of yeast from which the 10 examined were a random sample will differ according to whether 3, 6, or 9 hours of drying time are employed. The presence of a significant but relatively small interaction term indicates that the magnitude of this difference changes slightly from sample to sample. The sample means for 3-hour, 6-hour, and 9-hour drying times are (1) 6.98, (1) 6.73, and (1) 6.59, respectively. These, as might have been expected, show a systematic decrease in the percentage of total solids found as the drying time is increased, indicating the approach to constant weight. This decrease is plotted against drying time in Figure 7.1, where the points have been joined by a smooth curve. The insertion of the curve represents a technical inference and not a statistical one. In Section 7.74 we shall consider a further breakdown of these data which sheds more light on the nature of this dependence.

A special case of 2 crossed classifications arises when only 1 observation is available in each cell. In this case no estimate of σ^2 is available from the variation among replicate observations within a given cell, and the analysis of variance reduces to that in Table 7.21. The entire formulation for the general case is applicable if we remember that for $n = 1$ we can omit the subscript α, so that $\bar{x}_{ij} = T_{ij} = x_{ij}$, the single observation in the ijth cell. The variance ratio used in the test for row and column effects is distributed as F only if we assume that these effects were drawn from populations of size P and Q, large with respect to p and q. In the case where p/P or $q/Q \neq 0$ no exact tests are available. This model is frequently used with the a priori assumption that the λ_{ij} are zero; i.e., that $x_{ij} = \mu + \xi_i + \eta_j + \varepsilon_{ij}$ in which case the average values of the mean squares are $\sigma^2 + q\sigma_\xi^2$, $\sigma^2 + p\sigma_\eta^2$, and σ^2, respectively, and no distinction between the two cases is necessary.

TABLE 7.21

GENERAL ANALYSIS OF VARIANCE:

TWO CROSSED CLASSIFICATIONS—NO REPLICATION

Source of estimate	Sum of Squares	D.F.	Expected Value of Estimate
Between rows	$S_i = q\,\Sigma_i(\bar{x}_{i\cdot} - \bar{x})^2$	$p - 1$	$\sigma^2 + \left(1 - \dfrac{q}{Q}\right)\sigma_\lambda^2 + q\sigma_\xi^2$
Between columns	$S_j = p\,\Sigma_j(\bar{x}_{\cdot j} - \bar{x})^2$	$q - 1$	$\sigma^2 + \left(1 - \dfrac{p}{P}\right)\sigma_\lambda^2 + p\sigma_\eta^2$
Interaction	$S_{ij} = \Sigma_{ij}(x_{ij} - \bar{x}_{i\cdot} - \bar{x}_{\cdot j} + \bar{x})^2$	$(p - 1)(q - 1)$	$\sigma^2 + \sigma_\lambda^2$
Total	$S = \Sigma_{ij}(x_{ij} - \bar{x})^2$	$N - 1$	

As an example of this type of analysis Table 7.22 gives the results obtained in single determinations of total solids by each of 3 analysts on 10 different samples of wet brewers' yeast. The tests were performed by a method similar to that employed in the previous example. The experiment was designed to assess the reproducibility of the method when carried out by different analysts.

The ratio $F = 0.012/0.0076 = 1.58$ of the "between-analysts" to the "interaction" estimates is not significant at the 5% level ($F_{2,\,18,\,0.05} = 3.55$), indicating that no definitely detectable variation over and above that expected on the basis of the interactions has been added due to the use of different analysts.

TABLE 7.22

TOTAL SOLIDS IN WET BREWERS' YEAST
[Coding: Percentage of total solids −10]

Sample	1	2	3	4	5	6	7	8	9	10	Total
Analyst A	10.1	4.7	3.1	3.0	7.8	8.2	7.8	6.0	4.9	3.4	59.0
Analyst B	10.0	4.9	3.1	3.2	7.8	8.2	7.7	6.2	5.1	3.4	59.6
Analyst C	10.2	4.8	3.0	3.1	7.8	8.4	7.9	6.1	5.0	3.3	59.6
Total	30.3	14.4	9.2	9.3	23.4	24.8	23.4	18.3	15.0	10.1	178.2

$$\Sigma_{ij}x^2_{ij} = 1223.04; \quad \frac{\Sigma_i T_{i.}^2}{10} = 1058.523; \quad \frac{\Sigma_j T_{.j}^2}{3} = 1222.88; \quad \frac{T^2}{N} = 1058.508$$

The variation between samples is, of course, highly significant, since no effort was made to choose identical samples. In fact, the removal of this large variation between samples before making a test of the differences between chemists is one of the advantages of the above analysis. Note that this test for only 2 chemists would be identical with the test using

TABLE 7.23

ANALYSIS OF VARIANCE TABLE

Source of Estimate	Sum of Squares	D.F.	Estimate of Variance
Between analysts	0.024	2	0.012
Between samples	164.372	9	18.264
Interaction (error)	0.136	18	0.0076
Total	164.532	29	

paired differences given in Section 5.52, since in that case we should have 1 degree of freedom between analysts and 9 degrees of freedom for interaction, and our F-test would be equivalent to a t-test based on 9 degrees of freedom. The above analysis is the natural extension of paired differences to more than 2 means, just as the one-way classification with more than 2 means is the natural extension of the unpaired t-test between 2 means. In this example the between-samples variation represents the additional variation that would have been included in the error term if it had not been possible to pair or cross-classify.

The variance components in this case are:

Estimate of σ^2: 0.0076 $\quad (\sigma = 0.087)$

Estimate of σ_ξ^2 (variation due to analysts):

$$\frac{0.0120 - 0.0076}{10} = 0.00044 \qquad (\sigma_\xi = 0.021)$$

Estimate of σ_η^2 (variation between true values of samples):

$$\frac{18.264 - 0.0076}{3} = 6.09 \qquad (\sigma_\eta = 2.47)$$

In this case there is little question of the applicability of the interaction as the proper error estimate, since the 3 analysts are being considered as representative of a large number of analysts who might use the method, and the samples are obviously from a large population of possible samples. Also in this case the error estimate is comparable in size with the error estimate obtained on the previous example, even though these data were obtained by a slightly different and supposedly less precise method. This suggests that there is little, if any, contribution from the interactions in the residual mean square.

7.54. Missing Data

In considering the analysis of two-way classifications we have supposed the number of replicate determinations in each cell to be equal. If this is not true, it is usually impossible to separate the overall sum of squares into independent components due to separate treatment effects and interactions, and the data are said to be non-orthogonal. We consider the case of a $p \times q$ classification of a number of observations N such that n_{ij} observations occur in the ijth cell. If we regard each of the pq cells as separate classes, the data may be analyzed as a one-way classification, the sum of squares being subdivided as

$$\Sigma_{ij\alpha}(x_{ij\alpha} - \bar{x})^2 = \Sigma_{ij\alpha}(x_{ij\alpha} - \bar{x}_{ij})^2 + \Sigma_{ij}n_{ij}(\bar{x}_{ij} - \bar{x})^2$$

The separated components are independent, and, on the hypothesis that all class effects are zero, are distributed as $\dfrac{\sigma^2\chi^2}{\nu}$, with $\nu = N - pq$ and $pq - 1$ degrees of freedom. This hypothesis may be tested in the usual way by forming a variance ratio. If the class effects are real, a method of subdivision suitable to the non-orthogonal case must be employed.

We assume for this analysis that the observations are randomly and normally distributed with variance σ^2 and average values

$$\text{ave}\,(x_{ij\alpha}) = \mu + \xi_i + \eta_j$$

i.e., that in the $p \times q$ classification there are main effects due to classes, but no interaction terms. The sum of squares $\Sigma_{ij\alpha}(x_{ij\alpha} - \mu - \xi_i - \eta_j)^2$ will then be distributed as $\sigma^2\chi^2$ with N degrees of freedom. We wish to estimate the effects μ, ξ_i, and η_j, and as previously we take $\Sigma_i\xi_i = \Sigma_j\eta_j = 0$. To compute these estimates we may use the method of maximum likelihood which involves minimizing the sum of squares

$$Q = \Sigma_{ij\alpha}(x_{ij\alpha} - \mu - \xi_i - \eta_j)^2$$

subject to the condition $\Sigma_i\xi_i = \Sigma_j\eta_j = 0$, or, alternatively, minimizing the unrestricted sum

$$Q' = \Sigma_{ij\alpha}(x_{ij\alpha} - \mu - \xi_i - \eta_j)^2 + \lambda\Sigma_i\xi_i + \lambda'\Sigma_j\eta_j = 0$$

Differentiating with respect to μ, the ξ_i, and the η_j, we obtain

$$\frac{\partial Q'}{\partial \mu} = 0 = \Sigma_{ij\alpha}x_{ij\alpha} - N\hat{\mu} - \Sigma_{ij\alpha}\hat{\xi}_i - \Sigma_{ij\alpha}\hat{\eta}_j$$

$$\frac{\partial Q'}{\partial \xi_i} = 0 = \Sigma_{j\alpha}(x_{ij\alpha} - \hat{\mu} - \hat{\xi}_i - \hat{\eta}_j) - \lambda/2 \qquad (i = 1, 2, \cdots, p)$$

$$\frac{\partial Q'}{\partial \eta_j} = 0 = \Sigma_{i\alpha}(x_{ij\alpha} - \hat{\mu} - \hat{\xi}_i - \hat{\eta}_j) - \lambda'/2 \qquad (j = 1, 2, \cdots, q)$$

Summing the first set of equations over i and the second over j, we find that $\lambda = \lambda' = 0$, since $\Sigma_{ij\alpha}x_{ij\alpha} - N\hat{\mu} = 0$, $\Sigma_i\hat{\xi}_i = \Sigma_j\hat{\eta}_j = 0$. The equations may then be written

$$\begin{array}{l}
\hat{\xi}_1\Sigma_j n_{1j} + n_{11}\hat{\eta}_1 + n_{12}\hat{\eta}_2 + \cdots + n_{1j}\hat{\eta}_j + \cdots + n_{1q}\hat{\eta}_q = \Sigma_{j\alpha}(x_{1j\alpha} - \bar{x}) \\
\vdots \\
\hat{\xi}_i\Sigma_j n_{ij} + n_{i1}\hat{\eta}_1 + n_{i2}\hat{\eta}_2 + \cdots + n_{ij}\hat{\eta}_j + \cdots + n_{iq}\hat{\eta}_q = \Sigma_{j\alpha}(x_{ij\alpha} - \bar{x}) \\
\vdots \\
\hat{\xi}_p\Sigma_j n_{pj} + n_{p1}\hat{\eta}_1 + n_{p2}\hat{\eta}_2 + \cdots + n_{pj}\hat{\eta}_j + \cdots + n_{pq}\hat{\eta}_q = \Sigma_{j\alpha}(x_{pj\alpha} - \bar{x})
\end{array}$$

together with a similar set of q equations obtained from the partial derivatives $\partial Q'/\partial\eta_j$. In order to evaluate the $p + q$ values $\hat{\xi}_i$ and $\hat{\eta}_j$, these equations must be solved simultaneously. The procedure may be viewed as an application of the method of regression analysis in which each of the independent variables can take only the values 1 or 0, and in which a total of $p + q - 1$ independent variables is concerned. After fitting the coefficients, the residual sum of squares may be computed from the expression

$$\Sigma_{ij\alpha}(x_{ij\alpha} - \hat{\xi}_i - \hat{\eta}_j - \hat{\mu})^2 = \Sigma_{ij\alpha}x^2_{ij\alpha} - \Sigma_i\hat{\xi}_i\Sigma_{j\alpha}x_{ij\alpha} - \Sigma_j\hat{\eta}_j\Sigma_{i\alpha}x_{ij\alpha} - \hat{\mu}\Sigma_{ij\alpha}x_{ij\alpha}$$

The residual variance can also be estimated from the within-cells sum of squares $\Sigma_{ij\alpha}(x_{ij\alpha} - \bar{x}_{ij})^2$, so that, on the hypothesis that no real interaction effects exist,

$$\Sigma_{ij\alpha}(x_{ij\alpha} - \hat{\xi}_i - \hat{\eta}_j - \hat{\mu})^2 - \Sigma_{ij\alpha}(x_{ij\alpha} - \bar{x}_{ij})^2$$

may be shown to be distributed as $\sigma^2\chi^2$, with $pq - p - q + 1$ degrees of freedom. The hypothesis may be tested by forming the variance ratio

$$\frac{[\Sigma_{ij\alpha}(x_{ij\alpha} - \hat{\xi}_i - \hat{\eta}_j - \hat{\mu})^2 - \Sigma_{ij\alpha}(x_{ij\alpha} - \bar{x}_{ij})^2]\,[N - pq]}{\Sigma_{ij\alpha}(x_{ij\alpha} - \bar{x}_{ij})^2\,[pq - p - q + 1]}$$

If there is no evidence of significant interaction effects, the model selected is adequate to describe the data.

If the main effects ξ_i and η_j are to be tested separately, it must be remembered that the estimates $\hat{\xi}_i$ and $\hat{\eta}_j$ are not independent, so that the methods of the previous section are not available. We can, however, analyze the cell means $\bar{x}_{ij}$ to provide an unbiased estimate of σ^2. The variance of a given cell mean $\bar{x}_{ij}$ is σ^2/n_{ij}, so that, on the hypothesis that there are no real effects ξ_i, the unweighted marginal means defined as $\bar{x}_{i\cdot} = \dfrac{1}{q}\,\Sigma_j\bar{x}_{ij}$ will have variance $\dfrac{\sigma^2}{q^2}\left(\dfrac{1}{n_{i1}} + \dfrac{1}{n_{i2}} + \cdots + \dfrac{1}{n_{iq}}\right)$. We write $\dfrac{1}{q^2}\,\Sigma_j\,\dfrac{1}{n_{ij}} = \dfrac{1}{N_i}$ and may regard the marginal means as if they were obtained from N_i observations normally distributed with variance σ^2. The variance estimate "between marginal means" is then given by

$$s^2 = \frac{\Sigma_i N_i(\bar{x}_{i\cdot} - \Sigma_i N_i\bar{x}_{i\cdot}/\Sigma_i N_i)^2}{p - 1}$$

and this may be tested against the within-cells estimate of the residual variance $\Sigma_{ij\alpha}(x_{ij\alpha} - \bar{x}_{ij})^2/N - pq$ to determine the significance of the effects ξ_i. A similar analysis on the unweighted marginal means $\bar{x}_{\cdot j}$ would be employed to test the significance of the effects η_j.

As the power of the analysis of variance as a statistical method is considerably reduced when non-orthogonal data are to be examined, most experiments are designed to avoid these circumstances. Even with careful design, it often happens that a few observations are lost when the experiment is carried out, and, in order to avoid the non-orthogonal analysis, approximate methods have been developed. These lead to identical estimates of the treatment effects and the residual variance, and they provide significance tests which, although approximate, are unlikely to lead to misinterpretation of the results unless a substantial proportion (e.g., 20%) of the data is missing.

The method consists of inserting estimates of the missing observations, so chosen as to minimize the residual variance, and completing the analysis as though the original data were orthogonal. When a number of observations are available in each cell of the classification, this is clearly the equivalent of replacing the missing observation by the mean of the other observations of the cell. The inserted value makes no contribution to the residual sum of squares, and, provided that we are careful to include only true observations in computing the degrees of freedom available for the residual variance estimate, the latter is identical with that obtained by the non-orthogonal method. If the design requires a single observation in each cell, the estimation of the missing values is carried out by an extension of the method employed in the non-orthogonal case.

Suppose that in a $p \times q$ classification only $pq - r$ of the pq observations are available. Then, if the data are assumed to be randomly and normally distributed with variance σ^2 and average values ave $(x_{ij}) = \mu + \xi_i + \eta_j$, the maximum likelihood estimates $\hat{\mu}$, $\hat{\xi}_i$, and $\hat{\eta}_j$ are obtained by minimizing

$$Q' = \Sigma_{ij}(x_{ij} - \xi_i - \eta_j - \mu)^2 + \lambda\Sigma_i\xi_i + \lambda'\Sigma_j\eta_j$$

If for the r missing observations we insert X_{ij} this sum of squares may be written

$$Q' = \Sigma_{ij}(x_{ij} - \xi_i - \eta_j - \mu)^2 + \Sigma_{ij}(X_{ij} - \xi_i - \eta_j - \mu)^2 + \lambda\Sigma_i\xi_i + \lambda'\Sigma_j\eta_j$$

the first summation taking place over the $pq - r$ observations and the second over the r unknowns X_{ij}. Differentiating with respect to μ and the unknowns X_{ij}, we obtain

$$(a) \qquad (X_{ij} - \hat{\xi}_i - \hat{\eta}_j - \hat{\mu}) = 0$$

for the r missing values and

$$\Sigma_{ij}(x_{ij} - \hat{\xi}_i - \hat{\eta}_j - \hat{\mu}) + \Sigma_{ij}(X_{ij} - \hat{\xi}_i - \hat{\eta}_j - \hat{\mu}) = 0$$

In view of (*a*) this reduces to

$$(b) \qquad \Sigma_{ij}(x_{ij} - \hat{\xi}_i - \hat{\eta}_j - \hat{\mu}) = 0$$

Also from the partial derivatives with respect to ξ_i and η_j

$$\Sigma_j(x_{ij} - \hat{\xi}_i - \hat{\eta}_j - \hat{\mu}) + \Sigma_j(X_{ij} - \hat{\xi}_i - \hat{\eta}_j - \hat{\mu}) - \frac{\lambda}{2} = 0 \quad i = 1, \cdots, p$$

$$\Sigma_i(x_{ij} - \hat{\xi}_i - \hat{\eta}_j - \hat{\mu}) + \Sigma_i(X_{ij} - \hat{\xi}_i - \hat{\eta}_j - \hat{\mu}) - \frac{\lambda'}{2} = 0 \quad j = 1, \cdots, q$$

Since, from (*a*) the second summation vanishes in each of these $p + q$ equations, and since by summation over i and j we can show immediately using (*b*) that $\lambda = \lambda' = 0$, these equations reduce to

(*c*) $$\Sigma_j(x_{ij} - \hat{\xi}_i - \hat{\eta}_j - \hat{\mu}) = 0 \qquad i = 1, \cdots, p$$

(*d*) $$\Sigma_i(x_{ij} - \hat{\xi}_i - \hat{\eta}_j - \hat{\mu}) = 0 \qquad j = 1, \cdots, q$$

the summation taking place over the observed values only. Equations (*c*) and (*d*) are identical with those employed previously to determine $\hat{\xi}_i$, $\hat{\eta}_j$, and $\hat{\mu}$ in the non-orthogonal case, so that the estimates obtained by the two methods are identical. The estimates $\hat{\xi}_i$, $\hat{\eta}_j$, and $\hat{\mu}$ so obtained can be inserted in equations (*a*) to obtain the missing values X_{ij}. Furthermore, since the inserted values make no contribution to the residual sum of squares, and the degrees of freedom $(pq - p - q - r + 1)$ of this component are unchanged, we obtain the same residual variance estimate by the two methods.

TABLE 7.24

DETERMINATION OF CHLORINE IN COAL*

[Coding: (%Cl – 0.28) × 10^3]

		Muffle Temperature				Total
		475	500	550	600	
	2	28	20	11	10	69
	3	*a*	16	15	8	39 + *a*
Incineration time, hours	4	29	13	16	*b*	58 + *b*
	6	27	10	18	11	66
	8	28	11	15	10	64
Total		112 + *a*	70	75	39 + *b*	296 + *a* + *b*

* Data by courtesy of the Director, British Coke Research Association.

EXAMPLE. In order to illustrate the method we consider the following examplc of data of Table 7.24 obtained during an investigation of the effects of time of incineration and muffle temperature on the determination of chlorine in coal. The original data were complete, and 2 randomly selected observations have been omitted. We compute the components of the sum of squares in the usual way, obtaining

(*e*) Correction term: $1/20\,(296 + a + b)^2$

(*f*) Between temperatures: $1/5\,[(112 + a)^2 + 70^2 + 75^2 + (39 + b)^2] - 1/20\,[296 + a + b]^2$

(*g*) Between times: $1/4\,[69^2 + (39 + a)^2 + (58 + b)^2 + 66^2 + 64^2] - 1/20\,[296 + a + b]^2$

(*h*) Between observations: $[28^2 + a^2 + 29^2 + \cdots + b^2 + 11^2 + 10^2] - 1/20\,[296 + a + b]^2$

The residual component is computed as (*h*) — (*g*) — (*f*), and, neglecting purely numerical terms which will disappear on differentiation, we obtain

$$Q = a^2 + b^2 + \cdots + 1/20\,[592a + 592b + 2ab + a^2 + b^2 + \cdots]$$
$$- 1/4\,[78a + a^2 + 116b + b^2 + \cdots]$$
$$- 1/5\,[224a + a^2 + 78b + b^2 + \cdots]$$

Equating to zero the partial derivatives of Q with respect to a and b we obtain

$$\frac{\partial Q}{\partial a} = 0 = 2a + 29.6 + 0.1a + 0.1b - 19.5 - 0.5a - 44.8 - 0.4a$$

$$\frac{\partial Q}{\partial b} = 0 = 2b + 29.6 + 0.1a + 0.1b - 29 - 0.5b - 15.6 - 0.4b$$

or

$$12a + b = 347 \qquad a + 12b = 150$$

so that

$$a = 28.6 \qquad b = 10.2$$

These values may then be inserted in the table in place of a and b, and the analysis may be carried out as for the orthogonal case. This analysis of variance is given in Table 7.25.

The residual estimate is unbiased, but in deriving the estimates of the effects we have assumed that the row and column means are distributed with variance $\sigma^2/4$ and $\sigma^2/5$, respectively, whereas in those rows and columns for which estimates are missing this is not true. For this reason the average value of the row and column estimates is slightly greater than σ^2, if the null hypothesis is correct. This is unlikely to have any serious effect upon the conclusions in an example in which only 10% of the observations are missing, and we test these effects in the usual way, finding the temperature effect significant at the 0.001 level, and the between-times estimate significantly low at the 0.05 level.

It must be emphasized that the inserted values computed by this method do not represent observations, and the method gives results which approximate those obtained by a non-orthogonal analysis of the incomplete data. It is not possible from incomplete data to make any statements concerning the conclusions which might have been drawn if the data had been complete. In the example chosen, since the data were in fact complete, it is possible to state that the observed values of a and b were 23 and 24, and the reader may show that the analysis of the complete data results in a considerable change in the between-times estimate of the variance.

TABLE 7.25

Source of Estimate	Sum of Squares	D.F.	Estimate
Between temperatures	927	3	309
Between times	4	4	1
Residual	95	10	9.5
Total	1026	17	

7.6. Greater Numbers of Classifications

7.61. General Consideration of Three Crossed Classifications

The methods of the preceding sections are readily extended to any number of classifications, nested or crossed. In this section we shall develop a general formulation for a three-way crossed classification with replication and then consider an example of this case and a more complex one, including both nested and crossed classifications.

Suppose that we are given a set of $N = npqr \cdots$ observations $x_{ijk\cdots\alpha}$ divisible into $i = 1, \cdots, p$ classes according to a characteristic A, into $j = 1, \cdots, q$ classes according to a characteristic B, and into $k = 1, \cdots, r$ classes according to a characteristic C, etc., with $\alpha = 1, \cdots, n$ observations in each class, or cell, corresponding to a given combination of these characteristics. Then the analysis of variance for this set of data would consist of some or all of the following steps:

(1) *The assumption of a model.* We shall in general assume each observation to be normally distributed with a variance σ^2 common to all observations and an average value which is an additive combination of effects corresponding to the characteristics A, B, C, etc., and to all possible combinations of these characteristics. For example, in a three-way completely crossed classification we should assume each of the $N = npqr$ observations, $x_{ijk\alpha}$, $i = 1, \cdots, p$ rows, $j = 1, \cdots, q$ columns, $k = 1, \cdots, r$ groups and $\alpha = 1, \cdots, n$ replicates in each cell to be normally distributed with common variance σ^2 and

$$\text{ave}\,(x_{ijk\alpha}) = \mu + \xi_i + \eta_j + \zeta_k + \beta_{ij} + \gamma_{ik} + \delta_{jk} + \lambda_{ijk}$$

Here the ξ_i, η_j, and ζ_k represent the effects corresponding to the characteristics A, B, C, respectively, the β_{ij}, γ_{ik}, and δ_{jk} to the effects peculiar to the possible combinations of A and B, A and C, and B and C, respectively; and the λ_{ijk} represent the effects peculiar to the possible combinations of A, B, and C. The effects corresponding to a single characteristic are called main effects; those corresponding to a combination of 2 characteristics, two-factor interactions; those corresponding to a combination of 3 characteristics, three-factor interactions, etc. In a three-way completely crossed classification, there are 3 sets of main effects, 3 sets of two-factor interactions, and 1 set of three-factor interactions. Since the number of groups of effects and interactions is simply the number of combinations of the characteristics taken singly, 2 at a time, etc., they can be easily determined for a given number of characteristics, using the binomial coefficients, with the first omitted. The number of groups of main effects and interactions of various orders for up to 8 classifications are given in Table 7.26.

TABLE 7.26

NUMBER OF GROUPS OF EFFECTS IN VARIOUS CLASSIFICATIONS

Number of Factors (Classifications)	1	2	3	4	5	6	7	8
Groups of main effects	1	2	3	4	5	6	7	8
Groups of 2-factor interactions		1	3	6	10	15	21	28
Groups of 3-factor interactions			1	4	10	20	35	42
Groups of 4-factor interactions				1	5	15	35	70
Groups of 5-factor interactions					1	6	21	42
Groups of 6-factor interactions						1	7	28
Groups of 7-factor interactions							1	8
Groups of 8-factor interactions								1

(2) *The estimation of the effects.* In general this is done by minimizing the sum of squares of the observations from the parameter estimates, subject to the restriction that the two-factor interaction estimates make no average contribution to the main effects, the three-factor interaction estimates make no average contribution to the two-factor interactions, etc.

For the three-way case we should estimate the various main effects and interactions by determining those values $\hat{\mu}$, $\hat{\xi}_i$, $\hat{\beta}_{ij}$, etc., which minimize the sum of squares

$$Q = \Sigma_{ijk\alpha}(x_{ijk\alpha} - \mu - \xi_i - \eta_j - \zeta_k - \beta_{ij} - \gamma_{ik} - \delta_{jk} - \lambda_{ijk})^2$$

TABLE 7.27

Effect	Estimate
μ	$\bar{x}$
ξ_i	$\bar{x}_{i\cdot\cdot} - \bar{x}$
η_j	$\bar{x}_{\cdot j\cdot} - \bar{x}$
ζ_k	$\bar{x}_{\cdot\cdot k} - \bar{x}$
β_{ij}	$\bar{x}_{ij\cdot} - \bar{x}_{i\cdot\cdot} - \bar{x}_{\cdot j\cdot} + \bar{x}$
γ_{ik}	$\bar{x}_{i\cdot k} - \bar{x}_{i\cdot\cdot} - \bar{x}_{\cdot\cdot k} + \bar{x}$
δ_{jk}	$\bar{x}_{\cdot jk} - \bar{x}_{\cdot j\cdot} - \bar{x}_{\cdot\cdot k} + \bar{x}$
λ_{ijk}	$\bar{x}_{ijk} - \bar{x}_{ij\cdot} - \bar{x}_{i\cdot k} - \bar{x}_{\cdot jk} + \bar{x}_{i\cdot\cdot} + \bar{x}_{\cdot j\cdot} + \bar{x}_{\cdot\cdot k} - \bar{x}$

subject to the conditions that the sum of all effect estimates over either rows, columns, or groups individually should be zero. This procedure yields the estimates given in Table 7.27. The notation is a natural extension of that previously employed, i.e., $\bar{x}$ is the overall mean of all $N = npqr$ observations, $\bar{x}_{ijk}$ is the mean of the n observations in the ijkth cell, $\bar{x}_{\cdot jk}$ the mean of the np observations common to the jth column and the kth group, $\bar{x}_{\cdot j\cdot}$ the mean of the npr observations in the jth column, etc. Note that these estimates correspond symbolically to the quantities 1, $i - 1$, $j - 1$, $k - 1$, $(i - 1)$ $(j - 1)$, $(i - 1)$ $(k - 1)$, etc., where 1 indicates the mean of all observations, i the mean of the ith row, j the mean of the jth column, ij the mean of observations in both the ith row and the jth column, etc. Thus, since $(j-1)(k-1) = jk - j - k + 1$, the estimate of δ_{jk} would be $\hat{\delta}_{jk} = \bar{x}_{\cdot jk} - \bar{x}_{\cdot j \cdot} - \bar{x}_{\cdot \cdot k} - \bar{x}$. This symbolism is easily extended to higher classifications, since those subscripts not appearing in the required estimate are simply replaced by 1's in the symbolic product.

Since additivity is a fundamental part of our assumption, we can obtain the estimate of the sum or difference of 2 effects by adding or subtracting the corresponding estimates of the effects. Thus, to obtain the estimates of the column means $\mu + \eta_j$, we should add the estimates $\hat{\mu}$ and $\hat{\eta}_j$, and, to obtain the estimate of the difference $\xi_1 - \xi_2$ between the effects of rows 1 and 2, we should subtract the corresponding estimates. Note that, since all these estimates are linear combinations of the original observations, their average values are equal to the corresponding effects in the model, i.e., they are unbiased. Methods of obtaining the variance of the effects will be discussed in a later section.

(3) *The algebraic breakdown of the overall sum of squares into independent sums of squares corresponding to the effects in the model, and the computation of the degrees of freedom and average value for each of these component sums of squares.* For the case of the three-way classification, we have a breakdown of the sum of squares in the form

$$S = S_{\alpha(ijk)} + S_{ijk} + S_{ij} + S_{ik} + S_{jk} + S_i + S_j + S_k$$

The definitions of these sums of squares, their degrees of freedom, and the average values of the mean squares, assuming a model of type II, are given in Table 7.28. In general each sum of squares consists of the sum of the squares of the estimates of a given group of effects, multiplied by the number of times this group of effects appears. Thus, for example, the row-column interaction β_{ij} appears in n replicate observations in each of r groups; hence the corresponding sum of squares would be

$$S_{ij} = nr\Sigma_{ij}(\bar{x}_{ij\cdot} - \bar{x}_{i\cdot\cdot} - \bar{x}_{\cdot j\cdot} + \bar{x})^2$$

Similarly, each of the column effects η_j appears in n replicate observations in each of the p rows in each of the r groups, and the corresponding sum of squares would be $S_i = npr\Sigma_j(\bar{x}_{\cdot j\cdot} - \bar{x})^2$ since the η_j are estimated by the quantities $\bar{x}_{\cdot j\cdot} - \bar{x}$. Hence, given the estimate of any effect, the corresponding sum of squares is easily written. This, combined with the symbolic method given above for obtaining the estimate of a given effect, gives the breakdown of the overall sum of squares into the parts corresponding to the effects in a model having more than 2 factors of classification.

TABLE 7.28

GENERAL ANALYSIS OF VARIANCE:
THREE CROSSED CLASSIFICATIONS WITH REPLICATION

Source of Estimate	Sum of Squares	Definition of Sum of Squares	D.F.	Average Value of Mean Square*
Main effects:				
rows	S_i	$nqr\,\Sigma_i(\bar{x}_{i\cdot\cdot} - \bar{x})^2$	$p - 1$	$\sigma^2 + n\sigma_\lambda^2 + nq\sigma_\gamma^2 + nr\sigma_\beta^2 + nqr\sigma_\xi^2$
columns	S_j	$npr\,\Sigma_j(\bar{x}_{\cdot j\cdot} - \bar{x})^2$	$q - 1$	$\sigma^2 + n\sigma_\lambda^2 + nr\sigma_\beta^2 + np\sigma_\delta^2 + npr\sigma_\eta^2$
groups	S_k	$npq\,\Sigma_k(\bar{x}_{\cdot\cdot k} - \bar{x})^2$	$r - 1$	$\sigma^2 + n\sigma_\lambda^2 + nq\sigma_\gamma^2 + np\sigma_\delta^2 + npq\sigma_\zeta^2$
Two-factor interactions:				
rows-columns	S_{ij}	$nr\,\Sigma_{ij}(\bar{x}_{ij\cdot} - \bar{x}_{i\cdot\cdot} - \bar{x}_{\cdot j\cdot} + \bar{x})^2$	$(p - 1)(q - 1)$	$\sigma^2 + n\sigma_\lambda^2 + nr\sigma_\beta^2$
rows-groups	S_{ik}	$nq\,\Sigma_{ik}(\bar{x}_{i\cdot k} - \bar{x}_{i\cdot\cdot} - \bar{x}_{\cdot\cdot k} + \bar{x})^2$	$(p - 1)(r - 1)$	$\sigma^2 + n\sigma_\lambda^2 + nq\sigma_\gamma^2$
columns-groups	S_{jk}	$np\,\Sigma_{jk}(\bar{x}_{\cdot jk} - \bar{x}_{\cdot j\cdot} - \bar{x}_{\cdot\cdot k} + \bar{x})^2$	$(q - 1)(r - 1)$	$\sigma^2 + n\sigma_\lambda^2 + np\sigma_\delta^2$
Three-factor interaction:				
rows-columns-groups	S_{ijk}	$n\,\Sigma_{ijk}(\bar{x}_{ijk} - \bar{x}_{ij\cdot} - \bar{x}_{i\cdot k} - \bar{x}_{\cdot jk} + \bar{x}_{i\cdot\cdot} + \bar{x}_{\cdot j\cdot} + \bar{x}_{\cdot\cdot k} - \bar{x})^2$	$(p - 1)(q - 1)(r - 1)$	$\sigma^2 + n\sigma_\lambda^2$
Within replicates (error)	$S_{\alpha(ijk)}$	$\Sigma_{ijk\alpha}(x_{ijk\alpha} - \bar{x}_{ijk})^2$	$N - pqr = pqr(n - 1)$	σ^2
Total	S	$\Sigma_{ijk\alpha}(x_{ijk\alpha} - \bar{x})^2$	$N - 1$	

* All effects assumed drawn from infinite populations of such effects.

TABLE 7.29

Sum of Squares	Computational Form
S_i	$\frac{1}{nqr} \Sigma_i T^2_{i\cdot\cdot} - T^2/N$
S_j	$\frac{1}{npr} \Sigma_j T^2_{\cdot j\cdot} - T^2/N$
S_k	$\frac{1}{npq} \Sigma_k T^2_{\cdot\cdot k} - T^2/N$
S_{ij}	$\frac{1}{nr} \Sigma_{ij} T^2_{ij\cdot} - \frac{1}{nqr} \Sigma_i T^2_{i\cdot\cdot} - \frac{1}{npr} \Sigma_j T^2_{\cdot j\cdot} + T^2/N$
S_{ik}	$\frac{1}{nq} \Sigma_{ik} T^2_{i\cdot k} - \frac{1}{nqr} \Sigma_i T^2_{i\cdot\cdot} - \frac{1}{npq} \Sigma_k T^2_{\cdot\cdot k} + T^2/N$
S_{jk}	$\frac{1}{np} \Sigma_{jk} T^2_{\cdot jk} - \frac{1}{npr} \Sigma_j T^2_{\cdot j\cdot} - \frac{1}{npq} \Sigma_k T^2_{\cdot\cdot k} + T^2/N$
S_{ijk}	$\frac{1}{n} \Sigma_{ijk} T^2_{ijk} - \frac{1}{nr} \Sigma_{ij} T^2_{ij\cdot} - \frac{1}{nq} \Sigma_{ik} T^2_{i\cdot k} - \frac{1}{np} \Sigma_{jk} T^2_{\cdot jk} + \frac{1}{nqr} \Sigma_i T^2_{i\cdot\cdot} + \frac{1}{npr} \Sigma_j T^2_{\cdot j\cdot} + \frac{1}{npq} \Sigma_k T^2_{\cdot\cdot k} + T^2/N$
$S_{\alpha(ijk)}$	$\Sigma_{ijk\alpha} x^2_{ijk\alpha} - \frac{1}{n} \Sigma_{ijk} T^2_{ijk}$
S	$\Sigma_{ijk\alpha} x^2_{ijk\alpha} - T^2/N$

(4) *Calculation of sums of squares.* The sums of squares defined under (3) can be broken down algebraically into sums of squares of totals which are more easily computed from the original data. If, following the notation of the previous sections, we define T_{ijk} as the sum of the replicates in the ijkth cell, $T_{ij\cdot}$ the sum of the observations in the ith row and jth column, $T_{i\cdot\cdot}$ the sum of the observations in the ith row, T the sum of all observations, etc., then the computational forms for the sum of squares are given in Table 7.29. The 9 different quantities necessary to compute the sums of squares from this formulation are:

$$C = T^2/N \qquad C_i = (1/nqr)\,\Sigma_i T^2_{i\cdot\cdot} \qquad C_j = (1/npr)\,\Sigma_j T^2_{\cdot j\cdot}$$

$$C_k = (1/npq)\,\Sigma_k T^2_{\cdot\cdot k} \qquad C_{ij} = (1/nr)\,\Sigma_{ij} T^2_{ij\cdot}$$

$$C_{ik} = (1/nq)\,\Sigma_{ik} T^2_{i\cdot k} \qquad C_{jk} = (1/np)\,\Sigma_{jk} T^2_{\cdot jk}$$

$$C_{ijk} = (1/n)\,\Sigma_{ijk} T^2_{ijk} \qquad C_{ijk\alpha} = \Sigma_{ijk\alpha} x^2_{ijk\alpha}$$

These quantities can be obtained directly from tables of the original observations or from sums of the observations by accumulating squares and dividing by the appropriate number. Note again how the pattern of these computational formulae follows the symbolic formulation of the estimates given in (2). Thus for a given group of effects, say the two-factor interactions β_{ij}, we have

$$(i-1)(j-1) = ij - i - j + 1$$

$$\text{estimate of } \beta_{ij} = \bar{x}_{ij\cdot} - \bar{x}_{i\cdot\cdot} - \bar{x}_{\cdot j\cdot} + \bar{x}$$

$$S_{ij} \text{ (definition)} = nr\,\Sigma_{ij}(\bar{x}_{ij\cdot} - \bar{x}_{i\cdot\cdot} - \bar{x}_{\cdot j\cdot} + \bar{x})^2$$

$$S_{ij} \text{ (computation)} = C_{ij} - C_i - C_j + C$$

To obtain the C's in the case of 3 crossed classifications we need the following tables:

(*a*) The original observations. The sum and the sum of squares of these observations are T and $C_{ijk\alpha}$, respectively.

(*b*) A table giving the sums T_{ijk} of the replicates in each cell. The sum and the sum of squares of the entries in this table are T (check) and $\Sigma_{ijk}T^2_{ijk}$, respectively. From the latter we obtain C_{ijk} by dividing by n (without removing from the calculating machine).

(*c*) Three tables obtained from the table prepared in (*b*) by adding over rows, columns, and groups, respectively. Each of these tables will be classified in only 2 ways, and the individual entries will be the sums $T_{ij\cdot}$, $T_{i\cdot k}$, and $T_{\cdot jk}$. From the sum of squares of these entries, using the sum as a check and dividing by nr, nq, and np, respectively, we obtain C_{ij}, C_{ik}, and C_{jk}. The marginal totals of these 3 tables will be the sums

$T_{i..}$, $T_{.j.}$, and $T_{..k}$, each set occurring twice (they need be obtained only once, except as a check). From the sum of squares of these 3 sets of marginal totals, using T again as a check and dividing by the appropriate number, we obtain C_i, C_j, and C_k. Obtaining $C = T^2/N$ completes the evaluation of the quantities necessary to obtain the sums of squares in Table 7.28.

(5) *Tests for the existence of significant effects.* These should begin with the test for the highest-order interaction, and proceed toward proper tests for the main effects, at all times the average values of the estimates involved being taken into account. All too frequently an analysis of variance is concluded by taking the ratio of all the variances to the error term and declaring these ratios significant or not significant at a given level. Actually, we should begin at the bottom of the table with the highest-order interaction, and then work carefully up the table, identifying each significant source of variation, and taking it into account in succeeding tests.

Let us consider a hypothetical situation with respect to the three-classification case discussed thus far, using the average values for a model of type II given in Table 7.28. The first hypothesis tested would be $\sigma_\lambda^2 = 0$; this would be made using the ratio of the estimate based on the three-factor interactions to the error estimate. Let us suppose that this ratio is less than 2, and not significant at the 5% level. Then *for the purpose of future testing* we shall assume $\sigma_\lambda^2 = 0$, and pool the three-factor interaction and the error estimates to estimate σ^2. We could then test for the presence of significant two-factor interactions by examining the hypotheses $\sigma_\beta^2 = 0$, $\sigma_\gamma^2 = 0$, and $\sigma_\delta^2 = 0$, using this pooled estimate as the error estimate in each case. Now suppose that the row-column interactions β_{ij} are significant, but that for the other interactions the variance ratios are non-significant and less than 2. If these two non-significant interaction estimates are again pooled with the error estimate under the assumption $\sigma_\gamma^2 = \sigma_\delta^2 = 0$, the situation will now be reduced to that of Table 7.30. At this point the proper test for the hypothesis $\sigma_\zeta^2 = 0$, i.e., the absence of significant group effects, would be to compare the group estimate with the pooled error estimate; however, the proper test for significant main effects due to rows and columns would involve a comparison of the row and column estimates with the two-factor row-column interaction.

If more than one of the two-factor interactions are significant, the average value of the main effect mean square for that factor which is involved in both interactions (e.g., A) will be

$$\text{ave (M.S. } A) = \sigma^2 + nr\sigma_\beta^2 + nq\sigma_\gamma^2 + nqr\sigma_\xi^2$$

TABLE 7.30

Source of Estimate	D.F.	Average Value of Mean Square*
Main effects:		
rows	$p-1$	$\sigma^2 + nr\sigma_\beta^2 + nqr\sigma_\xi^2$
columns	$q-1$	$\sigma^2 + nr\sigma_\beta^2 + npr\sigma_\eta^2$
groups	$r-1$	$\sigma^2 + npq\sigma_\zeta^2$
Two-factor interactions:		
rows-columns	$(p-1)(q-1)$	$\sigma^2 + nr\sigma_\beta^2$
Pooled error: (row-group, column-group, row-column-group, and between replicates)	$N - pq - r + 1$	σ^2
Total	$N-1$	

* All populations of effects assumed infinite.

so that, for the denominator of an unbiased F-test, we require a mean square with average value $\sigma^2 + nr\sigma_\beta^2 + nq\sigma_\gamma^2$. This does not occur in the table, and we proceed by the approximate method of Section 7.32. In the two-factor classification the approximate method was required only when the populations were considered finite, but for classifications involving 3 or more factors it may be required even when all the populations involved are infinite.

(6) *Estimate of variance components.* Let us designate by s_i^2, s_j^2, s_k^2, s^2_{ij}, s^2_{ik}, s^2_{jk}, s^2_{ijk}, and s^2 the estimates obtained by dividing the sums of squares corresponding to the various effects by their respective degrees of freedom. Then, by combining these estimates in the manner dictated by the average values of Table 7.28, we can estimate the variances $\sigma_\xi^2, \sigma_\eta^2, \cdots, \sigma_\lambda^2, \sigma^2$ of the various main effects and interactions (frequently called the variance components). Starting, as usual, with the error and highest-order interaction and working toward the main effects, we have, for example,

Estimate of σ^2: s^2

Estimate of σ_λ^2: $\dfrac{s^2_{ijk} - s^2}{n}$

Estimate of σ_β^2: $\dfrac{s^2_{ij} - s^2_{ijk}}{nr}$

Estimate of σ_γ^2: $\dfrac{s^2_{ik} - s^2_{ijk}}{nq}$

Estimate of σ_ξ^2: $\dfrac{s_i^2 - s^2_{ij} - s^2_{ik} + s^2_{ijk}}{nqr}$

The other variance components can be similarly estimated.

These results relate to a model of type II, i.e., one in which all populations are infinite. If a model of type III is assumed, then the average values of the mean squares become

Mean Square	Average Value of Mean Square
s^2	σ^2
s^2_{ijk}	$\sigma^2 + n\sigma_\lambda^2$
s^2_{ij}	$\sigma^2 + n\left(1 - \frac{r}{R}\right)\sigma_\lambda^2 + nr\sigma_\beta^2$
s^2_{ik}	$\sigma^2 + n\left(1 - \frac{q}{Q}\right)\sigma_\lambda^2 + nq\sigma_\gamma^2$
s^2_{jk}	$\sigma^2 + n\left(1 - \frac{p}{P}\right)\sigma_\lambda^2 + np\sigma_\delta^2$
s_i^2	$\sigma^2 + n\left(1-\frac{r}{R}\right)\left(1-\frac{q}{Q}\right)\sigma_\lambda^2 + nr\left(1-\frac{q}{Q}\right)\sigma_\beta^2 + nq\left(1-\frac{r}{R}\right)\sigma_\gamma^2 + nqr\sigma_\xi^2$
s_j^2	$\sigma^2 + n\left(1-\frac{r}{R}\right)\left(1-\frac{p}{P}\right)\sigma_\lambda^2 + nr\left(1-\frac{p}{P}\right)\sigma_\beta^2 + np\left(1-\frac{r}{R}\right)\sigma_\delta^2 + npr\sigma_\eta^2$
s_k^2	$\sigma^2 + n\left(1-\frac{q}{Q}\right)\left(1-\frac{p}{P}\right)\sigma_\lambda^2 + nq\left(1-\frac{p}{P}\right)\sigma_\gamma^2 + np\left(1-\frac{q}{Q}\right)\sigma_\delta^2 + npq\sigma_\zeta^2$
$(\bar{x} - \mu_0)^2$	$\sigma^2 + n\left(1-\frac{r}{R}\right)\left(1-\frac{q}{Q}\right)\left(1-\frac{p}{P}\right)\sigma_\lambda^2 + nr\left(1-\frac{q}{Q}\right)\left(1-\frac{p}{P}\right)\sigma_\beta^2$ $+ nq\left(1-\frac{r}{R}\right)\left(1-\frac{p}{P}\right)\sigma_\gamma^2 + np\left(1-\frac{r}{R}\right)\left(1-\frac{q}{Q}\right)\sigma_\delta^2$ $+ nqr\left(1-\frac{p}{P}\right)\sigma_\xi^2 + npr\left(1-\frac{q}{Q}\right)\sigma_\eta^2 + npq\left(1-\frac{r}{R}\right)\sigma_\zeta^2$ $+ npqr(\mu - \mu_0)^2$

In this case the individual variance components are estimated as for the type II model except that not all the coefficients in the linear combination are equal. Thus, for example

$$\hat{\sigma}_\beta^2 = \frac{1}{nr}\left[s^2_{ij} - \left(1 - \frac{r}{R}\right)s^2_{ijk} - \frac{r}{R}s^2\right]$$

$$\hat{\sigma}_\xi^2 = \frac{1}{nqr}\left[s_i^2 - \left(1 - \frac{q}{Q}\right)\left[s^2_{ij} - \left(1 - \frac{r}{R}\right)s^2_{ijk} - \frac{r}{R}s^2\right]\right.$$
$$- \left(1 - \frac{r}{R}\right)\left[s^2_{ik} - \left(1 - \frac{q}{Q}\right)s^2_{ijk} - \frac{q}{Q}s^2\right]$$
$$\left. - \left(1 - \frac{r}{R}\right)\left(1 - \frac{q}{Q}\right)[s^2_{ijk} - s^2] - s^2\right]$$

Although we are estimating a variance component which is never negative, the estimate obtained in a particular instance may well be negative if the population value is small. If no additional information is available, the true value of the variance component may be estimated as zero. If repeated experiments are performed in each of which this component is estimated, the negative estimate should be used in obtaining a joint estimate. If we obtain a series of negative estimates of a variance component from repeated experiments, we infer that the residual effects in our data are not randomly distributed and that we have probably overlooked some important factor in our classification.

To place confidence limits upon the variance components in these more complicated examples, it is usually necessary to form a composite denominator for the F-ratio with an equivalent number of degrees of freedom by the method of Section 7.52 and then employ the approximation of Section 7.23 to obtain the approximate limits.

TABLE 7.31

DETERMINATION OF THE VOLATILE MATTER CONTENT OF COAL SAMPLES EMPLOYING 2 TYPES OF CRUCIBLES AND 3 TEMPERATURES

[Coding: $X + 23\% =$ Volatile Matter]

Coal Samples		*A*				*B*			
Crucible		Steel		Silica		Steel		Silica	
Temperatures	825	8.73	7.15	9.50	9.22	0.92	1.12	5.06	5.66
	875	9.76	9.47	9.57	9.91	2.22	1.66	6.08	6.33
	925	10.13	9.20	10.38	10.29	1.00	1.69	6.32	6.18

EXAMPLE. The above data were obtained in an investigation to determine the influence of muffle temperature and type of crucible upon the estimate of volatile matter content of 2 coal samples. In this instance the coals A and B were known to be of different composition and volatile matter content so that a highly significant main effect due to the coals was anticipated. The purpose of the experiment was the estimation of the crucible and temperature effects and the significance of the two-factor interactions.

In this example the coal samples may be considered as a random sample from the large number available to a particular survey laboratory. The choice of crucibles is restricted to 2 types, both of which have been examined, and the levels of the scale factor (temperature) were chosen on an arbitrary basis, and not by a random procedure. Thus inferences concerning the effects of temperature and crucible will be restricted to the classes actually examined, and the average values of the mean squares will be modified. For the purpose of the example we shall restate the assumptions that

(*a*) The average value of an observation in a particular cell corresponding to a temperature θ, a coal c, and a type of crucible t will be

$$\text{ave}\,(x_{\theta ct}) = \mu + \xi_\theta + \eta_c + \zeta_t + \beta_{\theta c} + \gamma_{\theta t} + \delta_{ct} + \lambda_{\theta ct}$$

where ξ_θ, η_c, and ζ_t represent the main effects upon the determined volatile content of the temperature, coal, and type of crucible, and the remaining terms represent the corresponding interactions. We assume additivity of these effects.

(*b*) The observed value $x_{\theta tc\alpha}$, where α is the subscript identifying the replicate determination, will be distributed normally about its average value with constant unknown variance σ^2.

Before estimating the various additive components it is convenient to compute the subtables which will be required for the analysis of variance. Summing over the replicate determinations, we obtain Subtable (*a*).

SUBTABLE (*a*)

TYPICAL ELEMENT $T_{\theta ct}$

Coals		*A*		*B*	
Crucibles		Steel	Silica	Steel	Silica
	825	15.88	18.72	2.04	10.72
Temperatures	875	19.23	19.48	3.88	12.41
	925	19.33	20.67	2.69	12.50

Summing Subtable (*a*) over the coal, crucible, and temperature classes, respectively, we obtain Subtables (*b*), (*c*), and (*d*).

SUBTABLE (*b*)

TYPICAL ELEMENT $T_{\theta \cdot t}$

Crucibles	Steel	Silica
825	17.92	29.44
Temperatures 875	23.11	31.89
925	22.02	33.17

SUBTABLE (*c*)

TYPICAL ELEMENT $T_{\theta c \cdot}$

Coals	*A*	*B*
825	34.60	12.76
Temperatures 875	38.71	16.29
925	40.00	15.19

SUBTABLE (*d*)

TYPICAL ELEMENT $T_{\cdot ct}$

Coals	*A*		*B*	
Crucibles	Steel	Silica	Steel	Silica
	54.55	58.87	8.61	35.63

From these tables we obtain the marginal tables for the 3 main effects Subtables (*e*), (*f*), and (*g*).

SUBTABLE (*e*)

TYPICAL ELEMENT

$T_{\cdot c \cdot}$

Coal	*A*	*B*
	113.31	44.24

SUBTABLE (*f*)

TYPICAL ELEMENT

$T_{\cdot \cdot t}$

Crucible	Steel	Silica
	63.05	94.50

SUBTABLE (*g*)

TYPICAL ELEMENT

$T_{\theta \cdot \cdot}$

Temperature	825	875	925
	47.36	55.00	55.19

Estimation. The overall total of the observations (T) is 157.55 so that μ is estimated as $\frac{157.55}{24} = 6.5646$

The values of ξ_θ, η_c, and ζ_t can now be estimated, using Subtables (*g*), (*e*), and (*f*), respectively. Since on the basis of the assumed model the interaction effects make no average contribution to these subtotals, then, for example,

$$\hat{\xi}_{825} = \frac{47.36}{8} - 6.5646 = -0.6446$$

Estimates of all the main effects may be computed in a similar fashion, and we obtain the following values:

$$\hat{\mu} = 6.5646$$

$$\hat{\xi}_{825} = -0.6446 \qquad \hat{\xi}_{875} = 0.3104 \qquad \hat{\xi}_{925} = 0.3342 \qquad \Sigma\hat{\xi} = 0$$

$$\hat{\eta}_A = 2.8779 \qquad \hat{\eta}_B = -2.8779 \qquad \Sigma\hat{\eta} = 0$$

$$\hat{\zeta}_{\text{steel}} = -1.3104 \qquad \hat{\zeta}_{\text{silica}} = +1.3104 \qquad \Sigma\hat{\zeta} = 0$$

Using these estimates of the main effects and referring to Subtable (*c*), we obtain for our estimate of the interaction $\beta_{825,\ A}$

$$\frac{34.60}{4} + 0.6446 - 2.8779 - 6.5646 = -0.1479$$

Estimates of all other two-factor interactions may be obtained in a similar way and are tabulated below.

$$\hat{\beta}_{825,A} = -0.1479 \quad \hat{\beta}_{825,B} = +0.1479 \quad \hat{\gamma}_{825,\text{steel}} = -0.1296 \quad \hat{\gamma}_{825,\text{silica}} = +0.1296$$

$$\hat{\beta}_{875,A} = -0.0754 \quad \hat{\beta}_{875,B} = +0.0754 \quad \hat{\gamma}_{875,\text{steel}} = +0.2129 \quad \hat{\gamma}_{875,\text{silica}} = -0.2129$$

$$\hat{\beta}_{925,A} = +0.2233 \quad \hat{\beta}_{925,B} = -0.2233 \quad \hat{\gamma}_{925,\text{steel}} = -0.0833 \quad \hat{\gamma}_{925,\text{silica}} = +0.0833$$

$$\hat{\delta}_{A,\text{steel}} = +0.9412 \qquad \hat{\delta}_{B,\text{steel}} = -0.9412$$

$$\hat{\delta}_{A,\text{silica}} = -0.9412 \qquad \hat{\delta}_{B,\text{silica}} = +0.9412$$

These tables illustrate clearly the conditions imposed on the interaction estimates. All the marginal totals are zero, and hence in one of the 2×3 tables the calculation of $1 \times 2 = 2$ individual estimates enables the other $(3 - 1) + (2 - 1) + 1 = 4$ estimates to be made without further reference to the original data. Examination of the tables indicates that the values of $\hat{\delta}$ are considerably greater than the values of $\hat{\beta}$ and $\hat{\gamma}$. In our subsequent significance tests we shall therefore expect a higher mean square for the coals-crucibles interaction.

We conclude the estimation process by evaluating the estimates of the three-factor interaction $\lambda_{\theta ct}$. In this case

$$\hat{\lambda}_{825,\ A,\ \text{steel}} = \frac{15.88}{2} + 0.1473 + 0.1296 - 0.9412 + 0.6446 + 1.310$$
$$- 2.8779 - 6.5646 = -0.2112$$

and the remaining values calculated in a similar fashion become

$\hat{\lambda}_{825,\,A,\,\text{steel}} = -0.2112$ $\hat{\lambda}_{825,\,A,\,\text{silica}} = +0.2112$ $\hat{\lambda}_{825,\,B,\,\text{steel}} = -0.2112$ $\hat{\lambda}_{825,\,B,\,\text{silica}} = +0.2112$

$\hat{\lambda}_{875,\,A,\,\text{steel}} = +0.0938$ $\hat{\lambda}_{875,\,A,\,\text{silica}} = -0.0938$ $\hat{\lambda}_{875,\,B,\,\text{steel}} = +0.0938$ $\hat{\lambda}_{875,\,B,\,\text{silica}} = -0.0938$

$\hat{\lambda}_{925,\,A,\,\text{steel}} = +0.1175$ $\hat{\lambda}_{925,\,A,\,\text{silica}} = -0.1175$ $\hat{\lambda}_{925,\,B,\,\text{steel}} = +0.1175$ $\hat{\lambda}_{925,\,B,\,\text{silica}} = -0.1175$

Our method of estimation imposes 10 linear restraints upon the values $\hat{\lambda}_{\theta ct}$, so that only 2 independent values are present in the above set.

Significance Testing. It would be possible, as we have estimated the individual terms of the main effects and interactions, to compute the variance estimates for significance tests from these values. Usually, however, the estimation process follows the significance tests, and we shall proceed as though this were true. The overall sum of squares may be broken down as

$$\Sigma_{\theta ct\alpha} x^2_{\theta ct\alpha} - \frac{T^2}{N} = S_{\alpha(\theta ct)} + S_{\theta ct} + S_{\theta c} + S_{\theta t} + S_{ct} + S_{\theta} + S_{c} + S_{t}$$

The relevant sums of squares from Table 7.31 and from Subtables (*a*)–(*g*) are:

From Table 7.31 $\quad S = \Sigma_{\theta ct\alpha} x^2_{\theta ct\alpha} - \dfrac{T^2}{N} = 1304.6609 - 1034.2501 = 270.4108$

From Subtable (*g*) $\quad S_\theta = \dfrac{1}{8}\Sigma_\theta T^2_{\theta\cdot\cdot} - \dfrac{T^2}{N} = 1039.2382 - 1034.2501 = 4.9881$

From Subtable (*f*) $\quad S_t = \dfrac{1}{12}\Sigma_t T^2_{\cdot\cdot t} - \dfrac{T^2}{N} = 1075.4627 - 1034.2501 = 41.2126$

From Subtable (*e*) $\quad S_c = \dfrac{1}{12}\Sigma_c T^2_{\cdot c\cdot} - \dfrac{T^2}{N} = 1233.0278 - 1034.2501 = 198.7777$

From Subtable (*d*) $\quad S_{ct} = \dfrac{1}{6}\Sigma_{ct} T^2_{\cdot ct} - \dfrac{1}{12}\Sigma T^2_{\cdot c\cdot} - \dfrac{1}{12}\Sigma_t T^2_{\cdot\cdot t} + \dfrac{T^2}{N}$

$$= 1295.5032 - 1233.0278 - 1075.4627 + 1034.2501 = 21.2628$$

From Subtable (*c*) $\quad S_{\theta c} = \dfrac{1}{4}\Sigma_{\theta c} T^2_{\theta c\cdot} - \dfrac{1}{12}\Sigma_c T^2_{\cdot c\cdot} - \dfrac{1}{8}\Sigma_\theta T^2_{\theta\cdot\cdot} + \dfrac{T^2}{N}$

$$= 1238.6355 - 1233.0278 - 1039.2382 + 1034.2501 = 0.6196$$

From Subtable (*b*) $\quad S_{\theta t} = \dfrac{1}{4}\Sigma_{\theta t} T^2_{\theta\cdot t} - \dfrac{1}{12}\Sigma_t T^2_{\cdot\cdot t} - \dfrac{1}{8}\Sigma_\theta T^2_{\theta\cdot\cdot} + \dfrac{T^2}{N}$

$$= 1081.0034 - 1075.4627 - 1039.2382 + 1034.2501 = 0.5526$$

From Subtable (*a*) $S_{\theta ct} = \frac{1}{2}\Sigma_{\theta ct}T^2_{\theta ct} - \frac{1}{4}\Sigma_{\theta t}T^2_{\theta\cdot t} - \frac{1}{4}\Sigma_{\theta c}T^2_{\theta c\cdot} - \frac{1}{6}\Sigma_{ct}T^2_{\cdot ct} + \frac{1}{8}\Sigma_{\theta}T^2_{\theta\cdot\cdot}$

$$+ \frac{1}{12}\Sigma_c T^2_{\cdot c\cdot} + \frac{1}{12}\Sigma_t T^2_{\cdot\cdot t} - \frac{T^2}{N}$$

$$= 1302.2012 - 1081.0034 - 1238.6355 - 1295.5032$$
$$+ 1039.2382 + 1233.0278 + 1075.4627 - 1034.2501$$
$$= 0.5377$$

$$S_{\alpha(\theta ct)} = \Sigma_{\theta ct\alpha}x^2_{\theta ct\alpha} - 1/2\ \Sigma_{\theta ct}T^2_{\theta ct} = 1304.6609 - 1302.2012 = 2.4597$$

Assembling the components into an analysis of variance table, we obtain Table 7.32. The average values of the mean squares have been modified in accordance with our assumption, the general values for a model of type III given in (6) of this section having been used.

TABLE 7.32

ANALYSIS OF VARIANCE

Source of Estimate	Sum of Squares	D.F.	Mean Square	Average Value of Estimate
Main effects				
between temperatures	4.9881	2	2.49	$\sigma^2 + 0 + 4\sigma^2_{\theta c} + 0 + 8\sigma_\theta^2$
between coals	198.7777	1	198.78	$\sigma^2 + 0 + 0 + 0 + 12\sigma_c^2$
between crucibles	41.2126	1	41.21	$\sigma^2 + 0 + 0 + 6\sigma^2_{ct} + 12\sigma_t^2$
Two-factor interactions				
temp. × coals	0.6196	2	0.310	$\sigma^2 + 0 + 4\sigma^2_{\theta c}$
temp. × crucibles	0.5526	2	0.276	$\sigma^2 + 2\sigma^2_{\theta ct} + 4\sigma^2_{\theta t}$
coals × crucibles	21.2628	1	21.263	$\sigma^2 + 0 + 6\sigma^2_{ct}$
Three-factor interaction				
temps. × coals × crucibles	0.5377	2	0.269	$\sigma^2 + 2\sigma^2_{\theta ct}$
Residual	2.4597	12	0.205	σ^2
Total	270.4180	23		

We proceed to test the following hypotheses:

(1) $\lambda_{\theta ct} = 0$. Under this hypothesis $\sigma^2_{\theta ct} = 0$, and we test the variance ratio

$$\frac{0.269}{0.205} = F_{2,\ 12} = 1.32 \qquad (F_{2,\ 12,\ 0.20} = 1.8)$$

This is less than 2 and is not significant even at the 0.20 level, so that we can certainly pool to increase the number of degrees of freedom of the variance estimate before proceeding to further tests, obtaining

$$s^2 = \frac{2.4597 + 0.5377}{14} = 0.213$$

(2) $\beta_{\theta c} = 0, \gamma_{\theta t} = 0, \delta_{ct} = 0$. These 3 hypotheses are tested separately, assuming $\sigma^2{}_{\theta ct} = 0$ by forming the variance ratios

$$\frac{0.310}{0.213} = 1.46 \qquad (F_{2,\,14,\,0.20} = 1.8)$$

$$\frac{0.276}{0.213} = 1.29 \qquad (F_{2,\,14,\,0.20} = 1.8)$$

$$\frac{21.2628}{0.213} = 100^{***} \qquad (F_{1,\,14,\,0.001} = 17.1)$$

employing the estimate of s^2 obtained by pooling. We conclude that $\beta_{\theta c} = 0$, $\gamma_{\theta t} = 0$, and that the δ_{ct} are certainly not zero. Before proceeding we can pool the sums of squares due to $\beta_{\theta c}$ and $\gamma_{\theta t}$ with the residual sums of squares. The new estimate of the variance is then

$$s^2 = \frac{2.9974 + 0.6196 + 0.5526}{18} = 0.232$$

(3) $\xi_\theta = 0$. In forming estimates of the temperature effect, the contribution of the interaction terms involving crucibles to the variance estimate is zero, since each has a coefficient $\left(1 - \frac{n_t}{N_t}\right) = \left(1 - \frac{2}{2}\right)$, as we have sampled all possible classes of crucible. The temperature $\times$ coals interaction which does not involve this coefficient is not significant, so that we test the temperature main effect by forming the variance ratio

$$\frac{2.49}{0.232} = 10.7^{***} \; (F_{2,\,18,\,0.001} = 10.4).$$

(4) $\eta_c = 0$. Similar considerations in this case lead to the use of the residual variance estimate in forming the variance ratio

$$\frac{198.78}{0.232} = 839^{***} \qquad (F_{1,\,18,\,0.001} = 15.4)$$

As anticipated, this is revealed as a highly significant main effect.

(5) $\zeta_t = 0$. In this instance the coefficients of the interactions are $2\left(1 - \frac{n_c}{N_c}\right)\left(1 - \frac{n_\theta}{N_\theta}\right)$, $4\left(1 - \frac{n_\theta}{N_\theta}\right)$, and $6\left(1 - \frac{n_c}{N_c}\right)$, and, since the classes of coals which we have employed represent a sample from an infinite population of such classes, while we are interested only in the given temperatures the coefficients reduce to 0, 0, and 6. The coals $\times$ crucibles interaction effects involved in the average value of the mean square term for the crucible effect, is highly significant, so that the variance ratio based on this estimate becomes

$$\frac{198.78}{21.263} = 9.35 \qquad (F_{1,\,1,\,0.20} = 9.5)$$

We draw the following conclusions from the significance tests:

(*a*) As anticipated, the difference between the two coals is highly significant.

(*b*) Variation of the temperature at which the determination is carried out from 825° to 925°C has a significant effect upon the estimate of the volatile matter content of a coal.

(*c*) The influence of the type of crucible upon the volatile matter content varies significantly from coal to coal, so that a constant correction term, or a modification of the temperature at which the determination is carried out, will not suffice to make the results obtained with different types of crucible comparable for all coals sampled.

(*d*) No significant consistent difference could be detected between the mean values of the volatile content of the coals as measured in silica and steel crucibles.

7.62. General Considerations for Nested Classifications

If a classification is to be considered completely crossed, it is implied that the effects corresponding to it are identical in all the other classification considered. In the example of Table 7.31 such a classification involving 3 factors was analyzed. The analysis of nested classifications considered in Section 7.52 may also be extended to cases in which many factors are involved, and each is nested within the previous classification. In a three-factor classification, we should assume that any individual observation $x_{ijk\alpha}$ was randomly and normally distributed with variance σ^2 and expected value

$$\text{ave}\,(x_{ijk\alpha}) = \mu + \xi_i + \beta_{j(i)} + \lambda_{k(ij)}$$

The classifications are nested, and there is no reason to include such terms as η_j, ζ_k, and the various interactions in the model, since such effects cannot exist in the data. The breakdown of the overall sum of squares takes the form

$$S = S_i + S_{j(i)} + S_{k(ij)} + S_{\alpha(ijk)}$$

and this may be presented in the form of an analysis of variance as in Table 7.33, where we have again assumed infinite populations of effects,

TABLE 7.33
ANALYSIS OF VARIANCE

Source of Estimate	Sum of Squares	D.F.	Average Values of Mean Square*
Between i	S_i	$p-1$	$\sigma^2 + n\sigma_\lambda^2 + rn\sigma_\beta^2 + qrn\sigma_\xi^2$
Between j within i	$S_{j(i)}$	$p(q-1)$	$\sigma^2 + n\sigma_\lambda^2 + rn\sigma_\beta^2$
Between k within i and j	$S_{k(ij)}$	$pq(r-1)$	$\sigma^2 + n\sigma_\lambda^2$
Between replicates within i, j, and k	$S_{\alpha(ijk)}$	$pqr(n-1)$	σ^2
Total	S	$pqrn-1$	

* Populations assumed infinite.

or a model of type II. If a model of type III is needed, the average values of the mean squares may be obtained as before. For example, to obtain the average value of the mean square associated with the main effects we have

$$\frac{1}{p-1} \text{ ave } \Sigma_i qrn\,(\bar{x}_{i..} - \bar{x})^2$$

$$= \frac{1}{p-1} \text{ ave } qrn\,\Sigma_i \left(\mu + \xi_i + \frac{1}{q}\Sigma_j\beta_{j(i)} + \frac{1}{qr}\Sigma_{jk}\lambda_{k(ij)} + \frac{1}{qrn}\Sigma_{jk\alpha}\varepsilon_{ijk\alpha} - \mu - \frac{1}{p}\Sigma_i\xi_i - \frac{1}{pq}\Sigma_{ij}\beta_{j(i)} - \frac{1}{pqr}\Sigma_{ijk}\lambda_{k(ij)} - \frac{1}{pqrn}\Sigma_{ijk\alpha}\varepsilon_{ijk\alpha}\right)^2$$

$$= \frac{qrn}{p-1}\left[\text{ave } \Sigma_i\left(\xi_i - \frac{1}{p}\Sigma_i\xi_i\right)^2 + \frac{1}{q^2}\text{ ave } \Sigma_i\left(\Sigma_j\beta_{j(i)} - \frac{1}{p}\Sigma_{ij}\beta_{j(i)}\right)^2 + \frac{1}{q^2r^2}\text{ ave } \Sigma_i\left(\Sigma_{jk}\lambda_{k(ij)} - \frac{1}{p}\Sigma_{ijk}\lambda_{k(ij)}\right)^2 + \frac{1}{q^2r^2n^2}\text{ ave } \Sigma_i\left(\Sigma_{jk\alpha}\varepsilon_{ijk\alpha} - \frac{1}{p}\Sigma_{ijk\alpha}\varepsilon_{ijk\alpha}\right)^2\right]$$

$$= qrn\,\sigma_\xi^2 + \frac{rn}{q}\text{ ave } (\Sigma_j\beta_{j(i)})^2 + \frac{n}{qr}\text{ ave } (\Sigma_{ij}\lambda_{k(ij)})^2 + \sigma^2$$

For models of type III it is shown in Appendix 7*A* that

$$\text{ave } (\Sigma_j \beta_{j(i)})^2 = q\left(1 - \frac{q}{Q}\right)\sigma_\beta^2$$

$$\text{ave } (\Sigma_{jk} \lambda_{k(ij)})^2 = qr\left(1 - \frac{r}{R}\right)\sigma_\lambda^2$$

Thus the average value of the mean square is

$$qrn\sigma_\xi^2 + rn\left(1 - \frac{q}{Q}\right)\sigma_\beta^2 + n\left(1 - \frac{r}{R}\right)\sigma_\lambda^2 + \sigma^2$$

The average values of all the mean squares for the three-way nested classifications, assuming a model of type III, are

Mean Square	Average Value
s^2	σ^2
$s^2_{k(ij)}$	$\sigma^2 + n\sigma_\lambda^2$
$s^2_{j(i)}$	$\sigma^2 + n\left(1 - \frac{r}{R}\right)\sigma_\lambda^2 + nr\sigma_\beta^2$
s_i^2	$\sigma^2 + n\left(1 - \frac{r}{R}\right)\sigma_\lambda^2 + nr\left(1 - \frac{q}{Q}\right)\sigma_\beta^2 + nrq\sigma_\xi^2$
$(\bar{x} - \mu_0)^2$	$\sigma^2 + n\left(1-\frac{r}{R}\right)\sigma_\lambda^2 + nr\left(1-\frac{q}{Q}\right)\sigma_\beta^2 + nrq\left(1-\frac{p}{P}\right)\sigma_\xi^2 + nrqp(\mu - \mu_0)^2$

For computational purposes the sums and squares in Table 7.33 are obtained as

$$S_i = C_i - C$$
$$S_{j(i)} = C_{ij} - C_i$$
$$S_{k(ij)} = C_{ijk} - C_{ij}$$
$$S_{\alpha(ijk)} = C_{ijk\alpha} - C_{ijk}$$

The number of subtables required in this case is fewer since a number of effects and interactions are absent.

EXAMPLE. As an example of a completely nested classification, consider the following data which were obtained as part of a statistical study on the variability of a number of properties of crude smoked rubber. The

Supplier	*A*								*B*								*C*								*D*							
Batch	I		II		III		IV		I		II		III		IV		I		II		III		IV		I		II		III		IV	
Mix	1	2	1	2	1	2	1	2	1	2	1	2	1	2	1	2	1	2	1	2	1	2	1	2	1	2	1	2	1	2	1	2

figures given in Table 7.34 represent measurements of the modulus at 700% elongation made according to the accompanying experimental design involving 32 samples. On each of these samples 3 determinations of the modulus at 700% elongation were made, resulting in a total of 96 observations. It should be noted that the allocation of identification numbers to the batches, samples, and replicate determinations was a random operation and does *not* represent the chronological order in which the operations were carried out.

TABLE 7.34

MODULUS OF ELASTICITY AT 700% ELONGATION OF 96 PREPARED SPECIMENS OF SMOKED SHEET RUBBER

Testing Mix A.C.S. Curing Time, 120 min

[Coding: Results × 10 = lb/sq in.]

Supplier	*A*		*B*		*C*		*D*	
Sample mix	1	2	1	2	1	2	1	2
Batch I	211 215 197	171 198 268	196 186 190	196 210 156	200 221 198	240 229 217	323 279 251	262 234 249
Batch II	229 196 200	234 210 226	209 193 204	200 186 196	191 189 186	196 198 175	255 235 223	249 247 239
Batch III	204 221 238	225 215 196	204 165 194	174 172 171	211 197 210	196 184 190	228 250 260	262 227 272
Batch IV	229 250 238	248 249 249	198 209 221	202 211 204	196 197 186	180 166 172	273 241 221	273 256 230

In analyzing the data we shall employ a notation such that $x_{ijk\alpha}$, $i = 1, \cdots, 4$, $j = 1, \cdots, 4$, $k = 1$ or 2, $\alpha = 1, \cdots, 3$ represent the αth replicate determination from sample k which in turn derives from batch j of supplier i. We assume that the observations $x_{ijk\alpha}$ are normally and randomly distributed with variance σ^2 and mean value

$$\text{ave}\,(x_{ijk\alpha}) = \mu + \xi_i + \beta_{j(i)} + \lambda_{k(ij)}$$

As in the previous example, we compute a series of subtables by summation. Summing Table 7.34 over α (the replicate determinations) we obtain Subtable (a).

SUBTABLE (a)
TYPICAL ELEMENT T_{ijk}

Supplier		*A*		*B*		*C*		*D*	
Sample mix		1	2	1	2	1	2	1	2
Batch	I	623	637	572	562	619	686	853	745
	II	625	670	606	582	566	569	713	735
	III	663	636	563	517	618	570	738	761
	IV	717	746	628	617	579	518	735	759

Summing Subtable (a) over k (the sample mixes) gives Subtable (b), and summation of this subtable over j (the batches) leads to the single-row

SUBTABLE (b)
TYPICAL ELEMENT $T_{ij\cdot}$

Supplier of shipment		*A*	*B*	*C*	*D*
Batch	I	1260	1134	1305	1598
	II	1295	1188	1135	1448
	III	1299	1080	1188	1499
	IV	1463	1245	1097	1494

table, Subtable (c). The grand total of 20,728 $= \Sigma_i T_{i\cdot\cdot} = T$ is identical for all the tables and serves as a check on the addition.

SUBTABLE (c)
TYPICAL ELEMENT $T_{i\cdot\cdot}$

Supplier of shipment	*A*	*B*	*C*	*D*
	5317	4647	4725	6039

Computing the sums of squares of the elements of the subtables, and the components of the overall sum of squares, we obtain

$$C_{ijk\alpha} = \Sigma_{ijk\alpha} x^2_{ijk\alpha} = 4{,}564{,}558.0 \qquad S = C_{ijk\alpha} - \frac{T^2}{N} = 89{,}037.3$$

$$C_i = \frac{1}{4.2.3} \Sigma_i T_i^2 = 4{,}527{,}510.2 \qquad S_i = C_i - \frac{T^2}{N} = 51{,}989.5$$

$$C_{ij} = \frac{1}{2.3} \Sigma_{ij} T^2_{ij} = 4{,}540{,}244.7 \qquad S_{j(i)} = C_{ij} - C_i = 12{,}734.5$$

$$C_{ijk} = \frac{1}{3} \Sigma_{ijk} T^2_{ijk} = 4{,}545{,}324.7 \qquad S_{k(ij)} = C_{ijk} - C_{ij} = 5{,}080.0$$

$$C = \frac{T^2}{N} = 4{,}475{,}520.7 \qquad S_{\alpha(ijk)} = C_{ijk\alpha} - C_{ijk} = 19{,}233.4$$

The components of the sums of squares are assembled into the analysis of variance Table 7.35.

TABLE 7.35

Source	Sum of Squares	D.F.	Mean Square	Average Value of Mean Square
Between suppliers	$S_i = 51{,}989.5$	3	17,329.8	$\sigma_D^2 + 3\sigma_M^2 + 6\sigma_B^2 + 24\sigma_S^2$
Between batches within suppliers	$S_{j(i)} = 12{,}734.5$	12	1,061.2	$\sigma_D^2 + 3\sigma_M^2 + 6\sigma_B^2$
Between mixes within batches	$S_{k(ij)} = 5{,}080.0$	16	317.5	$\sigma_D^2 + 3\sigma_M^2$
Between replicates	$S_{\alpha(ijk)} = 19{,}233.3$	64	300.5	σ_D^2
Total	$S = 89{,}037.3$	95		

In analyzing the data we shall suppose that in all cases except the suppliers' classification the formation of a nested class is achieved by sampling from a population which is sufficiently large to be considered infinite, so that factors of the type $\left(1 - \frac{p}{P}\right)$ which should appear in the estimates column become unity. The classification according to supplier is rather different, since the number of possible suppliers is not large. If we wish to determine the relative merits of shipments from various suppliers, then an estimate of σ_S^2 will not be of interest. If, however, we are obliged to accept material from all suppliers, we might wish to estimate σ_S^2 in order to assess the contribution of this factor to the overall variability of the determination. Such an estimate would in this case be based on only 3 degrees of freedom and would be correspondingly unreliable.

Estimation. The estimates of the components of the mean squares are obtained from the last two columns of the table and prove to be

Component	σ_D^2	σ_M^2	σ_B^2
Estimate	300.5	5.7	124

The effects of the various suppliers may be obtained from Subtable (*c*) as $\xi_i = \dfrac{T_{i..}}{2\cdot3\cdot4} - \dfrac{T}{2\cdot3\cdot4\cdot4}$.

Supplier	*A*	*B*	*C*	*D*
Estimated effects	5.62	−22.29	−19.04	35.71

Significance Tests.

(1) $\sigma_M^2 = 0$. We test this hypothesis by forming the variance ratio

$$F = \frac{317.5}{300.5} = 1.06 \qquad (F_{16,\,64,\,0.20} = 1.35)$$

Since this ratio is clearly non-significant we may accept the hypothesis for testing purposes and derive a better estimate of the variance σ_D^2 by pooling. This gives

$$s^2 = \frac{19{,}233.3 + 5080.0}{80} = 303.9$$

(2) $\sigma_B^2 = 0$. Testing this hypothesis we form the variance ratio

$$\frac{1061.2}{303.9} = 3.49^{***} \qquad (F_{12,\,80,\,0.001} = 3.2)$$

which is highly significant.

(3) $\sigma_S^2 = 0$. Since σ_B^2 is not zero the appropriate variance ratio in this instance is

$$\frac{17{,}329.8}{1061.2} = 16.3^{***} \qquad (F_{3,\,12,\,0.001} = 10.8)$$

which indicates that the differences between suppliers are highly significant. In order to determine which shipments gave rise to the differences we adopt the method of Section 7.35 and rank the class effects.

Class effects	−22.29		−19.04		5.62		35.71
Difference		3.25		24.66		30.09	

The estimated variance of a single observation on material from a given supplier is 1061.2, so that the variance of the difference of the means of 24 such observations is estimated as $s^2 = \dfrac{1061.2 \times 2}{24} = 88.43$. This estimate is based on 12 degrees of freedom, so that a significant separation of the class means is given by $t_{12,\,\alpha}\sqrt{s^2}$, which for $\alpha = 0.05$ becomes 20.5. The second and third differences are thus significant.

Conclusions. In terms of the original objectives of the investigation our conclusions might be

(1) In spite of variability introduced during the preparation and testing of the sample, significant differences can be detected between the performance of samples from the 4 shipments. The samples from shipment D give significantly greater moduli than those from shipment A, which in turn exceed the results from B and C. (Note: as we have no evidence concerning the variation from shipment to shipment of a given supplier, we do not know whether this ranking is a characteristic of the suppliers or of the particular shipments examined.)

(2) In the examination of a given shipment the contribution of the testing procedure to the overall variance is estimated to be $2\frac{1}{2}$ times as great as the component due to variation between batches of rubber. The contribution of mixing and subsampling procedure is negligible.

(3) If cost figures for the various operations are available, it is possible to devise a sampling procedure for a given shipment which for a given accuracy will minimize the sampling and testing costs. This aspect is considered in Section 8.38.

7.63. Analyses Involving Both Types of Classification

It is sometimes convenient to arrange an experiment in such a form that the analysis of variance will provide information on the main effects and interactions of a number of treatments, together with estimates of the variation introduced by the operations of the sampling and analytical procedure. Such a design involves crossed and nested classifications, and the method of analysis depends upon the sequence in which the criteria of classification occur.

We consider first the case in which the products of the various treatments are to be sampled, so that the classifications which correspond to the stages of the sampling procedure are nested within a single cell of the crossed treatment classifications. An example of this type would arise in an investigation of the effect of q treatments upon the sulphur content of p blends of material when the final determinations were made by first selecting r samples from each product, then crushing the individual samples and selecting n laboratory subsamples from each.

In analyzing the data from such an experiment we wish to estimate and test the main effects due to blend and treatment differences, and any interactions of these factors. Since the samples and subsamples would be selected by a random method, no systematic effects are anticipated in these classifications and we assume, as a model, that any observation $x_{ijk\alpha}$ is randomly and normally distributed with variance σ^2 and average value

$$\text{ave}\,(x_{ijk\alpha}) = \mu + \xi_i + \eta_j + \beta_{ij} + \lambda_{k(ij)}$$

where ξ_i and η_j represent the main effects of blends and treatments, β_{ij} represent the interaction effects of these factors, and $\lambda_{k(ij)}$ the random effects between the samples obtained from any cell ij. The breakdown of the sum of squares which provides variance estimates corresponding to these terms is

$$S = S_i + S_j + S_{ij} + S_{k(ij)} + S_{\alpha(ijk)}$$

The formal analysis of variance together with the average values of the variance estimates is given in Table 7.36, where the populations are assumed infinite. This differs from a sample $p \times q$ crossed classification with $N = rn$ replicates only in the composition of the residual variance estimate. The latter has been subdivided in order to provide separate estimates of the components due to bulk sampling and to laboratory methods, and we have

$$\text{estimate } \sigma^2 = s^2 = S_{\alpha(ijk)}/pqr(n-1)$$

$$\text{estimate } \sigma_\lambda^2 = s_\lambda^2 = \frac{1}{n}\,[S_{k(ij)}/pq(r-1) - S_{\alpha(ijk)}/pqr(n-1)]$$

If the sampling procedure in the above experiment had involved a greater number of stages of which the contributions to the experimental variance were to be determined, the model could be elaborated by the inclusion of additional effects (e.g., $\omega_{l(ijk)}$), and the component of the sum of squares obtained from observations within a given cell of the crossed classification would then be further subdivided to provide variance estimates corresponding to the new effects.

TABLE 7.36

Source of Estimate	Sum of Squares	D.F.	Average Value of Mean Square*
Between blends	S_i	$p - 1$	$\sigma^2 + n\sigma_\lambda^2 + rn\sigma_\beta^2 + prn\sigma_\xi^2$
Between treatments	S_j	$q - 1$	$\sigma^2 + n\sigma_\lambda^2 + rn\sigma_\beta^2 + qrn\sigma_\eta^2$
Blends × treatments interaction	S_{ij}	$(p-1)(q-1)$	$\sigma^2 + n\sigma_\lambda^2 + rn\sigma_\beta^2$
Between samples (within blends and treatments)	$S_{k(ij)}$	$pq(r-1)$	$\sigma^2 + n\sigma_\lambda^2$
Between replicates (within samples, blends, and treatments)	$S_{\alpha(ijk)}$	$pqr(n-1)$	σ^2
Total	S	$pqrn - 1$	

* Assuming infinite populations.

If the blends on which the above experiment was to be carried out were prepared from raw materials which were heterogeneous, the behavior of supposedly similar blends under fixed treatment conditions might be subject to considerable variations. If only a single sample of each blend were included in the tests, it would not be possible to determine whether a significantly high variance estimate between blends indicated that the blends differed systematically in behavior, or whether the effect could be explained by the variation between specimens of a given blend. To resolve this difficulty the experiment could be designed to provide an estimate of this latter variation by examining a number of specimens of each blend. We suppose that the previous investigation is modified, q

TABLE 7.37

Source of Estimate	Sum of Squares	D.F.	Average Value of Mean Square*
Between blends	S_i	$p - 1$	$\sigma^2 + n\sigma_\omega^2 + sn\sigma_\lambda^2 + qsn\sigma_\beta^2 + rsn\sigma_\eta^2 + qrsn\sigma_\xi^2$
Between specimens (within blends)	$S_{j(i)}$	$p(q - 1)$	$\sigma^2 + n\sigma_\omega^2 + sn\sigma_\lambda^2 + rsn\sigma_\eta^2$
Between treatments	S_k	$(r - 1)$	$\sigma^2 + n\sigma_\omega^2 + sn\sigma_\lambda^2 + qsn\sigma_\beta^2 + pqsn\sigma_\zeta^2$
Blends × treatment interaction	S_{ik}	$(p - 1)(r - 1)$	$\sigma^2 + n\sigma_\omega^2 + sn\sigma_\lambda^2 + qsn\sigma_\beta^2$
Specimens × treatment interaction (within blends)	$S_{jk(i)}$	$p(q - 1)(r - 1)$	$\sigma^2 + n\sigma_\omega^2 + sn\sigma_\lambda^2$
Between samples (within blends, specimens, and treatments)	$S_{l(ijk)}$	$pqr(s - 1)$	$\sigma^2 + n\sigma_\omega^2$
Between replicates (within blends, specimens, treatments, and samples)	$S_{\alpha(ijkl)}$	$pqrs(n - 1)$	σ^2
Total	S	$pqrsn - 1$	

* Assuming infinite populations.

specimens of each of p blends being examined after subjection to r treatments. From each of the pqr products s bulk samples are withdrawn and crushed, n laboratory determinations being carried out on each sample. We assume that the analytical determinations are randomly and normally distributed with variance σ^2 and average value,

$$\text{ave}\,(x_{ijkl}) = \mu + \xi_i + \eta_{j(i)} + \zeta_k + \beta_{ik} + \lambda_{jk(i)} + \omega_{l(ijk)}$$

The effects $\eta_{j(i)}$ correspond to differences between specimens of a given blend i. Since the specimens are randomly selected, there can be no relationship between the effects of the ith specimen in each of the p blends, and hence no main effect η_j is required in the model. The treatment factor is crossed with all the preceding classifications so that the model includes a main effect ζ_k and interactions β_{ik} and $\lambda_{jk(i)}$. The latter term corresponds to possible variations in the response of different specimens from the ith blend to the various treatments $1, \cdots, r$. To obtain variance estimates corresponding to each of these effects the breakdown of the sum of squares is

$$S = S_i + S_{j(i)} + S_k + S_{ik} + S_{jk(i)} + S_{l(ijk)} + S_{\alpha(ijkl)}$$

and from this the analysis of Table 7.37 may be derived.

For the purpose of significance testing the analysis of variance may be divided into 2 parts. Since all treatments are applied to each blend specimen those effects or interactions involving treatments may be estimated from the results obtained within specimens and are not affected by random variations η_j. The blend effects are estimated by determining the mean response over all specimens of a given blend, so that the average value of the corresponding variance estimate involves the component σ_η^2. In this type of design, which represents a special case of the split-plot experiment, to be considered in Chapter 8, the main effects of the 2 crossed factors are estimated with different levels of accuracy.

If the model selected for the mixed classification is of type III, it is often tedious to derive the average values of the mean squares by the full procedure of writing the observations in terms of the model. A useful device is illustrated by reference to the example of Table 7.37. This consists of a table on which the various populations involved are written symbolically and along which are written the components of the model, with the subscripts bracketed when nesting is involved. Thus for the example of Table 7.37 we have the accompanying table.

	i	j	k	l	α
$\varepsilon_{\alpha(ijkl)}$	1	1	1	1	$\left(1 - \frac{n}{N}\right)$
$\omega_{l(ijk)}$	1	1	1	$\left(1 - \frac{s}{S}\right)$	n
$\lambda_{jk(i)}$	1	$\left(1 - \frac{q}{Q}\right)$	$\left(1 - \frac{r}{R}\right)$	s	n
β_{ik}	$\left(1 - \frac{p}{P}\right)$	q	$\left(1 - \frac{r}{R}\right)$	s	n
ζ_k	p	q	$\left(1 - \frac{r}{R}\right)$	s	n
$\eta_{j(i)}$	1	$\left(1 - \frac{q}{Q}\right)$	r	s	n
ξ_i	$\left(1 - \frac{p}{P}\right)$	q	r	s	n

The rules for filling in the table are as follows:

(1) In any row write the number of classes under any column which corresponds to a letter not present in the subscript of the effect being considered.

(2) In any row write unity under any column which corresponds to a letter bracketed in the subscript of the row effect.

(3) In any row write in the remaining columns $\left(1 - \frac{\pi}{\Pi}\right)$ where π is the number of classes of the column heading and Π the number of classes in the corresponding population. In computing the average value of the mean square corresponding to the ξ_i effects, for instance, the components present will be all of those in the table corresponding to effects with i in the subscript. The coefficients for the variance components will be the product of all the numbers in the corresponding rows, excluding the numbers in column i. We obtain

$$qrsn\sigma_\xi^2 + rsn\left(1 - \frac{q}{Q}\right)\sigma_\eta^2 + qsn\left(1 - \frac{r}{R}\right)\sigma_\beta^2 + sn\left(1 - \frac{q}{Q}\right)\left(1 - \frac{r}{R}\right)\sigma_\lambda^2$$
$$+ n\left(1 - \frac{s}{S}\right)\sigma_\omega^2 + \left(1 - \frac{n}{N}\right)\sigma^2$$

Similarly for the average mean square corresponding to the $\eta_{j(i)}$ effects we exclude columns i and j and obtain

$$rsn\sigma_{\eta}^2 + sn\left(1 - \frac{r}{R}\right)\sigma_{\lambda}^2 + n\left(1 - \frac{s}{S}\right)\sigma_{\omega}^2 + \left(1 - \frac{n}{N}\right)\sigma^2$$

This method may be applied to crossed, nested, or mixed classifications in which the populations are of types I, II, or III.

If the number of factors of classification is increased, it is possible to arrange the sequence of nested and crossed classifications in a large number of ways, each of which involves a different arrangement of the analysis of variance. Examples of the use of designs of this type in chemical investigation have been published [44], [45], and we consider as an illustration a five-factor experiment carried out during an investigation of the physical properties of a number of samples prepared from crude smoked sheet rubber.

EXAMPLE. The data of Table 7.38 represent the sums of triplicate determinations of the 700% modulus of specimens prepared according to the experimental outline of Figure 7.2 from 4 shipments of rubber. From a given shipment 4 bales were selected, and, after cutting, 50-lb samples were withdrawn from each. A sample of this size was not convenient for laboratory treatment, and the weight was reduced by a representative sampling procedure before blending. The blending procedure was carefully controlled, and from each of the samples representing individual bales or "batches" 2 subsamples were withdrawn for processing. The latter were used to prepare standard mixes containing 0.5 parts per 100 of stearic acid, and further subdivided into portions which were cured in this form and portions receiving an additional 2.5 parts of stearic acid before curing. Alternative curing times of 60 and 120 min were employed for each type of mix, the product in each case being sufficient to enable 3 determinations of the 700% modulus to be made. The sequence of all operations was deliberately randomized so that any time trends during the investigation should not be confused with the effects which the experiment was designed to examine.

The "batch" classification in this experiment is nested within the main shipment classification, and "samples" are nested within "batches." The classification according to the proportion of stearic acid incorporated in the mix is crossed with "shipments," "batches," and "samples," and the "cures" classification is crossed with all preceding classifications. In writing down a model for these data we find it convenient to change our notation, using the symbol e to denote any effect and identifying particular effects by subscripts. Considering only the first 3 factors of classification,

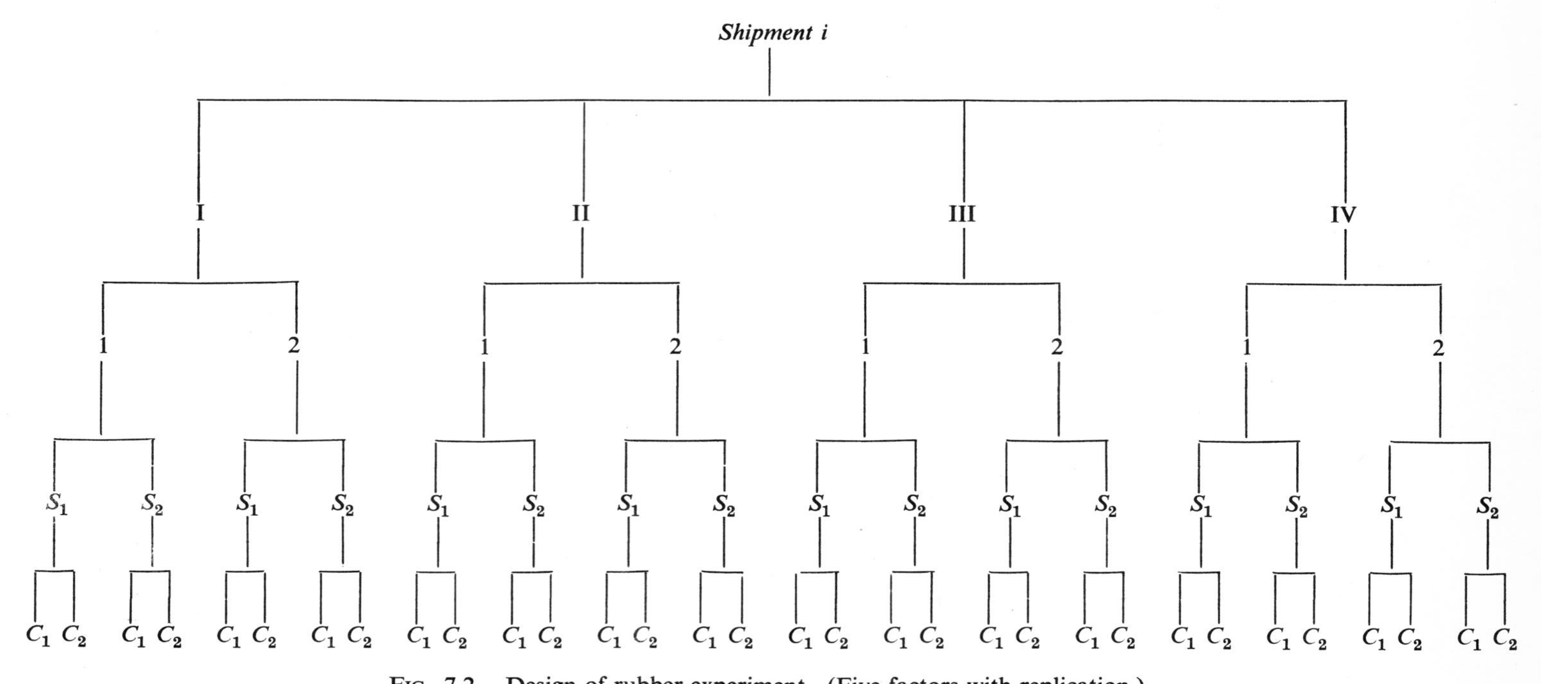

FIG. 7.2. Design of rubber experiment. (Five factors with replication.)
I, II, III, IV, batches (j); 1, 2 samples (k); S_1S_2 stearic levels (l); C_1C_2 cure levels (m).

namely shipments (i), batches (j), and samples (k), we describe the average value of an observation as

$$\text{ave}\,(\bar{x}_{ijk}\ldots) = e_0 + e_i + e_{j(i)} + e_{k(ij)}$$

Since the "stearic" classification (l) is crossed with the factors i, j, and k, we have

$$\text{ave}\,(\bar{x}_{ijkl}..) = e_0 + e_i + e_{j(i)} + e_{k(ij)} + e_l + e_{il} + e_{jl(i)} + e_{kl(ij)}$$

and for the complete model including the crossed classification "cures" (m)

$$\text{ave}\,(\bar{x}_{ijklm}.) = e_0 + e_i + e_{j(i)} + e_{k(ij)} + e_l + e_{il} + e_{jl(i)} + e_{kl(ij)} + e_m$$
$$+ e_{im} + e_{jm(i)} + e_{km(ij)} + e_{lm} + e_{ilm} + e_{jlm(i)} + e_{klm(ij)}$$

We assume that the 3 replicate determinations of the modulus for each cell are independently and normally distributed with the above average value and variance σ^2. The breakdown of the sums of squares takes the form

$$S = S_i + S_{j(i)} + S_{k(ij)} + S_l + S_{il} + S_{jl(i)} + S_{kl(ij)} + S_m + S_{im}$$
$$+ S_{jm(i)} + S_{km(ij)} + S_{lm} + S_{ilm} + S_{jlm(i)} + S_{klm(ij)} + S_{\alpha(ijklm)}$$

It is not possible to determine the overall sum of squares from the data of Table 7.38 since each figure recorded is actually the sum of 3 replicate observations. We quote from the original observations

$$C_{ijklm\alpha} = \Sigma_{ijklm\alpha} x^2{}_{ijklm\alpha} = 2{,}009{,}934{,}764$$

Bartlett's test does not indicate any heterogeneity in the residual variances.

As a first step in the analysis we compute the necessary subtables by summation. Since the batch and sample factors are nested within suppliers it is not necessary to compute all the possible subtables, the ones which need to be computed being determined by the proposed subdivision of the sum of squares. In this case the required subtables are those having the same subscripts as the components of the sums of squares, i.e., T_i, T_{ij}, T_{ijk}, T_l, T_{il}, T_{ijl}, T_{ijkl}, T_m, T_{im}, T_{ijm}, T_{ijkm}, T_{lm}, T_{ilm}, T_{ijlm}, T_{ijklm}, together with the grand total T. Since the list is extensive, only those subtables concerned with the effects and interactions of suppliers, stearic additions. and cures are reproduced.

TABLE 7.38

SUMS OF REPLICATE DETERMINATIONS

Shipment		A				B				C				D			
Sample		1		2		1		2		1		2		1		2	
Stearic Addition		S_1	S_2	S_1	S_2	S_1	S_2	S_1	S_2	S_1	S_2	S_1	S_2	S_1	S_2	S_1	S_2
Batch I	C_1	3810	6340	4080	6660	3004	5470	2833	6220	4210	7640	3910	6510	3810	8,100	4570	8,310
	C_2	6230	9050	6370	8650	5720	8390	5620	7300	6190	9720	6860	9960	8530	10,210	7450	10,650
Batch II	C_1	4350	6450	3800	6920	3380	6330	3308	6280	3310	6040	3302	6670	4380	8,010	4860	8,100
	C_2	6250	9090	6700	9000	6060	8770	5820	9180	5660	8540	5690	8720	7130	10,170	7350	9,430
Batch III	C_1	3690	6510	3590	6510	2758	5740	2945	6950	3445	6690	3273	5980	4160	7,370	4670	7,380
	C_2	6630	8680	6360	8440	5630	8200	5170	8310	6180	8060	5700	8210	7380	9,560	7610	9,400
Batch IV	C_1	4390	7470	4700	7210	3810	6830	3320	6020	2684	6010	3080	5630	5270	8,010	4640	7,540
	C_2	7170	9370	7460	8800	6280	9240	6170	8680	5790	7980	5180	7960	7350	10,510	7590	10,420

TYPICAL ELEMENT T_{ilm}

Supplier	A		B		C		D	
Stearic addition	S_1	S_2	S_1	S_2	S_1	S_2	S_1	S_2
Cure C_1	32,410	54,070	25,358	48,840	27,214	51,170	36,360	62,820
C_2	53,170	71,080	46,470	68,070	47,250	69,150	60,390	80,350

TYPICAL ELEMENT T_{il}

Supplier	A		B		C		D	
Stearic addition	S_1	S_2	S_1	S_2	S_1	S_2	S_1	S_2
Subtotal	85,580	125,150	71,828	117,910	74,464	120,320	96,750	143,170

TYPICAL ELEMENT T_{im}

Supplier	A		B		C		D	
Cure	C_1	C_2	C_1	C_2	C_1	C_2	C_1	C_2
Subtotal	86,480	124,250	75,198	114,540	78,384	116,400	99,180	140,740

TYPICAL ELEMENT T_{lm}

Stearic	S_1	S_2
Cure C_1	121,342	217,900
Cure C_2	207,280	288,650

TYPICAL ELEMENT T_i

Supplier	A	B	C	D
Subtotal	210,730	189,738	194,784	239,920

TYPICAL ELEMENT T_l

Stearic	S_1	S_2
Subtotal	328,622	506,550

TYPICAL ELEMENT T_m

Cure	C_1	C_2
Subtotal	339,242	495,930

Grand Total: 835,172

From the sums of squares of the subtotals in any table we obtain, on dividing by the number of observations which have been summed to provide these subtotals, Table 7.39.

TABLE 7.39

$\frac{T^2}{N} = C$	1,816,438,202	C_{im}	1,896,425,732
C_i	1,832,396,402	C_{ilm}	1,979,923,340
C_{ij}	1,837,266,513	C_{ijl}	1,920,568,355
C_{ijk}	1,837,688,829	C_{ijkl}	1,921,488,805
C_l	1,898,881,882	C_{ijm}	1,901,674,269
C_m	1,880,373,434	C_{ijkm}	1,902,920,302
C_{lm}	1,963,417,831	C_{ijlm}	1,986,250,397
C_{ilm}	1,915,176,868	C_{ijklm}	1,988,677,390

To compute the components of the sum of squares we use equations of the type

$$S_{il} = C_{il} - C_i - C_l - C$$

$$S_{j(i)} = C_{ij} - C_i$$

$$S_{kl(ij)} = C_{ijkl} - C_{ijk} - C_{ijl} + C_{ij}$$

In cases where some of the subscripts are bracketed, e.g., $S_{kl(ij)}$, the equations are easily obtained by first writing out the corresponding expression for the case when the bracketed subscripts are ignored, the ignored subscripts being subsequently introduced in the right-hand side of the equation without brackets, and in the left-hand side with brackets.

The components of the sum of squares are given in Table 7.40. The analysis of variance table is formed in the usual way, the degrees of

TABLE 7.40

$\frac{T^2}{N} = C$	1,816,438,202	S_{im}	94,098
S_i	15,958,200	S_{ilm}	116,425
$S_{j(i)}$	4,870,111	$S_{jl(i)}$	521,376
$S_{k(ij)}$	422,316	$S_{kl(ij)}$	498,134
S_l	82,443,680	$S_{jm(i)}$	378,426
S_m	63,935,232	$S_{km(ij)}$	823,717
S_{lm}	600,717	$S_{jlm(i)}$	557,144
S_{il}	336,786	$S_{klm(ij)}$	682,826

freedom associated with any component of the sum of squares being determined from the subscripts. Thus with $i = 1, \cdots, 4$, $j = 1, \cdots, 4$, $k = 1$ or 2, $l = 1$ or 2, $m = 1$ or 2, the degrees of freedom associated with $S_{j(i)}$ are $3 \cdot 4 = 12$, and with $S_{klm(ij)}$ are $1 \cdot 1 \cdot 1 \cdot 4 \cdot 4 = 16$.

TABLE 7.41

Source of Estimate	Sum of Squares		D.F.	Mean Square
Between suppliers	S_i	15,958,200	3	5.319×10^6
Between batches (within suppliers)	$S_{j(i)}$	4,870,111	12	4.058×10^5
Between samples (within suppliers and batches)	$S_{k(ij)}$	422,316	16	2.640×10^4
Between stearic levels	S_l	82,443,680	1	8.244×10^7
Between cures	S_m	63,935,232	1	6.394×10^7
Stearic × cures	S_{lm}	600,717	1	6.007×10^5
Suppliers × stearic	S_{il}	336,786	3	1.123×10^5
Suppliers × cures	S_{im}	94,098	3	3.137×10^4
Suppliers × stearic × cures	S_{ilm}	116,425	3	3.721×10^4
Stearic × batches (within suppliers)	$S_{jl(i)}$	521,376	12	4.345×10^4
Stearic × samples (within batches and suppliers)	$S_{kl(ij)}$	498,134	16	3.113×10^4
Cures × batches (within suppliers)	$S_{jm(i)}$	378,426	12	3.154×10^4
Cures × samples (within batches and suppliers)	$S_{km(ij)}$	823,717	16	5.148×10^4
Stearic × cures × batches (within suppliers)	$S_{jlm(i)}$	557,144	12	4.643×10^4
Stearic × cures × samples (within batches and suppliers)	$S_{klm(ij)}$	682,826	16	4.268×10^4
Residual	$S_{\alpha(ijklm)}$	9,117,374	255*	3.575×10^4
Total	S	181,356,562		

* One observation lost and replaced by cell mean.

In order to determine the average values of the mean square estimate it is necessary to consider the nature of the populations in our previous model. It is possible that the "suppliers" classes represent a random sample from a population, but the levels of stearic acid additions and curing times were certainly not obtained by a random process. It seems best therefore to treat these latter populations as if they consisted only of the levels examined. The repeated determinations, the samples, the batches, and the suppliers which were examined we shall regard as random samples from a population which is large in comparison with the sample size. Using the tabular method to determine the coefficients of the variance components, we have the accompanying table.

	i	j	k	l	m	α
$e_{\alpha(ijklm)}$	1	1	1	1	1	1
$e_{klm(ij)}$	1	1	1	0	0	3
$e_{jlm(i)}$	1	1	2	0	0	3
$e_{km(ij)}$	1	1	1	2	0	3
$e_{jm(i)}$	1	1	2	2	0	3
$e_{kl(ij)}$	1	1	1	0	2	3
$e_{jl(i)}$	1	1	2	0	2	3
e_{ilm}	1	4	2	0	0	3
e_{im}	1	4	2	2	0	3
e_{il}	1	4	2	0	2	3
e_{lm}	4	4	2	0	0	3
e_{m}	4	4	2	2	0	3
e_{l}	4	4	2	0	2	3
$e_{k(ij)}$	1	1	1	2	2	3
$e_{j(i)}$	1	1	2	2	2	3
e_{i}	1	4	2	2	2	3

TABLE 7.42

Source of Estimate	D.F.	Mean Square	Average Value of Mean Square
Between suppliers	3	(1) 5.319×10^6	$\sigma^2 + 12\sigma^2_{ijk} + 24\sigma^2_{ij} + 96\sigma_i^2$
Between batches (within suppliers)	12	(2) 4.058×10^5	$\sigma^2 + 12\sigma^2_{ijk} + 24\sigma^2_{ij}$
Between samples (within suppliers and batches)	16	(3) 2.640×10^4	$\sigma^2 + 12\sigma^2_{ijk}$
Between stearic levels	1	(4) 8.244×10^7	$\sigma^2 + 6\sigma^2_{ijlm} + 12\sigma^2_{ijl} + 48\sigma^2_{il} + 192\sigma_l^2$
Between cures	1	(5) 6.394×10^7	$\sigma^2 + 6\sigma^2_{ijkm} + 12\sigma^2_{ijm} + 48\sigma^2_{im} + 192\sigma_m^2$
Stearic × cures	1	(6) 6.007×10^5	$\sigma^2 + 3\sigma^2_{ijklm} + 6\sigma^2_{ijlm} + 24\sigma^2_{ilm} + 96\sigma^2_{lm}$
Suppliers × stearic	3	(7) 1.123×10^5	$\sigma^2 + 6\sigma^2_{ijkl} + 12\sigma^2_{ijl} + 48\sigma^2_{il}$
Suppliers × cures	3	(8) 3.137×10^4	$\sigma^2 + 6\sigma^2_{ijkm} + 12\sigma^2_{ijm} + 48\sigma^2_{im}$
Suppliers × stearic × cures	3	(9) 3.721×10^4	$\sigma^2 + 3\sigma^2_{ijklm} + 6\sigma^2_{ijlm} + 24\sigma^2_{ilm}$
Stearic × batches (within suppliers)	12	(10) 4.345×10^4	$\sigma^2 + 6\sigma^2_{ijkl} + 12\sigma^2_{ijl}$
Stearic × samples (within suppliers and batches)	16	(11) 3.113×10^4	$\sigma^2 + 6\sigma^2_{ijkl}$
Cures × batches (within suppliers)	12	(12) 3.154×10^4	$\sigma^2 + 6\sigma^2_{ijkm} + 12\sigma^2_{ijm}$
Cures × samples (within suppliers and batches)	16	(13) 5.148×10^4	$\sigma^2 + 6\sigma^2_{ijkm}$
Stearic × cures × batches (within suppliers)	12	(14) 4.643×10^4	$\sigma^2 + 3\sigma^2_{ijklm} + 6\sigma^2_{ijlm}$
Stearic × cures × samples (within suppliers and batches)	16	(15) 4.268×10^4	$\sigma^2 + 3\sigma^2_{ijklm}$
Residual	255	(16) 3.575×10^4	σ^2

As an illustration of the procedure, the average value of the mean square corresponding to the main effect of curing time may be computed. This will involve variance components due to all effects which include m in the subscript, and the coefficient of these components will be obtained by deleting the m column from the above table and taking the product of the remaining coefficients in the appropriate row. We thus have ave- [M.S. (effect of cures)] $= \sigma^2 + 6\sigma^2_{ijkm} + 12\sigma^2_{ijm} + 48\sigma^2_{im} + 192\sigma_m^2$ and, determining the average values of the remaining means squares in a similar way, we obtain Table 7.42.

Significance Testing. We first examine the hypotheses that the effect of stearic additions and curing time and the interaction effects stearic $\times$ cures are the same for all samples of a given batch and for all batches of a given supplier by testing mean squares (10) to (15) against the residual variance estimate. Individually these mean squares are not significant, and, if we take the total of the corresponding sums of squares, 3,461,623, and divide by the total of the degrees of freedom, we obtain a mean square of 4.121×10^4 and variance ratio of 1.15.

It is also clear that the suppliers $\times$ stearic $\times$ cures and the suppliers $\times$ cures interactions are not significant. The remaining interactions are tested as

SUPPLIERS $\times$ STEARIC

$$\frac{\text{M.S.}_{(7)}}{\text{M.S.}_{(16)}} = 3.14^* \qquad F_{3,\,\infty,\,0.05} = 2.6$$

STEARIC $\times$ CURES

$$\frac{\text{M.S.}_{(6)}}{\text{M.S.}_{(16)}} = 16.78^{***} \qquad F_{1,\,\infty,\,0.001} = 10.8$$

and both prove to be significant.

The main effect of cures must be tested against the residual estimate, and it is obviously highly significant. Since we regard the shipments examined as a random sample from a large population of possible shipments, and since the effects e_{il} are significant, the appropriate test for the significance of the stearic main effect is

$$\frac{\text{M.S.}_{(4)}}{\text{M.S.}_{(7)}} = 734^{***} \qquad F_{1,\,3,\,0.001} = 167$$

Effects between samples are clearly non-significant and the significance of the batch and supplier effects are tested as

$$\frac{\text{M.S.}_{(2)}}{\text{M.S.}_{(16)}} = 11.35^{***} \qquad F_{12,\,\infty,\,0.001} = 2.7$$

$$\frac{\text{M.S.}_{(1)}}{\text{M.S.}_{(2)}} = 13.11^{***} \qquad F_{3,\,12,\,0.001} = 10.8$$

Conclusions. The treatment factors, stearic additions, and curing time, are seen to be highly significant, and their effects can be evaluated from the tables of subtotals. There is an interaction effect in the response to the two treatments, the joint effect being some 9% less than the sum of the individual effects. This type of "diminished return" is a common occurrence where treatment factors designed to increase yields or physical properties of materials are concerned.

Of the interactions between treatment factors and the random effects due to suppliers, batches, or samples only the interaction between suppliers and stearic additions appears significant at the 5% level. If we consider the interactions between treatments and random factors jointly, this significance level is not very convincing, and the most we can say is that the effect of stearic additions may change between shipments from different suppliers. For the 4 shipments examined the differences expressed in terms of subtotals are

	A	*B*	*C*	*D*
Difference $S_2 - S_1$	39,570	46,082	48,738	46,420

and it appears that the material from suppliers *A* appreciates least from the extra stearic additions, whereas the remaining samples appreciate equally. The interaction is sufficiently small so that the ranking of the 4 shipments is the same at high and low levels of the stearic additions. The non-significance of the remaining interactions indicates that, with the possible exception of this interaction, the evaluation of the relative merits of the 4 shipments will be similar, regardless of which of the 4 treatment conditions is applied. It also indicates that the efficiency of a sampling procedure which might be designed on the basis of these results would not depend upon the test conditions selected, within the limits of the 4 test conditions examined.

For the design of such a procedure estimates of the variance components between shipments, batches, and samples are required. The last is negligible, indicating that the stratified sampling method employed in the tests was successful. The variance component between batches is estimated as 3.70×10^5, and, in view of the 255 degrees of freedom of the residual, it will be distributed very nearly as $\dfrac{\sigma^2_{ij}\chi^2_{[12]}}{12}$ so that the appropriate 5% confidence limits will be

$$\frac{s^2_{ij}\chi^2_{0.975}}{12} < \sigma^2_{ij} < \frac{s^2_{ij}\chi^2_{0.025}}{12}$$

or

$$1.4 \times 10^5 < \sigma^2_{ij} < 7.1 \times 10^5$$

$$3.74 \times 10^2 < \sigma_{ij} < 8.34 \times 10^2$$

The variance component between shipments is estimated as 4.91×10^6. If we wish to place confidence limits on this estimate, it is necessary to proceed by the method of Section 7.23, although with only 3 degrees of freedom this hardly seems worth while.

7.7. Application of the Analysis of Variance to Regression

7.71. Simple Linear Regression

The principles of the analysis of variance can be applied to the determination of the significance of the regression of a random variable y on a known variable x. The basis of this analysis for simple linear regression is the assumption (see Section 6.21) that, given n pairs of observations (x_i, y_i), the observations y_i are normally distributed with common variance σ^2 and ave $(y_i) = \alpha + \beta x_i$. This expression can be rewritten as

$$\text{ave}\,(y_i) = (\alpha - \beta\bar{x}) + \beta(x_i - \bar{x}) = \mu + \beta(x_i - \bar{x})$$

where $\mu = \alpha + \beta\bar{x}$ represents the overall mean and the term $\beta(x_i - \bar{x})$ represents the effects due to the linear dependence of ave (y_i) on x_i. The presence of significant effects of the latter type is equivalent to the presence of linear regression.

It was shown in Section 6.22 that the maximum likelihood estimates of β and μ are

$$b = \frac{\Sigma_i(y_i - \bar{y})\,(x_i - \bar{x})}{\Sigma_i(x_i - \bar{x})^2}$$

and

$$\bar{y} = a + b\bar{x}$$

Also, if $\tilde{y}_i$ are the predicted values of y_i, we have the algebraic identity

$$\begin{aligned}\Sigma_i(y_i - \bar{y})^2 &= \Sigma_i(y_i - \tilde{y}_i)^2 + \Sigma_i(\tilde{y}_i - \bar{y})^2 \\ &= \Sigma_i(y_i - \tilde{y}_i)^2 + b^2\,\Sigma_i(x_i - \bar{x})^2\end{aligned}$$

The sum $\Sigma_i(y_i - \bar{y})^2$ has $n - 1$ degrees of freedom, being made up of the squares of normal deviates subject to the single linear restraint $\Sigma_i(y_i - \bar{y}) = 0$. The term $\Sigma_i(y_i - \tilde{y}_i)^2$ is the sum of squares of the deviations from the estimated regression equation, and, since the terms $y_i - \tilde{y}_i$ are subject to 2 independent linear restraints, this expression has $n - 2$ degrees of freedom, and average value $(n - 2)\sigma^2$. Since the x_i are considered to be fixed, the quantity $b^2\,\Sigma_i(x_i - \bar{x})^2$ is the square of a linear combination of normally distributed variables and has a single degree of freedom. From Section 6.22 we know that the estimate b is normally distributed with average value β and variance

$$\sigma_b^2 = \frac{\sigma^2}{\Sigma_i(x_i - \bar{x})^2}$$

Hence, since ave $(b^2) = \sigma_b^2 + (\text{ave}\,(b))^2$,

$$\text{ave}\,[b^2\,\Sigma_i(x_i - \bar{x})^2] = \sigma_b^2\,\Sigma_i(x_i - \bar{x})^2 + \beta^2\,\Sigma_i(x_i - \bar{x})^2$$
$$= \sigma^2 + \beta^2\,\Sigma_i(x_i - \bar{x})^2$$

The corresponding analysis is shown in Table 7.43.

TABLE 7.43

Source of Estimate	Sum of Squares	D.F.	Average Value of Mean Square
Linear regression	$b^2\Sigma_i(x_i - \bar{x})^2$	1	$\sigma^2 + \beta^2\Sigma_i(x_i - \bar{x})^2$
Residual	$\Sigma_i(y_i - \tilde{y}_i)^2$	$n - 2$	σ^2
Total	$\Sigma_i(y_i - \bar{y})^2$	$n - 1$	

If $\beta = 0$, both estimates have average value σ^2, and under these circumstances their ratio is distributed as F, with $(1, n - 2)$ degrees of freedom. This property may be used to test the hypothesis $\beta = 0$. Since $\Sigma_i(y_i - \tilde{y}_i)^2/n - 2 = s^2_{y\cdot x}$ as defined in Section 6.22, the variance ratio $b^2\,\Sigma_i(x_i - \bar{x})^2/s^2_{y\cdot x}$ is the square of the statistic $t_{n-2} = (b - \beta)\sqrt{\Sigma_i(x_i - \bar{x})^2}/s_{y\cdot x}$ if $\beta = 0$. The distributions $F_{1,\,n-2}$ and t^2_{n-2} are, of course, identical so that the significance test may be made by either method. The use of the analysis of variance in regression analysis corresponds to the interpretation of a regression as "explanation" for part of the overall variability of the observations. The quantities required for Table 7.43 may be computed by the identities

$$\Sigma_i(y_i - \bar{y})^2 = \Sigma_i y_i^2 - \frac{T^2_{(y)}}{n}$$

$$b^2\,\Sigma_i(x_i - \bar{x})^2 = b^2\left[\Sigma_i x_i^2 - \frac{T^2_{(x)}}{n}\right] = b\left[\Sigma_i x_i y_i - \frac{T_{(x)}T_{(y)}}{n}\right]$$

and the term $\Sigma_i(y_i - \tilde{y}_i)^2$ may be obtained by difference. $T_{(y)}$ and $T_{(x)}$ are the sums of the y and x measurements, respectively, and

$$b = \frac{n\,\Sigma_i x_i y_i - T_{(x)}T_{(y)}}{n\,\Sigma_i x_i^2 - T^2_{(x)}} = \frac{\Sigma_i(x_i - \bar{x})\,(y_i - \bar{y})}{\Sigma_i(x_i - \bar{x})^2}$$

Alternately, we can compute $S = \Sigma_i(y_i - \bar{y})^2 = (n - 1)s_y^2$ in the usual manner and $\Sigma_i(y_i - \tilde{y}_i)^2 = (n - 2)s^2_{y\cdot x}$ by the method of Section 6.22, obtaining the sum of squares due to linear regression by subtraction.

It should be noted that the two methods of significance testing for the hypothesis $\beta = 0$ are equivalent to the t-test of the hypothesis $\rho = 0$ based on the estimate r of the correlation coefficient of a normal bivariate population.

EXAMPLE. As an illustration of the application of the analysis of variance to regression problems we shall examine the data of Table 6.4. From these data we compute

$$n = 9 \quad \Sigma y = 640.2 \quad \Sigma x = 0.91$$

$$\Sigma y^2 = 45{,}653.26 \quad \Sigma xy = 65.096$$

$$\Sigma x^2 = 0.0945$$

$$\Sigma(x_i - \bar{x})^2 = 0.002489$$

$$\Sigma(y_i - \bar{y})^2 = 113.70$$

$$\Sigma(x_i - \bar{x})(y_i - \bar{y}) = 0.36467$$

$$b^2 \Sigma(x_i - \bar{x})^2 = \frac{\{\Sigma(x_i - \bar{x})(y_i - \bar{y})\}^2}{\Sigma(x_i - \bar{x})^2} = 53.43$$

$$\Sigma(y_i - \tilde{y}_i)^2 = 113.70 - 53.43 = 60.27$$

The analysis of variance is shown in Table 7.44.

TABLE 7.44

Source of Estimate	Sum of Squares	D.F.	Mean Square
Linear regression	53.43	1	53.43
Residual	60.27	7	8.51
Total	113.70	8	

Testing the hypothesis $\beta = 0$ we form the variance ratio $\dfrac{53.43}{8.61} = 6.21^*$ ($F_{1,7,0.05} = 5.6$). Thus there is evidence, significant at the level $\alpha = 0.05$, of a linear regression of y upon x.

7.72. Analysis of Variance for Linear Regression in more than 1 Variable

A very similar analysis can be made in the case of linear regression on 2 dependent variables, x_1 and x_2. Given sets of observations (y_i, x_{1i}, x_{2i}) we assume that each y_i is normally distributed with variance σ^2 and that ave $(y_i) = \alpha + \beta_1 x_{1i} + \beta_2 x_{2i}$, or, equivalently, ave $(y_i) = \mu + \beta_1(x_{1i} - \bar{x}_1) + \beta_2(x_{2i} - \bar{x}_2)$, where $\mu = \alpha + \beta_1\bar{x}_1 + \beta_2\bar{x}_2$. Again, the best estimate of μ is $\bar{y}$, but the estimates b_1 and b_2 of β_1 and β_2 are more complicated, as was seen in Section 6.31. Using these estimates, we obtain for the breakdown of the overall sum of squares of the y values about their mean

$$\Sigma_i(y_i - \bar{y})^2 = \Sigma_i(y_i - \tilde{y}_i)^2 + \Sigma_i\{b_1(x_{1i} - \bar{x}_1) + b_2(x_{2i} - \bar{x}_2)\}^2$$

where the $\tilde{y}_i = \bar{y} + b_1(x_{1i} - \bar{x}_1) + b_2(x_{2i} - \bar{x}_2)$ are the predicted values of y for the given x_i, and, as before, the sums of squares on the right-hand side are those due to residual error and regression effects, respectively. In this case the residual sum of squares has $n - 3$ of the $n - 1$ degrees of freedom available after correction for the overall mean, and the sum of squares due to regression has 2 degrees of freedom corresponding to the estimates b_1 and b_2. The average value of the mean square based on residuals is σ^2; that of the regression can be shown to be

$$\sigma^2 + \tfrac{1}{2}\,\Sigma_i\{\beta_1(x_{1i} - \bar{x}_1) + \beta_2(x_{2i} - \bar{x}_2)\}^2$$

Hence we obtain the analysis of variance in Table 7.45, and can test for the significance of the regression in the same manner as in the previous section, where now the variance ratio has 2 and $n - 3$ degrees of freedom.

TABLE 7.45

GENERAL ANALYSIS OF VARIANCE:
REGRESSION WITH 2 INDEPENDENT VARIABLES

Source of Estimate	Sum of Squares	D.F.	Average Value of Mean Square
Regression	$\Sigma_i\{b_1(x_{1i} - \bar{x}_1) + b_2(x_{2i} - \bar{x}_2)\}^2$	2	$\sigma^2 + \frac{1}{2}\Sigma_i\{\beta_1(x_{1i} - \bar{x}_1) + \beta_2(x_{2i} - \bar{x}_2)\}^2$
Residual	$\Sigma_i(y_i - \tilde{y}_i)^2$	$n - 3$	σ^2
Total	$\Sigma_i(y_i - \bar{y})^2$	$n - 1$	

The computations in this case are most easily carried out by determining the overall sum of squares $S = \Sigma_i(y_i - \bar{y})^2$ in the usual manner, and $\Sigma_i(y_i - \tilde{y}_i)^2$ by the method given in Section 6.31, and then obtaining the regression sum of squares by subtraction. Note that the regression sum of squares breaks up into 2 parts corresponding to the separate effects of b_1 and b_2 only if $\Sigma_i(x_{1i} - \bar{x}_1)(x_{2i} - \bar{x}_2) = 0$, i.e., if the x_1's and x_2's are orthogonal as defined in Section 4.52. This means that we are testing the significance of the regression of y on x_1 and x_2 jointly and the test implies nothing concerning the relative importance of their individual contribution. Also in general the values b_1 and b_2 will not be the same as if we had determined them individually. However, we can first determine the sum of squares due to regression on x_1 alone as in the previous section. This sum of squares has 1 degree of freedom, and the corresponding residual sum of squares $n - 2$ degrees of freedom. Then the difference between the regression due to x_1 and x_2 together and the regression due to x_1 alone (or, alternatively, the reduction in the residual sum of squares due to the introduction of b_2), represents the additional reduction due to the introduction of x_2, and has 1 degree of freedom. This additional reduction is in fact that due to the use of a second orthogonal variable $x_{2\cdot1}$ where $x_{2\cdot1} = (x_{2i} - \bar{x}_2) - b_{21}(x_{1i} - \bar{x}_1)$. In the notation of Section 6.33 the sum of squares for regression becomes

$$\Sigma_i\{b_{y1}(x_{1i} - \bar{x}_1) + b_{y2\cdot1}(x_{2\cdot1})\}^2 = \Sigma_i b^2{}_{y1}(x_{1i} - \bar{x}_1)^2 + \Sigma_i b^2{}_{y2\cdot1}x^2{}_{2\cdot1}$$

the cross-product terms vanishing. This sum of squares can therefore be separated into 2 independent parts, each of which has a single degree of freedom and, on the hypothesis that the regression coefficient is zero, is an estimate of σ^2.

This type of analysis can be applied to regression problems in which k variables are involved, the sum of squares being separated into a portion due to regression with k degrees of freedom and a residual with $n - k - 1$ degrees of freedom. If we use as x variates $x_1, x_{2\cdot1}, x_{3\cdot12}, \cdots, x_{k\cdot123\cdots(k-1)}$, the sum of squares for regression can be further separated into k parts, each with a single degree of freedom and each with average value σ^2 if the appropriate regression coefficient is zero.

7.73. Analysis of Variance for Polynomial Regression

It was pointed out in Chapter 6 that the problem of determining the regression of y on the polynomial $a + b_1x + b_2x^2 + \cdots + b_kx^k$ is a special case of the regression problem of the preceding section where $x_1 = x$, $x_2 = x^2$, $x_k = x^k$. In this case the analysis given above would correspond to determining the reduction in the sum of squares due to linear regression, the additional reduction due to the use of a quadratic equation in x, etc.

If the x_i are equally spaced, as is usually possible and certainly advisable when the y_i are determined at points x_i chosen by the experimenter, then we can use the simple method of orthogonal polynomials considered in Section 6.32. In this case we have, rewriting equation (f) of that section,

$$\Sigma_i(y_i - \bar{y})^2 = \Sigma_i(y_i - \tilde{y}_i)^2 + b''{}_1{}^2D_1 + b''{}_2{}^2D_2 + \cdots + b''{}_k{}^2D_k$$

where the b''_i are given by equations (*b*) and (*e*) of Section 6.32 and the D_i are tabulated [13] for various n. This equation provides the required breakdown of the overall sum of squares into a residual component having $n - k - 1$ degrees of freedom and k *independent* components, each with 1 degree of freedom, the latter representing the independent contributions of each power of x to the regression. Hence we can test each contribution individually to assess the significance of this contribution to the overall regression. The analysis of variance is given in Table 7.46. Note that, if $n = k + 1$, the residual variance vanishes, since a polynomial of degree k can always be fitted exactly to $k + 1$ points. In this case we have separated the overall sum of squares into k independent components, each having a single degree of freedom.

TABLE 7.46

GENERAL ANALYSIS OF VARIANCE:
REGRESSION USING ORTHOGONAL POLYNOMIALS

Source of Estimate	Sum of Squares	D.F.	Average Value of Mean Square
Linear component	$b''^2_1 D_1$	1	$\sigma^2 + \beta''^2_1 D_1$
Quadratic component	$b''^2_2 D_2$	1	$\sigma^2 + \beta''^2_2 D_2$
	$\cdot$	$\cdot$	$\cdot$
	$\cdot$	$\cdot$	$\cdot$
kth degree component	$b''^2_k D_k$	1	$\sigma^2 + \beta''^2_k D_k$
Residual	$\Sigma_i (y_i - \tilde{y}_i)^2$	$n - k - 1$	σ^2
Total	$\Sigma_i (y_i - \bar{y})^2$	$n - 1$	

The most satisfactory method for computing the components of the sum of squares derives from equation (*e*) of Section 6.32.

$$b''^2_1 D_1 = \frac{(\Sigma_i \xi'_{1i} y_i)^2}{D_1^2} D_1 = \frac{(\Sigma_i \xi'_{1i} y_i)^2}{D_1}$$

Thus the first component of the sum of squares of Table 7.46 may be obtained by summing the product of the y_i and the corresponding ξ'_{1i}, squaring the total, and dividing by D_1. In many cases the regression coefficients $b''_1, \cdots, b''_k$ will be evaluated in order to estimate the regression line. The components of the sum of squares may then be determined directly as, e.g., $b''^2_j D_j$, although this method is less satisfactory from the viewpoint of computational accuracy.

EXAMPLE 1. Let us consider the example of Section 6.32. Corresponding to

$$\Sigma(y_i - \tilde{y}_i)^2 = \Sigma(y_i - \bar{y})^2 - b''^2_1 D_1 - b''^2_2 D_2 - b''^2_3 D_3 - b''^2_4 D_4 - b''^2_5 D_5$$

we have there

$$9983 = 699{,}151 - 607{,}504 - 19{,}983 - 60{,}116 - 618 - 947$$

TABLE 7.47

Source of Estimate	Sum of Squares	D.F.	Mean Square	Average Value of Mean Square
Linear component	607,504	1	607,504	$\sigma^2 + b''^2_1 D_1$
2nd-degree component	19,983	1	19,983	$\sigma^2 + b''^2_2 D_2$
3rd-degree component	60,116	1	60,116	$\sigma^2 + b''^2_3 D_3$
4th-degree component	618	1	618	$\sigma^2 + b''^2_4 D_4$
5th-degree component	947	1	947	$\sigma^2 + b''^2_5 D_5$
Residual component	9,983	7	1,426	σ^2
Total	699,151	12		

The analysis of variance is shown in Table 7.47. In order to estimate the significance of the component of any degree we test the hypothesis $\beta''_k = 0$ by forming the variance ratio corresponding to

$$\frac{\text{estimate } (\sigma^2 + \beta''^2_k D_k)}{\text{estimate } \sigma^2}$$

Except under special circumstances we should attempt to represent the data by a polynomial of low degree. Accordingly in this case we test the hypotheses in the following order:

(1) $\beta''_5 = 0$.

$$\text{variance ratio} = \frac{947}{1426} = 0.66 \quad \text{(not significant)}$$

and this component is pooled with the residual sum of squares to provide a new estimate of the residual variance

$$s^2 = \frac{9983 + 947}{8} = 1366$$

(2) $\beta''_4 = 0$.

$$\text{variance ratio} = \frac{618}{1366} = 0.45 \quad \text{(not significant)}$$

Pooling we obtain

$$s^2 = 1283$$

(3) $\beta''_3 = 0$.

$$\text{variance ratio} = \frac{60116}{1283} = 46.9^{***} \qquad F_{1,\,9,\,0.001} = 22.9$$

(4) $\beta_2'' = 0$.

$$\text{variance ratio} = \frac{19983}{1283} = 15.6^{**} \qquad F_{1,\,9,\,0.01} = 10.6$$

(5) $\beta_1'' = 0$.

$$\text{variance ratio} = \frac{607504}{1283} = 474^{***} \qquad F_{1,\,9,\,0.001} = 22.9$$

We conclude that the first, second, and third-degree terms are significant, and that the fourth and fifth-degree terms are not. In order to undertake the analysis we assumed that the terms of degree 6–12 were not significant.

In cases where n independent replicate observations are available at each of p fixed x values, a total of $N = np$ observations, we can form an independent estimate of the variance from the variation within classes

$$s^2 = \frac{\Sigma_{i\alpha}\,(y_{i\alpha} - \bar{y}_i)^2}{p(n-1)} = \frac{S_{\alpha(i)}}{n-p}$$

Under these circumstances each of $p - 1$ components of the between-classes sums of squares may be tested against this estimate of the residual variance. It is important to note that in so doing we assume that the replicate observations are distributed normally and independently about some average value, and then proceed to test hypotheses concerning the average value. If the replicate determinations are not independent, they do not provide an unbiased variance estimate for this test.

A variance analysis may be carried out on data which is to be represented by a polynomial, but in which the values of the independent variate are unequally spaced by regarding the variate values x_i, together with $x_i^2, x_i^3, \cdots$ as variates in a multiple regression analysis. It is not possible to determine the independent contributions to the sum of squares of the regressions on $x, x^2, x^3, \cdots, x^k$, since the variables are not orthogonal. We can, however, determine the independent contributions to the sum of squares by the regression on x_1, $x_{2\cdot 1}$, $x_{3\cdot 12}$, etc., where $x_1 = x$, $x_2 = x^2$, $x_3 = x^3$. If the required relationship is to be as simple as possible, we should normally proceed by fitting a linear regression to the data and testing for significance. Subsequently an expression involving linear and quadratic terms would be fitted, and the reduction in the residual component corresponding to the additional degree of freedom would be tested against the residual variance estimate. This procedure is continued until no significant reduction is achieved by the fitting of additional terms of higher degree. Since the sums of powers and products of degree $1, \cdots, 2k$ are required to fit a polynomial of degree k by this method, the computations rapidly become laborious.

EXAMPLE 2. Examination of Figure 6.3 suggests that the data might be better fitted by a curve than by a straight line. From Section 6.22 we note that for the linear curve fitted to these data we obtained a residual sum of squares 89.211, and hence the component due to linear regression alone is $533.647 - 89.211 = 444.436$. Computing the necessary additional sums of squares, cross products, and sums of squares of deviations from Table 6.3, and solving for the regression coefficients b_1 and b_2, we have for the residual sum of squares after fitting a quadratic regression

$$\begin{aligned}\Sigma(y_i - \tilde{y}_i)^2 &= S(y^2) - b_1 S(xy) - b_2 S(x^2 y)\\ &= 533.647 - 489.185 + 44.040 = 88.502\end{aligned}$$

TABLE 7.48

Source of Estimate	Sum of Squares	D.F.	Mean Square
Linear regression component	444.436	1	444.4
Difference between linear and (linear + quadratic) regression components	0.709	1	0.709
Total linear + quadratic regression	445.145	2	
Residual	88.502	33	2.682
Total	533.647	35	

On the hypothesis that, given a linear regression, there is no significant improvement on the addition of a quadratic term, the difference between the sum of squares due to linear and quadratic effects and the sum of squares due to linear effects only provides an estimate of σ^2. In this case the variance ratio $\dfrac{0.709}{2.682} = 0.264$ is not significant, and we accept the hypothesis that with these conditions there is no quadratic effect.

7.74. Subdivision of Scale Classifications

The classifications used in the analysis of variance can be subdivided into 2 major types; classification according to categories and classification according to scales. In the former the classification corresponds to a number of discrete objects or possibilities such as machines, analysts, methods, samples. In the latter the classification represents chosen levels of a continuous scale, such as a series of temperatures, pressures, reaction times, concentrations. With categories in addition to the overall variation present, we are interested in the estimation and testing of hypotheses concerning specific effects, or the classification of the objects tested, as described in the earlier sections of this chapter. With scales we are generally more interested in the dependence of the result on the scale used. Any of the methods of the previous sections can be used to subdivide the sum of squares corresponding to this particular classification into parts indicating the nature of this dependence. If the classifications chosen are equally spaced along the scale, then the methods of the previous section make it possible to subdivide the classification into parts corresponding to linear, quadratic, cubic, etc., and residual effects. If the values are not equally spaced, or if we are interested in other than polynomial dependence, then more complicated calculations are necessary.

In subdividing the sum of squares due to a particular classification, we generally wish to include this subdivision as part of the general analysis of variance. This means that it is necessary to express these sums of squares in the same basic units as the other sums of squares. For example, if we subdivide the sum of squares corresponding to row effects in a two-way analysis of variance, then, since the original overall sum of squares $\Sigma_i(\bar{x}_{i\cdot} - \bar{x})^2$ corresponding to row totals is multiplied by nq in the general analysis, each part of its breakdown must also be multiplied by nq. Similarly, if we work with row totals, as is more usual, then we must divide the usual expressions for the regression and residual sums of squares by nq, just as the sum of squares of row totals was divided by nq in the calculation of the overall sum of squares.

We consider the classification according to drying periods in the example of Section 7.53. Here the drying periods represent 3 choices from a continuous scale of drying periods that might have been used, and we are primarily interested in the dependence of the analytical results on the period chosen. Let us first consider the dependence of the average values for all samples on the drying period. In this case we have 3 equally spaced points, and the overall sum of squares for differences due to drying periods has 2 degrees of freedom.

We first subdivide this sum of squares into linear and quadratic components, each with 1 degree of freedom. Since the quadratic equation must fit 3 points exactly, these 2 components will account for the entire sum of squares due to differences in drying periods, and the comparison afforded is better described as a breakdown into linear and non-linear dependence. For 3 points, the orthogonal coefficients for the linear component are $-1, 0, 1$, and for the quadratic component $1, -2, 1$. Hence from Table 7.19 we obtain for the linear component of the drying period effects

$$-1(139.52) + 0(134.53) + 1(131.87) = -7.65$$

and for the corresponding sum of squares

$$\frac{(-7.65)^2}{40} = 1.4631$$

where the divisor is the number of individual results in each total, 20, multiplied by the sum of squares of the coefficients for the linear component, $(-1)^2 + (0)^2 + (1)^2 = 2$. Similarly, we obtain for the quadratic component

$$1(139.52) - 2(134.53) + 1(131.87) = 2.33$$

and for the corresponding sum of squares

$$\frac{(2.33)^2}{120} = 0.0452$$

Thus we have the breakdown of the sum of squares due to drying periods shown in Table 7.49.

TABLE 7.49

Source of Estimate	D.F.	Sum of Squares
Linear dependence	1	1.4631
Quadratic dependence (residual)	1	0.0452
Total	2	1.5083

This indicates that the significance of the effects due to drying period is largely attributable to a linear decrease of weight with drying time, as indicated by Figure 7.1. However, the quadratic, or residual, component gives a variance estimate which is still much greater than the error estimate of 0.0050, and appreciably greater than the interaction estimate of 0.0198, indicating some tendency towards constant weight.

It was noted in the original analysis of these data that a significant interaction existed between drying period and samples, possibly owing to different drying rates for different samples. This can be investigated by computing the linear and quadratic components of the drying period classification for each sample from the data of Table 7.19. For the first sample we compute the linear and quadratic components as :

$$-1(6.80) + 0(6.42) + 1(5.97) = -0.83,$$

and

$$1(6.80) - 2(6.42) + 1(5.97) = -0.07;$$

the complete results are given in Table 7.50.

TABLE 7.50

Sample	Linear Component	Quadratic Component (Residual)
1	−0.83	−0.07
2	−0.27	+0.07
3	−0.96	+0.56
4	−0.71	−0.13
5	−1.40	+0.34
6	−0.13	+0.07
7	−0.70	+0.28
8	−0.98	+0.34
9	−0.63	+0.25
10	−1.04	+0.62
Total	−7.65	2.33

The sum of squares for differences between linear components, after correction for the overall linear dependence previously obtained, is

$$\frac{(-0.83)^2 + (-0.27)^2 + \cdots + (-1.04)^2}{4} - \frac{(-7.65)^2}{40} = 0.3102$$

and is based on 9 degrees of freedom. We obtain the divisor 4 as before, remembering that in this instance each number used in computing the components was the total of 2 original observations. Similarly we obtain for the quadratic effects:

$$\frac{(-0.07)^2 + (0.07)^2 + \cdots + (0.62)^2}{12} - \frac{(2.33)^2}{120} = 0.0466$$

again with 9 degrees of freedom. The sum of these 2 parts is 0.3568 with 18 degrees of freedom, as was originally obtained for interaction. Hence the major portion of the interaction is apparently due to differences in the linear dependence of the analytical result on drying time. The final analysis of variance is shown in Table 7.51.

TABLE 7.51

ANALYSIS OF VARIANCE

Source of Estimate	Sum of Squares	D.F.	Estimate of Variance
Samples	242.0681	9	26.8965
Drying times			
linear dependence	1.4631	1	1.4631
quadratic dependence (residual)	0.0452	1	0.0452
Interaction			
samples × lin. dep.	0.3102	9	0.0345
samples × quad. dep.	0.0466	9	0.0052
Error	0.1495	30	0.0050
Total	244.0827	59	

Although the data have been employed to illustrate the practice of separating the sums of squares between levels of a scale factor into linear and quadratic components, such a division in this example is purely empirical and the statistical inferences apply only to the levels of 3, 6, and 9 hours at which observations have been made. It might be thought reasonable to interpolate in order to estimate the loss of weight at some intermediate time, and if this were necessary a transformation of the time scale might be employed to give a more convenient linear relationship.

As an empirical step the mean weights of the samples may be plotted against the logarithm of the drying time. This leads to an arrangement which is almost exactly linear, and which would therefore be convenient over the range in question. The logarithms of the drying periods are not equally spaced, and the breakdown of the sum of squares must now be obtained by fitting a linear regression. From Table 7.19 we obtain Table 7.52, and, working with the class totals for the sake of accuracy,

TABLE 7.52

Drying Time	x Log Drying Time	y Class Total
3	0.4771	139.52
6	0.7782	134.53
9	0.9452	131.87

we have $\Sigma y^2 = 54{,}953.8482$, $T_{(y)} = 405.92$, $\Sigma x^2 = 1.7437$, $T_{(x)} = 2.2095$, $\Sigma xy = 297.0866$; also $\dfrac{T^2_{(y)}}{3} = 54{,}923.6821$, $\dfrac{T^2_{(x)}}{3} = 1.6273$, and $\dfrac{T_{(x)}T_{(y)}}{3}$ $= 298.9601$. Dividing by 20 to obtain results in terms of the original analysis of variance, we obtain

Overall Sums of Squares:

$$\Sigma y^2 - \frac{T^2_{(y)}}{3} \div 20 = \frac{54953.8482 - 54923.6821}{20} = 1.5083$$

Sum of Squares Due to Regression:

$$\frac{\left(\Sigma xy - \dfrac{T_{(x)}T_{(y)}}{3}\right)^2}{\Sigma x^2 - \dfrac{T^2_{(x)}}{3}} \div 20 = \frac{(297.0866 - 298.9601)^2}{(1.7437 - 1.6273)20}$$

$$= \frac{(-1.8735)^2}{2.3280} = 1.5077$$

Residual: $1.5083 - 1.5077 = 0.0006$

Similarly by determining the individual regression of weight upon the logarithm of the drying time for each sample, computing the sum of squares, and subtracting the component due to the overall regression we can break down the sum of squares due to interactions into a part depending linearly upon the logarithm of the drying time and a residual. The resulting analysis is given in Table 7.53.

TABLE 7.53

ANALYSIS OF VARIANCE:
REGRESSION ON LOG DRYING TIME

Source of Estimate	Sum of Squares	D.F.	Estimate of Variance
Samples	242.0681	9	26.8965
Time of drying			
regression on log time	1.5077	1	1.5077
residual	0.0006	1	0.0006
Interaction			
regression on log time	0.3224	9	0.0358
residual	0.0344	9	0.0038
Between duplicates	0.1495	30	0.0050
Total	244.0827	59	

Analyzed in this form each of the residual mean squares is less than the mean square between duplicates, indicating that the selected relationship provides a very satisfactory description of the data. The interaction component due to linear regression is significantly high when compared with the variance estimate based on variation between duplicates, indicating that, although the loss of weight was linearly related to the log time, the rate varied from sample to sample.

It must be emphasized that, although the use of the logarithm of the drying time leads to a good agreement between model and data, it does not follow that this is the only model from which such close agreement could have been obtained, particularly since the experiment was restricted to 3 levels on the time scale. Theoretical considerations would indicate that the true relationship is likely to be much more complex, since it involves diffusion of moisture within the material and heat and mass transfer across a boundary layer. Furthermore, the logarithmic relationship, if extrapolated, leads to impossible values for very long or very short drying periods.

7.8. Analysis of Covariance

7.81. Introduction

The technique of analysis of covariance is designed for circumstances in which the data are in such a form that they must be analyzed by a combination of the methods of regression and analysis of variance. It frequently happens that we wish to perform an analysis on observations of the variable y, made at a series of fixed levels of one or more factors, in a crossed classification. If, however, some additional factor x varies during the period in which the observations were made, any dependence of y on x will tend to obscure and possibly to invalidate the results of an analysis of variance performed on the original data. If the situation is reversed and interest centers in the regression of y upon x, but the data are obtained from a number of sources or may be otherwise classified, our estimate of the regression coefficient may be influenced by class effects.

Probably the most familiar example of these circumstances occurs in the measurement of gas volumes, where in addition to variation due to the factors under investigation the results are influenced by the conditions of temperature and pressure under which the measurements are made. Under these conditions we normally assume a relationship between the temperature, pressure, and volume of a gas, and any analysis of the results is preceded by a correction to standard conditions of temperature and pressure. Frequently, however, a relationship based upon previous experience is not available as a basis for correction, but must be sought within the data of the experiment by a suitable analysis. It is usually impossible to follow what might appear to be the obvious procedure of determining the regression relationship, correcting the observations by means of this relationship to some standard condition, and examining the corrected values by the analysis of variance. The regression relationship which is determined from the data is only an estimate of the true relationship, and those observations which receive a large correction term will have a larger residual variance than observations for which the correction term is small.

7.82. Differences between Slopes

We shall first consider the case in which y depends linearly upon a single set of simultaneously observed variables x and may be classified into one of a set of p groups.

(*a*) *Comparisons for* 2 *Groups.* Suppose that a series of $n_1 + n_2$ pairs of observations on the variables y and x are made, n_1 pairs in the case where y belongs to the first class and n_2 when y belongs to the second class. Then the regression coefficients and intercepts within the two classes may be estimated independently as

$$a_1 = \frac{\Sigma_\alpha y_{1\alpha} - b_1 \Sigma_\alpha x_{1\alpha}}{n_1} \qquad a_2 = \frac{\Sigma_\alpha y_{2\alpha} - b_2 \Sigma_\alpha x_{2\alpha}}{n_2}$$

$$b_1 = \frac{\Sigma_\alpha(y_{1\alpha} - \bar{y}_1)(x_{1\alpha} - \bar{x}_1)}{\Sigma_\alpha(x_{1\alpha} - \bar{x}_1)^2} \qquad b_2 = \frac{\Sigma_\alpha(y_{2\alpha} - \bar{y}_2)(x_{2\alpha} - \bar{x}_2)}{\Sigma_\alpha(x_{2\alpha} - \bar{x}_2)^2}$$

Furthermore, if we assume that the residual variance within the two classes is the same, then

$$\frac{\Sigma_\alpha(y_{1\alpha} - \bar{y})^2 - b_1^2\,\Sigma_\alpha(x_{1\alpha} - \bar{x}_1)^2}{n_1 - 2}$$

and

$$\frac{\Sigma_\alpha(y_{2\alpha} - \bar{y}_2)^2 - b_2^2\,\Sigma_\alpha(x_{2\alpha} - \bar{x}_2)^2}{n_2 - 2}$$

are each estimates of σ^2, and the sum of squares of the residuals may be pooled to give an estimate of σ^2 based on $n_1 + n_2 - 4$ degrees of freedom, so that

$$s^2 = \frac{\Sigma_\alpha(y_{1\alpha} - \bar{y}_1)^2 + \Sigma_\alpha(y_{2\alpha} - \bar{y}_2)^2 - b_1^2\,\Sigma_\alpha(x_{1\alpha} - \bar{x}_1)^2 - b_2^2\,\Sigma_\alpha(x_{2\alpha} - \bar{x}_2)^2}{n_1 + n_2 - 4}$$

The estimates b_1 and b_2 of the regression coefficients are normally distributed each with 1 degree of freedom and with variance $\dfrac{\sigma^2}{\Sigma_\alpha(x_{1\alpha} - \bar{x}_1)^2}$ and $\dfrac{\sigma^2}{\Sigma_\alpha(x_{2\alpha} - \bar{x}_2)^2}$, respectively. The hypothesis that the true regression coefficient is the same in each class, i.e., $\beta_1 = \beta_2$, may be tested, since

$$\sigma^2{}_{b_1 - b_2} = \frac{\sigma^2}{\Sigma_\alpha(x_{1\alpha} - \bar{x}_1)^2} + \frac{\sigma^2}{\Sigma_\alpha(x_{2\alpha} - \bar{x}_2)^2}$$

so that under the given hypothesis

$$\frac{b_1 - b_2}{s\sqrt{\dfrac{1}{\Sigma_\alpha(x_{1\alpha} - \bar{x}_1)^2} + \dfrac{1}{\Sigma_\alpha(x_{2\alpha} - \bar{x}_2)^2}}}$$

is distributed as t with $(n_1 + n_2 - 4)$ degrees of freedom.

The hypothesis that the intercept is the same in each class, i.e., $\alpha_1 = \alpha_2 = \alpha$, may also be tested, since

$$\sigma^2{}_{a_1 - a_2} = \sigma^2\left(\frac{1}{n_1} + \frac{1}{n_2} + \frac{\bar{x}_1^2}{\Sigma_\alpha(x_{1\alpha} - \bar{x}_1)^2} + \frac{\bar{x}_2^2}{\Sigma_\alpha(x_{2\alpha} - \bar{x}_2)^2}\right)$$

and under this hypothesis

$$\frac{a_1 - a_2}{s\sqrt{\dfrac{1}{n_1} + \dfrac{1}{n_2} + \dfrac{\bar{x}_1^2}{\Sigma_\alpha(x_{1\alpha} - \bar{x}_1)^2} + \dfrac{\bar{x}_2^2}{\Sigma_\alpha(x_{2\alpha} - \bar{x}_2)^2}}}$$

is distributed as t with $(n_1 + n_2 - 4)$ degrees of freedom.

If the difference between the estimates of the regression coefficients is not significant, we may accept the hypothesis $\beta_1 = \beta_2 = \beta$ and employ

$$\frac{\Sigma_\alpha(y_{1\alpha} - \bar{y}_1)(x_{1\alpha} - \bar{x}_1) + \Sigma_\alpha(y_{2\alpha} - \bar{y}_2)(x_{2\alpha} - \bar{x}_2)}{\Sigma_\alpha(x_{1\alpha} - \bar{x}_1)^2 + \Sigma_\alpha(x_{2\alpha} - \bar{x}_2)^2}$$

as an estimate of β. We then have

$$a_1 = \frac{\Sigma_\alpha y_{1\alpha} - b\,\Sigma_\alpha x_{1\alpha}}{n_1} \qquad a_2 = \frac{\Sigma_\alpha y_{2\alpha} - b\,\Sigma_\alpha x_{2\alpha}}{n_2}$$

$$s^2 = \frac{\Sigma_\alpha(y_{1\alpha} - \bar{y}_1)^2 + \Sigma_\alpha(y_{2\alpha} - \bar{y}_2)^2 - b^2[\Sigma_\alpha(x_{1\alpha} - \bar{x}_1)^2 + \Sigma_\alpha(x_{2\alpha} - \bar{x}_2)^2]}{n_1 + n_2 - 3}$$

Using these values of a_1, a_2, and s, and, under the hypothesis that $\alpha_1 = \alpha_2 = \alpha$, we have

$$\frac{a_1 - a_2}{s\sqrt{\dfrac{1}{n_1} + \dfrac{1}{n_2} + \dfrac{\bar{x}_1^{\,2}}{\Sigma_\alpha(x_{1\alpha} - \bar{x}_1)^2} + \dfrac{\bar{x}_2^{\,2}}{\Sigma_\alpha(x_{2\alpha} - \bar{x}_2)^2}}}$$

distributed as t with $(n_1 + n_2 - 3)$ degrees of freedom.

The average values of y for the two regressions may be compared at any value of x, whether the hypothesis $\beta_1 = \beta_2 = \beta$ is accepted or rejected, but for a given value of x the variances of the estimates of these average values are unequal.

(*b*) *Comparison of More than* 2 *Groups.* If the series of observations of the pairs x_i, y_i can be grouped into p classifications with n_i pairs in the ith class, $\Sigma_i n_i = N$, the estimates $(b_1, \cdots, b_i, \cdots, b_p)$, $(a_1, \cdots, a_i, \cdots, a_p)$ should be tested as groups for heterogeneity. Unfortunately this cannot be achieved by a simple analysis of variance technique of the type considered in Section 7.21, since the variance of the estimates are unequal. This difficulty can be obviated by the use of a complete covariance analysis. For each of the p classes we compute

$$\Sigma_\alpha(x_{i\alpha} - \bar{x}_i)^2 = S_\alpha(x_i^2)$$
$$\Sigma_\alpha(y_{i\alpha} - \bar{y}_i)^2 = S_\alpha(y_i^2) \qquad b_i = \frac{S_\alpha(x_i y_i)}{S_\alpha(x_i^2)}$$
$$\Sigma_\alpha(y_{i\alpha} - \bar{y}_i)(x_{i\alpha} - \bar{x}_i) = S_\alpha(x_i y_i)$$

together with

$$\Sigma_{\alpha i}(x_{i\alpha} - \bar{x}_i)^2 = S_{\alpha(i)}(x^2) \qquad \Sigma_i n_i(\bar{x}_i - \bar{x})^2 = S_i(x^2)$$

$$\Sigma_{\alpha i}(y_{i\alpha} - \bar{y}_i)^2 = S_{\alpha(i)}(y^2) \qquad \Sigma_i n_i(\bar{y}_i - \bar{y})^2 = S_i(y^2)$$

$$\Sigma_{\alpha i}(y_{i\alpha} - \bar{y}_i)(x_{i\alpha} - \bar{x}_i) = S_{\alpha(i)}(xy) \qquad \Sigma_i n_i(\bar{y}_i - \bar{y})(\bar{x}_i - \bar{x}) = S_i(xy)$$

$$\Sigma_{\alpha i}(x_{i\alpha} - \bar{x})^2 = S(x^2) \qquad b_a = \frac{S_{\alpha(i)}(xy)}{S_{\alpha(i)}(x^2)}$$

$$\Sigma_{\alpha i}(y_{i\alpha} - \bar{y})^2 = S(y^2) \qquad b_m = \frac{S_i(xy)}{S_i(x^2)}$$

$$\Sigma_{\alpha i}(y_{i\alpha} - \bar{y})(x_{i\alpha} - \bar{x}) = S(xy) \qquad b_0 = \frac{S(xy)}{S(x^2)}$$

If we are concerned only to examine the class effect upon the individual regression coefficient, we can analyze the variance within sets as follows:

$$\begin{aligned}\Sigma_{\alpha i}(y_{i\alpha} - \bar{y}_i)^2 &= \Sigma_{\alpha i}[y_{i\alpha} - \bar{y}_i - b_i(x_{i\alpha} - \bar{x}_i)]^2 + \Sigma_{\alpha i} b_i^2(x_{i\alpha} - \bar{x}_i)^2 \\ &= \Sigma_{\alpha i}[y_{i\alpha} - \bar{y}_i - b_i(x_{i\alpha} - \bar{x}_i)]^2 + \Sigma_{\alpha i}(b_i - b_a)^2(x_{i\alpha} - \bar{x}_i)^2 \\ &\qquad + b_a^2\, \Sigma_{\alpha i}(x_{i\alpha} - \bar{x}_i)^2\end{aligned}$$

The first term of the right-hand side is the sum of the squares of $\Sigma n_i = N$ normally distributed variates of variance σ^2, subject to $2p$ linear restraints, and is distributed as $\sigma^2\chi^2_{N-2p}$. On the hypothesis $\beta_1 = \beta_2 = \cdots = \beta_p = \beta$, the second term is the sum of squares of p normally distributed variates of variance σ^2, subject to 1 linear restraint, and is distributed as $\sigma^2\chi^2_{p-1}$, whereas, on the hypothesis that there is no linear regression within the classes, i.e., that $\beta = 0$, the final term is distributed as $\sigma^2\chi_1^2$. Thus we can draw up the analysis of variance Table 7.54. To test the hypothesis $\beta_1 = \beta_2 = \cdots = \beta_p$ we form the variance ratio

$$\frac{\Sigma_i(b_i - b_a)^2 S_\alpha(x_i^2)}{p-1} \; \frac{N-2p}{\Sigma_{i\alpha}[y_{i\alpha} - \bar{y}_i - b_i(x_{i\alpha} - \bar{x}_i)]^2} = F_{p-1,\, N-2p}$$

TABLE 7.54

Source of Estimate	Sum of Squares	D.F.
Mean regression within classes	$b_a^2\, \Sigma_{\alpha i}(x_{i\alpha} - \bar{x}_i)^2 = b_a^2 S_{\alpha(i)}(x^2)$	1
Difference between class regressions	$\Sigma_{\alpha i}(b_i - b_a)^2(x_{i\alpha} - \bar{x}_i)^2 = \Sigma_i(b_i - b_a)^2 S_\alpha(x_i^2)$	$p - 1$
Residual corrected for class regressions	$\Sigma_{\alpha i}[y_{i\alpha} - \bar{y}_i - b_i(x_{i\alpha} - \bar{x}_i)]^2 = S_{\alpha(i)}(y^2) - \Sigma_i b_i^2 S_\alpha(x_i^2)$	$N - 2p$
Total	$\Sigma_{\alpha i}(y_{i\alpha} - \bar{y}_i)^2 = S_{\alpha(i)}(y^2)$	$N - p$

If this ratio is not significant, we may accept the hypothesis and pool the sums of squares due to difference between class regressions and the corrected residual to give an estimate of the residual variance based on $N - p - 1$ degrees of freedom. Using this estimate, we can test the significance of the mean regression coefficient within classes by forming the appropriate variance ratio.

7.83. Treatment Comparisons

The method of the previous section is appropriate if we are interested in the class effect upon the regression of y on x. To examine the class effects upon the constant term of the linear regression it is necessary to form an estimate of the class means, or of the variance between class means, and of the residual variance, all adjusted for variation in x. The residual variance has already been obtained as the residual of Table 7.54. Where the regression coefficients differ significantly from class to class the results of a comparison of the corrected class means will vary according to the value of x at which the comparison is made. Under these circumstances such a comparison is of limited value, but may occasionally be required.

An approximate comparison of the class means is obtained by using the actual means $\bar{y}_i$, $\bar{x}_i$ and the estimated regression coefficients b_i to compute the expected value $\tilde{y}_i$ at some fixed x, say $\bar{x}$. If the sum of squares $\Sigma_i n_i(\tilde{y}_i - \bar{y})^2$ is computed, it may be used, together with the residual sum of squares, to form a variance ratio. On the hypothesis that there is no class effect upon the regression constant, the average value of $\Sigma_i n_i(\tilde{y}_i - \bar{y})^2$ is greater than σ^2 unless all the $\bar{x}_i$ are equal, so that the method leads to an overestimate of the significance level of the variance ratio.

The exact test may be obtained by consideration of the variance-covariance Table 7.55. Each of the sums of squares under the heading y^2 may be further resolved into a part due to regression and a residual.

TABLE 7.55

Source of Estimate	y^2	x^2	xy	Regression Coefficient
Between classes	$S_i(y)^2$	$S_i(x)^2$	$S_i(xy)$	$b_m = S_i(xy)/S_i(x^2)$
Within classes	$S_{\alpha(i)}(y^2)$	$S_{\alpha(i)}(x^2)$	$S_{\alpha(i)}(xy)$	$b_a = S_{\alpha(i)}(xy)/S_{\alpha(i)}(x^2)$
Total	$S(y^2)$	$S(x^2)$	$S(xy)$	$b_0 = S(xy)/S(x^2)$

We first test the significance of the mean regression within classes by forming the variance ratio based on the within-classes regression and the residual

$$\frac{b_a S_{\alpha(i)}(xy)\,(N-p-1)}{S_{\alpha(i)}(y^2) - b_a S_{\alpha(i)}(xy)}$$

If this ratio indicates a significant regression within classes the hypothesis of equal class means implies that b_a, b_m, and b_0 are estimates of a common regression coefficient. We may employ these estimates to compute the residual sum of squares, as in Table 7.56. In this case the residual sums

TABLE 7.56

Source of Estimate	Residual Sum of Squares	D.F.
Between classes	$S_i(y^2) - b_m S_i(xy) = S_1$	$p - 2$
Within classes	$S_{\alpha(i)}(y^2) - b_a S_{\alpha(i)}(xy) = S_2$	$N - p - 1$
Overall	$S(y^2) - b_0 S(xy) = S_3$	$N - 2$

of squares, divided by the appropriate number of degrees of freedom, should on our hypothesis form an estimate of σ^2. If from the third-row sum of squares we subtract the second-row sum of squares, the difference $S_i(y^2) + b_a S_{\alpha(i)}(xy) - b_0 S(xy)$ is the reduction in the residual sum of squares due to the inclusion of $p - 1$ independent class constants in the model. Comparison of this term with the residual of the first row of Table 7.56 indicates a difference of 1 degree of freedom, owing to the fact that the latter is obtained by employing b_m as the estimate of the regression coefficient. The difference between these terms gives an estimate of the variance due to the difference between b_a and b_m, so that the residual sum of squares corrected for regression may be analyzed as in Table 7.57. The term $S_3 - S_2 - S_1$ may be rearranged, since b_0, b_a, and b_m are not independent, as follows:

$$S_3 - S_2 - S_1 = b_a S_{\alpha(i)}(xy) + b_m S_i(xy) - \frac{[S(xy)]^2}{S(x^2)}$$

$$= \frac{[S_i(x^2) + S_{\alpha(i)}(x^2)]\,[(b_a S_{\alpha(i)}(xy) + b_m S_i(xy)]}{S(x^2)}$$

$$- \frac{[S_{\alpha(i)}(xy)]^2 + 2S_{\alpha(i)}(xy)S_i(xy) + [S_i(xy)]^2}{S(x^2)}$$

which reduces to

$$S_3 - S_2 - S_1 = \frac{S_i(x^2)b_a S_{\alpha(i)}(xy) + S_{\alpha(i)}(x^2)b_m S_i(xy) - 2S_{\alpha(i)}(xy)S_i(xy)}{S(x^2)}$$

$$= \frac{(b_a - b_m)^2 S_{\alpha(i)}(x^2)S_i(x^2)}{S(x^2)}$$

Also, if on our hypothesis, b_a and b_m are estimates of β, then

$$\text{ave } (b_a - b_m) = 0$$

$$\sigma^2{}_{b_a - b_m} = \sigma^2 \left(\frac{1}{S_{\alpha(i)}(x^2)} + \frac{1}{S_i(x^2)}\right)$$

so that

$$\frac{(b_a - b_m)^2 S_{\alpha(i)}(x^2)S_i(x^2)}{S(x^2)}$$

is an estimate of σ^2. In order to test the hypothesis that there is no class effect upon the regression constant, we form the variance ratio $\dfrac{S_3 - S_2}{p - 1}\dfrac{(N - p - 1)}{S_2}$ which corresponds to the ratio

$$\frac{\text{corrected mean square between classes}}{\text{corrected mean square within classes}}$$

and accept or reject the hypothesis on this basis.

TABLE 7.57

Source of Estimate	Sum of Squares	D.F.
Deviation of class means from regression of class means b_m	$S_i(y^2) - b_m S_i(xy) = S_1$	$p - 2$
Difference between b_a and b_m	$-b_0 S(xy) + b_a S_{\alpha(i)}(xy) + b_m S_i(xy) = S_3 - S_2 - S_1$	1
Deviations from mean regression within classes b_a	$S_{\alpha(i)}(y^2) - b_a S_{\alpha(i)}(xy) = S_2$	$N - p - 1$
Total deviations from overall regression b_0	$S(y^2) - b_0 S(xy) = S_3$	$N - 2$

EXAMPLE. The series of tests from which the data of Table 6.3 derive were carried out on samples of ball clay from 3 regions. Sixty samples were examined in all, 8 from North Devon, 36 from South Devon, and 16 from Dorset.

Without retabulating the data of Table 6.3 which refers to the South Devon samples, the remaining figures are given in Table 7.58.

TABLE 7.58

CARBON CONTENT OF 24 SAMPLES OF BALL CLAY*

x = carbon content determined by combustion
y = carboniferous material from rational analysis

North Devon				Dorset							
y	x	y	x	y	x	y	x	y	x	y	x
1.09	0.74	0.50	0.52	0.28	0.13	1.09	0.48	0.68	0.41	3.88	2.05
0.06	0.13	0.53	0.27	0.63	0.25	0.01	0.26	0.67	0.37	2.21	1.30
−0.17	0.15	0.65	0.99	0.44	0.19	−0.14	0.17	0.87	0.98	0.36	0.26
2.33	1.47	0.72	0.53	0.56	0.33	0.53	0.20	6.55	3.90	1.61	0.72

*Private communication from the Director of the British Ceramic Research Association.

Using 1, 2, and 3 for South Devon, North Devon, and Dorset, we compute the various sums of squares and cross products shown in Table 7.55, and obtain Table 7.59.

TABLE 7.59

Source of Estimate	y^2	x^2	xy	Regression Coefficient
Within class 1	450.555164	220.814431	313.268403	1.418695
Within class 2	4.045788	1.472200	2.216200	1.505364
Within class 3	44.123793	14.545200	24.901900	1.712036
Total within classes	498.724745	236.831831	340.386513	1.437250
Between classes	46.844113	21.680027	31.667045	1.460656
Total	545.568858	258.511858	372.053558	1.439213

To test for differences in regression coefficients between classes, we form the residual sums of squares shown in Table 7.60, and proceed as follows:

TABLE 7.60

Source of Estimate	Corrected Sums of Squares	D.F.	Estimate of Variance
Class 1	$S_\alpha(y_1^2) - b_1 S_\alpha(x_1 y_1) = 6.122847$	34	0.18008
Class 2	$S_\alpha(y_2^2) - b_2 S_\alpha(x_2 y_2) = 0.709596$	6	0.11827
Class 3	$S_\alpha(y_3^2) - b_3 S_\alpha(x_3 y_3) = 1.490844$	14	0.10649
Sum over classes	$S_{\alpha(i)}(y^2) - \Sigma_i b_i S_\alpha(x_i y_i) = 8.323287$	54	0.1541
Difference between regression coefficients	$\Sigma_i (b_i - b_a)^2 S_\alpha(x_i^2) = 1.180957$	2	0.5905
Total within classes corrected for mean regression	$S_{\alpha(i)}(y^2) - b_a S_{\alpha(i)}(x^2) = 9.504244$	56	0.1697

(1) Since our hypothesis includes the assumption of equal residual variances in the 3 classes, we check this assumption by the use of Bartlett's test:

$$M = \sum_{i=1}^{3} \nu_i \ln \left(\frac{\sum_{i=1}^{3} \nu_i s_i^2}{\sum_{i=1}^{3} \nu_i} \right) - \sum_{i=1}^{3} \nu_i \ln s_i^2$$

$$= 54 \ln 0.1541 - 34 \ln 0.1801 - 6 \ln 0.1183 - 14 \ln 0.1065$$

$$= 1.46$$

The appropriate 5% point is approximately 6, so that there is no evidence of lack of homogeneity. This we should have suspected since the largest and smallest variances give a ratio of about 1.7, which on the two-sided F-test corresponds to $\alpha = 0.4$.

(2) Testing for class effects upon the regression coefficients we obtain

$$\frac{0.5095}{0.1541} = 3.83^* \qquad (F_{2,\,54,\,0.05} = 3.2)$$

Thus the difference between the individual class regression coefficients is significant at the 0.05 level. This would indicate either that the assumptions made in the rational analysis are not equally applicable to the 3 groups of clays, or that the carbonaceous matter differs in carbon content in the 3 groups. Since the 3 lines are not parallel, no useful purpose is served by examining the between-class sum of squares when corrected for the mean regression. We might, however, be interested in determining whether the intercepts of the 3 regression equations differed significantly from zero. These may be conveniently computed as $\dfrac{S_\alpha(y_i) - b_i S_\alpha(x_i)}{n_i}$, whence $a_1 = 0.1240$, $a_2 = 0.1895$, $a_3 = -0.0197$. The joint variance estimate may be employed to compute the variance of the predicted value of y at $x = 0$. Thus from Section 6.24

$$s^2_{a_1} = 0.1541\left[\frac{1}{36} + \frac{(1.4187)^2}{220.814}\cdot\frac{69.25}{36}\right] = 0.00948$$

$$s_{a_1} = 0.0974$$

Similarly

$$s^2_{a_2} = 0.1046 \qquad s^2_{a_3} = 0.0271$$

$$s_{a_2} = 0.323 \qquad s_{a_3} = 0.164$$

from which

$$t_1 = \frac{0.1240}{0.0974} = 1.27 \qquad t_2 = \frac{0.1895}{0.323} = 0.59 \qquad t_3 = \frac{-0.0197}{0.164} = -0.12$$

and there is no evidence of an intercept which differs significantly from zero. It should be noted that, in using a common variance estimate to compute t_1, t_2, and t_3, we forfeit the independence of these statistics.

7.84. Covariance with Multiple Classification

The methods of Section 7.83 may be applied when the results can be classified according to two or more criteria. This case will occur most frequently when the data are collected in accordance with an experimental design, and additional observations of a variate x are included in an attempt to increase the accuracy of the comparisons. In this instance there will normally be the same number of observations on each cell of the classification. These circumstances are not essential to the method, but with unequal numbers in the various subclasses the analysis is more difficult.

Suppose that the experimental design involves a two-criterion classification into p classes and q groups and that n replicates are carried out within each cell of this classification. With the notation $y_{ij\alpha}$ indicating the αth replicate measurement in the ith class and the jth group, the sums of squares and sums of products about the overall mean may be expressed in the form

$$\Sigma_{ij\alpha}(y_{ij\alpha} - \bar{y})^2 = \Sigma_{ij\alpha}(y_{ij\alpha} - \bar{y}_{ij})^2 + \Sigma_{ij\alpha}(\bar{y}_{ij} - \bar{y}_i - \bar{y}_j + \bar{y})^2 + \Sigma_{ij\alpha}(\bar{y}_i - \bar{y})^2 + \Sigma_{ij\alpha}(\bar{y}_j - \bar{y})^2$$

and can be tabulated as in Table 7.61.

TABLE 7.61

Source of Estimate	y^2	x^2	xy	Regression Coefficient
Between replicates, within cells	$S_{\alpha(ij)}(y^2)$	$S_{\alpha(ij)}(x^2)$	$S_{\alpha(ij)}(xy)$	$b_a = \dfrac{S_{\alpha(ij)}(xy)}{S_{\alpha(ij)}(x^2)}$
Between cells, within groups and classes	$S_{ij}(y^2)$	$S_{ij}(x^2)$	$S_{ij}(xy)$	$b_{mij} = \dfrac{S_{ij}(xy)}{S_{ij}(x^2)}$
Between groups	$S_j(y^2)$	$S_j(x^2)$	$S_j(xy)$	$b_{mj} = \dfrac{S_j(xy)}{S_j(x^2)}$
Between classes	$S_i(y^2)$	$S_i(x^2)$	$S_i(xy)$	$b_{mi} = \dfrac{S_i(xy)}{S_i(x^2)}$
Total	$S(y^2)$	$S(x^2)$	$S(xy)$	$b_0 = \dfrac{S(xy)}{S(x^2)}$

Each of the sums of y^2 may be further subdivided into residuals and parts due to regression, the significance of the regression within cells being tested by forming the variance ratio

$$\frac{b_aS_{\alpha(ij)}(xy)}{1} \frac{N - pq - 1}{S_{\alpha(ij)}(y^2) - b_aS_{\alpha(ij)}(xy)} = F_{1,\,N-pq-1}$$

If the ratio is significant and we are not concerned with the possibility of difference between regression in individual cells, then

$$\frac{S_{\alpha(ij)}(y^2) - b_aS_{\alpha(ij)}(xy)}{N - pq - 1}$$

may be employed as an estimate of the residual variance in the population.

If it is required to examine the variation between cells of the estimates of the regression coefficients within cells, this may be undertaken by a procedure identical with that of Table 7.54 except that the comparison will take up $pq - 1$ degrees of freedom since there are now pq cells. Table 7.62 presents the corrected sums of squares of the y variate together with the appropriate number of degrees of freedom.

TABLE 7.62

Source of Estimate	Sum of Squares	D.F.
Between replicates within cells corrected for regression within cells	$S_{\alpha(ij)}(y^2) - b_a S_{\alpha(ij)}(xy) = S_4$	$N - pq - 1$
Between cells within groups and classes corrected for regression within groups and classes	$S_{ij}(y^2) - b_{mij} S_{ij}(xy) = S_3$	$(p - 1)(q - 1) - 1$
Between groups corrected for regression between groups	$S_j(y^2) - b_{mj} S_j(xy) = S_2$	$q - 2$
Between classes corrected for regression between classes	$S_i(y^2) - b_{mi} S_i(xy) = S_1$	$p - 2$
Overall sum of squares corrected for overall regression	$S(y^2) - b_0 S(xy) = S_5$	$N - 2$

As in Table 7.56, each of these components of the sum of squares leads to an estimate of σ^2 on the hypothesis that there is no class effect, but in addition the hypothesis requires that b_a, b_{mij}, b_{mj}, and b_{mi} be estimates of a common regression coefficient β. The 3 independent comparisons between these 4 estimates take up the 3 remaining degrees of freedom.

In order to obtain a satisfactory joint test we employ the method of the previous section. When hypotheses concerning a number of effects and interactions are to be examined by this method, it is convenient to draw up a series of separate tables from which the required regression coefficients may be estimated. As an example we consider the interaction in a two-way classification as shown in Table 7.63.

TABLE 7.63

Source of Estimate	y^2	x^2	xy	Regression Coefficient
Between cells within groups and classes	$S_{ij}(y^2)$	$S_{ij}(x^2)$	$S_{ij}(xy)$	$b_{mij} = \dfrac{S_{ij}(xy)}{S_{ij}(x^2)}$
Between replicates within cells	$S_{\alpha(ij)}(y^2)$	$S_{\alpha(ij)}(x^2)$	$S_{\alpha(ij)}(xy)$	$b_a = \dfrac{S_{\alpha(ij)}(xy)}{S_{\alpha(ij)}(x^2)}$
Sum	$S_{ij}(y^2) + S_{\alpha(ij)}(y^2)$	$S_{ij}(x^2) + S_{\alpha(ij)}(x^2)$	$S_{ij}(xy) + S_{\alpha(ij)}(xy)$	$b_0' = \dfrac{S_{ij}(xy) + S_{\alpha(ij)}(xy)}{S_{ij}(x^2) + S_{\alpha(ij)}(x^2)}$

Denoting the regression coefficient in the sum of interaction and residual components by b_0' and the corrected sums of squares by S_3, S_4, and S_5' we obtain Table 7.64.

TABLE 7.64

Source of Estimate	Sum of Squares	D.F.
Between cells within groups and classes corrected for regression b_{mij}	S_3	$(p-1)(q-1)-1$
Between replicates within cells corrected for regression b_a	S_4	$N - pq - 1$
Sum corrected for regression b_0'	$S_{ij}(y^2) + S_{\alpha(ij)}(y^2) - b_0' S_{ij}(xy) - b_0' S_{\alpha(ij)}(xy) = S_5'$	$N - p - q$

It should be noted that $S_3 + S_4 \neq S_5'$, since the corrections for regression are based on different regression coefficients. In fact, by the method of the previous section we can show that

$$S_5' - S_4 - S_3 = (b_{mij} - b_a)^2 \frac{S_{\alpha(ij)}(x^2) S_{ij}(x^2)}{S_{\alpha(ij)}(x^2) + S_{ij}(x^2)} = S_3'$$

which, on the hypothesis that b_{mij} and b_a are estimates of a common regression coefficient β, is distributed as $\sigma^2\chi^2$ with 1 degree of freedom. In the usual type of problem we are interested in determining whether real class and interaction effects exist and only infrequently in determining whether these effects are to be explained by a difference in regression coefficients or by a non-linear variation of the class or cell means. Since S_3 and $S_5' - S_4 - S_3$ are distributed as $\sigma^2\chi^2$ with $(p - 1)(q - 1) - 1$ and 1 degree of freedom, the sum $S_5' - S_4$ is also distributed as $\sigma^2\chi^2$ with $(p - 1)(q - 1)$ degrees of freedom. Thus in order to make a combined test for the significance of interaction effects it is necessary to compute only the corrected sums of squares S_5' and S_4, the significance test being obtained from the variance ratio $(S_5' - S_4)(N - pq - 1)/S_4(p - 1)(q - 1)$. Since S_4 is common to all tests for interaction and effects, only 1 regression coefficient need be recomputed when a new component is to be tested. The formal analysis of variance for a $p \times q$ classification is given in Table 7.65. For the sake of completeness the corrected interaction and main effects have been separated into components $S_1 + S_1'$, $S_2 + S_2'$, $S_3 + S_3'$, although this elaboration is usually unnecessary.

As in a normal analysis of variance, we should first test the significance of the interaction effect $P \times Q$ by forming the variance ratio

$$\frac{S_3 + S_3'}{(p - 1)(q - 1)} \frac{N - pq - 1}{S_4}$$

If this is not significant we may pool $S_3 + S_3' + S_4$ to obtain the estimate of the residual variance based on $N - p - q$ degrees of freedom before testing the group and class effects by the variance ratios

$$\frac{S_2 + S_2'}{q - 1} \frac{N - p - q}{S_3 + S_3' + S_4}$$

and

$$\frac{S_1 + S_1'}{p - 1} \frac{N - p - q}{S_3 + S_3' + S_4}$$

If a significant interaction exists the group and class mean squares may be tested against either the corrected mean square obtained from the interaction or the residual mean square, depending on the type of model employed.

Although the method of covariance analysis has been illustrated by reference to a two-factor classification with replication, it is equally applicable to data from multiple classifications. In Chapter 8 where

TABLE 7.65

Source of Estimate	Corrected Sum of Squares	D.F.
Between classes corrected for regression of class means b_{mi}	$= S_1$	$p - 2$
Difference between b_{mi} and b_a	$(b_{mi} - b_a)^2 \dfrac{S_i(x^2)S_{\alpha(ij)}(x^2)}{S_i(x^2) + S_{\alpha(ij)}(x^2)} = S_1'$	1
Between groups corrected for regression of group means b_{mj}	$= S_2$	$q - 2$
Difference between b_{mj} and b_a	$(b_{mj} - b_a)^2 \dfrac{S_j(x^2)S_{\alpha(ij)}(x^2)}{S_j(x^2) + S_{\alpha(ij)}(x^2)} = S_2'$	1
Between cells, within groups and classes corrected for regression of cell means b_{mij}	$= S_3$	$(p - 1)(q - 1) - 1$
Difference between b_{mij} and b_a	$(b_{mij} - b_a)^2 \dfrac{S_{ij}(x^2)S_{\alpha(ij)}(x^2)}{S_{ij}(x^2) + S_{\alpha(ij)}(x^2)} = S_3'$	1
Between replicates, within cells corrected for regression within cells b_a	$= S_4$	$N - pq - 1$
Total sum of squares corrected for overall regression	$S_1 + S_1' + S_2 + S_2' + S_3 + S_3' + S_4 = S_5$	$N - 2$

experimental designs involving multifactor classifications are considered, the use of covariance analysis is illustrated by examples in Sections 8.54 and 8.63. In cases where it is desirable to break down the treatment sums of squares into a series of orthogonal comparisons, these may also be computed in a form corrected for the effect of regression. The correction of each comparison involves the computation of a regression coefficient based on the sum of the error and comparison covariances, which is time consuming. Finney [46] has proposed an approximate method whereby this may be avoided and which is satisfactory, if the number of degrees of freedom in the residual term is high.

7.85. Analysis of Covariance Involving Multiple Regression

The use of analysis of covariance to increase the accuracy of comparisons in experimental data may be extended to cases in which, in addition to the classification criteria, measurements are made of a number of fixed variables in order to eliminate from the comparison the effects of the regression of y upon these variates. The method is similar to that of Table 7.54 except that a series of regression coefficients for each of the x variates is estimated, resulting in a reduction in the number of degrees of freedom available for the final error estimate.

EXAMPLE. This method is illustrated by the following example based on data from a survey of the quality of coke as measured by the 2-inch shatter index at a number of British carbonizing units. The investigation was undertaken to determine whether the volatile matter content of the coal would provide an adequate index of the coal quality in relation to the 2-inch shatter index, or whether, after adjustment for this factor, there existed an inherent difference between the 4 classifications of coal A, B, C, and D. The factor of classification is the coalfield or area in which the coking plants were situated. Although some of the plants blend coals from more than one coalfield the classification may be taken to reflect the performance of coals from various coalfields. Since other operational factors such as oven width, carbonizing rate, and size distribution of the coal charge varied from plant to plant, it was desirable to eliminate the influence of these factors in addition to that of the volatile content of the coal before testing the significance of class differences. The problem may be viewed as a multiple regression problem in which we wish to determine whether the incorporation of a separate constant term for each class in our descriptive model will significantly reduce the estimate of the residual variance. Table 7.66 presents the coded data for this investigation, the variable oven width, which does not prove significant, being excluded.

TABLE 7.66

TWO-INCH SHATTER INDICES OF A NUMBER OF BRITISH COKES*

y = 2-inch shatter index
x_1 = coking rate (in./hr)
x_2 = volatile content of coal — 31%
x_3 = percentage of coal less than 1/8 inch
A, B, C, and D = coal classifications

Coal Class	x_1	x_2	x_3	y	Coal Class	x_1	x_2	x_3	y
A	0.884	+3.2	80	60.0	C	1.125	+4.0	67	58.0
A	0.923	+1.9	70	61.0	C	0.750	+2.0	75	62.3
A	0.883	−0.2	75	73.0	C	0.778	+1.4	75	65.6
A	0.662	−4.5	100	84.0	C	0.972	+5.4	21	48.7
A	0.914	+2.0	69	71.0	C	0.750	+1.5	54	68.0
A	0.812	+1.1	64	70.5	C	1.037	0.0	39	59.0
A	0.870	+0.4	75	73.0	C	1.040	0.0	68	55.0
A	0.859	+1.8	83	70.2	C	1.042	+2.2	50	58.6
A	0.875	−5.0	67	77.0	C	1.040	0.0	79	55.0
A	0.980	+0.7	79	62.0	C	1.037	+2.9	95	60.0
A	0.773	+1.3	82	68.0	C	1.000	+1.5	45	60.0
A	0.773	−1.4	90	69.0	C	1.003	+3.4	57	57.0
A	1.040	0.0	42	65.0	C	0.934	+4.0	50	68.0
A	0.950	+0.6	72	76.0	C	1.044	+1.6	64	56.5
A	1.116	−4.2	74	72.1	C	1.050	+1.4	80	56.0
A	0.550	+1.4	39	71.0	C	0.990	+4.5	63	65.0
B	0.792	−9.0	83	79.4	C	0.668	+1.0	95	79.0
B	0.900	−6.0	90	87.0	D	1.044	+6.2	56	46.0
B	0.528	−9.6	100	86.0	D	1.073	+2.0	80	62.0
B	0.930	−6.3	83	82.5	D	0.923	+5.5	70	44.0
B	0.947	−8.4	90	79.5	D	0.809	+2.1	38	50.0
B	0.500	−7.4	69	96.0	D	0.826	+0.2	73	80.0
B	0.580	−9.5	90	94.0	D	0.954	+4.7	35	53.2
B	0.469	−10.0	92	94.0	D	0.720	+4.4	80	69.5

* Data by courtesy of the Director, British Coke Research Association.

In order to proceed further the variance and covariance estimates are required. These are given in tabular form in Table 7.67, the within-class sums of squares being computed as

$$S_{\alpha(i)}(x_j x_k) = S(x_j x_k) - \Sigma_i n_i(\bar{x}_{ij} - \bar{x}_{\cdot j})(\bar{x}_{ik} - \bar{x}_{\cdot k})$$

TABLE 7.67

VARIANCES AND COVARIANCES

Source of Estimate \ Type of Product	x_1^2	x_2^2	x_3^2	x_1x_2	x_1x_3	x_2x_3	D.F.
Between classes	0.350031	716.954	3,681.9	15.0262	−34.6143	−1587.81	3
Within classes	0.965507	179.542	12,340.1	2.0384	−12.3161	−399.03	44
Total	1.315538	896.476	16,022.0	17.0646	−46.9314	−1986.84	47

	y^2	yx_1	yx_2	yx_3	D.F.
Between classes	4,612.54	38.4771	−1788.71	4,117.22	3
Within classes	2,685.99	27.0342	− 313.68	1,951.88	44
Total	7,298.53	65.5113	−2102.39	6069.10	47

From this table we can determine the multiple regression coefficients corresponding to the rows "within-classes" and "Total." The former gives the mean multiple regression within classes and the latter the overall regression obtained when the classification is ignored. The inverse matrices for the two solutions are given in the accompanying tabulations.

Within classes

+1.067,149	−0.010,503,5	+0.000,725,44
−0.010,503,5	+0.006,104,4	+0.000,186,9
+0.000,725,44	+0.000,186,9	+0.000,087,804

Total

+1.016,702	−0.017,585,49	+0.000,797,383
−0.017,585,49	+0.001,842,360	+0.000,176,954
+0.000,797,383	+0.000,176,954	+0.000,086,693,4

From these the corresponding regression coefficients, together with the variance of the regression coefficients, may be computed and tested for significance.

Within classes

$$\Sigma(y - \tilde{y})^2 = 1454.5 \qquad s^2 = 35.476$$

$$b_1 = -24.139, \quad s\sqrt{c_{11}} = 6.153, \quad t = 3.92^{***} \qquad (t_{41,\,0.001} = 3.55)$$

$$b_2 = -1.2661, \quad s\sqrt{c_{22}} = 0.4653, \quad t = 2.72^{**} \qquad (t_{41,\,0.01} = 2.70)$$

$$b_3 = +0.09314, \quad s\sqrt{c_{33}} = 0.0558, \quad t = 1.67 \qquad (t_{41,\,0.10} = 1.68)$$

Total

$$\Sigma(y - \tilde{y})^2 = 1589.9$$

$$b_1 = -24.836$$

$$b_2 = -0.647$$

$$b_3 = +0.1019$$

The effects of volatile content of the coal and coking rate are definitely significant for the regression within classes. The influence of the size distribution of the coal charge, as measured by the index "Percentage of coal less than $\frac{1}{8}$ inch," is of doubtful significance.

In this analysis we wish to determine whether class differences exist. It is not important to determine whether they correspond to deviations from the between-classes regression or to differences between this regression and the regression within classes. Accordingly, we compute the corrected sum of squares between classes as 1589.9 — 1454.5 = 135.4. The final analysis of variance is given in Table 7.68. In order to test the hypothesis that there

TABLE 7.68

Source of Estimate	Corrected Sum of Squares	D.F.	Estimate of Variance
Between classes	135.4	3	45.10
Within classes	1454.5	41	35.47
Total	1589.9	44	

are no significant differences between the class totals after correction for regression on the independent variates, we compare the mean square between classes with the residual variance estimate, obtaining

$$F = \frac{45.10}{35.47} = 1.27 \qquad (F_{3,\,41,\,0.20} = 1.6)$$

As this is not significant, we accept the hypothesis, and conclude that the overall regression equation provides a satisfactory description of the data for the 4 classes, and that no additional class effects can be detected. In this context satisfactory implies only that the description is not significantly improved by fitting individual class constants. The inference is conditional in the sense that it is theoretically restricted to the population of alternative observations which could have been made with the fixed values of the independent variables. In practice this restriction usually implies that we can expect the regression equation to describe additional observations in a given class, provided that the values of the independent variables associated with these new observations are within the range covered by previous measurements. Thus we might reasonably expect to predict the 2-inch shatter index from a coal of 24% volatile matter of class *B*, carbonized at 0.75 in./hr. We should proceed more cautiously in extrapolating our predictions to a coal from this class of 35% volatile matter carbonized at 1.10 in./hr, since we have no data from this class for this region.

7.86. Comments on the Application of Analyses of Covariance

Although this technique does not appear to have been widely used in chemical investigations either for increasing the efficiency of designed experiments or for comparison of regression relationships in a number of classes of materials, it is well suited for use in this kind of work. In investigations of relationships by regression methods 2 types of problem which can best be understood by this approach are a frequent source of difficulty.

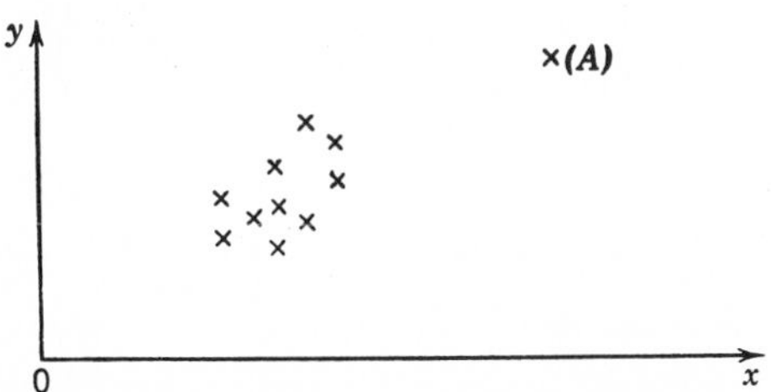

FIG. 7.3. Observations with one outlying point.

(*a*) *The problem of "outlier" points.* Figure 7.3 illustrates a hypothetical example of this problem in simple linear regression. If the data are analyzed as a sample from a homogeneous population, the regression of y upon x is significant at the 0.05 level. A similar analysis excluding the point A indicates no significant regression. The single point A and the remaining observations may thus be regarded as 2 classes showing no significant regression within classes. The between-classes regression is responsible for the overall regression in the data, and this is indistinguishable from any class effect upon the regression constants within the 2 classes. Thus, for practical purposes, if further data cannot be immediately acquired, it is necessary to examine the source of the data, to decide whether there is any basis for the subclassification other than the observed x and y values. If no such basis can be detected, we should accept the observed regression in the complete data, with reservation. In cases where the point A can be classified separately from the remainder of the data, we have no basis for supposing a significant regression of y upon x, since the observed effect might be due entirely to class differences.

(*b*) *The pooling of data.* Most chemical research projects are preceded by an examination of the results obtained by previous workers in the same field. In searching for relationships in these data it is often desirable to be able to pool the results of a number of workers who have employed different techniques, since by this means the range of the observations is increased and relationships are more easily detected. Difficulties frequently arise when this method is employed, if the range of the independent variates examined differs from worker to worker since we are in danger of accepting, as an indication of a significant regression on an independent variate, a set of data in which the non-random variations are due in fact to differences in experimental technique.

A similar example arises in attempting to determine the relationship between the Nusselt and Prandtl numbers under fixed flow conditions. In order to extend the range of Prandtl numbers it may be desirable to pool results obtained from a wide variety of fluids, say mercury, gases, water, and paraffin oils. If we undertake experiments on these 4 types of fluid, an analysis of covariance carried out on the results will indicate:

(1) Whether there is a significant relationship between the Nusselt and Prandtl numbers within the 4 classes of fluid.

(2) Whether there is any evidence that a different exponent of the Prandtl number should be used for different classes of fluid.

(3) Whether the relationship satisfactorily describes the observations, or whether some other factor causes additional variation between the 4 classes of fluid.

It may happen, since the range of Prandtl numbers within each class is limited, that, whereas we can reject the hypothesis that there is no overall regression in the observations, there is no evidence of a significant relationship under (1) and (2). The rejection of the first hypothesis implies only that, for the 4 classes of fluid examined, and with fixed mean values of the Prandtl numbers, there is a significant regression within the observations of the Nusselt number which could have been obtained.

The extension of the inference to cover the case of different fluids and different values of the mean Prandtl number can best be judged by comparing the component of the sum of squares due to regression with the component due to deviations of the class means from the regression line; departures from linearity are tested by comparing this latter estimate with the within-classes estimate of the variance.

Before undertaking an analysis of covariance, it is important to define the objectives for which the original data were accumulated in order to ensure that we do not eliminate a part of the effects which we wish to compare. Suppose, for example, that we have collected data from a number of steam-raising units and wish to compare their thermal efficiency in terms of "pounds of standard steam per pound of fuel consumed." It would be logical to use the method of covariance analysis to correct the data for such effects as the quality of fuel used, the shift during which the observation was made, and the output in proportion to the rated capacity of the unit. It would not be logical to correct the efficiency for the absolute value of the output, since the units of higher rated capacity will tend to be more efficient, and in making a correction to a standard absolute output we should eliminate a part of the difference between the various units which we are seeking to examine. If, however, we had wished to compare, not the efficiencies of the units, but the influence of a number of design modifications upon efficiency, it would then be logical to eliminate from the observations the effect of the rated capacity of the units to which the modifications were made, and we might well employ the absolute value of the output as an independent variate in the covariance analysis.

In general, if the range of values of an independent variate is a characteristic of the various classes between which comparisons are to be made, the objectives of the analysis should be considered carefully before steps are taken to eliminate the effect of this variable.

7.9. Relationship between Variates Each Including A Random Component

This subject, which was mentioned in the introduction to the previous chapter can be regarded as an example in variance component analysis, and has been deferred for this reason. Problems of this type can be reduced to simple regression problems by restricting the inference to observations in which the errors in the x variate or variates are held constant. The information obtained with such restrictions may be of use for prediction under special conditions, but cannot relate to the relationship between the variates.

7.91. Simple Regression with Known Slope

If we can assume that the slope of the regression line is known, as might be true if 2 direct analytical techniques were to be compared on a number of samples, our model takes the form

$$x_i = \theta_i + \varepsilon_i \qquad y_i = \alpha + \beta\theta_i + \eta_i \qquad i = 1, \cdots, n$$

where we assume that the ε_i and η_i are randomly distributed with zero mean, variance σ_ε^2 and σ_η^2, and covariance cov (ε, η). The slope β is known, and we require an estimate of α. Clearly

$$\begin{aligned} \text{ave } (y_i - \beta x_i) &= \text{ave } (\alpha + \beta\theta_i + \eta_i - \beta\theta_i - \beta\varepsilon_i) \\ &= \alpha + \text{ave } (\eta_i) - \beta \text{ ave } (\varepsilon_i) \\ &= \alpha \end{aligned}$$

so that $\hat{\alpha} = \frac{1}{n} \Sigma_i(y_i - \beta x_i)$ is an unbiased estimate of α. Also

$$\begin{aligned} \text{var } (y_i - \beta x_i) &= \text{var } (\alpha + \eta_i - \beta\varepsilon_i) \\ &= \sigma_\eta^2 + \beta^2\sigma_\varepsilon^2 - 2\beta \text{ cov } (\varepsilon, \eta) \end{aligned}$$

and, assuming the pairs (ε_i, η_i) to be independent, we have

$$\text{var } \hat{\alpha} = \frac{1}{n} \text{ var } (y_i - \beta x_i)$$

The quantity

$$\frac{1}{n-1} \Sigma_i(y_i - \beta x_i - \hat{\alpha})^2 = \frac{1}{n-1} \{\Sigma_i(y_i - \beta x_i)^2 - n\hat{\alpha}^2\}$$

is an unbiased estimate of var $(y_i - \beta x_i)$ based on $n - 1$ degrees of freedom, and can be used in the usual manner to test the significance of $\hat{\alpha}$ and obtain confidence intervals. If $\beta = 1$, the above procedure reduces to the simple paired comparison test of Section 5.52.

Making the additional assumption that the θ_i are independent of the ε_i and η_i, we have

$$\text{ave } \frac{1}{n-1} \Sigma_i(x_i - \bar{x})^2 = \text{var } (\theta) + \text{var } (\varepsilon)$$

$$\text{ave } \frac{1}{n-1} \Sigma_i(x_i - \bar{x})(y_i - \bar{y}) = \beta \text{ var } (\theta) + \beta \text{ cov } (\varepsilon, \eta)$$

$$\text{ave } \frac{1}{n-1} \Sigma_i(y_i - \bar{y})^2 = \beta^2 \text{ var } (\theta) + \text{var } (\eta)$$

and, if cov $(\varepsilon, \eta) = 0$ and β is known,

$$\hat{\sigma}_\varepsilon^{\,2} = \frac{1}{n-1}\left[\Sigma_i(x_i - \bar{x})^2 - \frac{1}{\beta}\,\Sigma_i(x_i - \bar{x})\,(y_i - \bar{y})\right]$$

$$\hat{\sigma}_\eta^{\,2} = \frac{1}{n-1}\,[\Sigma_i(y_i - \bar{y})^2 - \beta\,\Sigma_i(x_i - \bar{x})\,(y_i - \bar{y})]$$

7.92. Simple Regression with Slope Unknown

If β is not known, we cannot obtain estimates of $\sigma_\varepsilon^{\,2}$, $\sigma_\eta^{\,2}$, and β from the above equations, even if cov $(\varepsilon, \eta) = 0$, unless an additional relationship is available. The simplest additional relationship is one of the form $k\sigma_\varepsilon^{\,2} = \sigma_\eta^{\,2}$, and under these conditions we have

$$\text{ave}\,\frac{1}{n-1}\,\Sigma_i(x_i - \bar{x})^2 = \text{var}\,(\theta) + \text{var}\,(\varepsilon)$$

$$\text{ave}\,\frac{1}{n-1}\,\Sigma_i(x_i - \bar{x})\,(y_i - \bar{y}) = \beta\,\text{var}\,(\theta) + \text{cov}\,(\varepsilon, \eta)$$

$$\text{ave}\,\frac{1}{n-1}\,\Sigma_i(y_i - \bar{y})^2 = \beta^2\,\text{var}\,(\theta) + k\,\text{var}\,(\varepsilon)$$

so that, if cov $(\varepsilon, \eta) = 0$,

$$\frac{\beta^2 - k}{\beta} = \frac{1}{n-1}\,\frac{\text{ave}\,\Sigma_i(y_i - \bar{y})^2 - k\,\text{ave}\,\Sigma_i(x_i - \bar{x})^2}{\text{ave}\,\Sigma_i(x_i - \bar{x})\,(y_i - \bar{y})}$$

and we obtain an unbiased estimate of β by using the actual values $\Sigma_i(y_i - \bar{y})^2$, $\Sigma_i(x_i - \bar{x})^2$, and $\Sigma_i(y_i - \bar{y})\,(x_i - \bar{x})$ in the above expression. With an unbiased estimate of β we can estimate $\sigma_\varepsilon^{\,2}$, and hence $\sigma_\eta^{\,2}$. In the more general case where k is unknown some additional information is required to obtain a solution.

7.93. The Use of Instrumental Variates

The simplest case of the use of instrumental variates is that in which the data can be separated into two or more groups. The model to be considered is of the form

$$x_{ij} = \mu + \theta_i + \omega_{ij} + \varepsilon_{ij} \qquad j = 1, \cdots, n_i$$

$$y_{ij} = \alpha + \beta(\mu + \theta_i) + \beta_i\omega_{ij} + \eta_{ij} \qquad i = 1, \cdots, k$$

the number of observations within the ith group being n_i, $\Sigma n_i = N$, and the total number of groups k. We shall assume the average values of the ε_{ij} and η_{ij} to be zero and designate their variances and covariance by var (ε), var (η), and cov (ε, η) as usual. In addition we shall designate the variance of the ω_{ij} within the ith class by var (ω_i) and the variance of the θ_i by var (θ). Making the usual assumptions as to independence except in the case of the pairs $(\varepsilon_{ij}, \eta_{ij})$, we can compute the average values of the sums of squares needed in the analysis of the variances and covariances of x and y as follows:

Within groups

$$\text{ave } \Sigma_j(x_{ij} - \bar{x}_{i\cdot})^2 = (n_i - 1)\,[\text{var}\,(\omega_i) + \text{var}\,(\varepsilon)]$$

$$\frac{1}{N-k}\text{ ave } \Sigma_i\Sigma_j(x_{ij} - \bar{x}_{i\cdot})^2 = \sum_{i=1}^{k}\frac{(n_i - 1)}{N-k}\text{ var}\,(\omega_i) + \text{var}\,(\varepsilon)$$

$$\frac{1}{N-k}\text{ ave } \Sigma_i\Sigma_j(x_{ij} - \bar{x}_{i\cdot})\,(y_{ij} - \bar{y}_{i\cdot}) = \sum_{i=1}^{k}\frac{(n_i - 1)}{N-k}\beta_i\text{ var}\,(\omega_i) + \text{cov}\,(\varepsilon, \eta)$$

$$\frac{1}{N-k}\text{ ave } \Sigma_i\Sigma_j(y_{ij} - \bar{y}_{i\cdot})^2 = \sum_{i=1}^{k}\frac{(n_i - 1)}{N-k}\beta_i^2\text{ var}\,(\omega_i) + \text{var}\,(\eta)$$

Between groups

$$\text{ave } \Sigma_i n_i(\bar{x}_{i\cdot} - \bar{x})^2 = \text{ave } \Sigma_i n_i(\theta_i - \bar{\theta})^2 + \text{ave } \Sigma_i n_i(\bar{\omega}_{i\cdot} - \bar{\omega})^2 + \text{ave } \Sigma_i n_i(\bar{\varepsilon}_{i\cdot} - \bar{\varepsilon})^2$$

and, since

$$\text{ave } \Sigma_i n_i(\theta_i - \bar{\theta})^2 = \frac{N^2 - \Sigma n_i^2}{N}\text{ var}\,(\theta)$$

$$\text{ave } \Sigma_i n_i(\bar{\omega}_{i\cdot} - \bar{\omega})^2 = \sum_{i=1}^{k}\left(n_i - \frac{n_i^2}{N}\right)\text{ var}\,(\bar{\omega}_{i\cdot}) = \sum_{i=1}^{k}\left(1 - \frac{n_i}{N}\right)\text{ var}\,(\omega_i)$$

$$\text{ave } \Sigma_i n_i(\bar{\varepsilon}_{i\cdot} - \bar{\varepsilon})^2 = (k - 1)\text{ var}\,(\varepsilon)$$

we have

$$\frac{1}{k-1}\text{ ave } \Sigma_i n_i(\bar{x}_{i\cdot} - \bar{x})^2 = \frac{N^2 - \Sigma n_i^2}{N(k-1)}\text{ var}\,(\theta) + \sum_{i=1}^{k}\frac{\left(1 - \dfrac{n_i}{N}\right)}{k-1}\text{ var}\,(\omega_i) + \text{var}\,(\varepsilon)$$

and similarly

$$\frac{1}{k-1}\text{ ave } \Sigma_i n_i(\bar{x}_{i\cdot} - \bar{x})\,(\bar{y}_{i\cdot} - \bar{y}) = \frac{N^2 - \Sigma n_i^2}{N(k-1)}\beta\text{ var}\,(\theta) + \sum_{i=1}^{k}\frac{\left(1 - \dfrac{n_i}{N}\right)}{k-1}\beta_i\text{ var}\,(\omega_i) + \text{cov}\,(\varepsilon, \eta)$$

$$\frac{1}{k-1}\text{ ave } \Sigma_i n_i(\bar{y}_{i\cdot} - \bar{y})^2 = \frac{N^2 - \Sigma n_i^2}{N(k-1)}\beta^2\text{ var}\,(\theta) + \sum_{i=1}^{k}\frac{\left(1 - \dfrac{n_i}{N}\right)}{k-1}\beta_i^2\text{ var}\,(\omega_i) + \text{var}\,(\eta)$$

TABLE 7.69

Source of Estimate		Sum of Squares	D.F.	Average Value of M.S.
Between groups	(I)	$n\,\Sigma_i(\bar{x}_{i\cdot} - \bar{x})^2$	$k - 1$	$n \operatorname{var}(\theta) + \frac{1}{k}\sum_i^k \operatorname{var}(\omega_i) + \operatorname{var}(\varepsilon)$
	(II)	$n\,\Sigma_i(\bar{x}_{i\cdot} - \bar{x})(\bar{y}_{i\cdot} - \bar{y})$		$n\beta \operatorname{var}(\theta) + \frac{1}{k}\sum_i^k {}_i\beta \operatorname{var}(\omega_i) + \operatorname{cov}(\varepsilon, \eta)$
	(III)	$n\,\Sigma_i(\bar{y}_{i\cdot} - \bar{y})^2$		$n\beta^2 \operatorname{var}(\theta) + \frac{1}{k}\sum_i^k \beta_i^2 \operatorname{var}(\omega_i) + \operatorname{var}(\eta)$
Within groups	(IV)	$\Sigma_i\Sigma_j(x_{ij} - \bar{x}_{i\cdot})^2$	$N - k$	$\frac{1}{k}\sum_i^k \operatorname{var}(\omega_i) + \operatorname{var}(\varepsilon)$
	(V)	$\Sigma_i\Sigma_j(x_{ij} - \bar{x}_{i\cdot})(y_{ij} - \bar{y}_{i\cdot})$		$\frac{1}{k}\sum_i^k \beta_i \operatorname{var}(\omega_i) + \operatorname{cov}(\varepsilon, \eta)$
	(VI)	$\Sigma_i\Sigma_j(y_{ij} - \bar{y}_{i\cdot})^2$		$\frac{1}{k}\sum_i^k \beta_i^2 \operatorname{var}(\omega_i) + \operatorname{var}(\eta)$

If we can assume that var ω_i is constant for all k classes, or if the n_i are equal, these equations may be used to estimate β. We consider the second case, obtaining a variance-covariance analysis in the form shown in Table 7.69. From this table we have

$$\frac{1}{n} \text{ ave } [\text{M.S. (I)} - \text{M.S. (IV)}] = \text{var } (\theta)$$

$$\frac{1}{n} \text{ ave } [\text{M.S. (II)} - \text{M.S. (V)}] = \beta \text{ var } (\theta)$$

$$\frac{1}{n} \text{ ave } [\text{M.S. (III)} - \text{M.S. (VI)}] = \beta^2 \text{ var } (\theta)$$

so that β can be estimated from the ratios of these combinations of mean squares. If we also assume cov $(\varepsilon, \eta) = 0$ and $\beta_i = \beta$, we can estimate var (ε) and var (η), and form a confidence interval for β. To accomplish the latter we consider the analysis of $y - Bx$, where B is an arbitrary constant which we shall regard as a trial value for β. Under the hypothesis $B = \beta$ we have

$$\frac{\text{M.S. (III)} - 2\beta \text{ M.S. (II)} + \beta^2 \text{ M.S. (I)}}{\text{M.S. (VI)} - 2\beta \text{ M.S. (V)} + \beta^2 \text{ M.S. (IV)}}$$

distributed as $F_{(k-1),\,(n-k)}$. An $\alpha\%$ confidence interval for β is then obtained by putting

$$\frac{\text{M.S. (III)} - 2B \text{ M.S. (II)} + B^2 \text{ M.S. (I)}}{\text{M.S. (VI)} - 2B \text{ M.S. (V)} + B^2 \text{ M.S. (IV)}} = F_{(k-1),\,(n-k),\,\alpha}$$

and taking the solutions for B in this quadratic as the confidence limits.

The only assumptions concerning the groups which have been necessary in the foregoing analysis for β are

(1) All the θ_i are not zero.

(2) In any group ave (ε_{ij}) = ave $(\eta_{ij}) = 0$—that is, the covariance between groups and residuals is zero. To obtain confidence limits for β we also assumed zero covariance between the residuals ε_{ij} and η_{ij} for each group. It was also assumed at the outset that the relationship was truly linear so that $\beta_{yx} = \dfrac{1}{\beta_{xy}}$. The confidence interval for β which we obtain is conditional on this assumption.

We have

$$Pr(B_1 < \beta < B_2 \mid \beta_{\text{exists}}) = \alpha$$

where α is the confidence level of the F-ratio employed in solving for B. The alternative that β does not exist can be examined by determining the largest value of α for which the roots B_1 and B_2 are real. At any larger value of α the confidence limits for β are imaginary and β is non-existent. This is equivalent to the rejection of our original model for the linear relationship between x and y.

To deal with the case in which no information on grouping is available, Wald [104] and Bartlett [105] have proposed arbitrary subdivisions of the data. In the latter method the observations are arbitrarily divided into 3 groups on the basis of the values of one of the variates. As an alternative to natural or artificial groups to provide the additional information necessary, Geary [106] and Tukey [97] have shown that a third variate z may be so employed. In this case, if the covariances of z with the random components of x and y vanish, the sums of squares and cross products of x and y can be separated into components linearly dependent upon z and residuals which are linearly independent of z. These take the place of the between- and within-classes components, respectively, of this section.

APPENDIX 7A

The Average Values of Variance Components in Models of Type III

7A.1. Nested Classifications

In a two-way classification with 1 factor nested we have for a model

$$x_{ij\alpha} = \mu + \xi_i + \lambda_{j(i)} + \varepsilon_{ij\alpha}$$

$$i = 1, \cdots, p \qquad j = 1, \cdots, q \qquad \alpha = 1, \cdots, n$$

and we take the ξ_i to be random samples from a population of size P, the $\lambda_{j(i)}$ to be samples from a population of size PQ which are subject to the conditions $\sum_j^Q \lambda_{j(i)} = 0$ for all i, but are otherwise random, and the residual components $\varepsilon_{ij\alpha}$ to be a random sample from an infinite population. We define

$$\sigma_\xi^2 = \frac{1}{P-1} \sum_i^P \xi_i^2$$

$$\sigma_\lambda^2 = \frac{1}{P(Q-1)} \sum_{ij}^{PQ} \lambda^2{}_{j(i)}$$

$$\sigma_\varepsilon^2 = \sigma^2 = \int_{-\infty}^{+\infty} \varepsilon^2 f(\varepsilon) d\varepsilon$$

In determining the average values of the variance of the estimated means of subclasses, classes, and the population, we are required to determine

$$\text{ave}\left(\sum_\alpha^n \varepsilon_{ij\alpha}\right)^2 \qquad \text{ave}\left(\sum_{j\alpha}^{qn} \varepsilon_{ij\alpha}\right)^2 \qquad \text{ave}\left(\sum_{ij\alpha}^{pqn} \varepsilon_{ij\alpha}\right)^2$$

$$\text{ave}\left(\sum_j^q \lambda_{j(i)}\right)^2 \qquad \text{ave}\left(\sum_{ij}^{pq} \lambda_{j(i)}\right)^2$$

$$\text{ave}\left(\sum_i^p \xi_i\right)^2$$

Since the $\varepsilon_{ij\alpha}$ of the first row have zero covariance, this row is equivalent to

$$n\sigma^2 \qquad qn\sigma^2 \qquad pqn\sigma^2$$

For the first term of the second row we have

$$\begin{aligned}\text{ave}\,(\Sigma_j\lambda_{j(i)})^2 &= \text{ave}\,(\lambda_{1(i)} + \lambda_{2(i)} + \cdots + \lambda_{q(i)})^2 \\ &= q\left(1 - \frac{q}{Q}\right)\sigma_\lambda^2\end{aligned}$$

since the $\lambda_{j(i)}$, arising from a finite population of size Q, have (Section 3.44)

$$\text{cov}\,(\lambda_{j(i)},\lambda_{k(i)}) = -\frac{\sigma_\lambda^2}{Q}$$

Similarly for the term of the third row we have

$$\text{ave}\,(\Sigma_i\xi_i)^2 = p\left(1 - \frac{p}{P}\right)\sigma_\xi^2$$

The average value of $\left(\sum_{ij}^{pq}\lambda_{j(i)}\right)^2$ may be determined from combinatorial considerations. The total number of ways of forming $\sum_{ij}^{pq}\lambda_{j(i)}$ by taking q from each population Q and employing p such populations is

$$\binom{Q}{q}^p\binom{P}{p}$$

and any particular $\lambda_{s(r)}$ occurs in

$$\binom{Q-1}{q-1}\binom{Q}{q}^{p-1}\binom{P-1}{p-1}$$

of these cases. Hence in all possible squared combinations $\left(\sum_{ij}^{pq}\lambda_{j(i)}\right)^2$, $\lambda^2{}_{s(r)}$ will occur with this frequency, and thus over the whole population PQ of λ values we have each occurring on this number of occasions i.e., the direct square terms in every possible squared combination $\left(\sum_{ij}^{pq}\lambda_{j(i)}\right)^2$ in the whole population can be written

$$\binom{Q-1}{q-1}\binom{Q}{q}^{p-1}\binom{P-1}{p-1}\sum_{ij}^{PQ}\lambda^2{}_{j(i)}$$

The cross-product terms in the expression $\left(\sum_{ij}^{pq}\lambda_{j(i)}\right)^2$ may be divided into

2 classes, those of the type $2\lambda_{s(r)}\lambda_{u(r)}$ and those of the type $2\lambda_{s(r)}\lambda_{u(t)}$. Each term of the first type occurs

$$\binom{Q-2}{q-2}\binom{Q}{q}^{p-1}\binom{P-1}{p-1}$$

times when all possible combinations $\sum_{ij}^{pq}\lambda_{j(i)}$ are squared. Now for any class r we have

$$\lambda_{s(r)}\sum_{s\neq u}^{Q}\lambda_{u(r)} = \lambda_{s(r)}(-\lambda_{s(r)})$$

$$\left(\text{since } \sum_{u}^{P}\lambda_{u(r)} = 0\right)$$

$$= -\lambda^2_{s(r)}$$

and

$$\lambda_{u(r)}\sum_{u\neq s}^{Q}\lambda_{s(r)} = -\lambda^2_{u(r)}$$

and the term $2\lambda_{s(r)}\lambda_{u(r)}$ contributes $\lambda_{s(r)}\lambda_{u(r)}$ to each summation. Thus to form a summation leading to $-\lambda^2_{s(r)}$ we need $(Q-1)$ product terms. To form a summation leading to $-\sum_{s}^{Q}\lambda^2_{s(r)}$ we need $Q(Q-1)$ cross-product terms, and the total number of cross-product terms available is

$$2\binom{Q}{2}\binom{Q-2}{q-2}\binom{Q}{q}^{p-1}\binom{P-1}{p-1}$$

Thus in total

$$\sum_{s\neq u}^{Q}\lambda_{s(r)}\lambda_{u(r)} = -\frac{2\binom{Q}{2}\binom{Q-2}{q-2}\binom{Q}{q}^{p-1}\binom{P-1}{p-1}}{Q(Q-1)}\Sigma_s\lambda^2_{s(r)}$$

$$= -\binom{Q-2}{q-2}\binom{Q}{q}^{p-1}\binom{P-1}{p-1}\Sigma_s\lambda^2_{s(r)}$$

and, summing over all r, we obtain

$$\sum_{s\neq u}^{PQ}\lambda_{s(r)}\lambda_{u(r)} = \binom{Q-2}{q-2}\binom{Q}{q}^{p-1}\binom{P-1}{p-1}\sum_{rs}^{PQ}\lambda^2_{s(r)}$$

For the cross products of the type $2\lambda_{s(r)}\lambda_{u(t)}$ where $r \neq t$ we see that there are $Q^2(P-1)$ such products, each occurring

$$\binom{Q-1}{q-1}\binom{Q}{q}^{p-2}\binom{P-2}{p-2}$$

times. For any $\lambda_{s(r)}$ the products

$$\lambda_{s(r)}(\lambda_{1(t)} + \lambda_{2(t)} + \cdots + \lambda_{u(t)} + \cdots + \lambda_{Q(t)})$$

are zero since $\sum_{u}^{Q} \lambda_{u(t)} = 0$. Thus these cross products contribute nothing to the sum of the $\left(\sum_{ij}^{pq} \lambda_{j(i)}\right)^2$ when the sum is taken over all possible combinations. We have then

$$\sum_{a.c.} \left(\sum_{ij}^{pq} \lambda_{j(i)}\right)^2 = \binom{Q-1}{q-1} \binom{Q}{q}^{p-1} \binom{P-1}{p-1} \sum_{ij}^{PQ} \lambda^2_{j(i)}$$
$$- \binom{Q-2}{q-2} \binom{Q}{q}^{p-1} \binom{P-1}{p-1} \sum_{ij}^{PQ} \lambda^2_{j(i)}$$
$$= \binom{Q-1}{q-1} \binom{Q}{q}^{p-1} \binom{P-1}{p-1} \left[1 - \frac{q-1}{Q-1}\right] \sum_{ij}^{PQ} \lambda^2_{j(i)}$$

and the average value over all combinations of $\left(\sum_{ij}^{pq} \lambda_{j(i)}\right)^2$ is therefore

$$\frac{\binom{Q-1}{q-1}\binom{Q}{q}^{p-1}\binom{P-1}{p-1}\left[1-\frac{q-1}{Q-1}\right]}{\binom{Q}{q}^{p}\binom{P}{p}} \sum_{ij}^{PQ} \lambda^2_{j(i)} = \frac{pq}{PQ}\left[1-\frac{q-1}{Q-1}\right] \sum_{ij}^{PQ} \lambda^2_{j(i)}$$

Since we defined

$$\sigma_\lambda^2 = \frac{1}{P(Q-1)} \sum_{ij}^{PQ} \lambda^2_{j(i)}$$

we obtain finally

$$\text{ave} \left(\sum_{ij}^{pq} \lambda_{j(i)}\right)^2 = pq\left(1 - \frac{q}{Q}\right) \sigma_\lambda^2$$

and we have for the complete model

$\text{ave}\,(\Sigma_\alpha \varepsilon_{ij\alpha})^2 = n\sigma^2$ $\quad \text{ave}\,(\Sigma_{j\alpha} \varepsilon_{ij\alpha})^2 = qn\sigma^2$ $\quad \text{ave}\,(\Sigma_{ij\alpha} \varepsilon_{ij\alpha})^2 = pqn\sigma^2$

$\text{ave}\,(\Sigma_j \lambda_{j(i)})^2 = q\left(1 - \frac{q}{Q}\right)\sigma_\lambda^2$ $\quad \text{ave}\,(\Sigma_{ij} \lambda_{j(i)})^2 = pq\left(1 - \frac{q}{Q}\right)\sigma_\lambda^2$

$\text{ave}\,(\Sigma_i \xi_i)^2 = p\left(1 - \frac{p}{P}\right)\sigma_\xi^2$

The extension to a model of the type

$$x_{ijk\alpha} = \mu + \xi_i + \lambda_{j(i)} + \omega_{k(ij)} + \varepsilon_{ijk\alpha}$$

follows by a similar method from the result

$$\text{ave}\,(\Sigma_{ijk} \omega_{k(ij)})^2 = pqr\left(1 - \frac{r}{R}\right) \sigma_\omega^2$$

which is the only type of combination not represented in the two-way classification.

7A.2. Crossed Classifications

In the case of a two-way crossed classification we have for a model

$$x_{ij\alpha} = \mu + \xi_i + \eta_j + \lambda_{ij} + \varepsilon_{ij\alpha}$$

$$i = 1, \cdots, p \quad j = 1, \cdots, q \quad \alpha = 1, \cdots, n$$

and we take the ξ_i and η_j to be random samples from populations of size P and Q, respectively. With each of the PQ cells of the population is associated an interaction λ_{ij}, the values of λ_{ij} being subject to the conditions $\sum_j^Q \lambda_{ij} = 0$ for each i and $\sum_i^P \lambda_{ij} = 0$ for each j. The residual components are random samples from an infinite population. We define

$$\sigma_\xi^2 = \frac{1}{P-1} \sum_i^P \xi_i^2$$

$$\sigma_\eta^2 = \frac{1}{Q-1} \sum_j^Q \eta_j^2$$

$$\sigma_\lambda^2 = \frac{1}{(P-1)(Q-1)} \sum_{ij}^{PQ} \lambda^2_{ij}$$

$$\sigma_\varepsilon^2 = \sigma^2 = \int_{-\infty}^{+\infty} \varepsilon^2 f(\varepsilon)\, d\varepsilon$$

In determining the average values of the variance of the estimated cell, column, row, and population means, we are required to determine

$$\text{ave}\left(\sum_i^p \lambda_{ij}\right)^2 \quad \text{ave}\left(\sum_j^q \lambda_{ij}\right)^2 \quad \text{ave}\left(\sum_{ij}^{pq} \lambda_{ij}\right)^2$$

$$\text{ave}\left(\sum_i^p \xi_i\right)^2 \quad \text{ave}\left(\sum_j^q \eta_j\right)^2$$

The 4 cases in which a single summation is involved are treated by the method of Section 3.44, and we obtain

$$\text{ave}\left(\sum_i^p \lambda_{ij}\right)^2 = p\left(1 - \frac{p}{P}\right)\sigma_\lambda^2 \quad \text{ave}\left(\sum_j^q \lambda_{ij}\right)^2 = q\left(1 - \frac{q}{Q}\right)\sigma_\lambda^2$$

$$\text{ave}\left(\sum_i^p \xi_i\right)^2 = p\left(1 - \frac{p}{P}\right)\sigma^2_\xi \quad \text{ave}\left(\sum_j^q \eta_j^2\right) = q\left(1 - \frac{q}{Q}\right)\sigma_\eta^2$$

The double summation case may again be examined combinatorially.

In order to specify the selection of effects in this case it is only necessary to choose p columns and q rows. The number of possible selections of the λ_{ij} is then $\binom{P}{p}\binom{Q}{q}$, and of these selections $\binom{P-1}{p-1}\binom{Q-1}{q-1}$ include any particular element λ_{rs}. Thus in the summation $\Sigma_{a.c.}\left(\sum_{ij}^{pq}\lambda_{ij}\right)^2$ the coefficient of the direct square of each element is $\binom{P-1}{p-1}\binom{Q-1}{q-1}$. The cross-product terms may be separated into 3 groups—those involving terms in the same column, those involving terms in the same row, and a remainder. Any product $2\lambda_{rs}\lambda_{ru}$ of the first group will occur in the summation on $\binom{P-1}{p-1}\binom{Q-2}{q-2}$ occasions. The total number of such products for the rth column will then be $\binom{P-1}{p-1}\binom{Q-2}{q-2}Q\dfrac{(Q-1)}{2}$

By the method employed in the previous section we see that the sum of the products is equal to

$$-2\binom{P-1}{p-1}\binom{Q-2}{q-2}\frac{Q(Q-1)}{2Q(Q-1)}\sum_{s}^{Q}\lambda^2_{rs} = -\binom{P-1}{p-1}\binom{Q-2}{q-2}\sum_{s}^{Q}\lambda^2_{rs}$$

This is true for all values of r, and summing over r we have for the contribution of the first group of cross products

$$-\binom{P-1}{p-1}\binom{Q-2}{q-2}\sum_{rs}^{PQ}\lambda^2_{rs}$$

Since in the case of the crossed classification we also have $\sum_{r}^{P}\lambda_{rs} = 0$ for each s, the sum of the cross-product terms of the type $2\lambda_{rs}\lambda_{ts}$, i.e., involving 2 interaction effects from the same row, may be obtained by analogy with the above procedure as

$$-\binom{P-2}{p-2}\binom{Q-1}{q-1}\sum_{rs}^{PQ}\lambda^2_{rs}$$

The sum of the remaining cross-product terms involving 2 interactions which have neither a common row nor column does not vanish in this case. Each term of the type $2\lambda_{rs}\lambda_{tu}$ occurs on $\binom{P-2}{p-2}\binom{Q-1}{q-1}$ occasions, and, as there are $PQ(P-1)(Q-1)$ different terms of this type, there are in all $2\binom{P-2}{p-2}\binom{Q-1}{q-1}\dfrac{PQ(P-1)(Q-1)}{2}$ terms of the type $\lambda_{rs}\lambda_{tu}$.

Now

$$\begin{aligned}
\lambda_{rs}(\lambda_{11} &+ \lambda_{21} + \cdots + \lambda_{(r-1)1} + \lambda_{(r+1)1} + \cdots + \lambda_{P1} \\
+ \lambda_{12} &+ \lambda_{22} + \cdots + \lambda_{(r-1)2} + \lambda_{(r+1)2} + \cdots + \lambda_{P2} \\
&\cdot \\
&\cdot \\
&\cdot \\
+ \lambda_{1(s-1)} &+ \lambda_{2(s-1)} + \cdots + \lambda_{(r-1)\ (s-1)} + \lambda_{(r+1)\ (s-1)} + \cdots + \lambda_{P(s-1)} \\
+ \lambda_{1(s+1)} &+ \lambda_{2(s-1)} + \cdots + \lambda_{(r-1)\ (s+1)} + \lambda_{(r+1)\ (s+1)} + \cdots + \lambda_{P(s+1)} \\
&\cdot \\
&\cdot \\
&\cdot \\
+ \lambda_{1Q} &+ \lambda_{2Q} + \cdots + \lambda_{(r-1)Q} + \lambda_{(r+1)Q} + \cdots + \lambda_{PQ}) \\
&= -\lambda_{rs}(\lambda_{r1} + \lambda_{r2} + \cdots + \lambda_{r(s-1)} + \lambda_{r(s+1)} + \cdots + \lambda_{rQ}) \\
&= +\lambda_{rs}\lambda_{rs} = \lambda^2_{rs}
\end{aligned}$$

Thus we require $(P-1)(Q-1)$ cross-product terms to form λ^2_{rs} and $PQ(P-1)(Q-1)$ to form $\sum_{rs}^{PQ} \lambda^2_{rs}$, and hence from all such terms we obtain the contribution

$$+2\binom{P-2}{p-2}\binom{Q-2}{q-2}\frac{PQ}{2PQ}\frac{(P-1)(Q-1)}{(P-1)(Q-1)}\sum_{rs}^{PQ}\lambda^2_{rs}$$

$$= \binom{P-2}{p-2}\binom{Q-2}{q-2}\sum_{rs}^{PQ}\lambda^2_{rs}$$

The total coefficient of $\sum_{rs}^{PQ} \lambda^2_{rs}$ in $\sum_{a.c.}\left(\sum_{ij}^{pq} \lambda_{ij}\right)^2$ is therefore

$$\binom{P-1}{p-1}\binom{Q-1}{q-1}\left[1 - \frac{p-1}{P-1} - \frac{q-1}{Q-1} + \frac{(p-1)\ (q-1)}{(P-1)(Q-1)}\right]$$

The total number of combinations is $\binom{P}{p}\binom{Q}{q}$, so that the mean coefficient per combination is

$$\frac{pq}{PQ}\left(1 - \frac{p-1}{P-1}\right)\left(1 - \frac{q-1}{Q-1}\right)$$

Now

$$\sigma_\lambda^2 = \frac{1}{(P-1)(Q-1)}\sum_{ij}^{PQ}\lambda^2_{ij}$$

so that

$$\text{ave}\left(\sum_{ij}^{pq} \lambda_{ij}\right)^2 = pq\left(1 - \frac{p}{P}\right)\left(1 - \frac{q}{Q}\right)\sigma_\lambda^2$$

This method may be employed for more complicated models, such as

$$x_{ijk\alpha} = \mu + \xi_i + \eta_j + \zeta_k + \beta_{ij} + \gamma_{ik} + \delta_{jk} + \lambda_{ijk} + \varepsilon_{ijk\alpha}$$

where for example

$$\text{ave}\left(\sum_{ijk}^{pqr} \lambda_{ijk}\right)^2 = pqr\left(1 - \frac{p}{P}\right)\left(1 - \frac{q}{Q}\right)\left(1 - \frac{r}{R}\right)\sigma_\lambda^2$$

The general coefficient will be obvious from these examples.

CHAPTER 8

The Design of Experiments

8.1. Introduction

In the preceding chapter we have considered the application of the analysis of variance to data from a number of types of experiment. It appears that this type of analysis is a most satisfactory method of estimating and testing the effects and interactions of a number of factors, so that we should prefer to use it wherever possible. It is also apparent that these advantages depend upon the collection of the data according to a systematic scheme and upon certain assumptions concerning the independence and distribution of the experimental errors. In order to make the best use of experimental results it is desirable that the whole experiment should be so designed that the data, when collected, will be suitable for analysis by this method.

The formal development of the technique of experimental design was initiated by R. A. Fisher and his co-workers (notably F. Yates) at the Rothamstead Agricultural Experimental Station, and a large body of experience of the application of these methods to agricultural and biological work has been built up by these and other workers. The fundamental concepts of the subject are now firmly established, but new designs are still being developed for particular requirements. Designed experiments in industrial investigations are a more recent development, and sufficient experience is not always available in this field to indicate which of the many available designs is most suitable for a given purpose.

It is not surprising that the original fields of application of adequately designed experiments have been agricultural and biological, since this type of experiment is frequently subject to large variations in observed data, owing to natural causes beyond the control of the experimenter, and may be both time consuming and costly to carry out. Under these conditions a satisfactory design is required to:

(1) Eliminate as much as possible of the natural variation which causes uncertainty in the results.
(2) Ensure that the remaining variation is not confused with the effects which are to be tested.
(3) Detect existing effects with a minimum of experimental effort.

It can be seen that these requirements are not independent, since reduction in the residual error and the elimination of bias are the two principal means of increasing the sensitivity of a given design.

In principle the designs which have been developed for agricultural and biological work are applicable to the problems of chemical or engineering investigations and have already found considerable use in this field. The most important differences between the two fields of investigation are probably:

(1) In chemical experiments the tests are often carried out sequentially so that natural variations with time are of great importance.
(2) The source of many of the natural variations is frequently known, and their effects could often be eliminated by refinements in experimental technique. The economic factor in the selection of experimental designs is therefore likely to be more obvious in chemical investigation.
(3) Information from experience or theory which enables certain interaction effects to be declared non-existent may be available to the chemist. Although comparisons which would normally be employed to estimate these effects can always be employed to estimate the residual components, it is often desirable to use the comparison for some other purpose.

For these reasons the experimental designs which are likely to prove most satisfactory to the chemist are those which can be applied in a sequential form, and in which any prior information can be employed to reduce the expenditure of money and effort.

8.2. General Terminology

Experimental investigations are usually performed to determine the effect upon one or more results of a change in the experimental conditions under which the results were obtained. In such cases the complete experiment consists of obtaining numerical observations for a series of experimental conditions, the observations obtained under one set of conditions being described as an *experimental unit*. The experimental conditions may be grouped according to a series of factors, or criteria; the different classes within a given factor are described as *levels*, or *categories.* The combination of conditions obtaining in a particular experimental unit is often referred to as a *treatment.* As an illustration of the terminology, consider a two-factor experiment designed to investigate the effect of preliminary charring on a number of samples in an analytical determination involving a muffle operation. In this case charred and not charred would be the categories of one factor; those of the other factor would be the various samples.

In order to establish the effect of a given treatment more accurately and to provide an estimate of the variation between experimental units receiving the same treatment, the whole or part of the experiment may be repeated in independent tests. Such repetitions are described as *replication.* If all treatments are applied once, we refer to a *single replication.* If a proportion of the treatments are omitted, we have a *fractional replication.*

8.3. Collection of Data

8.31. General Considerations

The conclusions drawn from a statistical analysis are usually to be applied to a population of which the observations used in the analysis are members. This is possible only when the relationship between the whole population and the members which have been examined is known. In previous chapters we have assumed that these members constituted a random sample from the populations to be examined, since on this basis it was possible to develop the analyses and provide estimates and significance tests.

When the method of collection of the data is not under the control of the statistician the above considerations represent an important assumption which is not always subject to test. In Chapter 11 we shall consider methods which can sometimes be employed for detecting non-randomness in data, but these, at best, will only give assurance that some population was randomly sampled. If this was not the population to which the conclusions of the investigation are to be applied, the results of any statistical analysis may be biased. Here the use of advanced techniques in the analysis of the data is a waste of effort since any results obtained can only be interpreted in terms of previous knowledge or opinions on the extent of the bias. For example, a number of determinations of the composition of the product stream from a laboratory-scale Fischer-Tropsch synthesis unit might be examined to provide information on the population of such laboratory determinations. These same measurements provide much less information on the performance of a similar full-scale unit because in this instance we must depend upon previous experience to indicate the discrepancy in results which may be expected in moving from the laboratory to the plant scale of operation.

Thus, in addition to developing experimental designs which are suitable for the methods of analysis of Chapters 6 and 7, we must ensure that the data are obtained in such a way that the fundamental assumptions of these methods are satisfied. In the following sections we consider a number of aspects of this problem.

8.32. Sampling of Materials

The population to be sampled may consist of the intake of ore to a blast furnace, the density measurements which could have been obtained from a sample of material tested under standard conditions, or the analytical abilities of a group of technical chemists. Where material populations are concerned the types of sampling procedure involved fall into a number of broad classes having common features. If the characteristic of the sample in which interest centers is an attribute rather than a variable, the sampling is often simplified since such cases usually involve discrete similar articles and a logical sub-grouping is often apparent. This aspect of sampling is deferred to Chapter 9 where a further discussion of the treatment of data distributed according to the binomial, multinomial, or Poisson forms is presented. In the rest of Section 8.3 we shall be concerned with the sampling of variables.

In designing a sampling procedure it is usually required that the results obtained lead to unbiased estimates of the population parameters and that, for a given expenditure, the variance of these estimates be minimized.

8.33. Sampling of Discrete Articles

If the population to be sampled consists of milk bottles or automobile tires or similar units each with a standard content of material, or each of interest as a product rather than as an aggregation of material, then the size of a single increment of the sample is fixed. If a given number of increments is to be taken, it remains to ensure that the sampling is unbiased and efficient. We may ensure freedom from bias by identifying each member of the population with a number and then determine the units to be examined by making selections from a table of random numbers. Such a procedure is adopted because any systematic sampling scheme which we might adopt could coincide with natural subdivisions in the material and lead to bias in the results. If the articles were known to be perfectly mixed so that the average value of the variance between neighboring units was equal to the average value of the variance between widely separated units, it would not be necessary to use a random selection process to ensure an unbiased sample.

This random sampling procedure is well suited to a problem in which we have no information about the natural grouping of the units of the population, but improvements can be effected if it is known that the population may be divided into a series of subpopulations, strata, or regions containing units which on the average differ less than units randomly selected from the whole population.

8.34. Stratified Sampling

In the simplest examples these strata will be clearly defined as the output from individual machines, from one batch of raw material, or by some other logical grouping. If this is the case, and if the total number of units in each stratum is known, the means of the individual strata may be determined and combined, with suitable weighting, to determine the overall mean of the population. The distribution of the samples between strata so as to secure an unbiased estimate of the population means having minimum variance was considered in Section 3.5 where it was shown that

(1) The observations from a given stratum should be weighted equally in estimating the stratum mean.

(2) Stratum means should be weighted in direct proportion to the total number of articles in the stratum in determining the population mean.

(3) The number of samples selected from any stratum should be proportional to the product of the total number of articles in the stratum N_i and the standard deviation σ_i within the stratum.

Under these conditions we have

$$\text{var}\,(m) = \frac{1}{N^2}\,\frac{(\Sigma_i N_i\sigma_i)^2}{n} - \Sigma N_i\sigma_i^2$$

if the strata are finite or

$$\text{var}\,(m) = \frac{(\Sigma p_i\sigma_i)^2}{n}$$

where the strata can be considered infinite and the probability of drawing a value from the ith stratum is p_i. The samples within the strata are selected by a random method so that

(1) If the information on the number of members in the strata, or on p_i in the case of infinite strata, is correct, the stratified sample will provide an unbiased estimate of the population mean.

(2) If the information on the variances of the strata is correct the stratified sample will be at least as efficient as a completely random sample.

(3) If the means of the strata differ, the variance of the population mean estimate obtained by stratified sampling will be less than the variance of the mean estimate obtained from a completely random sample.

EXAMPLE. The intake to a unit producing a special foundry coke consists of fines from 4 collieries. It is required to control the phosphorus content of the coal blend at 0.008 %, and economic considerations indicate that this is best achieved by blending in the ratio 45 : 40 : 10 : 5 = A : B : C : D., the proportion of coal D being increased when this does not conflict with the maximum phosphorus content specification. The phosphorus contents of the 4 materials are approximately 0.006 %, 0.008 %, 0.011 %, and 0.020 %, and the variances of the estimates of the phosphorus contents are in each case proportional to the amount of phosphorus present. It is required to determine the best allocation of samples to the 4 shipments in order to minimize the variance of the estimated mean phosphorus content of the blend. We then have

$$W_A = \frac{N_A}{N} = 0.45 \quad W_B = \frac{N_B}{N} = 0.40 \quad W_C = \frac{N_C}{N} = 0.10$$

$$W_D = \frac{N_D}{N} = 0.05$$

$$\begin{aligned} n_A : n_B : n_C : n_D &= 0.45\sqrt{0.006} : 0.40\sqrt{0.008} : 0.10\sqrt{0.011} : 0.05\sqrt{0.020} \\ &= 0.0348 : 0.0358 : 0.0105 : 0.0071 \\ &\sim 10 : 10 : 3 : 2 \end{aligned}$$

Thus, if we are prepared to take 25 samples in all, a close approximation to the most desirable ratios may be achieved by taking $n_A = 10$, $n_B = 10$, $n_C = 3$, $n_D = 2$.

In many investigations the ratios of the variances of the subpopulations, strata, or regions will be unknown when the sampling scheme is considered. In these cases it is sometimes possible to undertake a pilot investigation in which these variances are estimated. By this means the number of samples in the major scheme to provide an estimate of a given accuracy can be determined, and the effectiveness of the scheme may be increased by a more efficient allocation of the samples. Furthermore, if the scheme is to be used repeatedly, it may be progressively refined in the light of experience obtained during its use.

This treatment is appropriate to the case in which the strata are clearly marked within the population. In many instances such a demarcation is not evident, although it is known that units from the same region of the population to be sampled tend to be more alike in properties than units from different regions. The region may consist of units which are close together in space or in time of production. Here the population may be divided arbitrarily into a series of regions, and a stratified sampling scheme employed. If, as is often true, the variance within regions decreases as the size of the region decreases, the most efficient estimate of the mean which can be obtained from n samples will result from a scheme in which the population is divided into n regions and a single unit is selected at random from each region. Unfortunately it is not then possible to place confidence limits on the mean estimate, so that it may sometimes be preferable to select 2 samples from each region and increase the size of the regions.

8.35. Sampling of Aggregates

In the previous section we considered the case in which the population consists of discrete articles such as packages of material, or automobile tires which are to be tested as tires and not as pieces of processed rubber. If we are interested in the properties of the rubber, the position is unchanged, provided that we are prepared to examine the whole of the tire and the results are to be expressed in units of tires. In many instances the material to be examined does not consist of uniform packages but rather of lumps of different sizes and content, and the property to be measured is to be expressed on a weight or volume basis. In addition, the property to be measured is often correlated with the particle size. If the material is to be screened before processing, a representative sampling procedure in which the screen fractions are regarded as strata may be employed; otherwise the increments withdrawn from unscreened material must consist of a number of particles selected in such a manner that the chance of a given particle occurring is proportional to the number of particles of that size in the population or stratum. This is usually considered to be achieved by selecting increments of weight or volume several times larger than the weight or volume of the largest piece in the aggregate. It is, of course, desirable to use small increments, other things being equal, since by this means the total volume of the sample is reduced. The optimum weight of increment and method of selecting the increment have been investigated for some materials, notably coal, but it does not appear possible to generalize the conclusions. The increment size being fixed, the increments can be regarded as units in a discrete population and stratified sampling techniques may be employed. Such methods are often advantageous with raw materials in which complete mixing is rare, and in which transport and handling often results in the partial segregation of materials of different sizes, shapes, and densities.

Since these factors are usually related to the chemical composition of particles of the aggregate, a stratified sampling procedure will often result in a considerable decrease in sampling errors.

A random sampling process, whether carried out on the complete population or on strata of this population, is frequently difficult to administer in a chemical plant. For this reason a systematic sampling scheme is often substituted, samples being selected at equal intervals of space or time, or if the materials are discrete at intervals of k units. If the distribution of the property to be measured is random with respect to the criterion on which the system is based, this procedure is in effect a random sampling scheme. More frequently in chemical units the fluctuations in quality of the products are made up of a random component and a regular periodic fluctuation caused by batchwise operation, shift changes, or regular alterations in control settings. If the sampling interval is small compared with the period of these fluctuations, a systematic sampling scheme results in improved accuracy in comparison with a completely random scheme. When the period and the interval are of similar size, serious bias in the mean estimate may result from a systematic procedure.

8.36. Sample Reduction

Most chemical tests are designed for application to a small quantity of material, whereas a satisfactory sampling procedure often results in a large bulk of sample. Unless a large number of tests are to be carried out, the sample must be reduced in bulk in such a way that the variance of the estimate of the mean of the bulk sample is minimized. This can usually be achieved by reducing the particle size of the material, mixing, and random sampling, although in the case of hard or sensitive materials care is required to prevent contamination or decomposition. In practice it is usually necessary to limit the size reduction, and as an indication of the additional variation introduced by the subsampling procedure the *size-weight ratio* [47, 48] is frequently employed. This is defined as

$$\frac{\text{weight of largest particle of impurity} \times 100}{\text{weight of reduced sample}} = b$$

The variance component due to subsampling is unlikely to exceed bx, where x represents the percentage proportion of impurity in the large sample. This follows since, if a gross sample of mass W is reduced in particle size so that the largest particle has mean linear dimension d and the density of the impurity is ρ, there will be a least $\dfrac{xW}{100\rho\, d^3}$ particles of impurity. Since this represents the worst case, we shall assume that all the impurity is present as large particles of this size. The average value of the number of particles of impurity in a subsample of mass w will then be $wx/100\rho d^3 = x/b$, and the actual number in any sample will be distributed in the Poisson form with this average value and hence with variance x/b. The proportion by mass of the impurity in the reduced sample will have average value $(x/b)\left(\frac{\rho\, d^3}{w}\cdot 100\right) = x$ and variance $(x/b)\left(\frac{\rho\, d^3}{w}\cdot 100\right)^2 = bx$.

This value usually provides an overestimate of the variance component due to subsampling, since the impurity is unlikely to consist of particles all of which are of the maximum size, and if the procedure is to be generally used a better approximation may be obtained by conducting a pilot sampling trial. Furthermore the assumption has been made that the constitution of the impurity remains fixed throughout the test, whereas, if the original sample were made up of particles some of which were "pure," and some of which contained laminations of impurity, the process of size reduction would almost certainly tend to separate the components of this laminated material, thus increasing the variance in the original sample.

8.37. Sampling Bias

In many of the major industries where difficult sampling problems arise, information is available on sources of sampling bias, and is frequently included in the recommendations of the national standardizing bodies such as the American Society for Testing Materials and the British Standards Institution. It would not be possible to consider all sources of bias in the sampling methods employed by the industrial chemist, but we shall select a few examples in order to illustrate the previous discussion of the relationship between the sample and the population to be sampled.

In the sampling of fluids the physical process of withdrawing the sample, or even of measurement in situ, may bias the observation. The measurement of local temperatures and velocities and the sampling of mixed vapors without immediate superheating provide many examples of this type of occurrence. In aggregates of solid materials differences in size, shape, or density of the particles are the factors most likely to lead to biased sampling. Any handling process to which the solid is subjected, and particularly those involving discharge and subsequent piling of the material, can be expected to produce segregation according to the above characteristics. For this reason it is usually advantageous to withdraw samples at a discharge point before any interruption in the fall has provided a basis for segregation, for instance at the outlet of a pipe or conveyor. Even under these conditions care is required to ensure that the full cross section of the stream is sampled, since on most conveyors both vertical and horizontal segregation is to be expected. Furthermore, the discharge should be divided into regions of time or volume, each randomly sampled in order to avoid systematic fluctuations introduced by previous segregation of the throughput.

In cases where the population sampled is of a more general nature the dangers of biased sampling are present, but may be less apparent [49]. Statistical investigations based on data from questionnaires completed by industrial concerns on a voluntary basis must often be carried out before data are available from all the units. It is then most improbable that returns actually received represent a random sample from the industry. A further example of this type arises when a number of industrial laboratories cooperate with a standardizing body in examining a new or modified test method. Laboratories equipped and prepared to undertake this type of work frequently provide an overestimate of the accuracy of the method in general use.

8.38. Replication

Experiments or observations are usually repeated in order to improve the accuracy of the estimate of some mean value and to provide an indication of the variance of this estimate. To do so efficiently it is necessary to specify the population of observations which is to be sampled and to ensure that the repeated observations from this population are independent. If the latter condition is satisfied, the measurements provide an unbiased estimate of the population variance and are termed replicate observations. It should be noted that replication can only be defined by reference to a population from which the observations are drawn and to which any inferences from the observations will be applied. In practice, although it is recognized that a number of sources contribute to the experimental variation, and that for a given number of tests the results will be most reliable when the contributions to different observations are independent, complete replications of observations may not provide the cheapest method of securing an answer of given accuracy. If some alternative method is to be used in order to minimize the cost of a sampling procedure, it is clear that information is required concerning the relative magnitudes of the variance contribution from the different sources and the relative costs of different sampling and testing procedures. For the purpose of illustration we consider the typical sequences of an analytical determination of the ash content of a shipment of coal shown in Figure 8.1. It will be seen that, if the stratified sample method is excluded, the maximum relative reduction in estimate variance which can be achieved by repeating an analytical determination is $\frac{1}{2}$, and this reduction only comes about when 2 batches are selected at random from the shipment

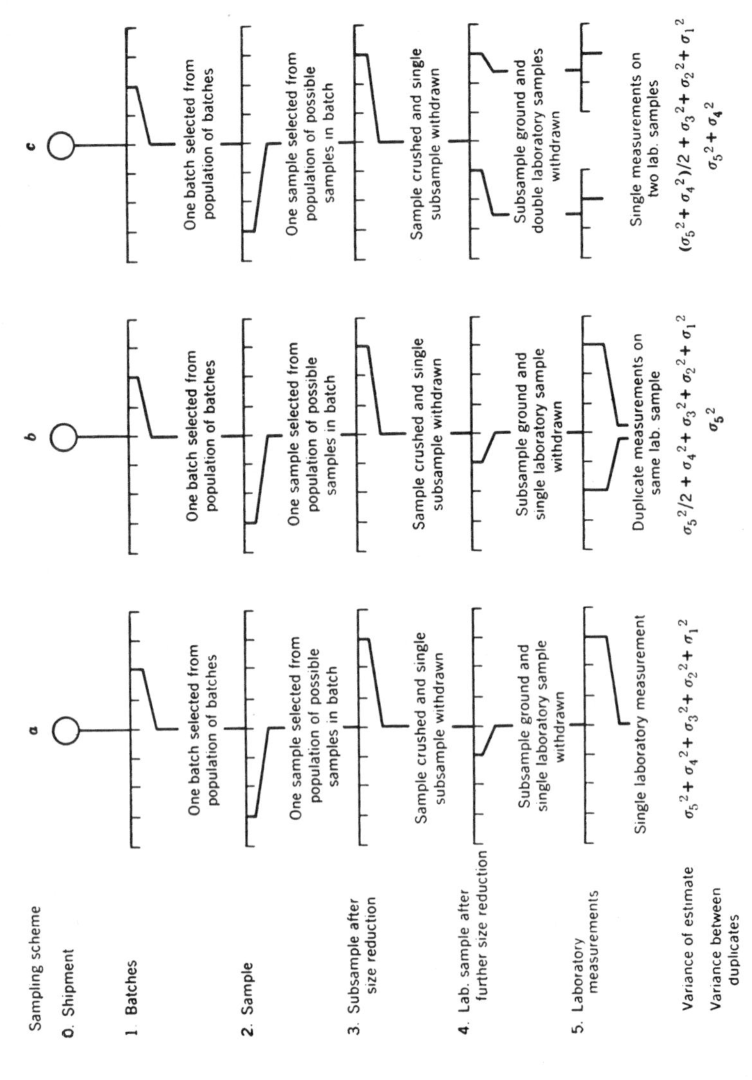

FIG. 8.1. Alternative sampling schemes.

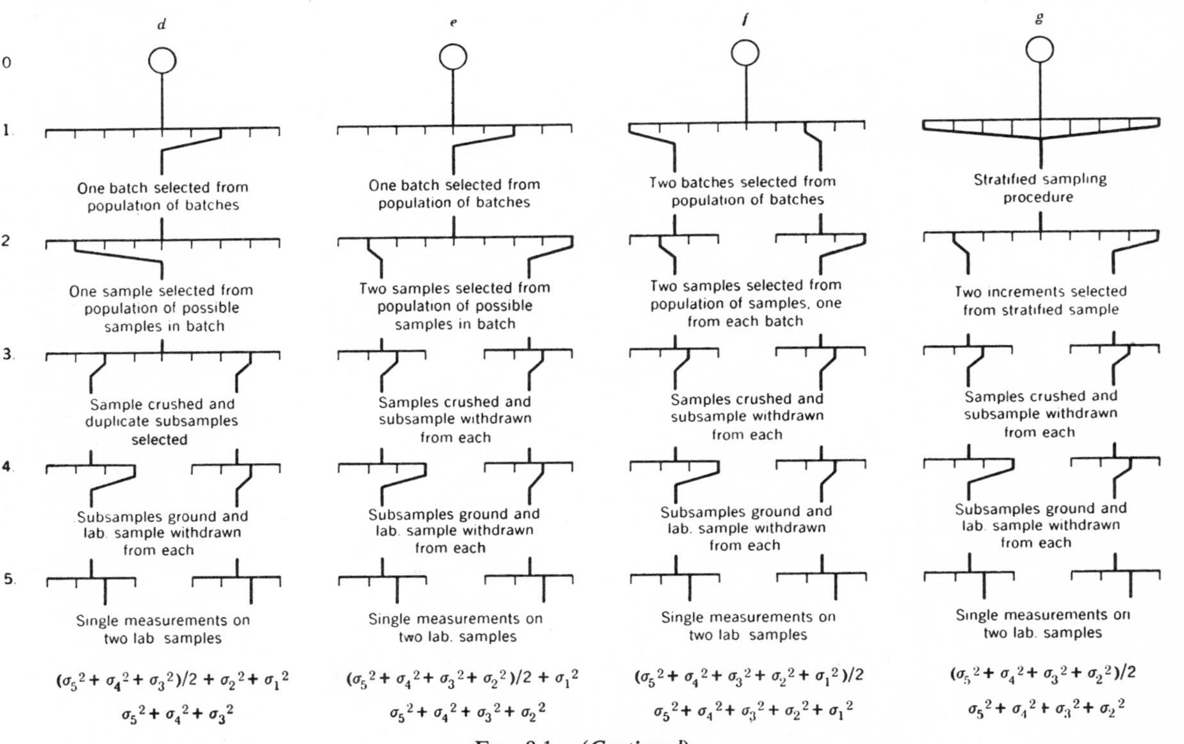

FIG. 8.1. *(Continued)*.

and the procedure of sample reduction and analysis is carried out independently for the samples so obtained. The ratio

$$M = \frac{\text{variance of estimate based on single determination}}{\text{variance of estimate based on } n \text{ determinations}}$$

has been called the improvement ratio [51]. If we consider the populations of effects at the various stages infinite, it is clear that this ratio will have a maximum value of n in the case of n determinations. If we suppose that, in the example of Fig. 8.1, σ_1^2, the variance component due to differences between batches or carloads, is small and the additional cost of moving to a separate car to withdraw the second sample is fairly high, it might be desirable to withdraw 2 samples from the same car. In this event the difference between the samples so obtained will no longer provide an unbiased estimate of the variance of the mean. However, if 4 samples are taken, 2 from each of 2 carloads, an unbiased estimate of the variance of the mean estimate can be obtained. For a more general treatment we may consider the case in which n_1 cars are sampled, n_2 samples being withdrawn from each. Each sample then provides n_3 subsamples and n_3n_4 laboratory samples, a total of $N_5 = n_1n_2n_3n_4n_5$ laboratory measurements being carried out. We assume that the variance components $\sigma_1^2, \cdots, \sigma_5^2$ have been estimated by a preliminary survey followed by an analysis of variance of the type considered in Section 7.62. If we suppose that the cost of n_5 independent laboratory determinations carried out on any laboratory sample is n_5c_5, the cost of producing n_4 laboratory samples from a single subsample is n_4c_4, and so on, we have for the total cost of the sampling procedure [107]

$$C = n_1c_1 + n_1n_2c_2 + n_1n_2n_3c_3 + n_1n_2n_3n_4c_4 + n_1n_2n_3n_4n_5c_5$$

and for the variance of the estimate of the shipment mean

$$\sigma_0^2 = \frac{\sigma_1^2}{n_1} + \frac{\sigma_2^2}{n_1n_2} + \frac{\sigma_3^2}{n_1n_2n_3} + \frac{\sigma_4^2}{n_1n_2n_3n_4} + \frac{\sigma_5^2}{n_1n_2n_3n_4n_5}$$

We may choose to minimize C for a fixed σ_0^2, or minimize σ_0^2 for a fixed C. In the first case, writing $N_1 = n_1$, $N_2 = n_1n_2$, $N_3 = n_1n_2n_3$, $N_4 = n_1n_2n_3n_4$, $N_5 = n_1n_2n_3n_4n_5$, we minimize C by making $\dfrac{\partial(C + \lambda\sigma_0^2)}{\partial N_i}$ vanish for $i = 1, \cdots, 5$. Thus

$$\frac{\partial}{\partial N_i}(C + \lambda\sigma_0^2) = c_i - \frac{\lambda\sigma_i^2}{N_i^2} = 0$$

or

$$N_i = \sqrt{\lambda}\sqrt{\sigma_i^2/c_i}$$

Now, if the fixed value of σ_0^2 is V, we have, substituting for the N_i

$$V = \frac{1}{\sqrt{\lambda}} \sum_{i=1}^{5} \sqrt{\sigma_i^2 c_i}$$

Thus

$$\sqrt{\lambda} = \frac{1}{V} \dot{\Sigma}_i \sqrt{\sigma_i^2 c_i}$$

and

$$N_i = \frac{\sqrt{\sigma_i^2/c_i}}{V} \Sigma_i \sqrt{\sigma_i^2 c_i}$$

or, in particular,

$$n_1 = N_1 = \frac{\sqrt{\sigma_1^2/c_1}}{V} \Sigma_i \sqrt{\sigma_i^2 c_i}$$

$$n_2 = \frac{N_2}{N_1} = \sqrt{\frac{\sigma_2^2 c_1}{\sigma_1^2 c_2}}$$

$$n_5 = \frac{N_5}{N_4} = \sqrt{\frac{\sigma_5^2 c_4}{\sigma_4^2 c_5}}$$

It is of interest to note that the optimum allocation of samples after the first stage is independent of V and for any stage n_i depends only on σ_i^2, σ^2_{i-1}, c_i, and c_{i-1}. Thus for different values of V the modification in the optimum sampling scheme will consist of changing the value n_1, the number of carloads sampled, while maintaining constant the treatment accorded to each selected carload. In the case where the most efficient sampling scheme for a fixed total cost C is required it may be shown by a procedure similar to the above that

$$n_1 = \frac{C\sqrt{\sigma_1^2/c_1}}{\Sigma_i \sqrt{\sigma_i^2 c_i}}, \quad n_2 = \sqrt{\frac{\sigma_2^2 c_1}{\sigma_1^2 c_2}}, \cdots, n_5 = \sqrt{\frac{\sigma_5^2 c_4}{\sigma_4^2 c_5}}$$

Similar relationships for cases in which C or σ_0^2 are fixed and certain of the n_i are fixed have been derived [107].

8.39. Randomization

The second requirement of a satisfactory experimental design, which was listed earlier in the chapter, may now be recognized as a statement that the estimates of the effects should be free from bias. The models which were assumed in discussing the analysis of variance required that the residual component be randomly distributed with average value zero. In practice this means that any effect, or combination of effects not under the control of the experimenter, should contribute randomly to the various experimental units, and that the average contribution to any group of these units should be zero. If this is not true, our method of analysis will lead us to consider a part of these contributions real effects, and the portions not so identified will lead to an underestimate of the residual variance. In order to guard against this possibility it is customary to randomize the contributions of the residual component. To do so we need to recognize the major sources of the residual variation, subsequently allocating the experimental units in a random fashion to the various combinations of effects due to these sources. If the residual effects do not arise from any major sources, but from a large number of independent small contributions, they will already be random.

As an illustration of the method of randomization consider the experiment of Table 7.31 in which a three-factor classification, coals $\times$ crucibles $\times$ temperature, was involved. In this case the tests were carried out during 3 separate days, the sequence of testing being completely randomized. Suppose that for convenience in manipulation the 3 temperature levels had been investigated separately, 1 on each day. It would then be impossible to distinguish between those effects of temperature changes which the experiment was designed to investigate and any random effects due to differences in daily conditions. By randomizing the order in which the tests were carried out, we ensure that any effects due to the change with time of uncontrolled factors will not on the average bias our results. The averaging process envisaged may be taken over the population of results which might be obtained if the experiment were repeated at different times, or over the population of results which might have been obtained during the particular days on which the experiment was carried out if all possible random arrangements (24!) had been employed. In the former case we are obliged to assume that the days on which the experiment was carried out were a random sample from all the days of the specified population. In the latter the inference is more restricted and we need not make this assumption. The theoretical aspects of randomization are extensively treated by Kempthorne [117]. Where randomization is to be carried out, tables of random numbers or some alternative device which does not involve human decisions should be employed. A description of the use of random number tables has been given by Fisher and Yates [13], p. 23.

Randomization of an experimental procedure is usually inconvenient in chemical investigations, since it involves additional control settings and interference with the sequence of analytical work. Furthermore, greater care is required in the identification of samples or specimens. It is important to note that the efficiency of the experimental procedure is to be measured, not in terms of the number of tests carried out in a given time, but rather in terms of the information produced by a given amount of effort. A large number of precise tests of uncertain bias are of little value as a basis for conclusions about the population. For a discussion of this subject, together with examples of randomized and non-randomized designs in analytical investigations, see [57]. One disadvantage of complete randomization is that information concerning the source of a part of the residual variance is frequently ignored, and the sensitivity of the experiment is reduced. In Section 8.5 we consider methods by which this information may be utilized without biasing the resulting estimates and significance tests, or by which the bias may be restricted to only a limited number of estimates which we are prepared to sacrifice.

8.4. Factorial Design

8.41. Classical and Factorial Designs

It is customary, in discussing the design of experiments, to distinguish between classical and factorial designs. We do so by considering an investigation in which the influence of n factors (each at p levels) upon some dependent variable is to be determined. In the classical system each factor in turn is selected for investigation at p levels, the remaining factors in any series of tests being maintained constant at standard values. The total number of experimental units involved in such a scheme is $n(p - 1) + 1$, corresponding to the unit in which all factors are at the standard levels and the $n(p - 1)$ units in which each factor in turn is examined. The alternative system requires that the effect on the dependent variable of each of the p^n possible treatment combinations be investigated.

The factorial design is frequently employed by workers who have no interest in statistical analysis, because it provides a picture of a complete field of investigation rather than a detailed examination of a few selected paths. By combining the factorial method of design with a statistical method of analysis of the results it is possible to obtain information on the effects and interactions of the factors involved, together with unbiased significance tests for these factors. The conclusions obtained are likely to be of wider application because the tests carried out cover a greater range of experimental conditions.

As an illustration of the application of the two methods consider an investigation of the efficiency of an extraction column. Suppose that 2 types of packing, helices and Raschig rings, were under consideration, and that 3 levels of the concentration of salting agent were to be examined. By the classical method we should carry out 4 tests, 2 to differentiate between the types of packing at a fixed concentration, and an additional 2 to examine the 2 remaining levels of salting agent for 1 type of packing.

A factorial scheme would involve 6 experimental units corresponding to the 6 possible combinations of treatments. If we suppose that each type of design were replicated and that the total number of observations by the 2 methods were equal (e.g., by 3 replicates of the classical method and 2 of the factorial method), then the advantages of the factorial design may be summarized as

(1) Greater efficiency in estimating the main effects.

(2) Greater scope, since the two-factor interaction effects may be estimated.

In order to generalize these conclusions we consider the case of an experiment involving 3 factors each at 2 levels. Denoting the levels by A_1A_2, B_1B_2, and C_1C_2, we have in all 8 experimental units. If 16 tests

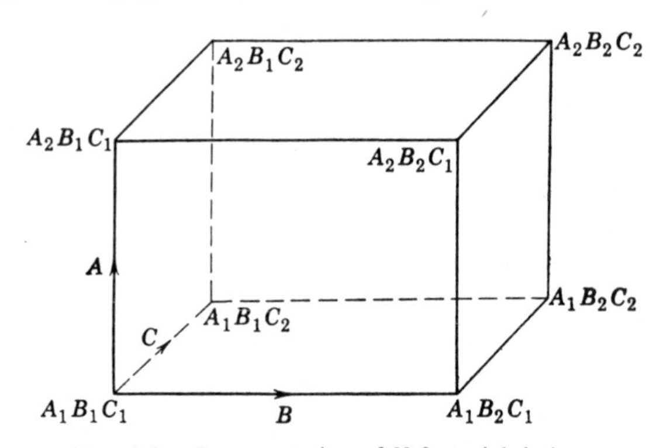

FIG. 8.2. Representation of 2^3 factorial design.

are to be carried out, we could replicate the design ($A_1B_1C_1$, $A_2B_1C_1$, $A_1B_2C_1$, $A_1B_1C_2$) 4 times, which would enable us to calculate the overall mean effect, the 3 differences between the means of given treatment combinations, and leave $(16 - 3 - 1) = 12$ degrees of freedom available for an estimate of the residual variance. The estimates of the differences between the means would in each case relate only to the first level of the remaining factors, and the means themselves would be based on 4 observations so that the variance of the estimate of a given effect would be (estimate of residual variance)/2.

Alternatively, the full design of 8 treatment combinations could be replicated twice, the resulting observations being illustrated schematically in Figure 8.2 where each vertex of the cube represents the two observations made under these conditions. The 8 observations of the upper face of the cube differ from the 8 of the lower face only in having factor A at level 2 in place of level 1. The difference between these means then provides an estimate of the effect due to increasing the level of A from 1 to 2, and all the observations are employed in this comparison. Similarly a comparison of the mean of the 8 observations constituting the right face of the cube with the corresponding mean for the left face gives the effect of change in the level of B; the effect of a change in the level of C may be determined from the front and rear faces of the cube. If we suppose that there are no real interaction effects, then the analysis of variance for this experiment takes the form of the accompanying table. The estimate of the residual variance is again based on 12 degrees of freedom, but each treatment effect is computed from means based on 8 observations, so that the estimated variance of any treatment effect is given by (estimated residual variance)/4.

Source of Estimate	D.F.	Average Value of Mean Square
Between treatments	3	$\sigma^2 + 8\sigma_\xi^2$
Residual	12	σ^2
Total	15	

In the event that real interactions do exist they can be estimated only from the factorial design. The existence of an interaction $A \times B$ implies that the magnitude of the effect of a change of level of A (e.g., from A_1 to A_2) depends upon the level of B at which this change is achieved. If we sum the terms in corresponding positions on the front and rear faces of the cube, we obtain

$$\begin{array}{ll}(A_2B_1C_1 + A_2B_1C_2) & (A_2B_2C_1 + A_2B_2C_2) \\ (A_1B_1C_1 + A_1B_1C_2) & (A_1B_2C_1 + A_1B_2C_2)\end{array}$$

The mean value for each vertex of the resultant square is then based on 4 observations. The difference in A effects with B at the first level is given by

$$(A_2B_1C_1 + A_2B_1C_2 - A_1B_1C_1 - A_1B_1C_2)$$

and with B at the second level by

$$(A_2B_2C_1 + A_2B_2C_2 - A_1B_2C_1 - A_1B_2C_2)$$

and the interaction $A \times B$ which is the difference of these two is given by

$$(A_2B_2C_1 + A_2B_2C_2 + A_1B_1C_1 + A_1B_1C_2 - A_1B_2C_1 - A_1B_2C_2 - A_2B_1C_1 - A_2B_1C_2)$$

If we interchange the subscripts on A and B in each term of the above expression, the value of the expression is unchanged, indicating that the interactions $A \times B$ and $B \times A$ are identical and that we could have reversed the roles of A and B in determining the interaction. Similar geometrical derivations can be given for the remaining two-factor interactions, and for the three-factor interaction ABC, but these will be deferred until an alternative nomenclature has been developed. If the interaction effects are real, the degrees of freedom available for the estimation of the residual variance are reduced by the number necessary to estimate these effects. The scope of the analysis is then increased, but the sensitivity is reduced.

In the analysis of the factorial design it is important to consider the types of population involved when significance tests are being made. Factorial experiments frequently involve scale factors and are carried out to investigate conditions at a number of fixed levels of these factors. Since the inference to be drawn in this case does not relate to a population of possible levels from which those selected represent a *random* sample, significant interactions should not be used for testing effects of lower order without careful consideration of the individual problem and of the average value of the mean squares involved.

8.42. The 2^3 Factorial Experiment

A factorial design involving k factors each at 2 levels will include a total of 2^k treatment combinations. In such an experiment each main effect, and every possible interaction between factors, will have 1 degree of freedom in the analysis of variance, a total of $2^k - 1$ degrees of freedom in all. The allocation of the degrees of freedom for values of k ranging from 2 to 8 has already been given in Table 7.26, where it was pointed out that in the general case this allocation may be derived from the coefficients of the binomial expansion. The analysis of the treatment effects in a factorial experiment can be made by breaking down the treatment sum of squares into components each with a single degree of freedom, and proceeding with an analysis of variance. In the case of the 2^k design it is possible to substitute a more direct method which gives a better understanding of the results of the experiment. We shall illustrate this method for the case of a 2^3 design, and for this purpose we adopt a notation which has been generally used in this type of experiment.

Let us designate by a, b, and c the presence of the factor A, B, and C at the second level, assigning the term arbitrarily in cases where 2 distinct categories are involved. Hence the treatment a would consist of factor A at the second level and factor B and C at the first level, and the treatment ab would consist of A and B at the second level and C at the first level.

If we denote the combination in which all factors are at the first level by (1), then the possible treatment combinations in a 2^3 factorial design are given by (1), a, b, c, ab, ac, bc, and abc. We also designate the result or total for a given experimental unit by these symbols. The letters A, B, C, AB, AC, BC, ABC are used to denote the totals of the effects due to these factors, and also the experimental estimates of these totals. Thus the estimate of the total effect of A would be the difference of the sum of observations from experimental units involving treatment a and the sum of observations from units not involving this treatment:

$$A = abc + ab + ac + a - bc - b - c - (1)$$

and similarly

$$B = abc + ab + bc + b - ac - a - c - (1)$$

$$C = abc + ac + bc + c - ab - a - b - (1)$$

Although these totals are generally described as, e.g., "The effect of $A \cdots$," they do not represent the mean effect on the observations of moving from the lower to the upper level of A, but 4 times this effect in a 2^3 design. In the 2^k design there are 2^{k-1} experimental units at the upper level of A and a similar number at the lower level. The difference between the sums of these two groups therefore represents 2^{k-1} times the mean effect of a change of level.

With the above notation the interaction effect AB is given by the difference between the effect of A at the upper level of B and the effect of A at the lower level of B, that is:

$$\begin{aligned} AB &= (abc - bc + ab - b) - (ac - c + a - (1)) \\ &= abc + ab + c + (1) - ac - bc - a - b \end{aligned}$$

and similarly

$$AC = abc + ac + b + (1) - ab - bc - a - c$$

$$BC = abc + bc + a + (1) - ab - ac - b - c$$

As in the previous treatment the same results are obtained if we interchange the roles of the factors involved in the interactions. To obtain the single three-factor interaction we first consider the estimate of the effect due to A for each of the 4 combinations of B and C. These are $(abc - bc)$, $(ab - b)$, $(ac - c)$, and $(a - (1))$. The AB interaction estimates computed separately for the two levels of C are $[(abc - bc) - (ac - c)]$ and $[(ab - b) - (a - (1))]$, and the difference between these expressions represents the effect of the level of C upon the AB interaction, i.e., the ABC interaction. Thus we have

$$\begin{aligned} ABC &= (abc - bc) - (ac - c) - (ab - b) + (a - (1)) \\ &= abc - ab - ac - bc + a + b + c - (1) \end{aligned}$$

Since the interchange of the symbols a, b, and c does not affect the estimate ABC, we should obtain the same result by interchanging the roles of A, B, and C in the derivation of this effect. This is generally true for interactions of any order. As with the main effects, the estimates of the interaction effects obtained by this method represent a comparison between observation totals, and in order to estimate the mean effect we divide these totals by 4 in the case of a 2^3 design, or by 2^{k-1} in the case of a 2^k design.

Table 8.1 gives the signs which must be employed in combining the observations from the experimental units to obtain estimates of the main effects and interactions. The numerical coefficient is unity in each

TABLE 8.1

Treatments	(1)	a	b	ab	c	ac	bc	abc
Effects								
I	+	+	+	+	+	+	+	+
A	−	+	−	+	−	+	−	+
B	−	−	+	+	−	−	+	+
AB	+	−	−	+	+	−	−	+
C	−	−	−	−	+	+	+	+
AC	+	−	+	−	−	+	−	+
BC	+	+	−	−	−	−	+	+
ABC	−	+	+	−	+	−	−	+

instance so that the comparisons are orthogonal, and therefore independent. The contribution of each effect to the total sum of squares may be obtained by squaring the effect estimates and dividing by the sum of squares of the coefficients involved in the comparison. In a 2^3 design this divisor is 8, since 8 experimental units are involved and each is employed in every comparison. More generally, for a 2^k design the divisor is 2^k.

If a design of this type is carried out in r replicates, the difference between replicates within experimental units provides an estimate of the inherent variability, and the effect estimates may be obtained from Table 8.1 by employing the sums of the observations for given experimental units as the factors in the comparison. In this case the overall mean will be estimated as $I/2^k r$, and the remaining effects and interactions will be estimated by dividing the total effect estimates by $2^{k-1}r$. The divisor for the components of the sum of squares will be equal to the total number of observations, i.e., $2^k r$.

EXAMPLE. The data of Table 8.2 represent results from a factorial experiment on a spinning band laboratory fractionating column. The factors involved in the design are:

A. Clearance between band and static tube (two levels).
B. Boil-up rate (two levels).
C. Rate of rotation of band (two levels).

The observations represent the number of equivalent theoretical plates as determined by computation from the refractive index of the still and condenser liquids and the known characteristics of the binary mixture examined.

TABLE 8.2

A. Band clearance	0.05 inch				0.1 inch			
B. Boil-up rate	1		2		1		2	
C. Band speed (rpm)	750	1500	750	1500	750	1500	750	1500
Number of eq. th. plates	11.8	20.9	8.5	16.2	9.9	18.3	8.1	16.0
Treatment combination	(1)	*c*	*b*	*bc*	*a*	*ac*	*ab*	*abc*

The main effects and interactions could be computed directly as, e.g.,

$$A = abc + ab + ac + a - bc - b - c - (1)$$

$$= 16.0 + 8.1 + 18.3 + 9.9 - 16.2 - 8.5 - 20.9 - 11.8$$

$$= -5.1$$

but a systematic tabular method due to Yates [52] is more convenient. The original data are listed in column (2) of Table 8.3 in the order indicated in column (1). The first 4 entries in column (3) are the sums [(i) + (ii)], [(iii) + (iv)], [(v) + (vi)], and [(vii) + (viii)], and second 4 are the differences [(ii) — (i)], [(iv) — (iii)], [(vi) — (v)], and [(viii) — (vii)], so that, for example, the term 16.6 represents $(ab + b)$. Operating on column (3)

TABLE 8.3

(1) Treatment Combination	(2) Observation		(3)	(4)	(5)	Effect	(Effect)^2/8
(1)	11.8	(i)	21.7	38.3	+109.7	*I*	1,504.261
a	9.9	(ii)	16.6	71.4	−5.1	*A*	3.251
b	8.5	(iii)	39.2	−2.3	−12.1	*B*	18.301
ab	8.1	(iv)	32.2	−2.8	+3.9	*AB*	1.901
c	20.9	(v)	−1.9	−5.1	+33.1	*C*	136.951
ac	18.3	(vi)	−0.4	−7.0	−0.5	*AC*	0.031
bc	16.2	(vii)	−2.6	+1.5	−1.9	*BC*	0.451
abc	16.0	(viii)	−0.2	+2.4	+0.9	*ABC*	0.101

in the same way gives column (4), of which the first entry is $[(1) + a + b + ab]$ and the second $[c + ac + bc + abc]$. On forming column (5) from (4) in the same way we obtain for the first entry

$$[c + ac + bc + abc] + [(1) + a + b + ab] = \mathrm{I}$$

and for the fifth

$$[c + ac + bc + abc] - [(1) + a + b + ab] = C$$

The other entries may be shown to correspond to the appropriate effects. The final column contains the contribution to the uncorrected sum of squares of each effect. Since the 8 comparisons are orthogonal, this may be obtained by squaring the effect total and dividing by the sum of squares of the coefficients. As a check on the results, the sum of squares of the original observations may be computed and compared with the total of column (7). The formal analysis of variance for the components, each with a single degree of freedom, is given in Table 8.4.

TABLE 8.4

Source of Estimate	Sum of Squares	D.F.
Effect *A*	3.25	1
B	18.30	1
C	136.95	1
Interaction *AB*	1.90	1
AC	0.03	1
BC	0.45	1
ABC	0.10	1
Total (corrected for mean effect)	160.98	7

In view of the small number of tests we cannot hope to learn much about the interactions. Since $F_{1,1,0.05} = 164$, there is no reason to suppose that any of them are disproportionately large, and we pool them to provide an estimate of the residual based on 4 degrees of freedom. This proves to be 0.62, and testing the main effect components of the sum of squares against it we obtain variance ratios:

$$A: \quad 3.25/0.62 = 5.2 \qquad (F_{1,4,0.1} = 4.5)$$
$$B: \quad 18.30/0.62 = 29.5^{**} \qquad (F_{1,4,0.01} = 21.2)$$
$$C: \quad 137/0.62 = 221^{***} \qquad (F_{1,4,0.001} = 74.1)$$

The data provide significant evidence of an improvement in fractionation performance with increase in band speed and decrease in boil-up rate. They suggest that within the range examined the column may be more efficient as the clearance is reduced.

8.43. The 2^k Factorial Experiment

The extension of the above formulation to the general case of k factors at 2 levels is somewhat simplified by extending the symbolic method of the previous section. In the above example, we can write the various effects as

$$A = (a - 1)(b + 1)(c + 1)$$
$$B = (a + 1)(b - 1)(c + 1)$$
$$AB = (a - 1)(b - 1)(c + 1)$$

Note that in this form the negative sign, or signs, always correspond to the main effects, or interactions, being studied. This rule can be immediately generalized to any number of factors. Thus in a two-factor experiment we should have

$$A = (a - 1)(b + 1)$$
$$B = (a + 1)(b - 1)$$
$$AB = (a - 1)(b - 1)$$

and, of course, $\mathrm{I} = (a + 1)(b + 1) =$ sum of all treatments. In a four-factor experiment we should have, for example,

$$A = (a - 1)(b + 1)(c + 1)(d + 1)$$
$$D = (a + 1)(b + 1)(c + 1)(d - 1)$$
$$AD = (a - 1)(b + 1)(c + 1)(d - 1)$$
$$BCD = (a + 1)(b - 1)(c - 1)(d - 1)$$
$$ABCD = (a - 1)(b - 1)(c - 1)(d - 1)$$

In all, there would be 15 such effects (16 including I). Each expansion would have 16 terms corresponding to the 16 treatments, and would indicate the method of combination necessary to obtain the desired main effect or interaction.

It can be seen that the number of treatment combinations increases rapidly with the number of factors involved, and hence even in a 2^5 factorial experiment, with 32 treatment combinations, a single replication requires 32 experimental units, and each succeeding replication an additional 32. For this reason higher-order factorial experiments are frequently carried out with only single replication, which does not allow a within-treatments estimate of error. Furthermore, because with such a large number of tests it is difficult to ensure that the experimental conditions are maintained constant throughout the testing period, it is usual to use one of the methods of local control considered in Section 8.5 in order to reduce residual variance. In the preceding chapter, tests were made in the absence of replication by assuming the highest-order interactions to be negligible, and using the estimate based on this interaction as an error estimate. In the 2^k design this interaction, like any other effect, has only 1 degree of freedom, and would not provide a satisfactory error estimate. However, in a 2^5 experiment, for example, there will be 5 four-factor interactions and 10 three-factor interactions, all or many of which may be known a priori, or can be assumed from the analysis, to be negligible. By assuming that, in addition to the single five-factor interaction, all 5 of the four-factor interactions were negligible, we could obtain 6 degrees of freedom for error estimation; on this basis some of the three-factor interactions might prove small enough to be pooled before testing the two-factor interactions and main effects.

In factorial experiments involving a large number of factors each at 2 or more levels, the number of experimental units required increases rapidly. Thus for a 2^8 design the effects of 256 different treatment combinations would be investigated, and estimates would be available for 1 eight-factor interaction, 8 seven-factor interactions, and 28 six-factor interactions. In such cases the number of interaction estimates which we are prepared to assume negligible is often more than adequate to provide a residual variance estimate, and it is desirable to reduce the number of experimental units in such a way that the important estimates are still available and the unimportant ones are sacrificed. Such modifications are considered in Section 8.7.

8.44. Designs Involving Factors at Other than Two Levels

The 2^k design is well suited for a preliminary investigation of a technical problem in which the available information suggests that a number of factors may be involved, and the relative importance of these factors is to be investigated. The selection of levels is largely a technical matter, but for commercial investigations it seems reasonable to select the difference between levels for the various factors so that they represent comparable intervals in production costs, or steps of comparable economic importance. If this is not done, we may find on analyzing the results that the effects of some of the factors which appear statistically significant are of no commercial importance and that a more exact estimate of the effects of other factors which are not statistically significant is required. When such an investigation has been completed, or if sufficient experience is available to render it unnecessary, those factors which appear important may be combined in a second experiment in order to study the nature of the relationships between factor and observations. For this purpose the number of levels at which the factors are tested must be increased. If the linearity of the relationship is to be tested, 3 levels may be adequate, but, if a maximum value is sought, 4 or 5 levels are desirable. In addition, where raw materials originate from a number of sources, or from a unit not under statistical control, it may be desirable to incorporate several samples of raw material in the design.

The general methods of analysis of results obtained from designs involving multilevel factors has been considered in Chapter 7, and references [52, 53, 55, 117] include a more detailed discussion of the many factorial designs which have been employed. In the following sections we consider a number of alternative methods of analysis which are available for special designs, and which often result in a better appreciation of the experimental data.

8.441. *Simplified Analysis when Some Factors are at Two Levels.* If in an experimental design one of the factors is to be investigated at only 2 levels, it is possible to simplify the analysis and the interpretation of the results by transforming the observations. Suppose that an experiment involves 4 factors A, B, C, D at levels a_1 and a_2, $b_1, \cdots, b_p$; $c_1, \cdots, c_q$, and $d_1, \cdots, d_r$, so that in all $2pqr$ treatment combinations are involved. If we consider the linear transformations of the observations from the cells $a_1b_lc_md_n$ and $a_2b_lc_md_n$

$$y_{1lmn} = x_{1lmn} + x_{2lmn}$$

$$y_{2lmn} = x_{1lmn} - x_{2lmn}$$

then y_{1lmn} and y_{2lmn} will be independently and normally distributed if the x's were so distributed. Separate analyses can then be performed on the $p \times q \times r$ classification for each of the y values. From the first, which corresponds to the subtable obtained in the usual analysis by summing over A, the effects I, B, C, D, BC, BD, CD, and BCD together with the corresponding sums of squares can be computed. The second analysis gives the interaction of these effects with the factor A since it is carried out on the difference between corresponding observations at the levels a_1 and a_2.

The method can be extended to cases in which more than one of the factors are to be investigated at 2 levels. If the experimental design is of the type $2^k \cdot p \cdot q \cdot r \cdot$, we can undertake the analysis as though the design involved a $p \times q \times r$ classification, each cell of which was a 2^k design. By transforming the observations in each of the pqr cells into independent estimates of the grand mean, the k main effects, $k(k-1)/2$ two-factor interactions, etc., of the 2^k classification and analyzing each of the transformed variates separately, the complete analysis can be synthesized.

EXAMPLE. In Section 7.61 the results of an investigation involving the determination of the volatile content of 2 coals in 2 types of crucible at each of 3 temperature levels was considered. The experiment in this instance was replicated, the corrected sum of squares of the individual observations being 270.4108. Arbitrarily we select coal B as the lower level of the coal factor B, and the steel crucible as the lower level of the crucible factor A. The table formed by summing over replicates may then be written as shown in Table 8.5.

TABLE 8.5

Treatment Combinations	(1)			(2)			(3)			Effect
	c_0	c_1	c_2	c_0	c_1	c_2	c_0	c_1	c_2	
(1)	2.04	3.88	2.69	12.76	16.29	15.19	47.36	55.00	55.19	I
a	10.72	12.41	12.50	34.60	38.71	40.00	11.52	8.78	11.15	A
b	15.88	19.23	19.33	8.68	8.53	9.81	21.84	22.42	24.81	B
ab	18.72	19.48	20.67	2.84	0.25	1.34	−5.84	−8.28	−8.47	AB

The sum-and-difference method for the analysis of a 2^2 factorial design is carried out at each of the 3 levels of c in sections (1), (2), and (3), and the entries in section (3) represent the estimates of the effects I, A, B, and BC at these levels.

In order to complete the computation so that a variance analysis may be undertaken, the estimates of effects at the 3 levels of c must be combined. Since factor C has been examined at 3 levels it is possible to separate the

TABLE 8.6

Effect	Level of C			Overall Effects			Components of Sum of Squares		
	c_0	c_1	c_2	I	C_L	C_Q	I	C_L	C_Q
I	47.36	55.00	55.19	157.55	7.83	−7.45	I(24) 1034.25	C_L(16) 3.83	C_Q(48) 1.16
A	11.52	8.78	11.15	31.45	−0.37	5.11	A(24) 41.21	AC_L(16) 0.0086	AC_Q(48) 0.544
B	21.84	22.42	24.81	69.07	2.97	1.81	B(24) 198.78	BC_L(16) 0.551	BC_Q(48) 0.068
AB	−5.84	−8.28	−8.47	−22.59	−2.63	−2.25	AB(24) 21.26	ABC_L(16) 0.437	ABC_Q(48) 0.105

temperature effects into linear and quadratic components by forming the orthogonal combinations

$$
\begin{aligned}
IX &= c_0X + c_1X + c_2X \\
C_LX &= -c_0X + c_2X \\
C_QX &= c_0X - 2c_1X + c_2X
\end{aligned}
$$

where X represents successively I, A, B, and AB. The contribution to the sum of squares of each of these final effects may then be computed by squaring and dividing by the sum of squares of the coefficients involved in the combination. The results of such a computation are given in Table 8.6, the figures in parentheses indicating the divisor for the sum of squares. If required a formal analysis of variance table may be drawn up from the final section of Table 8.6, and this procedure is recommended in cases where finite populations have been sampled since the correct significance tests are then more easily distinguished. The original analysis of the data was presented in Table 7.31, and will not be repeated. It is easily shown that the linear and quadratic or residual components of the temperature effect are significant at the 0.001 and 0.05 levels, respectively, and that the rest of the analysis is unchanged.

8.442. *Single Degrees of Freedom in the* 3^2 *and* 3^3 *Design.* In analyzing the example of Table 7.17 which was concerned with loss of weight, after 3 drying periods, of a number of samples of yeast, we were able to subdivide the sum of squares between drying periods into 2 independent components, corresponding to the linear and quadratic regressions of the sample weight on the drying time. The variation of the linear and quadratic components between samples was also examined by computing the interactions between these independent components and the 10 samples. In cases where the data may be classified according to a number of scale factors, each at 3 or more levels, each factor may be analyzed as a number of independent comparisons and the interactions between the independent comparisons of the various factors may be determined.

We first consider the case of a 3^2 design involving factors A and B and indicate the 9 treatment combinations by

$$
\begin{matrix}
a_0b_0 & a_1b_0 & a_2b_0 \\
a_0b_1 & a_1b_1 & a_2b_1 \\
a_0b_2 & a_1b_2 & a_2b_2
\end{matrix}
$$

The individual comparisons to be employed may be selected on the basis of any previous knowledge or theoretical considerations which would indicate the nature of the relationship. If the levels of the factor A and factor B were equally spaced, and if it were believed that the response to the treatments might be approximately proportional to the level of the factors involved, then the independent comparisons could be selected to correspond to linear and quadratic components. For the main effects of A and B these are simply

$$\begin{aligned} A_L &= (a_2b_2 + a_2b_1 + a_2b_0) - (a_0b_2 + a_0b_1 + a_0b_0) \\ &= (a_2 - a_0)(b_2 + b_1 + b_0) \end{aligned}$$

$$\begin{aligned} A_Q &= (a_2b_2 + a_2b_1 + a_2b_0) - 2(a_1b_2 + a_1b_1 + a_1b_0) + (a_0b_2 + a_0b_1 + a_0b_0) \\ &= (a_2 - 2a_1 + a_0)(b_2 + b_1 + b_0) \end{aligned}$$

$$\begin{aligned} B_L &= (a_2b_2 + a_1b_2 + a_0b_2) - (a_2b_0 + a_1b_0 + a_0b_0) \\ &= (a_2 + a_1 + a_0)(b_2 - b_0) \end{aligned}$$

$$\begin{aligned} B_Q &= (a_2b_2 + a_1b_2 + a_0b_2) - 2(a_2b_1 + a_1b_1 + a_0b_1) + (a_2b_0 + a_1b_0 + a_0b_0) \\ &= (a_2 + a_1 + a_0)(b_2 - 2b_1 + b_0) \end{aligned}$$

and for the interactions

$$\begin{aligned} A_LB_L &= (a_2b_2 - a_0b_2) - (a_2b_0 - a_0b_0) \\ &= (a_2 - a_0)(b_2 - b_0) \end{aligned}$$

$$\begin{aligned} A_LB_Q &= (a_2b_2 - a_0b_2) - 2(a_2b_1 - a_0b_1) + (a_2b_0 - a_0b_0) \\ &= (a_2 - a_0)(b_2 - 2b_1 + b_0) \end{aligned}$$

$$\begin{aligned} A_QB_L &= (a_2b_2 - 2a_1b_2 + a_0b_2) - (a_2b_0 - 2a_1b_0 + a_0b_0) \\ &= (a_2 - 2a_1 + a_0)(b_2 - b_0) \end{aligned}$$

$$\begin{aligned} A_QB_Q &= (a_2b_2 - 2a_1b_2 + a_0b_2) - 2(a_2b_1 - 2a_1b_1 + a_0b_1) \\ &\quad + (a_2b_0 - 2a_1b_0 + a_0b_0) \\ &= (a_2 - 2a_1 + a_0)(b_2 - 2b_1 + b_0) \end{aligned}$$

The coefficients for use with the 3×3 table of observations may be conveniently obtained as the product of the coefficients of the linear combinations of the two factors. Thus the main effect of A_L is given by $(-1, 0, 1) \times (1, 1, 1)$, the rule for multiplication being that the first row of the resultant is formed by multiplying the coefficients of the first bracket by the first coefficient of the second bracket, the second row by a similar multiplication using the next coefficient of the second bracket, and so on, leading to

$$\begin{matrix} -1 & 0 & 1 \\ -1 & 0 & 1 \\ -1 & 0 & 1 \end{matrix}$$

The 9 possible products which can be obtained in this way are given in Table 8.7, together with the corresponding divisors, and the effects to which they refer. The effects are then computed by multiplying the sum

TABLE 8.7

	(1, 1, 1) (1, 1, 1)			(−1, 0, 1) (1, 1, 1)			(1, −2, 1) (1, 1, 1)			(1, 1, 1) (−1, 0, 1)			(1, 1, 1) (1, −2, 1)		
	1	1	1	−1	0	1	1	−2	1	−1	−1	−1	1	1	1
	1	1	1	−1	0	1	1	−2	1	0	0	0	−2	−2	−2
	1	1	1	−1	0	1	1	−2	1	1	1	1	1	1	1
Divisor	9			6			18			6			18		
Effect	I			A_L			A_Q			B_L			B_Q		

	(−1, 0, 1) (−1, 0, 1)			(−1, 0, 1) (1, −2, 1)			(1, −2, 1) (−1, 0, 1)			(1, −2, 1) (1, −2, 1)		
	1	0	−1	−1	0	1	−1	2	−1	1	−2	1
	0	0	0	2	0	−2	0	0	0	−2	4	−2
	−1	0	1	−1	0	1	1	−2	1	1	−2	1
Divisor	4			12			12			36		
Effect	$A_L B_L$			$A_L B_Q$			$A_Q B_L$			$A_Q B_Q$		

of the results for each experimental unit by the corresponding coefficient of the appropriate table and summing. The contribution made by each effect to the overall sum of squares may be calculated by squaring the effect and dividing by r times the sum of squares of the coefficients in this table, r being the number of replicate measurements for each treatment combination.

If we had reason to suppose that the response to the various levels of one or both factors would be better described by linear combinations which did not represent the linear and quadratic portions of the regression of the observations on these factors, a similar method of analysis could be employed. Thus, taking the comparisons for factor A in the form (1, 1, 1) (−1, 0, 1) and (1, −2, 1) and those for B in the form I = (1, 1, 1), x = (2, −1, −1,), y = (0, 1, −1), the table of coefficients for the effect $A_Q B_x$ would be given by the product (1, −2, 1) (2, −1, −1) or

2	−4	2
−1	2	−1
−1	2	−1

An example of a complete analysis of a 3^2 experiment which was designed to investigate the effects of variation in temperature and mixed acid concentration on the nitration of cellulose has been published [44]. In the case of a 3^3 or similar experiment involving a number of scale factors at more than 2 levels the tables of coefficients may be obtained by a repetition of the multiplication process employed above. Alternatively the effects may be computed in stages, the independent comparisons for 2 of the factors being evaluated separately for each treatment combination of the remaining factors and subsequently combined into further comparisons. This procedure is more time consuming, but results in a lesser number of computational errors.

EXAMPLE. The data of Table 8.8 were reported by Fuell and Wagg [56]. The original experiment was designed to investigate the effects of the concentration of detergent, sodium carbonate, and sodium carboxymethyl cellulose upon the washing and suspending power of a solution. The data quoted refer to the washing tests, where improved performance is denoted by a higher figure. Each factor was investigated at 3 levels with equal spacing so that it is convenient to separate the linear and quadratic components of the main effects and interactions.

TABLE 8.8

Detergent Level		d_0			d_1			d_2		
Sodium carbonate		a_0	a_1	a_2	a_0	a_1	a_2	a_0	a_1	a_2
	c_0	106	197	223	198	329	320	270	361	321
Sodium C.M.C.	c_1	149	255	294	243	364	410	315	390	415
	c_2	182	259	297	232	389	416	340	406	387

The nature of the variation is obviously complex, and, although it will be necessary to pool a number of interactions in order to provide an estimate of the residual variance, it is advantageous to separate the sum of squares into components, each with a single degree of freedom.

We start by considering the design as though it were the result of 3 separate 3^2 experiments at different levels of d. For each of these experiments the linear and quadratic components of the main effects and interactions are computed. This leads to Table 8.9. By selecting linear combinations of any effect X at the 3 levels of factor D in the form

$$\begin{aligned} Xd_2 + Xd_1 + Xd_0 &= XI = X \\ Xd_2 \qquad\quad - Xd_0 &= XD_L \\ Xd_2 - 2Xd_1 + Xd_0 &= XD_Q \end{aligned}$$

the main effect X and its interactions with D_L and D_Q can be computed.

TABLE 8.9

Detergent level	d_0	d_1	d_2
Effect I	+1962	+2901	+3205
A_L	+377	+473	+198
A_Q	−171	−345	−266
C_L	+212	+190	+181
C_Q	−132	−150	−155
$A_L C_L$	−2	+62	−4
$A_L C_Q$	−58	−28	−102
$A_Q C_L$	+26	+10	+46
$A_Q C_Q$	+30	−120	−116

The effects so calculated together with the sum of squares of the coefficients of the combination (divisor) and the contribution of the effect to the overall sum of squares are given in Table 8.10.

The pooled mean square for all the three-factor interactions is 175, so that the mean square for $D_L A_L C_L$ is rather low, but since this represents the extreme case from 8 effects we shall use this pooled estimate to test the lower-order effects. The most satisfactory method of procedure is to combine the sums of squares corresponding to the single degrees of freedom of a given interaction etimate to provide a joint variance estimate based on 4 degrees of freedom. Testing these variance estimates against the pooled three-factor interaction estimate, we obtain

$$AC: \quad \frac{481}{175} = 2.60 \qquad (F_{4,\,8,\,0.05} = 3.8)$$

$$CD: \quad \frac{25}{175} = \frac{1}{7}^{*} \qquad (F_{8,\,4,\,0.05} = 6.1)$$

$$AD: \quad \frac{1844}{175} = 10.5^{**} \qquad (F_{4,\,8,\,0.01} = 7.0)$$

It appears that an examination of the individual components of the interactions should be undertaken for CD and AD. The low-variance estimate of the former cannot be attributed to any single component, but

TABLE 8.10

Source of Estimate	Effect	Divisor	Mean Square	Source of Estimate	Effect	Divisor	Mean Square
I	+8068	27	2,410,837	$A_L C_L$	+56	12	261
D_L	+1243	18	85,836	$A_L C_Q$	+188	36	982
D_Q	+635	54	7,467	$A_Q C_L$	+82	36	187
A_L	+1048	18	61,017	$A_Q C_Q$	−206	108	393
A_Q	−782	54	11,326	$D_L A_L C_L$	−2	8	0.5
C_L	+583	18	18,883	$D_L A_L C_Q$	−44	24	81
C_Q	−437	54	3,536	$D_L A_Q C_L$	+20	24	17
$D_L A_L$	−179	12	2,670	$D_L A_Q C_Q$	−146	72	296
$D_L A_Q$	−95	36	251	$D_Q A_L C_L$	−118	24	704
$D_Q A_L$	−371	36	3,823	$D_Q A_L C_Q$	−104	72	150
$D_Q A_Q$	−253	108	593	$D_Q A_Q C_L$	+52	72	38
$D_L C_L$	−31	12	80	$D_Q A_Q C_Q$	+154	216	110
$D_L C_Q$	−23	36	15				
$D_Q C_L$	−13	36	5				
$D_Q C_Q$	−13	108	1				

the D_Q effect is remarkably consistent for all levels of C. This suggests that in pooling the three-factor interaction components to provide a variance estimate we may have overestimated the true variance of a single observation. The high value of the variance estimate for the AD interaction appears to be due entirely to the components $D_L A_L$ and $D_Q A_L$, indicating that the linear response to sodium carbonate additions depends in a rather complicated way on the level of the detergent additions.

Since all the factors involved in the design are scale classifications the main effects may be tested against the residual estimate, the conclusions so obtained being restricted to the particular levels examined. On this basis all the main effects are significant.

8.443. *Dummy Comparisons.* Comparisons which estimate only random effects may arise when certain treatment combinations are identical so that the factors are not completely crossed. If, for example, a number of modifications to a process, or a number of promoters for a catalyst, were to be compared, we should probably wish to include the present standard method or the present type of catalyst as a control in the experiment. The design of the resulting experiment depends to a large extent on whether we wish to distinguish between the merits of the various modifications, or promoters, or to determine whether they represent an improvement on the standard procedure. In the first instance we should consider the standard method on an equal basis with any of the alternatives. If there were $p - 1$ modifications, the class would consist of p varieties, and as such would be crossed with any other classification of the design. Let us suppose for the moment that no other crossed classification were involved, but that the experiment was replicated r times. Then to separate from the sum of squares the component due to the difference between the standard condition and the modifications we should form the estimate of this effect $\dfrac{(p - 1)T_1 - T_2 - \cdots - T_p}{r(p - 1)}$, the component of the sum of squares being given by

$$\frac{[(p - 1)T_1 - T_2 - \cdots - T_p]^2}{rp(p - 1)}$$

and having, of course, a single degree of freedom. The analysis of variance would then take the form of the accompanying table. Since

Source of Estimate	D.F.
Mean effect of modifications	1
Difference between modifications	$p - 2$
Difference between all varieties	$p - 1$
Between replications	$(r - 1)p$
Total	$pr - 1$

there are only r results obtained under the standard conditions, and $r(p - 1)$ results from modifications, the first comparison is relatively weak, the variance of the estimate of the effect being $\sigma^2 p/r(p - 1)$.

If we regarded this comparison as more important than the comparisons between modifications, we might arrange that for every modification to be tested a control in the form of an experimental unit involving the standard conditions was included in the design. In this case if there were $p - 1$ modifications, we should require $2(p - 1)r$ experimental units. The mean effect of the modifications would then be estimated as

$$\frac{(T_1 + T_2 + \cdots + T_{p-1}) - (T_p + T_{p+1} + \cdots + T_{2p-2})}{r(p - 1)}$$

where $T_1, T_2, \cdots, T_{p-1}$ represent the totals from the controls and $T_p, T_{p+1}, \cdots, T_{2p-2}$ the totals from the $p - 1$ modifications. The component of the sum of squares corresponding to the comparison between controls and modifications would be

$$\frac{[(T_1 + T_2 + \cdots + T_{p-1}) - (T_p + T_{p+1} + \cdots + T_{p-2})]^2}{2r(p - 1)}$$

Of the remaining $2p - 2 - 2 = 2p - 4$ orthogonal comparisons between treatment totals, $p - 2$ would refer to differences between modifications and $p - 2$ to differences between controls. The latter are a measure of the variation in response to a constant treatment and may be combined with other random components between replicates and used to estimate the residual variance if such a procedure appears desirable. In more advanced designs the independent estimate of the residual variance provided by dummy comparisons may be retained in a separate form and employed to recover information. The analysis of variance takes the form of the accompanying table.

Source of Estimate	D.F.
Mean effect of modifications	1
Differences between modifications	$p - 2$
Differences between controls	$p - 2$
Total variety effects	$2p - 3$
Between replicates	$(r - 1)(2p - 2)$
Total	$r(2p - 2) - 1$

In the second design $(p - 2)$ additional experimental units have been introduced in order to strengthen the selected comparison between modifications and controls, and the variance of the estimate of this comparison is reduced to $2\sigma^2/r(p - 1)$. The precision of the other comparisons is unaffected except in so far as the residual variance estimate is improved by the inclusion of additional degrees of freedom from the dummy comparisons.

If the varietal classification which includes a number of dummy comparisons is crossed with a second classification, the interactions of these comparisons with those of the second factor may also be employed to estimate the residual variance. If in the example in which $(p - 1)$ plant modifications and $(p - 1)$ control tests were to be carried out the investigation were extended to examine 2 levels of output, the analysis of variance would take the form of the accompanying table. The estimates marked

Source of Estimate	D.F.
Main effects:	
Mean effect of modifications	1
Differences between modifications	$p - 2$
Differences between controls	$p - 2$*
Total between varieties	$2p - 3$
Between throughputs	1
Interactions:	
Mean effect of modification × throughput	1
Differences between modifications × throughput	$p - 2$
Differences between controls × throughput	$p - 2$*
Residual	$4(p - 1)(r - 1)$
Total	$4r(p - 1) - 1$

with an asterisk, which refer to random variations between observations from identical treatment combinations, may be used separately or in conjunction with the variations between replicates to estimate the residual variance.

8.5. Elimination of Experimental Error

8.51. General Considerations

Factorial designs are usually arranged so that all the observations made in connection with an experiment are used in computing each of the treatment effects. By this means the uncertainties in these estimates are minimized for a given number of observations and a fixed residual variance. When such a design is employed any further economies in the number of observations to be made can be achieved only by reducing the residual variance. This may be accomplished by:

(1) The careful control of all important factors, other than those to be

investigated, throughout the duration of the tests. This method suffers from the disadvantages that the necessary control may be costly to achieve, and, if human performances are involved, the application of the control may produce a systematic effect upon the results.

(2) The systematic elimination of the effects of known sources of error which can be identified but cannot economically be controlled. These effects may be estimated from associated measurements by the method of covariance analysis (Section 7.8), or the experiment may be so designed that the required comparisons are independent of the effects of the principal sources of error.

The first method is usual in scientific investigations, but it can often be used most effectively in conjunction with one or more of the statistical methods of (2). In the remainder of this section we shall consider a number of examples of the latter type of approach.

8.52. Replication in Blocks

One of the types of experimental error which can frequently be anticipated in industrial chemistry is that due to variation in raw material. Even where the raw material cannot be divided into logical sublots, samples of material from adjacent regions tend to differ less than those from widely separate regions. We shall consider first the case in which the sublot is sufficient to support a complete replicate of the experiment, and is chosen for this purpose from a compact region of the raw material when no logical basis for subdivision is available. The treatment combinations which make up the replicate will then be applied to random samples of raw material from the sublot in order to ensure that an unbiased estimate of the residual variance is obtained. The resulting replication is known as a *randomized complete block*, and any differences between the blocks of the complete experiment will not influence our estimates of the treatment effects, provided that the block and treatment effects do not interact.

As a standard example which we shall employ in later discussions, we consider an experiment designed to estimate the effects of variation in temperature and catalyst composition upon the yield of a particular fraction from a hydrocarbon synthesis unit, each factor being investigated at 2 levels. The raw material in this case is an H_2/CO mixture, the composition of which will fluctuate somewhat with time. Since a change in the H_2/CO ratio is known to influence the yields being measured, we consider the quantity of synthesis gas necessary to carry out the 4 experimental units of a single replicate as a sublot, carrying out the 4 units in random order. The synthesis gas required for the next 4 experimental units is considered the next sublot, and a complete, but randomized, replicate is performed upon this material. Supposing that 6 replicates of the test were to be carried out, the full design might appear as in the accompanying blocks.

b	a	ab	(1)	a	(1)
ab	b	b	b	(1)	ab
a	ab	a	ab	b	a
(1)	(1)	(1)	a	ab	b
Block 1	2	3	4	5	6

If we can assume that the residual variance within blocks is uniform, it might be reasonable to suppose that the average value of any observation could be represented by

$$\text{ave}\,(x_{ijk}) = e_0 + e_i + e_j + e_{ij} + e_k$$

i.e., by an overall mean, an effect due to the temperature level, an effect due to the type of catalyst, an effect due to the interaction of these two main effects, and an effect due to the block in which the test was undertaken. This is equivalent to the supposition that any variation in the synthesis gas between blocks has a similar constant effect upon all the experimental units in the succeeding block, or that there are no real interaction effects between blocks and the treatment combinations. Since all blocks include a complete replicate, the sums of squares due to blocks, treatments, and errors are each independent, and the analysis of variance takes the form of the accompanying table, if we suppose that

Source of Estimate	D.F.	Average Value of Mean Square
Between temperatures	1	$\sigma^2 + 6\sigma^2_{ij} + 12\sigma_i^2$
Between catalysts	1	$\sigma^2 + 6\sigma^2_{ij} + 12\sigma_j^2$
Interaction temperatures × catalyst	1	$\sigma^2 + 6\sigma^2_{ij}$
Total treatment effects	3	
Between blocks	5	$\sigma^2 + 4\sigma_k^2$
Residual	15	σ^2
Total	23	

the levels of temperature and catalyst represent random samples from an infinite population of possible levels. In this example we should probably restrict the statistical inference to the particular levels examined, and the underlined coefficients would then be zero. Since we assumed that there are no interactions between block effects and the other factors, the sums of squares which would normally lead to estimates of these effects, and which account for $5 + 5 + 5 = 15$ degrees of freedom, may be pooled to provide an estimate of the residual variance. If we consider the analysis of the same experimental data which would be made if we did not assume that the blocks $\times$ treatment interactions were zero, we obtain the accompanying analysis of variance, where the underlined coefficients are zero if we

Source of Estimate	D.F.	Average Value of Mean Square
Between temperatures	1	$\sigma^2 + \underline{\sigma}^2_{ijk} + \underline{6}\sigma^2_{ij} + 2\sigma^2_{ik} + 12\sigma_i^2$
Between catalysts	1	$\sigma^2 + \underline{\sigma}^2_{ijk} + \underline{6}\sigma^2_{ij} + 2\sigma^2_{jk} + 12\sigma_j^2$
Temperatures $\times$ catalysts	1	$\sigma^2 + \sigma^2_{ijk} + 6\sigma^2_{ij}$
Between blocks	5	$\sigma^2 + \underline{\sigma}^2_{ijk} + \underline{2}\sigma^2_{jk} + 4\sigma_k^2$
Blocks $\times$ temperatures	5	$\sigma^2 + \underline{\sigma}^2_{ijk} + 2\sigma^2_{ik}$
Blocks $\times$ catalysts	5	$\sigma^2 + \underline{\sigma}^2_{ijk} + 2\sigma^2_{jk}$
Blocks $\times$ temperatures $\times$ catalysts	5	$\sigma^2 + \sigma^2_{ijk}$
Total	23	

assume that the temperature and catalyst classifications are of type I and the blocks classification is of type II. The correct significance tests cannot now be made by pooling the block $\times$ treatment interactions as a single residual estimate.

If the block $\times$ treatment interactions are zero, the effectiveness of the method depends upon the existence and size of the effects e_k. If these are large, then the elimination of these terms from the residual variance improves the sensitivity of the experiment. If the block effects are negligible, then the experiment is weakened since the new estimate of the residual is based on 15 degrees of freedom in place of 20, and the variance ratio test is proportionately less sensitive. The general question of design efficiency and the requirements for the use of randomized blocks in place of complete randomization are discussed in [53], where additional references are given.

The method of paired comparisons of Section 5.52 is really the simplest example of a block design, each paired comparison constituting a block. In this case we assumed that the difference between pairs was independent of the block and estimated the variance of this difference from the within-block observations. If the residual variance component differs for different treatments, we may employ the paired-comparison method to ensure that the correct variance component is used in determining the significance of each treatment comparison. Thus, if (1), a, b, and ab had represented not a 2^2 factorial design but the tests in a trial of 4 different catalysts, we might wish to test the significance of the difference between a and b. If this difference is computed within each block, we can employ the method of Section 5.52 to ensure that an unbiased estimate of the variance of this difference is obtained. In so doing we are obliged to use a variance estimate based on only 5 degrees of freedom so that the significance test is less powerful.

TABLE 8.11

Furnace (Blocks)	Compacting Pressure	Size Range of Material					
		A	*B*	*C*	*D*	*E*	*F*
1	25	11.3	12.2	12.9	12.1	16.9	14.3
	50	21.1	21.1	21.7	24.4	23.6	23.5
2	25	11.9	10.4	12.4	13.9	14.9	15.0
	50	21.3	21.4	22.0	24.1	25.5	22.1
3	25	10.0	9.9	11.3	13.3	12.4	13.8
	50	18.8	19.5	21.6	23.8	23.3	20.5

EXAMPLE. The above data [59] were obtained during an investigation to determine the influence of the initial particle size and the compacting pressure upon the tensile strength of iron powder sinters. Sintering was carried out at a standard temperature in a dry hydrogen atmosphere on samples of materials of 6 size ranges A to F, material A having the greatest particle size and F the smallest. The compacting pressures employed were 25 and 50 tsi, and in order to expedite the treatment procedure 3 separate but similar furnaces were employed. The complete experiment was a triple replicate involving 48 experimental units, 12 being carried out in each furnace in a randomized complete block design. The allocation of the samples of each material to the blocks and treatments was randomized, subject to the condition that each furnace contain a complete replicate. The results obtained are given in Table 8.11.

The computational subtables may be obtained by summing over blocks,

Compacting Pressures	Size Range of Materials						Totals
	A	*B*	*C*	*D*	*E*	*F*	
25	33.2	32.5	36.6	39.3	44.2	43.1	228.9
50	61.2	62.0	65.3	72.3	72.4	66.1	399.3
Total	94.4	94.5	101.9	111.6	116.6	109.2	628.2

Block	1	2	3
Total	215.1	214.9	198.2

and computing in addition the marginal subtotals for pressures and size classes. The use of the usual computational methods leads to the analysis of variance in Table 8.12.

TABLE 8.12

Source of Estimate	Sum of Squares	D.F.	Mean Square
Between sizes	71.24	5	14.25
Between pressures	806.56	1	806.56
Size × pressure. Interact.	8.63	5	1.73
Treatment total	886.43	11	
Between blocks	15.68	2	7.84
Residual	16.62	22	0.775
Total	918.73	35	

The tests of significance are:

(1) *Interaction Size × Pressure.*

$$\text{variance ratio: } \frac{1.73}{0.755} = 2.29 \qquad (F_{5,\ 22,\ 0.05} = 2.7)$$

The interaction effect appears rather high, but is not significant at the 0.05 level. If we pool with the residual, we obtain a new variance estimate of 0.935, which will be used to test the main effects.

(2) *Size.*

$$\text{variance ratio: } \frac{14.25}{0.935} = 15.2^{***} \quad (F_{5,\ 27,\ 0.001} = 5.8)$$

(3) *Pressure.*

$$\text{variance ratio: } \frac{806.56}{0.935} = 863^{***} \quad (F_{1,\ 27,\ 0.001} = 13.6)$$

The use of randomized blocks in this example has contributed appreciably to the sensitivity of the experiment, the 2 degrees of freedom for block effects accounting for a component of the sums of squares almost as great as that due to the residual, which has 22 degrees of freedom.

In industrial experimentation the nature of the block will depend upon the circumstances of the experiment. Where the units of the design are carried out in chronological sequence a region of time may be considered to constitute the block. If certain definite time units such as days or weeks are involved, it is desirable that the sequence of the experiment be arranged so that these intervals constitute the blocks, since they represent logical subgroups. The use of time units as blocks tends to compensate for a large number of factors such as deterioration of plant, variation with time of raw material due to instability or due to the vagaries of a continuous unit in which the raw material is produced, and the change in human reactions over a period of time. Batches of raw materials, a number of experimenters, or a number of similar units of equipment also provide useful bases for the design of blocked experiments. Where the incidence of unwanted effects is under the control of the designer it is sometimes possible to arrange that a number of different effects may be removed in a single blocked design. Thus, if in the standard example a catalyst sample would serve for 2 experimental units without appreciable deterioration, but its subsequent performance was uncertain, it would be possible to use a separate catalyst sample for each block, differences between the performance of samples of catalyst of the same type being eliminated with the block effects. Where it is not possible to arrange matters in this way alternative designs may be used.

8.53. Latin Squares

One such class of designs includes the Latin square and the other orthogonal squares which derive from it. In Figure 8.3 the letters $A, B, C, \cdots$, are so arranged that each occurs once in each row and column of the square designs. The number of possible arrangements of a square increases rapidly as the length of side is increased, so that, whereas for a 2×2 square there are only 2 possible arrangements, obtained by interchanging A and B in Figure 8.3, there are 12 arrangements of the 3×3 square corresponding to the $3!\,2!$ ways of permuting the standard square, and in a 4×4 square there are 4 standard squares, each of which can be permuted in $4!\,3!$ ways, giving a total of 576 such squares. A standard square is one in which the first row and first column contain letters in alphabetical order. There is only one such square for the 2×2 and 3×3 arrangements. All the standard squares up to $r = 6$ and examples of standard squares up to $r = 12$ are given in reference [13].

A	B
B	A

2×2 Square

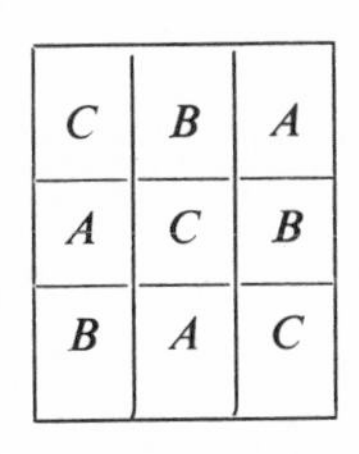

3×3 Square

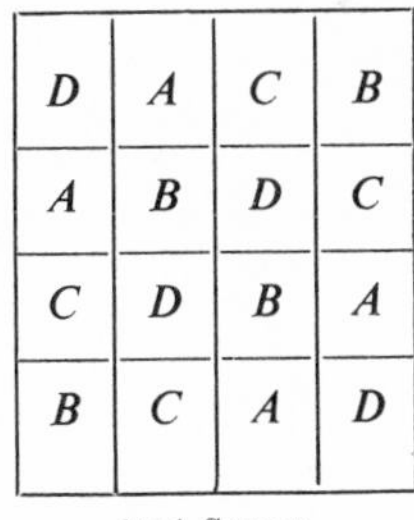

4×4 Square

FIG. 8.3.

As an illustration of the use of the Latin square, let us suppose that in our previous example it was desirable to carry out the synthesis tests in 4 similar pilot scale units. This might reasonably occur, if the results of the tests were required urgently. If we allocate the experimental units to the equipment in a random fashion for each replicate, we shall ensure that the test results will not in general be biased, but effects due to differences between the 4 synthesis units will increase the variability of the results. If we allocate the equipment to separate blocks, we cannot complete the experiment appreciably more rapidly by using 4 units than we could in the randomized block design in which only 1 unit was used. The solution is achieved by imposing on the experimental design a second condition, namely, that each of the experimental units shall be carried out once on each piece of equipment. This can be achieved, if we limit the number of replicates to 4, by using a Latin square design in which the 4 columns represent the first 4 time blocks of our previous example, and the

4 rows represent the different units of equipment. Subject to the conditions imposed by this square, the allocation of the time blocks to the columns, the reactors to the rows, and the treatment combinations to the letters *A*, *B*, *C*, and *D* may be achieved by a random process, in order to ensure that the experiment will be unbiased. Detailed instructions for the random selection of Latin squares are given in [13]. As a result of such a random process the accompanying experimental design was obtained. The analysis of the results obtained from this design depends on the

Time block		1	2	3	4
	I	b	ab	a	(1)
Synthesis unit	II	a	(1)	ab	b
	III	(1)	a	b	ab
	IV	ab	b	(1)	a

assumption that any observation is normally distributed with variance σ^2 and average value

$$\text{ave}\,(x_{kl}) = e_0 + e_i + e_j + e_{ij} + e_k + e_l$$

where e_i, e_j, and e_{ij} denote the temperature and catalyst effects and the associated interactions, and where e_k and e_l denote the effects associated with the experimental unit of block k and equipment l. In selecting this model we are assuming that there are no interactions between time blocks, synthesis units, and the 4 treatment combinations. Under these conditions the row, column, and treatment effects are mutually independent as may be seen by writing down any comparison in terms of the model. The analysis of variance then takes the form of the accompanying table.

Source of Estimate	D.F.	Average Value of Mean Square
Between temperatures	1	$\sigma^2 + 4\sigma^2_{ij} + 8\sigma_i^2$
Between catalysts	1	$\sigma^2 + 4\sigma^2_{ij} + 8\sigma_j^2$
Interaction temperature × catalyst	1	$\sigma^2 + 4\sigma^2_{ij}$
Total treatment effects	3	
Between blocks	3	$\sigma^2 + 4\sigma_k^2$
Between synthesis units	3	$\sigma^2 + 4\sigma_l^2$
Residual	6	
Total	15	

In general for a square of side n the sums of squares due to treatment, row, and column effects each account for $(n - 1)$ degrees of freedom, so that the residual is based on $n^2 - 1 - 3(n - 1) = (n - 1)(n - 2)$ degrees of freedom. In the 3×3 square the residual estimate is based on only 2 degrees of freedom which is rarely sufficient to render a single square of value as an experimental design. On the other hand, in an 8×8 square we are obliged to replicate the experimental units 8 times, which for many purposes is an excessive number. The Latin squares which are most useful for industrial designs are probably those of side 4, 5, or 6, the 3×3 square being employed with replication. Since in all Latin squares the presence of any real interactions between rows, columns, and treatments may bias the results in addition to increasing the residual error, it is important that the randomization process be properly carried out.

For certain standard squares it is possible to obtain various estimates for individual treatment comparisons. Thus for the 4×4 square 2

	I	II	III	IV
1	*A*	*B*	*C*	*D*
2	*B*	*A*	*D*	*C*
3	*C*	*D*	*A*	*B*
4	*D*	*C*	*B*	*A*

independent estimates of any comparison may be obtained, each unaffected by row or column effects. For the comparison $A - C$ we have

$$A\text{I}\,1 + A\text{III}\,3 - C\text{III}\,1 - C\text{I}\,3 = 2A - 2C$$

and

$$A\text{II}\,2 + A\text{IV}\,4 - C\text{IV}\,2 - C\text{II}\,4 = 2A - 2C$$

The sum of these estimates therefore provides an estimate of $4A - 4C$, and the difference of the estimates provides an estimate of the standard deviation of these comparisons. Similar arrangements are possible for the 6×6 square and the 8×8 square.

EXAMPLE. As an illustration of the use of a Latin square design we shall consider the following data taken from an investigation of the effect of preconditioning on the rate of abrasion of leather samples [54]. Small samples of leather of a suitable size for the testing machine were cut from a butt in a square pattern, and allocated to treatments in accordance with the symbols of a randomly chosen Latin square. The treatments consisted of conditioning in a desiccator at controlled humidities of 25.1, 37.8, 49.7, 61.8, 75.1, and 87.5%. A 6 × 6 Latin square and 36 samples were therefore required. Since the laboratory atmosphere was to be controlled during the subsequent abrasion testing at a value approaching the standard value at which the specimens had been conditioned, it was hardly practicable to randomize the order of treatment of the specimens. Accordingly only the 6 specimens to be treated at each humidity were examined in random order. Systematic differences between testing conditions would therefore be confused with treatment effects. The only likely source of such differences was the abrasive cloth to be employed. This was prepared in advance as 36 specimens which were allocated to the samples in a random fashion. The Latin square chosen for the investigation is given in Figure 8.4, where the letters *A*, *B*, *C*, *D*, *E*, and *F* indicate increasing relative humidity in the conditioning and testing process, and the geometrical position indicates the location of the original sample in the butt.

C 7.38	*D* 5.39	*F* 5.03	*B* 5.50	*E* 5.01	*A* 6.79
B 7.15	*A* 8.16	*E* 4.96	*D* 5.78	*C* 6.24	*F* 5.06
D 6.75	*F* 5.64	*C* 6.34	*E* 5.31	*A* 7.81	*B* 8.05
A 8.05	*C* 6.45	*B* 6.31	*F* 5.46	*D* 6.05	*E* 5.51
F 5.65	*E* 5.44	*A* 7.27	*C* 6.54	*B* 7.03	*D* 5.96
E 6.00	*B* 6.55	*D* 5.93	*A* 8.02	*F* 5.80	*C* 6.61

FIG. 8.4.

Before computing the various sums of squares we require an auxiliary table presenting the row, column, and treatment totals. The correspond-

Rows	3510	3735	3990	3783	3789	3891
Columns	4098	3763	3584	3661	3794	3798
Treatments	*A* 4610	*B* 4059	*C* 3956	*D* 3586	*E* 3223	*F* 3264

ing sums of squares may then be determined in the uncorrected form as

Between observations	1462.8916
Between rows	1433.2986
Between columns	1433.6831
Between treatments	1454.6389
Correction for mean	1431.1089

leading to the analysis of variance in Table 8.13. The treatment effects are clearly highly significant, since we have for the variance ratio

$$\frac{4.706}{0.174} = 27.0 \qquad (F_{5,\,20,\,0.001} = 6.5)$$

TABLE 8.13

Source of Estimate	Sum of Squares	D.F.	Mean Square
Between rows	2.1897	5	0.438
Between columns	2.5742	5	0.515
Between treatments	23.5300	5	4.706
Residual	3.4888	20	0.174
Total	31.7827	35	

In this particular example the results indicate that the treatment effects would have appeared significant even if the 6 replicates had been completely randomized. Nevertheless the 10 degrees of freedom which have been abstracted for row and column effects have taken with them a disproportionate amount of the residual sum of squares, both effects being significant at about the 0.05 level. The accuracy of the estimates of the treatment effects was therefore appreciably increased by the use of the Latin square design.

As the data provide an estimate of the residual variance based on 20 degrees of freedom, it is a simple matter to examine the linearity of the relationship between the treatment means and the humidity. From the previous table we obtain the treatment means in the accompanying table,

Relative humidity	25.1	37.8	49.7	61.8	75.1	87.5
Treatment mean	7.683	6.765	6.593	5.977	5.372	5.440

and a plot of these means versus the relative humidity is shown in Figure 8.5. Proceeding by the methods of Section 7.71 we obtain the analysis of variance of Table 8.14. Testing the residual estimate between means corrected for regression against the estimate from the original square, we obtain the variance ratio

$$\frac{0.389}{0.174} = 2.24 \qquad (F_{4,\,20,\,0.05} = 2.9)$$

indicating that there is insufficient evidence to confirm the significance of the reversal of the trend at the highest relative humidity suggested by Figure 8.5.

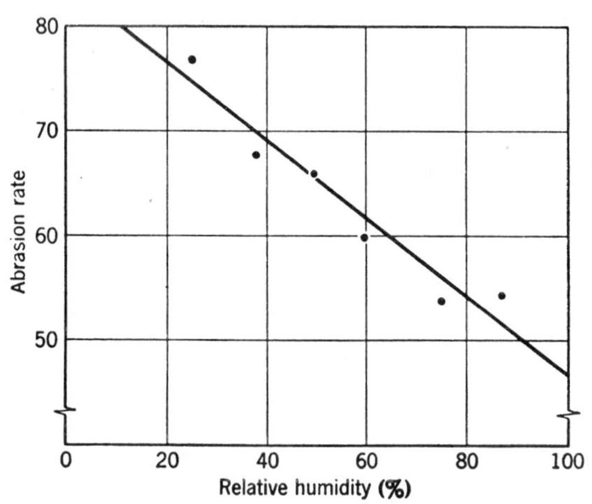

FIG. 8.5. Abrasion rates of leather at various humidities.

8.54. Graeco-Latin Squares

The assignment of the 2 sets of letters $A, B, C, D, \cdots$, and $\alpha, \beta, \gamma, \delta, \cdots$, to a square in such a manner that each of the sets forms a Latin square,

TABLE 8.14

Source of Estimate	Sum of Squares	D.F.	Mean Square
Linear regression of means	21.9730	1	21.9730
Deviations of means from linear regression	1.557	4	0.389
Total between means	23.5300	5	
Residual from Latin square	3.4888	20	0.174

and in addition each possible combination of a Latin letter and Greek letter occurs once and once only within the square, leads to a Graeco-Latin square. Examples of such squares for $n = 3$, $n = 4$, and $n = 5$ are given in Figure 8.6. Two Latin squares which can be superimposed

$C\beta$	$B\alpha$	$A\gamma$
$A\alpha$	$C\gamma$	$B\beta$
$B\gamma$	$A\beta$	$C\alpha$

3 × 3 Square

$A\alpha$	$B\beta$	$C\gamma$	$D\delta$
$B\gamma$	$A\delta$	$D\alpha$	$C\beta$
$C\delta$	$D\gamma$	$A\beta$	$B\alpha$
$D\beta$	$C\alpha$	$B\delta$	$A\gamma$

4 × 4 Square

$A\alpha$	$B\beta$	$C\gamma$	$D\gamma$	$E\varepsilon$
$B\delta$	$C\varepsilon$	$D\alpha$	$E\beta$	$A\gamma$
$C\beta$	$D\gamma$	$E\delta$	$A\varepsilon$	$B\alpha$
$D\varepsilon$	$E\alpha$	$A\beta$	$B\gamma$	$C\delta$
$E\gamma$	$A\delta$	$B\varepsilon$	$C\alpha$	$D\beta$

5 × 5 Square

FIG. 8.6.

to give a Graeco-Latin square are said to be orthogonal. Not all the standard squares of a given side length give rise to Graeco-Latin squares, and in particular such a square of side 6 is impossible [13]. Where such squares do exist and interactions between the 4 factors involved are non-existent, comparisons between the various levels of any factor are independent of the remaining factors.

Suppose that in our standard experiment it was required that the individual tests be started up in sequence, rather than simultaneously, in order to economize in supervision. We might elect to use the same sequence for the 4 units of equipment in each time block, in which case any systematic effect due to the order of starting the tests would be eliminated with the equipment effect in the Latin square design. By the use of the Graeco-Latin square we could separate these two effects without additional experimental work. In this case, after randomizing the selection of the square to be used, we should allocate the treatment combinations and the sequence of starting up the test to the Latin and Greek letters by a random method. The analysis of variance would then take the form of Table 8.15. Under these circumstances the number of degrees of freedom for residual estimate is reduced to 3, which is hardly satisfactory. In the general case $(n - 1)(n - 3)$ degrees of freedom remain for the residual estimate, so that Graeco-Latin squares of sides 5, 7, and 8 are usually the more suitable for industrial experiments, unless the residual variance can be estimated from additional data.

EXAMPLE. An example of the use of a Graeco-Latin square design in performance tests on 7 types of gasoline has been published [58]. Tests

TABLE 8.15

Source of Estimate	D.F.	Average Value of Mean Square
Between temperatures	1	$\sigma^2 + 4\sigma^2_{ij} + 8\sigma_i^2$
Between catalysts	1	$\sigma^2 + 4^2_{ij} + 8\sigma_j^2$
Interaction temperature × catalyst	1	$\sigma^2 + 4\sigma^2_{ij}$
	—	
Total treatment effects	3	
Between blocks	3	$\sigma^2 + 4\sigma_k^2$
Between synthesis units	3	$\sigma^2 + 4\sigma_l^2$
Between starting sequence	3	$\sigma^2 + 4\sigma_m^2$
Residual	3	σ^2
	—	
Total	15	

were carried out in a single automobile, at 7 distinct periods of the day, and for 7 days. In addition 7 drivers were employed. Such an experiment carried out as 49 experimental units of a 7 × 7 Graeco-Latin square will lead to a preliminary analysis of variance of the form of the accompanying table, which provides a residual variance estimate based on a satisfactory number of degrees of freedom.

Source of Estimate	D.F.
Between days	6
Between time of day	6
Between drivers	6
Between fuels	6
Residual	24
	—
Total	48

Although the speed of driving is an important variable if the performance of the fuel is to be evaluated in miles per gallon, it was not desirable to restrict the test to a tight driving schedule since

(1) Such a procedure would tend to prejudice the operational character of the test.

(2) By sacrificing an additional degree of freedom from the residual the influence of differences in mean driving speed could be evaluated, and the effects of such differences eliminated from the desired comparisons by covariance analysis.

TABLE 8.16

Period		Days						
		1	2	3	4	5	6	7
		A 2	*B* 6	*G* 1	*C* 3	*D* 7	*E* 4	*F* 5
1	mph	33.36	40.11	39.34	37.50	34.80	35.23	38.38
	mpg	35.74	31.45	30.65	30.84	31.98	33.28	31.76
		C 7	*E* 5	*D* 4	*A* 6	*F* 1	*B* 3	*G* 2
2	mph	34.03	37.40	36.36	41.40	42.40	36.14	25.22
	mpg	34.86	30.84	32.43	30.68	29.39	31.31	37.53
		D 6	*G* 4	*C* 2	*B* 1	*A* 5	*F* 7	*E* 3
3	mph	38.32	36.55	30.89	44.78	36.00	37.50	39.96
	mpg	31.05	31.73	36.61	29.26	33.26	32.36	31.83
		E 1	*F* 2	*A* 3	*D* 5	*C* 4	*G* 6	*B* 7
4	mph	34.75	32.49	38.71	36.46	33.49	42.58	36.49
	mpg	33.76	35.89	31.84	32.43	34.26	29.13	33.26
		B 4	*A* 7	*F* 6	*E* 2	*G* 3	*C* 5	*D* 1
5	mph	34.43	33.69	40.45	25.65	37.21	34.52	41.81
	mpg	34.17	33.02	31.25	38.32	32.18	34.01	29.22
		F 3	*C* 1	*B* 5	*G* 7	*E* 6	*D* 2	*A* 4
6	mph	37.89	34.53	35.89	36.64	42.15	25.26	39.67
	mpg	31.91	31.83	34.94	31.29	30.31	38.22	31.03
		G 5	*D* 3	*E* 7	*F* 4	*B* 2	*A* 1	*C* 6
7	mph	34.47	35.64	34.57	32.97	28.31	44.44	42.63
	mpg	32.64	29.92	33.85	32.97	38.87	27.88	28.92

A selection was made at random from the set of 7 × 7 Graeco-Latin squares, and the resulting experimental arrangement is shown in Table 8.16, where the letters *A–G* represent the types of fuel employed and the numbers 1–7 are employed in place of Greek letters to indicate the drivers. Before undertaking an analysis of variance it is convenient to determine the totals of replicate observations for days, periods, drivers, and fuels. Since it is proposed to include an analysis of covariance, the totals for the independent variate are also required. Both are given in Table 8.17.

TABLE 8.17

EFFECT TOTALS. MILES PER HOUR AND MILES PER GALLON

Days			Periods			Drivers			Fuels		
Day	mph	mpg	Period	mph	mpg	Driver	mph	mpg	Fuel	mph	mpg
1	247.25	234.13	1	258.72	225.70	1	282.05	211.99	*A*	267.27	223.45
2	250.41	224.68	2	252,95	227.04	2	201.18	261.18	*B*	256.15	233.26
3	256.21	231.57	3	264.00	226.10	3	263.05	219.83	*C*	247.59	231.33
4	255.40	225.79	4	254.97	230.57	4	248.70	229.87	*D*	248.65	225.25
5	254.36	230.25	5	247.76	322.17	5	253.12	229.88	*E*	249.71	232.19
6	255.67	226.19	6	252.03	229.53	6	287.64	212.79	*F*	262.08	225.53
7	264.16	223.55	7	253.03	225.05	7	247.72	230.62	*G*	252.01	225.15

In addition to the sums of squares of the observations (mpg), we require the sums of squares of the independent variate (mph) and the sums of cross products. Since all of these are additive (see Section 7.83) the terms for the residual are most conveniently determined by difference. Computing in the usual manner from the subtotals and correcting in each case for the mean, we obtain Table 8.18. It is apparent from Table 8.18 that the factor of greatest importance in determining the fuel consumption is the type of driver employed, and that an important difference between drivers is the mean speed of driving.

We first compute the regression in the residual, obtaining:

$b = -0.3948$

Sum of squares due to regression = 26.8348

Residual after correction for regression = 10.3315

TABLE 8.18

Source of Sum	Sum of Squares and Cross Products			D.F.
	x^2	xy	y^2	
Days	23.8828	−10.3161	13.4809	6
Periods	23.4032	−7.5219	6.4079	6
Drivers	693.4787	−402.1993	238.4597	6
Fuels	47.6882	−12.5840	14.1139	6
Residual	172.1469	−67.9671	37.1663	24
Total	960.5998	−500.5884	309.6287	48

The significance of the regression is conveniently tested by the variance ratio

$$\frac{26.8348}{1} \times \frac{23}{10.3315} = 59.6^{***} \quad (F_{1,\,23,\,0.001} = 14.2)$$

Since in this instance there is no technical reason for the separation of the sums of squares corrected for regression into components corresponding to the deviation of the class mean regression from the overall regression and to the deviation of the class means from the class mean regression, we may use the short method of analysis, pooling these components. This analysis for each of the 4 factors is presented in Table 8.19.

The variance ratios for the corrected mean squares become

Days

$$\frac{1.509}{0.449} = 3.36^{*}$$

Periods

$$\frac{0.684}{0.449} = 1.52$$

Drivers

$$\frac{1.654}{0.449} = 3.68^{*}$$

Fuels

$$\frac{1.906}{0.449} = 4.24^{**}$$

$$(F_{6,\,23,\,0.05} = 2.6) \quad (F_{6,\,23,\,0.01} = 3.8) \quad (F_{6,\,23,\,0.001} = 5.7)$$

so that the effect of period is non-significant, the days and drivers effects approach the 0.01 level of significance, and the effect of fuels is significant at about the 0.005 level. These results may be compared with those which would have been obtained from the experimental design without the use of the covariance analysis.

TABLE 8.19

ANALYSIS OF COVARIANCE

Source of Estimate	Sum of Squares and Cross Products			Portion Due to Regression	Corrected Sum of Squares	D.F.	Mean Square
	x^2	xy	y^2				
Between days	23.8828	−10.3161	13.4809		9.0538	6	1.509
Residual	172.1469	−67.9671	37.1663	26.8348	10.3315	23	0.449
Sum	196.0297	−78.2832	50.6472	31.2619	19.3853		
Between periods	23.4032	−7.5219	6.4079		4.1014	6	0.684
Residual	172.1469	−67.9671	37.1663	26.8348	10.3315	23	0.449
Sum	195.5501	−75.4890	43.5742	29.1413	14.4329		
Between drivers	693.4787	−402.1793	238.4597		9.9226	6	1.654
Residual	172.1469	−67.9671	37.1663	26.8348	10.3315	23	0.449
Sum	865.6256	−470.1664	275.6260	255.3719	20.2541		
Between fuels	47.6882	−12.5840	14.1139		11.4335	6	1.906
Residual	172.1469	−67.9671	37.1663	26.8348	10.3315	23	0.449
Sum	219.8351	−80.5511	51.2802	29.5152	21.7650		

In practice, of course, it is not sufficient merely to state that the fuels differ significantly; we should also determine the relative performance of the 7 classes of fuel. This may be done by correcting the mean mileage per gallon of each fuel for the effects of regression, i.e., bringing each to a common speed for the purposes of comparison, and leads to Table 8.20. The variances of the corrected estimates of the mean fuel consumption are unequal, and furthermore an overall estimate of the regression rather than an estimate derived for the individual fuels has been employed to

TABLE 8.20

CORRECTED VALUES OF MEAN FUEL CONSUMPTIONS

Fuel	A	B	C	D	E	F	G
Mean fuel consumption	31.92	33.32	33.05	32.18	33.17	32.22	32.16
Mean speed	38.18	36.59	35.37	35.52	35.67	37.44	36.00
Correction using $b = -0.395$	+0.70	+0.08	−0.41	−0.35	−0.29	+0.41	−0.15
Mean fuel consumption corrected to overall mean speed	32.62	33.40	32.64	31.83	32.88	32.63	32.01

make the correction. However, since the range of speeds in the tests is very much greater than the range of the mean speeds for the 7 classes of fuel, we shall not err seriously in assuming that the variance of the corrected means is simply σ^2/n, which we estimate as $0.449/7 = 0.06414$. This gives $s = 0.253$, and leads to a critical separation at the 0.05 level of $0.253 \times 2.07 \times \sqrt{2} = 0.741$. The 7 fuels cannot therefore be considered to separate into a number of distinct classes, but we should infer with some confidence that fuel B was better than fuel D if quality is measured in terms of miles per gallon at a mean speed of about 35 mph.

8.55. Hyper Graeco-Latin Squares

The principle of the Graeco-Latin square may be extended by superimposing a third orthogonal square to give an arrangement known as a *hypersquare*. In the previous example we might have found it convenient to employ 7 different automobiles of the same type to carry out the tests. If the automobiles were identified with the symbols of the third square, the resulting experimental design would have ensured that each car was used once to test each fuel, that the cars were employed once on each day, and during each period, and that each driver carried out tests in each car. Provided that the cars employed in carrying out the tests represented a random sample from a defined population, this procedure would enable the inferences concerning the performances of the various fuels to be applied to this population, and at the same time we should be able to estimate the variation in performance between similar cars. For these advantages it would be necessary to sacrifice 6 degrees of freedom from the residual estimate, and any real interaction effects between cars and other factors would tend to increase the residual sum of squares, resulting in an increased and less precise estimate of the residual variance.

In general we can study 4 independent effects, using 2 superimposed orthogonal 3×3 Latin squares (3×3 Graeco-Latin squares), 5 independent effects in a set of 3 superimposed orthogonal 4×4 Latin squares, and where such squares are available we can study $n + 1$ effects, using a superimposed set of $n - 1$ mutually orthogonal $n \times n$ Latin squares. Such orthogonal sets are given by Fisher and Yates [13] for $n = 2, 3, 4, 5, 7, 8$, and 9, and their existence has been proved for all primes and prime powers.

$A_1\alpha$	$B_2\beta$	$C_3\gamma$	$D_4\delta$
$B_4\gamma$	$A_3\delta$	$D_2\alpha$	$C_1\beta$
$C_2\delta$	$D_1\gamma$	$A_4\beta$	$B_3\alpha$
$D_3\beta$	$C_4\alpha$	$B_1\delta$	$A_2\gamma$

$A_1\alpha_1$	$B_2\beta_2$	$C_3\gamma_3$	$D_4\delta_4$	$E_5\varepsilon_5$
$B_3\delta_5$	$C_4\varepsilon_1$	$D_5\alpha_2$	$E_1\beta_3$	$A_2\gamma_4$
$C_5\beta_4$	$D_1\gamma_5$	$E_2\delta_1$	$A_3\varepsilon_2$	$B_4\alpha_3$
$D_2\varepsilon_3$	$E_3\alpha_4$	$A_4\beta_5$	$B_5\gamma_1$	$C_1\delta_2$
$E_4\gamma_2$	$A_5\delta_3$	$B_1\varepsilon_4$	$C_2\alpha_5$	$D_3\beta_1$

FIG. 8.7.

Examples of these so-called hyper Graeco-Latin squares are shown in Figure 8.7 for the case of 3 orthogonal sets in a 4×4 square, and 4 orthogonal sets in a 5×5 square, the additional sets being denoted by the subscripts on the Latin and Greek letters, respectively. Usually at least one of the $n + 1$ possible comparisons in an $n \times n$ square is omitted to provide for a residual estimate, although there are situations where an independent estimate of the error is available, or, where the square is to be replicated, where all $n + 1$ possible independent comparisons can be used.

The orthogonal squares can be regarded as the simplest case of an orthogonal figure in r dimensions having n^r cells. From each alphabet we select n letters, and in any direction the figure has n parallel hyperplanes so that each letter of a given alphabet will occur n^{r-2} times in a hyperplane. For the orthogonal squares and cubes we have 1 example of each letter per row and n examples of each letter per plane, respectively. The total number of alphabets which can be superimposed on the r-space hypercube of side n is

$$(n^r - 1)/(n - 1)$$

provided that n is a prime or prime power.

8.56. Other Uses of Latin and Graeco-Latin Squares

The smaller Latin squares are seldom used singly because they do not provide a sufficient number of degrees of freedom for error estimation, but they can sometimes be used with advantage as components of a randomized block design. For example, suppose that in a given routine laboratory an experiment were designed to test, in a 3×3 Latin square, the comparative deviations of the analyses of 3 analysts, using 3 different types of equipment, from the standard values of 3 different standards. Such a design is shown in Figure 8.8. One such experiment might tell

Standard solution.

Type of equipment	I	II	III
1	*A*	*B*	*C*
2	*B*	*C*	*A*
3	*C*	*A*	*B*

A, *B*, *C*: Analysts

FIG. 8.8.

us little, but a repetition could be made of the entire experiment on 5 different days of a given work week, 9 results being obtained each day, in a random order, corresponding to the 9 combinations of analyst, equipment, and standard solution. The analysis of variance would be as shown in the accompanying table.

Source of Estimate	Degrees of Freedom	
	General	Example
Between blocks (days)	$b - 1$	4
Between treatments (analysts)	$n - 1$	2
Between rows (equipment)	$n - 1$	2
Between columns (solutions)	$n - 1$	2
Interactions:		
block × treatments	$(b - 1)(n - 1)$	8
block × rows	$(b - 1)(n - 1)$	8
block × columns	$(b - 1)(n - 1)$	8
Residual	$b(n - 1)(n - 2)$	10
Total	$bn^2 - 1$	44

The particular example considered would require $3 \times 3 \times 3 \times 5 = 135$ experimental units, if it were carried out as a complete factorial design, and would provide no variance estimate unless the many factor interactions were negligible. By carrying out only one-third of this number of units we have apparently contrived to estimate the important effects and interactions and yet retained 10 degrees of freedom for variance estimation. This process is satisfactory provided that we can assume that the 3 factors in the individual Latin squares do not interact. When these interactions are present the design may be regarded as an example of fractional replication (Section 8.72). This implies that the comparisons which we take to estimate effects in fact estimate the sum or difference of these effects and certain interactions, whereas the residual comparisons actually estimate the sum or difference of pairs of interactions. The pairing of the main effects and interactions in a particular instance depends upon the particular Latin square which we employ, so that in order to obtain unbiased estimates and significance tests in the previous example we should choose the square to be employed on each day by a random method. It is not sufficient to select Monday's square by a random method and then use the same square for the remainder of the week unless the interaction effects in the square are known to be absent, and if this is the case even the first selection need not be random.

A slightly different extension of the use of Latin squares can be applied to the example of Section 8.53, where it was necessary to reduce the replications from 6 to 4 in order to use the Latin square design. We could equally well have increased the replications to 8 simply by repeating the process on 4 new lots. The resulting experiment would still have 4 treatments and 4 batches per lot, but 8 lots and 8 replications. The analysis would be as shown in the accompanying table. This type of

Source of Estimate	Degrees of Freedom
Between treatments	3
Between reactors	3
Between lots	7
Residual	18
Total	31

design can be applied in general when $n = kt$, i.e., when the number of replications is a multiple of the number of treatments. Notice that this is equivalent to applying the treatment to the successive batches in an order dictated by the columns of one or more Latin squares.

Latin square designs can be conveniently used to make preliminary studies of several sources of variation in a given process. For example, in developing a new analytical method involving a reduction, an extraction, and a colorimetric determination, we may be interested in determining the relative effect on the final result of the concentration of the reducing agent, the time allowed for extraction, and the time allowed for color development. Choosing, let us say, 4 levels of each of these factors (preferably at equal intervals) covering the ranges to be considered, and assigning them at random to the columns, rows, and treatments of a 4×4 Latin square, we should perform 16 analyses, using the 16 indicated combinations of the 3 factors. By analyzing the results in the usual manner we could determine which of the factors made a significant contribution to the overall variability.

A design of this type might be employed when it is required to estimate variance components rather than effects. Assuming that no interactions are present, the average values of the mean squares obtained in the analysis of an $n \times n$ Latin square are shown in the accompanying table and the

Source of Estimate	Average Value of Mean Square
Treatments	$\sigma^2 + n\sigma_\alpha^2$
Rows	$\sigma^2 + n\sigma_\eta^2$
Columns	$\sigma^2 + n\sigma_\xi^2$
Residual	σ^2

variance component due to each of the 3 factors is obtained by subtracting the residual estimate from the estimate corresponding to a given factor and dividing by n. In such experiments the residual effect may turn out to be the most important, indicating that the major source of variation is some factor not included in the experiment.

In square designs in which all the levels of the factors studied are chosen from continuous scales of such factors, the degrees of freedom for rows, columns, and treatments may be subdivided into linear, quadratic, and cubic components as in the example of Section 8.53. It is also possible to introduce additional sources of variation by using hyper Graeco-Latin squares. For example, using a 4×4 Graeco-Latin square, we could add another source of variation to the 3 studied above and still have 3 degrees of freedom remaining for the estimation of the residual error. This would not be sufficient for our purposes if we wished to estimate and test effects, but might be adequate if only the variance components were to be estimated. Similarly, by using a 5×5 square and performing 25 analyses, we could have compared 5 possible sources of variation and the residual, each with 4 degrees of freedom. The average values of the mean square estimates and the method of computing the variance components would be similar to those given for the Latin square.

8.6. Confounding

8.61. Introduction

In the previous section we have considered the use of factorial designs in conjunction with methods by means of which the effects of known sources of variation can be eliminated from the experimental comparisons. Frequently these two aspects of experimental design are at cross purposes, and it becomes necessary to find ways in which we can take full or partial advantage of one of these benefits at the least expense to the other.

A simple situation of this type arises when the number of treatments exceeds the capacity of the size of block which is desirable for the elimination of known sources of experimental error. As an example, let us suppose that in the standard experiment of Section 8.52 we wish to examine the influence of the H_2S content of the gas stream and are in a position to control this factor by making additions to the synthesis gas. Since 2 types of catalyst are involved we should anticipate that real interaction effects between this factor and the sulphur content would exist, so that the designs of the previous section would be unsuitable.

If for the purposes of the investigation 2 levels of the sulphur content sufficed, we might consider the 6 blocks of the proposed design in 3 groups each of 2 blocks, and each of which would accommodate the 8 units necessary for the experiment. A number of alternative methods of allocating the experimental units to a given group are then available:

(1) We may treat the group as a complete block and allocate the 8 treatment combinations in a completely random fashion. In this case the experiment will be less sensitive, if there are differences between the 2 four-unit blocks in a given group, since such differences will inflate the experimental error.

(2) To one of the blocks of each group we may allocate 4 treatment combinations involving the lower level of sulphur content, and to the other the remaining 4 combinations involving the higher level of sulphur content. In this case the design is known as a *split plot* experiment, and

it results in a high efficiency of estimation of all interactions and the main effects due to catalysts and temperatures, but a reduced efficiency in the estimation of the main effect due to sulphur.

(3) The allocation of treatments to the 2 blocks may be made in such a way that the comparison between blocks of a given group corresponds to one of the interactions, whereas the 6 remaining effects and interactions are independent of the block difference. The selected interaction is then said to be *confounded* with the block difference. Usually a high-order interaction is chosen for simple confounding, the split plot design corresponding to the case in which a main effect is confounded.

(4) The idea of grouping the blocks into units each of which contains a complete replicate of the experiment, may be rejected in favor of a design in which the experimental units occur together in a block on the same number of occasions. Such an arrangement is known as a *balanced incomplete block*, and it imposes conditions on the number of replicates which must be carried out [6].

8.62. Split Plots

Allocating the treatment combinations to the 6 available blocks by the method of (2) in the example of the Fischer-Tropsch synthesis we obtain Figure 8.9.

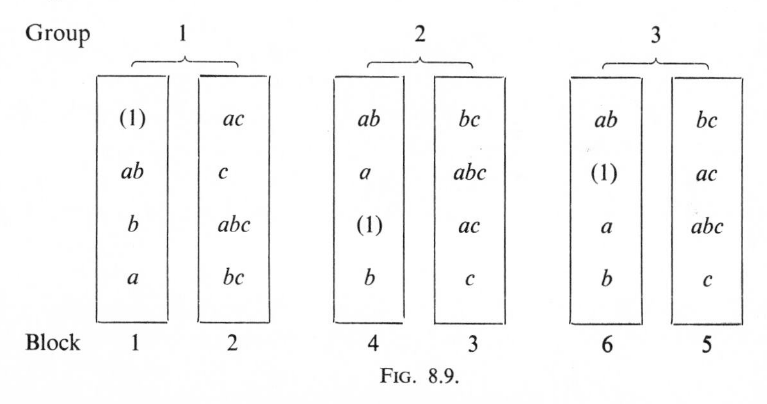

FIG. 8.9.

If the subscripts i, j, k denote the level of the factors A, B, and C and the subscripts l and m identify the group or plot and the block (or subplot) of a given plot, we assume that the observation x_{ijklm} is randomly and normally distributed with variance σ^2 and average value

$$\text{ave}\,(x_{ijklm}) = e_0 + e_i + e_j + e_k + e_{ij} + e_{ik} + e_{jk} + e_{ijk} + e_l + e_m$$

the values of k and m which an observation from a given group can take on being limited by the design. Alternatively, since the values of k and

m are interdependent, we may regard the effect of blocks (within groups and C treatments) as the interaction of groups and the 2 levels of C treatments. By arranging that all other comparisons utilize equal numbers of observations from each block of a group, and by assuming that there are no block $\times$ treatment interactions, we ensure that all other group $\times$ treatment interactions are zero. The model is then

$$\text{ave}\,(x_{ijkl\alpha}) = e_0 + e_i + e_j + e_k + e_{ij} + e_{ik} + e_{jk} + e_{ijk} + e_{kl}$$

Within a given group or replication the effects A, B, AB, AC, BC, and ABC are estimated as, e.g.,

$$A = abc + ab + ac + a - bc - b - c - (1)$$

$$AC = abc + ac - bc - c + ab + a - b - (1)$$

By substituting in the expressions for these 6 effects the average values of the various experimental units involved, it may be shown that these estimates are independent of the difference between the 2 blocks of the group. The estimate of the main effect C, however, involves the difference between the experimental units of the 2 blocks, and is not distinguishable in the case of a single replicate from this difference. If a number of replicates are available, we may suppose the e_{kl} to be randomly and normally distributed with average value zero (since any effect common to 2 blocks of a group would be regarded as a group effect), and our estimate of the main effect C in the various replicates will be subject to additional variation, owing to these random effects. The analysis of variance then takes the form shown in Table 8.21.

The significance test appropriate to the hypothesis that any particular effect is zero may be determined from the column of average values of the mean square estimate. These have been written for the case in which each of the factors is of type II. For the more general case, where the factors are of type III, the tabular method of Section 7.63 may be employed to determine the coefficients in the average values of the mean square.

The overall efficiency of the split plot design is no greater than that obtained by combining the subunits to produce complete blocks and allocating the experimental units to these blocks in a random manner. Whereas the use of randomized blocks leads to a uniform precision in estimating the effects, this alternative procedure gives higher precision in the estimate of interactions and all but one of the main effects, at the expense of the precision of estimate of this remaining effect. If an additional factor is introduced into an experiment in order to estimate the resulting interaction effects, a procedure which enhances the precision of the estimate of these effects may be advantageous. Thus in the example

TABLE 8.21

Source of Estimate	D.F.	Average Value of Mean Square*
Between groups	2	$\sigma^2 + 4\sigma_m^2 + 8\sigma_l^2$
Main effect C	1	$\sigma^2 + 4\sigma_m^2 + 3\sigma^2_{ijk} + 6\sigma^2_{ik} + 6\sigma^2_{jk} + 12\sigma_k^2$
Between blocks (within groups and C treatments)	2	$\sigma^2 + 4\sigma_m^2$
Total between blocks (within groups)	3	
Main effect A	1	$\sigma^2 + 3\sigma^2_{ijk} + 6\sigma^2_{ij} + 6\sigma^2_{ik} + 12\sigma_i^2$
Main effect B	1	$\sigma^2 + 3\sigma^2_{ijk} + 6\sigma^2_{ij} + 6\sigma^2_{jk} + 12\sigma_j^2$
Interactions $A \times B$	1	$\sigma^2 + 3\sigma^2_{ijk} + 6\sigma^2_{ij}$
$A \times C$	1	$\sigma^2 + 3\sigma^2_{ijk} + 6\sigma^2_{ik}$
$B \times C$	1	$\sigma^2 + 3\sigma^2_{ijk} + 6\sigma^2_{jk}$
$A \times B \times C$	1	$\sigma^2 + 3\sigma^2_{ijk}$
Residual	12	σ^2
Total	23	

* All populations of effects assumed infinite.

considered we might be certain from previous experience that increase in sulphur content of the gas stream would result in a lower conversion so that this effect would be of relatively little interest, but a knowledge of the difference between the 2 catalysts in sensitivity to sulphur poisoning might be of considerable value in suggesting lines for future investigation.

Although in this instance we have considered a subdivision of the main plot into two subunits, since the factor C was to be studied at 2 levels, this is not essential to the design. If 3 levels of sulphur content were to be investigated, requiring 12 experimental units, we might employ the design of Figure 8.10. The analysis of the results would then have the same form as that of Table 8.21, except that the number of degrees of freedom associated with effects involving the factor C would be doubled, and it would be possible to estimate the linear and quadratic components of these effects. An alternative arrangement for split plot designs is that in which the subunits are themselves subdivided so that the analysis of the results leads to 3 or more levels of precision in the estimates.

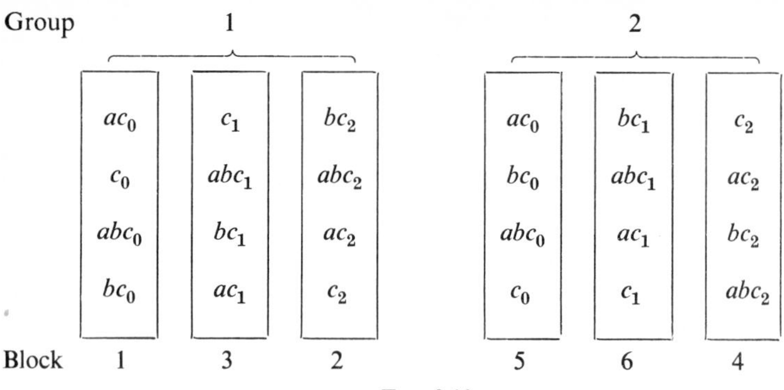

FIG. 8.10.

In cases where all the experimental units which are to receive a given treatment can be processed simultaneously, or without adjustment to the equipment, the split plot design is often convenient from the technical viewpoint. This is particularly true in experiments in which both pilot-scale and laboratory-scale treatments are involved in the same design. In this case the production of a separate pilot-scale batch for each experimental unit is wasteful, so that a number of samples or specimens to receive all alternative combinations of the laboratory treatments are usually withdrawn from each large batch. Comparisons involving laboratory treatments can then be made within batches, whereas the main effects of the pilot-scale treatments are confounded with random batch differences.

EXAMPLE. The following data [114] were obtained from a series of runs to determine the effect of 3 types of oil on the wear of 5 piston rings in a single-cylinder Waukesha engine under standard conditions. The figures quoted are the logarithms of the loss of weight in grams $\times$ 100 of the piston rings during 12-hour test runs. The logarithmic transformation was employed because it was known to stabilize the variance and to result in near normality of the residual components. The experiment consists of 5 replicates of a 3×5 factorial experiment, the 5 observations on each oil for different piston rings being obtained from a single engine test. At the end of each test the test engine was dismantled, cleaned, and reassembled with the same rings before charging with a new oil. Thus differences between oils are confounded with random effects introduced in stripping down and reassembling the engine.

The analysis of variance takes the form of Table 8.23.

The oils $\times$ rings interaction is clearly not significant, and we could pool with the residual before testing the mean square between rings, obtaining a new variance estimate of 0.005996. The effect of rings is

TABLE 8.22
DATA FOR OIL TESTS

Replicate		1			2			3			4			5		
Oil		*A*	*B*	*C*	*A*	*B*	*C*	*A*	*B*	*C*	*A*	*B*	*C*	*A*	*B*	*C*
Ring	1	1.782	1.568	1.570	1.642	1.539	1.562	1.682	1.616	1.630	1.654	1.680	1.740	1.496	1.626	1.558
	2	1.306	1.223	1.240	1.346	1.064	1.334	1.322	1.369	1.428	1.532	1.452	1.408	1.354	1.466	1.478
	3	1.149	1.029	1.068	1.090	0.778	1.136	1.176	1.053	1.202	1.233	1.193	1.228	1.038	1.167	1.330
	4	1.025	0.919	0.982	1.012	0.690	1.021	0.930	0.935	1.057	0.992	0.973	1.093	0.924	0.974	0.996
Oil ring		1.110	1.093	1.094	1.000	0.733	0.987	0.892	0.845	1.029	0.940	0.786	1.060	0.863	0.881	0.968

TABLE 8.23

Source of Estimate	Sum of Squares	D.F.	Mean Square	Average Value of Mean Square
Between replicates	0.140606	4	0.035151	$\sigma^2 + 5\sigma^2_{0r} + 5\sigma_t^2 + 15\sigma_r^2$
Between oils	0.138247	2	0.069123	$\sigma^2 + 5^2_{0r} + 5\sigma_t^2 + 25\sigma_0^2$
Between tests (within oils)	0.199457	8	0.024932	$\sigma^2 + 5\sigma^2_{0r} + 5\sigma_t^2$
Between rings	4.854990	4	1.213747	$\sigma^2 + 5\sigma^2_{0r} + 15\sigma_r^2$
Oils × rings	0.044848	8	0.005606	$\sigma^2 + 5\sigma^2_{0r}$
Residual	0.290933	48	0.006061	σ^2
Total	5.669081	74		

highly significant, but, since the object of the experiment was to obtain information on oils and oils × rings interaction, it is not necessary to separate the various inter-ring comparisons. The mean square for tests (within oils) is compared with the residual estimate,

$$F = 4.16^{***} \qquad (F_{8,\,56,\,0.001} = 3.9)$$

and is highly significant. The effect of oils is therefore properly tested against the mean square for tests within oils, so that

$$F = 2.77 \qquad (F_{2,\,8,\,0.10} = 3.11)$$

and no significant difference between oils can be distinguished. Note that the difference between replicates is also non-significant, and since the variance ratio is quite low we might have pooled this estimate with that from tests (within oils) to test the differences between oils. The conclusion would have been the same as above.

In this example the experimental comparison which is of greatest interest, that between oils, is estimated with low precision because of the split pilot design. In the circumstances, since the engine had to be stripped between tests in order to weigh the rings and remove the products of the previous test, it is difficult to see how the situation could have been avoided.

8.63. Confounding of Interactions

In factorial designs information on the main effects is generally considered of greater importance than a knowledge of the high-order interactions, so that an alternative arrangement, in which some high-order interaction is confounded, is desirable. Suppose that in the example of Figure 8.9 the three-factor interaction ABC were selected for confounding; then, if this effect is to be identified with the difference between blocks in a given replicate, we have

$$\begin{aligned} ABC &= (a - 1)(b - 1)(c - 1) \\ &= abc + a + b + c - ab - ac - bc - (1) \\ &= (abc + a + b + c) - (ab + ac + bc + (1)) \end{aligned}$$

If we assign the 4 treatment combinations of the first bracket to one block selected at random and the remaining combinations to the second block, the desired confounding will have been achieved. Furthermore, the estimates of the remaining effects will be independent of the block differences, as may be seen from Table 8.24, since each includes a given block effect with positive and negative signs on an equal number of occasions (i.e., they are orthogonal to the confounded comparison). It

TABLE 8.24

Block	1				2			
Treatment combinations	*abc*	*a*	*b*	*c*	*ab*	*ac*	*bc*	(1)
Effect								
ABC	+	+	+	+	−	−	−	−
A	+	+	−	−	+	+	−	−
B	+	−	+	−	+	−	+	−
C	+	−	−	+	−	+	+	−
AB	+	−	−	+	−	+	−	+
AC	+	−	+	−	−	+	−	+
BC	+	+	−	−	−	−	+	+

is easily seen that this would be true whichever interaction had been chosen for confounding.

In the 2^3 design which we have considered, replication would be necessary under most circumstances because of the small number of degrees of freedom involved. We shall suppose that 4 complete replicates were to be carried out, involving 8 blocks, each containing 4 experimental units. A design for this experiment is given in Figure 8.11. In each case the allocation of the treatments to positions within a block and the selection of the even- and odd-numbered blocks is achieved in a random manner. In the analysis of the design of Figure 8.11 we have a total of 32 results, and hence of 31 degrees of freedom, of which 7 correspond to differences between the 8 blocks. Of these 7 degrees of freedom 3 correspond to the

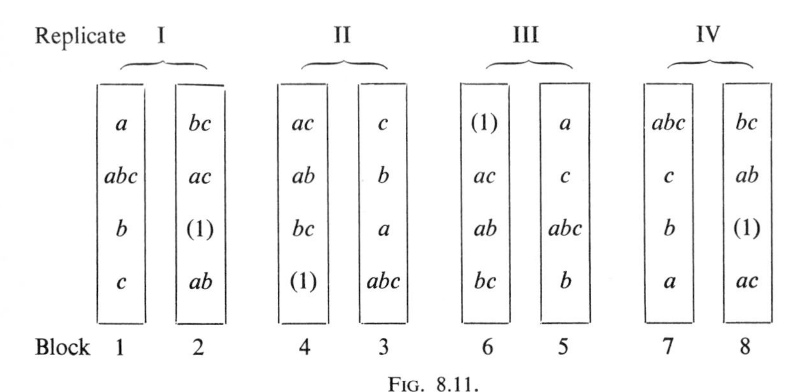

FIG. 8.11.

differences between replicates, 1 to the average difference between the pairs of blocks forming a replicate, and the remaining 3 to the variation in this block difference.

If the blocks in this design are regarded as a 2×4 classification where the two-level factor is in fact the upper and lower levels of the contrast *ABC*, then the block effects (within *ABC* and replicates) may be regarded as the interaction of groups with the levels of the *ABC* treatment. The model may then be written

$$\text{ave}\,(x_{ijkl\alpha}) = e_0 + e_i + e_j + e_k + e_{ij} + e_{ik} + e_{jk} + e_{ijk} + e_l + e_{ijkl}$$

As in the split plot design the average difference between blocks in the same replicate is an estimate of the confounded effect, although if the confounded interaction is of high order this information may not be recovered. The analysis of variance appropriate to a design of this type is given in Table 8.25.

TABLE 8.25

Source of Estimate	D.F.	Average Value of Mean Square*
Between replicates	3	$\sigma^2 + 4\sigma^2_{ijkl} + 8\sigma_l^2$
ABC interaction	1	$\sigma^2 + 4\sigma^2_{ijkl} + 4\sigma^2_{ijk}$
Between blocks (within replicates and *ABC* interaction)	3	$\sigma^2 + 4\sigma^2_{ijkl}$
Total between blocks	7	
Main effect *A*	1	$\sigma^2 + 4\sigma^2_{ijk} + 8\sigma^2_{ij} + 8\sigma^2_{ik} + 16\sigma_i^2$
B	1	$\sigma^2 + 4\sigma^2_{ijk} + 8\sigma^2_{ij} + 8\sigma^2_{jk} + 16\sigma_j^2$
C	1	$\sigma^2 + 4\sigma^2_{ijk} + 8\sigma^2_{ik} + 8\sigma^2_{jk} + 16\sigma_k^2$
Interaction *AB*	1	$\sigma^2 + 4\sigma^2_{ijk} + 8\sigma^2_{ij}$
AC	1	$\sigma^2 + 4\sigma^2_{ijk} + 8\sigma^2_{ik}$
BC	1	$\sigma^2 + 4\sigma^2_{ijk} + 8\sigma^2_{jk}$
Total treatment effects	13	
Residual	18	σ^2
Total	31	

* All populations of effects assumed infinite.

The hypothesis that the three-factor interaction is zero may first be tested by forming the appropriate variance ratio. If this hypothesis is acceptable, the remaining significance tests present no difficulties. In cases where the three-factor interaction is significant there is no exact test for the two-factor interactions or main effects unless the inferences are restricted. In many cases the three-factor interaction is assumed to be zero, and the corresponding sum of squares is not separated from that due to between-block effects. With this assumption the term $4\sigma^2_{ijk}$ vanishes from the average value of all the treatment mean squares, and exact significance tests are available in all cases. Alternatively, the approximate method of Section 7.31 may be employed.

EXAMPLE 1. The following data [65] are derived from an investigation of a test for the abrasion resistance of leather. Details of the testing procedure are given in the reference, the example in question being concerned with the effect of:

Applied load	A at 2 levels	16.41 and 32.24 psi
Speed	B at 2 levels	12.8 and 25.4 strokes per min
Air rate	C at 2 levels	0 and 10 arbitrary units

Abrasion was carried out on standard abrasive cloth which moved at a controlled rate normal to the direction of motion of the specimens, the air blast serving to clean the abrasive surface of abraded leather. Since the rate of progress of the abrasive cloth was constant for all tests any significant effects due to the factor B might be caused by the more intensive service of a given portion of the cloth under the high-speed conditions. Seven replicates were carried out, each in 2 blocks of 4 experimental units. In view of the high variability of leather, the blocks were based on the region of the hide from which the specimens were obtained.

TABLE 8.26

Block	1				2			
Treatment combination	*abc*	*a*	*b*	*c*	*ab*	*ac*	*bc*	(1)
Replicate i	11.32	7.41	4.64	9.70	6.69	14.64	8.84	5.92
ii	12.52	8.60	4.96	9.89	5.58	14.60	9.63	5.69
iii	12.42	8.13	4.57	10.03	6.33	14.61	9.09	5.89
iv	12.50	7.94	4.95	9.56	6.38	13.39	7.94	5.14
v	11.57	9.11	4.96	10.91	6.06	13.12	9.05	6.62
vi	14.84	8.45	5.44	10.60	7.45	14.82	9.43	6.50
vii	12.53	7.81	5.00	9.86	6.56	14.56	9.83	6.24

The analysis is expedited by forming subtables of the sums of replicates for each treatment combination, and of the sums of treatment combinations in each block. For the first we obtain column (2) of Table 8.27 and

TABLE 8.27

Treatment Combination	Sum of Observations	(3)	(4)	(5)	Effect	(Effect)^2/56
(1)	42.00	99.45	179.02	+500.82	I	4478.941
a	57.45	79.57	321.80	+79.06	*A*	111.616
b	34.52	170.29	25.98	−38.66	*B*	26.689
ab	45.05	151.51	53.08	−10.22	*AB*	1.865
c	70.55	15.45	−19.88	+142.78	*C*	364.038
ac	99.74	10.53	−18.78	+27.10	*AC*	13.114
bc	63.81	29.19	−4.92	+1.20	*BC*	0.026
abc	87.70	23.89	−5.30	−0.38	*ABC*	0.0026

proceed to estimate the effects by the sum-and-difference method. The components of the sum of squares due to treatment effects are obtained in each case by squaring the effect totals of column (5) and dividing by the sum of squares of the coefficients involved in the comparison times the number of replications, in this case 56.

The second subtable, giving the block and replicate totals is Table 8.28, and from it we compute the total sum of squares between blocks, the sum of squares between replicates, and, as a check, the component of the sum of

TABLE 8.28

Replicate	i	ii	iii	iv	v	vi	vii	Total
Block 1	33.07	35.97	35.15	34.95	36.55	39.33	35.20	250.22
Block 2	36.09	35.50	35.92	32.85	34.85	38.20	37.19	250.60
Total	69.16	71.47	71.07	67.80	71.40	77.53	72.39	500.82

squares due to the confounded interaction *ABC*. The first of these less the second and third gives the portion of the sum of squares due to variation in the block difference between replicates. After computing the overall sum of squares from the original data, the full analysis is as shown in Table 8.29.

TABLE 8.29

Source of Estimate	Sum of Squares	D.F.	Mean Square
Between replicates	7.0629	6	1.177
ABC interaction	0.0026	1	0.0026
Between block differences	2.8061	6	0.4677
Total between blocks	9.8716	13	
Main effect *A*	111.6158	1	111.6
B	26.6892	1	26.69
C	364.0380	1	364.0
Interaction *AB*	1.8651	1	1.865
AC	13.1145	1	13.11
BC	0.0216	1	0.0216
Residual	10.4281	36	0.2897
Total	537.6439	55	

Testing the confounded interaction ABC against the estimate of the variance derived from the between-block differences component of the sum of squares, we find that the former is small, but not significantly so, and we accept the hypothesis that there are no real three-factor interaction effects. The two-factor interactions are properly tested against the residual variance estimate, giving

$$AB\text{: } 1.865/0.290 = 6.44^{*} \qquad (F_{1,\,36,\,0.01} = 7.4)$$

$$AC\text{: } 13.11/0.290 = 45.3^{***} \qquad (F_{1,\,36,\,0.001} = 12.9)$$

BC: Not significant

The variance estimate to be employed in testing the main effects depends upon the hypothesis to be tested. In this case it seems most profitable to restrict our conclusions to the actual levels of the factors which have been employed in the experiment, in which case the average value of the mean square for each of the main effects does not contain a component due to the interaction effects. The main effect variance estimates are therefore tested against the residual, giving the variance ratios

$$A\text{: } 111.6/0.290 = 384^{***}$$

$$B\text{: } 26.7/0.290 = 92^{***} \qquad (F_{1,\,36,\,0.001} = 12.9)$$

$$C\text{: } 364/0.290 = 1255^{***}$$

If there are no three-factor interaction effects the variance estimate derived from the component of the sum of squares between blocks (within replicates) is $2.8087/7 = 0.401$, which is not significantly greater than the residual estimate even at the 0.20 level. The estimated gain in efficiency due to the reduction in block size may be shown to be a little over 20%, barely justifying the confounding.

The technical conclusions from the analysis, which relate only to the levels examined, are

(1) Increase of loading, reduction of speed, and increase in air flow all serve to increase the rate of abrasion.

(2) The rate of abrasion at the higher load and speed is less than would be anticipated from the main effects alone, probably because under these circumstances the abrasive cloth in a given area is used more intensively for a greater number of strokes, and is appreciably blunted before traversing the specimen.

(3) The rate of abrasion at the higher loads and air flow is greater than would be anticipated from the main effects alone. This would be expected since higher loads result in higher abrasion rates and hence increased "clogging" of the cloth. It would be anticipated that the air blast would be most effective under these conditions.

EXAMPLE 2. Let us consider the data of Bainbridge [108] from 2 replicates of a 2^3 factorial experiment in which the ABC interaction was confounded between blocks. In addition, the method of covariance analysis was employed to improve the efficiency of the investigation. The data relate to the strength of the product solution in a recirculation synthesis unit with a fixed proportion, in arbitrary units, of purge gas and 2 levels of reactor temperature (a), gas throughput (b), and concentration of active constituent in the make-up gas (c). The covariance variate (x) was the proportion in arbitrary units of a harmful impurity in the gas stream. The data are given in Table 8.30, where the factor D refers to the replication involved.

It is required to separate estimates of the effects A, B, AB, C, AC, BC, ABC of the residual variance and of the components of the sums of squares corresponding to the difference between replicates and the variation in the difference between blocks. It is also required to estimate the coefficient of linear regression of solution strength upon impurities and to correct all estimates for this regression. The situation is rather different from the cases of covariance analysis considered in Chapter 7, since we have no replicates within cells but only replication in randomized blocks and confounding of the three-factor interaction.

TABLE 8.30

Treatment	Strength	Impurity	Treatment	Strength	Impurity
	y	x		y	x
(1)	99	10	d	46	11
a	18	16	ad	18	13
b	51	13	bd	62	12
ab	52	10	abd	−47	16
c	108	10	cd	104	13
ac	42	12	acd	22	11
bc	95	17	bcd	67	12
abc	35	14	$abcd$	36	11

To analyze the results without a covariance correction we employ Yates' method of sums and differences, and operate as though the two levels of replication were a fourth factor, obtaining the data of Table 8.31.

TABLE 8.31

Treatment	y	(1)	(2)	(3)	(4)	Effect	$\frac{\text{Effect}}{16}$	$\frac{(\text{Effect})^2}{16}$
(1)	99	117	220	500	808	I^+	50.5	40,804
a	18	103	280	308	−456	A	−28.5	12,996
b	51	150	79	−206	−106	B	−6.625	702.25
ab	52	130	229	−250	58	AB	3.625	210.25
c	108	64	−80	−34	210	C	13.125	2,756.25
ac	42	15	−126	−72	−22	AC	−1.375	30.25
bc	95	126	−137	88	20	BC	1.250	25
abc	35	103	−113	−30	56	ABC^+	3.5	196
d	46	−81	−14	60	−192	D^+	−12.0	2,304
ad	18	1	−20	150	−44	AD†	−2.75	121
bd	62	−66	−49	−46	−38	BD†	−2.375	90.25
abd	−47	−60	−23	24	−118	ABD†	−7.375	870.25
cd	104	−28	82	−6	90	CD†	5.625	506.25
acd	22	−109	6	26	70	ACD†	4.375	306.25
bcd	67	−82	−81	−76	32	BCD†	2.0	64
$abcd$	36	−31	51	132	208	$ABCD^+$	13.0	2,704

The effect ABC^+ estimates the average difference between blocks, and the effect $ABCD^+$ estimates the variability of this difference. The effect D^+ measures the difference between replicates. On the hypothesis that the effects and interaction of factors A, B, and C are the same in each replicate (i.e., that there are no factor $\times$ replicate interactions) those effects marked † are non-existent and the corresponding mean squares are estimates of the residual variance. We thus obtain

$$\frac{1}{16}\,\Sigma(\text{effects}\dagger)^2 = 1958.0 \qquad s^2 = \frac{1}{16}\,\frac{\Sigma(\text{effects}\dagger)^2}{6} = 326.33$$

The interaction effects AB, AC, and BC are not significant when tested separately, and although the corresponding mean squares appear low a joint test indicates that they are not significantly so, since

$$\frac{210.25 + 30.25 + 25}{3} = 88.5 \qquad \frac{326.33}{88.5} = 3.7 \qquad (F_{6,\,3,\,0.05} = 8.9)$$

We may therefore pool these estimates to give a new variance estimate of $(1958.0 + 265.5)/9 = 247$ before testing the main effects as

$$A\colon\ \frac{12{,}996}{247} = 52.6^{***} \qquad (F_{1,\,9,\,0.001} = 22.9)$$

$$B\colon\ \frac{702.25}{247} = 2.84 \qquad (F_{1,\,9,\,0.05} = 5.1)$$

$$C\colon\ \frac{2{,}756.25}{247} = 11.2^{**} \qquad (F_{1,\,9,\,0.01} = 10.6)$$

It appears from this analysis that the effect of throughput (b) is quite small but that the other factors, temperature and make-up gas concentration, are important.

We now proceed to consider the analysis with covariance correction. If we suppose that the regression on the x variate is linear so that ave $(y) = \mu_y + \beta(x - \mu_x)$, all the observations should be corrected before the analysis is undertaken, and, if β and the average values of x and y are known, this can be done. In the present case β is unknown, but it has been assumed that the treatment $\times$ replicate interactions are zero, so that the average values of the linear combinations which estimate these effects are also zero. We can thus estimate β by taking b to minimize the sums of squares

$$\phi = \Sigma_i(\eta_i - b\xi_i)^2$$

where the η_i are those linear combinations of the y's which have average value zero and the ξ_i are the corresponding combinations of the x's. Differentiating with respect to b and equating to zero, we obtain

$$b = \frac{\Sigma_i \eta_i \xi_i}{\Sigma_i \xi_i^{\,2}}$$

and

$$\text{var}\,(b) = \frac{\Sigma_i \xi_i^{\,2}\,\text{var}\,(\eta_i)}{[\Sigma_i \xi_i^{\,2}]^2}$$

Since the η_i are orthogonal linear combinations of 16 independent observations of variance σ^2, var $(\eta_i) = 16\sigma^2$, so that

$$\text{var}\ (b) = \frac{16\sigma^2}{\Sigma_i \xi_i^2}$$

If we now use the sum-and-difference procedure on both the y and x observations, we obtain the data of Table 8.32.

TABLE 8.32

Effect	η	ξ	$\eta^2/16$	$\eta\xi/16$	$\xi^2/16$	Estimate of b
I	808	201	40,804	10,150.5	2,525.0625	—
A	−456	5	12,996	−142.5	1.5625	−91.20
B	−106	9	702.25	−59.625	5.0625	−11.78
AB	58	−11	210.25	−39.875	7.5625	−5.273
C	210	−1	2,756.25	−13.125	0.0625	−210.0
AC	−22	−13	30.25	17.875	10.5625	1.692
BC	20	7	25	8.750	3.0625	2.857
ABC	56	3	196	10.500	0.5625	—
D	−192	−3	2,304	36.000	0.5625	—
AD	−44	1	121	−2.750	0.0625	
BD	−38	−3	90.25	7.125	0.5625	
ABD	−118	17	870.25	−125.375	18.0625	
CD	90	−9	506.25	−50.625	5.0625	−5.245 (AD through BCD)
ACD	70	−5	306.25	−21.875	1.5625	
BCD	32	−17	64	−34.000	18.0625	
$ABCD$	208	−5	2,704	−65.000	1.5625	−41.60

If a joint estimate of β is based on the sums of the squares $\xi^2/16$ and cross products $\eta\xi/16$ for the 6 effects AD through BCD, for which the average value would be zero provided that no covariance correction were proposed, we have

$$b = \frac{\Sigma\eta\xi}{\Sigma\xi^2} \qquad s^2 = 1/5\ (\Sigma\eta^2/16 - b\ \Sigma\eta\xi/16)$$

the summation taking place over the 6 effects AD through BCD only. On the hypothesis that any of the effects A through BC are zero, the corresponding values of the square and cross product should also lead to an independent estimate of β. Thus for these effects

$$b_j = \frac{\eta_j\xi_j}{\xi_j^2} = \frac{\eta_j}{\xi_j}$$

and ave $(b - b_j) = 0$. Under these conditions

$$\text{var}\ (b - b_j) = 16\sigma^2 \left[\frac{1}{\Sigma\xi^2} + \frac{1}{\xi_j^2}\right]$$

or

$$\text{ave } (b - b_j)^2 \cdot \frac{\xi_j^2 \, \Sigma\xi^2}{16(\xi_j^2 + \Sigma\xi^2)} = \sigma^2$$

In this case, since only a single degree of freedom is available for each of the effects A through BC, there is no question of testing the departure from linearity in the regression of the class means and the significance of an effect may be determined by comparing the estimate of σ^2 based on the difference between the regression coefficients with the residual variance estimate s^2. Also an equivalent test may be obtained by correcting the effect means, using the regression coefficient estimated from the effects AD through BCD. We then have

$$[\text{corrected value } \eta_j/16] = \eta_j/16 - b\xi_j/16$$

and, on the hypothesis that the corrected effect is zero,

$$\text{ave } (\eta_j/16 - b\xi_j/16) = 0$$

$$\text{var } (\eta_j/16 - b\xi_j/16) = \left(\frac{\sigma^2}{16} + \frac{\xi_j^2}{16}\,\frac{\sigma^2}{\Sigma\xi^2}\right)$$

so that

$$\frac{(\eta_j/16 - b\xi_j/16)}{\dfrac{s}{4}\sqrt{1 + \xi_j^2/\Sigma\xi^2}} = t_{n-1}$$

where n is the number of terms over which the summation sign extends. Squaring and rearranging we have

$$\frac{(b_j - b)^2}{16s^2}\,\frac{\xi_j^2 \, \Sigma\xi^2}{(\xi_j^2 + \Sigma\xi^2)} = t^2_{n-1} = F_{1,\, n-1}$$

which is the same expression for the variance ratio as that obtained by the previous argument.

For the data of Table 8.32 we have

$$b = -5.245 \qquad s^2 = [1958 - (-5.245)(-227.5)]/5 = 153$$

$$\frac{(-5.245)(-227.5)}{153} = 7.8^* \qquad (F_{1,\,5,\,0.05} = 6.6)$$

so that the estimate of the regression coefficient differs significantly from zero. In this example the true regression coefficient β is known to be negative from past experience, and the significance test serves rather to indicate that, with the range of observed values of the x variate, a correction for regression will be worth while. Using the estimates of the regression, we can test the significance of differences between blocks in order to determine whether such differences are due simply to within-block variations and changes in proportion of impurities or whether additional factors are involved. Since any consistent difference between blocks within replicates may be due to the *ABC* interaction we test the effect *ABCD*, obtaining

$$\frac{(41.60 - 5.245)^2 (1.5625)(43.375)}{153(1.5625 + 43.375)} = 13^* \qquad (F_{1,\,5,\,0.01} = 16.3)$$

This ratio approaches the 0.01 significance level, and it may be supposed that there are additional sources of variation between blocks. It is also apparent by inspection that the *ABC* interaction, which is properly tested against the *ABCD* interaction, is negligible.

The effects *A* through *BC* are tested in a similar fashion to that employed in testing the *ABCD* effect. For the interactions we obtain the variance ratios

$$AB\colon 0.000033 \qquad AC\colon 2.67 \qquad BC\colon 1.23$$

The *AC* and *BC* interactions are not significant, but the *AB* interaction is significantly low at the 0.01 level when the covariance correction is made. This may well be attributed to chance effects. For the main effects the variance ratios are

$$A\colon 72.8^{***} \qquad B\colon 1.27 \qquad C\colon 17.10^{**}$$

so that the effects of temperature and concentration of active component of the feed are marked as clearly significant as in the previous analysis. The additional information gained from the analysis of covariance consists of an estimate of the regression coefficient β and a reduction in the residual variance estimate. Following the method of the previous treatment, we might conclude that the interaction effects *AB*, *AC*, and *BC* were not only non-significant but actually zero. In this event these estimates might be used in conjunction with the estimates *AD* through *BCD* to determine a new value of b and obtain a residual variance estimate based on 8 degrees of freedom.

In the previous examples the same effect was confounded in each replicate, but this is not an essential feature of the method. Had we so chosen, the main effects and interactions might have been confounded successively, 7 replicates being required, so that all the final estimates could be made with proportionate precision. Alternatively, if 4 replicates only were to be undertaken, we might choose to confound each of the interactions in turn, according to the design of Figure 8.12. In estimating the interaction effects only those replicates in which the interaction in question is not confounded are employed, so that in effect each interaction

Replicate	1		2		3		4	
	a	(1)	*c*	*bc*	*abc*	*bc*	*abc*	*b*
	c	*bc*	*abc*	*ac*	(1)	*c*	*bc*	*c*
	abc	*ab*	(1)	*a*	*ac*	*ab*	(1)	*ab*
	b	*ac*	*ab*	*b*	*b*	*a*	*a*	*ac*
Confounded	*ABC*		*AB*		*AC*		*BC*	

FIG. 8.12.

is estimated with the precision which would have been obtained if only 3 replicates had been carried out with some other interaction confounded. The analysis of variance resulting from this type of design is given formally in Table 8.33. The variation in the numerical coefficients of the components of the variance estimate should be noted.

When factorial designs involving a greater number of experimental units are undertaken so that replication is impracticable, it is still possible to confound a main effect or interaction in order to reduce the block size. In this case it is not possible to recover information on the confounded effect since no estimate of the variability of the difference between blocks is available, so that a split plot design is usually unsuitable. Since the higher-order interaction estimates will in any case be employed to estimate the residual variance, on the hypothesis that no real high-order interactions exist, little is lost by selecting one such estimate for confounding. In the following example we consider the case of a 2^4 design in which the three- and four-factor interactions are to be pooled to yield a variance estimate. The interaction *ABC* was selected for confounding.

EXAMPLE 3. The data presented in the second column of Table 8.34 were obtained [66] during an investigation of the economics of the Fleissner process as applied to lignitic coals of the Latrobe Valley, Australia. The object of the process is to reduce the moisture content of the lignite by treating under pressure with superheated steam, the water being removed partly by evaporation and partly by the direct draining which follows the reduction of viscosity and surface tension at high temperatures. The factors investigated were:

Steam pressure	*A* at 1.9 atm and 28.6 atm
Time of treatment	*B* at 10 min and 40 min
Source of coal	*C* Yallovin seam or Latrobe seam
Size of coal	*D* screened 3 in.–2 in. and $1\frac{1}{2}$ in.–1 in.

TABLE 8.33

Source of Estimate	D.F.	Average Value of Mean Square*
Between replicates	3	$\sigma^2 + 4\sigma_m^2 + 8\sigma_l^2$
Between blocks (within replicates)	4	$\sigma^2 + 4\sigma_m^2 + (1/4)\sigma^2_{ijk} + (1/2)\sigma^2_{ij} + (1/2)\sigma^2_{ik} + (1/2)\sigma^2_{jk}$
Total between blocks	7	
Main effect *A*	1	$\sigma^2 + 4\sigma^2_{ijk} + 8\sigma^2_{ij} + 8\sigma^2_{ik} + 16\sigma_i^2$
B	1	$\sigma^2 + 4\sigma^2_{ijk} + 8\sigma^2_{ij} + 8\sigma^2_{jk} + 16\sigma_j^2$
C	1	$\sigma^2 + 4\sigma^2_{ijk} + 8\sigma^2_{ik} + 8\sigma^2_{jk} + 16\sigma_k^2$
Interactions *AB*	1	$\sigma^2 + 3\sigma^2_{ijk} + 6\sigma^2_{ij}$
AC	1	$\sigma^2 + 3\sigma^2_{ijk} + 6\sigma^2_{ik}$
BC	1	$\sigma^2 + 3\sigma^2_{ijk} + 6\sigma^2_{jk}$
ABC	1	$\sigma^2 + 3\sigma^2_{ijk}$
Total block and treatment effects	14	
Residual	17	σ^2
Total	31	

*All populations of effects assumed infinite.

In order to avoid sampling difficulties the complete charge was weighed before and after treatment, the recorded figures being an improvement ratio defined as (mass before treatment)/(mass after treatment).

The results of an analysis by the tabular method given in Table 8.34, are summarized in Table 8.35. Of the two-factor interactions only *AC* gives a mean square estimate significantly greater than the residual. The

TABLE 8.34

Treatment Combination	Improvement Ratio	(1)	(2)	(3)	(4)	Effect	Component of Sum of Squares
(1)	1.60	3.63	7.40	13.68	27.37	I	46.81981
a	2.03	3.77	6.28	13.69	2.45	*A*	0.37516
b	1.69	3.12	7.38	1.24	0.37	*B*	0.00856
ab	2.08	3.16	6.31	1.21	0.17	*AB*	0.00181
c	1.47	3.63	0.82	0.18	−2.19	*C*	0.29976
ac	1.65	3.75	0.42	0.19	−0.75	*AC*	0.03516
bc	1.46	3.12	0.78	0.02	−0.15	*BC*	0.00141
abc	1.70	3.19	0.43	0.15	0.09	*ABC*	0.00051*
d	1.64	0.43	0.14	−1.12	0.01	*D*	0.000006
ad	1.99	0.39	0.04	−1.07	−0.03	*AD*	0.000056
bd	1.66	0.18	0.12	−0.40	0.01	*BD*	0.000006
abd	2.09	0.24	0.07	−0.35	0.13	*ABD*	0.001060†
cd	1.47	0.35	−0.04	−0.10	0.05	*CD*	0.000155
acd	1.65	0.43	0.06	−0.05	0.05	*ACD*	0.000155†
bcd	1.47	0.18	0.08	0.10	0.05	*BCD*	0.000155†
abcd	1.72	0.25	0.07	−0.01	−0.11	*ABCD*	0.00076†
						Total	47.54453

* Confounded. † Pooled as error estimate.

TABLE 8.35

Source of Estimate	Sum of Squares	D.F.	Mean Square
Between blocks (*ABC*)	0.00051	1	0.00051
Main effect *A*	0.37516	1	0.37516
B	0.00856	1	0.00856
C	0.29976	1	0.29976
D	0.00001	1	0.00001
Interactions *AB*	0.00181	1	0.00181
AC	0.03516	1	0.03516
AD	0.00006	1	0.00006
BC	0.00141	1	0.00141
BD	0.00001	1	0.00001
CD	0.00016	1	0.00016
Pooled residual	0.00213	4	0.00053
Total	0.72474	15	

variance ratio in this case is 66.3** ($F_{1,\,4,\,0.001} = 74.1$), which is almost significant at the 0.001 level. Pooling the remaining two-factor interactions leads to a new estimate 0.00062 for the residual variance. Tested against this residual the estimate due to main effect D is not significant; that due to the main effect B gives a variance ratio of 13.8** ($F_{1,\,9,\,0.01} =$ 10.6). The two most important main effects A and C are highly significant if tested against the residual. We conclude that there is a difference in response between the 2 types of lignite, and that with these 2 types the process is more effective with the higher steam pressure. The increase in steam pressure is not equally effective on the 2 coals. Increase in time of treatment also results in a higher improvement ratio.

In this example the reduction in block size has produced no change in efficiency and a decrease in sensitivity, since the size of the residual is unaffected, but it has 1 less degree of freedom. Nevertheless the smaller blocks have the advantage in this type of experiment that, if, owing to failure of equipment, it becomes impossible to complete the full design as 1 undertaking, only the incompleted block or blocks need be repeated when the equipment is replaced. Furthermore, by regarding the completed block or blocks of the interrupted design as a partial replication it may be possible to reach interim conclusions which will indicate whether the completion of the experiment is worthwhile.

If the size of block which is desirable in order to eliminate a known source of experimental error is insufficient to accommodate a half replicate of the design, then more than one effect must be confounded. Because the effects to be confounded cannot be selected independently it is necessary to examine the relationship between them in order to produce a design in which the lost information will relate, as far as possible, to unimportant effects.

In Table 8.1 the signs to be allocated to the observations of a 2^3 factorial design in order to estimate the 8 effects have been given. If corresponding coefficients in any 2 of these linear forms are multiplied, then (1) the sum of the products is zero, i.e., the forms are orthogonal; (2) the new coefficients lead to an estimate of some other effect of the design. We take as an example the effects A and B for which the appropriate coefficients are

$$A = +(abc) + (ab) + (ac) + (a) - (bc) - (b) - (c) - (1)$$

$$B = +(abc) + (ab) - (ac) - (a) + (bc) + (b) - (c) - (1)$$

The new expression formed by the product of corresponding coefficients is

$$+(abc) + (ab) - (ac) - (a) - (bc) - (b) + (c) + (1)$$

which is the combination leading to an estimate of the AB interaction. If the effects AB and BC had been chosen for illustration, the product of corresponding coefficients in the comparisons would lead to the comparison estimating the AC interaction. Under this operation the effect comparisons form a group, the product of any 2 elements of which is the *generalized interaction* of these elements. If we adopt a system of multiplication for the effect symbols in which $A^2 = B^2 = C^2 = \mathrm{I}$ (a *unit element*), then the symbol for the generalized interaction of 2 effects is the product of the effect symbols.

$$A \times B = AB \quad AB \times BC = AB^2C = A\mathrm{I}C = AC$$

It is found that if 2 effects are selected for confounding in a given design then their generalized interaction must also be confounded. This may be seen to be true by considering a 2^3 design confounded in 4 blocks of 2 experimental units, and by selecting 2 effects, e.g., AB and BC, for confounding. The 8 effects in the analysis are estimated by 8 linear combinations of the observations in the form of Table 8.1, which are mutually orthogonal. The linear combinations of the block effects W, X, Y, and Z are given by the 4 comparisons

$$W + X + Y + Z = \mathrm{I}$$
$$W + X - Y - Z$$
$$W - X - Y + Z$$
$$W - X + Y - Z$$

the coefficients having properties similar to those of the effect comparisons. If 4 of the treatment effect estimates are to be independent of, and hence orthogonal to, the block effects, the remaining treatment effects must be identified with the block comparisons. If we choose to identify AB and AC as

$$AB = W + X - Y - Z$$
$$BC = W - X - Y + Z$$

then, since

$$AB = abc + ab + c + (1) - ac - bc - a - b$$
$$BC = abc - ab - c + (1) - ac + bc + a - b$$

this can only be achieved by arranging the blocks as

$$W = abc + (1) \quad X = ab + c \quad Y = ac + b \quad Z = bc + a$$

The third comparison between blocks is then obtained by taking the product of corresponding signs in the coefficients of the other two comparisons. Since these also correspond to the effect AB and BC the third comparison between blocks must correspond to the generalized interaction of these two effects AC, so that the latter is also confounded. The relationship between confounding and the theory of finite groups has been discussed by R. A. Fisher [72] and O. Kempthorne [117].

In a 2^k factorial design carried out in 2^r blocks the overall effect I and $2^r - 1$ additional treatment comparisons must be confounded. Of the $2^r - 1$ only r are independent, the remaining $2^r - r - 1$ being obtained as the generalized interactions of the selected effects. The confounded interactions therefore form a subgroup of the main group. For the case $k = 3$, 3 types of confounding in 4 blocks are available, and are exemplified (A, B, AB), (A, BC, ABC), and (AC, BC, AB). The main effects can only be preserved by sacrificing the 3 two-factor interactions, so that the design is of little practical value unless many replications are to be carried out. If $k = 4$ and the design is carried out in 4 blocks, the arrangements in which main effects are not confounded are exemplified by (AB, BC, AC), (AB, CD, $ABCD$), and (AB, BCD, ACD), so that 1 two-factor interaction must be sacrificed in the best arrangement. It is easily shown that the 2^5 and 2^6 designs may be arranged in 4 blocks by confounding interactions of 3 or more factors, and that the latter design may be arranged in 8 blocks without losing any of the two-factor interactions. As it is usually desirable to preserve the two-factor interactions, a block size of at least 8 experimental units is required in all cases except that of the 2^3 design with ABC confounded. If, on the other hand, some of the two-factor interactions are known to be zero, the block size may be further reduced.

When the effects to be confounded have been selected it is necessary to allocate the treatment combinations so as to achieve the desired arrangement. When the number of blocks is not more than 8 this may be done most conveniently by writing down the linear combinations leading to the maximum number of independent confounded effects, and allocating the treatment combinations to blocks, according to the arrangement of signs in the selected comparisons. Thus for a 2^5 design in 8 blocks we might confound (I, AB, CD, ACE, ADE, BCE, BDE, $ABCD$), selecting AB, CD, and ACE for expansion since none of these 3 is the generalized interaction of the remaining 2. Expanding these effects, we obtain Table 8.36. The treatment combinations can then be separated as

	(+ + +)	(+ + −)	(+ − +)	(− + +)	(− − +)	(− + −)	(+ − −)	(− − −)
	abcde	*abcd*	*abce*	*acde*	*ace*	*bcde*	*abde*	*ade*
	ab	*abe*	*abd*	*bed*	*bde*	*acd*	*abc*	*bce*
	cd	*cde*	*de*	*be*	*ad*	*ae*	*ce*	*bd*
	e	(1)	*c*	*e*	*bc*	*b*	*d*	*ac*
Block	1	2	3	4	5	6	7	8

TABLE 8.36

Effect	Treatment Combination							
	abcde	*abcd*	*abce*	*abde*	*abc*	*abd*	*abe*	*cde*
AB	+	+	+	+	+	+	+	+
CD	+	+	−	−	−	−	+	+
ACE	+	−	+	−	−	+	−	−
	ab	*cd*	*ce*	*de*	*c*	*d*	*e*	(1)
AB	+	+	+	+	+	+	+	+
CD	+	+	−	−	−	−	+	+
ACE	+	+	−	+	+	−	+	−
	acde	*bcde*	*acd*	*ace*	*ade*	*bcd*	*bce*	*bde*
AB	−	−	−	−	−	−	−	−
CD	+	+	+	−	−	+	−	−
ACE	+	−	−	+	−	+	−	+
	ac	*ad*	*ae*	*bc*	*bd*	*be*	*a*	*b*
AB	−	−	−	−	−	−	−	−
CD	−	−	+	−	−	+	+	+
ACE	−	+	−	+	−	+	+	−

An alternative method of allocating treatments which is more convenient when the experiment is large and the block capacity small is to place in the *principal block* those treatments which have an even number of letters in common with each of the confounded effects. Since zero is an even number the principal block always includes the control (1), and consideration of the factorized expansions for the treatment effects shows that it includes all other treatment combinations with a similar sign arrangement. The principal block includes combinations which form a subgroup of the main group, and by generating cosets from this subgroup the design is obtained. This is achieved in practice by multiplying the elements of the subgroup by any element of the group which is neither a member of the subgroup nor of any additional set already generated.

In the previous example the principal block is 2, and we generate block 1 by multiplying each element of 2 by e, subject to the condition $a^2 = b^2 = \cdots = e^2 = 1$. Since c has not occurred in either of the blocks we use this as multiplier to form block 3. Similarly, with a, bc, b, d, and ac as multipliers we can generate the remaining blocks. It may be noted that the multipliers are not restricted to the low-order effects, a block can be generated by the use of any one of the treatment combinations as multiplier.

EXAMPLE. 4 In an experiment designed to investigate the effect of trace elements on the growth of a number of crops [67] 6 factors, each at 2 levels, were to be investigated. The crops were grown in large pots on greenhouse benches, and in order to reduce the effect of variations due to the location of the pots the experiment was subdivided into 8 blocks. In this case it is possible to preserve all the two-factor interactions by selecting for confounding three-factor interactions, any pair of which has a single common letter. There was no reason to suppose that any of the three-factor interactions would be particularly important, and ACE, BDE, and BCF, ADF, $ABCD$, $CDEF$, and $ABEF$ were chosen for confounding, the principal block becoming (1), *ade*, *acf*, *bdf*, *bce*, *abcd*, *cdef*, and *abef*. The remaining blocks are most easily generated by multiplying by a, b, c, d, e, f, and ab. The resulting design together with the yields obtained in 1 experiment are given in Table 8.37, the symbols referring to the following additions:

a: mixed fertilizer N.P.K.	100	d: H_3BO_3	2
b: $MgSO_4 \cdot 7H_2O$	75	e: $MnSO_4 \cdot 4H_2O$	15
c: $CuSO_4 \cdot 5H_2O$	5	f: $ZnSO_4 \cdot 7H_2O$	5

and the yields to weights of air-dried soybean in grams per pot.

The data are most conveniently analyzed by the tabular method, all 64 effects being computed. The five- and six-factor interactions together with the unconfounded four-factor interactions may be pooled to provide a residual variance estimate based on 19 degrees of freedom. The analysis is left as an example to the reader.

8.64. Confounding in 3^k and Other Factorial Designs

Confounded arrangements suitable for the case in which the number of levels of the factors is more than 2 have been published [52]. If the number of levels of the various factors differ, the design may be rather complicated, and a number of replicates may be required in order to balance the degree of confounding of the interactions of a given order. For the 3^k design the confounding procedure is still relatively simple and may be understood by first considering the 3×3 Graeco-Latin square for the factors A, B, C, and D at levels 0, 1, and 2.

If A and B are the two factors in a 3×3 factorial design, the row and column effects each represent 2 independent comparisons which are the main effects of A and B, respectively. In addition the square contains 4 other independent comparisons, corresponding to the effects of C and D.

TABLE 8.37

Block Number

1		2		3		4		5		6		7		8	
(1)	1.18	*f*	2.48	*d*	3.73	*a*	4.18	*c*	3.58	*b*	2.65	*e*	5.45	*ab*	5.10
ade	2.50	*adef*	2.83	*ae*	3.43	*de*	2.73	*acde*	3.50	*abde*	2.78	*ad*	5.35	*bde*	2.60
acf	3.18	*ac*	2.98	*acdf*	3.50	*cf*	3.98	*af*	3.80	*abcf*	4.45	*acef*	3.70	*bcf*	3.33
bdf	2.73	*bd*	4.50	*bf*	2.33	*abdf*	3.88	*bcdf*	3.60	*df*	3.45	*bdef*	2.35	*adf*	3.85
bce	3.83	*bcef*	2.33	*bcde*	2.33	*abce*	4.48	*be*	3.13	*ce*	3.90	*bc*	3.20	*ace*	4.75
abcd	4.25	*abcdf*	4.18	*abc*	4.38	*bcd*	2.58	*abd*	3.03	*acd*	3.38	*abcde*	4.30	*cd*	4.80
abef	5.65	*abe*	4.48	*abdef*	2.55	*bef*	2.03	*abcef*	2.68	*aef*	3.05	*abf*	3.20	*ef*	2.75
cdef	2.88	*cde*	2.98	*cef*	4.05	*acdef*	3.90	*def*	2.60	*bcdef*	3.98	*cdf*	3.93	*abcdef*	4.58

These must be equivalent to the interaction effects $A \times B$ since they represent the remaining independent comparisons in the data.

	b_0	b_1	b_2
a_0	c_0d_0	c_2d_1	c_1d_2
a_1	c_1d_1	c_0d_2	c_2d_0
a_2	c_2d_2	c_1d_2	c_0d_1

FIG. 8.13.

Unfortunately this arrangement of the interaction effect comparisons does not represent a type in which we should normally be interested, such as A_LB_L, A_LB_Q, A_QB_L, and A_QB_Q so that if we confound some of these interaction effects the remainder will not have any recognizable meaning in terms of the types of comparison which we usually prefer to make. If we rearrange the Graeco-Latin square so that C and D are the row and column effects, we have

	d_0	d_1	d_2
c_0	a_0b_0	a_2b_2	a_1b_1
c_1	a_1b_2	a_0b_1	a_2b_0
c_2	a_2b_1	a_1b_0	a_0b_2

If we then arrange that the experimental units of d_0 are placed in the first block, of d_1 in the second block, and of d_2 in the third block, the two independent comparisons of d_0, d_1, and d_2 will be completely confounded with block comparisons, but the three sets of 2 comparisons between a_0, a_1, and a_2, b_0, b_1 and b_2, and c_0, c_1, and c_2 will be unaffected by blocks. This means that the comparisons between main effects are preserved together with 2 interaction comparisons. If we are prepared to assume that the latter are negligible, we can employ the 2 degrees of freedom for an error estimate. In most instances this would be insufficient, and we should be obliged to replicate the experiment, in which case the rows rather than the columns would be chosen to constitute the blocks. Each of the 4 interaction comparisons would then be equally confounded, and we should have the equivalent of 1 determination of each interaction effect, and 2 of each main effect. The analysis of variance would take the form of the accompanying table.

Source of Estimate	D.F.	Average Value of Mean Square
Between replicates	1	
Between A	2	$\sigma^2 + 2\sigma^2{}_{AB} + 6\sigma_A{}^2$
Between B	2	$\sigma^2 + 2\sigma^2{}_{AB} + 6\sigma_B{}^2$
Between blocks	4	$\sigma^2 + \sigma^2{}_{AB} + 3\sigma^2{}_{Blocks}$
$A \times B$ interaction	4	$\sigma^2 + \sigma^2{}_{AB}$
Residual	4	σ^2
Total	17	

This method may be extended to the 3^3 and in general to the 3^k design. To comply with established notation [52] we remark that the 2 groups of 3 interaction sums which lead to the 4 independent comparisons for interaction are usually identified as in Figure 8.14, the subscripts identifying the

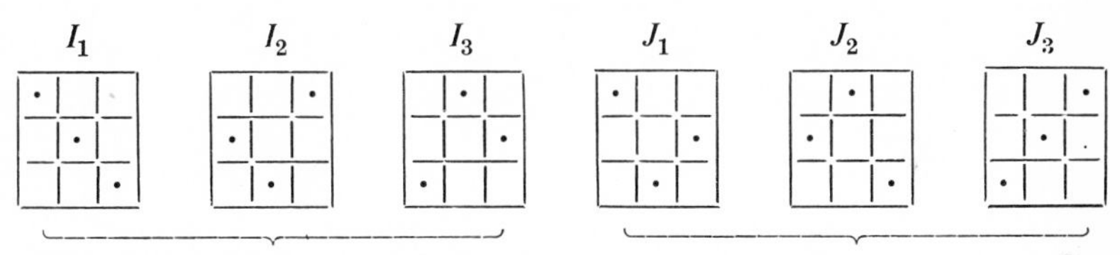

FIG. 8.14.

cell of the first column which is employed. To obtain the three-factor interaction which could be confounded in the 3^3 design we consider the sums I_1, I_2, and I_3, and J_1, J_2, and J_3 to be evaluated for each level of the third factor C, giving 2 sets of 9 sums as shown in the accompanying figure. Now in each set

(1) The sums over rows represent the effect of C at the 3 levels.

(2) The sums over the columns represent the I or J effects of the interactions A and B.

The three-factor interactions may be obtained by selecting either the I or J sets from each of these sets of 9. By this procedure we obtain 4 sets, each of 3 totals. The 3 totals represent 2 independent comparisons each, and

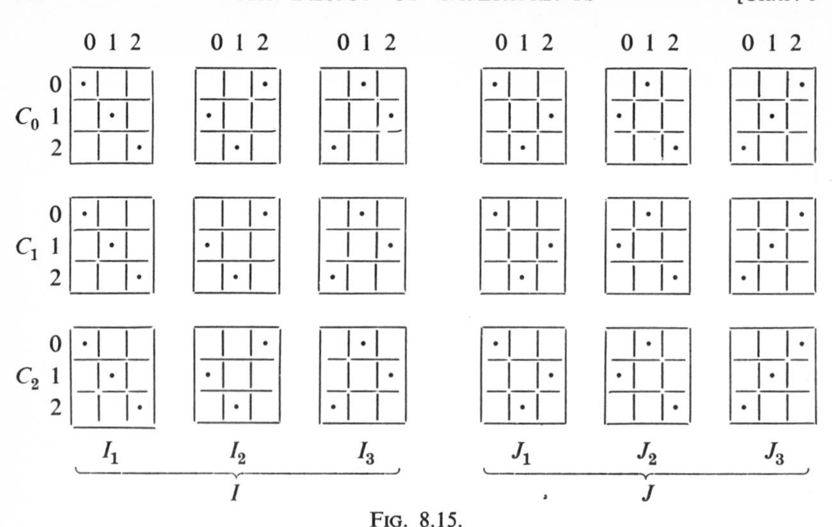

FIG. 8.15.

overall we have the 8 independent comparisons of the three-factor interactions. These may be sketched out in the manner shown above, but for convenience we write, e.g., 021 to describe the experimental unit $a_0b_2c_1$ (a is the vertical direction of the diagram) and obtain Table 8.38.

TABLE 8.38

W_1	W_3	W_2	X_1	X_2	X_3	Y_1	Y_3	Y_2	Z_1	Z_2	Z_3
I_1I	I_2I	I_3I	J_1I	J_2I	J_3I	I_1J	I_2J	I_3J	J_1J	J_2J	J_3J
000	001	002	000	001	002	000	001	002	000	001	002
110	111	112	110	111	112	210	211	212	210	211	212
220	221	222	220	221	222	120	121	122	120	121	122
101	102	100	102	100	101	101	102	100	102	100	101
211	212	210	212	210	211	011	012	010	012	010	011
021	022	020	022	020	021	221	222	220	222	220	221
202	200	201	201	202	200	202	200	201	201	202	200
012	010	011	011	012	010	112	110	111	111	112	110
122	120	121	121	122	120	022	020	021	021	022	020

The W, X, Y, Z nomenclature is also due to Yates [52]. To confound the 27 experiments from a single replication in blocks of 9 we can arrange to use either of the above sets, e.g., W, and can place each of the subsets

W_1, W_2, and W_3 in separate blocks. The remaining 6 independent three-factor comparisons, together with the two-factor comparisons and main effects, are then unaffected by block differences. As in the 3^2 design the confounded interaction comparisons are of a complicated nature, and an interaction component of the form $A_L B_L C_L$ cannot be recovered. If 4 replicates are carried out, the W, X, Y, and Z components of the three-factor interaction being confounded successively, then three-quarters of the information will be available on all the three-factor interaction components, which may be assembled in any required orthogonal linear combinations.

Factors at 4 levels may be confounded by regarding the factor as the formal equivalent of 2 two-level factors. The 3 orthogonal comparisons between observations at the 4 levels of the factor may be taken in the form

$$a_1 + a_2 - a_3 - a_4 = A'$$
$$a_1 - a_2 - a_3 + a_4 = A''$$
$$a_1 - a_2 + a_3 - a_4 = A'''$$

of which

$$2A' + A''' = \text{linear component}$$
$$A'' = \text{quadratic component}$$
$$2A''' - A' = \text{cubic component}$$

If we choose to confound A''' and suppose the cubic component to be negligible, then A' may be used to estimate the linear component of the regression and A'' to estimate the quadratic component. A''' is obviously the generalized interaction of these factors. If the experiment involved a 4×2 classification, it would be convenient to confound the $A'''B$ interaction

$$a_1b_2 - a_2b_2 + a_3b_2 - a_4b_2 - a_1b_1 + a_2b_1 - a_3b_1 + a_4b_1 = A'''B$$

as

$$(a_1b_2 + a_3b_2 + a_2b_1 + a_4b_1) - (a_2b_2 + a_4b_2 + a_1b_1 + a_3b_1)$$

Alternatively with three replicates we could confound in turn each of the effects $A'''B$, $A''B$, and $A'B$ to obtain a balanced design. Similar methods can be applied to a 4^2 or 4^k design or to designs of the type 4^k2^l. Designs for confounding experiments of the type 3×2^k, $3^2 \times 2$ are given by Yates [52], and Kempthorne [117] gives an account of the theory of confounding in the more complicated cases.

It is sometimes convenient to arrange that an experiment should be confounded with respect to 2 sets of limitations. Thus, if we were carrying out a 2^4 experiment involving a process which could be carried out only once per day on a given piece of equipment, and we had available

4 such sets of equipment, it would be reasonable to use a Latin square arrangement for the confounding. If we decide to confound $ABCD$, AB, and CD between rows (days), we obtain (i), where the 4 blocks have been generated in the usual manner. It will not now be possible to confound any of the interactions $ABCD$, AB, and CD between columns (equipment) in view of the requirement which we have placed on the rows. We might choose to confound the interactions ABC, BCD, and AD, which are independent of the first group, obtaining for the principal block (1), *bc*, *abd*, *acd*. If we produce such a column by rearranging the

Day				
1	(1)	*ab*	*cd*	*abcd*
2	*a*	*b*	*acd*	*bcd*
3	*c*	*abc*	*d*	*abd*
4	*bc*	*ac*	*bd*	*ad*

(i)

Equipment I	II	III	IV
(1)	*ab*	*cd*	*abcd*
acd	*bcd*	*a*	*b*
abd	*d*	*adc*	*c*
bc	*ac*	*bd*	*ad*

(ii)

FIG. 8.16.

rows, the remaining terms in the rows may also be arranged to give the columns (*ab*, *bcd*, *d*, and *ac*), (*cd*, *a*, *abc*, and *bd*), and (*abcd*, *b*, *c*, *ad*) which are required for the selected confounding. The remaining 9 independent comparisons in such a square provide the estimates of the unconfounded effects of the 2^4 design. It is not necessary for double confounding that a square arrangement should be used; we could for example confound a 2^5 design in 4 blocks of 8 and simultaneously in 8 blocks of 4, provided that we were prepared to forgo the groups of comparisons involved in the confounding process.

8.65. Incomplete Blocks

The confounding techniques of the previous section represent examples of incomplete blocks used in conjunction with factorial experiments. In these examples the interaction effects were usually chosen for confounding because it was believed that they were likely to be less important than the main effects. In a factorial design this is frequently a reasonable assumption, but, in a case where the 8 treatment combinations are replaced by 8 varieties so that the relative importance of the various possible comparisons cannot be determined in advance, such a procedure might result in a low level of precision for a comparison which, after the experiment, appeared to be the most important. To avoid this situation a series of designs known as balanced incomplete blocks have been developed. In such a design the treatment conditions or varieties are allocated to the blocks so that all possible pairs of varieties occur an equal number of times within the blocks. Simple examples of such balanced designs are shown in Figure 8.17. Designating the number of blocks by b, the number

(1)

Treatment \ Block	I	II	III
1	X		X
2	X	X	
3		X	X

(2)

Treatment \ Block	I	II	III	IV	V	VI
1	X	X	X			
2	X			X	X	
3		X		X		X
4			X		X	X

(3)

Treatment \ Block	I	II	III	IV
1	X	X	X	
2	X	X		X
3	X		X	X
4		X	X	X

FIG. 8.17.

of treatments by t, the number of treatments per block by k, and the number of replicates of each treatment by r, it is apparent that the total number of observations is $N = rt = bk$ and that the number of times any treatment pair occurs in a block will be

$$\frac{r(k-1)}{t-1} = \lambda$$

If we assume no interactions between treatment and blocks, we have for our model

$$x_{ij} = e_0 + e_i + e_j + \varepsilon_{ij} \qquad i = 1, \cdots, t \quad j = 1, \cdots, b$$

The sum of the observations on treatment p will therefore be given by

$$\text{(a)} \qquad \sum_j^r x_{pj} = re_0 + re_p + \sum_j^r e_j + \sum_j^r \varepsilon_{pj}$$

Similarly the sum of the observations in a block j containing the treatment p will be

$$\text{(b)} \qquad \sum_i^k x_{ij} = ke_0 + e_p + \sum_i^{k-1} e_i + ke_j + \varepsilon_{pj} + \sum_i^{k-1} \varepsilon_{ij}$$

and the sum for all blocks of this type will be

$$\text{(c)} \qquad \sum_j^r \sum_i^k x_{ij} = rke_0 + re_p + \sum_j^r \sum_i^{k-1} e_i + k \sum^r e_j + \sum^r \varepsilon_{pj} + \sum_j^r \sum_i^{k-1} \varepsilon_{ij}$$

If we multiply (*a*) by k and subtract (*c*), we obtain

$$k \sum_j^r x_{pj} - \sum_j^r \sum_i^k x_{ij} = r(k-1)e_p - \sum_j^r \sum_i^{k-1} e_i + (k-1) \sum_j^r \varepsilon_{pj} - \sum_j^r \sum_i^{k-1} \varepsilon_{ij}$$

Now in the r blocks containing the treatment p any pair of observations (x_p, x_i) occurs $\lambda = \dfrac{r(k-1)}{t-1}$ times so that since x_p occurs once in every block every observation $x_i (i \neq p)$ occurs λ times. Thus the above equation may be written

$$k \sum_j^r x_{pj} - \sum_j^r \sum_i^k x_{ij} = [r(k-1) + \lambda]e_p - \lambda \sum_i^t e_i + (k-1) \sum_j^r \varepsilon_{pj} - \sum_j^r \sum_i^{k-1} \varepsilon_{ij}$$

Since e_0 is defined such that over the population of possible treatments and blocks $\Sigma_i e_i = \Sigma_j e_j = 0$ and since the sums of the random components have average value zero,

$$\begin{aligned} \text{ave } (k \sum_j^r x_{pj} - \sum_j^r \sum_i^k x_{ij}) &= [r(k-1) + \lambda]e_p \\ &= \frac{rt(k-1)}{t-1} e_p \end{aligned}$$

If $\sum_j^r x_{pj} = T_p$ and $\sum_i^k x_{ij} = B_j$, we have

$$\text{ave } \frac{(t-1)}{N(k-1)} (kT_p - \Sigma_{(p)} B_j) = e_p$$

when the summation takes place over all blocks containing treatment p.

In general we wish to compare the effects of 2 treatments, and we obtain an unbiased estimate of $e_p - e_q$ by subtracting the corresponding estimates $\hat{e}_p$ and $\hat{e}_q$. The variance of the comparison may be obtained as

$$\frac{2k(t-1)}{N(k-1)} \sigma^2$$

In the limited number of cases where the design is balanced with respect to blocks, the block effects may be estimated by the same procedure and the block differences will have the same variance as the treatment differences. The first and third examples of Figure 8.17 illustrate this case. When balance with respect to blocks is not present the treatment effects may be estimated and employed to correct the block totals. In the latter instance the variance of a comparison between block totals having c common treatments, and therefore $(k - c)$ pairs of treatments for which correction must be made, is

$$2 \left[k + \frac{(k-c)\, k(t-1)}{k-1} \right] \sigma^2$$

and the corresponding variance of a comparison of block effects is

$$\frac{2}{k}\left[1 + \frac{(k-c)(t-1)}{(k-1)}\right]\sigma^2$$

The analysis of variance takes the form of the accompanying table.

Source of Estimate	D.F.	Average Value of Mean Square
Between treatments	$t-1$	$\sigma^2 + \frac{b(k-1)}{(t-1)}\sigma_t^2$
Between blocks	$b-1$	$\sigma^2 + \sigma_t^2 + k\sigma_b^2$
Residual	$N-t-b+1$	σ^2
Total	$N-1$	

TABLE 8.39

Treatment		1	2	3	4	5	Total
	1	35	16				51
	2	20		10			30
	3	13			26		39
	4	25				21	46
Bale number	5		16	5			21
	6		21		24		45
	7		27			16	43
	8			20	37		57
	9			15		20	35
	10				31	17	48
Total		93	80	50	118	74	415

EXAMPLE. As an example of the use of such a design we consider the data of Table 8.39 which relate to the results obtained from 5 series of tests on 10 bales of rubber. We shall regard the bales of rubber as the blocks and the 5 series of tests as treatments. The test measurements recorded are (percentage elongation—300) of specimens of rubber stressed at 400 psi.

With the notation of this section we have

$$t = 5 \quad b = 10 \quad r = 4 \quad k = 2 \quad N = 20$$

$$\lambda = \frac{r(k-1)}{t-1} = 1$$

The sum of squares between bales, treatments being ignored, is simply

$$\tfrac{1}{2}[(51)^2 + (30)^2 + \cdots + (48)^2] - \tfrac{1}{20}(415)^2 = 504.25$$

The treatment effects are estimated as

$$\hat{e}_p = \frac{t-1}{N(k-1)}(kT_p - \Sigma_{(p)}B_j)$$

and we have for treatment 1

$$\hat{e}_1 = \tfrac{1}{5}[2(35 + 20 + 13 + 25) - (51 + 30 + 39 + 46)]$$
$$= \tfrac{1}{5}[20] = 4$$

Similarly for the other treatments we have

$$\hat{e}_2 = \tfrac{1}{5}[0] = 0 \qquad \hat{e}_4 = \tfrac{1}{5}[47] = 9.4$$
$$\hat{e}_3 = \tfrac{1}{5}[-43] = -8.6 \qquad \hat{e}_5 = \tfrac{1}{5}[-24] = -4.8$$

The sum of squares for treatments is most conveniently computed by squaring the treatment effect totals and multiplying by

$$\frac{t-1}{Nk(k-1)} = \frac{1}{10}$$

Doing this, we obtain

$$\tfrac{1}{10}[(20)^2 + (0)^2 + (-43)^2 + (47)^2 + (-24)^2] = 503.4$$

The residual sum of squares is computed by determining the overall sum of squares

$$[(35)^2 + (16)^2 + (20)^2 + (10)^2 + \cdots + (31)^2 + (17)^2] - \tfrac{1}{20}(415)^2 = 1207.75$$

and subtracting the treatment and bales components.

The analysis of variance may then be set out as in Table 8.40. The variance ratio

$$\frac{125.85}{33.35} = 3.77 \qquad (F_{4,\,6,\,0.05} = 4.5)$$

indicates that the treatment effects are not significant at the 5% level.

If the same experiment had been carried out in randomized complete blocks, involving in all N observations, then the average value of the

TABLE 8.40

Source of Estimate	Sum of Squares	D.F.	Mean Square	Average Value of Mean Square
Between bales	504.25	9	56.02	$\sigma^2 + \sigma_t^2 + 2\sigma_b^2$
Between treatments	503.4	4	125.85	$\sigma^2 + \frac{5}{2}\sigma_t^2$
Residual	200.1	6	33.35	σ^2
Total	1207.75	19		

mean square between treatments would have been $\sigma^2 + r\sigma_t^2$. The ratio $\frac{b(k-1)}{r(t-1)} = \frac{1 - 1/k}{1 - 1/t} = \frac{t\lambda}{rk}$ is defined as the efficiency of the experimental design. The rest of the information on treatment effects is to be found in the between-blocks mean square, which includes the term σ_t^2 in its average value. The recovery of this information has been discussed by Yates [60, 109].

A large number of possible designs for balanced incomplete blocks has been given by Fisher and Yates [13]. One of the principal difficulties with this type of design from the viewpoint of chemical experimentation is the large number of replications required if a balanced design is to be employed.

8.66. Incomplete Latin Squares or Youden Squares

For certain arrangements of balanced incomplete block design it is possible to arrange the treatments so than an incomplete Latin square [110] is obtained. Such arrangements are generally known as Youden

(i)

A	*B*	*C*	*D*
B	*C*	*D*	*A*
C	*D*	*A*	*B*

(ii)

A	*B*	*C*	*D*	*E*	*F*	*G*
B	*C*	*D*	*E*	*F*	*G*	*A*
D	*E*	*F*	*G*	*A*	*B*	*C*

FIG. 8.18.

squares [111, 112]. The simplest examples of this type of design are obtained by removing 1 row (or column) from a complete Latin square as in Figure 8.18 (i), but more useful designs when the number of treatments is large are obtained by omitting more than 1 line from larger squares as shown in (ii). For 4, 5, 6, 8, 9, and 10 treatments only squares of type (i) are available; for 7 and 11 treatments 2 squares of type (ii) and 1 of type (i) are available.

Since each row of the above designs constitutes a complete replication it is possible to eliminate a component of variation between replications, or, provided that the absence of interactions between the factors concerned can be assumed, these replicates may be used to examine 3 levels of a single factor. Alternatively, the design may be employed as a means of controlling 2 sources of error. The method of analysis is similar to that for the usual balanced incomplete block, except that a component of the sum of squares between replicates may be separated and that, in addition to the balance of treatment comparisons with respect to blocks, the block comparisons are balanced with respect to treatments.

TABLE 8.41

(Data coded as (Degrees C − 26.8700) × 10^4)

Thermometer		1	2	3	4	5	6	7	Totals
Sample *A*	I	$19_{(3)}$				$24_{(2)}$		$35_{(1)}$	78
	II	$31_{(1)}$	$22_{(3)}$				$21_{(2)}$		74
	III		$41_{(1)}$	$15_{(3)}$				$29_{(2)}$	85
Sample *B*	IV	$61_{(2)}$		$71_{(1)}$	$68_{(3)}$				200
	V		$75_{(2)}$		$76_{(1)}$	$68_{(3)}$			219
	VI			$66_{(2)}$		$66_{(1)}$	$56_{(3)}$		188
Sample *C*	VII				$18_{(2)}$		$14_{(1)}$	$14_{(3)}$	46
Totals		111	138	152	162	158	91	78	890

EXAMPLE. As an example of this type we consider the data obtained during tests on the triple point of diphenyl ether [113]. The tests were carried out on 7 samples of ether, each contained in a separate cell. Observations were made with 7 thermometers, 7 tests being made on each of 3 days. The design and data are presented in Table 8.41, where the subscripts refer to the day on which the tests were undertaken. The totals for days are

Day	1	2	3
Total	334	294	262

If the data are to be described by a model

$$x_{ijk} = e_0 + e_i + e_j + e_k + \varepsilon_{ijk}$$

where i, j, and k refer to days, cells, and thermometers, respectively, then for the estimate of the day effects we have

$$\textstyle\sum_{jk} x_{ijk} = 7e_0 + 7e_i + \sum_j^7 e_j + \sum_k^7 e_k + \sum_{jk}^7 \varepsilon_{ijk}$$

$$\text{ave}\left(\tfrac{1}{7}\textstyle\sum_{jk} x_{ijk}\right) = e_0 + e_i$$

Similarly for the sum of squares due to days

$$\frac{1}{7}\sum_i \left(\sum_{jk} x_{ijk} - \frac{1}{3}\sum_{ijk} x_{ijk}\right)^2 = \frac{1}{7}\sum_i \left(7e_0 + 7e_i + \sum_j^7 e_j + \sum_k^7 e_k + \sum_{jk}^7 \varepsilon_{ijk}\right.$$

$$\left. - 7e_0 - \frac{7}{3}\sum_i e_i - \sum_j^7 e_j - \sum_k^7 e_k - \frac{1}{3}\sum_{ijk}^{21} \varepsilon_{ijk}\right)^2$$

$$= \frac{1}{7}\sum_i \left[\left(7e_i - \frac{7}{3}\sum_i e_i\right)^2 + \left(\sum_{jk}^7 \varepsilon_{ijk} - \frac{1}{3}\sum_{ijk}^{21} \varepsilon_{ijk}\right)^2\right]$$

The average value of this expression is simply

$$14\sigma_i^2 + 2\sigma^2$$

so that the mean square between days has average value $7\sigma_i^2 + \sigma^2$. To obtain the corrected cell or thermometer effects we may proceed by the method used for balanced incomplete blocks, taking for example:

$$\text{Cell I total} = 3(78) - 111 - 158 - 78 = -113$$

$$\text{Cell I effect} = \tfrac{1}{7}(-113) = -16.14$$

$$\text{Thermometer I total} = 3(111) - 78 - 74 - 200 = -19$$

$$\text{Thermometer I effect} = \tfrac{1}{7}(-19) = -2.71$$

Repeating the process for all cells and thermometers, we may form Table 8.42. Since the cell and thermometer effects are non-orthogonal the analysis of variance may be computed in two ways:

(*a*) In terms of thermometer totals and cell totals corrected for thermometers.

(*b*) In terms of cell totals and thermometer totals corrected for cells.

TABLE 8.42

Number	1 (I)	2 (II)	3 (III)	4 (IV)	5 (V)	6 (VI)	7 (VII)
Cell Totals	78	74	85	200	219	188	46
Corrected total	−113	−118	−113	175	199	163	−193
Corrected means	−16.14	−16.86	−16.14	25.00	28.43	23.29	−27.57
Thermometer totals	111	138	152	162	158	91	78
Corrected totals	−19	36	−17	21	−11	−35	25
Corrected means	−2.71	5.14	−2.43	3.00	−1.57	−5.00	3.57

Using method (*a*) we have

$$\text{S.S. thermometers} = \tfrac{1}{3}[(111)^2 + (138)^2 + \cdots + (78)^2] - \tfrac{1}{21}(890)^2$$

$$= 40{,}014.00 - 37{,}719.05$$

$$= 2294.95$$

$$\text{S.S. cells (corrected for thermometers)} = \frac{6}{(21)(3)(2)}[(113)^2 + (118)^2 + \cdots + (193)^2]$$

$$= 8262.19$$

and using (*b*)

$$\text{S.S. thermometers (corrected for cells)} = \frac{6}{(21)(3)(2)}[(19)^2 + (36)^2 + \cdots + (25)^2]$$

$$= 207.52$$

$$\text{S.S. cells} = \frac{1}{3}[(78)^2 + (74)^2 + \cdots + (46)^2] - \frac{1}{21}(890)^2$$

$$= 48{,}068.67 - 37{,}719.05$$

$$= 10{,}349.62$$

For either method the total of the two sums of squares is 10,557.14. The corrected sum of squares of the individual observations is

$$\text{Total S.S.} = [(19)^2 + (24)^2 + (35)^2 + \cdots + (14)^2 + (14)^2] - \frac{1}{21}(890)^2$$

$$= 11{,}050.95$$

and for the day effects

$$\text{S.S. days} = \frac{1}{7}[(334)^2 + (294)^2 + (262)^2] - \frac{1}{21}(890)^2$$

$$= 38{,}090.86 - 37{,}719.05$$

$$= 371.81$$

The residual sum of squares is therefore

$$11{,}050.95 - 371.81 - 10{,}349.62 = 279.52$$

The analysis of variance is shown in Tables 8.43 and 8.44.

TABLE 8.43

Source of Estimate	Sum of Squares	D.F.	Mean Square
Between days	371.81	2	185.6
Between thermometers	2,294.95	6	382.5
Between cells (corrected)	8,262.19	6	1377.6
Residual	279.52	6	46.59
Total	11,050.95	20	

TABLE 8.44

Source of Estimate	Sum of Squares	D.F.	Mean Square
Between days	371.81	2	185.6
Between thermometers (corrected)	207.52	6	34.59
Between cells	10,349.62	6	1724.9
Residual	279.52	6	46.59
Totals	11,050.95	20	

It is clear from the analysis that the differences between cells are the principal source of variation, and for the variance ratio between cells (corrected)/residual we have

$$\frac{1377.6}{46.59} = 29.6^{***} \qquad (F_{6,\,6,\,0.001} = 20)$$

which is highly significant. The day effect is almost significant at the 5% level, the variance ratio being

$$\frac{185.6}{46.59} = 4.0 \qquad (F_{2,\,6,\,0.05} = 5.1)$$

The uncorrected mean square between thermometers is also significant, but this is probably caused by the component of the cell effects which is present in this term.

The variance for a comparison of any two class means is

$$\frac{2k(t-1)\sigma^2}{N(k-1)} = \frac{6}{7}\sigma^2$$

so that the least significant difference between corrected class means is $t_{6,\,0.05}\sqrt{39.93} = 15.5$ and the cells B(IV, V, and VI) are clearly differentiated as of higher purity than the remainder. Cell C(VII) is probably less pure than the A or B groups. The variation between the 3 cells of either the A or B groups is so small as to indicate that the difference in the cell construction makes no appreciable contribution to the variation in the results.

8.7. Limiting the Number of Experimental Units

8.71. Introduction

To reduce the number of experimental units necessary for the investigation of a problem it is necessary to forego some of the information which the complete factorial experiment would have provided. If this information is already available, or if the corresponding effects are known to be negligible, it is usually possible to utilize a design in which either the effect is not estimated, or the estimate is used for some other purpose. The simplest examples of this approach have already been considered in the use of replicated randomized blocks, where the factors which produced differences between blocks were assumed to have no interactions with the factors involved within the blocks. Those comparisons which could have been employed to estimate these interactions were therefore employed as an estimate of the residual variance. A similar technique was employed in multifactor factorial designs, where the higher-order interactions of all factors were assumed absent and the corresponding estimates were used to estimate the residual variance. Although it is desirable in most experimental designs to obtain an estimate of the residual variance, it may sometimes happen that the residual variance is small, or can be assumed known, or that much larger effects are required for economic significance than for statistical significance. In this event a design may be used which provides no estimate of the residual variance.

If all the higher interactions of all factors seem likely to be small, it may happen that the corresponding comparisons are more numerous than we require for a useful estimate of the residual variance. The possibility of employing a fraction of the factorial design so chosen that all the data can be employed to estimate each of the effects which are believed to be present must then be considered.

8.72. Fractional Replication

The theory of fractional replication has been developed for experiments of the type p^k where p is a prime or prime power. We shall restrict our discussion of the subject to the 2^k design. The reader is referred to Kempthorne [117] for a more complete treatment.

A fractional replicate may consist of any block of a confounded 2^k design, but for the purpose of illustration we consider first a 2^4 experiment to be undertaken in 2 blocks by confounding the four-factor interaction. The signs for the coefficients in the 16 linear orthogonal combinations by means of which the effects are estimated are given in Table 8.45. The experimental units have been arranged in accordance with the proposed scheme of confounding, i.e., so that those experimental units occurring in a given block have a common sign in the combination leading to the estimate of $ABCD$. In the principal block all these signs are positive so that the generalized interaction of $ABCD$ with any effect has the same signs as the effect. In the second block the signs of the generalized interaction of an effect with $ABCD$ are the reverse of the corresponding signs

TABLE 8.45

Block	I								II							
Treatment Combination	*abcd*	*ab*	*ac*	*ad*	*bc*	*bd*	*cd*	(1)	*abc*	*abd*	*acd*	*bcd*	*a*	*b*	*c*	*d*
Effect I	+	+	+	+	+	+	+	+	+	+	+	+	+	+	+	+
A	+	+	+	+	−	−	−	−	+	+	+	−	+	−	−	−
B	+	+	−	−	+	+	−	−	+	+	−	+	−	+	−	−
C	+	−	+	−	+	−	+	−	+	−	+	+	−	−	+	−
D	+	−	−	+	−	+	+	−	−	+	+	+	−	−	−	+
AB	+	+	−	−	−	−	+	+	+	+	−	−	−	−	+	+
AC	+	−	+	−	−	+	−	+	+	−	+	−	−	+	−	+
AD	+	−	−	+	+	−	−	+	−	+	+	−	−	+	+	−
BC	+	−	−	+	+	−	−	+	+	−	−	+	+	−	−	+
BD	+	−	+	−	−	+	−	+	−	+	−	+	+	−	+	−
CD	+	+	−	−	−	−	+	+	−	−	+	+	+	+	−	−
ABC	+	−	−	+	−	+	+	−	+	−	−	−	+	+	+	−
ABD	+	−	+	−	+	−	+	−	−	+	−	−	+	+	−	+
ACD	+	+	−	−	+	+	−	−	−	−	+	−	+	−	+	+
BCD	+	+	+	+	−	−	−	−	−	−	−	+	−	+	+	+
ABCD	+	+	+	+	+	+	+	+	−	−	−	−	−	−	−	−

in the effect estimates. Thus, if the data are examined when only 1 block has been completed, each pair of effects which have $ABCD$ (the confounded effect) for their generalized interaction will be estimated by the same linear combination of treatment results. These pairs are then said to be *aliased*. For example, estimates of the main effects in the 2 blocks, considered separately, will be aliased with a three-factor interactions, thus

$$\text{Block I} \quad A + BCD = abcd + ab + ac + ad - bc - bd - cd - (1)$$

$$\text{Block II} \quad A - BCD = abc + abd + acd - bcd + a - b - c - d$$

If information is available from both blocks, then the sum and difference of the separate block comparisons lead to unaliased estimates of A and BCD.

In a 2^k experimental design in which any single effect is confounded, and one of the possible blocks is carried out, any two effects having the confounded effect for their generalized interaction will be aliased. It will usually be most convenient in this case to select the k factor interaction for confounding, since under these circumstances the main effects are aliased with the $(k - 1)$ factor interactions and the latter can frequently be assumed negligible. The advantages of fractional replication become apparent when 6 or more factors are involved. Under these circumstances, if the higher-order interactions are negligible, it is usually possible to estimate the main effects and two-factor interactions with some confidence, pooling the remaining degrees of freedom to provide a variance estimate. The analysis of variance for such an experiment might take the form of the accompanying table.

HALF REPLICATE OF 2^6 FACTORIAL DESIGN

Source of Estimate	D.F.
Main effects (aliased with 5-factor interactions)	6
Two-factor interactions (aliased with 4-factor interactions)	15
Residual (pooled 3-factor, aliased interactions)	10
Total	31

In experimental work where the tests will be carried out sequentially, and the blocks represent regions of time, a preliminary analysis after the completion of the first block might be used to indicate the approximate importance of the main effects and two-factor interactions. The following alternatives are then available:

(1) If none of the two-factor interactions is significant, but certain main effects are large, the test may be considered to have provided the desired information and be discontinued.

(2) If the results are insufficient to lead to satisfactory conclusions the remaining half replicate may be completed.

This procedure may have considerable advantages for technical work in which a preliminary indication of the conclusion from the first part of an experimental design could be used to determine the structure of the rest of the design. It is natural to attempt to extend the analysis of fractional replicates to cases in which a number of effects are confounded, and one-quarter or one-eighth of the total replicate is available for analysis.

In the case of a 2^4 design confounded in blocks of 4 we may suppose that the four-factor interaction and 2 two-factor interactions are to be confounded, e.g., $ABCD$, AB, and CD, and that the preliminary analysis is to be carried out when a half replicate is available. Any two of the blocks may be used for the half replicate, but it is most convenient to include those containing the experimental units which occur with a common sign in the $ABCD$ effect comparison. $ABCD$ is then termed the *defining contrast*, and, as previously, any 2 effects of which the generalized interaction is $ABCD$ are aliased. In particular, the two-factor interactions AB and CD are aliased and also confounded between blocks. The loss of the two-factor interactions is no great disadvantage in this instance since the remaining interactions of this order must be pooled as a variance estimate. A quarter replicate in this instance would be of little value since it involves only 4 tests, but it will be examined as an illustration of the method. The quarter replicate may be chosen to consist of any of the 4 blocks, the 3 confounded effects representing the defining contrasts. Any two effects of which the generalized interaction is one of these confounded effects will be aliased. It is easily seen that the effects are aliased in groups of 4 and that in the previous example these groups consist of (A, B, ACD, BCD), (C, D, ABC, ABD), (AC, BC, AD, BD), and (I, AB, CD, $ABCD$) so that main effects are aliased with main effects and no useful information can be recovered. If, however, we consider a 2^6 factorial experiment carried out in 4 blocks, it is possible to confound 3 four-factor interactions, or the six-factor interaction and 2 three-factor interactions. Adopting the first alternative and choosing $ABCD$, $CDEF$, and $ABEF$ as the defining contrasts, we find that the aliases are

(A, BCD, BEF, $ACDEF$) (B, ACD, AEF, $BCDEF$) (C, ABD, DEF, $ABCEF$) (D, ABC, CEF, $ABDEF$)
(E, CDF, ABF, $ABCDE$) (F, CDE, ABE, $ABCDF$) (AB, CD, EF, $ABCDEF$) (AC, BD, $ADEF$, $BCEF$)
(AD, BC, $ACEF$, $BDEF$) (AE, BF, $BCDE$, $ACDF$) (AF, BE, $BCDF$, $ACDE$) (BC, AD, $BDEF$, $ACEF$)
(CF, DE, $ABDF$, $ABCE$) (ACE, BDE, ADF, BCF) (BDF, ACF, BCE, ADE) (I, $ABCD$, $CDEF$, $ABEF$)

All the two-factor interactions are then aliased with at least one other two-factor interaction, but the lowest-order aliases of the main effects are three-factor interactions. If all interactions involving 3 or more factors can be assumed to be negligible, the main effects may be estimated, but only 2 degrees of freedom will be available for the residual estimate. In order to provide satisfactory significance tests for the main effects it is necessary to assume that the two-factor interactions are also negligible.

EXAMPLE. An example of the analysis of a fractionally replicated experiment carried out in a penicillin production unit has been published [44]. As a further illustration we consider the information available from 4 blocks of a complete 2^6 experiment originally confounded in 8 blocks of 8. The defining contrast is selected as $ABCD = \mathrm{I}$, and the other confounded effects are $ABEF$, $CDEF$, ACE, BDE, BCF, and ADF. We may select the 32 treatment combinations to be considered so that they have an even number of letters in common with $ABCD$, and derive the principal block by determining the 8 treatments which have also an even number of letters in common with the 6 other confounded effects. The resulting design is given in Table 8.46, the remaining blocks being obtained by multiplying the treatment combinations of the first block by e, f, and ef, respectively. The location of the experimental units within a given block are, of course, randomized.

TABLE 8.46

HALF REPLICATE 2^6 FACTORIAL IN BLOCKS OF EIGHT

Block 1		Block 2		Block 3		Block 4	
(1)	1.25	*e*	0.85	*f*	0.20	*ef*	0.20
ade	2.40	*ad*	1.50	*adef*	2.15	*adf*	1.85
bce	1.35	*bc*	0.80	*bcef*	0.70	*bcf*	1.25
bdf	0.15	*bdef*	1.15	*bd*	1.90	*bde*	1.40
acf	1.90	*acef*	1.90	*ac*	2.00	*ace*	2.10
abcd	1.95	*abcde*	2.35	*abcdf*	2.60	*abcdef*	1.55
cdef	1.00	*cdf*	1.55	*cde*	1.70	*cd*	1.30
abef	2.65	*abf*	1.90	*abe*	2.30	*ab*	1.75

The data quoted in Table 8.46 refer to the yields of alfalfa obtained during the investigation of the effects of trace elements on plant growth [67]. The factors involved are those considered in the previous discussion of these data in example 4 of Section 8.63. The analysis is carried out by ignoring the effects of 1 factor and computing the effects of the remaining 5 factors as though the experiment were a completely replicated 2^5 design. In this case any of the factors *a*, *b*, *c*, or *d* may be ignored and we select the last. The resulting analysis is given in Table 8.47.

In addition to the confounded effect there are 10 groups of aliases which contain only three-, four-, and five-factor interactions. If these are assumed negligible, the corresponding components of the sum of squares may be pooled to provide a residual variance estimate 0.22575. Since the defining contrast is a four-factor interaction, 3 two-factor interactions are aliased with two-factor interactions and cannot be examined. For the remainder a variance estimate above 1.12 or below 0.00097 is required to indicate significance at the 5% level ($F_{1,\,10,\,0.05} = 5.0$) ($F_{10,\,1,\,0.05} = 235$). The upper value is not approached by any of the estimates, and only the *DE* interaction estimate is significantly low. Since this represents 1 estimate from 11 in what is in fact a two-sided test giving an effective probability level of 10%, we may reasonably conclude that those two-factor interaction estimates which can be examined show no evidence of the existence of real effects. If it appeared desirable, we might assume these effects to be zero and pool the estimates with the residual before proceeding to test the main effects. In this instance the variance estimate was so obtained and the critical values at the 5% level for the main effect mean square estimates are 0.846 and 0.00078. Only the factor *A* is marked as significant.

The technique of fractional replication is a post World War II development in experimental design, and few examples of its use are available in

TABLE 8.47

Treatment	Observation	3	4	5	6	7	Effect	Alias	(Effect)²/32
(1)	1.25	2.75	6.40	12.45	26.90	49.60	I	*ABCD*†	76.88000
a(*d*)	1.50	3.65	6.05	14.45	22.70	16.10	*A*	*BCD*	8.10031
b(*d*)	1.90	3.30	6.95	11.40	5.80	1.90	*B*	*ACD*	0.11281
ab	1.75	2.75	7.50	11.30	10.30	0.60	*AB*	*CD*	0.01125
c(*d*)	1.30	3.25	4.10	1.95	0.70	2.40	*C*	*ABD*	0.18000
ac	2.00	3.70	7.30	3.85	1.20	−2.70	*AC*	*BD*	0.22781
bc	0.80	3.80	6.15	5.10	0.00	−3.70	*BC*	*AD*	0.42781
abc(*d*)	1.95	3.70	5.15	5.20	0.60	3.40	*ABC*	*D*	0.36125
e	0.85	2.05	0.10	0.35	0.20	1.90	*E*	*ABCDE*	0.11281
a(*d*)*e*	2.40	2.05	1.85	0.35	2.20	2.00	*AE*	*BCDE*	0.12500
b(*d*)*e*	1.40	3.45	2.45	0.40	0.70	0.40	*BE*	*ACDE*	0.00500
abe	2.30	3.85	1.40	0.80	−3.40	−1.70	*ABE*	*CDE*	0.09031 *
c(*d*)*e*	1.70	2.35	3.40	0.05	−2.00	−3.30	*CE*	*ABDE*	0.34031
ace	2.10	3.80	1.70	−0.05	−1.70	−2.80	*ACE*‡	*BDE*‡	0.24500
bce	1.35	2.90	3.45	1.10	2.10	−1.60	*BCE*	*ADE*	0.08000 *
abc(*d*)*e*	2.35	2.25	1.75	−0.50	1.30	−0.10	*ABCE*	*DE*	0.00031
f	0.20	0.25	0.90	−0.35	2.00	−4.20	*F*	*ABCDF*	0.55125
a(*d*)*f*	1.85	−0.15	−0.55	0.55	−0.10	4.50	*AF*	*BCDF*	0.63281
b(*d*)*f*	0.15	0.70	0.45	3.20	1.90	0.50	*BF*	*ACDF*	0.00781
abf	1.90	1.15	−0.10	−1.00	0.10	0.60	*ABF*	*CDF*	0.01125 *
c(*d*)*f*	1.55	1.55	0.00	1.75	0.00	2.00	*CF*	*ABDF*	0.12500
acf	1.90	0.90	0.40	−1.05	0.40	−4.10	*ACF*	*BDF*	0.52531 *
bcf	1.25	0.40	1.45	−1.70	−0.10	0.30	*BCF*‡	*ADF*‡	0.00281
abc(*d*)*f*	2.60	1.00	−0.65	−1.70	−1.60	−0.80	*ABCF*	*DF*	0.02000
ef	0.20	1.65	−0.40	−1.45	0.90	−2.10	*EF*	*ABCDEF*	0.13781
a(*d*)*ef*	2.15	1.75	0.45	−0.55	−4.20	−1.80	*AEF*	*BCDEF*	0.10125 *
b(*d*)*ef*	1.15	0.35	−0.65	0.40	−2.80	0.40	*BEF*	*ACDEF*	0.00500 *
abef	2.65	1.35	0.60	−2.10	0.00	−1.50	*ABEF*‡	*CDEF*‡	0.07031
c(*d*)*ef*	1.00	1.95	0.10	0.85	0.90	−5.10	*CEF*	*ABDEF*	0.81281 *
acef	1.90	1.50	1.00	1.25	−2.50	2.80	*ACEF*	*BDEF*	0.24500 *
bcef	0.70	0.90	−0.45	0.90	0.40	−3.40	*BCEF*	*ADEF*	0.36125 *
abc(*d*)*ef*	1.55	0.85	−0.05	0.40	−0.50	−0.90	*ABCEF*	*DEF*	0.02531 *

* Pooled as variance estimate. † Defining contrast. ‡ Confounded.

the literature. It appears better suited for industrial investigations in which the time and duration of the tests are under the control of the experimenter than for agricultural work where a decision to complete the replicate would often involve a delay of a year. Finney [68] [69] has described methods for designing fractional replicate experiments of the 2^k and 3^k types with additional confounding to reduce the size of block involved; and Kempthorne [70] has extended the treatment to designs of the type p^k, p being any prime. In a second paper by this author [71] the relationship between fractional replication and confounded complete replicates in which block $\times$ treatment interactions occur is considered.

It is easily seen that a 2^4 design in which the $ABCD$ interaction is confounded between blocks consists of the two sets

$$(1) \quad ab \quad ac \quad ad \quad bc \quad bd \quad cd \quad abcd$$

$$ae \quad be \quad ce \quad de \quad abce \quad abde \quad acde \quad bcde$$

if e represents the "upper" level of the block factor. The aliased effects between the two blocks are therefore $ABCD = E$, and the whole experiment may be regarded as a half replicate of a 2^5 experiment, the other half of which could be completed by interchanging the blocks with respect to the tests. In the available half replicate we have aliased

$I = ABCDE$	$AB = CDE$	$ABC = DE$
	.	.
$E = ABCD$	.	.
	.	.
$A = BCDE$	.	.
. .	.	.
. .	.	.
. .	.	.
$D = ABCE$	$CD = ABE$	$BCD = AE$

Thus a large value of 1 of the three-factor interactions may be caused by the interaction of blocks and a single factor. This interpretation of the data should be considered when the technology of the experiment suggests that interactions between blocks and factors may be present. In cases where the original experiment was confounded in 4 blocks we may employ a similar method of interpretation, the original 2^k experiment being regarded as a quarter replicate of a 2^{k+2} design.

8.73. The Use of Latin Squares and Hypersquares

If we consider the 2×2 Latin square in which the rows, columns, and treatments each correspond to the two levels of a different factor, we obtain

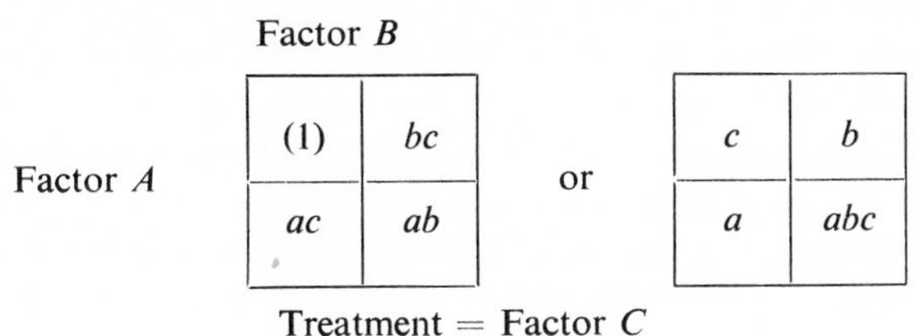

It is apparent that either of these squares constitutes a half replicate of the 2^3 experiment involving factors A, B, and C and that in any one of the Latin squares we have the aliased effects $A = BC$, $B = AC$, $C = AB$, corresponding to the 3 available degrees of freedom. If we assume that the treatment effects do not interact, these comparisons are unequivocal. In the case of the 3×3 Latin square the design can be shown to be the equivalent of a one-third replicate of a 3^3 design, in which the 2 comparisons of the W factor are confounded between blocks, and, only 1 of the blocks is carried out, so that we have the aliases

$$A = J_{BC} = Z \qquad J_{AB} = J_{AC} = \mathrm{I}_{BC}$$

$$B = \mathrm{I}_{AC} = Y \qquad W = \mathrm{I}$$

$$C = \mathrm{I}_{AB} = X$$

The corresponding Graeco-Latin square is the equivalent of a one-ninth replicate of a 3^4 design. It will be seen, therefore, that in using a Graeco-Latin square or hypersquare design we are carrying out the equivalent of the smallest fractional replication in which the main effects are not aliased one with another. If we use a square of side n to study less than $n + 1$ different treatments, the components which we employ as estimates of the residual correspond to a series of aliased interactions. Since there is no Graeco-Latin square of side 2 it follows that a quarter replicate of a 2^4 factorial design will either have main effects aliased with main effects, e.g., by selecting (I, AB, CD, $ABCD$) as the confounded subgroup, or will have 1 or more main effects confounded. In the latter case the design will in effect be a half replicate of a 2^3 design or a full replicate of a 2^2 design.

If we consider the orthogonal cube of side 2, $(2^3 - 1)/(2 - 1) = 7$ alphabets can be superimposed so that 2^4, 2^5, 2^6, and 2^7 factorials can be fractionally replicated in 8 experimental units without aliasing main effects with main effects. For the 3^k series 4 alphabets can be superimposed on the 3×3 square so that a one-third replicate of a 3^3 factorial or a one-ninth replicate of a 3^4 factorial will not involve mutual aliasing of main effects. Since $(3^3 - 1)/(3 - 1) = 13$ alphabets can be inserted in a cube of side 3 it follows that all the 3^k designs from $k = 5$ to $k = 13$ can be fractionally replicated in 27 experimental units with the main effects preserved. Thus in employing a Latin, Graeco-Latin, or hyper square in order to reduce the number of experimental units necessary for the experiment we must (*a*) randomize the selection of the square; (*b*) appreciate that apparent main effects may be attributable to interactions of other factors.

In spite of the considerable number of aliases which arise from hyper squares of sides 4, 5, and 7, these designs may be useful in practice where the effects of the factors involved are known to be almost additive. In Section 8.56 the possibility of employing a 3×3 Latin square with replication to study the effects of analysts, equipment, and standards on an analytical method was considered. In this instance the replications were to be carried out on different days, and the interaction effects of days with analysts, equipment, and solutions could be isolated if this seemed desirable. Thus in place of replications we might have substituted any other factor that we wished to investigate at a number of levels. If this factor interacted with some or all of those of the Latin square, the corresponding components could be isolated. If no interactions were present, these components might be pooled with the residual. In experiments where low-order interactions between certain factors appear possible, and the factors are to be tested at 2 levels, it is possible to use a 4×4 Graeco-Latin or hyper square, pairing the factors between which two-factor interactions seem likely, and repeating the square at different levels of any factor for which the three-factor interactions may be anticipated. Such an application depends upon the possibility of selecting factors, or groups of factors, for which the interaction effects may be assumed negligible, so that we can safely identify an effect among its aliases.

A situation which differs slightly arises when most of the factors must be considered to interact so that a factorial design is essential, but 1 or more factors can be considered additive in effects. In this case it may be possible to confound certain of the higher-order interactions in the design with the different levels of the additive factors. Thus in a 3^3 design where the three-factor interactions could be assumed negligible the W set of these interactions could be confounded with a fourth three-level factor, and if necessary multiple confounding could be used to introduce additional three-level factors in place of the X, Y, and Z sets of three-factor interactions.

EXAMPLE. The following data were obtained in connection with an investigation of the errors in a new analytical determination. The determination consisted of a catalytic reduction to a standard valency condition in which the quantity of catalyst and reducing agent and the time for reduction were each controlled at 4 levels. The material was oxidized to a higher valency state, 4 levels of oxidizing agent being employed, and then extracted by 4 concentrations of a standard reactant. It was felt that interactions between the first 3 factors were possible, although the three-factor interactions could be neglected. The fourth and fifth factors were considered to affect the results additively, and were introduced in place of 2 of the three-factor interactions. This was achieved in this case by superimposing a 4 × 4 Graeco-Latin square on the 4 × 4 design at the level T_1, and using a cyclic permutation of the rows of the square for each additional level of T. This results in the design shown in Table 8.48. Before carrying out the experiment, the subscripts on the C's, R's, and T's should have been randomized, but,

TABLE 8.48

	T_1				T_2			
	C_1	C_2	C_3	C_4	C_1	C_2	C_3	C_4
R_1	O_1E_1 345	O_2E_2 2004	O_3E_3 2118	O_4E_4 2086	O_2E_3 2111	O_1E_4 1917	O_4E_1 905	O_3E_2 2071
R_2	O_2E_3 1663	O_1E_4 1805	O_4E_1 402	O_3E_2 2023	O_3E_4 2095	O_4E_3 2138	O_1E_2 540	O_2E_1 314
R_3	O_3E_4 2123	O_4E_3 2124	O_1E_2 204	O_2E_1 424	O_4E_2 1938	O_3E_1 1387	O_2E_4 1896	O_1E_3 861
R_4	O_4E_2 1979	O_3E_1 984	O_2E_4 1889	O_1E_3 1337	O_1E_1 86	O_2E_2 672	O_3E_3 2061	O_4E_4 2051

	T_3				T_4			
	C_1	C_2	C_3	C_4	C_1	C_2	C_3	C_4
R_1	O_3E_4 2111	O_4E_3 2040	O_1E_2 734	O_2E_1 1464	O_4E_2 1998	O_3E_1 1441	O_2E_4 2091	O_1E_3 1744
R_2	O_4E_2 1976	O_3E_1 1334	O_2E_4 1908	O_1E_3 1269	O_1E_1 574	O_2E_2 1709	O_3E_3 2110	O_4E_4 2111
R_3	O_1E_1 318	O_2E_2 1263	O_3E_3 2079	O_4E_4 1956	O_2E_3 1856	O_1E_4 1158	O_4E_1 1032	O_3E_2 1926
R_4	O_2E_3 1803	O_1E_4 1672	O_4E_1 1027	O_3E_2 2051	O_3E_4 1955	O_4E_3 2120	O_1E_2 396	O_2E_1 146

owing to a misunderstanding, this procedure was omitted. The significance tests were therefore invalidated, if real interaction effects exist where they have been assumed absent. For the sake of illustration the data will be analyzed on the assumption that all interaction effects except $C \times R$, $R \times T$, and $C \times T$ are zero.

In the table the subscripts are proportional to the amount or concentration of the factor, and it was anticipated that the treatment effects, if significant, would show a plateau at the higher levels, so that the 3 independent comparisons for each factor were to be taken in the form $(3x_1 - x_2 - x_3 - x_4)$, $(2x_2 - x_3 - x_4)$ and $(x_3 - x_4)$. The analysis of variance is carried out in the usual way, subtables being composed by summing over C, R, and T and then over CR, TR, and CT. The sums of squares due to the main effects of C, R, and T and the interactions CR, TR, and CT may then be computed and the residual sum of squares obtained by difference. From this residual, 6 comparisons corresponding to the effects of O and E may be extracted by forming the sums over O_1, O_2, O_3 and O_4, and E_1, E_2, E_3, E_4 in Table 8.48. These sums are then squared, corrected for the mean value, divided by 16, and entered in the sums of squares column of the table. The true residual based on $27 - 6 = 21$ degrees of freedom may then be determined by difference. The analysis of variance is shown in Table 8.49 where the individual comparisons for the main effects have not been separated. The effect of oxidant quantities and extractant concentrations is highly significant.

TABLE 8.49

ANALYSIS OF VARIANCE

Source of Estimate	Sum of Squares	D.F.	Mean Square
Between catalyst concentrations (C)	676,275	3	225,425
Between reducing times (T)	143,704	3	47,901
Between reductant quantities (R)	960,331	3	320,110
$C \times R$	532,694	9	59,188
$T \times R$	1,232,531	9	136,948
$C \times T$	413,449	9	59,188
Between oxidant quantities (O)	8,241,401	3	2,747,134
Between extractant concentrations (E)	13,500,104	3	4,500,035
Residual	1,856,348	21	88,398
Total	27,556,837	63	

$(F_{9,\,21,\,0.05} = 2.37)$ $(F_{3,\,21,\,0.05} = 3.07)$

The effect of reductant quantity is significant at the 0.05 level, and the data suggests that some small effects due to catalyst concentration may be present. The totals for these effects are:

C_1	C_2	C_3	C_4	R_1	R_2	R_3	R_4
24931	25768	21392	23834	27180	23971	22545	22229
O_1	O_2	O_3	O_4	E_1	E_2	E_3	E_4
14960	23213	29869	27883	12183	23484	29434	30824

and the analysis in terms of comparisons corresponding to individual degrees of freedom is left as an example for the reader.

8.74. Other Types of Design

By the use of the techniques of fractional replication which have been considered in this section it is often possible to arrive at a design which appears suitable for a particular investigation. When the standard types of design appear unsuitable it may be possible to formulate a series of alternative designs which, although not belonging to any general class, appear to meet the experimental requirements in a particular instance. It is then advisable to write out a model for the data which it is proposed to obtain, incorporating in the model any assumptions which are justified by technical experience in the field concerning the effects or interactions. The average values of the various experimental comparisons which are to be made can then be considered in advance to ensure that unbiased estimates of those model parameters which are of interest will be available from the data.

EXAMPLE. As an example we consider an experiment [115] in which it was required to determine the lead content of a series of 18 tin castings which were to be employed as standard samples. The specimens consisted of bars, cast vertically, and the experimental determinations, which were to be spectrographic, were to be designed to detect differences in lead content between the top and bottom of the specimens, and between the 18 standard specimens. In particular, since the specimen numbers represent the order of pouring, it seemed possible that some systematic trend might be observed when the relationship between lead content and specimen number was considered. Appreciable variations among the photographic plates were anticipated, and it was not possible to obtain more than 6 spectra on a single plate. It was considered reasonable to assume that interaction effects between specimens, positions, and plates were absent.

If all the specimens are to be analyzed, and the mean difference between top and bottom of the specimens with respect to lead content is to be determined, 36 appears to be the minimum possible number of experimental units. This will involve 6 plates, and since differences between plates must be eliminated 30 degrees of freedom will then be available for the various comparisons and the residual variance estimate. Table 8.50 gives the selected design together with the experimental determinations

TABLE 8.50

Plate / Casting	I	II	III	IV	V	VI
1	0.064*T*			0.062*B*		
2		0.068*T*			0.065*B*	
3			0.063*T*			0.090*B*
4	0.058*B*			0.065*T*		
5		0.069*B*			0.065*T*	
6			0.061*B*			0.095*T*
7	0.065*T*			0.059*B*		
8		0.071*T*			0.062*B*	
9			0.059*T*			0.089*B*
10	0.063*B*			0.061*T*		
11		0.067*B*			0.065*T*	
12			0.066*B*			0.092*T*
13	0.060*T*			0.067*B*		
14		0.070*T*			0.065*B*	
15			0.064*T*			0.089*B*
16	0.063*B*			0.059*T*		
17		0.066*B*			0.070*T*	
18			0.060*B*			0.092*T*

Sampling point: T = Top; B = Bottom.

obtained. It is apparent that we could elect to carry out determinations on samples from the top and bottom of 3 specimens on one plate, or that 1 observation from each of 6 specimens could be recorded on a single plate. Since we are not concerned with the position $\times$ specimens interaction but wish to retain as many comparisons between specimens as possible, the second alternative seems preferable. If we represent the data by a model of the form

$$x_{ijk} = e_0 + e_i + e_j + e_k + \varepsilon_{ijk}$$

$$i = 1, \cdots, 18 \quad j = b \text{ or } t \quad k = \text{I}, \cdots, \text{VI}$$

assuming ε_{ijk} to be randomly and normally distributed with variance σ^2, we have

$$\frac{1}{6} \text{ ave } (\Sigma_{ij} x_{ij\text{I}} - \Sigma_{ij} x_{ij\text{IV}}) = e_{\text{I}} - e_{\text{IV}}$$

together with similar expressions for $e_{\text{II}} - e_{\text{V}}$ and $e_{\text{III}} - e_{\text{VI}}$. Also

$$\frac{1}{6} \text{ ave } (\Sigma_i x_{it\text{I}} + \Sigma_i x_{it\text{IV}} - \Sigma_i x_{ib\text{I}} - \Sigma_i x_{ib\text{IV}}) = e_t - e_b$$

and, if we take the sum of the determinations on a particular casting, unbiased comparisons between castings within the groups (1, 4, 7, 10, 13, 16), (2, 5, 8, 11, 14, 17), and (3, 6, 9, 12, 15, 18) may be made but differences between these groups are confounded with differences between the groups of plates (I, IV), (II, V), and (III, VI). The existence of a fluctuation of so short a period in the casting purity is extremely unlikely, so that the lost information is of little consequence.

The analysis of variance may be carried out separately for the 3 pairs of plates, and the results may then be collected in a joint analysis. These are shown in Tables 8.51 and 8.52. In the latter the first mean square includes a component of differences between castings, owing to the confounding of certain casting comparisons with the differences between the sets of plates. The second mean square derived from comparisons between plates which are independent of casting or positional effects is significant at the 0.001 level. The remaining mean squares are not significantly high, but the differences between castings are significantly low at the 0.01 level. This suggests that the assumption of independent random components may have been in error. If interaction effects between position and casting were present, the metal being uniform when poured, but segregating to different extents in different molds, the residual estimate would include a component due to this interaction, which would not be present in the mean square between castings.

TABLE 8.51

Source of Estimate	Sum of Squares			D.F.	Mean Square		
	I and IV	II and V	III and VI		I and IV	II and V	III and VI
Between plates	0.0	30.1	2523.0	1	0.0	30.1	2523.0
Between positions	0.4	18.7	8.4	1	0.4	18.7	8.4
Between castings							
Linear	0.9	4.1	0.2	1	0.9	4.1	0.2
Remainder	7.8	1.3	29.5	4	1.95	0.33	7.4
Total	8.7	5.4	29.7	5	1.74	1.08	5.94
Residual	78.6	26.7	23.6	4	19.6	6.7	5.9
Total	87.7	80.9	2584.7	11			

TABLE 8.52
JOINT ANALYSIS OF VARIANCE

Source of Estimate	Sum of Squares	D.F.	Mean Square
Between sets of plates	1311.6	2	656.8
Between plates (within sets)	2553.1	3	851.0
Between positions (all sets)	20.3	1	20.3
Between positions (within sets)	7.1	2	3.55
Between castings			
Linear components	0.1	1	0.1
Between linear components	5.1	2	2.6
Remainder	38.6	12	3.2
Total	43.8	15	2.9
Residual	128.9	12	10.7
Total	4064.8	35	

The number of observations can also be reduced in those instances where the objective of the investigation is the determination of a combination of conditions for which the dependent variable has a maximum or minimum value. Subsequently we shall suppose that the value sought is a maximum, and that the objective is to locate this value and to determine the nature of the response surface in the immediate neighborhood. This calls for a technique for proceeding from an arbitrary starting point towards a region of increased response until the maximum can be located somewhere within a subregion of suitable size. The subregion can then be further investigated. Such a procedure has the advantage of being sequential in operation, but since the whole experimental region is not investigated, when applied to a complicated response surface having more than one local maximum, it might lead to an investigation of the wrong peak. In most chemical investigations this is probably not a serious criticism.

Two techniques have been proposed, that formalized by Friedman and Savage [126] in which the independent variables are changed in level one by one and the maximum slope technique due to Box and Wilson [127]. In the former with all but one of the factors held constant the conditional maximum for this factor is located experimentally and in subsequent tests the factor is maintained at the level associated with this maximum. When all factors have been investigated in this way the cycle may be repeated. In the maximum slope method the procedure depends upon the nature of the yield surface in the neighborhood of the starting point. If this point is far removed from a stationary point a linear approximation is often adequate and the line of maximum slope is determined from a fractional 2^k experiment in which either main effects or main effects and 2 factor interactions are not mutually aliased. A new experimental region is then selected some way along the line so determined, and a further fractional replicate is carried out to determine the line of maximum slope in the new region. One of the technical problems associated with this procedure is that of choosing suitable scales for the factors involved since the direction of maximum slope depends upon these scales although under many conditions costs could be employed to provide a common scale. As the maximum is approached the linear approximation will become progressively worse and the value of the slope will be reduced so that random errors in the observations will lead to increased uncertainty in the determination of the maximum slope direction. When this stage is reached composite designs, not of the factorial type, are proposed for determining interactions and quadratic components of main effects. For further details of these methods the reader is referred to [127] and to [128] where a discussion of fractional experiments, not of the factorial form, which are suitable for maximum slope determination will be found.

8.8. Combined Experiments

8.81. General Discussion

In agricultural experimentation the results of a number of similar experimental designs carried out in different counties or states and in different years have been collected and subjected to a joint or combined analysis. Where the use of different land areas and different seasons was an integral part of the overall experimental design we may speak of a combined experiment. We have already seen that the range of validity of experimental conclusions depends to a large extent upon the methods employed in selecting those classes to be examined in the experiment from the population of classes which might have been examined. It is understandable that the agricultural experimenter would wish to increase the range of validity of his results by examining the effects of different soils and seasons, but it must be noted that, if this extension is to have statistical validity, the population must be defined and the sampling procedure must involve a random element.

In the chemical industry raw materials have a role similar to that of the soil in agriculture, and, although climatic conditions are less important, experience shows that a number of other factors which change with time may contribute to differences in experimental results. If the conclusions drawn from such results are to be applied to a population of possible raw materials, all of which are available for sampling at the time of the experiment, it will be desirable to draw a random sample from the population for examination. Such a procedure may not be feasible in practice, and we may be able from previous experience to examine what we believe to be a representative group of raw materials. In that event the experimental conclusions from a statistical analysis apply only to the materials examined, and the extension to the population is entirely based on personal judgment. Such an arrangement may represent a practical solution to the problem, but the alternative of random sampling is theoretically available.

The effect of unidentified factors which change with time cannot be offset by a random sampling procedure since the period in which the experimental conclusions are to be applied cannot be sampled. We are therefore obliged to depend upon experience in applying the results of past experiments to future or present problems. Such experience indicates that as a greater proportion of the time-dependent factors are identified and controlled the residual variations tend to behave more and more as independent random components. It also suggests that in most cases where the residual variations are not independent successive effects will be positively correlated so that today's results give a more accurate indication of the results which might be obtained tomorrow than of those which might be obtained a year or a decade hence. Observations which are widely spaced in time will tend to behave as independent measurements and to give a more accurate estimate of the true population variance. The methods of statistical quality control to be considered in Chapter 10 are for the most part designed to detect important fluctuations caused by variations in time-dependent variables in a continuous production unit, and to distinguish between such effects and the host of minor effects which are acceptable as "random" variations. In research or development work an equivalent system would lead to a constant repetition of experimental measurements which could rarely be justified.

In considering the design of combined experiments it is convenient to examine separately the populations which are available for sampling by a random method and those, such as the effects of time-dependent variables, which cannot be randomly sampled.

8.82. Raw Materials

We shall consider the raw materials of a process as an example of the first type. If the production of raw materials at the various sources which constitute the population to be sampled is subject to statistical quality control, then the population will not change with time; otherwise we may regard the population as that existing at the time of the experiment. The actual specimens of raw material to be investigated may be selected by a completely random process, or by a stratified sampling scheme such as that adopted in the example of Table 7.38, which may be regarded as a 2^2 design carried out on 32 different samples. In either event the combined analysis should be carried out as though the raw materials classification were completely crossed with the remaining factors of the design, since materials $\times$ treatments interactions may be present, and the whole of the experimental program should be randomized. If the experiments on the different classes of raw material are carried out in sequence, variations due to raw materials will not be distinguished from variations due to time effects.

In considering the population of effects due to raw materials in the combined analysis, it is important to define the population which is to be sampled. Other things being equal, it is usual to find that those who operate a plant will obtain the best and most consistent results with the raw material which they are accustomed to process. Thus, if raw materials are tested in different places by different operators, we may be considered to have sampled the population of effects due to the combination (raw material + consumer skill). If, recognizing this difficulty, we design an experiment in which raw materials and consumers are treated as separate crossed classifications, these factors may be expected to interact, and the analysis may be rendered difficult by a complicated heterogeneity of the residual variance.

8.83. Time Effects and Sequential Experiments

In Section 8.71 it was pointed out that a complete replicate of an experiment confounded in blocks could be considered a fractional replicate of a large experiment if treatment $\times$ block interactions were present. In many experiments where a period of time constitutes the experimental block the resultant aliases lead to some uncertainty unless these interactions are known to be negligible. The experimental program which involves a series of experiments or parts of experiments not carried out in a completely random order will always be subject to this uncertainty. The many-factor experiment in which all possible variables are considered appears to offer the only logical solution to this problem, but in practice it must be recognized that such an experiment, even if carried out as a fractional replicate, is often too costly, and presupposes the ability of the experimenter to list the important variables from previous experience. What appears to be required is a sequence of experiments, each involving a small number of factors. On the basis of the results obtained from these experiments the more important variables would be identified and their interactions determined in a single factorial experiment. In order to minimize the incidence of time effects in the preliminary work the factors not examined in a particular experiment should be controlled, or their effects eliminated by covariance analysis. Gore [116] has described a sequence of experiments carried out to determine reaction rates in polymerization. In this instance some 13 variables were listed as possibly important, and, of these, those combinations of 3 between which interaction effects were considered probable were incorporated in half-replicate 2^3 designs (2×2 Latin squares). The Latin squares were carried out in duplicate, and those cases where 2 of the factors involved had significant effects the remaining half replicate of the 2^3 factorial was carried out in duplicate so that interaction effects could be examined. After a series of experiments of this type those of the original 13 variables which appeared important were to be incorporated in a multifactor design.

This method of procedure serves to illustrate a general difficulty which arises in sequences of experiments. The experimenter requires flexibility of design so that preliminary results may be used to determine the pattern of subsequent work, as in the decision to carry out the second half replicate of the 2^3 design. At the same time he requires that, when the experimental program is completed, the data obtained shall be analyzed collectively. Unfortunately these properties appear incompatible, since the significance tests in the joint analysis must be biased if the decision to carry out the second half replicate is based on results obtained in the earlier work.

CHAPTER 9

Analysis of Counted Data

9.1. Introduction

The preceding chapters have dealt primarily with the analysis of data obtained by the measurement of continuously variable quantities which could be represented mathematically by continuous distribution functions. Situations frequently arise—less often, perhaps, in chemical than in other applications—where characteristics can only be evaluated qualitatively in terms of certain attributes which the objects in the sample do or do not possess. The predominant characteristic of such data is that we count rather than measure. Typical examples of data of this type are the enumeration of such characteristics as eye color, or the qualitative evaluation of color development in terms of a series of standards. Often, as in the latter case, it would be possible to obtain a quantitative measurement by some more refined technique, and in this case that technique is generally to be preferred, unless it is impracticable for economic or technical reasons.

The distributions usually used to represent qualitative data, which are by nature discrete, are the binomial and the associated multinomial and Poisson distributions. Hence the first part of our study of the sampling of attributes will concern problems of estimation and significance testing with respect to these distributions. We shall then consider the application of the "goodness of fit" criterion, based on the χ^2-distribution, developed by Karl Pearson. Finally, we shall touch upon the problem of acceptance sampling by attributes, which is one of the most common applications of statistical analysis to counted data.

9.2. Transformation of Counted Data

Although the tables of the binomial and Poisson distributions referred to in Sections 4.61 and 4.63, and extensive tables of $n!$ and $\binom{n}{x}$, are now available, the computations involved in exact tests of counted data based on the binomial and Poisson distributions may become tedious when n and x are large. Also we have seen that in using analysis of variance techniques one of the important assumptions is that the variance be constant for all observations, regardless of the variation in the average values of the observations. One way of overcoming both these difficulties is to develop transformations which will give approximate percentage points of the binomial and Poisson distributions, and which will stabilize the variance of data with these distributions.

The limiting forms of the binomial and Poisson distributions given in Section 4.64 can be used to approximate the $Pr(x \leq x_0)$, where x has the binomial or Poisson distribution, when λ, or np, $p < q$, is large (say 50 or more). A transformation which will give satisfactory approximations for values of λ, or np, $p < q$, as low as 5 has been given by Freeman and Tukey [62, 88]. For the binomial distribution this approximation is

$$(a) \qquad Pr(x \leq x_0) \sim Pr(t < 2[\sqrt{(x_0 + 1)(1 - p)} - \sqrt{(n - x_0)p}])$$

where t has the unit normal distribution; for the Poisson distribution this reduces to

$$(b) \qquad Pr(x \leq x_0) \sim Pr(t < 2[\sqrt{x_0 + 1} - \sqrt{\lambda}])$$

Tukey and Mosteller [61] have developed a special type of graph paper based on (*a*) which enables the quick graphical solution of problems involving simple dichotomies and other related problems.

For the stabilization of variance many transformations have been suggested (see Section 7.45). Freeman and Tukey (see above references) have also shown that, as long as λ, or np, $p < q$, is greater than unity, the averaged angular transformation

$$(c) \qquad \sin^{-1}\sqrt{\frac{x}{n+1}} + \sin^{-1}\sqrt{\frac{x+1}{n+1}}$$

will have a variance very nearly $\dfrac{1}{n + \frac{1}{2}}$ (in radians); for Poisson data this is replaced by

$$(d) \qquad \sqrt{x} + \sqrt{x+1}$$

with variance very nearly equal to unity. If the transformation (*c*) is used, the average value of the transformed data will be approximately $2 \sin^{-1}\sqrt{p}$; if transformation (*d*) is used, the average value will be approximately $\sqrt{4\lambda + 1}$.

EXAMPLE 1. For the first example given in Section 4.64, we have, using (a),

$$Pr(x \leq 830) = Pr(t < 2\,[\sqrt{(831)\tfrac{1}{2}} - \sqrt{(770)\tfrac{1}{2}}])$$

$$= Pr(t < 2\,[20.3838 - 19.6214])$$

$$= Pr(t < 1.525)$$

and in the second, using (b),

$$Pr(x \leq 10{,}100) = Pr(t < 2\,[\sqrt{10{,}101} - \sqrt{10{,}000}])$$

$$= Pr(t < 2\,[100.5037 - 100])$$

$$= Pr(t < 1.007)$$

As would be expected for these large values of np and λ, the computed probabilities agree quite closely with the exact values.

Examples of the use of (c) and (d) will be given in Section 9.34.

9.3. Inferences from Counted Data

9.31. Estimation of the Probability p of Success

Situations frequently arise in which data are classified in one of 2 classes on the basis of some characteristic or attribute. We have already seen that such a classification or dichotomy leads to results which are distributed as repeated samples of size n from a binomial distribution if the probability p is constant for all n objects in the sample. For a single sample of n the maximum likelihood estimate (see Appendix 5A) of p is x/n, where x is the number of successes observed. From Section 4.61, we have

(a) $$\text{ave}\,(x/n) = p$$

hence this estimate is unbiased. Also from this section we have

(b) $$\text{var}\,(x/n) = \frac{p(1-p)}{n}$$

It follows that, given 2 samples of n_1 and n_2 with x_1 and x_2 successes, respectively, each from a population for which the probability of success is p, an unbiased estimate of p is obtained by considering the two samples as 1 large sample of $n_1 + n_2$ with $x_1 + x_2$ successes, i.e., by using the estimate

(c) $$\hat{p} = \frac{x_1 + x_2}{n_1 + n_2}$$

The variance of this estimate would be

$$(d) \qquad \text{var}\left(\frac{x_1 + x_2}{n_1 + n_2}\right) = \frac{p(1 - p)}{n_1 + n_2}$$

Note that, if we had averaged the two separate ratios, i.e., taken as our estimate

$$\hat{p} = \frac{1}{2}\left(\frac{x_1}{n_1} + \frac{x_2}{n_2}\right)$$

which is also unbiased, the variance would have been

$$\text{var}(\hat{p}) = \frac{1}{4}\left(\frac{1}{n_1} + \frac{1}{n_2}\right) p(1 - p)$$

which is larger than that given by (d) except when $n_1 = n_2$, in which case the two estimates are identical and the variance equal.

The problem of obtaining confidence intervals for estimating the unknown proportion p is more difficult than the corresponding problem for the mean μ in samples from a normal distribution, since in this case the variation of the sample ratio x/n depends on the quantity p to be estimated. For a given n and p we can compute, or obtain from tables of the binomial distribution, limits within which the ratio x/n will fall with a probability of at least $1 - \alpha$. For example, for $n = 10$, Figure 9.1 shows 2 step-like boundaries such that for a given p the probability is at least $1 - \alpha$ that the ratio x/n will fall inside or on these boundaries, at most $\alpha/2$ that it will be to the left, and at most $\alpha/2$ that it will be to the right. Hence the probability of an arbitrary point $(p, x/n)$ falling in the central portion is at least $1 - \alpha$. Then for a given observed proportion x_0/n of successes the probability is also at least $1 - \alpha$ that the unknown value of p will be within the interval between the points C and D. In practice this means that we determine the smallest value of p, say p_2, such that

$$(e) \qquad Pr(x \leq x_0) = \sum_{x=0}^{x_0} \binom{n}{x} p^x(1 - p)^{n-x} \leq \frac{\alpha}{2}$$

and the largest value of p, say p_1, such that

$$(f) \qquad Pr(x \geq x_0) = \sum_{x=x_0}^{n} \binom{n}{x} p^x(1 - p)^{n-x} \leq \frac{\alpha}{2}$$

from which it follows that the statement

$$(g) \qquad p_1 < p < p_2$$

will be correct with probability at least $1 - \alpha$ if x_0 successes have been observed in n trials. One-sided limits can be obtained by replacing $\alpha/2$ by α in either (*e*) or (*f*), whichever is applicable.

EXAMPLE 1. For $n = 10$, $x_0 = 4$, and $\alpha = 0.05$ we can obtain directly from tables of the binomial distribution, or from tables of the incomplete beta function, using the identity (*d*) of Section 4, Appendix 4*A*, $p_1 = 0.12$ and $p_2 = 0.74$. This means that on the basis of having observed 4 successes in 10 trials, we can state with 95% confidence that the unknown p is between 0.12 and 0.74. Similarly, for $\alpha = 0.01$ we obtain $0.07 < p < 0.81$.

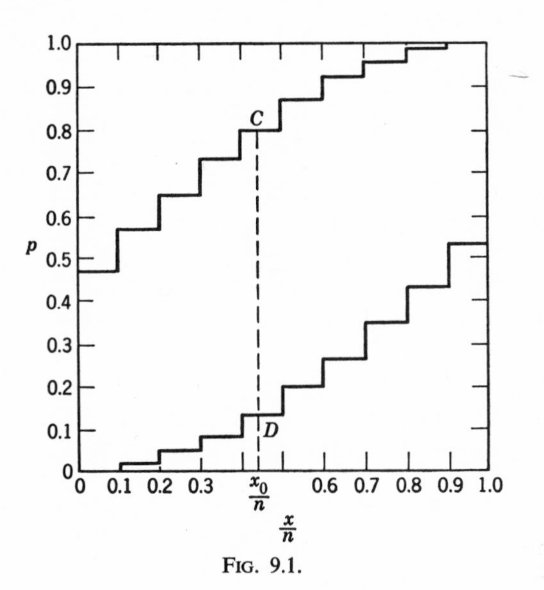

FIG. 9.1.

For sample sizes n beyond those tabled, and in cases where tables are not available and the observed number of successes x_0 (or failures $n - x_0$) is not too small, (*a*) of Section 9.2 can be used to obtain approximate confidence limits. By using this approximation directly on (*e*) above, we find that p_2 can be obtained from the equation

$$(h) \qquad 2\left(\sqrt{(x_0 + 1)(1 - p_2)} - \sqrt{(n - x_0)p_2}\right) = t_\alpha$$

where t_α is the unit normal deviate exceeded in absolute value with probability α. Since $Pr(x \geq x_0) = 1 - Pr(x \leq x_0 - 1)$, it follows directly from (*f*) that p_1 can be obtained from the equation

$$(i) \qquad 2\left(\sqrt{x_0(1 - p_1)} - \sqrt{(n - x_0 + 1)p_1}\right) = t_\alpha$$

After some algebraic manipulation we find that p_2 is the larger of the two solutions of the equation

$$(j) \qquad (n+1)^2 p_2^2 - 2p_2\left[(n+1)(x_0+1) + \frac{t_\alpha^2}{4}(n - 2x_0 - 1)\right] + \left(x_0 + 1 - \frac{t_\alpha^2}{4}\right)^2 = 0$$

and p_1 the smaller of the solutions of the same equation with x_0 replaced by $x_0 - 1$.

EXAMPLE 2. For $n = 10$, $x_0 = 4$, and $\alpha = 0.05$, equation (j) becomes

$$121p_2^2 - 2(55.96)p_2 + 16.32 = 0$$

and the larger of the two solutions is $p_2 = 0.744$. Putting $x_0 = 3$ in (j) we obtain the equation

$$121p_1^2 - 2(46.88)p_1 + 9.24 = 0$$

for which the smaller solution is $p_1 = 0.116$. Rounded to 2 decimal places, these solutions agree exactly with those given in Example 1.

EXAMPLE 3. Of a test lot of 1253 pieces produced by a new process, 75 were rejected. The best estimate of the proportion of pieces which would be rejected if the process used were adopted is $75/1253 = 0.0598$, or 6.0%. 95% confidence limits for the unknown proportion of rejects can be obtained by solving equation (j) for $n = 1253$, $x_0 = 75$ and 74, and $\alpha = 0.05$ as in the previous example; the equations are

$$1{,}572{,}516p_2^2 - 2(96{,}362.88)p_2 + 5631.0016 = 0$$

for which the larger root is $p_2 = 0.0745$, and

$$1{,}572{,}516p_1^2 - 2(95{,}110.80)p_1 + 5481.9216 = 0$$

for which the smaller root is $p_1 = 0.0474$.

A test of the hypothesis $p = p_0$ on the basis of a sample of n can be made by determining whether p_0 falls within the confidence limits for p based on the observed number of successes x_0. However, to avoid solving (j) it is simpler in this case to substitute x_0 and p_0 in the left-hand side of (h) if $x_0 < np_0$, or (i) if $x_0 > np_0$, and reject the hypothesis if the resulting value of t is greater than t_α in absolute value. For the one-sided test of the hypothesis $p \geq p_0$ against the alternatives $p < p_0$, we should compute the left-hand side of (h), regardless of the value of x_0, and reject the hypothesis if the value obtained was less than $-t_{2\alpha}$; similarly (i) would always be used to test the hypothesis $p \leq p_0$ against the alternatives $p > p_0$.

EXAMPLE 4. Suppose that the data of Example 3 had been collected to test the hypothesis that the new process would produce *less than* 7% rejects. Then for $n = 1253$, $x_0 = 75$, and $p_0 = 0.07$ we have from the left-hand side of (h)

$$
\begin{aligned}
t &= 2(\sqrt{76(0.93)} - \sqrt{(1178)(0.07)}) \\
&= 2(8.407 - 9.081) \\
&= -1.35
\end{aligned}
$$

Since this is greater than the value $-t_{0.10} = -1.645$, we should not reject the hypothesis at the 5% level on the basis of this evidence; i.e., we cannot be sure that the process will produce less than 7% rejects.

Although the evidence was not conclusive, these data seemed to indicate some improvement in the process and a second test lot of 1165 pieces was produced, of which 54 were rejected. Combining these results with those from the previous lot we have a total of 2418 test pieces, of which 129 were rejected. For the combined samples we have $n = 2418$, $x_0 = 129$, and

$$
\begin{aligned}
t &= 2(\sqrt{130(0.93)} - \sqrt{2289(0.07)}) \\
&= 2(1.100 - 1.266) \\
&= -3.32
\end{aligned}
$$

Since at the 0.001 significance level the critical value for a one-sided test is $-t_{0.002} = 3.09$, the combined samples indicate conclusively that the process will produce less than 7% defects.

9.32. Estimation of the Expected Value λ

Given a series of observations from a distribution known to be approximately represented by a Poisson distribution, we may wish to estimate the value of the single parameter λ representing the expected or average number of occurrences. We have seen (Appendix 5A) that the maximum likelihood estimate of λ based on a sample of n observations $x_1, \cdots, x_n$ is $\bar{x}$, the sample mean. This estimate is unbiased, since from Section 4.63

$$\text{(a)} \qquad \text{ave}(\bar{x}) = \lambda$$

and has variance

$$\text{(b)} \qquad \text{var}(\bar{x}) = \frac{\lambda}{n}$$

Note that to compute the estimate $\bar{x}$ we need know only the total number of occurrences, Σx_i, and the number of observations n. In this case the quantity λ is also estimated by the sample variance s^2. However, it is not so good an estimate of λ as the mean. As in the case of estimating p in Section 9.31, if 2 samples of size n_1 and n_2 are given, both from a Poisson distribution with expected value λ, then the best estimate of λ is the sample mean

$$(c) \qquad \frac{n_1\bar{x}_1 + n_2\bar{x}_2}{n_1 + n_2} = \frac{\Sigma x_{1i} + \Sigma x_{2i}}{n_1 + n_2}$$

of the combined sample.

Confidence intervals for λ[102] can be obtained in a fashion very similar to that used for obtaining confidence limits for p, except that in this instance there is no upper limit to the possible values of λ. Table 9.1 gives the 95% and 99% confidence intervals for $n\lambda$ when $n\bar{x} = \Sigma x_i \leq 50$. Confidence intervals for λ may be obtained by dividing both limits by n. Actually it makes no difference in obtaining limits whether we consider Σx_i as the sum of n observations with average value λ or as a single observation with average value $n\lambda$; it is merely a matter of the units in which λ is to be expressed, i.e., accidents per month or accidents per year, counts per minute or counts per hour.

As in the case of the binomial, the limits in the table were obtained by determining for each total count x_0 the largest value λ_2 of λ such that

$$(d) \qquad Pr(x \leq x_0) = \sum_{x=0}^{x_0} \frac{e^{-\lambda}\,\lambda^x}{x!} \leq \frac{\alpha}{2}$$

and the smallest value λ_1 of λ such that

$$(e) \qquad Pr(x \geq x_0) = \sum_{x=x_0}^{\infty} \frac{e^{-\lambda}\,\lambda^x}{x!} \leq \frac{\alpha}{2}$$

For observed counts x_0 beyond those in the table we can use the approximation given by (b) of Section 9.2 for the above probabilities, and obtain the equations

$$(f) \qquad 2(\sqrt{x_0 + 1} - \sqrt{\lambda_2}) = -t_\alpha$$

$$2(\sqrt{x_0} - \sqrt{\lambda_1}) = t_\alpha$$

which can be solved for λ_1 and λ_2.

EXAMPLE 1. It is well known that the number of disintegrations observed in a given quantity of radioactive material in equal intervals of time is approximately represented by a Poisson distribution under the assumptions of a long half-life and the absence of daughter decay. This distribution will also apply approximately, except at high counting rates where coincidence may become an appreciable factor, to the counts

TABLE 9.1

CONFIDENCE LIMITS FOR POISSON FREQUENCY DISTRIBUTIONS*

The probability is at least $1 - \alpha$ that $n\lambda$ will be between the limits given when Σx_i is the total of n observations on a Poisson variable with average value λ.

Total Observed Count $x_0 = \Sigma x_i$	Significance Level				Total Observed Count $x_0 = \Sigma x_i$	Significance Level			
	$\alpha = 0.01$		$\alpha = 0.05$			$\alpha = 0.01$		$\alpha = 0.05$	
	Lower Limit	Upper Limit	Lower Limit	Upper Limit		Lower Limit	Upper Limit	Lower Limit	Upper Limit
0	0.0	5.3	0.0	3.7					
1	0.0	7.4	0.1	5.6	26	14.7	42.2	17.0	38.0
2	0.1	9.3	0.2	7.2	27	15.4	43.5	17.8	39.2
3	0.3	11.0	0.6	8.8	28	16.2	44.8	18.6	40.4
4	0.6	12.6	1.0	10.2	29	17.0	46.0	19.4	41.6
5	1.0	14.1	1.6	11.7	30	17.7	47.2	20.2	42.8
6	1.5	15.6	2.2	13.1	31	18.5	48.4	21.0	44.0
7	2.0	17.1	2.8	14.4	32	19.3	49.6	21.8	45.1
8	2.5	18.5	3.4	15.8	33	20.0	50.8	22.7	46.3
9	3.1	20.0	4.0	17.1	34	20.8	52.1	23.5	47.5
10	3.7	21.3	4.7	18.4	35	21.6	53.3	24.3	48.7
11	4.3	22.6	5.4	19.7	36	22.4	54.5	25.1	49.8
12	4.9	24.0	6.2	21.0	37	23.2	55.7	26.0	51.0
13	5.5	25.4	6.9	22.3	38	24.0	56.9	26.8	52.2
14	6.2	26.7	7.7	23.5	39	24.8	58.1	27.7	53.3
15	6.8	28.1	8.4	24.8	40	25.6	59.3	28.6	54.5
16	7.5	29.4	9.4	26.0	41	26.4	60.5	29.4	55.6
17	8.2	30.7	9.9	27.2	42	27.2	61.7	30.3	56.8
18	8.9	32.0	10.7	28.4	43	28.0	62.9	31.1	57.9
19	9.6	33.3	11.5	29.6	44	28.8	64.1	32.0	59.0
20	10.3	34.6	12.2	30.8	45	29.6	65.3	32.8	60.2
21	11.0	35.9	13.0	32.0	46	30.4	66.5	33.6	61.3
22	11.8	37.2	13.8	33.2	47	31.2	67.7	34.5	62.5
23	12.5	38.4	14.6	34.4	48	32.0	68.9	35.3	63.6
24	13.2	39.7	15.4	35.6	49	32.8	70.1	36.1	64.8
25	14.0	41.0	16.2	36.8	50	33.6	71.3	37.0	65.9

* "The Concept of Confidence or Fiducial Limits Applied to the Poisson Frequency Distribution," by W. E. Ricker. Published by permission of the *Journal of the American Statistical Association*, Vol. *32* (1937), pp. 349–386, Washington, D.C.

recorded by a counterscaler unit with fixed geometry (percent of disintegrations recorded). Hence, if 11,286 counts were recorded during a 20-minute count of a radioactive sample, the average counts per minute, 564.3, would be the best estimate of the true counts per minute, λ, to be expected of this sample. Note that to obtain this estimate it is not necessary to know the individual counts for each of the 20 1-minute intervals.

Since the observed number of counts is quite large, we can obtain confidence limits for this estimate from (f) above. Using $\alpha = 0.05$, we have for 95% confidence limits for $n\lambda$

$$2(\sqrt{11,287} - \sqrt{\lambda_2}) = -1.96$$

$$\sqrt{\lambda_2} = -106.24 - 0.98$$

$$\lambda_2 = 11,496$$

$$2(\sqrt{11,286} - \sqrt{\lambda_1}) = 1.96$$

$$\sqrt{\lambda_1} = -106.24 + 0.98$$

$$\lambda_1 = 11,080$$

and hence, dividing each limit by $n = 20$,

$$554.0 < \lambda < 574.8$$

EXAMPLE 2. During a 15 minute background count on a given counter 45 counts were observed. In this case the best estimate of the expected number of counts due to background is $45/15 = 3.0$ counts per min. The 99% confidence limits can be obtained by the use of Table 9.1, from which we have

$$29.6 < 15\,\lambda < 65.3$$

$$1.97 < \lambda < 4.35$$

If we had used equations (f) and (g), we should have obtained

$$2(\sqrt{46} - \sqrt{\lambda_2}) = -2.576$$

$$\sqrt{\lambda_2} = -6.78 - 1.29$$

$$\lambda_2 = 65.1$$

$$2(\sqrt{45} - \sqrt{\lambda_1}) = 2.576$$

$$\sqrt{\lambda_1} = -6.71 + 1.29$$

$$\lambda_1 = 29.4$$

in excellent agreement with the limits of the table.

As in the case of binomial data, the hypothesis $\lambda = \lambda_0$ can be tested by computing the left-hand side of (f) or (g), depending on whether $x_0 < \lambda_0$ or $x_0 > \lambda_0$, and by rejecting the hypothesis if the value of t obtained is greater than t_α in absolute value. A one-sided test of the hypothesis $\lambda \leq \lambda_0$ against the alternatives $\lambda < \lambda_0$ is obtained by computing the left-hand side of (g), regardless of the values of x_0, and rejecting the hypothesis when the actual value of t obtained is greater than $t_{2\alpha}$. A similar one-sided test of $\lambda \geq \lambda_0$ against $\lambda < \lambda_0$ can be obtained using (f).

EXAMPLE 3. In low background α-counting, the counters are to be decontaminated when the background exceeds 0.25 c/m. To test the hypothesis that the background λ, in counts per minute, is less than 0.25 c/m against the alternative that $\lambda > 0.25$ c/m and decontamination is required, overnight background counts of 16 hr (960 min) are obtained, and the counter is decontaminated if the hypothesis is rejected at the 5% level on the basis of the observed count. For one such count the observed number was 308; since the expected number under the hypothesis to be tested is (960) (0.25) = 240, we have from the left-hand side of (g)

$$\begin{aligned} t &= 2(\sqrt{308} - \sqrt{240}) \\ &= 2(17.55 - 15.49) \\ &= 4.12 \end{aligned}$$

Since this is far beyond the critical value $t_{0.10} = 1.645$ for the one-sided test being used, there is no doubt that decontamination is indicated in this case.

9.33. Comparison of Counts

The methods of the preceding sections can also be used in the comparison of counts. Let us first consider the comparison of 2 binomial counts; the procedure in the Poisson case differs only in the use of different transformations.

Given 2 observed counts x_1 and x_2 of the number of successes in samples of n_1 and n_2, respectively, we wish to determine whether these observed counts are consistent with the hypothesis that both were obtained from a single binomial population with probability of success p. Under the hypothesis to be tested the best estimate of p is given by (c) of Section 9.31. Assuming $x_1 < n_1\hat{p}$ and $x_2 > n_2\hat{p}$, which is always possible since $\hat{p}$ must lie between the two observed proportions, we can obtain, using the left-hand side of (h) in Section 9.31

$$t_1 = 2(\sqrt{(x_1 + 1)(1 - \hat{p})} - \sqrt{(n - x_1)\hat{p}}) \tag{a}$$

and similarly from (i) of the same section,

$$(b) \qquad t_2 = 2(\sqrt{x_2(1-\hat{p})} - \sqrt{(n - x_2 + 1)\hat{p}})$$

The difference $t_2 - t_1$ is approximately distributed as the range of 2 normally distributed variables for which the upper 5% and 1% points are 2.77 and 3.64, respectively. Hence we can compare the magnitude of this difference with these percentage points to test the hypothesis that the two observed counts came from the same distribution.

Similarly, given 2 observed total counts x_1 and x_2 from a Poisson distribution, where ave $(x_1) = n_1\lambda_1$ and ave $(x_2) = n_2\lambda_2$, we can test the hypothesis $\lambda_1 = \lambda_2$ by obtaining $\hat{\lambda} = \dfrac{x_1 + x_2}{n_1 + n_2}$ and computing, assuming $x_1 < n_1\hat{\lambda}$ and $x_2 > n_2\hat{\lambda}$

$$(c) \qquad t_1 = 2(\sqrt{x_1 + 1} - \sqrt{n_1\hat{\lambda}})$$

and

$$(d) \qquad t_2 = 2(\sqrt{x_2} - \sqrt{n_2\hat{\lambda}})$$

again comparing the difference $t_2 - t_1$ with the percentage points given above.

The above procedures can be immediately generalized to the case of more than 2 counts, for, considering the binomial case first, we can from a series of observed counts $x_1, x_2, \cdots, x_k$ on samples of $n_1, n_2, \cdots, n_k$ obtain an estimate $\hat{p} = \dfrac{x_1 + x_2 + \cdots + x_k}{n_1 + n_2 + \cdots + n_k}$ of the probability p assumed to be common to all k samples. Using this value of $\hat{p}$, we can compute a series of values $t_1, t_2, \cdots, t_k$, using (a) if $x_i < n_i\hat{p}$ and (b) if $x_i > n_i\hat{p}$. The hypothesis that all the samples came from the same population could again be tested by comparing the range of these k values with the percentage points of the range of k normally distributed variables given in Table 10.7; however, unless k is small, it is better to compute Σt_i^2, which will have approximately the χ^2-distribution with $k - 1$ degrees of freedom, and compare it with the percentage points of this distribution obtained from Table II.

A similar procedure could be followed for a series of observed counts x_i, $i = 1, \cdots, k$, from Poisson distributions such that ave $(x_i) = n_i\lambda_i$. To test the hypothesis that $\lambda_i = \lambda$ for all i, we should obtain an estimate $\hat{\lambda} = \dfrac{\Sigma x_i}{\Sigma n_i}$, and compute a series of values t_i, using (c) and (d). Σt_i^2 could then be compared with the percentage points of the χ^2-distribution for $k - 1$ degrees of freedom as in the binomial case.

EXAMPLE 1. During a given period of time, 134 analyses of a given type were performed in one laboratory of a large industrial plant, and 110 in a second laboratory. In the first laboratory it was necessary to rerun 15 analyses, owing to evidence of gross error in one or more of the quadruplicate determinations performed; in the second laboratory 19 such reruns were necessary. To test whether the difference in proportion of reruns necessary in the two laboratories is significant, we compute

$$\hat{p} = \frac{15 + 19}{134 + 110} = 0.139$$

and

$$\begin{aligned} t_1 &= 2\left(\sqrt{16(0.861)} - \sqrt{119(0.139)}\right) \\ &= 2(3.71 - 4.07) \\ &= -0.72 \\ t_2 &= 2\left(\sqrt{19(0.861)} - \sqrt{92(0.139)}\right) \\ &= 2(4.04 - 3.58) \\ &= 0.92 \\ t_2 - t_1 &= 1.64 \end{aligned}$$

Since this is below the 5% level of significance, these data give no indication of any difference in the probability of a rerun for the 2 laboratories.

EXAMPLE 2. In making 10-minute check counts on 4 alpha counters, the same source being used, the following total counts were obtained: 18,289, 18,197, 18,290, 18,095. From these we obtain $\hat{\lambda} = 1821.8$ for the estimated counts per minute of the source, or $10\hat{\lambda} = 18{,}218$ for the average value of a 10-minute count. Using (*c*) and (*d*) of this section we obtain for t_1 through t_4

$$0.54, \quad -0.14, \quad 0.54, \quad -0.90$$

and hence $\Sigma t_i^2 = 1.41$. Since this is less than the average value of 3 for the χ^2-distribution with 3 degrees of freedom, the check shows no evidence of any difference between the counters tested.

EXAMPLE 3. Table 9.2 gives, for 12 lots of a given product, the number of acceptable and unacceptable pieces produced as determined by 100% inspection. Considering these lots as samples, we wish to determine whether the observed variation in the proportion of rejects from lot to lot is consistent with the hypothesis that the process being used has a constant probability of producing a reject. Dividing the total number of rejects by the total number of pieces, we obtain an estimate $\hat{p} = 0.0832$

for the probability of a reject. By using this, the values t_i given in the table were computed from (*a*) and (*b*) of this section. Several typical computations are

$$\begin{aligned} t_5 &= 2\left(\sqrt{229(0.9168)} - \sqrt{1459(0.0832)}\right) \\ &= 2(14.49 - 11.02) \\ &= +6.94 \end{aligned}$$

$$\begin{aligned} t_{11} &= 2\left(\sqrt{135(0.9168)} - \sqrt{1525(0.0832)}\right) \\ &= 2(11.125 - 11.26) \\ &= -0.27 \end{aligned}$$

TABLE 9.2

Lot Number	Pieces in Lot	Acceptable Pieces	Unacceptable Pieces	t
1	1,781	1,610	171	+1.88
2	1,685	1,546	139	−0.02
3	1,606	1,485	121	−1.08
4	1,622	1,527	95	−3.78
5	1,687	1,458	229	+6.94
6	1,638	1,529	109	−2.48
7	1,650	1,494	156	+1.62
8	1,580	1,474	106	−2.34
9	1,611	1,463	148	+1.22
10	1,774	1,612	162	+1.22
11	1,659	1,525	134	−0.27
12	1,750	1,652	98	−4.38
Total	20,043	18,375	1668	

Since $\Sigma t_i^2 = 103.6$, which is far beyond even the 0.1% point for χ^2 with 11 degrees of freedom, there is no doubt that the probability of producing a reject, or the stringency of the inspection procedure, or some such factor, was varying during the production of these lots.

An exact test of the hypothesis $p_1 = p_2$ when $n_1 = n_2 = n$ is afforded by a test similar to the sign test for paired differences discussed in Section 5.52. For, if we consider the two sets of observations in the order obtained, we have n ordered pairs of observations, consisting of 1 observation on an object from the first population and 1 observation on an object from the second. If we designate a success by 1 and a failure by 0, we can thus obtain for different pairs: (0, 0), (0, 1), (1, 0), and (1, 1). Let us discard all pairs of the type (0, 0) and (1, 1), and suppose that of the remaining pairs there are m_1 of the type (0, 1) and m_2 of the type (1, 0), $m_1 + m_2 = m \leq n$. The probability of obtaining a pair of type (0, 1), i.e., a failure on the observation from the first population and a success on the observation from the second, given that it is already either of type (0, 1) or (1, 0), is

$$p = \frac{p_2(1 - p_1)}{p_2(1 - p_1) + p_1(1 - p_2)}$$

and similarly the probability of a pair of type (1, 0) is

$$1 - p = \frac{p_1(1 - p_2)}{p_2(1 - p_1) + p_1(1 - p_2)}$$

Hence the hypothesis $p_1 = p_2$ is equivalent to the hypothesis $p = \frac{1}{2}$, which can easily be tested exactly on the basis of m_2 (or m_1) occurrences in a sample of m by determining from Table 5.2 of Section 5.34 whether m_2 is outside the percentage point of the binomial distribution for $n = m$ and $p = \frac{1}{2}$.

Although it seems strange to discard observations completely, this test in common with other paired tests, has the advantage that variations in the probability from pair to pair do not affect the results. Also a sequential probability ratio test (see Section 5.7) of the hypothesis $p_1 > p_2$ can be defined on the basis of the above formulation [92] or [93].

9.34. Analysis of Variance for Counted Data

In the preceding sections we have considered the analysis of counted data either by exact methods or by transforming the observed counts to approximately standard normal deviates, and assessing the significance of the transformed data by comparison with the percentage points of the normal or χ^2-distributions. In more complicated situations it is natural to attempt to use the analysis of variance techniques in conjunction with the transformation to normality. Two objections arise: (1) in the analysis of variance, we are more interested in constant variance than approximate normality (see Section 7.45), and (2) the methods of the last section involve obtaining an estimate of the parameters p or λ on which to base the transformation, and evaluating the effect of this estimation procedure on the independence of the transformed deviates. Hence we are led to use the preparatory transformations (*c*) and (*d*) of Section 9.2 which stabilize the variance and do not involve any unknown parameters.

Note that the variance of transformed binomial data using (*c*) is $\dfrac{1}{n + 1/2}$

(in radians), which, although independent of p, does depend on the sample size n. Hence the variance is stabilized in this case only if the observed counts are on samples of equal size.

EXAMPLE 1. Let us consider the 4 counts in Example 2 of the preceding section. The transformed counts, obtained by use of (*d*) of Section 9.2, are 270.48, 269.80, 270.48, 269.04, and for these values $\bar{x} = 269.95$, $\Sigma(x_i - \bar{x})^2 = 1.4124$, and $s^2 = 0.47$. Since s^2 is less than the expected variance of 1, we would immediately conclude that these counts were not excessively variable. To show the relationship with the previous method, we compute

$$\frac{(n-1)s^2}{\sigma^2} = \frac{\Sigma(x_i - \bar{x})^2}{\sigma^2} = 1.41$$

which, under the additional assumption that the transformed values are approximately normally distributed, has the χ^2-distribution, with 3 degrees of freedom. This value is identical with that previously obtained, as would be expected with the large counts involved. The only essential difference in procedure is that here we obtained the sum of squares of the deviations from the mean of the transformed data, whereas, in the previous case, we computed deviations from the transformed mean. To obtain an estimate $\hat{\lambda}$ in this case we have from Section 9.2

$$\sqrt{4\hat{\lambda} + 1} = \bar{x} = 269.95$$
$$4\hat{\lambda} + 1 = 72{,}873$$
$$\hat{\lambda} = 1821.8$$

which is identical with the estimate previously obtained.

EXAMPLE 2. W. L. Gore [64] has reported the results of an experiment designed to test the effect of injection temperature, injection cycle, and mould temperature on the brittleness of nylon bars. Two levels were used for each factor, as shown in Table 9.3. For each of the 8 treatment combinations, 100 bars were examined, and the number of brittle bars was determined.

TABLE 9.3

Factor	Symbol	Low Level	High Level
Injection temperature	A	450°F	530°F
Injection cycle	B	30 sec	90 sec
Mould temperature	C	45°F	250°F

TABLE 9.4

Treatment	x	$\sqrt{\frac{x}{101}}$	$\sqrt{\frac{x+1}{101}}$	$y = \sin^{-1}\sqrt{\frac{x}{101}} + \sin^{-1}\sqrt{\frac{x+1}{101}}$
(1)	2	0.14075	0.17235	0.3146
a	30	0.54500	0.55401	1.1636
b	10	0.31466	0.33002	0.6564
ab	52	0.71753	0.72440	1.6105
c	10	0.31466	0.33002	0.6564
ac	35	0.58867	0.59702	1.2692
bc	25	0.49752	0.50737	1.0528
abc	60	0.77075	0.77715	1.7701

These results are given in Table 9.4. Also given in this table are the transformed values $y = \sin^{-1}\frac{x}{101} + \sin^{-1}\frac{x+1}{101}$, to which the methods of factorial analysis described in Chapter 8 can be applied. The effects calculated from the transformed values, and the corresponding mean squares, are given in Table 9.5, the I effect, or sum, being omitted. If the interactions are pooled, we obtain the analysis of variance in Table 9.6, which shows that, compared with the pooled interactions, the effect of injection temperature is highly significant, the effect of injection cycle time significant at the 0.01 level, and the effect of mould temperature significant at the 0.05 level. Note that the estimate 0.0088 based on the pooled interactions is quite close to the expected variance of $\frac{1}{n + 1/2} = \frac{1}{100.5} = 0.00995$.

TABLE 9.5

Effect	Calculated Value	Sum of Squares
A	3.1332	1.2271
B	1.6860	0.3553
C	1.0034	0.1259
AB	0.2096	0.0055
AC	−0.4730	0.0280
BC	0.1086	0.0015
ABC	−0.0006	0.0000

TABLE 9.6

Source of Estimate	Sum of Squares	Degrees of Freedom	Mean Square	F
Injection temperatures	1.2271	1	1.2271	139.4***
Injection cycles	0.3553	1	0.3553	40.4**
Mould temperatures	0.1259	1	0.1259	14.3*
Pooled interactions	0.0350	4	0.0088	
Total	1.7433			

EXAMPLE 3. The data in Table 9.7, given by Davies [63], p. 184, are the piston-ring failures observed over the same period of time in each of

TABLE 9.7
PISTON-RING FAILURES

Compressor \ Leg	North	Center	South	Total
1	17	17	12	46
2	11	9	13	33
3	11	8	19	38
4	14	7	28	49
Total	53	41	72	166

TABLE 9.8
TRANSFORMED VALUES: $y = \sqrt{x} + \sqrt{x+1}$

Compressor \ Leg	North	Center	South	Total
1	8.37	8.37	7.07	23.81
2	6.78	6.16	7.35	20.29
3	6.78	5.83	8.83	21.44
4	7.61	5.47	10.68	23.76
Total	29.54	25.83	33.93	89.30

3 legs of 4 apparently identical compressors. The values $y = \sqrt{x} + \sqrt{x+1}$ for these data are given in Table 9.8. The analysis of variance for the transformed data is shown in Table 9.9.

TABLE 9.9

ANALYSIS OF VARIANCE

Source of Estimate	Sum of Squares	Degrees of Freedom	Mean Square
Between legs	8.221	2	4.110
Between compressors	3.063	3	1.021
Interaction	12.032	6	2.005
Total	23.318	12	

In the usual two-factor analysis of variance with only 1 observation in each category, we can test for differences between rows and columns only if we assume that there are no real interactions present, and use the interaction mean square as an estimate of the error variance. However, if we are willing to assume in this case that the observed counts are independent observations from Poisson distributions whose *transformed* mean values are additive combinations of leg and compressor effects and leg $\times$ compressor interactions, then we can use the expected variance of unity for the transformed data in testing for the presence of interactions. The variance ratio is $\frac{2.005}{1} = 2.005$, which for 6 and ∞ degrees of freedom is just below the 5% level. Table 9.10 shows

TABLE 9.10

INDIVIDUAL INTERACTIONS

Compressor \ Leg	North	Center	South
1	+0.49	+1.42	−1.91
2	+0.07	+0.38	−0.45
3	−0.31	−0.33	+0.64
4	−0.25	−1.47	+1.72

the individual interactions, none of which seem to stand out; the major contribution to the interaction mean square arises from the reversal of behavior of center and south on compressors 1 and 4.

To make the proper tests for variations between legs and between compressors, we must consider the populations involved. Clearly we are dealing here with the entire population of legs; hence the mean square between compressors is properly tested against the error variance, leading to a variance ratio 1.021 which is clearly non-significant. If we are interested only in the 4 compressors tested, then the mean square between legs is tested against the error variance, leading to a variance ratio 4.110 which for 2 and ∞ degrees of freedom is significant at the 5% level. However, if we consider these compressors a sample of a much larger number of similar compressors, the mean square between legs is properly compared with the interaction mean square, leading to a variance ratio 2.05 based on 2 and 6 degrees of freedom, which is clearly non-significant. Hence, although there is some evidence for the existence of a systematic difference between legs in these 4 compressors, there is no basis, in view of the interactions present, for extending this inference to a larger number of compressors of this type.

9.4. Chi-Square Tests

9.41. Chi-Square Test of Goodness of Fit

Situations frequently arise where data can be classified in one of k classes, with probabilities $p_1, p_2, \cdots, p_k$, $\Sigma p_i = 1$, of falling into each class. If these probabilities are assumed constant for each member of a sample of n, then the numbers $x_1, x_2, \cdots, x_k$, $\Sigma_i x_i = n$, of observations falling into each class can be expected to have the multinomial distribution discussed in Section 4.62.

In order to be able to test hypotheses concerning relationships between the probabilities p_i, it can be shown, by methods beyond the scope of this book, that the quantity

$$(a) \qquad \chi^2 = \Sigma_i \frac{(x_i - np_i)^2}{np_i}$$

where the x_i, n, and p_i are defined as above, has approximately the χ^2-distribution with degrees of freedom $\nu = k - r$, where r is the number of linear or nearly linear restrictions imposed on the differences $x_i - np_i$ when the p_i are estimated from the observations. Note that, even if all the p_i are specified, we still lose 1 degree of freedom, since we must have $\Sigma_i x_i = \Sigma_i np_i = n\,\Sigma_i p_i = n$. Since, from Section 4.62, np_i is simply the average value of x_i. this form for χ^2 corresponds to summing, over all classes, the squares of the deviations of the observed x_i from their average values divided by their average values.

In particular, if there are $k = 2$ classes, χ^2 has 1 degree of freedom, and, if we replace x_1 and x_2 by x and $n - x$ and p_1 and p_2 by p and $1 - p$, we have

$$\begin{aligned}
\chi^2 &= \frac{(x_1 - np_1)^2}{np_1} + \frac{(x_2 - np_2)^2}{np_2} \\
&= \frac{(x - np)^2}{n} + \frac{[n - x - n(1 - p)]^2}{n(1 - p)} \\
&= \frac{(x - np)^2}{n}\left[\frac{1}{p} + \frac{1}{1 - p}\right] \\
&= \frac{(x - np)^2}{n}\left[\frac{1 - p + p}{p(1 - p)}\right] \\
&= \frac{(x - np)^2}{np(1 - p)}
\end{aligned}$$

and hence, since χ^2 with 1 degree of freedom is the square of a normally distributed variable with zero mean and unit variance

$$(b) \qquad t = \sqrt{\chi^2} = \frac{x - np}{\sqrt{np(1 - p)}}$$

i.e., for 2 classes this form for χ^2 reduces to the limiting form of a single binomial variable given in Section 4.64. This is to be expected, since the derivation of the above criterion is based on an approximation of the multinomial distribution by a multivariate normal distribution similar to the approximation of the binomial by the single variate normal distribution. It emphasizes the fact that the test is only approximate, especially when the smallest expectation np_i is, say <50.

This criterion was developed by Karl Pearson for testing the "goodness of fit" of an observed to a theoretical distribution, but it has had wide application to other problems, especially in connection with contingency tables. Note that, if the probabilities p_i of the observations falling in a particular class are completely specified by the given distribution function, without the use of parameters estimated from the data, this test is non-parametric, and is probably the first such test developed.

EXAMPLE 1. W. L. Gore [64] gives an example of the determination of goodness of fit in connection with testing for brittleness in nylon test bars, using a mandrel bend test. Each test bar could be tested in 5 places. It was required to determine whether the brittleness was randomly distributed throughout a set of test bars moulded under similar conditions from the same batch of nylon molding powder, or whether it was concentrated in particular test bars. In the first case the number of breaks per bar would be expected to have a binomial distribution, with $n = 5$ and $p =$ probability of a break in an individual test assumed to be constant throughout all bars. In the second case we should expect the probability to change from bar to bar, resulting in a situation in which a given bar would either produce several breaks or none.

TABLE 9.11

Breaks per per Bar	Observed Frequency		Computed Average Frequency		Difference Observed − Computed
0	157		130.2		26.8
1	69		107.7		−38.7
2	35		35.7		−0.7
3	17		5.9		
4	1	19	0.5	6.4	12.6
5	1		0.0		
Total	280		280.0		

Table 9.11 gives the distribution of the number of breaks observed in 280 test bars. The average number of breaks per bar is 0.7107. Equating this to ave $(x) = np$, we have

$$5p = 0.7107$$

$$p = 0.1421$$

Using this estimated value of p, the computed numbers of breaks given in Table 9.11 are obtained by multiplying the successive terms of the binomial distribution

$$p(x) = \binom{5}{x} (0.1421)^x (0.8579)^{5-x} \quad x = 0, \cdots, 5$$

by 280. From these observed and computed values, grouping the last 3 classes to obtain a sufficiently large computed frequency, we obtain

$$\chi^2 = \frac{(26.8)^2}{130.2} + \frac{(-38.7)^2}{107.7} + \frac{(-0.7)^2}{35.7} + \frac{(12.6)^2}{6.4}$$

$$= 5.52 + 13.91 + 0.01 + 24.81$$

$$= 44.25$$

In this case, in addition to requiring the total of observed and computed values to be the same, we have estimated the probability p from the data; hence the χ^2 value has in this case $4 - 2 = 2$ degrees of freedom. Since $\chi^2_{2, 0.001} = 13.82$, this value is highly significant, indicating that the data are not properly fitted by a binomial distribution. Looking at the differences between observed and computed numbers, we see that there are more cases of 0 breaks, less cases of 1 break, and more cases of 3 or more breaks than would be expected if the binomial distribution applied. This is the situation that might be expected if some bars were brittle throughout, so that one break would be followed by others, whereas other bars were tough throughout, showing no breaks at all.

9.42. Contingency Tables

The cross classification of a series of observations according to 2 characteristics, or attributes, is referred to as a contingency table. The simplest case is that in which each classification is a simple dichotomy, resulting in a table with 4 classes, corresponding to all possible combinations of the presence or absence of each characteristic. For example, Table 9.12 indicates the frequency of failure due to cracking of specimens in 30-day tests on a number of large industrial boilers. The observations are classified as cracked and uncracked, and also according to the addition or non-addition of tannin to the feedwater. In cases such as this we are generally interested in testing the independence of the 2 criteria of classification, i.e., in determining whether the presence or absence of one characteristic is "contingent" upon the presence or absence of the other. Thus in the above example we should like to know whether the failure of the test specimens was influenced by the addition of tannin to the boiler feedwater.

TABLE 9.12

Tannin	Uncracked	Cracked	Total
Added	69	17	86
Not added	88	61	149
Total	157	78	235

The general 2×2 contingency table is of the form shown in Table 9.13, where n_{11}, n_{12}, n_{21}, and n_{22} are the observed numbers in each of the cross

TABLE 9.13

n_{11}	n_{12}	$n_{1\cdot}$
n_{12}	n_{22}	$n_{2\cdot}$
$n_{\cdot 1}$	$n_{\cdot 2}$	n

classifications, $n_{1\cdot} = n_{11} + n_{12}$, $n_{2\cdot} = n_{21} + n_{22}$, $n_{\cdot 1} = n_{11} + n_{21}$, and $n_{\cdot 2} = n_{12} + n_{22}$ are the subtotals for rows and columns, respectively, and n is the total number of observations. The best estimate of the probability of an observation falling into the first row, regardless of the column, is $n_{1\cdot}/n$. Hence in the case of independence, we should expect that of the $n_{\cdot 1}$ observations in the first column, $n_{\cdot 1}(n_{1\cdot}/n)$ would fall in the first row. By similar arguments, or simply by subtraction from the row and column totals (frequently called the marginal totals), estimated average values for the remaining classes can be determined. These are

TABLE 9.14

$\dfrac{n_{\cdot 1}\, n_{1\cdot}}{n}$	$\dfrac{n_{\cdot 2}\, n_{1\cdot}}{n}$	$n_{1\cdot}$
$\dfrac{n_{\cdot 1}\, n_{2\cdot}}{n}$	$\dfrac{n_{\cdot 2}\, n_{2\cdot}}{n}$	$n_{2\cdot}$
$n_{\cdot 1}$	$n_{\cdot 2}$	n

summarized in Table 9.14. Note that each of these is the product of the corresponding marginal totals divided by the grand total.

Based on these two tables giving the observed and average values for the 4 classes, we can compute

$$(a) \qquad \chi^2 = \frac{(n_{11} - a_{11})^2}{a_{11}} + \frac{(n_{12} - a_{12})^2}{a_{12}} + \frac{(n_{21} - a_{21})^2}{a_{21}} + \frac{(n_{22} - a_{22})^2}{a_{22}}$$

where the a's represent the computed average values of Table 9.14. Although there are 4 classes, this value has only 1 degree of freedom, since, in addition to requiring the average values to total to the number of observations n, we have used in their computation the estimates $n_{1\cdot}/n$ and $n_{\cdot 1}/n$ of the probabilities of falling in the first row and first column, respectively, assuming only the independence of these two estimates.

A second way of looking at this is to note that for the fixed marginal totals required of both observed and average values, only 1 class frequency can be chosen at random, the remaining 3 being automatically fixed by this choice. Also in this case the magnitude of the deviation of the observed value from the average value is the same for each class, and this common difference d can be considered as representing the single degree of freedom. In terms of this common difference (*a*) can be written

$$\chi^2 = d^2 \left(\frac{1}{a_{11}} + \frac{1}{a_{12}} + \frac{1}{a_{21}} + \frac{1}{a_{22}}\right) \tag{b}$$

Since in this case we are again approximating a discrete distribution by a continuous one, it is necessary to make a correction for continuity similar to that used in Section 4.64 in connection with the normal distribution of the limiting forms of the single variate binomial and Poisson distributions. This is accomplished in this case merely by decreasing the common difference d by one half in absolute value prior to the computation of χ^2. Hence we have for the computation of χ^2 from a 2×2 table

$$\chi^2 = (d - \tfrac{1}{2})^2 \left(\frac{1}{a_{11}} + \frac{1}{a_{12}} + \frac{1}{a_{21}} + \frac{1}{a_{22}}\right) \tag{c}$$

EXAMPLE 1. For the data of Table 9.12 we have

$$a_{11} = \frac{(86) \cdot (157)}{235} = 57.46 \quad a_{12} = \frac{(86) \cdot (78)}{235} = 28.54$$

$$a_{21} = \frac{(149) \cdot (157)}{235} = 99.54 \quad a_{22} = \frac{(149) \cdot (78)}{235} = 49.46$$

Thus $d = \pm 11.54$ and

$$\begin{aligned} \chi^2 &= (11.04)^2 \left(\frac{1}{57.46} + \frac{1}{28.54} + \frac{1}{99.54} + \frac{1}{49.46}\right) \\ &= (121.88)\,(0.0825) \\ &= 10.06 \end{aligned}$$

Since $\chi^2_{1,\,0.01} = 6.64$, this value is significant, and when the signs of the differences are considered it is apparent that the association is between uncracked tubes and tannin treatment of the feedwater.

When the numbers in some or all of the classes of a 2×2 contingency table are small it may be desirable to consider an exact test rather than the approximate tests treated above, although except for very small frequencies the calculations are quite laborious. A description of the exact test, which is independent of the true probability, can be found in [12], Section 21.02. It is also possible to extend the above approximate treatment to the case of more than 2 classes in each classification. However, except where we are really testing the association between 2 qualitative characteristics of a series of observations, it is preferable to use the methods of Sections 9.33 and 9.34, because the transformations used are more accurate for small observed counts than the limiting form on which the χ^2-test is based, and because the analysis of variance techniques are more directly applicable when we are merely classifying observed counts.

9.43. Tests of Homogeneity of Observed Counts

The χ^2-test can be used to test the homogeneity of a series of observed counts by methods which are frequently more convenient than those given in Section 9.33, although they are possibly not so accurate if small average values are involved.

Let us consider first a series of observed counts $x_1, \cdots, x_k$ in samples of $n_1, \cdots, n_k$. If we assume that all these samples came from the same binomial population, the estimated average values are given by $n_1\hat{p}$, $n_2\hat{p}, \cdots, n_k\hat{p}$, subject to the single linear restriction that $\Sigma n_i\hat{p} = \Sigma x_i$. Hence

$$(a) \qquad \chi^2 = \frac{\Sigma(x_i - n_i\hat{p})^2}{n_i\hat{p}}$$

has approximately the χ^2-distribution with $k - 1$ degrees of freedom, and excessive values of χ^2 would indicate the non-homogeneity of the samples obtained.

Similarly, for the Poisson distribution, the average value of a series of observed counts $x_1, \cdots, x_k$ is best estimated by $\bar{x}$, if all counts are assumed to have the same distribution, and hence

$$(b) \qquad \chi^2 = \frac{\Sigma(x_i - \bar{x})^2}{\bar{x}}$$

has approximately the χ^2-distribution with $k - 1$ degrees of freedom. Note that this same result is obtained if we remember that for the Poisson distribution ave $(s^2) = \lambda$, and hence, when λ is sufficiently large so that the Poisson distribution can be considered approximately normal,

$$(c) \qquad \chi^2 = \frac{(k - 1)s^2}{\lambda} = \frac{\Sigma(x_i - \bar{x})^2}{\lambda}$$

will have approximately the χ^2-distribution with $k - 1$ degrees of freedom. Replacing λ by the estimate $\bar{x}$, we obtain (b).

EXAMPLE 1. For the data of Example 3 of Section 9.33, the observed number of rejects, the computed average value, the difference, and the difference squared divided by the average value are given for each lot in Table 9.15.

TABLE 9.15

Lot	Observed Rejects	Computed Rejects	Difference	Contribution to χ^2
1	171	148.2	22.8	3.36
2	139	140.2	−1.2	0.00
3	121	133.7	−12.7	1.11
4	95	135.0	−40.0	11.56
5	229	140.4	88.6	55.28
6	109	136.3	−27.3	5.27
7	156	137.3	18.7	2.41
8	106	131.5	−25.5	4.75
9	148	134.1	13.9	1.34
10	162	147.6	14.4	1.31
11	134	138.1	−4.1	0.09
12	98	145.6	−47.6	15.24
	1668	1668		101.72

In computing the contribution to χ^2, one-half was subtracted from the absolute value of each difference to correct for continuity. This value of χ^2 is highly significant, as was the value of 103.6 previously obtained.

EXAMPLE 2. For the 4 counts of Example 2 of Section 9.33, we have

$$\bar{x} = 18{,}218$$

$$\Sigma(x_i - \bar{x})^2 = 25{,}795$$

Hence we obtain

$$\chi^2 = \frac{25{,}795}{18{,}218} = 1.42$$

which is in good agreement with the value previously obtained.

9.5. Acceptance Sampling by Attributes

One of the most important uses of statistical inference with respect to discrete data arises in acceptance sampling by attributes. The most common situation is that in which we are faced with lots of N items, the quality of which is measured by the number which do or do not possess a given characteristic, i.e., by the number of "defectives" present, and we are asked to determine the sampling scheme which will best enable us to sort the lots presented in such a fashion that only acceptable lots are allowed to pass.

It is obvious that the sampling scheme will depend on the definition of an "acceptable lot." In the first place, if an acceptable lot is one which contains no defectives (as might be true of gas masks, safety belts, or such items), then only 100% inspection (and probably not that) will enable us to screen out the defectives present. However, if we are willing to accept somewhat less than perfection, it will be possible to base our acceptance or rejection of particular lots on the number x of defectives in a sample of n of the N items.

For a lot of size N, a single sampling plan will consist of a sample size n and an acceptance number c such that whenever the number of defectives in the sample of size n chosen at random from the lot is less than or equal to c we accept the lot, but whenever the number of defectives is greater than c we reject the lot. In a double sampling plan 2 sample sizes n_1 and n_2 and 2 acceptance numbers c_1 and c_2 are given; a sample of size n_1 is chosen at random from the lot, and, if the number if defectives is less than or equal to c_1, the lot is accepted, whereas, if the number of defectives is greater than c_2, the lot is rejected. If the number of defectives is greater than c_1 but less than or equal to c_2, a second sample of n_2 is chosen and the total number of defectives in the combined sample of n_1 and n_2 is determined. If this number is less than or equal to c_2, the lot is accepted, and, if it exceeds c_2, the lot is rejected. In either single or double sampling, inspection may be terminated when a decision is reached, but, if the data from the sampling plan are used for estimation and process control in addition to inspection, it may not be desirable to make decisions on less than the full sample number. In a sequential sampling plan successive lots of size n are inspected, and acceptance and rejection numbers given for the total number of rejects at any stage; inspection is terminated when a decision is reached. Single, double, or sequential sampling plans can be constructed with essentially the same properties with regard to the height of the barrier that they erect against poor quality material; for plans equivalent in this respect, single sampling plans will require the most inspection, and sequential plans, on the average, the least. Double and sequential sampling plans are psychologically more desirable in the sense that they do not seem so "cut and dried" but "give the lot another chance," although such plans are frequently more difficult to administer. For any of these plans the rejected lots may either be accepted after 100% inspection and replacement of defective items by non-defective ones, or they may be returned to the producer.

The choice of a sampling plan must be based on a balance between the risks of accepting poor lots and either the amount of inspection to be done or the risk of rejecting a good lot. The consumer who receives the accepted material is ordinarily protected by one of the two following methods:

(*a*) The risk α of accepting lots which contain more than a certain fraction p_1 of defectives is fixed at some suitably small level. α is called consumers' risk and p_1 is called the lot tolerance fraction defective.

(*b*) In plans calling for 100% inspection and replacement of defective items in rejected lots, we can compute for any fraction defective $p = X/N$ in the lot submitted the average proportion of defectives which will be present in the accepted material. This is called the average outgoing quality (AOQ); it will be zero if the material presented for inspection is perfect, i.e., contains no defectives, and it will approach zero as the material submitted becomes very bad, in which case almost all material will be inspected 100%. Between these two extremes will be a point where the AOQ has its maximum value; this maximum value, known as the average outgoing quality limit (AOQL), is frequently specified to determine the sampling plan, since the consumer is sure that the product he receives can be no worse than the AOQL on the average, regardless of the nature of the lots submitted.

Similarly, there are 2 ways which are commonly used to protect the producer. These are:

(*a*) In plans calling for the acceptance of all lots after 100% inspection and the replacement of defective items, we can minimize the amount (and hence usually the cost) of inspection. The fraction defective $\bar{p}$ which, for a given n and c, minimizes the average amount of inspection, is called the process average (as we shall see later, this nomenclature arises because in practice this procedure is reversed).

(*b*) In cases where lots are rejected and returned to the producer, we may wish to require that the probability β of rejecting a lot with a fraction defective less than p_0 be some small value; i.e., we do not want good material to be rejected very often. β is called the producers' risk, and p_0 the acceptable quality level.

For a given sampling plan we can determine the behavior of the sampling plan, in the sense of how frequently it will accept the lots submitted, for lots of a given quality p. The curve relating the probability of acceptance of a lot by a given plan to the proportion of defectives in the lot submitted is called the operating characteristic curve of the sampling plan. The specification of consumer risk for a given lot tolerance percent defective, or of producer risk for a given acceptable quality level, is equivalent to choosing a point through which this curve must pass. The specification of an AOQL or the requirement of minimum sampling for a given process average does not directly involve the operating characteristic curve.

The choice of the method of specifying a plan will depend on the circumstances in which it is to be used. Single and double sampling plans which emphasize protection to the consumer by fixing the AOQL or the lot tolerance fraction defective for a consumer risk of 0.10 (10%) have been developed by Dodge and Romig [17]. The plans tabulated also assure that for a given process average the average inspection required be a minimum, assuming 100% inspection and replacement of defective items; in most cases this minimum is appreciably less for double sampling. A complete treatment of acceptance sampling is given in [18], with the emphasis on single, double, and sequential plans having a given acceptable quality level at a producer risk of 0.05 (5%) and operating characteristic curves which are almost equivalent; the additional conditions required involve both the steepness of the operating characteristic curve (essentially defining the consumer risk) and the amount of inspection required. Both [17] and [18] give many details required for effectively introducing and operating acceptance sampling plans.

It should be noted that a sampling plan based on producer and consumer risk is simply a test of the statistical hypothesis that a particular lot has a proportion of defectives $\leq p_0$ against the alternative hypothesis that the proportion of defectives present is $\geq p_1$. The producer and consumer risks are the errors of type I and II, respectively, and the operating characteristic curve of the plan is that of the test. In the case of single sampling, the determination of n and c for a specified α, β, p_0, and p_1 is very simple; for, if $p(x \mid n, p)$ represents the probability of obtaining x defectives in a sample of n from a lot containing a proportion $p = X/N$ of defectives, then we simply require the minimum value of n and the corresponding value of c such that

$$\sum_{x=0} p(x \mid n, p_0) \geq 1 - \beta$$

(a)

$$\sum_{x=0}^{c} p(x \mid n, p_1) \leq \alpha$$

The exact probability $p(x \mid n, p)$ is given for lots of size n by the hypergeometric distribution; in practical situations the proportion of defectives is usually small and the lots are large, so that the binomial or even the Poisson distribution gives a sufficiently good approximation. A similar formulation can be written for double sampling, the third condition imposed being that the average sample size be minimized in some fashion. The development of multiple and sequential sampling plans which minimize in some fashion the average sample size for a given producer and consumer risk is more complex.

CHAPTER 10

Control Charts

10.1. Introduction

Of all the applications of statistical methods in industrial work, the statistical control of quality has been the most widespread. Introduced by W. A. Shewhart in 1926, these techniques have been successfully applied in a wide variety of industrial situations, and in particular they were, and are, extensively used by the Army, Navy, and other government agencies in connection with the production of material. Since there is much available literature on this subject, including the famous treatise *The Economic Control of Quality of a Manufactured Product* by Shewhart [26], we shall give here only a brief resumé of the principles underlying the use of the control chart and some of the methods of obtaining control limits which are most frequently applied.

10.2. Statistical Control

10.21. The Concept of Statistical Control

It has become a commonly accepted fact that observations or measurements of a given characteristic, obtained either under carefully controlled experimental conditions or from routine plant or laboratory operations, vary to a greater or lesser degree. We do not expect the most precise analytical method to give identical results when repeated determinations are made, nor do we expect the most carefully controlled process to produce all material with identical properties. What we do hope to attain is an analytical method which gives very nearly the same answer in a large percentage of cases, and a process which will produce batches of material of very nearly the same properties.

The variation in an observed or measured characteristic is always due to a large variety of causes. Thus it would be impossible to enumerate the possible reasons why repeated determinations differ, or why successive batches of material fail to be identical. Some of these causes of variation can be identified and removed, and these will be called assignable causes, but there will remain other causes inherent in the process or method which it is impossible, or economically unfeasible, to remove. Such causes are known as chance causes, or as a system of chance causes.

In any particular case there is no sharp dividing line between these two types of cause; as some of the more apparent assignable causes are removed, others may become assignable, until finally we are left with a system in which we are willing to accept the remaining causes as chance causes. We should not conclude that in a practical situation all assignable causes can be economically removed, or that we never know the source of the chance causes which remain.

A process from which the assignable causes of variation have been removed is said to be controlled, or to exhibit a state of control. The observations or measurements obtained from such a process will vary, owing to chance causes only, and should approximately satisfy our concept of randomness. On the other hand, the mass behavior of such variations is predictable; hence we can predict by statistical methods the limits within which repeated measurements may be expected to fall. These qualities of randomness, and prediction within limits, are characteristic of controlled data. They are also the essential characteristics on which much of the theory of the previous chapters has been based.

The statistical concepts used in obtaining control are similar to those used in preceding chapters to test significance or estimate confidence limits, but the aim is completely different. The methods previously developed have with few exceptions been aimed at testing the significance of, or estimating the effect of, *known* sources, or causes, of variation. Thus we have tested the differences between groups of observations on objects *known* to differ in a certain respect, or estimated the parameters of an *expected* relationship between variables. The problem of establishing control is that of deciding whether any such sources, or causes, of variation, over and above the inherent variation, *do* exist. When possible assignable causes of variation are so indicated, the methods of the preceding chapters may be of immense help in identifying these causes and facilitating their removal.

W. A. Shewhart, to whom these concepts of systems of chance causes, assignable causes, and state of control are due, has given 5 criteria for determining whether a state of control exists. One of these, the control chart, based on the concept of control limits, is widely used in industry to facilitate the removal of assignable causes of variation in product quality and, along with other criteria, to detect assignable causes, or nonrandomness, in experimental data.

10.22. The Concept of Control Limits

As stated in the preceding section, the variation in a quality characteristic of a product produced by a controlled process is predictable within limits; i.e., we can by statistical methods establish limits which within repeated measurements of the given characteristics should almost always fall if no assignable causes are present. Such limits are known as *control limits*. Variations beyond the control limits will be considered an indication of the presence of assignable causes which should be investigated and, if possible, eliminated.

Control limits are conceptually different from tolerance limits or specification limits, which express the desired or specified quality to be expected or to be attained. All chemists are familiar with the specification of the quality of a chemical in terms of the impurities present, or the specification of the quality of a pipette or burette in terms of the accuracy of its calibration. Generally these specifications also take the form of limits within which the given quality characteristic is expected to fall. Thus a given compound may be specified as containing less than 1% impurity, having a specific gravity between 1.047 and 1.049, and a distillation range from 227° to 235° under standard conditions. Similarly a pipette may be specified as delivering between 4.998 and 5.002 cc of liquid at 20°C. Such limits are known as specification limits, or tolerance limits, and express the quality that a product has, or must possess. They are the logical expression of the quality of, or required of, a given product. Control limits, on the other hand, merely indicate the desirability of investigative action. When, by this and other means, we have established a state of control, we can determine if the process is capable of producing a product of which a sufficiently large percentage will fall within specified limits, or compute tolerance limits which reflect the variations in quality of the product produced.

The control limits most frequently used are 3 standard deviations on either side of the average value of the characteristic being controlled. If the process is in control, these limits should contain approximately 997 values out of every 1000 in the case of a normal distribution, and a high fraction of the values regardless of the distribution. In many instances it is preferable to establish the control limits for the mean of several values, since these means will be more nearly normally distributed than the original distribution. Limits 2 standard deviations on either side of the average value are often used as warning limits. Limits 1.96, 2.57, and 3.09 standard deviations on either side of the average value, corresponding to normal probabilities of 1 in 20, 1 in 100, and 1 in 500, respectively, are also frequently used; in Great Britain 1.96σ warning limits and 3.09σ action limits are standard. The correct choice of limits should be a matter of economic analysis, but it is often a matter of convention. In general, the wider the limits, the more likely is the presence of an assignable cause if a measurement is outside the limits; on the other hand, there is also a greater possibility that assignable causes will be present in measurements within limits, and hence they will go undetected. If we narrow the limits, the number of cases in which measurements will exceed the limits due to the inherent variability may serve to confuse the identification of those variations actually due to assignable causes. One point should be emphasized: *in practice, the limits should be set so as to make the complete investigation of the reasons for all points out of limits practicable.* The benefits of the establishment of control limits are in the identification and elimination of the assignable causes indicated by the presence of measurements beyond the control limits. Without such action we are merely performing a lip service and ignoring the real benefits of statistical control.

The control chart (see Figure 10.1) is the graphical expression of the concept of control limits. The measurements are plotted in the order obtained, and the central line and control limits are indicated on the graph. Measurements outside the limits, indicating a lack of control, are immediately apparent. In addition, this procedure of plotting the points in the order obtained may also give a first indication of some of the other types of non-randomness, such as trends or cyclic effects which are to be discussed in Chapter 11.

10.3. Control Charts for Variables

10.31. Rational Subgroups

We shall consider first the preparation of control charts for a quality characteristic which is measured on a continuous scale, such as specific gravity, boiling point, or some other physical measurement. Control charts for qualitative quality characteristics, or attributes, will be considered in the next section.

In order to establish control limits and prepare control charts we must first obtain estimates of the average value μ and the standard deviation σ of the characteristic to be controlled. This must generally be done on the basis of available data, or data specifically obtained for the purpose, from the process to be controlled. *If we were sure* that the process was in control, then the overall mean and standard deviation of these data could be used as the desired estimates. However, if assignable causes of variation are present, as is almost always true with the initial data, the overall standard deviation will not be a good estimate of the inherent variability of the process and its use should be avoided. It is customary to divide the data into *rational subgroups*, within which we have reason to believe the variation is more likely to be due to chance causes, but between which assignable causes may be present. If we obtain an estimate s of σ from the variation within these groups, we can expect that it will be a better estimate of the inherent variability than the overall standard deviation for the measurements. (Note the similarity between this procedure and the use of blocks in experimental design.) If the fundamental distribution of the measurements is in question, we can expect the means of these subgroups to behave more nearly normally than the measurements themselves, and, what is more important, to reflect more quickly the presence of assignable causes of variation between groups.

The constitution of a rational subgroup will depend on the conditions in the application in question. In controlling the production of manufactured items, measurements on samples of items taken from the process at, or very nearly at, the same time are frequently taken as subgroups. In chemical batch processes, a sample of the batches processed during a given day, or a given shift, or some other time period may be the logical choice. If no other basis seems indicated, division of the measurements into subgroups in the order received seems the best procedure, on the basis that items or batches produced consecutively will be more nearly homogeneous. It has been shown [27] that where the size of the subgroup is not determined by any technical consideration subgroups of 4 or 5 measurements are to be preferred. The subgroups should, if possible, contain equal numbers, since the computations are then greatly simplified, but this requirement should be subject to the two important points in the formation of subgroups; namely, that (1) the variations within each subgroup reflect as nearly as possible the inherent variability of the process, and (2) the variation between subgroups reflect the existence of assignable causes. The latter point almost invariably requires that the subgroups retain the natural order in which the measurements were taken.

10.32. Control Charts for Specified Quality Levels

One of the simplest applications of the control chart is to determine whether a process is in control with a specified average quality level μ and a specified variability σ. To determine whether the variations of the subgroup means $\bar{x}$ are in control, we prepare a control chart for these values with the average value and control limits based on the given values μ and σ. Since ave $(\bar{x}) = \mu$ and $\sigma_{\bar{x}} = \sigma/\sqrt{n}$, where n is the size of the subgroup, here assumed constant, the central line will be drawn at μ and the limits will be $\mu \pm 3\sigma/\sqrt{n}$. In order to facilitate the computation of these limits, the quantity $A = 3/\sqrt{n}$ has been tabulated for $n \leq 25$, and these values are given in Table 10.1. Thus the control limits in this case can be written in the form $\mu \pm A\sigma$, where the choice of A depends on the size of the subgroup. The essential features of this chart, which will be common to almost all charts used in this chapter, are shown in Figure 10.1. The designations UCL and LCL are commonly used abbreviations for upper control limit and lower control limit, respectively.

TABLE 10.1*

Number in Subgroup n	Factors for $\bar{x}$ Chart		Factors for R Chart					Factors for s Chart		
	A	A_2	d_2	D_1	D_2	D_3	D_4	c_2'	B_1'	B_2'
2	2.121	1.880	1.128	0.000	3.686	0.000	3.267	0.798	0.000	2.298
3	1.732	1.023	1.693	0.000	4.358	0.000	2.575	0.886	0.000	2.111
4	1.500	0.729	2.059	0.000	4.698	0.000	2.282	0.921	0.000	1.982
5	1.342	0.577	2.326	0.000	4.918	0.000	2.115	0.940	0.000	1.889
6	1.225	0.483	2.534	0.000	5.078	0.000	2.004	0.951	0.085	1.817
7	1.134	0.419	2.704	0.205	5.203	0.076	1.924	0.960	0.158	1.762
8	1.061	0.373	2.847	0.387	5.307	0.136	1.864	0.965	0.215	1.715
9	1.000	0.337	2.970	0.546	5.394	0.184	1.816	0.969	0.262	1.676
10	0.949	0.308	3.078	0.687	5.469	0.223	1.777	0.973	0.302	1.644
11	0.905							0.976	0.336	1.616
12	0.866							0.977	0.365	1.589
13	0.832							0.980	0.392	1.568
14	0.802							0.981	0.414	1.548
15	0.775							0.982	0.434	1.530
16	0.750							0.984	0.454	1.514
17	0.728							0.984	0.469	1.499
18	0.707							0.986	0.486	1.486
19	0.688							0.986	0.500	1.472
20	0.671							0.987	0.513	1.461
21	0.655							0.988	0.525	1.451
22	0.640							0.988	0.536	1.440
23	0.626							0.989	0.546	1.432
24	0.612							0.989	0.556	1.422
25	0.600							0.990	0.566	1.414

* The factors A, A_2, d_2, D_1, D_2, D_3, D_4 are reproduced from, and the factors B_1' and B_2' derived from, Table $B2$ of the *ASTM Manual on Quality Control of Materials*, American Society for Testing Materials, Philadelphia, January, 1951.

Where n differs for the different subgroups, there are 2 possible procedures. If the differences are small, it is usually a sufficiently close approximation to use the average $\bar{n}$ of the numbers in the subgroups to obtain the limits, but, if any appreciable discrepancies occur, it is better to compute individual control limits for each value of $\bar{x}$, rather than use continuous control limits as in Figure 10.1.

To determine whether the variability within subgroups is in control, we can prepare a control chart for s, the standard deviation of a subgroup, or, if $n \leq 10$, for the range R which is more easily computed. Since, as we have seen in Section 5.41, the average value of s is not in general equal to σ, the central line of the chart for s is drawn at $c_2'\sigma$, where c_2' is the required

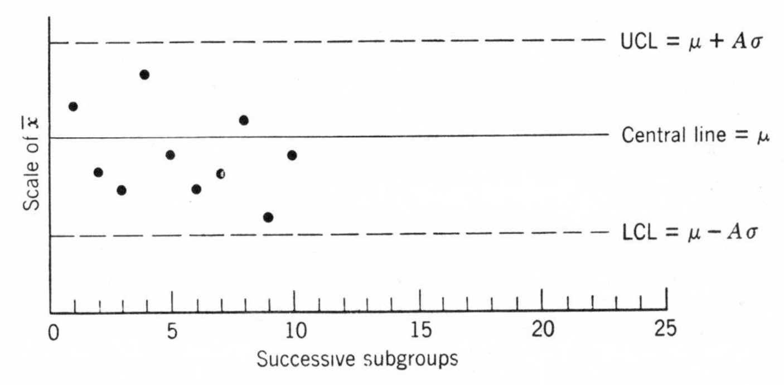

FIG. 10.1. Control chart for averages. Plotted points are subgroup averages.

correction for bias. Values of c_2' for $n \leq 25$ are given in Table 10.1. For control chart purposes s is generally assumed to be symmetrically distributed, and $\sigma/\sqrt{2n}$ is used for its standard deviation. Hence the control limits for s are $c_2'\sigma \pm 3\sigma/\sqrt{2n}$. Values of $B_1' = c_2' - 3/\sqrt{2n}$ and $B_2' = c_2' + 3/\sqrt{2n}$ for $n \leq 25$ are given in Table 10.1 to facilitate the direct computation of these limits. Using these values, the lower control limit, central line, and upper control limit for s are $B_1'\sigma$, $c_2'\sigma$, and $B_2'\sigma$, respectively.

In a similar fashion the average value, or central line, for the control chart for R is given by $d_2\sigma$, where the values of d_2 are those given in Section 5.41, and retabulated in Table 10.1 for convenience. The standard deviation of R, σ_R, was originally computed for various n by Tippett, and is given in the tables of the range referred to in Section 4.56. From these values factors D_1 and D_2 have again been computed so that $D_1\sigma$ and $D_2\sigma$ give the lower control limits and upper control limits for R directly, and these are tabulated for $n \leq 10$ in Table 10.1. In practice, control charts for the range are almost always used in preference to those for the standard deviation. This is due to the fact that the natural occurrence of true rational subgroups with $n > 10$ is infrequent (as we have mentioned, $n = 4$ or 5 is to be preferred) and for $n \leq 10$ the more easily computed and understood range is to be preferred.

The preparation of control charts for s and R is less simple when the subgroups are of unequal size since both the control limits and the central value vary with sample size. For values of $n > 10$, where s charts would be employed for control purposes, the variation in the central value is unimportant since c_2' differs little from unity in this region. If the discrepancy in subgroup sizes is not too great, the use of an average number $\bar{n}$ for the computation of the control limits and the central value is usually satisfactory; otherwise we can compute individual limits. For values of $n \leq 10$ the variation of the factor d_2 with sample size is appreciable, and control charts based on the range are not easily interpreted. In this case it is preferable to plot R/d_2, for which the average value is σ, regardless of sample size, and compute individual limits $D_1\sigma/d_2$ and $D_2\sigma/d_2$ for each value plotted.

10.33. Control Charts When no Levels are Specified

The more usual situation with respect to a quality characteristic to be controlled is that the average quality level μ and the inherent variability σ must be estimated from available process data, or from measurements obtained for this specific purpose. The best estimate of μ is the overall sample mean $\bar{\bar{x}}$, which may be computed from the subgroup means $\bar{x}$, weighted by the method of Section 5.32 when the subgroups are of unequal size. The method of estimating σ, the variability within groups, depends upon the type of control chart to be used. If a control chart for s is to be used, the standard deviation must be computed in any event and little additional labor is involved in recording the subgroup variance estimates s^2 and computing the best available estimate of σ by the method of Section 5.42. If the subgroup sizes are equal, this is equivalent to computing the square root of the average subgroup variance. For small subgroups where a control chart based on the sample range is employed the above method may be used, but it involves considerable additional labor, and the loss in efficiency is small if the average of the subgroup ranges $\bar{R}$ is used to estimate σ. It must be remembered that this estimate is biased, since ave $(R) =$ ave $(\bar{R}) = d_2\sigma$, where d_2 depends upon the subgroup size, and therefore $\bar{R}/d_2$ provides an unbiased estimate of σ as stated in Section 5.41. If the subgroup sizes are unequal, the division by the proper value of d_2 must be performed before these values are averaged.

The preparation of control charts in this instance is identical with the procedure of the preceding section except that the estimates $\bar{\bar{x}}$ and s' or

$\bar{R}/d_2$ are used in place of μ and σ. Hence the central line and control limits for $\bar{x}$ are given by $\bar{\bar{x}}$ and $\bar{\bar{x}} \pm As'$, or if the estimate $\bar{R}/d_2$ of σ has been used, by $\bar{\bar{x}}$ and $\bar{\bar{x}} \pm A\bar{R}/d_2$. The values $A_2 = A/d_2$ have been tabulated in Table 10.1 for $n \leq 10$, so that the control limits in the latter case can be computed directly in the form $\bar{\bar{x}} \pm A_2\bar{R}$, assuming, as above, that the subgroup sizes are equal.

Similarly, a control chart for s is prepared by simply replacing σ by s'; hence the central line and control limits are $c_2's'$, $B_1's'$, and $B_2's'$, respectively. If a control chart for R is to be used, the value $\bar{R}$ is a direct

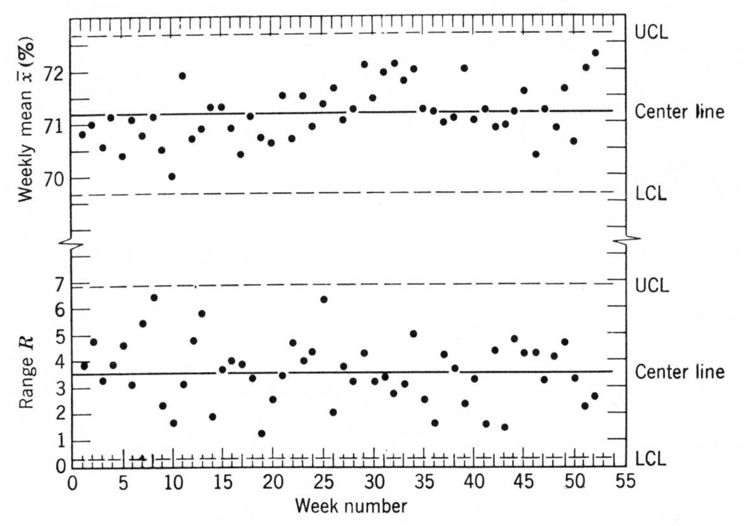

FIG. 10.2. Means and ranges of weekly subgroups.

estimate of the position of the central line, and the control limits will be $\bar{R} \pm 3\sigma_R$. Since the values of σ_R referred to in the previous section can be determined in terms of $\bar{R}$, factors D_3 and D_4 have been computed so that these limits are given directly as $D_3\bar{R}$ and $D_4\bar{R}$, respectively. Values of D_3 and D_4 for $n \leq 10$ are included in Table 10.1. Again these factors are conveniently applied only when the subgroup sizes are equal.

EXAMPLE 1. Let us consider the data of Table 2.1, giving the average daily yield of coke for 52 weeks. In this case it seems natural to choose a week as a subgroup. Table 10.2 gives the mean, variance, standard deviation, and range for each of the 52 weeks. From these we obtain $\bar{\bar{x}} = 71.186$, $s'^2 = 1.8231$ (in this case s'^2 is merely the mean of the variance estimates s^2, since the numbers in all subgroups are equal), $s' = 1.350$, and $\bar{R} = 3.60$. Note that the estimates $s' = 1.350$ and $\bar{R}/d_2 = 3.60/2.704 = 1.33$ are quite close to the overall standard deviation 1.352 for the data of Table 2.1 as a whole, indicating that there must be very little additional variability between subgroups in this case.

Figure 10.2 shows the control charts for subgroup means and ranges. In this case the central lines are $\bar{\bar{x}} = 71.19$ for the means, and $\bar{R} = 3.60$ for the ranges. The limits for the means are given by

$$\text{UCL} = \bar{\bar{x}} + A_2\bar{R} = 71.19 + 0.419(3.60) = 72.70$$

$$\text{LCL} = \bar{\bar{x}} - A_2\bar{R} = 71.19 - 0.419(3.60) = 69.68$$

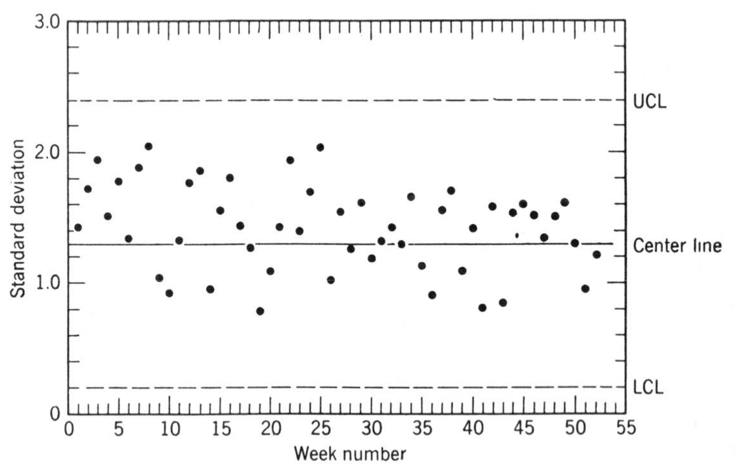

FIG. 10.3. Standard deviations of weekly subgroups.

and for the ranges by

$$\text{UCL} = D_4\bar{R} = 1.924(3.60) = 6.91$$

$$\text{LCL} = D_3\bar{R} = 0.076(3.60) = 0.29$$

In this case we could also have used s' in place of the range, to fix the control limits for the subgroup means and should have obtained slightly wider limits in each instance. Also, in place of the range chart, we could have used the control chart for s' shown in Figure 10.3, the central line for this chart is given by $c_2's' = 0.960(1.350) = 1.30$ and the limits by

$$\text{UCL} = B_2's' = 1.762(1.350) = 2.38$$

$$\text{LCL} = B_1's' = 0.158(1.350) = 0.21$$

Neither the range chart nor the standard deviation chart gives any indication of lack of control of variability within subgroups. On the mean chart, although there are no points outside the limits, there is a suspiciously large number of the earlier values below the central line, followed by what may be a slight trend upward ending with the thirty-fourth week. These features will be discussed later in this chapter.

TABLE 10.2

Week	$\bar{x}$	s^2	s	R	Week	$\bar{x}$	s^2	s	R
1	70.84	1.636	1.28	3.9	27	71.06	2.086	1.44	3.8
2	71.03	2.769	1.66	4.8	28	71.26	1.190	1.09	3.3
3	70.57	3.746	1.94	3.3	29	72.07	2.326	1.53	4.3
4	71.17	1.996	1.41	3.9	30	71.46	0.996	1.00	3.3
5	70.40	2.987	1.73	4.6	31	71.97	1.349	1.16	3.4
6	71.10	1.300	1.14	3.0	32	72.09	1.661	1.29	2.8
7	70.79	3.455	1.86	5.5	33	71.81	1.291	1.14	3.1
8	71.16	4.300	2.07	6.4	34	72.00	2.507	1.58	5.0
9	70.49	0.665	0.82	2.3	35	71.27	0.872	0.93	2.6
10	70.03	0.452	0.67	1.7	36	71.20	0.417	0.65	1.7
11	71.94	1.320	1.15	3.1	37	70.97	2.139	1.46	4.3
12	70.71	2.945	1.72	4.8	38	71.14	2.713	1.65	3.7
13	70.90	3.373	1.84	5.8	39	72.03	0.779	0.88	2.4
14	71.30	0.503	0.71	1.9	40	71.04	1.643	1.28	3.4
15	71.31	2.118	1.46	3.7	41	71.24	0.290	0.54	1.6
16	70.93	3.062	1.75	4.0	42	70.91	2.218	1.49	4.4
17	70.40	1.723	1.31	3.9	43	70.96	0.326	0.57	1.5
18	71.17	1.242	1.11	3.4	44	71.21	2.058	1.43	4.8
19	70.74	0.230	0.48	1.3	45	71.59	2.305	1.52	4.3
20	70.63	0.739	0.86	2.6	46	70.40	1.993	1.41	4.3
21	71.49	1.685	1.30	3.5	47	71.23	1.449	1.20	3.3
22	70.71	3.728	1.93	4.7	48	70.91	1.938	1.39	4.1
23	71.49	1.565	1.25	4.0	49	71.64	2.353	1.53	4.7
24	70.93	2.629	1.62	4.4	50	70.63	1.326	1.15	3.4
25	71.37	4.199	2.05	6.3	51	72.03	0.499	0.71	2.3
26	71.64	0.623	0.79	2.1	52	72.29	1.088	1.04	2.7

EXAMPLE 2. In a refining process, samples are taken periodically for analysis to determine the amount of various impurities present. Table

10.3 gives the results of the analyses for silica content (ppm) for a 9-month period. For these data we have

$$\bar{\bar{x}} = \frac{12(65.8) + 9(36.9) + \cdots + 6(59.0)}{12 + 9 + \cdots + 6}$$

$$= \frac{5032.8}{79} = 63.71$$

$$s'^2 = \frac{11(1027.30) + 8(441.61) + \cdots + 5(187.20)}{11 + 8 + \cdots + 5}$$

$$= \frac{31770.79}{70} = 453.87$$

$$s' = 21.3$$

TABLE 10.3

Month	Number of Analyses	$\bar{x}$	s^2	s
1	12	65.8	1027.30	32.1
2	9	36.9	441.61	21.0
3	10	49.4	605.82	24.6
4	11	74.4	317.65	17.8
5	7	59.3	229.57	15.2
6	11	76.6	295.65	17.2
7	8	79.5	307.43	17.5
8	5	70.2	221.70	14.9
9	6	59.0	187.20	13.7

Control charts for the mean $\bar{x}$ and standard deviation s, with individual limits computed for each month, are shown in Figure 10.4. Although the central line, as well as the limits of the s chart should change with n, in this case this change is so slight that there is little error in using a constant line based on the average number of analyses in each subgroup.

A lack of control is indicated by the low value obtained for the average silica content during the second month. Equally interesting is the fact that, although all the individual standard deviations are within the control limits, the variation in the analyses appears to have decreased steadily during the period studied.

10.34. Discussion of the Use of Control Charts

It will be noted that, in computing control limits for the preparation of control charts, we have in many instances used approximations where more precise limits from a statistical viewpoint might be obtained by the methods of Chapter 5. It should be remembered that control limits are not being used in order to make precise probability statements as in the case of confidence limits but as a basis for action, and in this situation experience has shown the more easily computed limits given above to be completely satisfactory. If the control of a process is thought to justify

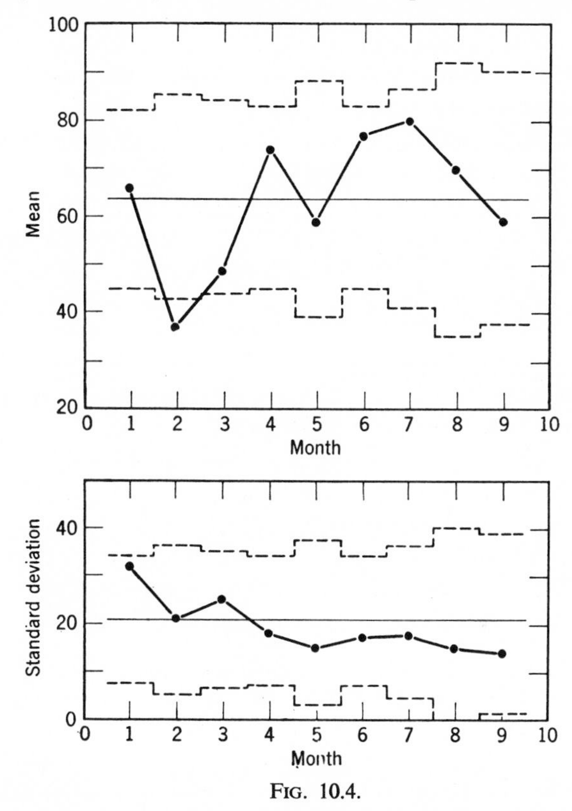

FIG. 10.4.

the additional labor involved, "exact" probability limits may be introduced by basing the limits for the standard deviation upon the χ^2-distribution, and those for the range upon the distribution mentioned in Section 4.56. It should be remembered that the resulting probability statements are exact only if the assumptions concerning the parent population are satisfied, and that the real objective of the control chart is to provide a basis for plant control rather than an exact probability statement. For these reasons the simpler techniques are almost always satisfactory.

If it is necessary to estimate μ and σ from data already available or obtained for the purpose, it is very likely that there will be assignable causes present, and these may affect the estimates of $\bar{\bar{x}}$ and s' or $\bar{R}$ obtained. As these assignable causes are detected and removed it will be necessary to revise our estimates on the basis of data obtained from the system as a condition of statistical control is approached. A knowledge of the source of a part of the variation in the earlier data may enable us to exclude some of this material in computing new control limits, but observations should not be excluded because they represent points which are out of control unless investigation has revealed an assignable cause for this divergence. This latter condition should also be observed in the preparation of initial limits from available data.

If μ and σ are specified, and the data appear out of control on the basis of the limits computed from these values, it is advisable to determine limits based on the estimates $\bar{\bar{x}}$ and s' or $\bar{R}$ to see if the data are in control at any level. If the data are in control on the latter basis, the process is producing a consistent product, but is incapable of meeting the specified quality levels without modification.

10.4. Control Charts for Attributes

10.41. Control Charts for Fraction Defective

The quality of a product is frequently measured in terms of some characteristic which the product does or does not possess. Items produced which are unsatisfactory are generally referred to as defectives. If we assume that for a process in a state of control each item produced has an equal probability p of being defective, then we should expect the fraction defective x/n in successive rational subgroups to vary in a binomial distribution with expected value p and standard deviation $\sqrt{\dfrac{p(1-p)}{n}}$, as was shown in Section 4.61. Control charts based on this distribution can be used with variables by considering all measurements outside certain limits as defectives. This procedure is less effective than those of the previous section in the sense that a larger number of observations is required to provide the same control, but it may have economic advantages if the cost of classification is appreciably less than the cost of measurement. Examples of this situation are provided by the use of "go, no-go" gauges in the engineering industry or color standards in the measurement of H_2S by the staining of lead acetate paper.

If the fraction defective p at which the process is to be controlled is specified, the central line and control limits for x/n are given by p and $p \pm 3\sqrt{\dfrac{p(1-p)}{n}}$. If the subgroup sizes are unequal, we can as before use the average number $\bar{n}$ in computing the limits if the discrepancies are not too great, but, if there are large variations in subgroup size, individual limits should be computed. For example, if we choose a day's production of 475 to 500 items as a subgroup, there would certainly be little loss in using the average daily production to compute constant limits for the fraction of each day's production defective.

If the expected fraction defective p is not specified, we must again obtain an estimate $\hat{p}$ from available data and use it in place of p for the central line and in computing the control limits. As shown in Section 9.31, the correct estimate is obtained by dividing the total number of defectives by the total number of items. The estimate $\hat{p}$ will in this case be affected by assignable causes present, and should be adjusted as these are identified and removed.

In order to justify the use of symmetric limits the number in the subgroups should be fairly large, say $n \geq 100$, and the expected number of defectives per subgroup $np \geq 50$. If control limits are desired for smaller subgroups, they can be obtained by computing 0.997, or possibly 0.99 or 0.998, probability limits for the binomial by the exact methods of Section 4.61.

EXAMPLE. In Example 3 of Section 9.33, the number of unacceptable pieces produced in 12 consecutive lots of a given product were tabulated. Table 10.4 shows these data, along with the fraction defective for each lot and for the combined lots. In this case the numbers in each lot are sufficiently close that the use of constant limits based on the average number in each lot should be satisfactory. Hence, using $\bar{n} = 1670$ and $\hat{p} = 0.0832$, we have

$$\begin{aligned}
\text{UCL} &= \hat{p} + 3\sqrt{\frac{\hat{p}(1-\hat{p})}{n}} \\
&= 0.0832 + 3\sqrt{\frac{(0.0832)\,(0.9168)}{1670}} \\
&= 0.0832 + 3(0.0066) \\
&= 0.1030
\end{aligned}$$

$$\text{Central Line} = \hat{p} = 0.0832$$

$$\begin{aligned}
\text{LCL} &= 0.0832 - 3(0.0066) \\
&= 0.0634
\end{aligned}$$

TABLE 10.4

Lot	Number in Lot	Number Unacceptable	Fraction Unacceptable
1	1,781	171	0.096
2	1,685	139	0.082
3	1,606	121	0.075
4	1,622	95	0.059
5	1,687	229	0.136
6	1,638	109	0.067
7	1,650	156	0.095
8	1,580	106	0.067
9	1,611	148	0.092
10	1,774	162	0.091
11	1,659	134	0.081
12	1,750	98	0.056
Total	20,043	1668	0.0832

The control chart is shown in Figure 10.5. Of the 12 points 3 are out of limits and the lack of control indicated by the previous analysis of these data is confirmed. Since 2 of the 3 points are below expectation, it seems probable that the overall fraction defective 0.0832 indicated by these 12 lots could be appreciably lowered if assignable causes were removed.

10.42. Control Charts for Number Defective

If the numbers in each subgroup are equal, it is sometimes preferable to prepare a control chart for x, the actual number of defectives, rather than the fraction defective. This is not convenient in unequal subgroups since the average number of defectives np will vary with n. For fixed n and a specified p we can use this average value for the central line and the limits $np \pm 3\sqrt{np(1-p)}$ to construct the control chart. If p is small, say <0.05, and n sufficiently large so that $np > 50$, we can assume the Poisson approximation to the binomial and replace these limits by $np \pm 3\sqrt{np}$.

If p is not specified, we can again replace it by $\hat{p}$. In this case $n\hat{p}$ is equivalent to $\bar{x}$, the average number of defectives since all subgroups are assumed to be of size n. If the Poisson approximation is applicable, the limits become approximately $\bar{x} \pm 3\sqrt{\bar{x}}$, and are independent of the subgroup size n. These limits can be used where the subgroups are indeterminately large but of approximately equal size, and only the average number of defectives can be determined. Such limits are particularly useful in connection with the control of counting instruments commonly used in radiochemistry.

10.5. Control Charts for Errors of Measurement

10.51. The Need for the Control of Measurement Errors

An application of control charts which is of great importance in the chemical industry is the control of errors of measurement, particularly

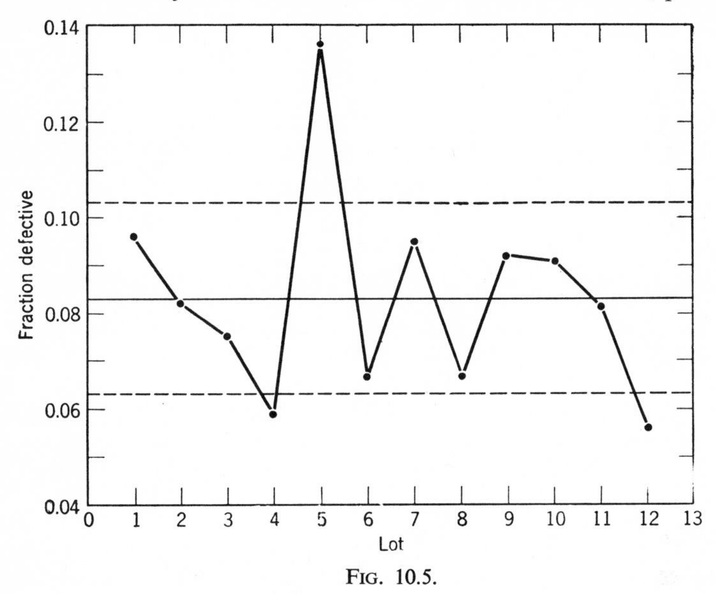

FIG. 10.5.

those associated with chemical analyses. Although in general the observed variation between measurements of a given quality characteristic will consist of two parts—the actual variation from item to item in the characteristic being controlled and the variation due to errors in measurement—the latter variation in some engineering and production applications is quite small and not worthy of separate consideration. In the chemical industry many measurements, both of quality characteristics and of other items essential to production control or for inventory purposes, are the result of chemical analyses or other indirect methods of measurement which contribute appreciably to the overall variation. This is especially true where the analytical procedures and production procedures differ only in scale, as in the use of essentially the same extraction and precipitation procedure both for processing the batch and for analyzing the resulting product. It should be noted that this problem can also arise in the case of attributes where the defectiveness or non-defectiveness of an item involves judgment on the part of the operator. However, we shall not consider this phase of measurement error, since the control of attributes as such occurs only infrequently in chemical problems.

10.52. Control Charts for Accuracy

The $\bar{x}$ chart can easily be adapted to control the accuracy of measure-

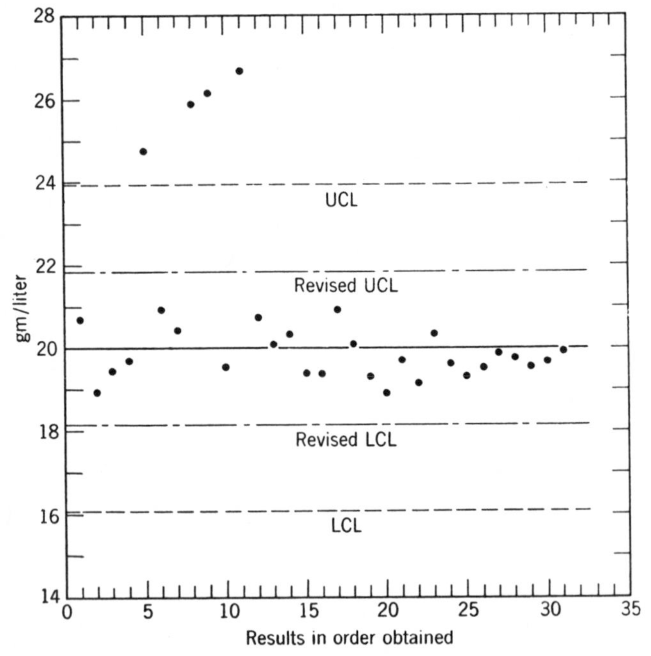

FIG. 10.6. Control analysis results.

ments. If repeated measurements are made of a standard so that the average value μ is known, these measurements can be divided into rational subgroups, and a control chart prepared for the subgroup means $\bar{x}$ based on the expected value μ and the estimated variability s' or $\bar{R}/d_2$ within subgroups. The limits are computed exactly as in Section 10.33 except that the value μ of the standard replaces $\bar{\bar{x}}$. Points out of control on this chart would indicate a lack of accuracy for the measurements in question.

Where the subgrouping is arbitrarily chosen we are often more interested in the individual determinations than in the means of the subgroups. Unless there is some indication of lack of normality in the underlying distribution, it may be more desirable in this case to prepare a control chart for individual determinations, the estimate of the inherent variability based on the subgrouping being used. In this case the controls limits would be simply $\mu \pm 3s'$, or $\mu \pm 3\bar{R}/d_2$.

EXAMPLE. The data of Table 10.5 represent 31 determinations of the uranium content of a standard sample with a concentration of 20 g/l, reported in the order in which they were obtained. To obtain an estimate of the inherent variability these have been arbitrarily grouped in subgroups of 4. From the ranges of the complete subgroups we obtain $\bar{R} = 2.72$, and hence, using the standard value $\mu = 20.0$ g/l in place of $\bar{\bar{x}}$, we have for the control limits for an individual value

$$\text{UCL} = \mu + 3\bar{R}/d_2 = 23.96$$

$$\text{LCL} = \mu - 3\bar{R}/d_2 = 16.04$$

Figure 10.6 shows the control chart for the individual values, with these limits.

The fifth, eighth, ninth, and eleventh points are beyond the control limits, and they are obviously discrepant from the remaining determinations. An investigation was made of the source of the discrepancies, and

TABLE 10.5

Determination	1	2	3	4	R
1–4	20.62	18.95	19.40	19.70	1.67
5–8	24.76	20.87	20.41	25.90	5.49
9–12	26.12	19.52	26.73	20.73	7.21
13–16	20.04	20.27	19.38	19.36	0.91
17–20	20.92	20.08	19.26	18.94	1.98
21–24	19.72	19.13	20.30	19.59	1.17
25–28	19.30	19.50	19.88	19.76	0.58
29–31	19.58	19.63	19.93	—	—

for purposes of future control the limits were revised, a variance estimate $\bar{R}/d_2 = 0.61$, based on those subgroups not involving the discrepant values, being used. These limits are also shown in Figure 10.6, where there is no further evidence of lack of control. However, it should be noted that 12 of the last 13 points are below the standard value, seemingly owing to a downward trend in the results obtained, and also that the variation in the results seems to be decreasing. The first observation indicates that possible causes of systematic error in the analysis should be investigated, and the latter indicates the possibility of further revision of the control limits if the trend continues.

Since measurements for accuracy control must be made on a known standard, they represent non-productive measurements on the part of the laboratory, inspector, or instrument being controlled. The number of such control measurements which can be obtained will depend on the economic factors involved; for example, checking a micrometer on a test block is a different matter from checking a 3-hour analysis on a standard solution. One convenient method of determining the number of control measurements to be made is to consider the fraction of the total productive time which can be devoted to such measurements, taking into account the seriousness of undetected deviations of the measurements from the true value due to an assignable cause. Thus we may decide that perhaps 1% or 5% of the capacity of the instrument, inspector, or laboratory can be economically devoted to the control of accuracy, and accordingly we obtain approximately 1 measurement of a standard for every 100 or 20 routine measurements. It may be possible to combine such limits with some obvious choice of rational subgroups for these measurements. For example, if a laboratory is performing approximately 30 analyses of a given type per shift, it might be practicable to perform 1 control analysis at the beginning of each shift, the 3 analyses during each work day forming a subgroup, or, if an inspector were gauging approximately 500 items per shift, 5 test block measurements made throughout the shift might be considered a rational subgroup.

A second factor that must be considered in the routine control of accuracy is the psychological fact that increased care will frequently be taken by operators or analysts who realize that the results are to be used for accuracy control purposes, and hence these measurements may not be truly representative of the accuracy of the measurements as a whole. This is less important in situations where the measurement errors are primarily instrumental, as in the case of scales, spectrophotometers, gauges, etc., than in analytical measurements depending to a great extent on the skill of the analyst. One possible method of overcoming this difficulty is to submit the standard samples to be analyzed in such a manner that their nature is unknown, but this procedure often has disadvantages from an administrative viewpoint.

10.53. Control Charts for Precision

Although it is impossible to control the accuracy of routine measurements except through control analyses performed specifically for the purpose, it is possible where more than 1 routine measurement is made on a single item to estimate the variability or precision of these measurements. Here the measurements made on a single item will constitute a subgroup. The control charts used for this purpose are identical with the control charts for s and $\bar{R}$ discussed in Section 10.32 and 10.33. Since chemical analyses are usually made in duplicate, and less frequently in triplicate or quadruplicate (instances of greater numbers are rare), the control chart for the range is almost invariably used for this purpose.

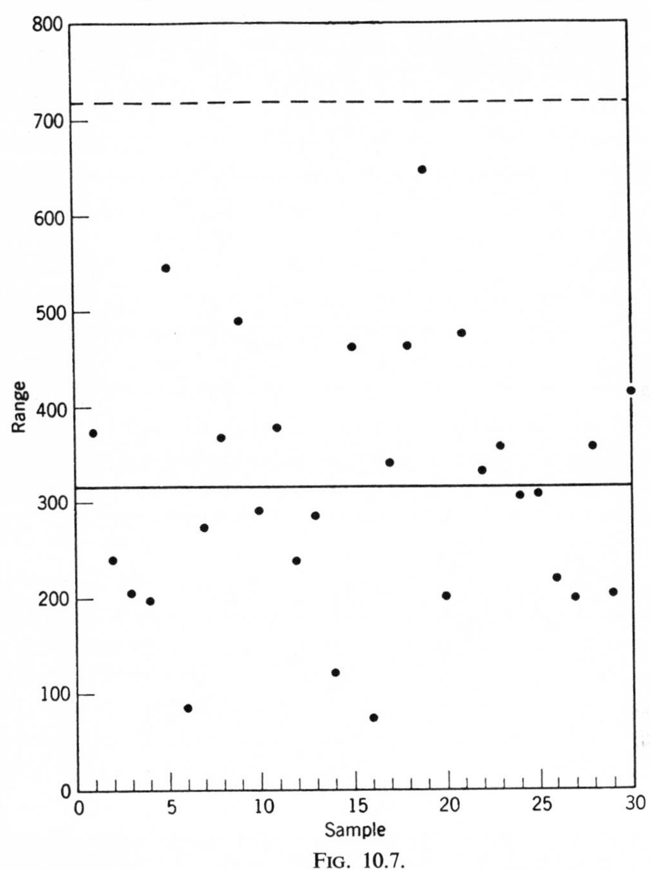

FIG. 10.7.

EXAMPLE. Table 10.6 gives the results in counts/minute, of 4 replicate determinations on each of 30 samples analyzed for plutonium by a radio-assay method. The mean and range of the results on each sample are also given. To determine whether the precision of these analyses is in control, we construct a range chart. From the tabulated ranges we have $\bar{R} = 314.5$, and hence

$$\text{UCL} = D_4\bar{R} = (2.282)(314.5) = 717.7$$

$$\text{Central line} = \bar{R} = 314.5$$

$$\text{LCL} = D_3\bar{R} = (0)(314.5) = 0$$

The control chart for ranges is given in Figure 10.7, and shows no evidence of lack of control.

In many cases control limits of this type are used as a basis for determining whether the given measurement will be repeated. For example, if in a chemical analysis the range of duplicate determinations is above the UCL, we may decide that an assignable cause was present, regardless of whether it can be identified, and repeat the analysis, or perform an additional determination to discover which of the original determinations

TABLE 10.6

Sample	Determination				Mean	Range
	1	2	3	4	$\bar{x}$	R
1	14,002	14,134	13,771	13,761	13,917	373
2	13,925	13,730	13,970	13,901	13,882	240
3	13,923	13,908	13,729	13,934	13,874	205
4	14,028	14,065	14,108	14,226	14,107	198
5	13,826	13,723	13,280	13,390	13,555	546
6	14,449	14,478	14,519	14,533	14,495	84
7	13,889	13,624	13,616	13,642	13,693	273
8	14,095	14,123	13,792	13,755	13,941	368
9	14,066	14,007	13,639	13,575	13,821	491
10	14,171	14,037	13,881	14,021	14,028	290
11	14,127	14,088	13,914	13,749	13,970	378
12	13,865	13,901	14,070	14,103	13,985	238
13	13,397	13,330	13,254	13,539	13,380	285
14	14,046	14,079	14,119	14,166	14,103	120
15	13,829	13,882	14,291	14,091	14,023	462
16	14,299	14,335	14,263	14,321	14,305	72
17	15,442	15,222	15,501	15,562	15,432	340
18	14,731	15,000	15,029	15,194	14,989	463
19	14,646	14,551	15,199	14,846	14,811	648
20	14,815	14,702	14,802	14,902	14,805	200
21	15,428	15,647	15,241	15,714	15,508	473
22	15,620	15,656	15,952	15,882	15,778	332
23	15,645	15,677	15,343	15,320	15,496	357
24	15,550	15,558	15,681	15,856	15,661	306
25	15,414	15,480	15,655	15,347	15,474	308
26	15,564	15,476	15,409	15,347	15,449	217
27	14,881	14,912	14,974	15,079	14,962	198
28	14,990	14,814	15,090	14,734	14,907	356
29	15,500	15,517	15,577	15,702	15,574	202
30	14,623	14,687	14,880	15,034	14,806	411

was in error. In a sense we are then using the control chart as a test of the hypothesis that the variation found between this pair of duplicate analyses is consistent with that observed in previous analyses. In such cases it may be desirable, especially where the cost of the analysis is small compared with the consequences of the possible incorrect result, to use 95% or 99% limits in preference to the 3σ limits customarily used, on the basis that the additional protection afforded is worth the "useless" repetition of 1% to 5% of the analyses. However, it must be remembered that these are not control limits in the usual sense, and we may still wish to base our investigation of assignable causes on the more conservative 3σ limits. Factors are given in Table 10.7 for $n \leq 10$ which, when multiplied by $\bar{R}$ or either s' or σ, will give the 95% and 99% upper limits for R. The limits so obtained are one-sided, since when using control limits in the manner described above we usually are interested only in detecting those cases where the variation is too large. Occasionally it may be desirable to have a lower limit to detect some feature, such as lack of independence, which might be indicated by an unusually small range. Factors for obtaining such limits can be found in the more complete table of percentage points for the range given in [16, Table 2].

It should be re-emphasized that, if control of this type is to be effective, the repeated measurements in a subgroup should represent completely

TABLE 10.7

PERCENTAGE POINTS FOR RANGE*

Number of Determinations	Upper 5% Points		Upper 1% Points	
	Based on R	Based on σ	Based on R	Based on σ
2	2.45	2.77	3.23	3.64
3	1.95	3.31	2.43	4.12
4	1.76	3.63	2.14	4.40
5	1.66	3.86	1.98	4.60
6	1.59	4.03	1.88	4.76
7	1.54	4.17	1.80	4.88
8	1.51	4.29	1.75	4.99
9	1.48	4.39	1.71	5.08
10	1.45	4.47	1.68	5.16

* Reproduced by permission of Professor E. S. Pearson from Table 2 of: "The Probability Integral of the Range in Samples of n Observations from a Normal Population," *Biometrika*, *32* (1942), 301–310, by E. S. Pearson and H. O. Hartley.

independent measurements. Thus duplicate analytical determinations on a single sample may be used to control analytical precision, but they give no control at all over the variations due to sampling, which may be an important part of the measurement error, and it would usually be preferable to have the duplicate determinations made on completely independent samples, so that the control would include both sources of error.

10.54. Special Control Charts for Duplicate Analyses

It is a common practice in most routine analytical laboratories to perform duplicate analyses to provide a control over the presence of gross analytical errors. A special form of control chart which requires no computation on the part of the analyst using the chart has been successfully used in control based on duplicate analyses and provides an additional control based on the order in which the two measurements are obtained.

For duplicate measurements x and y we have $R = |x - y|$. However, if instead of the range we consider the difference $d = x - y$, where x and y are the observations *in the order obtained*, we have the additional advantage of controlling not only the size of the difference but also its direction. Thus in controlling d we might recognize a tendency for the second result to be lower than the first, as might be the case with colorimetric determinations with somewhat unstable color development, whereas on a control chart for R this might go unnoticed. We have seen in Section 4.52 that, if x and y are independently distributed with ave $(x) =$ ave $(y) = \mu$ and var $(x) =$ var $(y) = \sigma^2$, then ave $(d) =$ ave $(x - y) =$ ave $(x) -$ ave $(y) = 0$, and var $(d) =$ var $(x) +$ var $(y) = 2\sigma^2$. Hence 3σ control limits for d would be given by $\pm 3\sqrt{2}\sigma$. When σ is unspecified the estimate $s' = \Sigma d^2/2k$, where k is the number of subgroups, or pairs of duplicate analyses, available for estimation, or the estimate $\bar{R}/d_2$, where $\bar{R} = \dfrac{\Sigma |d|}{k}$, can be used. In this case, if a modern calculating machine is available, there is little difference in ease of calculation, and hence the estimate s' is to be preferred.

EXAMPLE 1. Table 10.8 gives the values of each of 20 duplicate plutonium determinations made routinely on plant batches. From the differences we obtain $\Sigma |d| = 66.00$, and $\Sigma d^2 = 313.1954$. Hence

$$\bar{R} = \frac{66.00}{20} = 3.30$$

$$s'^2 = \frac{313.1954}{40} = 7.8299$$

$$s' = 2.80$$

TABLE 10.8

Batch	1st Analysis	2nd Analysis	Difference
1	263.36	269.81	−6.45
2	245.68	253.82	−8.14
3	248.64	252.05	−3.41
4	272.68	267.53	+5.15
5	261.10	266.08	−4.98
6	287.33	282.90	+4.43
7	266.41	272.52	−6.11
8	287.26	286.30	+0.96
9	276.32	275.08	+1.24
10	243.65	245.26	−1.61
11	256.42	252.01	+4.41
12	282.65	283.85	−1.20
13	250.97	253.75	−2.78
14	284.27	284.55	−0.28
15	258.25	254.54	+3.71
16	291.05	286.53	+4.52
17	294.34	297.80	−3.46
18	261.02	260.37	+0.65
19	267.87	268.14	−0.27
20	280.37	282.61	−2.24

The estimate of σ based on $\bar{R}$ is $3.30/1.128 = 2.93$, in good agreement with the value of s'.

Figure 10.8 shows control charts for both differences and ranges. The limits for the difference chart were obtained as $\pm 3\sqrt{2}s' = \pm 11.88$, the central line being 0; the limit for the range chart was obtained as $D_4\bar{R} = 10.78$; and the central line was placed at $\bar{R}$ in the usual manner. Neither of these charts shows any evidence of lack of control. The difference chart indicates no systematic tendency for the first analysis to be lower than the second, or vice versa. This information is not available from the range chart.

Another form of control chart for this case is obtained merely by plotting the duplicate measurements x and y, *in the order obtained.* These plotted points would be expected to fall on the 45° line $d = x - y = 0$, which corresponds to the central line of the usual chart, and within the limits given by the lines $d = x - y = \pm 3\sqrt{2}\sigma$. The chart thus obtained is illustrated in Figure 10.9. Note that the perpendicular distance from the central line to the control limits is exactly 3σ. Any consistent tendency for the first measurement to be greater than the second, or vice versa, would be indicated by points falling consistently above or below the central line.

The advantages of this chart over the usual control chart for d are that the duplicate results can be plotted directly without numerical calculations and that the magnitude of the determinations themselves is retained. One disadvantage is that there is no convenient way to preserve the order of the pairs of measurements, but this can be partly overcome by plotting

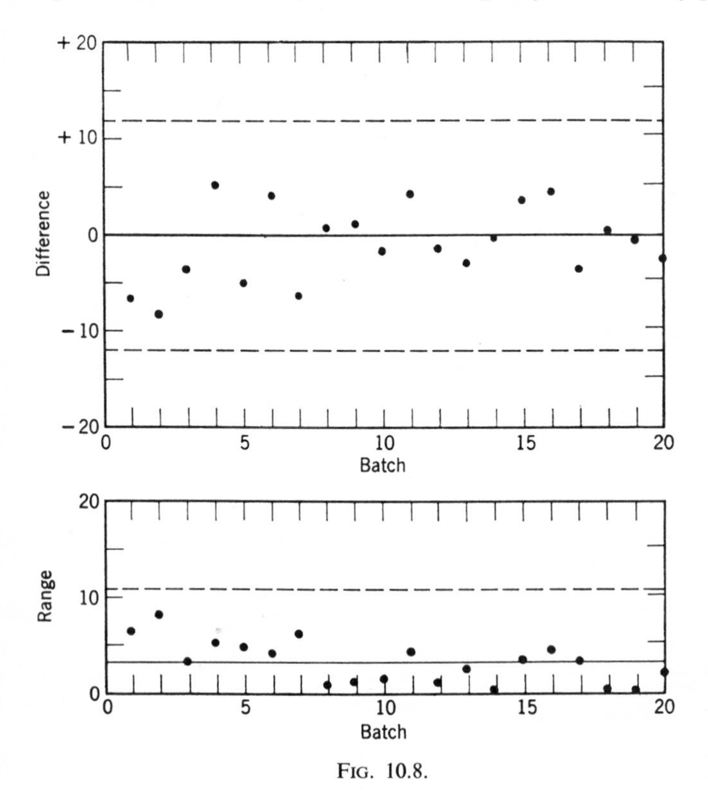

FIG. 10.8.

an auxiliary chart indicating *in the order obtained* whether each point is above or below the central line. The number of runs above and below the central line can then be used as a means of detecting non-randomness. The techniques involved are considered in Section 11.3.

If a standard in the same concentration range as the routine samples is being analyzed at intervals, control limits for this standard can be included on the same chart. We assume the pair of independent measurements on the standard to have a bivariate normal distribution with $\rho = 0$, ave (x) = ave $(y) = \mu$, the concentration of the standard, and var (x)

$= \text{var}(y) = \sigma^2$. Hence the point (x, y) would, from (f) of Appendix 4B, have a probability α of falling within the circle

$$(x - \mu)^2 + (y - \mu)^2 = \lambda_\alpha^2\sigma^2$$

with center at the point (μ, μ) on the 45° central line and radius $\lambda_\alpha\sigma$, where λ_α is chosen so that $Pr[\chi_2^2 < \lambda_\alpha^2] = \alpha$. The value of λ_α which gives a circular limit equivalent to the usual 3σ limit is 3.44. Points falling outside this limit would indicate an assignable cause of inaccuracy.

EXAMPLE 2. The data of Table 10.9 represent the duplicate results

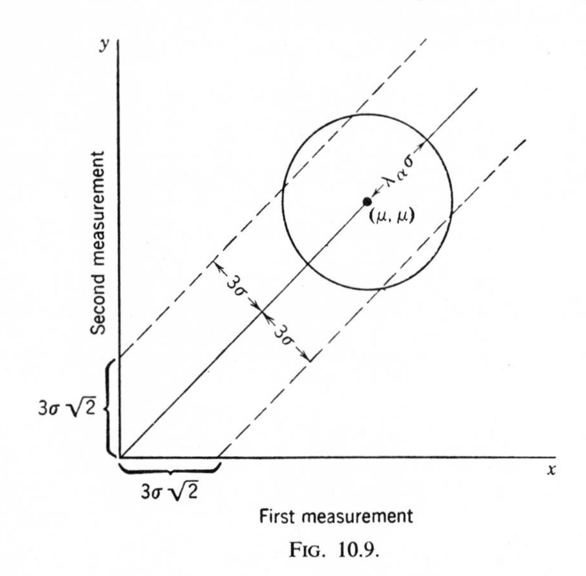

FIG. 10.9.

obtained on a standard sample containing 29.27% iron by 38 students in the chemistry laboratory of a large university. From the differences we compute the values

$$\bar{R} = \frac{\Sigma |d|}{k} = \frac{3.47}{38} = 0.0913$$

$$s' = \sqrt{\frac{\Sigma d^2}{2k}} = \sqrt{\frac{0.4661}{76}} = 0.0783$$

From the first of these values we obtain as an estimate of σ

$$\frac{\bar{R}}{d_2} = \frac{0.0913}{1.128} = 0.0809$$

which agrees well, as it should in this case, with that obtained from s'.

TABLE 10.9

(% iron found − 25)

Student	Results 1	Results 2	Diff.	Student	Results 1	Results 2	Diff.
1	4.41	4.32	0.07	20	4.27	4.25	0.02
2	4.27	4.39	−0.12	21	4.09	3.98	0.11
3	3.87	3.87	0.00	22	3.92	3.99	−0.07
4	4.04	4.03	0.01	23	4.26	4.15	0.11
5	4.73	4.96	−0.23	24	4.00	3.90	0.10
6	3.79	3.89	−0.10	25	4.40	4.23	0.17
7	4.30	4.20	0.10	26	3.96	4.04	−0.08
8	4.13	4.13	0.00	27	3.95	4.10	−0.15
9	4.21	4.05	0.16	28	3.90	3.97	−0.07
10	4.09	4.22	−0.13	29	4.19	4.20	−0.01
11	3.76	3.76	0.00	30	4.09	4.23	−0.14
12	4.33	4.24	0.09	31	4.07	4.10	−0.03
13	4.66	4.58	0.08	32	4.24	4.13	0.11
14	4.49	4.60	−0.11	33	4.27	4.20	0.07
15	4.42	4.39	0.03	34	4.85	4.80	0.05
16	4.32	4.25	0.07	35	4.35	4.45	−0.10
17	4.21	4.25	−0.04	36	4.32	4.11	0.21
18	4.30	4.08	0.22	37	4.20	4.42	−0.22
19	4.67	4.62	0.05	38	4.40	4.36	0.04

For $\sigma = s' = 0.0783$, Figure 10.10 shows a control chart, similar to that in Figure 10.9, on which the pairs of results from the above table have been plotted. The control lines shown are

$$x - y = \pm 3\sqrt{2}(0.0783)$$
$$= \pm 0.332$$

and the circle has center (4.27, 4.27), corresponding to the standard value of 29.27% iron, and radius $\lambda\sigma = 3.44(0.0783) = 0.269$. From this chart it is evident that (1) as measures of precision the results are in control; (2) as measures of accuracy the results are hopelessly out of control; (3) the lack of control in (2) is not restricted to a few of the pairs of measurements but is general. The high precision and low accuracy of the results may be due to selection or "cooking" by the students in reporting the results, to consistent errors in the analytical techniques, or to consistent sampling errors if both determinations were carried out on the same sample. Under the conditions of the test the last possibility is unlikely, and it seems probable that serious and consistent errors in analytical techniques are present and that most, if not all, of the students are responsible for such errors.

Although in this case there is no real significance to the order in which the students have been arranged, the usual supplementary chart indicating the position of each point in order of occurrence has been included.

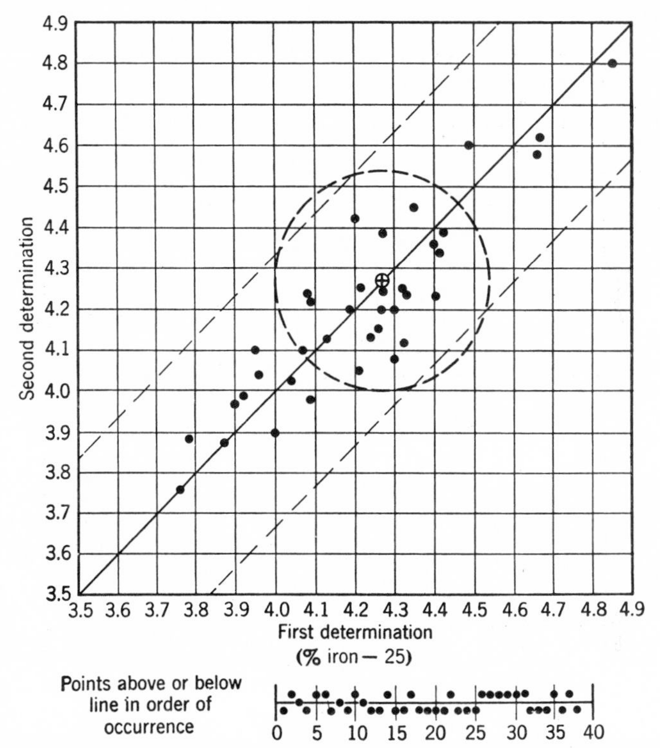

FIG. 10.10. Control chart for duplicate determinations on a given standard.

Three points for which the duplicate determinations were identical within the number of significant figures given have been plotted directly on the line. The tests for non-randomness to be considered in Chapter 11 do not indicate that the order of occurrence of the observations is in any way unusual.

10.55 Control Charts for Counting Instruments

Another special case in which the use of control charts has been particularly beneficial is in the control of radio-assay counting instrument.

Owing to the statistical nature of the emission of radioactive particles, the preparation of control charts for this purpose is quite direct. We have already indicated that the number of radio-active particles emitted in equal time intervals by a substance undergoing decay has a Poisson distribution if it is assumed (1) that the half-life of the decay is sufficiently long so that the probability of an emission in a given time interval remains constant, and (2) that any daughter decay present is negligible. It can

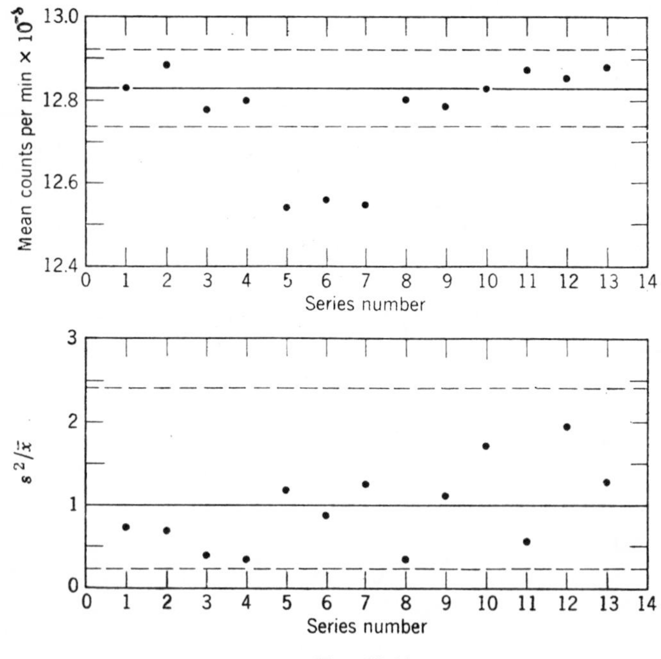

FIG. 10.11.

also be shown that, unless the counting rate or the resolution time is quite large, the actual counts recorded by a counterscaler unit will also have very nearly a Poisson distribution. Since the Poisson distribution is very closely approximated by the normal distribution for ave $(x) \geq 50$, we can for all practical purposes assume that successive counts of equal length are normally distributed if we choose our time interval so that the total count in each interval will be of the order of 100 or greater.

The point of particular interest in this situation is that not only will the observations be very nearly normally distributed but also their variance will be equal to their average value, as in the Poisson distribution. Hence the control limits for the means $\bar{x}$ of subgroups of n counts can be estimated completely from the overall mean $\bar{\bar{x}}$, and are given by

$$\bar{\bar{x}} \pm \frac{3\sqrt{\bar{\bar{x}}}}{\sqrt{n}} = \bar{\bar{x}} \pm A\sqrt{\bar{\bar{x}}}$$

The presence of assignable causes has comparatively little effect on these limits. If the counts are made on a known standard with expected value μ, the limits $\mu \pm A\sqrt{\mu}$ can be used to obtain a control chart for the means $\bar{x}$ of the subgroups. Note that in preparing these charts for $\bar{x}$ a knowledge of the individual counts in each subgroup is unnecessary.

Another type of control chart commonly used in counting situations is based on the fact (see Section 9.43) that in this case $\dfrac{(n-1)s^2}{\bar{x}}$, where $\bar{x}$ and s^2 are the mean and variance, respectively, of a subgroup of n counts on the same sample, has approximately the χ^2-distribution with $n - 1$ degrees of freedom. It follows that the quantity $s^2/\bar{x}$ has expected value unity, and control limits for this quantity can easily be obtained from tabulated values of χ^2 by dividing by $n - 1$. Usually in this case limits are used which have a probability of 0.001 or 0.01 on either side in place of the usual 3σ limits. Notice that these limits depend only on n and that the central line or average value is always unity.

EXAMPLE. Table 10.10 gives the mean, variance, and $s^2/\bar{x}$ for 13 series of 10 1-minute counts obtained on a low geometry alpha counter. These series of counts were obtained routinely on a standard with an actual counting rate of 12,828 counts/min as a means of controlling the performance of the counter and checking the geometry factor. In this case limits based on a probability of 0.01 rather than 3σ limits are used for both $\bar{x}$ and $s^2/\bar{x}$ charts, since it is comparatively easy to recheck any series indicating assignable causes. The control charts are shown in Figure 10.11. The limits for the $\bar{x}$ chart are given by

$$\text{UCL} = 12{,}828 + 2.576\sqrt{\frac{12{,}828}{10}} = 12{,}920$$

$$\text{LCL} = 12{,}828 - 2.576\sqrt{\frac{12{,}828}{10}} = 12{,}736$$

and for the $s^2/\bar{x}$ chart by

$$\text{UCL} = \frac{\chi^2_{9,\,0.01}}{9} = 2.41$$

$$\text{LCL} = \frac{\chi^2_{9,\,0.99}}{9} = 0.23$$

TABLE 10.10

Series	$\bar{x}$	s^2	$s^2/\bar{x}$
1	12,826	9,356	0.72
2	12,884	8,616	0.67
3	12,776	4,747	0.37
4	12,796	4,226	0.33
5	12,531	14,108	1.13
6	12,554	10,920	0.87
7	12,542	15,435	1.23
8	12,799	4,092	0.32
9	12,782	13,810	1.08
10	12,830	21,869	1.70
11	12,869	6,839	0.53
12	12,853	25,037	1.95
13	12,885	16,458	1.28

The mean of the fifth series was out of control, and series 6 and 7 represent immediate rechecks confirming this behavior. The difficulty was traced to a defective mica window, which was replaced; thereafter the remaining series showed no lack of control. The chart for $s^2/\bar{x}$ shows no evidences of abnormal short-term behavior on the part of the instrument, even in those cases for which the mean was low.

CHAPTER 11

Some Tests for Randomness

11.1. The Concept of Randomness

As we have frequently noted, all mathematical statistical inference is based in some way on the assumption that the available data represent one or more random samples from a specified population or set of populations. In Chapter 8, great emphasis was placed on the design of statistical experiments in such a fashion that this assumption would be valid, but there are many occasions on which it is necessary to analyze data which were not obtained from a properly randomized experiment, and we may wish to test the validity of this assumption. Also we can consider tests for randomness as criteria used to detect the presence or absence of statistical control; in this sense they are serving the primary purpose of indicating the presence of assignable causes, rather than the secondary purpose of checking the validity of an assumption.

Testing for randomness is difficult because there is no positive definition of what is meant by randomness which can be conveniently used for this purpose, but we have mentioned and can consider tests for certain indications of non-randomness. Thus, although we shall never be able to make, with any specified degree of confidence, a positive statement to the effect that a given series of observations *is* random, we can conclude in certain instances with any desired degree of confidence that the data *are not* random. For the most part these conclusions are based on certain intuitive concepts of non-randomness, which can be summed up in the 4 following characteristics of a series of observations: (1) the presence of extreme variations, (2) the presence of trends, (3) the presence of periodic fluctuations, (4) the presence of discontinuities. Extreme examples of these forms of non-randomness are shown in Figure 11.1 (*a*) through 11.1 (*e*); Figure 11.1 (*f*) shows a series of data which have been chosen as nearly as possible to represent random variation. It should be noted that the first of these is somewhat different from the last 3, in that whether an extreme variation is considered non-random depends on our definition

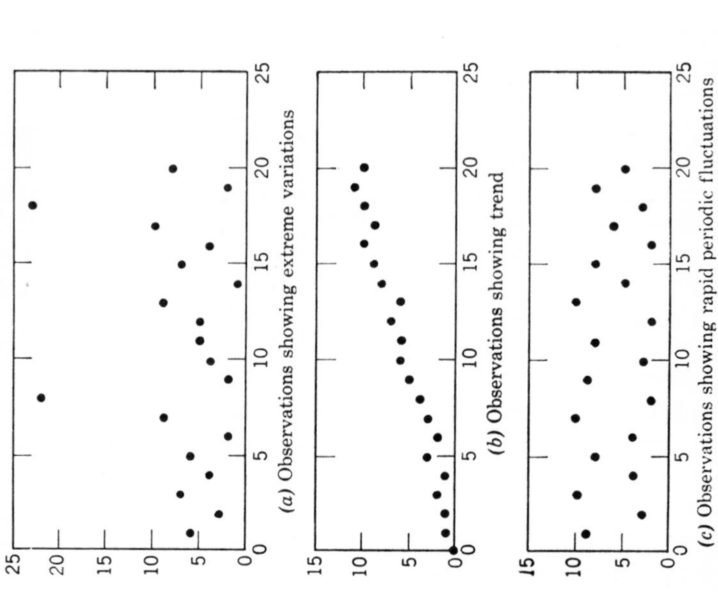

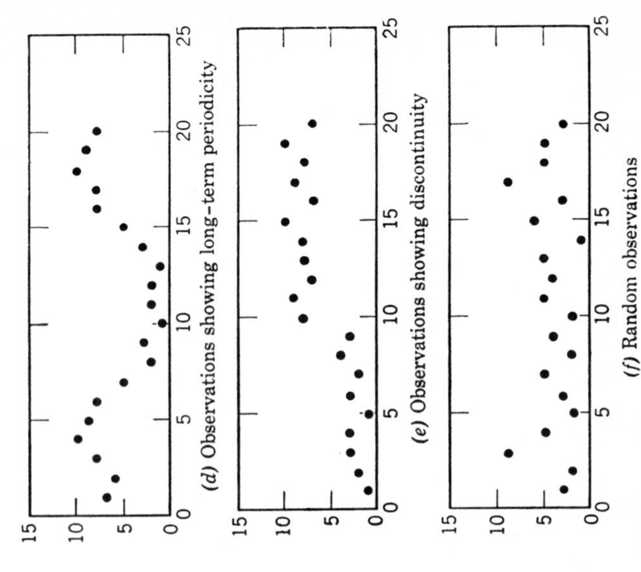

FIG. 11.1. Types of non-randomness.

of the population sampled; a series of data with a seemingly discrepant observation may be a perfectly random sample from a contaminated population with long tails, and it is our concept of this population, not the randomness of the sampling, which we are testing.

The last 3 of the 4 concepts of non-randomness are based on the order in which the observations occur. For example, if we were presented with the numbers 1, 2, 3, 4, 5, 6, 7, 8, 9, 10 in this order, we should not be inclined to consider them random; however, if we were presented with the same numbers in the order 3, 6, 2, 9, 10, 4, 7, 1, 5, 8, we should have a completely different opinion. Hence any tests for non-randomness of these types must somehow involve the order of occurrence of the observed values as well as their group behavior.

The two most common tests for the non-randomness of a series of observations are the control chart, used to detect extreme variations, and *runs*, which are based on the order of the observations and which frequently detect non-randomness of the last 3 types mentioned above. Other tests which may be more sensitive in the latter cases, but which are more difficult to compute and interpret are the *mean square successive difference* and *the serial correlation.*

11.2. Extreme Variations

11.21. Use of Control Charts to Detect Extreme Variations

The discussion of control charts in the previous chapter was primarily directed to their use in the detection and elimination of assignable causes of variation in the quality of routine production. In that case the emphasis was on the routine plotting of the subgroup means and ranges or standard deviation, and the immediate investigation of any indication of lack of control. The control chart can also be used as a means of looking at a series of observations or measurements in retrospect to determine whether these particular observations or measurements themselves can be considered a controlled or random series. A control chart used in this manner may detect any of the types of non-randomness mentioned in the preceding section, but it is particularly effective in detecting the presence of extreme variation on the part of individual measurements.

As the average value and variance of groups of experimental observations or measurements will very seldom be specified, the control chart must be based on the overall mean $\bar{\bar{x}}$ of the series of measurements and an estimate s' or $\bar{R}/d_2$ of the variability within suitably chosen subgroups. Since in this situation there will only infrequently be any obvious basis for the choice of subgroups, it is generally best to follow the procedure previously mentioned dividing the data into subgroups of 4 or 5 *in the order in which the observations were taken.*

11.22. Rejection of Observations

When seemingly extreme observations are obtained, the question always arises whether these observations should be considered discrepant and rejected. Points of view range from the flat statement that "data should never be rejected" to the view that it is not "practical" to include any suspicious measurement, regardless of the fact that the suspicions arose as a result of the examination of the data. Both points of view have their merits; often the single discrepant observation contains the most significant information of the group; on the other hand, this information may be completely irrelevant to the question that the data were originally supposed to answer. For example, one discrepant analysis in a group of 5 may give pertinent information with regard to difficulties present in the analytical method, but may also bias the analytical result if included in its calculation.

Many criteria have been suggested for the rejection of outlying observations when we wish our conclusions to apply only to the uncontaminated distribution. An extensive study of the behavior of these criteria with respect to several types of contamination has been made by W. J. Dixon [89].

When σ^2 can be considered known, or an independent estimate s^2 based on a comparatively large number of degrees of freedom is available, the most satisfactory test seems to be the ratio of the range of the observations (including the outlier) to the known standard deviation σ, or the studentized range discussed in Section 5.53. Percentage points for these criteria are given in Table 10.7 (use values based on σ) and Table 5.8; the outlying observation (or observations) should be rejected if the desired significance level is exceeded. When our test must be based on the observations themselves, the ratio

$$(a) \qquad r_{10} = \frac{x_{(n)} - x_{(n-1)}}{x_{(n)} - x_{(1)}}$$

of the difference between the two largest values and the range is a satisfactory criterion for small samples; for larger samples the ratio

$$(b) \qquad r_{22} = \frac{x_{(n)} - x_{(n-2)}}{x_{(n)} - x_{(3)}}$$

which excludes the extreme values (except for the suspected one) is to be preferred. If the smallest value $x_{(1)}$ rather than the largest value $x_{(n)}$ is suspected of being discrepant, the above ratios become

$$(c) \qquad r_{10} = \frac{x_{(2)} - x_{(1)}}{x_{(n)} - x_{(1)}}$$

and

$$(d) \qquad r_{22} = \frac{x_{(3)} - x_{(1)}}{x_{(n-2)} - x_{(1)}}$$

For symmetric populations the distributions of (*c*) and (*d*) are identical

TABLE 11.1

Percentage Points for Ratios Involving Extreme Values*

(For samples of n from a normal distribution, the probability of r_{10} or r_{22} exceeding the value tabulated is α.)

Test Criteria	n \ α	0.10	0.05	0.02	0.01
	3	0.886	0.941	0.976	0.988
	4	0.679	0.765	0.846	0.889
	5	0.557	0.642	0.729	0.780
r_{10}	6	0.482	0.560	0.644	0.698
	7	0.434	0.507	0.586	0.637
	8	0.399	0.468	0.543	0.590
	9	0.370	0.437	0.510	0.555
	10	0.349	0.412	0.483	0.527
	11	0.578	0.637	0.703	0.745
	12	0.543	0.600	0.661	0.704
	13	0.515	0.570	0.628	0.670
	14	0.492	0.546	0.602	0.641
	15	0.472	0.525	0.579	0.616
	16	0.454	0.507	0.559	0.595
	17	0.438	0.490	0.542	0.577
r_{22}	18	0.424	0.475	0.527	0.561
	19	0.412	0.462	0.514	0.547
	20	0.401	0.450	0.502	0.535
	21	0.391	0.440	0.491	0.524
	22	0.382	0.430	0.481	0.514
	23	0.374	0.421	0.472	0.505
	24	0.367	0.413	0.464	0.497
	25	0.360	0.406	0.457	0.489

* Reproduced by permission from Tables I and VI of "Ratios Involving Extreme Values," *Annals Math. Stat.*, *XXII* (1951), 68–78, by W. J. Dixon.

with (*a*) and (*b*), respectively. In any case the suspected value is rejected if the ratios are too large. If more than one value is suspected, the test can be repeated on the reduced sample. Extensive tables of percentage points of these and other similar ratios are given by Dixon [90] for samples of n from a normal population, and the upper 10%, 5%, 2%, and 1% points of r_{10} for $n \leq 10$, and of r_{22} for $11 \leq n \leq 25$ are reproduced in Table 11.1. A complete discussion of the rejection of observations based on optimum ratios of the above type is given in [10].

11.3. Use of Runs to Detect Non-Randomness

11.31. Relationship of Runs to Non-Randomness

A convenient method of testing for non-randomness in a series of observations or measurements is given by the theory of runs. The procedure is particularly effective in detecting non-randomness which takes the form of trends, periodicities, or discontinuities in the data.

Let us consider the simple operation of tossing an unbiased coin. Anyone who has matched pennies with a friend knows that the behavior of the individual tosses will be somewhat erratic; 20 tosses might, for example, give the results

(*a*) $\quad T\,T\,T\,T\,H\,H\,T\,T\,H\,T\,H\,T\,T\,H\,T\,H\,T\,T\,H\,T$

and, even though tails has occurred a total of 13 out of 20 tosses, including 4 consecutive times at the beginning, we should be willing to accept such behavior as random, since occasional occurrences of consecutive heads and tails of this type do occur. However, if 20 tosses had given the results

(*b*) $\quad H\,T\,T\,H\,T\,T\,H\,T\,T\,H\,T\,T\,H\,T\,T\,H\,T\,T\,H\,T$

we should immediately have suspected something unusual; similarly the series

(*c*) $\quad T\,T\,T\,T\,T\,T\,T\,H\,H\,H\,H\,H\,H\,H\,T\,T\,T\,T\,T\,T$

would be considered very improbable. Actually each series contains 7 heads and 13 tails, and hence is equally probable as a group; what differentiates the first from the last two is not the number of occurrences but the order in which they occur. One way of assessing this order systematically is to count the number of consecutive occurrences, or runs, of heads and tails. Thus in (*a*) we have runs of 4, 2, 1, 2, 1, 2, and 1 tails and runs of 2, 1, 1, 1, 1, and 1 heads. In (*b*) there are 7 runs of tails, 6 of length 2, and 7 runs of heads, all of length 1; it is this consistency in place of erratic behavior, on the part of the runs, that we consider improbable. In (*c*) we have 2 runs of 6 and 7 tails, and 1 run of 7 heads; here it is the length of the runs, and the small number, which is improbable.

In order to apply this concept of runs to a series of measurements, we must somehow convert the measurements to a series of "heads" or "tails" such as those considered above. One way of doing this is to consider the direction of change between each pair of measurements, i.e., whether the difference is positive or negative. Consecutive positive differences are then called runs up and consecutive negative differences, runs down. In this case many short runs would indicate a more or less sawtooth pattern on the part of the measurements, whereas a few long runs would indicate either trends or periodic movements of greater length. This type of analysis is not sensitive to discontinuities of the type illustrated in Figure 11.1 (*e*). A second method is to consider the observations above or below a given value, indicating them by a and b, respectively. For example, in using control charts in routine production, long runs of values above or below the central line might indicate the presence of an assignable cause, even if no points were actually out of control. In a particular series of observations the median is frequently used, since this ensures equal numbers of a's and b's (if the number of observations is odd, the median itself is excluded) and simplifies the mathematical study of the behavior of the runs. In this case a few long runs are a reflection of trends, discontinuities, or long-term periodicities; many short runs are again characteristic of sawtooth behavior or short-term periodicity.

11.32. Distribution of Number of Runs

We have noted, more or less intuitively, that many short runs, or a few long runs, are usually indicative of certain types of non-randomness. In order to formalize this intuitive concept, it is necessary to consider what type of behavior we expect from these runs, i.e., what the expected distribution of runs will be in a given instance.

One simple approach to the problem of the length of an individual run above or below a chosen value is to assume there is a constant probability p of obtaining a single measurement above the chosen value. Then, if in addition we assume consecutive measurements to be independent, the probability of obtaining n consecutive measurements above the chosen value would be p^n. Thus, for example, in the use of control charts for routine control we might assume that the probabilities of points above or below the central line were equal, and therefore equal to $1/2$. Then the probability of n consecutive points above the central line would be $1/2^n$, and hence a group of 6 points above or below the central line would be significant at the 0.05 probability level, a run of 8 points significant at the

0.01 level, and a run of 10 points significant at the 3σ probability level for a normal distribution. Such a run would be just as great an indication of lack of control as a single point outside the limits. Similarly, if in a group of 9 points, 8 were on one side and 1 on the other, this would be significant at the 0.05 level, and for a group of 12 points, 11 on one side and 1 on the other would be significant at the 0.01 level. Similar values for other numbers of points can be obtained from Table 5.2.

These considerations are not applicable to groups of points deliberately chosen from larger groups, and this approach to the length of a run is not easily extended to the length of the runs in a given series of observations. In the discussion thus far we have consistently associated few runs with long runs, and many runs with short runs, which suggests that an equivalent but simpler approach would be to consider the number of runs rather than their length, either too many or too few runs being indicative of non-randomness. The chance distribution of the number of runs above and below a given value can be directly determined, and depends only on the number of observations in the series and the total numbers above or below the given value.

Let us consider n measurements of which n_1 are designated as a's and n_2 as b's, $n_1 + n_2 = n$. We wish to determine the probability of obtaining exactly u different groups of a's and b's. To do this we must divide the total number of arrangements of a's and b's containing exactly u groups by the total number of arrangements possible. The latter is easily seen to be $\binom{n}{n_1} = \binom{n}{n_2} = \dfrac{n!}{n_1!\,n_2!}$ (Appendix IVB), but to compute the former is more difficult.

We begin by considering the number of ways in which the n_1 a's can be divided into exactly r_1 groups containing one or more a's. This is equal to the coefficient of t^{n_1} in the multinomial expansion of

$$(t + t^2 + t^3 + \cdots)^{r_2}$$

i.e., the number of ways r_1 of the exponents $1, 2, 3, \cdots, n_1$ can be combined to total n_1, including all distinct permutations of each. This can be written as

$$\left(\frac{1}{1-t} - 1\right)^{r_1} = \left(\frac{t}{1-t}\right)^{r_1} = t^{r_1}(1-t)^{-r_1}$$

and hence the coefficient of t^{n_1} is the coefficient of $t^{n_1-r_1}$ in the expansion of $(1-t)^{-r_1}$. From the binomial expansion with a negative exponent this is found to be

$$\frac{(n_1-1)!}{(n_1-r_1)!\,(r_1-1)!} = \binom{n_1-1}{r_1-1}$$

Similarly, the number of ways in which b's can be broken up into r_2

groups of one or more is $\binom{n_2 - 1}{r_2 - 1}$. Now in any complete arrangement of a's and b's, r_1 and r_2 can differ by at most 1; i.e., either $r_1 = r_2$ or $r_1 = r_2 \pm 1$, since the arrangement must either begin and end with the same letter, in which case $r_1 = r_2 \pm 1$, or begin and end with a different letter, in which case $r_1 = r_2$. Hence, if $r_1 = r_2$, the number of possible arrangements containing r_1 groups of a's and r_2 groups of b's is twice the product of the numbers of individual arrangements, since either letter can occur first; however, if $r_1 = r_2 \pm 1$, it is simply the product of the numbers of arrangements, since the letter having the larger number of groups must occur first. We can combine all these considerations into the following statement: the probability of exactly r_1 runs of a's and r_2 runs of b's in an arrangement containing n_1 a's and n_2 b's is

$$p(r_1, r_2) = K \frac{\binom{n_1 - 1}{r_1 - 1}\binom{n_2 - 1}{r_2 - 1}}{\binom{n}{n_1}}$$

where $K = 2$ if $r_1 = r_2$ and $K = 1$ if $r_1 = r_2 \pm 1$.

Now let us consider the probability of a total of exactly u runs. This is obtained by summing over all possible combinations of r_1 and r_2 such that $r_1 + r_2 = u$. If u is even, then r_1 and r_2 must be either both even or both odd, and, since they can differ by at most 1, this means that we must have $r_1 = r_2$. Hence in this case we have

$$p(u) = 2 \frac{\binom{n_1 - 1}{u/2 - 1}\binom{n_2 - 1}{u/2 - 1}}{\binom{n}{n_1}} \qquad u \text{ even}$$

If u is odd, we can have either $r_1 = r_2 + 1$ or $r_1 = r_2 - 1$; in the first case $r_1 = \frac{u+1}{2}$ and $r_2 = \frac{u-1}{2}$, and in the second, $r_1 = \frac{u-1}{2}$ and $r_2 = \frac{u+1}{2}$; hence

$$p(u) = p\left(\frac{u+1}{2}, \frac{u-1}{2}\right) + p\left(\frac{u-1}{2}, \frac{u+1}{2}\right)$$

$$= \frac{\binom{n_1 - 1}{\frac{u-1}{2}}\binom{n_2 - 1}{\frac{u-3}{2}} + \binom{n_1 - 1}{\frac{u-3}{2}}\binom{n_2 - 1}{\frac{u-1}{2}}}{\binom{n}{n_1}} \qquad u \text{ odd}$$

These two formulae give the discrete probability distribution of u.

Tables of the cumulative probability $P(u \leq u')$ are given by Swed and Eisenhart [28] for $n_1 \leq n_2 \leq 20$ ($n_1 = m$ and $n_2 = n$ in their notation) and all possible values of u'. They also give for the same range of n_1 and n_2 the exact critical values u' such that $P(u \leq u') \leq \alpha$ for $\alpha = 0.005, 0.01, 0.025$, and 0.05, and $P(u \leq u') \geq \alpha$ for $\alpha = 0.95, 0.975, 0.99$, and 0.995. Similar values are obtained for $n_1 = n_2 = 20$ to 100 by using a normal approximation due to Wald and Wolfowitz [29] which states that, for sufficiently large $n_1 = n_2 = m$, u is normally distributed with ave $(u) = m + 1$ and var $(u) = m/2$.

EXAMPLE 1. Let us consider the series of heads and tails obtained in the previous section. In the third series we might wish to test whether the 3 runs observed were a significantly small number. In this case $n = 20$, $n_1 = 7$, $n_2 = 13$, and either by direct calculation or from the tables referred to above we obtain $P(u \leq 3) = p(3) + p(2) = 0.000258$. Since the observed number of runs is quite improbable, we conclude that this test has given a definite indication of non-randomness. We could also have obtained from the tables the value $u'_{0.01} = 5$, and, since the observed number of runs was less than this value, we should reject the hypothesis of randomness. In the second series we are interested in whether an unusually large number of runs was obtained. A total of 14 runs was observed, and $P(u \geq 14) = p(14) + p(15) = 1 - P(u \leq 13) = 0.034$. Hence in this case the test has not given nearly so strong an indication of non-randomness, in spite of the regularity of the occurrence of heads and tails. In the third series, which actually represents a series of 20 tosses, we obtained 13 runs, and $P(u \geq 13) = p(13) + p(14) + p(15) = 1 - P(u \leq 12) = 0.116$; hence we have about 1 chance in 9 of obtaining as many or more than 13 runs, which is not a very unlikely event.

In applying this distribution to the number of runs above and below the median in a series of n observations, we have the special case $n_1 = n_2 = m$, where $m = n/2$ if n is even and $m = \dfrac{n-1}{2}$ if n is odd, since in the latter case the median is excluded. The above distribution in this case reduces to

$$p(u) = \frac{2\binom{m-1}{u/2-1}^2}{\binom{2m}{m}} \qquad n \text{ even}$$

$$p(u) = \frac{2\binom{m-1}{\frac{u-1}{2}}\binom{m-1}{\frac{u-3}{2}}}{\binom{2m}{m}} \qquad n \text{ odd}$$

where the range of u is $2 \leq u \leq 2m$ and the distribution is symmetric with the maximum value occurring at $u = m + 1$. Table 11.2, abstracted from the tables referred to above, gives values of u_α' and u_α'' such that $P(u \leq u_\alpha') \leq \alpha$ and $P(u \geq u_\alpha'') \leq \alpha$ for $\alpha = 0.05$ and 0.01 and $n_1 = n_2 =$ 5 to 30. An observed number of runs $\leq u_\alpha'$ or $\geq u_\alpha''$ will be reason to conclude at the given significance level that the data are not random. It should be noted that this test is completely independent of any assumption concerning the type of distribution sampled, depending only on the

TABLE 11.2*

$m = n_1 = n_2$	Lower Percentage Points, u_α'		Upper Percentage Points, u_α''	
	$\alpha = 0.05$	$\alpha = 0.01$	$\alpha = 0.05$	$\alpha = 0.01$
5	3	2	9	10
6	3	2	11	12
7	4	3	12	13
8	5	4	13	14
9	6	4	14	16
10	6	5	16	17
11	7	6	17	18
12	8	7	18	19
13	9	7	19	21
14	10	8	20	22
15	11	9	21	23
16	11	10	23	24
17	12	10	24	26
18	13	11	25	27
19	14	12	26	28
20	15	13	27	29
21	16	14	28	30
22	17	14	29	32
23	17	15	31	33
24	18	16	32	34
25	19	17	33	35
26	20	18	34	36
27	21	19	35	37
28	22	19	36	39
29	23	20	37	40
30	24	21	38	41

* Reproduced by permission from Tables II and III of: "Tables for Testing Randomness of Grouping in a Sequence of Alternatives," *Annals Math. Stat.*, *XIV* (1943), 66–87, by Freda S. Swed and C. Eisenhart.

order of the observations. On the other hand, in one of the above series of tosses an obviously systematic arrangement was significant only at the 5% level, so that what we gain in the sense of increased applicability we may lose in power to detect non-randomness.

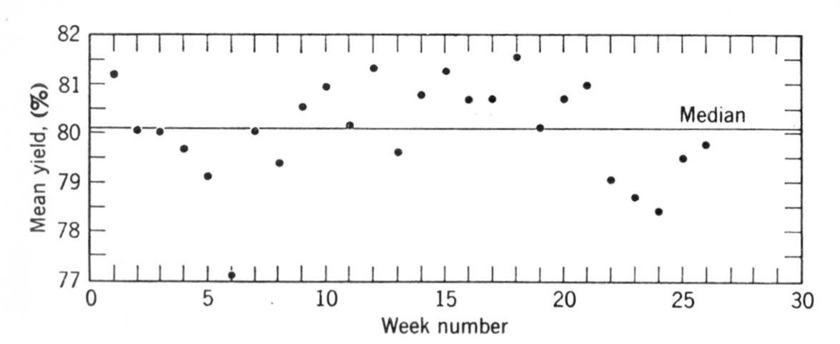

FIG. 11.2. Mean weekly yields, plant 1.

EXAMPLE 2. The data in Table 11.3, taken from a British Coke Research Association report [124], give the mean weekly yield in percent for 2 coke oven plants for a 26-week period. The medians, chosen as the midpoint of the two middle values, are 80.12 and 69.36, respectively, and each value in Table 11.3 is classified as above or below the median. The data are plotted in Figures 11.2 and 11.3, and they seem to give evidence in both instances of the presence of some type of periodicity of a fairly long-term nature, which might be expected to result in too few runs.

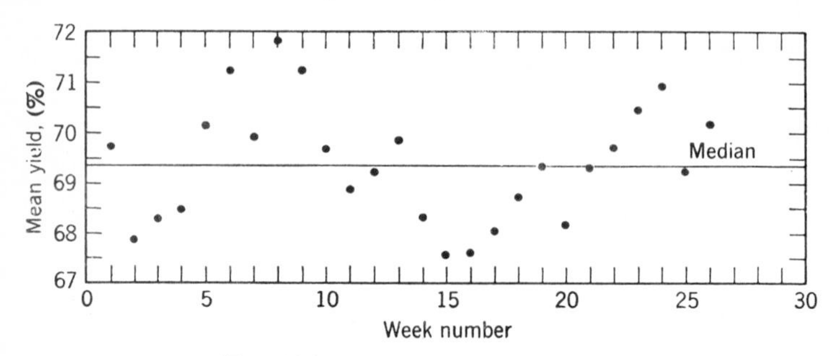

FIG. 11.3. Mean weekly yields, plant 4.

In this case $m = 13$; hence the expected number of runs is 14; and the lower 0.05 and 0.01 critical values are 9 and 7, respectively. For plant 1, there is a total of 6 runs, which is significantly low at the 0.01 level. For plant 4 there are 11 runs, which, although smaller than expected, is not significant at either level. This is a somewhat surprising result, since the periodicity is seemingly more apparent for plant 4.

Much more work has been done on the distribution theory of runs. Mood [31] has considered the general distribution of runs of a given length, and Mosteller [32] has given the probability of the largest run above or below the median in a series of observations exceeding a given value, and also the critical values at the 0.05 and 0.01 level, which are directly applicable in control chart work. Olmstead [33], who is in part responsible

TABLE 11.3

Week	Plant 1			Plant 4		
	Mean Yield Percent	Run	a = above Median b = below Median	Mean Yield Percent	Run	a = above Median b = below Median
1	81.02	1	a	69.76	1	a
2	80.08	2	b	67.88	2	b
3	80.05		b	68.28		b
4	79.70		b	68.48		b
5	79.13		b	70.15	3	a
6	77.09		b	71.25		a
7	80.09		b	69.94		a
8	79.40		b	71.82		a
9	80.56	3	a	71.27		a
10	80.97		a	69.70		a
11	80.17		a	68.89	4	b
12	81.35		a	69.24		b
13	79.64	4	b	69.86	5	a
14	80.82	5	a	68.35	6	b
15	81.26		a	67.61		b
16	80.75		a	67.64		b
17	80.74		a	68.06		b
18	81.59		a	68.72		b
19	80.14		a	69.37	7	a
20	80.75		a	68.18	8	b
21	81.01		a	69.35		b
22	79.09	6	b	69.72	9	a
23	78.73		b	70.46		a
24	78.45		b	70.94		a
25	79.56		b	69.26	10	b
26	79.80		b	70.20	11	a

for the tables given in [32], has given similar tables for runs up and down, and Wolfowitz [34] has given the approximate distribution of runs up and down of a given length.

EXAMPLE 3. The mean weekly yields for the data of Table 2.1 were tabulated and given in control chart form in Example 1 of Section 10.33. The median of these subgroup means is 71.165. The longest run above or below this value is 9 (weeks 28–36). From [32] a run of 10 on either side of the median in a group of 52 values is required to be significant at the 0.05 level.

An analysis of the number of runs of a given length both above and below the median is given in Table 11.4 for the original data of Table 2.1.

TABLE 11.4

Length	Number of Runs		
	Above	Below	Expected
1	44	42	46
2	19	24	23
3	15	11	11
4	5	4	6
5	1	2	3
> 5	4	5	2.5
Total	88	88	91.5

In this case $m = n_1 = n_2 = 364/2 = 182$, and the total number of runs $u = 88 + 88 = 176$. Using the normal approximation given above, we have

$$\text{ave}(u) = m + 1 = 183$$

$$\text{var}(u) \sim \frac{m}{2} = 91$$

$$\sigma_u \sim \sqrt{91} = 9.54$$

$$t = \frac{176 - 183}{9.54} = -0.73$$

which is far from being significant. Also note the agreement of the distribution of lengths with the expected numbers for each length, which were computed from results given in [31]. For all practical purposes, when n is as large as in this example and $m = n_1 = n/2$, the average number of runs both above and below the median of length l is given by $\frac{m+1}{2^l}$, and either above or below by $\frac{m+1}{2^{l+1}}$.

11.33. Other Uses of the Theory of Runs

The above test for non-randomness can be adapted to other problems similar to those considered in earlier chapters, but with less restrictive assumptions. As one example, let us suppose that we are given 2 samples of n_1 and n_2 observations, respectively. We assume only that each sample is a random sample from some distribution, and wish to test the null hypothesis that these two distributions are identical. Under this assumption, the two samples can be considered a single sample consisting of the $n = n_1 + n_2$ values $x_1, \cdots, x_{n_1}$ and $y_1, \cdots, y_{n_2}$.

If we arrange these values in order of magnitude, we should expect that under this hypothesis the x's and y's would fall into some random arrangement such as $x, x, y, x, y, y, x, x, x, y$. Any tendency for the x's to be consistently less than the y's, or vice versa, would be reflected in arrangements such as $x, x, x, x, y, x, y, y, y, y$ containing a few long runs. Thus the occurrence of too few runs could be considered evidence of a difference between the two distributions.

EXAMPLE. Let us consider the two series of chlorine determinations given in Example 1 of Section 5.51. Arranging these in order of magnitude, enclosing those from the first series in parentheses to differentiate them from the second, we have 55.71, 56.21, 56.58, 56.65, 56.72, (57.03), 57.08, 57.13, (57.33), 57.56, (57.80), 57.92, (58.00), (58.04), 58.27, (58.41), (58.45), (58.59), (58.64), (59.64), and the number of runs of consecutive elements from the same series is 10. Since the lower critical value for the 0.05 significance level for $n_1 = n_2 = 10$ is 6, we cannot conclude from this test that a significant difference exists.

The above use of the distribution of the number of runs is an example of a non-parametric test, since no specific assumptions are made concerning the nature of the distributions involved. Other examples of this type are given in [28].

11.34. Use of Mean Square Successive Difference to Detect Non-Randomness

Two other statistics frequently used to detect non-randomness are the *mean square successive difference* and the *serial correlation*, which will be discussed in the next section. Although these statistics are less easily computed and used than those of the previous sections, they are also more sensitive in many instances to non-random fluctuations.

The mean square successive difference, which was developed [36] for use in the examination of ballistics data, is particularly sensitive in detecting long-term trends, periodic or otherwise, or excessively rapid oscillation in an observed series. As its name implies, this statistic is obtained by taking the mean of the squares of the $n - 1$ successive differences between the observations, i.e.,

$$\delta^2 = \frac{\sum_{i=1}^{n-1} (x_{i+1} - x_i)^2}{n - 1}$$

As is usual with any statistic used to test randomness, the value of δ^2 depends on the order of the observations. Where the observations are a random sample from a normal distribution ave $(\delta^2) = 2\sigma^2$, and hence $\delta^2/2$ is an unbiased estimate of σ^2. Although its asymptotic efficiency is 2/3, it is less biased than the usual estimate s^2 by trends of a gradual nature in the mean of the distribution sampled, and in this case may be a better estimate of the basic variance σ^2. For example, it might be preferable in estimating the variation in a series of observations made under gradually changing experimental conditions.

In using the mean square successive difference to test for randomness, we are interested in comparing the values of δ^2 and s^2, since it is the disparity between these estimates that will indicate trends or short period oscillations. In the first case the ratio $\eta = \delta^2/s^2$ will be small, since, as we have noted above, the value of δ^2 will not be increased by the trend to as great an extent as s^2. In the second case both δ^2 and s^2 will increase, but this time the increase in δ^2 will be proportionately greater. The distribution of the quantity $\varepsilon = 1 - \eta/2 = 1 - \delta^2/2s^2$, which is symmetrical with average value zero for random samples from a normal distribution, was studied by von Neumann [37], and Hart [38] has computed percentage points k_α and k_α' such that $Pr(\eta < k_\alpha) = \alpha$ and $Pr(\eta > k_\alpha') = \alpha$ for $\alpha = 0.05$ and 0.01. These values, corrected for the fact that in [38] η was defined with $s^2 = \frac{1}{n} \Sigma(x_i - \bar{x})^2$ rather than $\frac{1}{n-1} \Sigma(x_i - \bar{x})^2$ as we have done throughout this book, are given in Table 11.5. For values of $n > 25$, ε is very nearly normally distributed with average value zero and variance $\frac{n-2}{(n-1)(n+1)}$, so that we can use $t = \varepsilon/\sigma_\varepsilon$ and the percentage points for a standard normal deviate in testing the significance of ε for large n. High negative values of t would be expected if long-term trends were present, and high positive values of t if short rapid oscillations were present.

TABLE 11.5*

Sample Size n	Lower Percentage Points		Upper Percentage Points	
	$\alpha = 0.05$	$\alpha = 0.01$	$\alpha = 0.05$	$\alpha = 0.01$
4	0.78	0.63	3.22	3.37
5	0.82	0.54	3.18	3.46
6	0.89	0.56	3.11	3.44
7	0.94	0.61	3.06	3.39
8	0.98	0.66	3.02	3.34
9	1.02	0.71	2.98	3.29
10	1.06	0.75	2.94	3.25
11	1.10	0.79	2.90	3.21
12	1.13	0.83	2.87	3.17
15	1.21	0.92	2.79	3.08
20	1.30	1.04	2.70	2.96
25	1.37	1.13	2.63	2.87

* Reproduced by permission from: "Significance Levels for the Ratio of the Mean Square Successive Difference to the Variance," *Annals Math. Stat.*, *XIII* (1942), 445–447, by B. I. Hart.

In testing for non-randomness, particularly in the latter case, it is frequently found that, although no single series gives a significantly high value of ε, there may be a tendency on the part of successive series to give positive values. In this case we can make use of the fact that the average $\bar{\varepsilon}$ of k values of ε, each based on a series of n observations, would also be approximately normally distributed with variance σ_ε^2/k, and hence we use $t = \dfrac{\bar{\varepsilon}}{\sigma_\varepsilon}\sqrt{k}$ to test for the significance of this behavior.

EXAMPLE 1. Table 11.6 lists the successive differences for the yields of plants 1 and 4 given in Table 11.3. For plant 1 we have from Table 11.3

$$\Sigma x_i = 2081.94 \quad \Sigma x_i^2 = 166{,}736.9454 \quad \Sigma(x_i - \bar{x})^2 = 26.4006$$

and from Table 11.6

$$\sum_{i=1}^{25} (x_{i+1} - x_i)^2 = 31.7348$$

TABLE 11.6

Week	Successive Differences, $x_{i+1} - x_i$	
	Plant 1	Plant 4
1	−0.94	−1.88
2	−0.03	0.40
3	−0.35	0.20
4	−0.57	1.67
5	−2.04	1.10
6	3.00	−1.31
7	−0.69	1.88
8	1.16	−0.55
9	0.41	−1.57
10	−0.80	−0.81
11	1.18	0.35
12	−1.71	0.62
13	1.18	−1.51
14	0.44	−0.74
15	−0.51	0.03
16	−0.01	0.42
17	0.85	0.66
18	−1.45	0.65
19	0.61	−1.19
20	0.26	1.17
21	−1.92	0.37
22	−0.36	0.74
23	−0.28	0.48
24	1.11	−1.68
25	0.24	0.94
26	—	—

Hence

$$s^2 = \frac{\Sigma(x_i - \bar{x})^2}{n - 1} = \frac{26.4006}{25} = 1.0506$$

$$\delta^2 = \frac{\sum_{i=1}^{25} (x_{i+1} - x_i)^2}{n - 1} = \frac{31.7348}{25} = 1.2694$$

and

$$\eta = \delta^2/s^2 = 1.21^*$$

which is well below the lower 5% point for $n = 25$, and close to the 1% point. Similarly for plant 4 we have from Table 11.3

$$\Sigma x_i = 1804.38 \quad \Sigma x_i^2 = 125{,}256.7532 \quad \Sigma(x_i - \bar{x})^2 = 34.1692$$

and from Table 11.6

$$\sum_{i=1}^{25} (x_{i+1} - x_i)^2 = 28.1832$$

Hence

$$s^2 = \frac{34.1692}{25} = 1.3668$$

$$\delta^2 = \frac{28.1832}{25} = 1.1273$$

$$\eta = 0.82^{**}$$

which is far below the 1% point. Thus it is indicated that non-random fluctuations are present in both these sets of data, especially, as we should suspect from the graph, in the latter.

The estimates of the basic variance σ^2 of a mean weekly yield obtained from the mean square successive differences are

$$\text{Plant 1:} \quad \delta^2/2 = \frac{1.2694}{2} = 0.6347$$

$$\text{Plant 4:} \quad \delta^2/2 = \frac{1.1273}{2} = 0.5636$$

They are in good agreement and probably represent fairly closely the variation in mean weekly yield exclusive of non-random fluctuations with time, although they may still be somewhat biased on the high side. Note the agreement between these values and the standard deviation 0.516, given in Section 4.51, for the weekly means of Table 10.2, which are similar to those of Table 11.5.

EXAMPLE 2. (Young, [39]). Table 11.7 gives the percentages of defective product turned out daily, over a period of 24 days, by a single workman, and the successive differences. From these values we obtain

$$\Sigma x_i = 242.6 \qquad \Sigma x_i^2 = 2495.82$$

$$\Sigma(x_i - \bar{x})^2 = 2495.82 - 2452.28 = 43.54$$

$$\sum_{i=1}^{23} (x_{i+1} - x_i) = 55.42$$

and hence

$$s^2 = \frac{43.54}{23} = 1.893$$

$$\delta^2 = \frac{55.42}{23} = 2.410$$

$$\eta = \delta^2/s^2 = 1.273*$$

This is well below the 5% point for $n = 25$, but above the 1% point, and the randomness of this series of observations is definitely questionable.

It is interesting to note that, if we use the χ^2-test for homogeneity discussed in Section 9.43, to test the hypothesis that these observations represent random fluctuations about a constant average proportion of defects, we obtain $\chi^2 = 21.52$ for 23 degrees of freedom, which is about what would be expected and far from significant. Thus it is not the size of the variations which is questionable in this instance, but the order in

TABLE 11.7

Days n	% Defective x_i	Successive Differences $x_{i+1} - x_i$
1	7.4	1.4
2	8.8	2.6
3	11.4	−1.1
4	10.3	1.6
5	11.9	0.3
6	12.2	−2.2
7	10.0	−1.6
8	8.4	1.0
9	9.4	1.5
10	10.9	−1.0
11	9.9	1.9
12	11.8	−1.8
13	10.0	−1.1
14	8.9	0.8
15	9.7	−0.4
16	9.3	2.7
17	12.0	0.3
18	12.3	−2.0
19	10.3	−1.7
20	8.6	1.8
21	10.4	0.7
22	11.1	−1.7
23	9.4	−1.2
24	8.2	—

which they occur, and it seems probable that some factor which causes periodic fluctuations in the quality of the man's product is present.

EXAMPLE 3. Let us again consider the data of Table 2.1. In this case there are several ways in which the mean square successive difference can be applied. For the overall set of 364 observations we have

$$\sum_{i=1}^{363} (x_{i+1} - x_i)^2 = 1276.41$$

$$\delta^2 = 3.5163$$

and, from previous considerations, $s^2 = 1.8231$. Hence

$$\eta = \frac{3.5163}{1.8231} = 1.9287$$

Since $n = 364$ is far beyond our tables, we compute

$$\varepsilon = 1 - \eta/2 = 1 - 0.964 = 0.036$$

$$\sigma_\varepsilon^2 = \frac{n-2}{(n-1)(n+1)} = \frac{362}{(363)(365)} = 0.002732$$

$$\sigma_\varepsilon = 0.052$$

and from these we obtain the approximate standardized normal deviate

$$t = \frac{\varepsilon}{\sigma_\varepsilon} = \frac{0.036}{0.052} = 0.69$$

which is definitely not significant. Also we can consider the weekly means given in Table 10.2, for which

$$\sum_{i=1}^{51} (x_{i+1} - x_i)^2 = 22.0247$$

$$\delta^2 = \frac{22.0247}{51} = 0.4319$$

and, from Section 4.51,

$$s^2 = 0.2663$$

and hence

$$\eta = \frac{0.4319}{0.2663} = 1.62$$

Again, since n is beyond the table, we compute

$$\varepsilon = 1 - \eta/2 = 1 - 0.81 = 0.19$$

$$\sigma_\varepsilon^2 = \frac{n-2}{(n-1)(n+1)} = \frac{50}{(51)(53)} = 0.0189$$

$$\sigma_\varepsilon = 0.136$$

$$t = \frac{\varepsilon}{\sigma_\varepsilon} = \frac{0.19}{0.136} = 1.40$$

which is not significant.

As a final test, the mean square successive difference was computed for each subgroup. These values are plotted in Figure 11.4, the limits shown being based on the 5% and 1% points obtained from Table 11.5. The values seem to show a slight tendency to be low, although the single value outside the 98% limits (since the percentage points given in Table

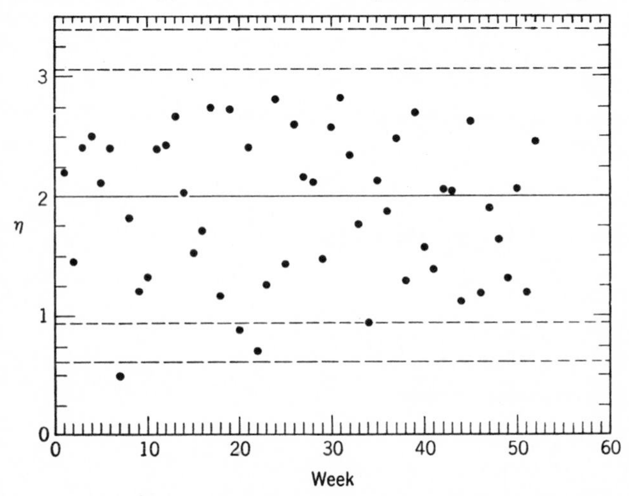

FIG. 11.4. Ratio of mean square successive difference to variance for weekly subgroups.

11.5 are one sided) and the 4 values on or below the 90% limits do not represent anything abnormal in a group of 52 points. The average η for the $k = 52$ values is $\bar{\eta} = 1.88$, and hence, since

$$\bar{\varepsilon} = 1 - \bar{\eta}/2 = 0.056$$

and for $n = 7$

$$\sigma_{\varepsilon}^2 = \frac{5}{(6)(8)} = 0.1042$$

$$\sigma_{\varepsilon} = 0.323$$

we have

$$t = \frac{\bar{\varepsilon}}{\sigma_{\varepsilon}}\sqrt{k} = \frac{0.056}{0.323}\sqrt{52} = 1.25$$

which indicates no significant average tendency on the part of the 52 subgroups. However, the distribution of η's is not what might be expected, owing to a lack of symmetry and a somewhat peculiar lack of values of η near the average value of 2.00.

11.35. Use of the Serial Correlation to Detect Non-Randomness

The serial correlation is the correlation between pairs of equally spaced observations from the same sample. Thus for a series of observations

$x_1, \cdots, x_n$ the serial correlation would be the correlation between the pairs x_i and x_{i+h}, where h is called the lag. For $i + h > n$, we define $x_{i+h} = x_{i+h-n}$. For example, for $n = 10$ and $h = 3$, the value paired with x_9 would be $x_{9+3-10} = x_2$, and the 10 pairs of measurements in this case would be (x_1, x_4), (x_2, x_5), (x_3, x_6), (x_4, x_7), (x_5, x_8), (x_6, x_9), (x_7, x_{10}), (x_8, x_1), (x_9, x_2) and (x_{10}, x_3). For such pairs (x, y) we have in general $\Sigma y_i = \Sigma x_i$ and $\Sigma y_i^2 = \Sigma x_i^2$, since each involves the sum and sum of squares of the same values $x_1, \cdots, x_n$, and $\Sigma x_i y_i = \Sigma_i x_i x_{i+h}$, with x_{i+h} defined as above. Hence, using the definition of the sample correlation coefficient given in Section 2.53, we have for the serial correlation of lag h

$$R_h = \frac{\Sigma_i x_i x_{i+h} - \dfrac{(\Sigma_i x_i)^2}{n}}{\Sigma_i x_i^2 - \dfrac{(\Sigma_i x_i)^2}{n}}$$

R_0 is identically 1, since it merely represents the correlation of each value with itself. Also it is easily seen that $R_h = R_{h+n}$, so that there are $n - 1$ serial correlations for each sample of n, representing lags of from 1 to $n - 1$. As with the usual correlation coefficient we must have $-1 < R_h < +1$.

The serial correlation as defined here is particularly useful in the detection of periodic effects in series of observations, such as that exhibited in the second series of tosses, and hence has been much used in the analysis of fluctuations in economic time series. For example, if in a chemical production process the production of a batch of high purity tended to decrease the purity of the batch immediately following, and vice versa, we should expect to find a negative serial correlation between the purity of successive batches. The serial correlation coefficient can also be expected to detect trends, although in this case it is preferable to modify the previous definition, which refers to the "circular" serial correlation, by omitting those pairs depending on using the definition $x_{i+h} = x_{i+h-n}$ for $i + h > n$. Thus in the above example we should omit the pairs (x_8, x_1), (x_9, x_2), and (x_{10}, x_3) and compute the correlation between the remaining pairs.

To use the serial correlation as a test for non-randomness, we must know something about its distribution when computed from random samples. The distribution of the circular serial correlation was studied by R. L. Anderson [40] under the assumption that $x_1, x_2, \cdots, x_n$ are a random sample from a normal distribution. He found the exact distributions for lag 1, and from these obtained the one-sided upper and lower percentage points of R_1 for the 0.05 and 0.01 significance levels, given in

Table 11.8 for $n = 5$–30. He also showed that for large n, R_1 is approximately normally distributed with

(a)
$$\text{ave}(R_1) \sim \frac{-1}{n-1}$$

$$\text{var}(R_1) \sim \frac{n-2}{(n-1)^2}$$

Hence for $n > 30$ we can compute

(b)
$$t = \frac{R_1 - \text{ave}(R_1)}{\sigma_{R_1}}$$

and compare it with the percentage points for a standard normal deviate. This approximation is conservative in that the approximate percentage points are somewhat larger in absolute value than the exact ones.

TABLE 11.8*

n	Lower Percentage Points		Upper Percentage Points	
	$\alpha = 0.05$	$\alpha = 0.01$	$\alpha = 0.05$	$\alpha = 0.01$
5	−0.753	−0.798	0.253	0.297
6	−0.708	−0.863	0.345	0.447
7	−0.674	−0.799	0.370	0.510
8	−0.625	−0.764	0.371	0.531
9	−0.593	−0.737	0.366	0.533
10	−0.564	−0.705	0.360	0.525
11	−0.539	−0.679	0.353	0.515
12	−0.516	−0.655	0.348	0.505
13	−0.497	−0.634	0.341	0.495
14	−0.479	−0.615	0.335	0.485
15	−0.462	−0.597	0.328	0.475
20	−0.399	−0.524	0.299	0.432
25	−0.356	−0.473	0.276	0.398
30	−0.325	−0.433	0.257	0.370

* Reproduced by permission from Table I of: "Distribution of the Serial Correlation Coefficient," *Annals Math. Stat.*, *XIII* (1942), 1–13, by R. L. Anderson.

The study of the serial correlation for lags other than 1 is simplified by the fact that, if h and n have no common factors, the sampling distributions of R_h and R_1 are identical. In particular, if n is a prime number, all $n - 1$ serial correlations will have identical distributions. The case where h and n have a common factor, and $h > 1$, is considered in the above reference, but in practical situations it is generally simpler to drop several observations so as to obtain one of the first two conditions. For example, if we had 60 measurements, then by dropping 1 measurement so as to obtain a prime number, the serial correlation with any lag could be tested, using the above approximation for R_1.

A later paper by Wald and Wolfowitz [41] studies the distribution of R_h, assuming only that the values $x_1, \cdots, x_n$ are a random sample from a distribution with a continuous cumulative density function, a condition almost always satisfied in dealing with experimental results. This is done by considering the distribution of values of R_h given by all possible permutations of the sample values, a method similar to that by which the distribution of runs was considered. Since the quantities $\Sigma_i x_i$ and $\Sigma_i x_i^2$ are not affected by permuting the sample values, this is equivalent to studying the distribution of the values $R_h' = \Sigma_i x_i x_{i+h}$. For R_1' it is shown that

$$\text{ave}\,(R_1') = \frac{1}{n-1}(S_1^2 - S_2)$$

(c)

$$\text{var}\,(R_1') = \frac{S_2^2 - S_4}{n-1} + \frac{S_1^2 - 4S_1^2S_2 + 4S_1S_3 + S_2^2 - 2S_4}{(n-1)(n-2)} - \frac{(S_1^2 - S_2)^2}{(n-1)^2}$$

where $S_k = \Sigma x_i^k$ is the kth power sum of the observations. The distribution is shown to approach normality as n increases. As before, the distribution of R_1' and R_h' are identical, if h and n have no common factor; hence in this case we can replace R_1' by R_h', and test the significance of R_h' for sufficiently large n by considering

(d)
$$t = \frac{R_h' - \text{ave}\,(R_h')}{\sigma_{R_h'}}$$

as a normally distributed variable.

In practice the serial correlations are computed only for small values of h, since these will in general be sufficient to indicate any important lack of non-randomness. Frequently the successive values of R_h (or R_h') are plotted as a function of h, the resulting chart being called a correlogram. The presence of trends or patterns on such a chart may indicate the presence of non-randomness even when individual correlations are not significant. In practice R_h is more easily used than R_h' since for the latter S_3, S_4, ave (R_1'), and var (R_1') must be evaluated, although the test so obtained is based on assumptions more directly applicable to tests for non-randomness. It should be noted that tests based on either the circular or non-circular definition of R_1 are essentially equivalent to those based on the mean square successive difference of the preceding section.

EXAMPLE. Let us consider the weekly yields for plant 4, given in Table 11.3, which seemed to show a marked periodicity. For these data we have

$$\Sigma x_i = 1804.38$$

$$\Sigma x_i^2 = 125{,}256.7532$$

and

$$\Sigma(x_i - \bar{x})^2 = \Sigma x_i^2 - \frac{(\Sigma x_i)^2}{n} = 34.1692$$

For $h = 1$, we obtain

$$\Sigma x_{i+1}x_i = 125{,}242.5648$$

$$\Sigma x_{i+1}x_i - \frac{(\Sigma x_i)^2}{n} = 19.9808$$

Hence for these data

$$R_1 = \frac{19.9808}{34.1692} = 0.584760$$

From Table 11.8, the upper critical values for $n = 25$ are 0.276 at the 0.05 level and 0.398 at the 0.01 level. Hence the observed correlation is highly significant.

11.36. Choice of Tests for Non-Randomness

It will be observed that the sensitivity of the various tests for non-randomness of observations depends upon the type of non-randomness which is present. In cases where this can be anticipated from technical considerations the appropriate test may be selected in advance, but, if the choice of tests is based on an examination of the results, the significance levels are clearly biased. An obvious alternative, that of subjecting all data to the same series of tests, is not free from criticism since many of the tests are not independent, although the degree of dependence is not yet known. It would appear that, without technical guidance, the best procedure is probably to use a series of tests, interpreting the significance levels as a general indication rather than a specific prediction.

APPENDIX

TABLE I

ORDINATES AND AREAS OF THE NORMAL DISTRIBUTION*

t	$f(t)$	$F(t)$	$1-F(t)$	t	$f(t)$	$F(t)$	$1-F(t)$
0.00	0.39894	0.500000	0.500000	0.40	0.36827	0.655422	0.344578
0.01	0.39892	0.503989	0.496011	0.41	0.36678	0.659097	0.340903
0.02	0.39886	0.507978	0.492022	0.42	0.36526	0.662757	0.337243
0.03	0.39876	0.511966	0.488034	0.43	0.36371	0.666402	0.333598
0.04	0.39862	0.515953	0.484047	0.44	0.36213	0.670031	0.329969
0.05	0.39844	0.519939	0.480061	0.45	0.36053	0.673645	0.326355
0.06	0.39822	0.523922	0.476078	0.46	0.35889	0.677242	0.322758
0.07	0.39797	0.527903	0.472097	0.47	0.35723	0.680822	0.319178
0.08	0.39767	0.531881	0.468119	0.48	0.35553	0.684386	0.315614
0.09	0.39733	0.535856	0.464144	0.49	0.35381	0.687933	0.312067
0.10	0.39695	0.539828	0.460172	0.50	0.35207	0.691462	0.308538
0.11	0.39654	0.543795	0.456205	0.51	0.35029	0.694974	0.305026
0.12	0.39608	0.547758	0.452242	0.52	0.34849	0.698468	0.301532
0.13	0.39559	0.551717	0.448283	0.53	0.34667	0.701944	0.298056
0.14	0.39505	0.555670	0.444330	0.54	0.34482	0.705401	0.294599
0.15	0.39448	0.559618	0.440382	0.55	0.34294	0.708840	0.291160
0.16	0.39387	0.563559	0.436441	0.56	0.34105	0.712260	0.287740
0.17	0.39322	0.567495	0.432505	0.57	0.33912	0.715661	0.284339
0.18	0.39253	0.571424	0.428576	0.58	0.33718	0.719043	0.280957
0.19	0.39181	0.575345	0.424655	0.59	0.33521	0.722405	0.277595
0.20	0.39104	0.579260	0.420740	0.60	0.33322	0.725747	0.274253
0.21	0.39024	0.583166	0.416834	0.61	0.33121	0.729069	0.270931
0.22	0.38940	0.587064	0.412936	0.62	0.32918	0.732371	0.267629
0.23	0.38853	0.590954	0.409046	0.63	0.32713	0.735653	0.264347
0.24	0.38762	0.594835	0.405165	0.64	0.32506	0.738914	0.261086
0.25	0.38667	0.598706	0.401294	0.65	0.32297	0.742154	0.257846
0.26	0.38568	0.602568	0.397432	0.66	0.32086	0.745373	0.254627
0.27	0.38466	0.606420	0.393580	0.67	0.31874	0.748571	0.251429
0.28	0.38361	0.610261	0.389739	0.68	0.31659	0.751748	0.248252
0.29	0.38251	0.614092	0.385908	0.69	0.31443	0.754903	0.245097
0.30	0.38139	0.617911	0.382089	0.70	0.31225	0.758036	0.241964
0.31	0.38023	0.621720	0.378280	0.71	0.31006	0.761148	0.238852
0.32	0.37903	0.625516	0.374484	0.72	0.30785	0.764238	0.235762
0.33	0.37780	0.629300	0.370700	0.73	0.30563	0.767305	0.232695
0.34	0.37654	0.633072	0.366928	0.74	0.30339	0.770350	0.229650
0.35	0.37524	0.636831	0.363169	0.75	0.30114	0.773373	0.226627
0.36	0.37391	0.640576	0.359424	0.76	0.29887	0.776373	0.224627
0.37	0.37255	0.644309	0.355691	0.77	0.29659	0.779350	0.220650
0.38	0.37115	0.648028	0.351972	0.78	0.29431	0.782305	0.217695
0.39	0.36973	0.651732	0.348268	0.79	0.29200	0.785236	0.214764

* A description of this table and its use is given in Section 4.41.

TABLE I (*continued*)

t	$f(t)$	$F(t)$	$1 - F(t)$	t	$f(t)$	$F(t)$	$1 - F(t)$
0.80	0.28969	0.788145	0.211855	1.20	0.19419	0.884930	0.115070
0.81	0.28737	0.791030	0.208970	1.21	0.19186	0.886860	0.113140
0.82	0.28504	0.793892	0.206108	1.22	0.18954	0.888768	0.111232
0.83	0.28269	0.796731	0.203269	1.23	0.18724	0.890651	0.109349
0.84	0.28034	0.799546	0.200454	1.24	0.18494	0.892512	0.107488
0.85	0.27798	0.802337	0.197663	1.25	0.18265	0.894350	0.105650
0.86	0.27562	0.805105	0.194895	1.26	0.18037	0.896165	0.103835
0.87	0.27324	0.807850	0.192150	1.27	0.17810	0.897958	0.102402
0.88	0.27086	0.810570	0.189430	1.28	0.17585	0.899727	0.100273
0.89	0.26848	0.813267	0.186733	1.29	0.17360	0.901475	0.098525
0.90	0.26609	0.815940	0.184060	1.30	0.17137	0.903200	0.096800
0.91	0.26369	0.818589	0.181411	1.31	0.16915	0.904902	0.095098
0.92	0.26129	0.821214	0.178786	1.32	0.16694	0.906582	0.093418
0.93	0.25888	0.823814	0.176186	1.33	0.16474	0.908241	0.091759
0.94	0.25647	0.826392	0.173608	1.34	0.16256	0.909877	0.090123
0.95	0.25406	0.828944	0.171056	1.35	0.16038	0.911492	0.088508
0.96	0.25164	0.831472	0.168528	1.36	0.15822	0.913085	0.086915
0.97	0.24923	0.833977	0.166023	1.37	0.15608	0.914657	0.085343
0.98	0.24681	0.836457	0.163543	1.38	0.15395	0.916207	0.083793
0.99	0.24439	0.838913	0.161087	1.39	0.15183	0.917736	0.082264
1.00	0.24197	0.841345	0.158655	1.40	0.14973	0.919243	0.080757
1.01	0.23955	0.843752	0.156248	1.41	0.14764	0.920730	0.079270
1.02	0.23713	0.846136	0.153864	1.42	0.14556	0.922196	0.077804
1.03	0.23471	0.848495	0.151505	1.43	0.14350	0.923641	0.076359
1.04	0.23230	0.850830	0.149170	1.44	0.14146	0.925066	0.074934
1.05	0.22988	0.853141	0.146859	1.45	0.13943	0.926471	0.073529
1.06	0.22747	0.855428	0.144572	1.46	0.13742	0.927855	0.072145
1.07	0.22506	0.857690	0.142310	1.47	0.13542	0.929219	0.070781
1.08	0.22265	0.859929	0.140071	1.48	0.13344	0.930563	0.069437
1.09	0.22025	0.862143	0.137857	1.49	0.13147	0.931888	0.068112
1.10	0.21785	0.864334	0.135666	1.50	0.12952	0.933193	0.066807
1.11	0.21546	0.866500	0.133500	1.51	0.12758	0.934478	0.065522
1.12	0.21307	0.868643	0.131357	1.52	0.12566	0.935745	0.064255
1.13	0.21069	0.870762	0.129238	1.53	0.12376	0.936992	0.063008
1.14	0.20831	0.872857	0.127143	1.54	0.12188	0.938220	0.061780
1.15	0.20594	0.874928	0.125072	1.55	0.12001	0.939429	0.060571
1.16	0.20357	0.876976	0.123024	1.56	0.11816	0.940620	0.059380
1.17	0.20121	0.879000	0.121000	1.57	0.11632	0.941792	0.058208
1.18	0.19886	0.881000	0.119000	1.58	0.11450	0.942946	0.057054
1.19	0.19652	0.882977	0.117023	1.59	0.11270	0.944082	0.055918

TABLE I (*continued*)

t	$f(t)$	$F(t)$	$1 - F(t)$	t	$f(t)$	$F(t)$	$1 - F(t)$
1.60	0.11092	0.945201	0.054799	2.00	0.05399	0.977250	0.022750
1.61	0.10915	0.946301	0.053699	2.01	0.05292	0.977784	0.022216
1.62	0.10741	0.947384	0.052616	2.02	0.05186	0.978308	0.021692
1.63	0.10567	0.948449	0.051551	2.03	0.05082	0.978822	0.021178
1.64	0.10396	0.949497	0.050503	2.04	0.04980	0.979325	0.020675
1.65	0.10226	0.950528	0.049472	2.05	0.04879	0.979818	0.020182
1.66	0.10059	0.951543	0.048457	2.06	0.04780	0.980301	0.019699
1.67	0.09893	0.952540	0.047460	2.07	0.04682	0.980774	0.019226
1.68	0.09728	0.953521	0.046479	2.08	0.04586	0.981237	0.018763
1.69	0.09566	0.954491	0.045509	2.09	0.04491	0.981691	0.018309
1.70	0.09405	0.955434	0.044566	2.10	0.04398	0.982136	0.017864
1.71	0.09246	0.956367	0.043633	2.11	0.04307	0.982571	0.017429
1.72	0.09089	0.957284	0.042716	2.12	0.04217	0.982997	0.017003
1.73	0.08933	0.958185	0.041815	2.13	0.04128	0.983414	0.016586
1.74	0.08780	0.959070	0.040930	2.14	0.04041	0.983823	0.016177
1.75	0.08628	0.959941	0.040059	2.15	0.03955	0.984222	0.015778
1.76	0.08478	0.960796	0.039204	2.16	0.03871	0.984614	0.015386
1.77	0.08329	0.961636	0.038364	2.17	0.03788	0.984997	0.015003
1.78	0.08183	0.962462	0.037538	2.18	0.03706	0.985371	0.014629
1.79	0.08038	0.963273	0.036727	2.19	0.03626	0.985738	0.014262
1.80	0.07895	0.964070	0.035930	2.20	0.03547	0.986096	0.013904
1.81	0.07754	0.964852	0.035148	2.21	0.03470	0.986447	0.013553
1.82	0.07614	0.965620	0.034380	2.22	0.03394	0.986791	0.013209
1.83	0.07477	0.966375	0.033625	2.23	0.03319	0.987126	0.012847
1.84	0.07341	0.967116	0.032884	2.24	0.03246	0.987454	0.012546
1.85	0.07206	0.967843	0.032157	2.25	0.03174	0.987776	0.012224
1.86	0.07074	0.968557	0.031443	2.26	0.03103	0.988089	0.011911
1.87	0.06943	0.969258	0.030742	2.27	0.03034	0.988396	0.011604
1.88	0.06814	0.969946	0.030054	2.28	0.02965	0.988696	0.011304
1.89	0.06687	0.970621	0.029379	2.29	0.02898	0.988989	0.011011
1.90	0.06562	0.971283	0.028717	2.30	0.02833	0.989276	0.010724
1.91	0.06438	0.971933	0.028067	2.31	0.02768	0.989556	0.010444
1.92	0.06316	0.972571	0.027429	2.32	0.02705	0.989830	0.010170
1.93	0.06195	0.973196	0.026804	2.33	0.02643	0.990097	0.009903
1.94	0.06077	0.973810	0.026190	2.34	0.02582	0.990358	0.009642
1.95	0.05959	0.974412	0.025588	2.35	0.02522	0.990613	0.009387
1.96	0.05844	0.975002	0.024998	2.36	0.02463	0.990862	0.009138
1.97	0.05730	0.975581	0.024419	2.37	0.02406	0.991106	0.008894
1.98	0.05618	0.976148	0.023852	2.38	0.02349	0.991344	0.008656
1.99	0.05508	0.976705	0.023295	2.39	0.02294	0.991576	0.008424

TABLE I (*continued*)

t	$f(t)$	$F(t)$	$1 - F(t)$	t	$f(t)$	$F(t)$	$1 - F(t)$
2.40	0.02239	0.991802	0.008198	2.80	0.00792	0.997445	0.002555
2.41	0.02186	0.992024	0.007976	2.81	0.00770	0.997523	0.002477
2.42	0.02134	0.992240	0.007760	2.82	0.00748	0.997599	0.002401
2.43	0.02083	0.992450	0.007550	2.83	0.00727	0.997673	0.002327
2.44	0.02033	0.992656	0.007344	2.84	0.00707	0.997744	0.002256
2.45	0.01984	0.992857	0.007143	2.85	0.00687	0.997814	0.002186
2.46	0.01936	0.993053	0.006947	2.86	0.00668	0.997882	0.002118
2.47	0.01889	0.993244	0.006756	2.87	0.00649	0.997948	0.002052
2.48	0.01842	0.993431	0.006569	2.88	0.00631	0.998012	0.001988
2.49	0.01797	0.993613	0.006387	2.89	0.00613	0.998074	0.001926
2.50	0.01753	0.993790	0.006210	2.90	0.00595	0.998134	0.001866
2.51	0.01709	0.993963	0.006037	2.91	0.00578	0.998193	0.001807
2.52	0.01667	0.994132	0.005868	2.92	0.00562	0.998250	0.001750
2.53	0.01625	0.994297	0.005703	2.93	0.00545	0.998305	0.001695
2.54	0.01585	0.994457	0.005543	2.94	0.00530	0.998359	0.001641
2.55	0.01545	0.994614	0.005386	2.95	0.00514	0.998411	0.001589
2.56	0.01506	0.994766	0.005234	2.96	0.00499	0.998462	0.001538
2.57	0.01468	0.994915	0.005085	2.97	0.00485	0.998511	0.001489
2.58	0.01431	0.995060	0.004940	2.98	0.00471	0.998559	0.001441
2.59	0.01394	0.995201	0.004799	2.99	0.00457	0.998605	0.001395
2.60	0.01358	0.995339	0.004661	3.00	0.00443	0.998650	0.001350
2.61	0.01323	0.995473	0.004527	3.01	0.00430	0.998694	0.001306
2.62	0.01289	0.995604	0.004396	3.02	0.00417	0.998736	0.001264
2.63	0.01256	0.995731	0.004269	3.03	0.00405	0.998777	0.001223
2.64	0.01223	0.995855	0.004145	3.04	0.00393	0.998817	0.001183
2.65	0.01191	0.995975	0.004025	3.05	0.00381	0.998856	0.001144
2.66	0.01160	0.996093	0.003907	3.06	0.00370	0.998893	0.001107
2.67	0.01130	0.996207	0.003793	3.07	0.00358	0.998930	0.001070
2.68	0.01100	0.996319	0.003681	3.08	0.00348	0.998965	0.001035
2.69	0.01071	0.996427	0.003573	3.09	0.00337	0.998999	0.001001
2.70	0.01042	0.996533	0.003467	3.10	0.00327	0.999032	0.000968
2.71	0.01014	0.996636	0.003364	3.11	0.00317	0.999064	0.000936
2.72	0.00987	0.996736	0.003264	3.12	0.00307	0.999096	0.000904
2.73	0.00961	0.996833	0.003167	3.13	0.00298	0.999126	0.000874
2.74	0.00935	0.996928	0.003072	3.14	0.00288	0.999155	0.000845
2.75	0.00909	0.997020	0.002980	3.15	0.00279	0.999184	0.000816
2.76	0.00885	0.997110	0.002890	3.16	0.00271	0.999211	0.000789
2.77	0.00861	0.997197	0.002803	3.17	0.00262	0.999238	0.000762
2.78	0.00837	0.997282	0.002718	3.18	0.00254	0.999264	0.000736
2.79	0.00814	0.997365	0.002635	3.19	0.00246	0.999289	0.000711

TABLE I (*continued*)

t	$f(t)$	$F(t)$	$1 - F(t)$	t	$f(t)$	$F(t)$	$1 - F(t)$
3.20	0.00238	0.999313	0.000687	3.65	0.00051	0.999869	0.000131
3.21	0.00231	0.999336	0.000664	3.66	0.00049	0.999874	0.000126
3.22	0.00224	0.999359	0.000641	3.67	0.00047	0.999879	0.000121
3.23	0.00216	0.999381	0.000619	3.68	0.00046	0.999883	0.000117
3.24	0.00210	0.999402	0.000598	3.69	0.00044	0.999888	0.000112
3.25	0.00203	0.999423	0.000577	3.70	0.00042	0.999892	0.000108
3.26	0.00196	0.999443	0.000557	3.71	0.00041	0.999896	0.000104
3.27	0.00190	0.999462	0.000538	3.72	0.00039	0.999900	0.000100
3.28	0.00184	0.999481	0.000519	3.73	0.00038	0.999904	0.000096
3.29	0.00178	0.999499	0.000501	3.74	0.00037	0.999908	0.000092
3.30	0.00172	0.999516	0.000484	3.75	0.00035	0.999912	0.000088
3.31	0.00167	0.999534	0.000466	3.76	0.00034	0.999915	0.000085
3.32	0.00161	0.999550	0.000450	3.77	0.00033	0.999918	0.000082
3.33	0.00156	0.999566	0.000434	3.78	0.00031	0.999922	0.000078
3.34	0.00151	0.999581	0.000419	3.79	0.00030	0.999925	0.000075
3.35	0.00146	0.999596	0.000404	3.80	0.00029	0.999928	0.000072
3.36	0.00141	0.999610	0.000390	3.81	0.00028	0.999930	0.000070
3.37	0.00136	0.999624	0.000376	3.82	0.00027	0.999933	0.000067
3.38	0.00132	0.999638	0.000362	3.83	0.00026	0.999936	0.000064
3.39	0.00127	0.999650	0.000350	3.84	0.00025	0.999938	0.000062
3.40	0.00123	0.999664	0.000336	3.85	0.00024	0.999941	0.000059
3.41	0.00119	0.999675	0.000325	3.86	0.00023	0.999943	0.000057
3.42	0.00115	0.999687	0.000313	3.87	0.00022	0.999946	0.000054
3.43	0.00111	0.999698	0.000302	3.88	0.00021	0.999948	0.000052
3.44	0.00107	0.999709	0.000291	3.89	0.00021	0.999950	0.000050
3.45	0.00104	0.999720	0.000280	3.90	0.00020	0.999952	0.000048
3.46	0.00100	0.999730	0.000270	3.91	0.00019	0.999954	0.000046
3.47	0.00097	0.999740	0.000260	3.92	0.00018	0.999956	0.000044
3.48	0.00094	0.999749	0.000251	3.93	0.00018	0.999958	0.000042
3.49	0.00090	0.999758	0.000242	3.94	0.00017	0.999959	0.000041
3.50	0.00087	0.999767	0.000233	3.95	0.00016	0.999961	0.000039
3.51	0.00084	0.999776	0.000224	3.96	0.00016	0.999962	0.000038
3.52	0.00081	0.999784	0.000216	3.97	0.00015	0.999964	0.000036
3.53	0.00079	0.999792	0.000208	3.98	0.00014	0.999966	0.000034
3.54	0.00076	0.999800	0.000200	3.99	0.00014	0.999967	0.000033
3.55	0.00073	0.999807	0.000193	4.00	0.00013	0.999968	0.000032
3.56	0.00071	0.999815	0.000185				
3.57	0.00068	0.999822	0.000178	4.25	0.00005	0.999989	0.000011
3.58	0.00066	0.999828	0.000172				
3.59	0.00063	0.999835	0.000165	4.50	0.00002	0.999997	0.000003
3.60	0.00061	0.999841	0.000159				
3.61	0.00059	0.999847	0.000153	4.75	0.00001	0.999999	0.000001
3.62	0.00057	0.999853	0.000147				
3.63	0.00055	0.999858	0.000142	5.00	0.00000	1.000000	0.000000
3.64	0.00053	0.999864	0.000136				

TABLE II

PERCENTAGE POINTS OF THE χ^2-DISTRIBUTION*

ν \ α	0.995	0.990	0.975	0.950	0.900	0.750
1	$392{,}704 \times 10^{-10}$	$157{,}088 \times 10^{-9}$	$982{,}069 \times 10^{-9}$	$393{,}214 \times 10^{-8}$	0.0157908	0.1015308
2	0.0100251	0.0201007	0.0506356	0.102587	0.210720	0.575364
3	0.0717212	0.114832	0.215795	0.351846	0.584375	1.212534
4	0.206990	0.297110	0.484419	0.710721	1.063623	1.92255
5	0.411740	0.554300	0.831211	1.145476	1.61031	2.67460
6	0.675727	0.872085	1.237347	1.63539	2.20413	3.45460
7	0.989265	1.239043	1.68987	2.16735	2.83311	4.25485
8	1.344419	1.646482	2.17973	2.73264	3.48954	5.07064
9	1.734926	2.087912	2.70039	3.32511	4.16816	5.89883
10	2.15585	2.55821	3.24697	3.94030	4.86518	6.73720
11	2.60321	3.05347	3.81575	4.57481	5.57779	7.58412
12	3.07382	3.57056	4.40379	5.22603	6.30380	8.43842
13	3.56503	4.10691	5.00874	5.89186	7.04150	9.29906
14	4.07468	4.66043	5.62872	6.57063	7.78953	10.1653
15	4.60094	5.22935	6.26214	7.26094	8.54675	11.0365
16	5.14224	5.81221	6.90766	7.96164	9.31223	11.9122
17	5.69724	6.40776	7.56418	8.67176	10.0852	12.7919
18	6.26481	7.01491	8.23075	9.39046	10.8649	13.6753
19	6.84398	7.63273	8.90655	10.1170	11.6509	14.5620
20	7.43386	8.26040	9.59083	10.8508	12.4426	15.4518
21	8.03366	8.89720	10.28293	11.5913	13.2396	16.3444
22	8.64272	9.54249	10.9823	12.3380	14.0415	17.2396
23	9.26042	10.19567	11.6885	13.0905	14.8479	18.1373
24	9.88623	10.8564	12.4001	13.8484	15.6587	19.0372
25	10.5197	11.5240	13.1197	14.6114	16.4734	19.9393
26	11.1603	12.1981	13.8439	15.3791	17.2919	20.8434
27	11.8076	12.8786	14.5733	16.1513	18.1138	21.7494
28	12.4613	13.5648	15.3079	16.9279	18.9392	22.6572
29	13.1211	14.2565	16.0471	17.7083	19.7677	23.5666
30	13.7867	14.9535	16.7908	18.4926	20.5992	24.4776
40	20.7065	22.1643	24.4331	26.5093	29.0505	33.6603
50	27.9907	29.7067	32.3574	34.7642	37.6886	42.9421
60	35.5346	37.4848	40.4817	43.1879	46.4589	52.2938
70	43.2752	45.4418	48.7576	51.7393	55.3290	61.6983
80	51.1720	53.5400	57.1532	60.3915	64.2778	71.1445
90	59.1963	61.7541	65.6466	69.1260	73.2912	80.6247
100	67.3276	70.0648	74.2219	77.9295	82.3581	90.1332
t_α	−2.5758	−2.3263	−1.9600	−1.6449	−1.2816	−0.6745

* Reproduced by permission of Professor E. S. Pearson from "Tables of the Percentage Points of the χ^2-Distribution," *Biometrika, 32* (1941), pp. 188–189, by Catherine M. Thompson.

TABLE II (*continued*)

ν \ α	0.500	0.250	0.100	0.050	0.025	0.010	0.005
1	0.454937	1.32330	2.70554	3.84146	5.02389	6.63490	7.87944
2	1.38629	2.77259	4.60517	5.99147	7.37776	9.21034	10.5966
3	2.36597	4.10835	6.25139	7.81473	9.34840	11.3449	12.8381
4	3.35670	5.38527	7.77944	9.48773	11.1433	13.2767	14.8602
5	4.35146	6.62568	9.23635	11.0705	12.8325	15.0863	16.7496
6	5.34812	7.84080	10.6446	12.5916	14.4494	16.8119	18.5476
7	6.34581	9.03715	12.0170	14.0671	16.0128	18.4753	20.2777
8	7.34412	10.2188	13.3616	15.5073	17.5346	20.0902	21.9550
9	8.34283	11.3887	14.6837	16.9190	19.0228	21.6660	23.5893
10	9.34182	12.5489	15.9871	18.3070	20.4831	23.2093	25.1882
11	10.3410	13.7007	17.2750	19.6751	21.9200	24.7250	26.7569
12	11.3403	14.8454	18.5494	21.0261	23.3367	26.2170	28.2995
13	12.3398	15.9839	19.8119	22.3621	24.7356	27.6883	29.8194
14	13.3393	17.1170	21.0642	23.6848	26.1190	29.1413	31.3193
15	14.3389	18.2451	22.3072	24.9958	27.4884	30.5779	32.8013
16	15.3385	19.3688	23.5418	26.2962	28.8454	31.9999	34.2672
17	16.3381	20.4887	24.7690	27.5871	30.1910	33.4087	35.7185
18	17.3379	21.6049	25.9894	28.8693	31.5264	34.8053	37.1564
19	18.3376	22.7178	27.2036	30.1435	32.8523	36.1908	38.5822
20	19.3374	23.8277	28.4120	31.4104	34.1696	37.5662	39.9968
21	20.3372	24.9348	29.6151	32.6705	35.4789	38.9321	41.4010
22	21.3370	26.0393	30.8133	33.9244	36.7807	40.2894	42.7956
23	22.3369	27.1413	32.0069	35.1725	38.0757	41.6384	44.1813
24	23.3367	28.2412	33.1963	36.4151	39.3641	42.9798	45.5585
25	24.3366	29.3389	34.3816	37.6525	40.6465	44.3141	46.9278
26	25.3364	30.4345	35.5631	38.8852	41.9232	45.6417	48.2899
27	26.3363	31.5284	36.7412	40.1133	43.1944	46.9630	49.6449
28	27.3363	32.6205	37.9159	41.3372	44.4607	48.2782	50.9933
29	28.3362	33.7109	39.0875	42.5569	45.7222	49.5879	52.3356
30	29.3360	34.7998	40.2560	43.7729	46.9792	50.8922	53.6720
40	39.3354	45.6160	51.8050	55.7585	59.3417	63.6907	66.7659
50	49.3349	56.3336	63.1671	67.5048	71.4202	76.1539	79.4900
60	59.3347	66.9814	74.3970	79.0819	83.2976	88.3794	91.9517
70	69.3344	77.5766	85.5271	90.5312	95.0231	100.425	104.215
80	79.3343	88.1303	96.5782	101.879	106.629	112.329	116.321
90	89.3342	98.6499	107.565	113.145	118.136	124.116	128.299
100	99.3341	109.141	118.498	124.342	129.561	135.807	140.169
t_α	0.0000	+0.6745	+1.2816	+1.6449	+1.9600	+2.3263	+2.5758

For $30 < \nu < 100$, linear interpolation where necessary will give 4 significant figures.
For $\nu > 100$ take $\chi^2_{\nu,\alpha} = \frac{1}{2}(t_\alpha + \sqrt{2\nu - 1})^2$.
A description of this table is given in Section 4.46.

TABLE III

Percentage Points of the t-Distribution*

ν \ α	0.50	0.25	0.10	0.05	0.025	0.01	0.005
1	1.00000	2.4142	6.3138	12.706	25.452	63.657	127.32
2	0.81650	1.6036	2.9200	4.3027	6.2053	9.9248	14.089
3	0.76489	1.4226	2.3534	3.1825	4.1765	5.8409	7.4533
4	0.74070	1.3444	2.1318	2.7764	3.4954	4.6041	5.5976
5	0.72669	1.3009	2.0150	2.5706	3.1634	4.0321	4.7733
6	0.71756	1.2733	1.9432	2.4469	2.9687	3.7074	4.3168
7	0.71114	1.2543	1.8946	2.3646	2.8412	3.4995	4.0293
8	0.70639	1.2403	1.8595	2.3060	2.7515	3.3554	3.8325
9	0.70272	1.2297	1.8331	2.2622	2.6850	3.2498	3.6897
10	0.69981	1.2213	1.8125	2.2281	2.6338	3.1693	3.5814
11	0.69745	1.2145	1.7959	2.2010	2.5931	3.1058	3.4966
12	0.69548	1.2089	1.7823	2.1788	2.5600	3.0545	3.4284
13	0.69384	1.2041	1.7709	2.1604	2.5326	3.0123	3.3725
14	0.69242	1.2001	1.7613	2.1448	2.5096	2.9768	3.3257
15	0.69120	1.1967	1.7530	2.1315	2.4899	2.9467	3.2860
16	0.69013	1.1937	1.7459	2.1199	2.4729	2.9208	3.2520
17	0.68919	1.1910	1.7396	2.1098	2.4581	2.8982	3.2225
18	0.68837	1.1887	1.7341	2.1009	2.4450	2.8784	3.1966
19	0.68763	1.1866	1.7291	2.0930	2.4334	2.8609	3.1737
20	0.68696	1.1848	1.7247	2.0860	2.4231	2.8453	3.1534
21	0.68635	1.1831	1.7207	2.0796	2.4138	2.8314	3.1352
22	0.68580	1.1816	1.7171	2.0739	2.4055	2.8188	3.1188
23	0.68531	1.1802	1.7139	2.0687	2.3979	2.8073	3.1040
24	0.68485	1.1789	1.7109	2.0639	2.3910	2.7969	3.0905
25	0.68443	1.1777	1.7081	2.0595	2.3846	2.7874	3.0782
26	0.68405	1.1766	1.7056	2.0555	2.3788	2.7787	3.0669
27	0.68370	1.1757	1.7033	2.0518	2.3734	2.7707	3.0565
28	0.68335	1.1748	1.7011	2.0484	2.3685	2.7633	3.0469
29	0.68304	1.1739	1.6991	2.0452	2.3638	2.7564	3.0380
30	0.68276	1.1731	1.6973	2.0423	2.3596	2.7500	3.0298
40	0.68066	1.1673	1.6839	2.0211	2.3289	2.7045	2.9712
60	0.67862	1.1616	1.6707	2.0003	2.2991	2.6603	2.9146
120	0.67656	1.1559	1.6577	1.9799	2.2699	2.6174	2.8599
∞	0.67449	1.1503	1.6449	1.9600	2.2414	2.5758	2.8070

* Computed by Maxine Merrington from "Tables of Percentage Points of the Incomplete Beta Function," *Biometrika*, *32* (1941), pp. 168–181, by Catherine M. Thompson, and reproduced by permission of Professor E. S. Pearson.

A description of this table is given in Section 4.54. Where necessary, interpolation should be carried out using the reciprocals of the degrees of freedom. The function $120/\nu$ is convenient for this purpose.

TABLE IV

Percentage Points of the F-Distribution*

$\alpha = 0.50$

ν_2 \ ν_1	1	2	3	4	5	6	7	8	9
1	1.0000	1.5000	1.7092	1.8227	1.8937	1.9422	1.9774	2.0041	2.0250
2	0.66667	1.0000	1.1349	1.2071	1.2519	1.2824	1.3045	1.3213	1.3344
3	0.58506	0.88110	1.0000	1.0632	1.1024	1.1289	1.1482	1.1627	1.1741
4	0.54863	0.82843	0.94054	1.0000	1.0367	1.0617	1.0797	1.0933	1.1040
5	0.52807	0.79877	0.90715	0.96456	1.0000	1.0240	1.0414	1.0545	1.0648
6	0.51489	0.77976	0.88578	0.94191	0.97654	1.0000	1.0169	1.0298	1.0398
7	0.50572	0.76655	0.87095	0.92619	0.96026	0.98334	1.0000	1.0126	1.0224
8	0.49898	0.75683	0.86004	0.91464	0.94831	0.97111	0.98757	1.0000	1.0097
9	0.49382	0.74938	0.85168	0.90580	0.93916	0.96175	0.97805	0.99037	1.0000
10	0.48973	0.74349	0.84508	0.89882	0.93193	0.95436	0.97054	0.98276	0.99232
11	0.48644	0.73872	0.83973	0.89316	0.92608	0.94837	0.96445	0.97661	0.98610
12	0.48369	0.73477	0.83530	0.88848	0.92124	0.94342	0.95943	0.97152	0.98097
13	0.48141	0.73145	0.83159	0.88454	0.91718	0.93926	0.95520	0.96724	0.97665
14	0.47944	0.72862	0.82842	0.88119	0.91371	0.93573	0.95161	0.96360	0.97298
15	0.47775	0.72619	0.82569	0.87830	0.91073	0.93267	0.94850	0.96046	0.96981
16	0.47628	0.72406	0.82330	0.87578	0.90812	0.93001	0.94580	0.95773	0.96705
17	0.47499	0.72219	0.82121	0.87357	0.90584	0.92767	0.94342	0.95532	0.96462
18	0.47385	0.72053	0.81936	0.87161	0.90381	0.92560	0.94132	0.95319	0.96247
19	0.47284	0.71906	0.81771	0.86987	0.90200	0.92375	0.93944	0.95129	0.96056
20	0.47192	0.71773	0.81621	0.86830	0.90038	0.92210	0.93776	0.94959	0.95884
21	0.47108	0.71653	0.81487	0.86688	0.89891	0.92060	0.93624	0.94805	0.95728
22	0.47033	0.71545	0.81365	0.86559	0.89759	0.91924	0.93486	0.94665	0.95588
23	0.46965	0.71446	0.81255	0.86442	0.89638	0.91800	0.93360	0.94538	0.95459
24	0.46902	0.71356	0.81153	0.86335	0.89527	0.91687	0.93245	0.94422	0.95342
25	0.46844	0.71272	0.81061	0.86236	0.89425	0.91583	0.93140	0.94315	0.95234
26	0.46793	0.71195	0.80975	0.86145	0.89331	0.91487	0.93042	0.94217	0.95135
27	0.46744	0.71124	0.80894	0.86061	0.89244	0.91399	0.92952	0.94126	0.95044
28	0.46697	0.71059	0.80820	0.85983	0.89164	0.91317	0.92869	0.94041	0.94958
29	0.46654	0.70999	0.80753	0.85911	0.89089	0.91241	0.92791	0.93963	0.94879
30	0.46616	0.70941	0.80689	0.85844	0.89019	0.91169	0.92719	0.93889	0.94805
40	0.46330	0.70531	0.80228	0.85357	0.88516	0.90654	0.92197	0.93361	0.94272
60	0.46053	0.70122	0.79770	0.84873	0.88017	0.90144	0.91679	0.92838	0.93743
120	0.45774	0.69717	0.79314	0.84392	0.87521	0.89637	0.91164	0.92318	0.93218
∞	0.45494	0.69315	0.78866	0.83918	0.87029	0.89135	0.90654	0.91802	0.92698

* Reproduced by permission of Professor E. S. Pearson from "Tables of Percentage Points of the Inverted Beta (F) Distribution," *Biometrika, 33* (1943), pp. 73–88, by Maxine Merrington and Catherine M. Thompson.

TABLE IV (*continued*)

ν_2 \ ν_1	10	12	15	20	24	30	40	60	120	∞
1	2.0419	2.0674	2.0931	2.1190	2.1321	2.1452	2.1584	2.1716	2.1848	2.1981
2	1.3450	1.3610	1.3771	1.3933	1.4014	1.4096	1.4178	1.4261	1.4344	1.4427
3	1.1833	1.1972	1.2111	1.2252	1.2322	1.2393	1.2464	1.2536	1.2608	1.2680
4	1.1126	1.1255	1.1386	1.1517	1.1583	1.1649	1.1716	1.1782	1.1849	1.1916
5	1.0730	1.0855	1.0980	1.1106	1.1170	1.1234	1.1297	1.1361	1.1426	1.1490
6	1.0478	1.0600	1.0722	1.0845	1.0907	1.0969	1.1031	1.1093	1.1156	1.1219
7	1.0304	1.0423	1.0543	1.0664	1.0724	1.0785	1.0846	1.0908	1.0969	1.1031
8	1.0175	1.0293	1.0412	1.0531	1.0591	1.0651	1.0711	1.0771	1.0832	1.0893
9	1.0077	1.0194	1.0311	1.0429	1.0489	1.0548	1.0608	1.0667	1.0727	1.0788
10	1.0000	1.0116	1.0232	1.0349	1.0408	1.0467	1.0526	1.0585	1.0645	1.0705
11	0.99373	1.0052	1.0168	1.0284	1.0343	1.0401	1.0460	1.0519	1.0578	1.0637
12	0.98856	1.0000	1.0115	1.0231	1.0289	1.0347	1.0405	1.0464	1.0523	1.0582
13	0.98421	0.99560	1.0071	1.0186	1.0243	1.0301	1.0360	1.0418	1.0476	1.0535
14	0.98051	0.99186	1.0033	1.0147	1.0205	1.0263	1.0321	1.0379	1.0437	1.0495
15	0.97732	0.98863	1.0000	1.0114	1.0172	1.0229	1.0287	1.0345	1.0403	1.0461
16	0.97454	0.98582	0.99716	1.0086	1.0143	1.0200	1.0258	1.0315	1.0373	1.0431
17	0.97209	0.98334	0.99466	1.0060	1.0117	1.0174	1.0232	1.0289	1.0347	1.0405
18	0.96993	0.98116	0.99245	1.0038	1.0095	1.0152	1.0209	1.0267	1.0324	1.0382
19	0.96800	0.97920	0.99047	1.0018	1.0075	1.0132	1.0189	1.0246	1.0304	1.0361
20	0.96626	0.97746	0.98870	1.0000	1.0057	1.0114	1.0171	1.0228	1.0285	1.0343
21	0.96470	0.97587	0.98710	0.99838	1.0040	1.0097	1.0154	1.0211	1.0268	1.0326
22	0.96328	0.97444	0.98565	0.99692	1.0026	1.0082	1.0139	1.0196	1.0253	1.0311
23	0.96199	0.97313	0.98433	0.99558	1.0012	1.0069	1.0126	1.0183	1.0240	1.0297
24	0.96081	0.97194	0.98312	0.99436	1.0000	1.0057	1.0113	1.0170	1.0227	1.0284
25	0.95972	0.97084	0.98201	0.99324	0.99887	1.0045	1.0102	1.0159	1.0215	1.0273
26	0.95872	0.96983	0.98099	0.99220	0.99783	1.0035	1.0091	1.0148	1.0205	1.0262
27	0.95779	0.96889	0.98004	0.99125	0.99687	1.0025	1.0082	1.0138	1.0195	1.0252
28	0.95694	0.96802	0.97917	0.99036	0.99598	1.0016	1.0073	1.0129	1.0186	1.0243
29	0.95614	0.96722	0.97835	0.98954	0.99515	1.0008	1.0064	1.0121	1.0177	1.0234
30	0.95540	0.96647	0.97759	0.98877	0.99438	1.0000	1.0056	1.0113	1.0170	1.0226
40	0.95003	0.96104	0.97211	0.98323	0.98880	0.99440	1.0000	1.0056	1.0113	1.0169
60	0.94471	0.95566	0.96667	0.97773	0.98328	0.98884	0.99441	1.0000	1.0056	1.0112
120	0.93943	0.95032	0.96128	0.97228	0.97780	0.98333	0.98887	0.99443	1.0000	1.0056
∞	0.93418	0.94503	0.95593	0.96687	0.97236	0.97787	0.98339	0.98891	0.99445	1.0000

A description of these tables is given in Section 4.55. Where necessary, interpolation should be carried out using the reciprocals of the degrees of freedom. The function $120/\nu$ is convenient for this purpose.

TABLE IV (*continued*)

PERCENTAGE POINTS OF THE F-DISTRIBUTION*

$\alpha = 0.25$

ν_2 \ ν_1	1	2	3	4	5	6	7	8	9
1	5.8285	7.5000	8.1999	8.5810	8.8198	8.9833	9.1021	9.1922	9.2631
2	2.5714	3.0000	3.1534	3.2320	3.2799	3.3121	3.3352	3.3526	3.3661
3	2.0239	2.2798	2.3555	2.3901	2.4095	2.4218	2.4302	2.4364	2.4410
4	1.8074	2.0000	2.0467	2.0642	2.0723	2.0766	2.0790	2.0805	2.0814
5	1.6925	1.8528	1.8843	1.8927	1.8947	1.8945	1.8935	1.8923	1.8911
6	1.6214	1.7622	1.7844	1.7872	1.7852	1.7821	1.7789	1.7760	1.7733
7	1.5732	1.7010	1.7169	1.7157	1.7111	1.7059	1.7011	1.6969	1.6931
8	1.5384	1.6569	1.6683	1.6642	1.6575	1.6508	1.6448	1.6396	1.6350
9	1.5121	1.6236	1.6315	1.6253	1.6170	1.6091	1.6022	1.5961	1.5909
10	1.4915	1.5975	1.6028	1.5949	1.5853	1.5765	1.5688	1.5621	1.5563
11	1.4749	1.5767	1.5798	1.5704	1.5598	1.5502	1.5418	1.5346	1.5284
12	1.4613	1.5595	1.5609	1.5503	1.5389	1.5286	1.5197	1.5120	1.5054
13	1.4500	1.5452	1.5451	1.5336	1.5214	1.5105	1.5011	1.4931	1.4861
14	1.4403	1.5331	1.5317	1.5194	1.5066	1.4952	1.4854	1.4770	1.4697
15	1.4321	1.5227	1.5202	1.5071	1.4938	1.4820	1.4718	1.4631	1.4556
16	1.4249	1.5137	1.5103	1.4965	1.4827	1.4705	1.4601	1.4511	1.4433
17	1.4186	1.5057	1.5015	1.4873	1.4730	1.4605	1.4497	1.4405	1.4325
18	1.4130	1.4988	1.4938	1.4790	1.4644	1.4516	1.4406	1.4312	1.4230
19	1.4081	1.4925	1.4870	1.4717	1.4568	1.4437	1.4325	1.4228	1.4145
20	1.4037	1.4870	1.4808	1.4652	1.4500	1.4366	1.4252	1.4153	1.4069
21	1.3997	1.4820	1.4753	1.4593	1.4438	1.4302	1.4186	1.4086	1.4000
22	1.3961	1.4774	1.4703	1.4540	1.4382	1.4244	1.4126	1.4025	1.3937
23	1.3928	1.4733	1.4657	1.4491	1.4331	1.4191	1.4072	1.3969	1.3880
24	1.3898	1.4695	1.4615	1.4447	1.4285	1.4143	1.4022	1.3918	1.3828
25	1.3870	1.4661	1.4577	1.4406	1.4242	1.4099	1.3976	1.3871	1.3780
26	1.3845	1.4629	1.4542	1.4368	1.4203	1.4058	1.3935	1.3828	1.3737
27	1.3822	1.4600	1.4510	1.4334	1.4166	1.4021	1.3896	1.3788	1.3696
28	1.3800	1.4572	1.4480	1.4302	1.4133	1.3986	1.3860	1.3752	1.3658
29	1.3780	1.4547	1.4452	1.4272	1.4102	1.3953	1.3826	1.3717	1.3623
30	1.3761	1.4524	1.4426	1.4244	1.4073	1.3923	1.3795	1.3685	1.3590
40	1.3626	1.4355	1.4239	1.4045	1.3863	1.3706	1.3571	1.3455	1.3354
60	1.3493	1.4188	1.4055	1.3848	1.3657	1.3491	1.3349	1.3226	1.3119
120	1.3362	1.4024	1.3873	1.3654	1.3453	1.3278	1.3128	1.2999	1.2886
∞	1.3233	1.3863	1.3694	1.3463	1.3251	1.3068	1.2910	1.2774	1.2654

* Reproduced by permission of Professor E. S. Pearson from "Tables of Percentage Points of the Inverted Beta (F) Distribution," *Biometrika*, *33* (1943), pp. 73-88, by Maxine Merrington and Catherine M. Thompson.

TABLE IV (*continued*)

ν_2 \ ν_1	10	12	15	20	24	30	40	60	120	∞
1	9.3202	9.4064	9.4934	9.5813	9.6255	9.6698	9.7144	9.7591	9.8041	9.8492
2	3.3770	3.3934	3.4098	3.4263	3.4345	3.4428	3.4511	3.4594	3.4677	3.4761
3	2.4447	2.4500	2.4552	2.4602	2.4626	2.4650	2.4674	2.4697	2.4720	2.4742
4	2.0820	2.0826	2.0829	2.0828	2.0827	2.0825	2.0821	2.0817	2.0812	2.0806
5	1.8899	1.8877	1.8851	1.8820	1.8802	1.8784	1.8763	1.8742	1.8719	1.8694
6	1.7708	1.7668	1.7621	1.7569	1.7540	1.7510	1.7477	1.7443	1.7407	1.7368
7	1.6898	1.6843	1.6781	1.6712	1.6675	1.6635	1.6593	1.6548	1.6502	1.6452
8	1.6310	1.6244	1.6170	1.6088	1.6043	1.5996	1.5945	1.5892	1.5836	1.5777
9	1.5863	1.5788	1.5705	1.5611	1.5560	1.5506	1.5450	1.5389	1.5325	1.5257
10	1.5513	1.5430	1.5338	1.5235	1.5179	1.5119	1.5056	1.4990	1.4919	1.4843
11	1.5230	1.5140	1.5041	1.4930	1.4869	1.4805	1.4737	1.4664	1.4587	1.4504
12	1.4996	1.4902	1.4796	1.4678	1.4613	1.4544	1.4471	1.4393	1.4310	1.4221
13	1.4801	1.4701	1.4590	1.4465	1.4397	1.4324	1.4247	1.4164	1.4075	1.3980
14	1.4634	1.4530	1.4414	1.4284	1.4212	1.4136	1.4055	1.3967	1.3874	1.3772
15	1.4491	1.4383	1.4263	1.4127	1.4052	1.3973	1.3888	1.3796	1.3698	1.3591
16	1.4366	1.4255	1.4130	1.3990	1.3913	1.3830	1.3742	1.3646	1.3543	1.3432
17	1.4256	1.4142	1.4014	1.3869	1.3790	1.3704	1.3613	1.3514	1.3406	1.3290
18	1.4159	1.4042	1.3911	1.3762	1.3680	1.3592	1.3497	1.3395	1.3284	1.3162
19	1.4073	1.3953	1.3819	1.3666	1.3582	1.3492	1.3394	1.3289	1.3174	1.3048
20	1.3995	1.3873	1.3736	1.3580	1.3494	1.3401	1.3301	1.3193	1.3074	1.2943
21	1.3925	1.3801	1.3661	1.3502	1.3414	1.3319	1.3217	1.3105	1.2983	1.2848
22	1.3861	1.3735	1.3593	1.3431	1.3341	1.3245	1.3140	1.3025	1.2900	1.2761
23	1.3803	1.3675	1.3531	1.3366	1.3275	1.3176	1.3069	1.2952	1.2824	1.2681
24	1.3750	1.3621	1.3474	1.3307	1.3214	1.3113	1.3004	1.2885	1.2754	1.2607
25	1.3701	1.3570	1.3422	1.3252	1.3158	1.3056	1.2945	1.2823	1.2689	1.2538
26	1.3656	1.3524	1.3374	1.3202	1.3106	1.3002	1.2889	1.2765	1.2628	1.2474
27	1.3615	1.3481	1.3329	1.3155	1.3058	1.2953	1.2838	1.2712	1.2572	1.2414
28	1.3576	1.3441	1.3288	1.3112	1.3013	1.2906	1.2790	1.2662	1.2519	1.2358
29	1.3541	1.3404	1.3249	1.3071	1.2971	1.2863	1.2745	1.2615	1.2470	1.2306
30	1.3507	1.3369	1.3213	1.3033	1.2933	1.2823	1.2703	1.2571	1.2424	1.2256
40	1.3266	1.3119	1.2952	1.2758	1.2649	1.2529	1.2397	1.2249	1.2080	1.1883
60	1.3026	1.2870	1.2691	1.2481	1.2361	1.2229	1.2081	1.1912	1.1715	1.1474
120	1.2787	1.2621	1.2428	1.2200	1.2068	1.1921	1.1752	1.1555	1.1314	1.0987
∞	1.2549	1.2371	1.2163	1.1914	1.1767	1.1600	1.1404	1.1164	1.0838	1.0000

A description of these tables is given in Section 4.55. Where necessary, interpolation should be carried out using the reciprocals of the degrees of freedom. The function $120/\nu$ is convenient for this purpose.

TABLE IV (*continued*)

PERCENTAGE POINTS OF THE F-DISTRIBUTION*

$\alpha = 0.10$

ν_2 \ ν_1	1	2	3	4	5	6	7	8	9
1	39.864	49.500	53.593	55.833	57.241	58.204	58.906	59.439	59.858
2	8.5263	9.0000	9.1618	9.2434	9.2926	9.3255	9.3491	9.3668	9.3805
3	5.5383	5.4624	5.3908	5.3427	5.3092	5.2847	5.2662	5.2517	5.2400
4	4.5448	4.3246	4.1908	4.1073	4.0506	4.0098	3.9790	3.9549	3.9357
5	4.0604	3.7797	3.6195	3.5202	3.4530	3.4045	3.3679	3.3393	3.3163
6	3.7760	3.4633	3.2888	3.1808	3.1075	3.0546	3.0145	2.9830	2.9577
7	3.5894	3.2574	3.0741	2.9605	2.8833	2.8274	2.7849	2.7516	2.7247
8	3.4579	3.1131	2.9238	2.8064	2.7265	2.6683	2.6241	2.5893	2.5612
9	3.3603	3.0065	2.8129	2.6927	2.6106	2.5509	2.5053	2.4694	2.4403
10	3.2850	2.9245	2.7277	2.6053	2.5216	2.4606	2.4140	2.3772	2.3473
11	3.2252	2.8595	2.6602	2.5362	2.4512	2.3891	2.3416	2.3040	2.2735
12	3.1765	2.8068	2.6055	2.4801	2.3940	2.3310	2.2828	2.2446	2.2135
13	3.1362	2.7632	2.5603	2.4337	2.3467	2.2830	2.2341	2.1953	2.1638
14	3.1022	2.7265	2.5222	2.3947	2.3069	2.2426	2.1931	2.1539	2.1220
15	3.0732	2.6952	2.4898	2.3614	2.2730	2.2081	2.1582	2.1185	2.0862
16	3.0481	2.6682	2.4618	2.3327	2.2438	2.1783	2.1280	2.0880	2.0553
17	3.0262	2.6446	2.4374	2.3077	2.2183	2.1524	2.1017	2.0613	2.0284
18	3.0070	2.6239	2.4160	2.2858	2.1958	2.1296	2.0785	2.0379	2.0047
19	2.9899	2.6056	2.3970	2.2663	2.1760	2.1094	2.0580	2.0171	1.9836
20	2.9747	2.5893	2.3801	2.2489	2.1582	2.0913	2.0397	1.9985	1.9649
21	2.9609	2.5746	2.3649	2.2333	2.1423	2.0751	2.0232	1.9819	1.9480
22	2.9486	2.5613	2.3512	2.2193	2.1279	2.0605	2.0084	1.9668	1.9327
23	2.9374	2.5493	2.3387	2.2065	2.1149	2.0472	1.9949	1.9531	1.9189
24	2.9271	2.5383	2.3274	2.1949	2.1030	2.0351	1.9826	1.9407	1.9063
25	2.9177	2.5283	2.3170	2.1843	2.0922	2.0241	1.9714	1.9292	1.8947
26	2.9091	2.5191	2.3075	2.1745	2.0822	2.0139	1.9610	1.9188	1.8841
27	2.9012	2.5106	2.2987	2.1655	2.0730	2.0045	1.9515	1.9091	1.8743
28	2.8939	2.5028	2.2906	2.1571	2.0645	1.9959	1.9427	1.9001	1.8652
29	2.8871	2.4955	2.2831	2.1494	2.0566	1.9878	1.9345	1.8918	1.8568
30	2.8807	2.4887	2.2761	2.1422	2.0492	1.9803	1.9269	1.8841	1.8490
40	2.8354	2.4404	2.2261	2.0909	1.9968	1.9269	1.8725	1.8289	1.7929
60	2.7914	2.3932	2.1774	2.0410	1.9457	1.8747	1.8194	1.7748	1.7380
120	2.7478	2.3473	2.1300	1.9923	1.8959	1.8238	1.7675	1.7220	1.6843
∞	2.7055	2.3026	2.0838	1.9449	1.8473	1.7741	1.7167	1.6702	1.6315

* Reproduced by permission of Professor E. S. Pearson from "Tables of Percentage Points of the Inverted Beta (F) Distribution," *Biometrika*, *33* (1943), pp. 73–88, by Maxine Merrington and Catherine M. Thompson.

TABLE IV (*continued*)

ν_2 \ ν_1	10	12	15	20	24	30	40	60	120	∞
1	60.195	60.705	61.220	61.740	62.002	62.265	62.529	62.794	63.061	63.328
2	9.3916	9.4081	9.4247	9.4413	9.4496	9.4579	9.4663	9.4746	9.4829	9.4913
3	5.2304	5.2156	5.2003	5.1845	5.1764	5.1681	5.1597	5.1512	5.1425	5.1337
4	3.9199	3.8955	3.8689	3.8443	3.8310	3.8174	3.8036	3.7896	3.7753	3.7607
5	3.2974	3.2682	3.2380	3.2067	3.1905	3.1741	3.1573	3.1402	3.1228	3.1050
6	2.9369	2.9047	2.8712	2.8363	2.8183	2.8000	2.7812	2.7620	2.7423	2.7222
7	2.7025	2.6681	2.6322	2.5947	2.5753	2.5555	2.5351	2.5142	2.4928	2.4708
8	2.5380	2.5020	2.4642	2.4246	2.4041	2.3830	2.3614	2.3391	2.3162	2.2926
9	2.4163	2.3789	2.3396	2.2983	2.2768	2.2547	2.2320	2.2085	2.1843	2.1592
10	2.3226	2.2841	2.2435	2.2007	2.1784	2.1554	2.1317	2.1072	2.0818	2.0554
11	2.2482	2.2087	2.1671	2.1230	2.1000	2.0762	2.0516	2.0261	1.9997	1.9721
12	2.1878	2.1474	2.1049	2.0597	2.0360	2.0115	1.9861	1.9597	1.9323	1.9036
13	2.1376	2.0966	2.0532	2.0070	1.9827	1.9576	1.9315	1.9043	1.8759	1.8462
14	2.0954	2.0537	2.0095	1.9625	1.9377	1.9119	1.8852	1.8572	1.8280	1.7973
15	2.0593	2.0171	1.9722	1.9243	1.8990	1.8728	1.8454	1.8168	1.7867	1.7551
16	2.0281	1.9854	1.9399	1.8913	1.8656	1.8388	1.8108	1.7816	1.7507	1.7182
17	2.0009	1.9577	1.9117	1.8624	1.8362	1.8090	1.7805	1.7506	1.7191	1.6856
18	1.9770	1.9333	1.8868	1.8368	1.8103	1.7827	1.7537	1.7232	1.6910	1.6567
19	1.9557	1.9117	1.8647	1.8142	1.7873	1.7592	1.7298	1.6988	1.6659	1.6308
20	1.9367	1.8924	1.8449	1.7938	1.7667	1.7382	1.7083	1.6768	1.6433	1.6074
21	1.9197	1.8750	1.8272	1.7756	1.7481	1.7193	1.6890	1.6569	1.6228	1.5862
22	1.9043	1.8593	1.8111	1.7590	1.7312	1.7021	1.6714	1.6389	1.6042	1.5668
23	1.8903	1.8450	1.7964	1.7439	1.7159	1.6864	1.6554	1.6224	1.5871	1.5490
24	1.8775	1.8319	1.7831	1.7302	1.7019	1.6721	1.6407	1.6073	1.5715	1.5327
25	1.8658	1.8200	1.7708	1.7175	1.6890	1.6589	1.6272	1.5934	1.5570	1.5176
26	1.8550	1.8090	1.7596	1.7059	1.6771	1.6468	1.6147	1.5805	1.5437	1.5036
27	1.8451	1.7989	1.7492	1.6951	1.6662	1.6356	1.6032	1.5686	1.5313	1.4906
28	1.8359	1.7895	1.7395	1.6852	1.6560	1.6252	1.5925	1.5575	1.5198	1.4784
29	1.8274	1.7808	1.7306	1.6759	1.6465	1.6155	1.5825	1.5472	1.5090	1.4670
30	1.8195	1.7727	1.7223	1.6673	1.6377	1.6065	1.5732	1.5376	1.4989	1.4564
40	1.7627	1.7146	1.6624	1.6052	1.5741	1.5411	1.5056	1.4672	1.4248	1.3769
60	1.7070	1.6574	1.6034	1.5435	1.5107	1.4755	1.4373	1.3952	1.3476	1.2915
120	1.6524	1.6012	1.5450	1.4821	1.4472	1.4094	1.3676	1.3203	1.2646	1.1926
∞	1.5987	1.5458	1.4871	1.4206	1.3832	1.3419	1.2951	1.2400	1.1686	1.0000

A description of these tables is given in Section 4.55. Where necessary, interpolation should be carried out using the reciprocals of the degrees of freedom. The function $120/\nu$ is convenient for this purpose.

TABLE IV (*continued*)

PERCENTAGE POINTS OF THE F-DISTRIBUTION*

$\alpha = 0.05$

ν_2 \ ν_1	1	2	3	4	5	6	7	8	9
1	161.45	199.50	215.71	224.58	230.16	233.99	236.77	238.88	240.54
2	18.513	19.000	19.164	19.247	19.296	19.330	19.353	19.371	19.385
3	10.128	9.5521	9.2766	9.1172	9.0135	8.9406	8.8868	8.8452	8.8123
4	7.7086	6.9443	6.5914	6.3883	6.2560	6.1631	6.0942	6.0410	5.9988
5	6.6079	5.7861	5.4095	5.1922	5.0503	4.9503	4.8759	4.8183	4.7725
6	5.9874	5.1433	4.7571	4.5337	4.3874	4.2839	4.2066	4.1468	4.0990
7	5.5914	4.7374	4.3468	4.1203	3.9715	3.8660	3.7870	3.7257	3.6767
8	5.3177	4.4590	4.0662	3.8378	3.6875	3.5806	3.5005	3.4381	3.3881
9	5.1174	4.2565	3.8626	3.6331	3.4817	3.3738	3.2927	3.2296	3.1789
10	4.9646	4.1028	3.7083	3.4780	3.3258	3.2172	3.1355	3.0717	3.0204
11	4.8443	3.9823	3.5874	3.3567	3.2039	3.0946	3.0123	2.9480	2.8962
12	4.7472	3.8853	3.4903	3.2592	3.1059	2.9961	2.9134	2.8486	2.7964
13	4.6672	3.8056	3.4105	3.1791	3.0254	2.9153	2.8321	2.7669	2.7144
14	4.6001	3.7389	3.3439	3.1122	2.9582	2.8477	2.7642	2.6987	2.6458
15	4.5431	3.6823	3.2874	3.0556	2.9013	2.7905	2.7066	2.6408	2.5876
16	4.4940	3.6337	3.2389	3.0069	2.8524	2.7413	2.6572	2.5911	2.5377
17	4.4513	3.5915	3.1968	2.9647	2.8100	2.6987	2.6143	2.5480	2.4943
18	4.4139	3.5546	3.1599	2.9277	2.7729	2.6613	2.5767	2.5102	2.4563
19	4.3808	3.5219	3.1274	2.8951	2.7401	2.6283	2.5435	2.4768	2.4227
20	4.3513	3.4928	3.0984	2.8661	2.7109	2.5990	2.5140	2.4471	2.3928
21	4.3248	3.4668	3.0725	2.8401	2.6848	2.5727	2.4876	2.4205	2.3661
22	4.3009	3.4434	3.0491	2.8167	2.6613	2.5491	2.4638	2.3965	2.3419
23	4.2793	3.4221	3.0280	2.7955	2.6400	2.5277	2.4422	2.3748	2.3201
24	4.2597	3.4028	3.0088	2.7763	2.6207	2.5082	2.4226	2.3551	2.3002
25	4.2417	3.3852	2.9912	2.7587	2.6030	2.4904	2.4047	2.3371	2.2821
26	4.2252	3.3690	2.9751	2.7426	2.5868	2.4741	2.3883	2.3205	2.2655
27	4.2100	3.3541	2.9604	2.7278	2.5719	2.4591	2.3732	2.3053	2.2501
28	4.1960	3.3404	2.9467	2.7141	2.5581	2.4453	2.3593	2.2913	2.2360
29	4.1830	3.3277	2.9340	2.7014	2.5454	2.4324	2.3463	2.2782	2.2229
30	4.1709	3.3158	2.9223	2.6896	2.5336	2.4205	2.3343	2.2662	2.2107
40	4.0848	3.2317	2.8387	2.6060	2.4495	2.3359	2.2490	2.1802	2.1240
60	4.0012	3.1504	2.7581	2.5252	2.3683	2.2540	2.1665	2.0970	2.0401
120	3.9201	3.0718	2.6802	2.4472	2.2900	2.1750	2.0867	2.0164	1.9588
∞	3.8415	2.9957	2.6049	2.3719	2.2141	2.0986	2.0096	1.9384	1.8799

* Reproduced by permission of Professor E. S. Pearson from "Tables of Percentage Points of the Inverted Beta (F) Distribution," *Biometrika*, *33* (1943), pp. 73–88, by Maxine Merrington and Catherine M. Thompson.

TABLE IV (*continued*)

ν_2 \ ν_1	10	12	15	20	24	30	40	60	120	∞
1	241.88	243.91	245.95	248.01	249.05	250.09	251.14	252.20	253.25	254.32
2	19.396	19.413	19.429	19.446	19.454	19.462	19.471	19.479	19.487	19.496
3	8.7855	8.7446	8.7029	8.6602	8.6385	8.6166	8.5944	8.5720	8.5494	8.5265
4	5.9644	5.9117	5.8578	5.8025	5.7744	5.7459	5.7170	5.6878	5.6581	5.6281
5	4.7351	4.6777	4.6188	4.5581	4.5272	4.4957	4.4638	4.4314	4.3984	4.3650
6	4.0600	3.9999	3.9381	3.8742	3.8415	3.8082	3.7743	3.7398	3.7047	3.6688
7	3.6365	3.5747	3.5108	3.4445	3.4105	3.3758	3.3404	3.3043	3.2674	3.2298
8	3.3472	3.2840	3.2184	3.1503	3.1152	3.0794	3.0428	3.0053	2.9669	2.9276
9	3.1373	3.0729	3.0061	2.9365	2.9005	2.8637	2.8259	2.7872	2.7475	2.7067
10	2.9782	2.9130	2.8450	2.7740	2.7372	2.6996	2.6609	2.6211	2.5801	2.5379
11	2.8536	2.7876	2.7186	2.6464	2.6090	2.5705	2.5309	2.4901	2.4480	2.4045
12	2.7534	2.6866	2.6169	2.5436	2.5055	2.4663	2.4259	2.3842	2.3410	2.2962
13	2.6710	2.6037	2.5331	2.4589	2.4202	2.3803	2.3392	2.2966	2.2524	2.2064
14	2.6021	2.5342	2.4630	2.3879	2.3487	2.3082	2.2664	2.2230	2.1778	2.1307
15	2.5437	2.4753	2.4035	2.3275	2.2878	2.2468	2.2043	2.1601	2.1141	2.0658
16	2.4935	2.4247	2.3522	2.2756	2.2354	2.1938	2.1507	2.1058	2.0589	2.0096
17	2.4499	2.3807	2.3077	2.2304	2.1898	2.1477	2.1040	2.0584	2.0107	1.9604
18	2.4117	2.3421	2.2686	2.1906	2.1497	2.1071	2.0629	2.0166	1.9681	1.9168
19	2.3779	2.3080	2.2341	2.1555	2.1141	2.0712	2.0264	1.9796	1.9302	1.8780
20	2.3479	2.2776	2.2033	2.1242	2.0825	2.0391	1.9938	1.9464	1.8963	1.8432
21	2.3210	2.2504	2.1757	2.0960	2.0540	2.0102	1.9645	1.9165	1.8657	1.8117
22	2.2967	2.2258	2.1508	2.0707	2.0283	1.9842	1.9380	1.8895	1.8380	1.7831
23	2.2747	2.2036	2.1282	2.0476	2.0050	1.9605	1.9139	1.8649	1.8128	1.7570
24	2.2547	2.1834	2.1077	2.0267	1.9838	1.9390	1.8920	1.8424	1.7897	1.7331
25	2.2365	2.1649	2.0889	2.0075	1.9643	1.9192	1.8718	1.8217	1.7684	1.7110
26	2.2197	2.1479	2.0716	1.9898	1.9464	1.9010	1.8533	1.8027	1.7488	1.6906
27	2.2043	2.1323	2.0558	1.9736	1.9299	1.8842	1.8361	1.7851	1.7307	1.6717
28	2.1900	2.1179	2.0411	1.9586	1.9147	1.8687	1.8203	1.7689	1.7138	1.6541
29	2.1768	2.1045	2.0275	1.9446	1.9005	1.8543	1.8055	1.7537	1.6981	1.6377
30	2.1646	2.0921	2.0148	1.9317	1.8874	1.8409	1.7918	1.7396	1.6835	1.6223
40	2.0772	2.0035	1.9245	1.8389	1.7929	1.7444	1.6928	1.6373	1.5766	1.5089
60	1.9926	1.9174	1.8364	1.7480	1.7001	1.6491	1.5943	1.5343	1.4673	1.3893
120	1.9105	1.8337	1.7505	1.6587	1.6084	1.5543	1.4952	1.4290	1.3519	1.2539
∞	1.8307	1.7522	1.6664	1.5705	1.5173	1.4591	1.3940	1.3180	1.2214	1.0000

A description of these tables is given in Section 4.55. Where necessary, interpolation should be carried out using the reciprocals of the degrees of freedom. The function $120/\nu$ is convenient for this purpose.

TABLE IV (*continued*)

PERCENTAGE POINTS OF THE F-DISTRIBUTION*

$\alpha = 0.025$

ν_2 \ ν_1	1	2	3	4	5	6	7	8	9
1	647.79	799.50	864.16	899.58	921.85	937.11	948.22	956.66	963.28
2	38.506	39.000	39.165	39.248	39.298	39.331	39.355	39.373	39.387
3	17.443	16.044	15.439	15.101	14.885	14.735	14.624	14.540	14.473
4	12.218	10.649	9.9792	9.6045	9.3645	9.1973	9.0741	8.9796	8.9047
5	10.007	8.4336	7.7636	7.3879	7.1464	6.9777	6.8531	6.7572	6.6810
6	8.8131	7.2598	6.5988	6.2272	5.9876	5.8197	5.6955	5.5996	5.5234
7	8.0727	6.5415	5.8898	5.5226	5.2852	5.1186	4.9949	4.8994	4.8232
8	7.5709	6.0595	5.4160	5.0526	4.8173	4.6517	4.5286	4.4332	4.3572
9	7.2093	5.7147	5.0781	4.7181	4.4844	4.3197	4.1971	4.1020	4.0260
10	6.9367	5.4564	4.8256	4.4683	4.2361	4.0721	3.9498	3.8549	3.7790
11	6.7241	5.2559	4.6300	4.2751	4.0440	3.8807	3.7586	3.6638	3.5879
12	6.5538	5.0959	4.4742	4.1212	3.8911	3.7283	3.6065	3.5118	3.4358
13	6.4143	4.9653	4.3472	3.9959	3.7667	3.6043	3.4827	3.3880	3.3120
14	6.2979	4.8567	4.2417	3.8919	3.6634	3.5014	3.3799	3.2853	3.2093
15	6.1995	4.7650	4.1528	3.8043	3.5764	3.4147	3.2934	3.1987	3.1227
16	6.1151	4.6867	4.0768	3.7294	3.5021	3.3406	3.2194	3.1248	3.0488
17	6.0420	4.6189	4.0112	3.6648	3.4379	3.2767	3.1556	3.0610	2.9849
18	5.9781	4.5597	3.9539	3.6083	3.3820	3.2209	3.0999	3.0053	2.9291
19	5.9216	4.5075	3.9034	3.5587	3.3327	3.1718	3.0509	2.9563	2.8800
20	5.8715	4.4613	3.8587	3.5147	3.2891	3.1283	3.0074	2.9128	2.8365
21	5.8266	4.4199	3.8188	3.4754	3.2501	3.0895	2.9686	2.8740	2.7977
22	5.7863	4.3828	3.7829	3.4401	3.2151	3.0546	2.9338	2.8392	2.7628
23	5.7498	4.3492	3.7505	3.4083	3.1835	3.0232	2.9024	2.8077	2.7313
24	5.7167	4.3187	3.7211	3.3794	3.1548	2.9946	2.8738	2.7791	2.7027
25	5.6864	4.2909	3.6943	3.3530	3.1287	2.9685	2.8478	2.7531	2.6766
26	5.6586	4.2655	3.6697	3.3289	3.1048	2.9447	2.8240	2.7293	2.6528
27	5.6331	4.2421	3.6472	3.3067	3.0828	2.9228	2.8021	2.7074	2.6309
28	5.6096	4.2205	3.6264	3.2863	3.0625	2.9027	2.7820	2.6872	2.6106
29	5.5878	4.2006	3.6072	3.2674	3.0438	2.8840	2.7633	2.6686	2.5919
30	5.5675	4.1821	3.5894	3.2499	3.0265	2.8667	2.7460	2.6513	2.5746
40	5.4239	4.0510	3.4633	3.1261	2.9037	2.7444	2.6238	2.5289	2.4519
60	5.2857	3.9253	3.3425	3.0077	2.7863	2.6274	2.5068	2.4117	2.3344
120	5.1524	3.8046	3.2270	2.8943	2.6740	2.5154	2.3948	2.2994	2.2217
∞	5.0239	3.6889	3.1161	2.7858	2.5665	2.4082	2.2875	2.1918	2.1136

* Reproduced by Permission of Professor E. S. Pearson from "Tables of Percentage Points of the Inverted Beta (F) Distribution," *Biometrika, 33* (1943), pp. 73–88, by Maxine Merrington and Catherine M. Thompson.

TABLE IV (*continued*)

ν_2 \ ν_1	10	12	15	20	24	30	40	60	120	∞
1	968.63	976.71	984.87	993.10	997.25	1001.4	1005.6	1009.8	1014.0	1018.3
2	39.398	39.415	39.431	39.448	39.456	39.465	39.473	39.481	39.490	39.498
3	14.419	14.337	14.253	14.167	14.124	14.081	14.037	13.992	13.947	13.902
4	8.8439	8.7512	8.6565	8.5599	8.5109	8.4613	8.4111	8.3604	8.3092	8.2573
5	6.6192	6.5246	6.4277	6.3285	6.2780	6.2269	6.1751	6.1225	6.0693	6.0153
6	5.4613	5.3662	5.2687	5.1684	5.1172	5.0652	5.0125	4.9589	4.9045	4.8491
7	4.7611	4.6658	4.5678	4.4667	4.4150	4.3624	4.3089	4.2544	4.1989	4.1423
8	4.2951	4.1997	4.1012	3.9995	3.9472	3.8940	3.8398	3.7844	3.7279	3.6702
9	3.9639	3.8682	3.7694	3.6669	3.6142	3.5604	3.5055	3.4493	3.3918	3.3329
10	3.7168	3.6209	3.5217	3.4186	3.3654	3.3110	3.2554	3.1984	3.1399	3.0798
11	3.5257	3.4296	3.3299	3.2261	3.1725	3.1176	3.0613	3.0035	2.9441	2.8828
12	3.3736	3.2773	3.1772	3.0728	3.0187	2.9633	2.9063	2.8478	2.7874	2.7249
13	3.2497	3.1532	3.0527	2.9477	2.8932	2.8373	2.7797	2.7204	2.6590	2.5955
14	3.1469	3.0501	2.9493	2.8437	2.7888	2.7324	2.6742	2.6142	2.5519	2.4872
15	3.0602	2.9633	2.8621	2.7559	2.7006	2.6437	2.5850	2.5242	2.4611	2.3953
16	2.9862	2.8890	2.7875	2.6808	2.6252	2.5678	2.5085	2.4471	2.3831	2.3163
17	2.9222	2.8249	2.7230	2.6158	2.5598	2.5021	2.4422	2.3801	2.3153	2.2474
18	2.8664	2.7689	2.6667	2.5590	2.5027	2.4445	2.3842	2.3214	2.2558	2.1869
19	2.8173	2.7196	2.6171	2.5089	2.4523	2.3937	2.3329	2.2695	2.2032	2.1333
20	2.7737	2.6758	2.5731	2.4645	2.4076	2.3486	2.2873	2.2234	2.1562	2.0853
21	2.7348	2.6368	2.5338	2.4247	2.3675	2.3082	2.2465	2.1819	2.1141	2.0422
22	2.6998	2.6017	2.4984	2.3890	2.3315	2.2718	2.2097	2.1446	2.0760	2.0032
23	2.6682	2.5699	2.4665	2.3567	2.2989	2.2389	2.1763	2.1107	2.0415	1.9677
24	2.6396	2.5412	2.4374	2.3273	2.2693	2.2090	2.1460	2.0799	2.0099	1.9353
25	2.6135	2.5149	2.4110	2.3005	2.2422	2.1816	2.1183	2.0517	1.9811	1.9055
26	2.5895	2.4909	2.3867	2.2759	2.2174	2.1565	2.0928	2.0257	1.9545	1.8781
27	2.5676	2.4688	2.3644	2.2533	2.1946	2.1334	2.0693	2.0018	1.9299	1.8527
28	2.5473	2.4484	2.3438	2.2324	2.1735	2.1121	2.0477	1.9796	1.9072	1.8291
29	2.5286	2.4295	2.3248	2.2131	2.1540	2.0923	2.0276	1.9591	1.8861	1.8072
30	2.5112	2.4120	2.3072	2.1952	2.1359	2.0739	2.0089	1.9400	1.8664	1.7867
40	2.3882	2.2882	2.1819	2.0677	2.0069	1.9429	1.8752	1.8028	1.7242	1.6371
60	2.2702	2.1692	2.0613	1.9445	1.8817	1.8152	1.7440	1.6668	1.5810	1.4822
120	2.1570	2.0548	1.9450	1.8249	1.7597	1.6899	1.6141	1.5299	1.4327	1.3104
∞	2.0483	1.9447	1.8326	1.7085	1.6402	1.5660	1.4835	1.3883	1.2684	1.0000

A description of these tables is given in Section 4.55. Where necessary, interpolation should be carried out using the reciprocals of the degrees of freedom. The function $120/\nu$ is convenient for this purpose.

TABLE IV (*continued*)

PERCENTAGE POINTS OF THE F-DISTRIBUTION*

$\alpha = 0.01$

ν_2 \ ν_1	1	2	3	4	5	6	7	8	9
1	4052.2	4999.5	5403.3	5624.6	5763.7	5859.0	5928.3	5981.6	6022.5
2	98.503	99.000	99.166	99.249	99.299	99.332	99.356	99.374	99.388
3	34.116	30.817	29.457	28.710	28.237	27.911	27.672	27.489	27.345
4	21.198	18.000	16.694	15.977	15.522	15.207	14.976	14.799	14.659
5	16.258	13.274	12.060	11.392	10.967	10.672	10.456	10.289	10.158
6	13.745	10.925	9.7795	9.1483	8.7459	8.4661	8.2600	8.1016	7.9761
7	12.246	9.5466	8.4513	7.8467	7.4604	7.1914	6.9928	6.8401	6.7188
8	11.259	8.6491	7.5910	7.0060	6.6318	6.3707	6.1776	6.0289	5.9106
9	10.561	8.0215	6.9919	6.4221	6.0569	5.8018	5.6129	5.4671	5.3511
10	10.044	7.5594	6.5523	5.9943	5.6363	5.3858	5.2001	5.0567	4.9424
11	9.6460	7.2057	6.2167	5.6683	5.3160	5.0692	4.8861	4.7445	4.6315
12	9.3302	6.9266	5.9526	5.4119	5.0643	4.8206	4.6395	4.4994	4.3875
13	9.0738	6.7010	5.7394	5.2053	4.8616	4.6204	4.4410	4.3021	4.1911
14	8.8616	6.5149	5.5639	5.0354	4.6950	4.4558	4.2779	4.1399	4.0297
15	8.6831	6.3589	5.4170	4.8932	4.5556	4.3183	4.1415	4.0045	3.8948
16	8.5310	6.2262	5.2922	4.7726	4.4374	4.2016	4.0259	3.8896	3.7804
17	8.3997	6.1121	5.1850	4.6690	4.3359	4.1015	3.9267	3.7910	3.6822
18	8.2854	6.0129	5.0919	4.5790	4.2479	4.0146	3.8406	3.7054	3.5971
19	8.1850	5.9259	5.0103	4.5003	4.1708	3.9386	3.7653	3.6305	3.5225
20	8.0960	5.8489	4.9382	4.4307	4.1027	3.8714	3.6987	3.5644	3.4567
21	8.0166	5.7804	4.8740	4.3688	4.0421	3.8117	3.6396	3.5056	3.3981
22	7.9454	5.7190	4.8166	4.3134	3.9880	3.7583	3.5867	3.4530	3.3458
23	7.8811	5.6637	4.7649	4.2635	3.9392	3.7102	3.5390	3.4057	3.2986
24	7.8229	5.6136	4.7181	4.2184	3.8951	3.6667	3.4959	3.3629	3.2560
25	7.7698	5.5680	4.6755	4.1774	3.8550	3.6272	3.4568	3.3239	3.2172
26	7.7213	5.5263	4.6366	4.1400	3.8183	3.5911	3.4210	3.2884	3.1818
27	7.6767	5.4881	4.6009	4.1056	3.7848	3.5580	3.3882	3.2558	3.1494
28	7.6356	5.4529	4.5681	4.0740	3.7539	3.5276	3.3581	3.2259	3.1195
29	7.5976	5.4205	4.5378	4.0449	3.7254	3.4995	3.3302	3.1982	3.0920
30	7.5625	5.3904	4.5097	4.0179	3.6990	3.4735	3.3045	3.1726	3.0665
40	7.3141	5.1785	4.3126	3.8283	3.5138	3.2910	3.1238	2.9930	2.8876
60	7.0771	4.9774	4.1259	3.6491	3.3389	3.1187	2.9530	2.8233	2.7185
120	6.8510	4.7865	3.9493	3.4796	3.1735	2.9559	2.7918	2.6629	2.5586
∞	6.6349	4.6052	3.7816	3.3192	3.0173	2.8020	2.6393	2.5113	2.4073

* Reproduced by permission of Professor E. S. Pearson from "Tables of Percentage Points of the Inverted Beta (F) Distribution," *Biometrika*, *33* (1943), pp. 73-88, by Maxine Merrington and Catherine M. Thompson.

TABLE IV (*continued*)

ν_2 \ ν_1	10	12	15	20	24	30	40	60	120	∞
1	6055.8	6106.3	6157.3	6208.7	6234.6	6260.7	6286.8	6313.0	6339.4	6366.0
2	99.399	99.416	99.432	99.449	99.458	99.466	99.474	99.483	99.491	99.501
3	27.229	27.052	26.872	26.690	26.598	26.505	26.411	26.316	26.221	26.125
4	14.546	14.374	14.198	14.020	13.929	13.838	13.745	13.652	13.558	13.463
5	10.051	9.8883	9.7222	9.5527	9.4665	9.3793	9.2912	9.2020	9.1118	9.0204
6	7.8741	7.7183	7.5590	7.3958	7.3127	7.2285	7.1432	7.0568	6.9690	6.8801
7	6.6201	6.4691	6.3143	6.1554	6.0743	5.9921	5.9084	5.8236	5.7372	5.6495
8	5.8143	5.6668	5.5151	5.3591	5.2793	5.1981	5.1156	5.0316	4.9460	4.8588
9	5.2565	5.1114	4.9621	4.8080	4.7290	4.6486	4.5667	4.4831	4.3978	4.3105
10	4.8492	4.7059	4.5582	4.4054	4.3269	4.2469	4.1653	4.0819	3.9965	3.9090
11	4.5393	4.3974	4.2509	4.0990	4.0209	3.9411	3.8596	3.7761	3.6904	3.6025
12	4.2961	4.1553	4.0096	3.8584	3.7805	3.7008	3.6192	3.5355	3.4494	3.3608
13	4.1003	3.9603	3.8154	3.6646	3.5868	3.5070	3.4253	3.3413	3.2548	3.1654
14	3.9394	3.8001	3.6557	3.5052	3.4274	3.3476	3.2656	3.1813	3.0942	3.0040
15	3.8049	3.6662	3.5222	3.3719	3.2940	3.2141	3.1319	3.0471	2.9595	2.8684
16	3.6909	3.5527	3.4089	3.2588	3.1808	3.1007	3.0182	2.9330	2.8447	2.7528
17	3.5931	3.4552	3.3117	3.1615	3.0835	3.0032	2.9205	2.8348	2.7459	2.6530
18	3.5082	3.3706	3.2273	3.0771	2.9990	2.9185	2.8354	2.7493	2.6597	2.5660
19	3.4338	3.2965	3.1533	3.0031	2.9249	2.8442	2.7608	2.6742	2.5839	2.4893
20	3.3682	3.2311	3.0880	2.9377	2.8594	2.7785	2.6947	2.6077	2.5168	2.4212
21	3.3098	3.1729	3.0299	2.8796	2.8011	2.7200	2.6359	2.5484	2.4568	2.3603
22	3.2576	3.1209	2.9780	2.8274	2.7488	2.6675	2.5831	2.4951	2.4029	2.3055
23	3.2106	3.0740	2.9311	2.7805	2.7017	2.6202	2.5355	2.4471	2.3542	2.2559
24	3.1681	3.0316	2.8887	2.7380	2.6591	2.5773	2.4923	2.4035	2.3099	2.2107
25	3.1294	2.9931	2.8502	2.6993	2.6203	2.5383	2.4530	2.3637	2.2695	2.1694
26	3.0941	2.9579	2.8150	2.6640	2.5848	2.5026	2.4170	2.3273	2.2325	2.1315
27	3.0618	2.9256	2.7827	2.6316	2.5522	2.4699	2.3840	2.2938	2.1984	2.0965
28	3.0320	2.8959	2.7530	2.6017	2.5223	2.4397	2.3535	2.2629	2.1670	2.0642
29	3.0045	2.8685	2.7256	2.5742	2.4946	2.4118	2.3253	2.2344	2.1378	2.0342
30	2.9791	2.8431	2.7002	2.5487	2.4689	2.3860	2.2992	2.2079	2.1107	2.0062
40	2.8005	2.6648	2.5216	2.3689	2.2880	2.2034	2.1142	2.0194	1.9172	1.8047
60	2.6318	2.4961	2.3523	2.1978	2.1154	2.0285	1.9360	1.8363	1.7263	1.6006
120	2.4721	2.3363	2.1915	2.0346	1.9500	1.8600	1.7628	1.6557	1.5330	1.3805
∞	2.3209	2.1848	2.0385	1.8783	1.7908	1.6964	1.5923	1.4730	1.3246	1.0000

A description of these tables is given in Section 4.55. Where necessary, interpolation should be carried out using the reciprocals of the degrees of freedom. The function $120/\nu$ is convenient for this purpose.

TABLE IV (*continued*)

PERCENTAGE POINTS OF THE *F*-DISTRIBUTION*

$\alpha = 0.005$

ν_2 \ ν_1	1	2	3	4	5	6	7	8	9
1	16211	20000	21615	22500	23056	23437	23715	23925	24091
2	198.50	199.00	199.17	199.25	199.30	199.33	199.36	199.37	199.39
3	55.552	49.799	47.467	46.195	45.392	44.838	44.434	44.126	43.882
4	31.333	26.284	24.259	23.155	22.456	21.975	21.622	21.352	21.139
5	22.785	18.314	16.530	15.556	14.940	14.513	14.200	13.961	13.772
6	18.635	14.544	12.917	12.028	11.464	11.073	10.786	10.566	10.391
7	16.236	12.404	10.882	10.050	9.5221	9.1554	8.8854	8.6781	8.5138
8	14.688	11.042	9.5965	8.8051	8.3018	7.9520	7.6942	7.4960	7.3386
9	13.614	10.107	8.7171	7.9559	7.4711	7.1338	6.8849	6.6933	6.5411
10	12.826	9.4270	8.0807	7.3428	6.8723	6.5446	6.3025	6.1159	5.9676
11	12.226	8.9122	7.6004	6.8809	6.4217	6.1015	5.8648	5.6821	5.5368
12	11.754	8.5096	7.2258	6.5211	6.0711	5.7570	5.5245	5.3451	5.2021
13	11.374	8.1865	6.9257	6.2335	5.7910	5.4819	5.2529	5.0761	4.9351
14	11.060	7.9217	6.6803	5.9984	5.5623	5.2574	5.0313	4.8566	4.7173
15	10.798	7.7008	6.4760	5.8029	5.3721	5.0708	4.8473	4.6743	4.5364
16	10.575	7.5138	6.3034	5.6378	5.2117	4.9134	4.6920	4.5207	4.3838
17	10.384	7.3536	6.1556	5.4967	5.0746	4.7789	4.5594	4.3893	5.2535
18	10.218	7.2148	6.0277	5.3746	4.9560	4.6627	4.4448	4.2759	4.1410
19	10.073	7.0935	5.9161	5.2681	4.8526	4.5614	4.3448	4.1770	4.0428
20	9.9439	6.9865	5.8177	5.1743	4.7616	4.4721	4.2569	4.0900	3.9564
21	9.8295	6.8914	5.7304	5.0911	4.6808	4.3931	4.1789	4.0128	3.8799
22	9.7271	6.8064	5.6524	5.0168	4.6088	4.3225	4.1094	3.9440	3.8116
23	9.6348	6.7300	5.5823	4.9500	4.5441	4.2591	4.0469	3.8822	3.7502
24	9.5513	6.6610	5.5190	4.8898	4.4857	4.2019	3.9905	3.8264	3.6949
25	9.4753	6.5982	5.4615	4.8351	4.4327	4.1500	3.9394	3.7758	3.6447
26	9.4059	6.5409	5.4091	4.7852	4.3844	4.1027	3.8928	3.7297	3.5989
27	9.3423	6.4885	5.3611	4.7396	4.3402	4.0594	3.8501	3.6875	3.5571
28	9.2838	6.4403	5.3170	4.6977	4.2996	4.0197	3.8110	3.6487	3.5186
29	9.2297	6.3958	5.2764	4.6591	4.2622	3.9830	3.7749	3.6130	3.4832
30	9.1797	6.3547	5.2388	4.6233	4.2276	3.9492	3.7416	3.5801	3.4505
40	8.8278	6.0664	4.9759	4.3738	3.9860	3.7129	3.5088	3.3498	3.2220
60	8.4946	5.7950	4.7290	4.1399	3.7600	3.4918	3.2911	3.1344	3.0083
120	8.1790	5.5393	4.4973	3.9207	3.5482	3.2849	3.0874	2.9330	2.8083
∞	7.8794	5.2983	4.2794	3.7151	3.3499	3.0913	2.8968	2.7444	2.6210

* Reproduced by permission of Professor E. S. Pearson from "Tables of Percentage Points of the Inverted Beta (*F*) Distribution," *Biometrika, 33* (1943), pp. 73–88, by Maxine Merrington and Catherine M. Thompson.

TABLE IV (*continued*)

ν_2 \ ν_1	10	12	15	20	24	30	40	60	120	∞
1	24224	24426	24630	24836	24940	25044	25148	25253	25359	25465
2	199.40	199.42	199.43	199.45	199.46	199.47	199.47	199.48	199.49	199.51
3	43.686	43.387	43.085	42.778	42.622	42.466	42.308	42.149	41.989	41.829
4	20.967	20.705	20.438	20.167	20.030	19.892	19.752	19.611	19.468	19.325
5	13.618	13.384	13.146	12.903	12.780	12.656	12.530	12.402	12.274	12.144
6	10.250	10.034	9.8140	9.5888	9.4741	9.3583	9.2408	9.1219	9.0015	8.8793
7	8.3803	8.1764	7.9678	7.7540	7.6450	7.5345	7.4225	7.3088	7.1933	7.0760
8	7.2107	7.0149	6.8143	6.6082	6.5029	6.3961	6.2875	6.1772	6.0649	5.9505
9	6.4171	6.2274	6.0325	5.8318	5.7292	5.6248	5.5186	5.4104	5.3001	5.1875
10	5.8467	5.6613	5.4707	5.2740	5.1732	5.0705	4.9659	4.8592	4.7501	4.6385
11	5.4182	5.2363	5.0489	4.8552	4.7557	4.6543	4.5508	4.4450	4.3367	4.2256
12	5.0855	4.9063	4.7214	4.5299	4.4315	4.3309	4.2282	4.1229	4.0149	3.9039
13	4.8199	4.6429	4.4600	4.2703	4.1726	4.0727	3.9704	3.8655	3.7577	3.6465
14	4.6034	4.4281	4.2468	4.0585	3.9614	3.8619	3.7600	3.6553	3.5473	3.4359
15	4.4236	4.2498	4.0698	3.8826	3.7859	3.6867	3.5850	3.4803	3.3722	3.2602
16	4.2719	4.0994	3.9205	3.7342	3.6378	3.5388	3.4372	3.3324	3.2240	3.1115
17	4.1423	3.9709	3.7929	3.6073	3.5112	3.4124	3.3107	3.2058	3.0971	2.9839
18	4.0305	3.8599	3.6827	3.4977	3.4017	3.3030	3.2014	3.0962	2.9871	2.8732
19	3.9329	3.7631	3.5866	3.4020	3.3062	3.2075	3.1058	3.0004	2.8908	2.7762
20	3.8470	3.6779	3.5020	3.3178	3.2220	3.1234	3.0215	2.9159	2.8058	2.6904
21	3.7709	3.6024	3.4270	3.2431	3.1474	3.0488	2.9467	2.8408	2.7302	2.6140
22	3.7030	3.5350	3.3600	3.1764	3.0807	2.9821	2.8799	2.7736	2.6625	2.5455
23	3.6420	3.4745	3.2999	3.1165	3.0208	2.9221	2.8198	2.7132	2.6016	2.4837
24	3.5870	3.4199	3.2456	3.0624	2.9667	2.8679	2.7654	2.6585	2.5463	2.4276
25	3.5370	3.3704	3.1963	3.0133	2.9176	2.8187	2.7160	2.6088	2.4960	2.3765
26	3.4916	3.3252	3.1515	2.9685	2.8728	2.7738	2.6709	2.5633	2.4501	2.3297
27	3.4499	3.2839	3.1104	2.9275	2.8318	2.7327	2.6296	2.5217	2.4078	2.2867
28	3.4117	3.2460	3.0727	2.8899	2.7941	2.6949	2.5916	2.4834	2.3689	2.2469
29	3.3765	3.2111	3.0379	2.8551	2.7594	2.6601	2.5565	2.4479	2.3330	2.2102
30	3.3440	3.1787	3.0057	2.8230	2.7272	2.6278	2.5241	2.4151	2.2997	2.1760
40	3.1167	2.9531	2.7811	2.5984	2.5020	2.4015	2.2958	2.1838	2.0635	1.9318
60	2.9042	2.7419	2.5705	2.3872	2.2898	2.1874	2.0789	1.9622	1.8341	1.6885
120	2.7052	2.5439	2.3727	2.1881	2.0890	1.9839	1.8709	1.7469	1.6055	1.4311
∞	2.5188	2.3583	2.1868	1.9998	1.8983	1.7891	1.6691	1.5325	1.3637	1.0000

A description of these tables is given in Section 4.55. Where necessary, interpolation should be carried out using the reciprocals of the degrees of freedom. The function $120/\nu$ is convenient for this purpose.

REFERENCES

1. Arley, N., and K. R. Buch. *Introduction to the Theory of Probability and Statistics.* John Wiley & Sons, New York, 1950.
2. Kendall, M. G. *The Advanced Theory of Statistics, I and II.* Griffin & Co., Ltd., London, 1946.
3. Wilks, S. S. *Mathematical Statistics.* Princeton University Press, Princeton, 1943.
4. *Handbook of Chemistry and Physics.* Chemical Rubber Publishing Co., Cleveland, Ohio.
5. *Tables of Probability Functions, Vol. II.* Prepared by Federal Works Agency, W.P.A., Mathematical Tables Project, New York, 1942.
6. Fry, T. C. *Probability and Its Engineering Uses.* Van Nostrand Co., New York, 1928.
7. Molina, E. C. *Tables of Poisson's Exponential Limit.* Van Nostrand Co., New York, 1942.
8. *Tables of the Incomplete Gamma Function.* Edited by Karl Pearson. H.M. Stationery Office, London, 1922.
9. *Tables of the Incomplete Beta Function.* Edited by Karl Pearson. *Biometrika* Office, University College, London, 1934.
10. Dixon, W. J. "Processing Data for Outliers." *Biometrics*, *9* (1953), 74–88.
11. Birge, R. T. "Least Squares' Fitting of Data by Means of Polynomials," *Review of Modern Physics*, *19* (1947), 298–347.
12. Fisher, R. A. *Statistical Methods for Research Workers.* 11th ed. Oliver and Boyd, Edinburgh, 1950.
13. Fisher, R. A., and F. Yates. *Statistical Tables for Biological, Agricultural, and Medical Research.* 3rd ed. Oliver and Boyd, Edinburgh, 1948.
14. Pearson, K. *Tables for Statisticians and Biometricians, Part II.* Cambridge University Press, 1931.
15. David, F. N. *Tables of the Correlation Coefficient.* *Biometrika* Office, University College, London, 1938.
16. Pearson, E. S., and H. O. Hartley. "The Probability Integral of the Range in Samples of n Observations from a Normal Population," *Biometrika*, *32* (1942), 301–310.
17. Dodge, H. F., and H. G. Romig. *Sampling Inspection Tables. Single and Double Sampling.* John Wiley & Sons, New York, 1944.
18. *Sampling Inspection.* Statistical Research Group, Columbia University. McGraw-Hill, New York, 1948.
19. *Techniques of Statistical Analysis.* Statistical Research Group, Columbia University. McGraw-Hill, New York, 1947.

20. Dixon, W. J., and F. J. Massey. *Introduction to Statistical Analysis.* McGraw-Hill, New York, 1951.
21. Wilks, S. S. "Order Statistics," *Bull. A.M.S., 54* (1948), 6 ff.
22. Welch, B. L. "The Generalization of 'Student's' Problem when Several Different Population Variances are Involved," *Biometrika, 34* (1947), 28–35.
23. Olds, E. G. "The 5% Significance Levels for Sums of Squares of Rank Differences and a Correction," *Annals Math. Stat., XX* (1949), 117–118.
24. Olds, E. G. "Distributions of Sums of Squares of Rank Differences for Small Numbers of Individuals," *Annals Math. Stat., IX* (1938), 133–148.
25. Dwyer, P. S. *Linear Computations.* John Wiley & Sons, New York, 1951.
26. Shewart, W. A. *The Economic Control of Quality of a Manufactured Product.* D. Van Nostrand Co., New York, 1931.
27. Shewart, W. A. *Statistical Method from the Viewpoint of Quality Control.* Edited by W. E. Deming. The Graduate School, Dept. of Agriculture, Washington, D.C., 1939.
28. Swed, Freda S., and C. Eisenhart. "Tables for Testing Randomness of Grouping in a Sequence of Alternatives," *Annals Math. Stat., XIV* (1943), 66–87.
29. Wald, A., and J. Wolfowitz. "On a Test Whether Two Samples are from the Same Population," *Annals Math. Stat., XI* (1940), 147–162.
30. Dwyer, P. S. "Recent Developments in Correlation Technique," *Journal of the American Statistical Association, 37* (1942), 441–460.
31. Mood, A. M. "The Distribution Theory of Runs," *Annals Math. Stat., XI* (1940), 367–392.
32. Mosteller, F. "Note on an Application of Runs to Quality Control Charts," *Annals Math. Stat., XII* (1941), 228.
33. Olmstead, P. S. "Distribution of Sample Arrangements for Runs Up and Down," *Annals Math. Stat., XVII* (1946), 24–33.
34. Wolfowitz, J. "Asymptotic Distribution of Runs Up and Down," *Annals Math. Stat., XV* (1944), 163–172.
35. Dwyer, P. S. "The Evaluation of Determinants," *Psychometrika, 6* (1941), 191–204.
36. Bellinson, H. R., J. von Neumann, R. H. Kent, and B. I. Hart. "The Mean Square Successive Difference," *Annals Math. Stat., XII* (1941), 153–162.
37. von Neumann, J. "Distribution of the Ratio of the Mean Square Successive Difference to the Variance," *Annals Math. Stat., XII* (1941), 307–395.
38. Hart, B. I. "Significance Levels for the Ratio of the Mean Square Successive Difference to the Variance," *Annals Math. Stat., XIII* (1942), 445–447.
39. Young, L. C. "Randomness in Ordered Sequences," *Annals Math. Stat., XII* (1941), 299.
40. Anderson, R. L. "Distribution of the Serial Correlation Coefficient," *Annals Math. Stat., XIII* (1942), 1–13.
41. Wald, A., and J. Wolfowitz. "An Exact Test for Randomness in the Non-Parametric Case Based on Serial Correlation," *Annals Math. Stat., XIV* (1943), 378–388.
42. Tukey, J. W. "Comparing Individual Means in the Analysis of Variance," *Biometrics, 5* (1949), 99

43. Yates, F. "The Principles of Orthogonality and Confounding in Replicated Experiments," *J. Agr. Sci.*, *23* (1933), 108–145.

44. Brownlee, K. A. *Industrial Experimentation.* 3rd American ed. Chemical Publishing Co., New York, 1949.

45. Gould, C. E., and W. M. Hampton. "Statistical Methods Applied to the Manufacture of Spectacle Glasses," *J. Royal Stat. Soc.*, *Supp. III* (1936), 137.

46. Finney, D. J. "Standard Errors of Yields Adjusted for Regression on an Independent Measurement," *Biometrics Bull.*, *2* (1946), 53–55.

47. Bailey, E. G. "Accuracy in Sampling Coal," *Ind. Eng. Chem.*, *1* (1909), 161.

48. Grumell, E. S. "The Sampling of Coal with Special Reference to the Size-weight Ratio Theory," *British Standard Specification 763* (1937).

49. Yates, F. "Some Examples of Biased Sampling," *Ann. Eugenics*, *6* (1935), 202.

50. "Student." "On the distribution of the Means of Samples which are not Drawn at Random," *Biometrika*, *7* (1909), 210–214.

51. Mandel, J. "Improvement of Precision by Repeated Measurements," *Ind. Eng. Chem.* (*Anal. ed.*), *18* (1946), 280.

52. Yates, F. "The Design and Analysis of Factorial Experiments," *Imp. Bureau Soil Science*, *Tech. Communication 35* (1937).

53. Cochran, W. G., and G. M. Cox. *Experimental Designs.* Mathematical Statistics Series, John Wiley & Sons, New York, 1950.

54. Mitton, R. G., and T. R. G. Lewis. "The Abrasion of Leather," Part IV, *J. Intern. Soc. Leather Trades' Chemists*, *30* (1946), 287.

55. Fisher, R. A. *The Design of Experiments.* 4th ed. Oliver and Boyd, Edinburgh, 1947.

56. Fuell, H. J., and R. E. Wagg. "Statistical Methods in Detergency Investigations," *Research*, *2* (1949), 334.

57. Bray, G. T., *et alia.* "The Determination of the Factor for Pyrethrin I in the Mercury Method," *J. Soc. Chem. Ind. London*, *66* (1947), 275.

58. Davies, H. M. "The Application of Variance Analysis to Some Problems of Petroleum Technology," *J. Inst. Petroleum*, *32* (1946), 465.

59. Lesser, A. "The Statistical Approach in Industrial Research," *Iron Age*, *158* (1946), 50.

60. Yates, F. "Incomplete Randomised Blocks," *Annals Eugenics*, *7* (1936), 121.

61. Tukey, J. W., and F. Mosteller. "The Uses and Usefulness of Binomial Probability Paper," *Jour. A.S.A.*, *44* (1949), 174.

62. Freeman, M. F., and J. W. Tukey. "Transformations Related to the Angular and the Square Root," *Memorandum Report 24*, Statistical Research Group, Princeton University, 1949.

63. Davies, O. L. *Statistical Methods in Research and Production.* Oliver and Boyd, Edinburgh, 1949.

64. Gore, W. L. "Quality Control in the Chemical Industry, IV," *Industrial Quality Control*, *IV* (1947), 5.

65. Mitton, R. G., and T. R. G. Lewis. "The Abrasion of Leather," Part I, *J. Intern. Soc. Leather Trades' Chemists*, *30* (1946).

66. Bainbridge, J. R., and K. Satchwell. "Experiments in Fleissner Drying Victorian Brown Coal," *Fuel in Science and Practice*, *26* (1947), 28.

67. Hams, H. C. "The Effect of Minor Elements on the Growth of Certain Crops," *Soil Science Soc. Am. Proc.*, *7* (1942), 345.

68. Finney, D. J. "The Fractional Replication of Factorial Arrangements," *Annals Eugenics*, *12* (1945), 291–301.

69. Finney, D. J. "Fractional Replication," *J. Agr. Sci.*, *36* (1946), 184–191.

70. Kempthorne, O. "A Simple Approach to Confounding and Fractional Replication in Factorial Experiments," *Biometrika*, *34* (1947), 255–272.

71. Kempthorne, O. "A Note on Differential Response in Blocks," *J. Agr. Sci.*, *37* (1947), 245–248.

72. Fisher, R. A. "The Theory of Confounding in Factorial Experiments in Relation to the Theory of Groups," *Annals Eugenics*, *11* (1942), 341–353.

73. Tippett, L. H. C. "Statistical Methods in Industry," *Iron and Steel Indus. Research Council Publication* (1943), 39.

74. Wernimont, G. "Quality Control in the Chemical Industry, II," *Industrial Quality Control*, *III* (1947), 5.

75. Tukey, J. W. "Some Sampling Simplified," *Jour. A.S.A.*, *45* (1950), 501–519.

76. Feller, William. *An Introduction to Probability Theory and Its Application*. John Wiley & Sons, New York, 1950.

77. Deming, W. E. *Some Theory of Sampling*. John Wiley & Sons, New York, 1950.

78. "Student." "Errors of Routine Analysis," *Biometrika*, *19* (1927), 151. Included in [79].

79. *"Student's" Collected Papers*. Edited by E. S. Pearson and John Wishart. *Biometrika* Office, University College, London. See particularly No. 2, "The Probable Error of a Mean," originally published in *Biometrika*, *6* (1908), 1.

80. Pearson, E. S., and H. O. Hartley. "Tables of the Probability Integral of the 'Studentized' Range," *Biometrika*, *33* (1943), 89–99.

81. *Tables of the Binomial Probability Distribution*. National Bureau of Standards, Applied Mathematics Series 6. U.S. Government Printing Office, Washington, D.C., 1950.

82. Mosteller, F. "Some Useful 'Inefficient' Statistics," *Annals Math. Stat.*, *XVII* (1946), 377.

83. Meier, P. "Weighted Means and Lattice Designs." Unpublished thesis, Princeton University, October, 1951.

84. Scheffé, H., and J. W. Tukey. *Another Beta-Function Approximation. Memorandum Report 28*, Statistical Research Group, Princeton University, 1949.

85. Aspin, Alice A. "An Examination and Further Development of a Formula Arising in the Problem of Comparing Two Mean Values," *Biometrika*, *35* (1948), 88–96.

86. Aspin, Alice A. "Tables for Use in Comparisons whose Accuracy Involves Two Variances, Separately Estimated," *Biometrika*, *36* (1949), 290–296.

87. Wilcoxon, Frank. *Some Rapid Approximate Statistical Procedures*. Insecticide and Fungicide Section, Stamford Research Laboratory, American Cyanamid Company, Stamford, Conn.

88. Freeman, M. F., and J. W. Tukey. "Transformations Related to the Angular and the Square Root," *Annals Math. Stat.*, *XXI* (1950), 607–611.

89. Dixon, W. J. "Analysis of Extreme Values," *Annals Math. Stat.*, *XXI* (1950), 488–506.

90. Dixon, W. J. "Ratios Involving Extreme Values," *Annals Math. Stat.*, *XXII* (1951), 68–78.

91. Ezekiel, M. *Methods of Correlation Analysis.* John Wiley & Sons, New York, 1941.

92. Wald, A. *Sequential Analysis.* John Wiley & Sons, New York, 1952.

93. Statistical Research Group, Columbia University. *Sequential Analysis of Statistical Data: Applications.* Columbia University Press, 1945.

94. Youden, W. J. "Statistics in Analytical Chemistry." From a Conference on "The Place of Statistical Methods in Biological and Chemical Experimentation." *Ann. N.Y. Acad. Sci.*, *52*, Art. 6 (1950), 815–819.

95. Geary, R. C. "The Distribution of 'Student's' Ratio for Non-Normal Samples," *J. Royal Stat. Soc.* (*Suppl.*), *3* (1936), 178–184.

96. Bross, I. "Fiducial Intervals for Variance Components," *Biometrics*, *6* (1950), 136.

97. Tukey, J. W. "Components in Regression," *Biometrics*, *7* (1951), 33.

98. Nair, K. R. "The Distribution of the Extreme Deviation from the Sample Mean and Its Studentized Form," *Biometrika*, *35* (1948), 118.

99. Cochran, W. G. "Some Consequences when the Assumptions for Analysis of Variance are not Satisfied," *Biometrics*, *3* (1947), 22.

100. Youden, W. J. "A Note on the Four by Four Latin Squares," *Biometrics*, *6* (1950), 289.

101. Bartlett, M. S. "The Use of Transformations," *Biometrics*, *3* (1947), 39.

102. Ricker, W. E. "The Concept of Confidence or Fiducial Limits Applied to the Poisson Frequency Distribution," *Jour. A.S.A.*, *32* (1937), 349–386.

103. Satterthwaite, F. E. "An Approximate Distribution of Estimates of Variance Components," *Biometrics Bull.*, *2* (1946), 110–114.

104. Wald, A. "The Fitting of Straight Lines if Both Variables are Subject to Error," *Annals Math. Stat.*, *XI* (1940), 284.

105. Bartlett, M. S. "Fitting a Straight Line when Both Variables are Subject to Error," *Biometrics*, *5* (1949), 207.

106. Geary, R. C. "Determination of Linear Relations between Systematic Parts of Variables," *Econometrica*, *17* (1949), 30.

107. Marcuse, S. "Optimum Allocation in Nested Sampling," *Biometrics*, *5* (1949), 189.

108. Bainbridge, J. R. "Factorial Experiments in Pilot Plant Studies," *Ind. Eng. Chem.*, *43* (1951), 1300.

109. Yates, F. "The Recovery of Interblock Information in Balanced Incomplete Block Designs," *Ann. Eugenics*, *10* (1940), 317.

110. Yates, F. "The Analysis of Latin Squares when Two or More Rows are Missing," *J. Royal Stat. Soc.* (*Suppl.*), *6* (1939), 67.

111. Youden, W. J. "The Use of Incomplete Block Replications in Estimating Tobacco Mosaic Virus," *Contribs. Boyce Thompson Inst.*, *9* (1937), 41.

112. Youden, W. J. "Experimental Designs to Increase the Accuracy of Greenhouse Studies," *Contribs. Boyce Thompson Inst.*, 11 (1939), 219.

113. Minor, J. E. Pure Substances Division, National Bureau of Standards, Washington, D.C. Private communication.

114. Frazier, D. The Standard Oil Company (Ohio), Cleveland, Ohio. Private communication.

115. Scrivener, B. F. National Bureau of Standards, Washington, D.C. Private communication.

116. Gore, W. L. "Statistical Design in Chemical Experimentation," *Ind. Eng. Chem.*, *43* (1951), 2327.

117. Kempthorne, O. *The Design and Analysis of Experiments.* John Wiley & Sons, New York, 1952.

118. Geary, R. C., and E. S. Pearson. "Tests of Normality," *Biometrika* Office, University College, London, 1938.

119. Lord, E. "The Use of the Range in Place of the Standard Deviation in the *t*-Test," *Biometrika*, *34* (1947), 41.

120. Walsh, J. E. "On the Range-Midrange Test and Some Tests with Bounded Significance Levels," *Annals Math. Stat.*, *XX* (1949), 257–267.

121. Tukey, J. W. "The Simplest Signed Rank Tests," *Memorandum Report 17*, Statistical Research Group, Princeton University, 1949.

122. Merrington, Maxine, and Catherine M. Thompson. "Tables for Testing the Homogeneity of a Set of Estimated Variances," *Biometrika*, *33* (1946), 296–304.

123. "Extended and Corrected Tables of the Upper Percentage Points of the 'Studentized' Range." Computed by Joyce M. May. *Biometrika*, *39* (1952), 192–193.

124. "The Measurement of Coke Yields at Coking Plants." Report of Panel No. 2, British Coke Research Association.

125. Welch, B. L. "The Specification of Rules for Rejecting too Variable a Product," *J. Royal. Stat. Soc.* (*Supplement*), *3* (1936), 29.

126. Friedman, M., and L. T. Savage, *Selected Techniques of Statistical Analysis.* McGraw Hill Book Co., New York, 1947.

127. Box, G. E. P., and K. B. Wilson, "On the Experimental Attainment of Optimum Conditions," *J. Royal. Stat. Soc.* (*Series B*), *13* (1951), 1.

128. Box, G. E. P. "Multi-Factor Designs of First Order," *Biometrika*, *39* (1952), 49.

INDEX

Acceptance sampling, *see* Sampling, acceptance
Analysis of covariance, 5, 441
  difference between slopes, 442
  effect of outlier points, 461
  for multiple classifications, 451
  pooling of data, 462
  treatment comparisons, 446
  with more than one covariance variable, 457
Analysis of variance, 4, 319
  application to regression, 427
    linear regression in more than one variable, 429
    polynomial regression, 431
    simple linear regression, 427
  classification of means, 341
  combination of nested and crossed classifications, 410
  comparison of class means, 340
  confidence limits for class means, 339
  covariance between residuals, 351
  failure in the assumptions concerning the model, 351
  general consideration of nested classifications, 402
  general linear comparisons, 335
  independence of linear comparisons, 336
  many-way classifications, 350
  non-additivity of effects, 353
  non-randomness of residuals, 351
  one-way classification, 319
    comparison of class effects, 333
    computations for, 323
    estimation of class effects, 331
    hypotheses concerning the overall mean, 344
    interpretation of class effects, 330
    model for, 321, 348
    with equal numbers, 321
    with unequal numbers, 327
Analysis of variance (*cont.*)
  pooling of variance estimates, 392
  single degrees of freedom, 335, 436
  subdivision of scale classifications, 436
  three crossed classifications, 385
    breakdown of sum of squares, 388
    calculation of sum of squares, 391
    estimation of effects, 387
    estimation of variance components, 393
    model for, 386
    tests of significance, 392
  transformations, 355
    to obtain additivity, 355
    to stabilize variance, 356
  two crossed classifications, 368
  two nested classifications, 358
  two-way classification, 349
    crossed with no interactions, 349
    interactions, 353, 368
    missing data, 379
    models for, 349
    nested, 350
  unequal variance of residuals, 351
Arcsin transformation, *see* Transformations, of counted data
Arithmetic mean, *see* Mean, sample
Average outgoing quality limit, 629
Average value, definition of, 47

Bartlett's test, 197
Batching factors, 191
Behrens-Fisher problem, 177
Beta distribution, 127
Beta function, 125
  incomplete, 126
Binomial distribution, 111
  limiting form, 118

Cell boundaries, 11
Cell midpoints, 11
Central limit theorem, 89

Central line, 634
Chi-square distribution, 95
  application to contingency tables, 623
  to test goodness of fit, 620
Combinations, 59, 66
Conditional distributions, 130
Confidence levels, 148
Confidence limits, 136
  for a population mean, 137, 154
  for an expected number of occurrences, 608
  for an unknown proportion, 604
  for class means in the analysis of variance, *see* Analysis of variance
  for difference of two means, 176, 181
  for intercept in linear regression, 231
  for median based on order statistics, 157
  for predicted values in linear regression, 229
  for several means, 187
  for slope, 228
  for variance, 172
Confounding, 538
  in 2^k designs, 545
  in 3^k and other factorial designs, 564
  of interactions, 545
  of main effects, 539
Consistency, 209
Consumer risk, 629
Contingency tables, *see* Chi-square distribution
Contour ellipse, 129
Control charts, 634
  for accuracy, 648
  for attributes, 644
  for counting instruments, 659
  for duplicate analyses, 654
  for errors of measurement, 647
  for fraction defective, 644
  for number defective, 646
  for precision, 650
  for specified levels, 635
  for variables, 634
  use of, 642
  when no levels are specified, 638
Control limits, 633
Correlation, 273
  partial, 287
  rank, 283
  three or more variables, 286
Correlation coefficient, 37
  distribution of, 132
  estimation of, 273
  multiple, 287
  partial, 287
  transformation of, 132
Correlation table, 35
Counted data, 601
  analysis of variance for, 615
  comparison of, 611
  tests of homogeneity for, 626
Covariance, 35
  analysis of, *see* Analysis of covariance

Data coding of, 23
  collection of, 480
  grouping of, 10
  ordered, 7
  organization of, 6
  raw, 6
Defective, 627
Degrees of freedom, 27
  single, in the analysis of variance, *see* Analysis of variance
    in factorial design, *see* Factorial design
Design of experiments, 4, 478 (*see also* Factorial design)
  combination of experiments, 597
  effect of raw materials, 598
  elimination of experimental error, 514
  fractional replication, 581
  Graeco-Latin squares, 526, 537
  hyper-squares, 533
  incomplete blocks, 570
  Latin squares, 521, 535
  Latin squares and hyper-squares considered as fractional replications, 587
  miscellaneous designs, 592
  replication in blocks, 515
  sequential experiments, 599
  Youden squares, 575
Discrete observations, *see* Counted data
Discrete probability distribution, 71
  of two variables, 74
Discriminant function, 288
Distribution function, 41, 68
  cumulative, 68
    more than two variables, 76

Distribution function (*cont.*)
of two variables, 73
relationship with probabilities, 70
marginal, 74

Efficiency, 136
asymptotic, 209
Error of Type I, 141
Error of Type II, 141
Errors of measurement, 206
Estimate, 133
best, 135
best unbiased, 135
interval, 133
joint, 166
from duplicate measurements, 168
maximum likelihood, 209
point, 134
unbiased, 135
Estimation, 4
of a linear combination, 152
of an average value, 149
of an expected number of occurrences, 607
of an unknown proportion, 603
of regression coefficients, 280
of slope, 224, 232, 465
of the intercept, 224
of the standard deviation, 164
of the variance, 164
Evaluation of determinants, 300
Experimental unit, 479
Extreme variations, 665
detection by control charts, 665
rejection of, 666

F-distribution, 108
Factorial design, 493
2^3 experiments, 496
2^k experiments, 501
dummy comparisons, 512
factors with more than two levels, 502
generalized interaction, 561
main effects and interactions, 494
computation in 2^k designs, 497, 501
relationship to classical design, 493
single degrees of freedom, 496, 502, 506
Factors, relationships between, 4
Fraction defective, 627, 629

Frequency, diagram, 7
cumulative, 7
distribution, 11
cumulative, 13
grouped, 11
histogram, 13

Gamma distribution, 125
Gamma function, 122
incomplete, 125
Graeco-Latin squares, *see* Design of experiments

Hypergeometric distribution, 120
limiting form, 121
Hyper-squares, *see* Design of experiments
Hypothesis testing, 140

Incomplete blocks, *see* Design of experiments
Interactions, *see* Analysis of variance, two-way classification, *and* Factorial design
Intercept, *see* Regression, linear
Inverse matrix, computation of, *see* Solution of simultaneous equations

Kurtosis, 82

Latin squares, *see* Design of experiments
Least squares, relation to maximum likelihood, 214
Levels, 479
Linear combinations, independence of, 101
in the analysis of variance, *see* Analysis of variance
Lot tolerance fraction defective, 629

Maximum likelihood, *see* Estimate
Mean, geometric, definition of, 18
Mean, population, *see* Average value
Mean, sample, 16
approximate computation from grouped data, 24
average value of, 55
computation of, 23
distribution of, 99
limits for, 26, 28 (*see also* Confidence limits)
variance of, 26, 56

Mean deviation, 18
bias in, 165
computation of, 22
efficiency of, 165
Mean square successive difference, 665, 677
distribution of, 678
used to detect non-randomness, 678
used to estimate variance, 678
Means, comparison of, 176
independence of comparisons, 340
in the analysis of variance, *see* Analysis of variance
paired, 180
several, 185
two, 176
substitute t-ratio, 177
Measurements, pairs of, 31, 216
reliability of, 3
Median, 17
distribution of, 110
efficiency of, 149
limits for, 30 (*see also* Confidence limits)
Midrange, 18
efficiency of, 150
Moments, 76
about the mean, 77
Moment generating functions, 78
Multinomial distribution, 114

Non-additivity, *see* Analysis of variance
Non-randomness, of residuals in the analysis of variance, *see* Analysis of variance
tests for, *see* Tests of significance
Normal distribution, bivariate, 128
Normal distribution function, 83
moment generating function of, 85
properties of, 84
tables of, 86
Normal equations, 224
general linear case, 245
Normally distributed variables, 83, 88
distribution of a linear combination, 88
mean of a linear combination, 88
variance of a linear combination, 88
Null hypothesis, 141

Operating characteristic curve, 142
Operating characteristic curve (*cont.*)
for tests of significance concerning variances, 174
for tests of significance for means, 155
for the comparison of two variances, 194
in acceptance sampling, 629
Order statistics, 17 (*see also* Confidence limits, Tests of significance, *and* Tolerance limits)
Orthogonal polynomials, 256
Outlier points in the analysis of covariance, *see* Analysis of covariance

Permutations, 66
Point estimate, *see* Estimate
Poisson distribution, 115
limiting form, 119
Polygon, cumulative, 13
Pooling, of data, *see* Analysis of Covariance
of variance estimates, *see* Analysis of variance
Population, 41
definition of, 42
relationship to sample, 54
Probability, 43
definition of, 43
laws of, 45
Probability density function, 71
of two variables, 74
Probability element, 73
Probability paper, 15
logarithmic, 91
Producer risk, 629

Quantiles, 7
deciles, 7
percentiles, 7
quartiles, 7

Random selection, 43
Random variables, 43
average value of, 47
average value of a linear combination, 49
average value of an arbitrary function of, 52
average value of sum, 48
average value of the products, 49
independence of, 45
variance of, 47

Random variables (*cont.*)
variance of a linear combination, 50
variance of an arbitrary function of, 52
Randomization, 491
Randomized blocks, *see* Design of experiments, replication in blocks
Randomness, 41
tests for, 663
Range, 7, 18
bias of, 165
distribution of, 110
efficiency of, 165
"Studentized," distribution of, 111
Rational subgroups, 634
Regression, 5, 36
linear, intercept,
confidence limits for, 231
estimation of, 224
tests of significance for, 231
variance of, 230
more than one independent variable, 245
one dependent variable, 223
orthogonal combinations of independent variables, 265
predicted values, 224
confidence limits for, 229
tests of significance for, 229
slope, *see* Slope
statistical model for, 223
with both variables subject to error, 463
with unequal residual variances, 243
multiple, 245
polynomial, 251
on equally spaced observations, 255
Regression coefficients, 247
estimation of, 280
linear, 36
variance of, 249
Regression line, 36, 131
estimation of, 280
Relationship between variables, *see* Regression
Replication, 480, 487
fractional, 480
in nested experiments, 487
with optimum allocation, 490
Risk, 148
consumer, *see* Consumer risk
Risk (*cont.*)
producer, *see* Producer risk
Runs, 665, 667
distribution of length of, 675
distribution of number of, 669
other uses for, 677
relationship to non-randomness, 667

Sample, 41
chemical, reduction of, 485
random, 55
relationship to population, 54
Sample mean, *see* Mean, sample
Sample variance, *see* Variance, sample
Sampling, acceptance, by attributes, 627
bias in, 486
from finite populations, 57
of aggregates, 484
of discrete articles, 481
of materials, 481
random, 55
stratified, 61, 482
efficiency of, 64
estimating the mean, 61
from infinite subpopulations, 64
minimizing the variance, 62
without replacement, 58
Scatter diagram, 33
Semi-invariants, 79
calculation of, 80
interpretation of, 81
properties of, 80
sample, 81
variance of, 95
Semi-invariant generating function, 79
Serial correlation, 665, 684
distribution of, 685
in non-parametric case, 687
used to detect non-randomness, 685
Sign test, 160
Signed rank tests, 182
Significance level, 141, 148
of a series of tests, 688
Size-weight ratio, 485
Skewness, 81
Slope, 37, 131
comparison of, *see* Analysis of covariance, difference between slopes
confidence limits for, 228
estimation of, 224

Slope (*cont.*)
with both variables subject to error, 465
with zero intercept, 232
limits for, 38
tests of significance for, 228
variance of, 39, 227
Solution of linear equations, 296
by the abbreviated Doolittle method, 302
exact, 296
method of single division, 302
sets of equations, 315
to obtain inverse matrix, 315
Split plots, *see* Confounding, of main effects
Square root transformation, *see* Transformations, of counted data
Standard curve, 217
Standard deviation, approximate computation from grouped data, 22
computation of, 21
definition of, 19
interpretation of, 26
sample, bias of, 165
Standard normal deviate, 86
Statistical approach, 2
Statistical control, 631
Statistical methods, misapplication of, 5
nature of, 2
place of, 1
Statistics, 16 (*see also* Estimate)
relative merits of, 20
to determine dispersion, 18
to determine location, 16
Stirling's approximation, 123
Stochastic convergence, 44
Strata, 61
"Student," 1

t-distribution, 105
t-tests, 154
substitute, 160
Tests for non-normality, 92
Tests of significance, 4, 140
based on order statistics, 157
for correlation coefficient, 278
for intercept in linear regression, 231
for means, 154
for non-randomness, using mean square successive difference, 678
using runs, 667
for population mean, 142
for predicted values in linear regression, 229
for rank correlation, 284
for slope, 228
for variance, 174
sequential, 200
Time as a variable, 221
Tolerance limits, 162
based on order statistics, 162
Transformations, logarithmic, 91
of counted data, 601
orthogonal, 102
to approximate normality, 91
to obtain additivity, *see* Analysis of variance
to obtain linear regressions, 234
to orthogonal polynomials, 256
to stabilize variance in the analysis of variance, *see* Analysis of variance
Treatment, 480

Variance, population, 47
ratio, 108, 192
interpretation of, 329
sample, 19
average value of, 57
distribution of, 102
estimation of, 164
Variances, comparison of several, 196
comparison of two, 192
Variance-covariance matrix, 288

Youden squares, *see* Design of experiments